Biotechnology & Genetic Engineering Reviews

Volume 12

BIOTECHNOLOGY AND GENETIC ENGINEERING REVIEWS

Biotechnology & Genetic Engineering Reviews

Volume 12

Editor:
MICHAEL P. TOMBS

Special Professor, University of Nottingham

Intercept

Andover

British Library Cataloguing in Publication Data
Biotechnology & genetic engineering reviews.—
Vol. 12
1. Biotechnology—Periodicals

A CIP catalogue record for this book is available from the British Library
ISBN 0–946707–99–5
ISSN 0264–8725

Published in December1994 by Intercept Limited,
PO Box 716, Andover, Hants. SP10 1YG, England

Typeset in Times by
Ann Buchan (Typesetters), Shepperton, Middlesex.
Printed by Athenaeum Press Ltd, Gateshead, Tyne & Wear.

Contributors

P. ADLERCREUTZ, *Department of Biotechnology, Chemical Centre, Lund University, P.O. Box 124, S-221 00 Lund, Sweden*

MUKUL KUMAR BASU, *Biomembrane Division, Indian Institute of Chemical Biology, 4, Raja S.C. Mullick Road, Calcutta 700 032, India*

W.S. BAWDEN, *School of Biochemistry and Molecular Genetics, University of New South Wales, Sydney, Australia*

J.A. BEELEY, *Department of Oral Science, University of Glasgow Dental Hospital and School, 378 Sauchiehall Street, Glasgow G2 3JZ, UK*

A.W. BUNCH, *Research School of Biosciences, Biological Laboratory, The University, Canterbury, Kent CT2 7NJ, UK*

P. CHARLIER, *Centre d'Ingénierie des Protéines et Laboratoire d'Enzymologie, Université de Liège, Institut de Chimie B6 Sart-Tilman, B-4000 Liège, Belgium*

O. DIDEBERG, *Laboratoire de Cristallographie Macromoléculaire, Institut de Biologie Structurale, Avenue des Martyrs 41, F-38027, Grenoble, France*

N.V. EFREMOVA, *A.N. Bach Institute of Biochemistry, Russian Academy of Science, Leninsky Prospect 33, Moscow 117071, Russia*

P. ELOMAA, *Institute of Biotechnology, P.O. Box 45, Karvaamokuja 3 A, FIN-00014 University of Helsinki, Finland*

E. FONZÉ, *Centre d'Ingénierie des Protéines et Laboratoire d'Enzymologie, Université de Liège, Institut de Chimie B6 Sart-Tilman, B-4000 Liège, Belgium*

G. FORTIER, *Laboratoire d'Enzymologie Appliquée, Groupe de recherche en Biothérapeutique Moléculaire, Départment de Chimie-Biochimie, Université du Québec à Montréal, C.P. 8888, Succ. Centre-Ville, Montréal (Québec), Canada H3C 3P8*

J.-M. FRÈRE, *Centre d'Ingénierie des Protéines et Laboratoire d'Enzymologie, Université de Liège, Institut de Chimie B6 Sart-Tilman, B-4000 Liège, Belgium*

J. GIBSON, *Department of Oral Surgery/Oral Medicine, University of Glasgow Dental Hospital and School, 378 Sauchiehall Street, Glasgow G2 3JZ, UK*

C. GRAHAM, *Section on Molecular Structure and Function, LMDB, National Eye Institute, Building 6 Room 222, National Institutes of Health, Bethesda, MD 20892, USA*

T. HOLTON, *Florigene Pty Ltd, ACN 006 455 616, 16 Gipps Street, Collingwood, Victoria 3066, Australia*

Y. INOUE, *Research Institute for Food Science, Kyoto University, Kyoto 611, Japan*

C. JAWORSKI, *Section on Molecular Structure and Function, LMDB, National Eye Institute, Building 6 Room 222, National Institutes of Health, Bethesda, MD 20892, USA*

J.A. KELLY, *Department of Molecular and Cell Biology, University of Connecticut, STORRS, CT 06269-3125, USA*

J. KNOX, *Department of Molecular and Cell Biology, University of Connecticut, STORRS, CT 06269-3125, USA*

J. LAMOTTE-BRASSEUR, *Centre d'Ingénierie des Protéines et Laboratoire d'Enzymologie, Université de Liège, Institut de Chimie B6 Sart-Tilman, B-4000 Liège, Belgium*

G.M. LEE, *Department of Biotechnology, Korea Advanced Institute of Science and Technology, 373-1, Kusong-Dong, Yusong-Gu, Taejon 305-701, Korea*

M.J. LENN, *Research School of Biosciences, University of Kent, Canterbury, Kent CT2 7NJ, UK*

A.G. MACKINLAY, *School of Biochemistry and Molecular Genetics, University of New South Wales, Sydney, Australia*

C. OLDFIELD, *Energy Biosystems Corporation, 4200 Research Forest Drive, The Woodlands, Texas TX 77381, USA*

B.O. PALSSON, *Department of Chemical Engineering, University of Michigan, Ann Arbor, MI 48109, USA*

R.J. PASSEY, *School of Biochemistry and Molecular Genetics, University of New South Wales, Sydney, Australia*

J. RICHARDSON, *Section on Molecular Structure and Function, LMDB, National Eye Institute, Building 6 Room 222, National Institutes of Health, Bethesda, MD 20892, USA*

G.K. ROBINSON, *Research School of Biosciences, University of Kent, Canterbury, Kent CT2 7NJ, UK*

D.Y. SCHAFHAUSER-SMITH, *Department of Environmental Science, Virginia Institute of Marine Science, The College of William and Mary, Gloucester Point, Virginia, USA*

L. SEGOVIA, *Section on Molecular Structure and Function, LMDB, National Eye Institute, Building 6 Room 222, National Institutes of Health, Bethesda, MD 20892, USA*

R. SHARON-FRILING, *Section on Molecular Structure and Function, LMDB, National Eye Institute, Building 6 Room 222, National Institutes of Health, Bethesda, MD 20892, USA*

K.B. STOREY, *Department of Biology, Carleton University, Ottawa, Ontario, Canada K1S 5B6*

I.N. TOPCHIEVA, *Department of Chemistry, Moscow State University, Lenin Hills, Moscow 119899, Russia*

G. WISTOW, *Section on Molecular Structure and Function, LMDB, National Eye Institute, Building 6 Room 222, National Institutes of Health, Bethesda, MD 20892, USA*

Contents

1

Crystallins: The Over-expression of Functional Enzymes and Stress Proteins in the Eye Lens

GRAEME WISTOW, JILL RICHARDSON, CYNTHIA JAWORSKI, CAROLINE GRAHAM, RONIT SHARON-FRILING AND LORENZO SEGOVIA

Section on Molecular Structure and Function, LMDB, National Eye Institute, National Institutes of Health, Bethesda, MD 20892, USA

Introduction

The ocular lens in vertebrates, and even some invertebrates, is a transparent, avascular, cellular tissue whose principal function is light refraction. The necessary refractive index is achieved by the accumulation of a few soluble proteins which, from their role in the 'crystalline' (clear) lens, are known as crystallins (Bloemendal, 1981). In some species crystallins may account for 90% of total lens protein, which in turn may make up 60–70% of the wet weight of the tissue. The lens grows throughout life, with new layers of cells continually overlaying the old (Piatigorsky, 1981). As the differentiated lens fibre cells mature they lose cell nuclei and other organelles, in a process resembling early stages of apoptosis (Gao, Bassnett and Zelenka, 1994). There is thus very little protein turnover in the mature fibre cells. Since the lens must serve throughout life, crystallins must be maintained in a stable, benign environment.

Remarkably, in spite of their unusual role in maintaining lens transparency for years on end, crystallins are not highly specialized structural proteins. Instead they are generally proteins with non-lens functions which have acquired a new role in the lens by a novel process of gene recruitment through tissue-specific enhancement of gene expression (Wistow, 1993b). Some crystallins are quite ancient; their gene recruitment must have occurred in a common ancestor of all vertebrates and they are

Abbreviations: ALDH, aldehyde dehydrogenase; ASL, argininosuccinate lyase; ATPase, adenosine triphosphatase; cAMP, cyclic adenosine monophosphate; CAT, chloramphenicol acetyltransferase; CRE, cAMP-responsive element; DT-A, diphtheria toxin A; EMSA, electrophoretic mobility shift assays; αENO, α-enolase/τ-crystallin; GluTR, glutamyl-tRNA reductases; GST, glutathione S-transferase; hGH, human growth hormone; hsp, heat-shock protein; LDHB, lactate dehydrogenase B; LSR, lens specific region; NADPH, reduced nicotinamide adenine dinucleotide phosphate; NMR, nuclear magnetic resonance; OCD, ornithine cyclodeaminases; SDS-PAGE, sodium dodecyl sulphate polyacrylamide gel electrophoresis; shsp, small heat-shock protein; TK, thymidine kinase; uPA, urokinase-type plasminogen activator; ZPE, ζ protected element.

Biotechnology and Genetic Engineering Reviews — Vol. 12, December 1994
0264–8725/94/12/1–38 $20.00 + $0.00 © Intercept Ltd, P.O. Box 716, Andover, Hampshire SP10 1YG, UK

essentially ubiquitous in their modern distribution. These crystallins are either identical to, or derived from, proteins with a role in stress responses. Some of these proteins, the a-crystallins, are members of the small heat-shock protein family (Klemenz *et al.*, 1991b; de Jong, Leunissen and Voorter, 1993) and have chaperone-like functions (Horwitz, 1992) which could contribute to the protection and solubilization of other lens proteins. Other crystallins (*Figure 1*) have arisen more recently as evolutionary pressures acted on the visual systems of various species (Wistow, 1993b). These taxon-specific crystallins are enzymes, and most of them are products of a single gene which has enormously increased expression in lens while still serving the ancestral catalytic role with low expression in many tissues. Because of this phenomenon, the lenses of some natural species are already very rich sources of active enzymes. For example, crude soluble extract of duck lens has fully one-quarter the argininosuccinate lyase (ASL) activity of the purified human liver enzyme (Piatigorsky *et al.*, 1988), and as such is considerably more active than most commercial preparations of this enzyme.

Recent advances in the molecular biology of crystallins has led to the identification of tissue-specific or strongly tissue-preferred gene promoters, which are capable of directing the expression of transgenes in various species (see below and Piatigorsky and Zelenka, 1992). Although determination of the basis of tissue-specific expression has proved to be a more complex problem than expected, the targeting of genes to the lens is a practical reality. This article reviews the nature of those proteins which already serve as crystallins, the tools available for manipulation of gene expression in transgenic animals and considers the implications for biotechnology and gene therapy in this system.

Ubiquitous stress-protein crystallins

α-CRYSTALLINS: STRESS PROTEINS AND CHAPERONES

α-Crystallins are abundant in the lenses of all vertebrates (de Jong, Leunissen and Voorter, 1993). Indeed, in some species, including humans, the αA-crystallin gene may be the single largest contributor of protein in the lens. There are two similar α-crystallin genes, αA- and αB-crystallin which are unlinked in those species in which they have been localized (Wistow and Piatigorsky, 1988). The first suggestion of a non-lens role for α-crystallins arose when the sequences of *Drosophila* small heat-shock proteins (shsp) revealed close similarities with αB- (Ingolia and Craig, 1982) (and hence with the closely related αA-) crystallin. This relationship was later extended to the whole shsp family, which is ubiquitous in eukaryotes. The significance of this became clearer when αB-crystallin was detected in a variety of non-lens tissues, showing that it has a non-lens role (de Jong, Leunissen and Voorter ,1993). It was first detected in scrapie-infected hamster brain (Duguid, Rohwer and Seed, 1988) and since then elevated levels of αB-crystallin have been shown to be associated with a number of neurological disorders, including Alexander's disease (Iwaki *et al.*, 1989), Lewy body disease (Lowe *et al.*, 1990) and Creutzfeldt–Jakob disease (Renkawek *et al.*, 1992). Indeed, αB-crystallin is now a valuable neural-tissue marker and antisera to αB-crystallin are commercially available. Lower amounts of αB-crystallin are also expressed constitutively in many tissues in both mammals and

Table 1. Crystallins

	Related or **identical**	Activity	Cofactor	Species
Ubiquitous stress-protein crystallins				
αA, αB	**Small heat-shock protein**	Solubilize heat-stressed protein	None	All vertebrates
βA1–4, βB1–3, γA–F, γS	*Micro-organism spore proteins,* M. xanthus *protein S,* P. polycephalum *spherulin 3a*	?	None	All vertebrates: γA–F not in birds
Taxon-specific enzyme crystallins				
δ1, δ2	**Argininosuccinate lyase**	δ2 has ASL activity	None	Most birds, reptiles
ε	**Lactate dehydrogenase B**	LDH activity	NADH	Many birds, crocodiles
ζ	**NADPH: quinone oxidoreductase**	Quinone reductase activity	NADPH	Hystricomorphs, camels
η	**Cytosolic aldehyde dehydrogenase**	?	NAD+ ?	Elephant shrews
λ	*Hydroxyacyl-CoA dehydrogenase*	?	NADH ?	Rabbits, hares
μ	*Ornithine cyclodeaminase glutamyl-tRNA reductase*	?	NADPH	Some marsupials
ρ	*Aldo-keto reductases*	?	NADPH	Frogs (*Rana*)
τ	**α-enolase**	Low enolase activity	None	Some reptiles, birds, several other species
Invertebrates				
S	*Glutathione-S transferase*	Weak GST activity in some in some subunits	GSH	Octopus, squid
Ω	*ALDH*	No ALDH activity detected	NAD+ ?	Octopus

birds (see, for example, Dubin, Wawrousek and Piatigorsky, 1989; Iwaki, Kume-Iwaki and Goldman, 1990; Kato *et al.*, 1991b; de Jong, Leunissen and Voorter, 1993; Lee, Kim and Wistow, 1993). Most significantly, it is induced by heat-shock in NIH 3T3 cells (mouse fibroblasts) (Klemenz *et al.*, 1991b), showing that it is in fact a small heat-shock protein; it can also be induced by other insults such as osmotic stress (Dasgupta, Hohman and Carper, 1992), ischaemia (Chiesi, Longoni and Limbruno, 1990) and by some proto-oncogenes (Klemenz *et al.*, 1991a). Inducibility of αA-crystallin has not yet been demonstrated, but the protein is present, albeit at very low levels in non-lens tissues, particularly in spleen and thymus (Kato *et al.*, 1991a), hinting at a possible role in the immune system.

The function of the shsp family is not well understood. Members of other heat-shock protein (hsp) families are known as chaperones and have dynamic roles in directing protein-folding pathways and the transport and assembly of multisubunit complexes (Hartl, Hlodan and Langer, 1994). The ubiquitous HSP70 proteins are ATPases with structural similarity to actin (Flaherty *et al.*, 1991). A chaperone- like role for the shsp family has now been suggested by observations that α-crystallins are capable of preventing the insolubilization of heat-denatured proteins (Horwitz, 1992). There is also some evidence to suggest that shsps, which have been identified as actin-depolymerizing proteins, and α-crystallins may have a role in assembly/disassembly of cytoskeleton in lens and in other cell types (Miron *et al.*, 1991; Nicholl and Quinlan, 1994).

α-Crystallins undergo a number of post-translational modifications, including cAMP-dependent phosphorylation (Voorter *et al.*, 1986; Chiesa, Gawinowicz-Kolks and Spector, 1987; Chiesa *et al.*, 1987; Voorter *et al.*, 1989), autophosphorylation (Kantorow and Piatigorsky, 1994) and cytoplasmic glycosylation (Roquemore *et al.*, 1992) which could influence their states of aggregation and other interactions. The quaternary structure of α-crystallins continues to be controversial with many competing models (see de Jong, Leunissen and Voorter, 1993 for references). It has been suggested that, given their apparent role as chaperones, α-crystallins and other shsp family members may have multiple modes of assembly and dynamic interchanges between alternative structures (Wistow, 1993c). This could explain some of the apparent contradictions in the biochemical and biophysical data. In spite of the growing interest in this family, there is little tertiary structural data. CD and ORD spectroscopy both suggest that α-crystallins have predominantly β-sheet structure (Li and Spector, 1967, 1974). Sequence comparison has recently suggested a region of similarity between shsp and HSP70 including a conserved α-helix (Lee *et al.*, 1993). However an X-ray structure determination is urgently needed.

β- AND γ-CRYSTALLINS: OSMOTIC-STRESS PROTEINS?

The other two recognized classes of ubiquitous crystallins, the β- and γ-crystallins, belong to the same protein superfamily, the βγ-crystallins (Driessen *et al.*, 1980; Wistow *et al.*, 1981). All vertebrates seem to have six or seven β-crystallin polypeptides (Ostrer and Piatigorsky, 1980; de Jong, 1981; Berbers *et al.*, 1984; Wistow and Piatigorsky, 1988). They form dimers and higher aggregates of mixed composition (Harding and Crabbe, 1984). Most species have a similar number of γ-crystallins which, in contrast, are strictly monomeric (Harding and Crabbe, 1984). The best-defined γ-crystallins are those found in mammals, which are principally expressed early in development so that they predominate in the central (or 'nuclear') regions of the lens consisting of cells of embryonic origin. This group of γ-crystallins is apparently completely absent from birds (Treton *et al.*, 1984). However bird lenses do contain one ubiquitous protein, now called γS-crystallin, which is a more distantly related member of the γ-crystallin family and which, in mammals, is expressed only later in development in the less-dense cortical regions of the lens (Harding and Crabbe, 1984; Quax-Jeuken *et al.*, 1985; van Rens, de Jong and Bloemendal, 1991). The confusion of identity for this protein is such that it was originally called βS-crystallin. Like β-crystallins, and unlike classic γ-crystallins, this protein has an N-terminal extension and is N-terminally acetylated.

The tertiary structures of β- and γ-crystallins are known in considerable detail thanks to a number of X-ray crystallographic studies. These were first carried out on bovine γII-crystallin (Blundell *et al.*, 1981; Wistow *et al.*, 1983; Summers *et al.*, 1984) (which now has the systematic name γB-crystallin). The protein has a remarkably symmetrical structure of two domains, each containing a further twofold repeat for a total of four characteristic structural motifs, modified 'greek-keys', with a distinctive pattern of required amino-acid residues. This same pattern is seen in the β-crystallins for which X-ray structures are now also available (Bax *et al.*, 1990). The existence of the structural signature in the sequence of this superfamily allowed the identification of two non-lens members. Both these proteins are expressed specifically in micro-organisms which are undergoing spore or cyst formation in response to stress, typically osmotic stress or dehydration. The first to be noticed was protein S of the bacterium *Myxococcus xanthus* which has the same four-motif, two-domain structure as the β- and γ-crystallins (Wistow, Summers and Blundell, 1985). The predicted structure of protein S has now been confirmed by an elegant NMR study (Bagby *et al.*, 1994a,b). The major difference between this structure and that of a γ-crystallin is that the interdomain contact in the bacterial protein is unsymmetrical and principally involves motifs 2 and 3, while that in γ-crystallin is highly symmetrical and involves motifs 2 and 4. The difference in interaction may relate to the calcium-binding properties of protein S. The second protein, spherulin 3a, is found in a eukaryotic slime mould, *Physarum polycephalum* (Wistow, 1990). Interestingly, this eukaryotic protein represents a more primitive stage of molecular evolution than its bacterial relative and the crystallins. Instead of four motifs and two domains it has only two motifs and one domain and seems to have missed out on the last internal duplication of sequence in the evolution of this superfamily. Intriguingly, an unusual feature links prokaryotic *M. xanthus* and eukaryotic *P. polycephalum*. Unlike more closely related sporulating species, both these, otherwise very distantly related, micro-organisms sporulate in response to osmotic stress induced by polyols, sugar alcohols. Thus both organisms share an unusual stress response and both express members of the βγ-crystallin superfamily under such stress. This has led to the suggestion that the ancestral non-lens role of the β- and γ-crystallins may have been as osmotic stress proteins (Wistow, 1990, 1993b).

This may also point to a continuing non-lens function of one or more of the modern β- or γ-crystallins. Recently, very sensitive RNA protection methods have detected non-lens expression of γ-crystallin family genes in *Xenopus* larvae, although there has been no detection of non-lens protein (Smolich *et al.*, 1994). There is stronger evidence of significant expression of β-crystallins, particularly βB2-crystallin, in non-lens tissues of newly hatched chickens, especially in retina, brain and kidney (Head, Peter and Clayton, 1991). The possibility of a non-lens function for some β-crystallins, perhaps as constitutive stress proteins in certain tissues, is therefore quite real.

In contrast, the main role of the γ-crystallins seems to be as specialized lens proteins. Their fundamental role in the lens seems to be the maintenance of a stable, low-water, high protein concentration environment suitable for the densest, highest refractive index regions of the lens. They are most abundant in very hard lenses, such as those of fish or nocturnal, burrowing rodents such as mice and rats (de Jong, 1981; Harding and Crabbe, 1984). They are much less abundant, or even absent, from softer lenses of diurnal, terrestrial species such as birds. Decreased γ-crystallin content may be due to reduced gene expression, as in humans (Russell *et al.*, 1987; Brakenhoff *et al.*, 1990), or gene loss, as in birds (Treton *et al.*, 1984). However, γ-crystallin content may also be reduced by another mechanism, namely the 'dilution' of γ-crystallins in the lens fibre cells by overexpression of other proteins through the direct gene recruitment of metabolic enzymes to a new structural role.

Gene recruitment: taxon-specific enzyme crystallins

Vertebrates with eyes originally evolved under water. This optically dense medium necessitates lenses with high refractive indices, just as is observed in lenses of fish and those produced by convergent evolution in cephalopods such as squid and octopus. However, when vertebrates emerged from water on to the land during the course of evolution there would have been strong pressure to modify the properties of their hard aquatic lenses for this new, low refractive index environment. Lenses with the high refractive index most useful under water would not be able to focus more distant objects, such as food or predators, in air. This has been hypothesized to be the basis of the remarkable diversity of taxon-specific crystallins in vertebrates (Wistow, 1993b) (*Table 1, Figure 1*). Different lineages of modern vertebrates must have made multiple independent modifications to their repertoire of crystallins in response to selective pressures of environment and habit.

Frequently this adaptation was achieved by the direct gene recruitment of enzymes as crystallins (*Figure 2*). In this novel evolutionary process, genes already being expressed in the lens for metabolic or other reasons underwent modification in their expression specifically in the lens. This led to a huge increase in the concentration of the recruited protein in the lens and thereby to the acquisition of a new structural role. It is important to recognize that after recruitment the same gene continues its normal pattern of expression in other tissues. This means that a recruited enzyme crystallin becomes a dual functional protein, retaining its well-adapted, ancient, catalytic role while, at the same time, acting as a crystallin, interacting with other proteins at high concentration to maintain lens transparency. These dual roles subject the protein to two different sets of evolutionary selective pressures. In some cases, as for example seen in lactate dehydrogenase B/ε-crystallin (Wistow, Mulders and de Jong, 1987; Hendriks *et al.*, 1988; Wistow, Anderson and Piatigorsky, 1990), the new role in the lens may drive unusual protein sequence modifications in the enzyme. It has been suggested that such competing selective pressures can lead to an adaptive conflict, in

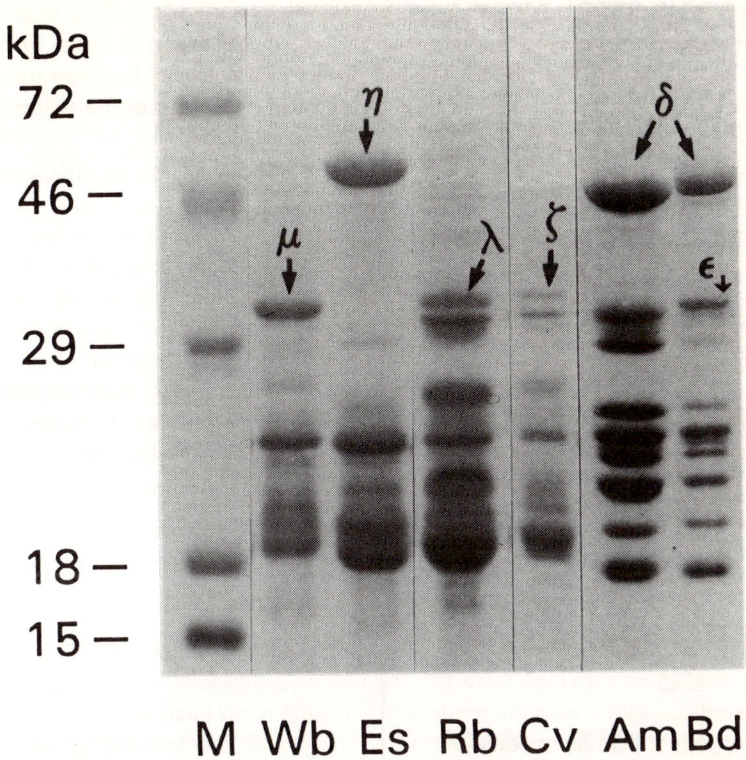

Figure 1. Diversity and similarity of abundant lens crystallins. SDS-PAGE of total soluble protein (Wistow, 1993a) from lenses of several species. Taxon specific crystallins are marked (see *Table 1*). Species are: Wb, tamar wallaby; Es, elephant shrew; Rb, rabbit; Cv, rock cavy; Am, American merganser; Bd, black duck.

which changes beneficial for one role are deleterious for the other (Wistow, Anderson and Piatigorsky, 1990; Wistow, 1993b). This can be solved in various ways, including loss of the recruited function (revision) or, more rarely, by gene duplication followed by specialization and separation of enzyme and crystallin functions.

RECRUITED ENZYMES

The enzymes which have so far been identified as crystallins are listed in *Table 1*. The archetype of this phenomenon is ε-crystallin. This protein is extremely abundant in the lenses of many birds and crocodiles, two groups which are closely related (Stapel *et al.*, 1985). ε-Crystallin accounts for as much as half the total protein in the lens of a humming-bird (*Calypte ana*) (Wistow, Anderson and Piatigorsky, 1990), and is 10–25% of total protein in other species, particularly sea-birds, waterfowl and others that hunt their food in bright light (Stapel *et al.*, 1985; Wistow, Mulders and de Jong, 1987). When ε-crystallin from duck (*Anas platyrhynchos*) lens was partially sequenced, it became clear that it was very closely related to lactate dehydrogenase B (LDHB), the predominant heart isoform of this glycolytic enzyme (Stapel *et al.*, 1985; Wistow, Mulders and de Jong, 1987). Indeed, peptide sequences of authentic heart LDHB showed identity to the lens protein (Wistow, Mulders and de Jong, 1987) and, later,

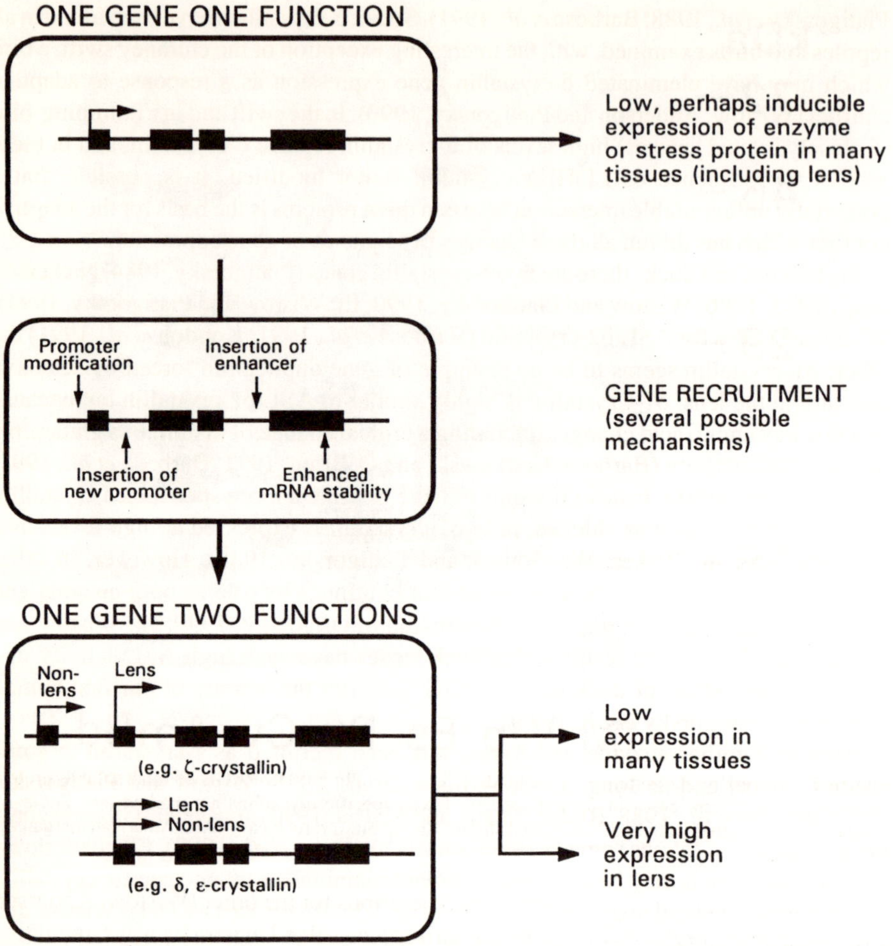

Figure 2. The gene recruitment of crystallins. A schematic gene is shown. Boxes are exons.

cloning experiments confirmed that the heart and lens proteins are the products of the same single gene (Hendriks *et al.*, 1988). In its new role in lens, LDHB/ε-crystallin must survive without turnover for very long periods, even years, while almost continually exposed to light and while interacting closely with other lens components. Peptide sequencing reveals that there is some post-translational modification of ε-crystallin in the lens as two asparagine residues become deamidated, probably as result of ageing in the lens (Wistow, Mulders and de Jong, 1987; Hendriks *et al.*, 1988). However, in spite of all this, duck lens LDHB/ε-crystallin is almost as enzymatically active *in vitro* as the purified heart enzyme (Wistow, Mulders and de Jong, 1987).

Duck lens is an even more impressive source of another enzyme, argininosuccinate lyase (ASL). This enzyme is usually associated with the urea cycle in mammals but is also expressed in non-ureotelic tissues such as those of the eye (Rao and Cotlier, 1984) and may be involved in various pathways such as nitric oxide synthesis (Hecker *et al.*, 1990). In birds, ASL is identical to δ2-crystallin (Wistow and Piatigorsky, 1987;

Piatigorsky *et al.*, 1988; Barbosa *et al.*, 1991). So far, δ-crystallins have been found in all reptiles and birds examined, with the interesting exception of the chimney swift, a bird which may have eleminated δ-crystallin gene expression as a response to adaptive conflict (Wistow, Anderson and Piatigorsky, 1990). In the swift and in a humming-bird (a closely related species) high levels of ε-crystallin replace δ-crystallin, and in these species the sequence of LDHB/ε-crystallin is not modified. It is possible that a potentially unfavourable interaction between these proteins is the basis for the adaptive conflict which has driven all these changes in sequence or gene expression.

In chickens and ducks there are two δ-crystallin genes (Piatigorsky, 1984; Nickerson *et al.*, 1985, 1986; Wistow and Piatigorsky, 1990; Li, Wistow and Piatigorsky, 1994). While one codes for ASL/δ2-crystallin (Barbosa *et al.*, 1991; Kondoh *et al.*, 1991) the other, δ1-crystallin seems to be an example of gene duplication forced by selective pressure in the lens. δ1-Crystallin is highly similar to ASL/δ2-crystallin but because of a few key amino acid changes, including a critical change of histidine to glutamine, has lost ASL activity (Barbosa, Cialkowski and O'Brien, 1991; Barbosa *et al.*, 1991; Kondoh *et al.*, 1991). Instead it seems to have become a more specialized crystallin. In some species, such as chicken, only δ1-crystallin is expressed at high levels as a structural protein (Parker, Wawrousek and Piatigorsky, 1988). However, in other birds, for example ducks, ASL/δ2-crystallin continues its role as both enzyme and crystallin (Piatigorsky *et al.*, 1988; Wistow and Piatigorsky, 1990; Li, Wistow and Piatigorsky, 1994). The result is that duck lenses have such high ASL activity that crude soluble extract of duck lens has fully a quarter the activity of purified human liver ASL (Piatigorsky *et al.*, 1988).

Another familiar enzyme, α-enolase, has been recruited as τ-crystallin in some species (Stapel and de Jong, 1983; Williams *et al.*, 1985; Wistow and Piatigorsky, 1987; Wistow *et al.*, 1988; Rudner, Katar and Maisel, 1990) and, at least in embryonic turtle lenses, also retains some enzyme activity (Wistow *et al.*, 1988). Elephant shrew η-crystallin, which may be the most abundant mammalian taxon-specific crystallin, is apparently identical to cytoplasmic aldehyde dehydrogenase (ALDH) (Wistow and Kim, 1991). However, no significant ALDH activity could be measured in lens extracts of elephant shrews obtained *post mortem* from zoos. Yet this could easily be due to the relative lack of freshness of these samples and the low inherent activity of the enzyme which, furthermore, has an active-site cysteine residue which could be sensitive to oxidation (see Weiner, 1990). Now that elephant shrew η-crystallin has been cloned (Graham, Hodin and Wistow, in preparation), the recombinant protein is being expressed in *Escherichia coli* to see if ALDH activity can be detected in freshly synthesized protein. It is also possible that η-crystallin, like δ1-crystallin, is the result of gene duplication and specialization, with subsequent loss of enzyme activity. However, sequence analysis shows that, unlike δ1-crystallin, all the residues required for enzyme activity (see Weiner, 1990) are present in η-crystallin (Graham, Hodin and Wistow, in preparation).

NOVEL ENZYMES

Although some taxon-specific crystallins were easily identified as familiar enzymes, others seem to be novel enzymes discovered for the first time as crystallins. ρ-Crystallin (originally named ε-crystallin) which is found in the lenses of frogs of the genus *Rana* (Tomarev *et al.*, 1984), belongs to an aldo-keto reductase superfamily which includes aldehyde and aldose reductase, prostaglandin F synthase and several detoxification enzymes (Carper *et al.*, 1987, 1989; Watanabe *et al.*, 1988; Bohren *et al.*, 1989). ρ-Crystallin does bind NADPH, like other members of the superfamily, but its particular enzymatic specificity is not known (Carper *et al.*, 1989). Similarly, λ-crystallin, found in rabbits and hares, is another previously unknown enzyme. It is distantly related to hydroxyacyl- and hydroxybutyryl-CoA dehydrogenase and conserves the NADH-binding site, but its activity, if any, has not been determined (Mulders *et al.*, 1988).

A particularly interesting example of a novel enzyme discovered first as a crystallin is μ-crystallin. This protein may account for as much as 25% of total protein in the lenses of some Australian marsupials, specifically in macropods (kangaroos) and the one dasyurid (*Dasyurops maculata*) examined (Wistow and Kim, 1991). When the complete sequence of kangaroo lens μ-crystallin was determined by cDNA cloning (Kim, Gasser and Wistow, 1992) it became apparent that it has no close relatives among eukaryotic protein sequences. However, it is significantly similar (over 30% identical in amino-acid sequence) to ornithine cyclodeaminases (OCD) of the bacterium *Agrobacterium tumefaciens*. The similarity includes a putative ornithine binding site (Sans, Schindler and Schröder, 1988). OCD catalyses the conversion of ornithine to proline in one step in the presence of NAD^+ (Sans, Schindler and Schröder, 1988). Kangaroo μ-crystallin also is capable of binding pyridine dinucleotide cofactors but its highest affinity is for NADPH (Wistow and Segovia, 1994), in contrast to bacterial OCDs (Sans, Schindler and Schröder, 1988). *In vitro* assays of μ-crystallin with ornithine, proline and all four pyridine dinucleotide cofactors have so far revealed no enzymatic activity. However, μ-crystallin seems to be the product a single gene which is expressed in other tissues, presumably in an enzymatic role (Kim, Gasser and Wistow, 1992). In kangaroos the primary non-lens tissues containing μ-crystallin mRNA are retina and brain (Kim, Gasser and Wistow, 1992). μ-Crystallin is also present at enzymatic levels in other species. It has been cloned from human retina (Kim, Gasser and Wistow, 1992; Segovia and Wistow, in preparation) and seems to be expressed preferentially in photoreceptors in humans, rats and chickens (Wistow and Segovia, 1994). We have recently noticed that μ-crystallins and OCDs are both related to another family of enzymes involved in unusual amino-acid metabolism, the glutamyl-tRNA reductases (GluTR) (Segovia and Wistow, in preparation). These enzymes convert glutamyl-tRNA to a glutamate semialdehyde (Jahn, Verkamp and Söll, 1992) and use NADPH as cofactor. Since glutamate is the major neurotransmitter of the photoreceptors (Barnstable, 1993) there is the interesting possibility that μ-crystallin could be involved in the metabolism of glutamate or its derivatives and could play an important role in the visual process.

Finally, there is one example of a taxon-specific crystallin whose novel enzymatic function has been uncovered. ζ-Crystallin is found in guinea-pigs, some related animals and also in camels and llamas (presumably a rare example of independent recruitment of the same gene in two different lineages) (Huang *et al.*, 1987; Garland *et al.*, 1991; Lee *et al.*, 1993). Sequence analysis showed that ζ- crystallin is a distantly related member of the alcohol dehydrogenase superfamily (Borras, Persson and Jornvall, 1989; Rodokanaki, Holmes and Borras, 1989). Cofactor-binding studies showed further that ζ- crystallin binds NADPH, suggesting a role as a reductase (Rao and Zigler, 1990). This led to an heroic effort to test substrates for the putative enzyme. Remarkably, an activity was found. ζ-Crystallin turns out to be a novel NADPH:quinone oxidoreductase (Rao, Krishna and Zigler, 1992). Unlike the familiar quinone reductases, ζ-crystallin operates through a single electron transfer mechanism.

INVERTEBRATE CRYSTALLINS

Lenses are also found in a variety of invertebrate eyes. In cephalopods, such as octopus and squid, the lenses are cellular and superficially similar to those of vertebrates. The principal structural proteins of these lenses, S-crystallins, are related to the enzyme glutathione S-transferase (GST) (Wistow and Piatigorsky, 1987; Tomarev and Zinovieva, 1988). Most do not have GST activity, but some minor subunits in the lens do seem to have detectable activity (Tomarev *et al.*, 1994). The recruitment events which led to cephalopod lenses may have been much more ancient than those in vertebrates. Consequently, there may have been more tissue-specific specialization over time. A minor component of octopus lenses has been named Ω-crystallin (Chiou, 1988). This protein has been cloned (Tomarev, Zinovieva and Piatigorsky, 1991) and shown to be related to aldehyde dehydrogenases. This draws a parallel with η-crystallin in elephant shrews, although Ω-crystallin has not conserved all the residues thought to be essential for enzymatic function (Tomarev, Zinovieva and Piatigorsky, 1991). This may be another example of greater specialization and more ancient recruitment in cephalopods. Remarkably, another ALDH-related protein has been recruited to an apparent structural role in another lens-like tissue. Some squid have an organ which emits light, probably for signalling or camouflage purposes. This organ has a diffusing lens derived from muscle tissue. The principal component of this tissue has been named L-crystallin and like η- and Ω-crystallins belongs to the ALDH superfamily (Montgomery and McFall-Ngai, 1992). The coincidence of multiple recruitment events for ALDH-like proteins is interesting. It suggests that these enzymes are easy to recruit under the conditions which give rise to lenses. As suggested below, this could relate to a role for ALDH in elongating cells, perhaps involving osmoregulation or interactions with cytoskeleton.

Some jellyfish also have cellular lenses. The major lens proteins of some cubomedusan jellyfish, named J-crystallins, have been cloned (Piatigorsky *et al.*, 1989). Sequencing revealed no obvious identity with known proteins, although there is a weak but intriguing similarity to the HSP60 family of molecular chaperones (Wistow, unpublished). N-terminal protein sequence has also been obtained for the major proteins of the eye lens from *Aplysia californica* (Cox, Glick and Strumwasser, 1991). The limited sequence available does not reveal the relatedness, if any, of these proteins. Finally, a protein of the extracellular corneal lens of *Drosophila* has been partially characterized and named drosocrystallin (Komori, Usukura and Matsumoto, 1992). Again, it is not clear what the superfamily relationships of this protein might be.

A common origin for crystallins?

While crystallins are clearly quite diverse, it is striking that they fall into only two rather narrow categories (Wistow, 1993b). Even the taxon-specific crystallins are restricted to 35–54 kDa subunit metabolic enzymes, all with small molecule substrates, products and cofactors. Many of the enzymes bind NAD(P)H (*Table 1*). There are probably functional constraints on what kind of proteins will work well as crystallins. Size, thermodynamic stability and an ability to interact with other proteins to maintain short-range order (Delaye and Tardieu, 1983) without superaggregating (thereby creating centres for light-scattering) are all important characteristics. However, it is also possible that existing crystallins share common origins in some basic process of lens development.

It has been pointed out that all cellular lenses undergo enormous increases in cell volume and elongation (Wistow, 1993b). This must involve massive reorganization of cytoskeleton and either proactive or reactive control of osmolarity. Both groups of stress-protein crystallins have plausible connections with osmoregulatory events or with manipulation of cytoskeleton. αB-crystallin is induced by osmotic stress (Dasgupta, Hohman and Carper, 1992) and the βγ-crystallins are related to proteins which are also induced under similar conditions (Wistow, Summers and Blundell, 1985; Wistow, 1990). A protein isolated from turkey gizzard as an actin depolymerizing protein was found to be a small heat-shock protein related to α-crystallins (Miron *et al*., 1991), suggesting that proteins of this class may have active roles in manipulating cytoskeleton. Indeed, there have been many observations of α-crystallin association with cytoskeleton and membrane fractions in lens, and it has recently been shown that α-crystallins exert a chaperone-like effect in assembly of cytoskeleton in lens (Nicholl and Quinlan, 1994).

In addition, several of the enzymes recruited as crystallins are able to bind substrates which are known to be potent osmolytes in certain systems, such as lactate and proline (see Kim, Gasser and Wistow, 1992 for references). ρ-Crystallin is related to aldose reductase, an enzyme which is expressed at high levels in the kidney and in the differentiating (and elongating) cells of the rat eye lens (Bondy and Lightman, 1989), which may both be subject to osmotic stress. There are also a number of examples in which enzymes, particularly glycolytic enzymes, are associated with cytoskeleton, forming a cytomatrix (Clarke *et al*., 1985; Knull and Walsh, 1992). Indeed, in *Tetrahymena*, intracellular structures described as 14 nm filaments were found to be composed of the enzyme citrate synthase (Numata *et al*., 1991), again showing that enzymes can serve as structural proteins. All this suggests the possibility that crystallins may represent proteins involved in important processes of cell elongation. Even in the absence of recruitment as crystallins, lens cells would need to express these proteins in order to be able to differentiate into fibre cells. The genes for such proteins with a fundamental role in an essential cellular process would necessarily be active in the lens and should thus be relatively facile targets for enhanced expression as crystallins. Higher expression of proteins capable of, for example, protecting cytoskeleton or exerting other protective effects, could indeed by beneficial for the long-term stability of the lens.

Gene expression: targeting the lens

Whatever their origin, all crystallins share the characteristic of very high gene expression in the lens. Several crystallin gene promoters and enhancers have been analysed in some detail. What follows is a summary of promoter elements which have so far been identified in the expression of crystallin genes and a review of the transgenic animals which have been produced using these promoters.

αA-CRYSTALLIN

The αA-crystallin gene is expressed predominantly in the lens of all species examined. Promoter fragments from mouse (Chepelinsky, Sommer and Piatigorsky, 1987; Jaworski, Chepelinsky and Piatigorsky, 1991), hamster (van den Heuvel *et al.*, 1985), mole rat (Hendriks *et al.*, 1987), chicken (Thompson, Hawkins and Piatigorsky, 1987) and human (Jaworski, Chepelinsky and Piatigorsky, 1991) have been cloned and sequenced, and extensive regions of 5'- flanking regions have been examined in the mouse, human and chicken (1.8, 1.8 and 2.6 kbp, respectively).

The mechanisms regulating αA-crystallin gene expression in these different species appear to be similar, since mouse and chicken promoters show appropriate expression in a variety of heterologous systems. αA-Crystallin transcription has been studied using chick lens epithelial cells, an SV40-transformed cell line derived from mouse lens (αTN4-1 cells), a rabbit lens cell line (N1003A) and in transgenic mice. In general, the same promoter regions are implicated for function regardless of whether the transcription system is derived from rabbit, mouse or chicken. Despite this apparent evolutionary conservation of transcriptional control mechanisms, analysis of the promoters has revealed striking differences (*Figure 3*).

Global alignments of promoter sequences for mouse, human and chicken display only limited similarity; only four regions of significant sequence matches can be aligned in all three species (Jaworski, Chepelinsky and Piatigorsky, 1991; Sax and Piatigorsky, 1994). A number of short sequence motifs are common to all three species, but these are distributed differently with regard to location, orientation and copy number in each. An unexpected similarity among the αA-crystallin promoters is the presence of a repetitive element, characteristic of each species, at approximately 1 kbp upstream of the coding regions (Jaworski, Chepelinsky and Piatigorsky, 1991).

Definition of minimal promoter sequences for the mouse and chicken genes by transfection studies have established that –111 bp are sufficient for activity of the mouse promoter (Chepelinsky, Sommer and Piatigorsky, 1987), but a longer region, extending to –162, is necessary for functional activity of the chicken promoter (Yasuda *et al.*, 1988; Klement, Wawrousek and Piatigorsky, 1989). Other methods confirm the need for a greater extent of distal sequences in the chicken (Cooper, Constable and Davidson, 1984; Wawrousek *et al.*, 1990).

Functional elements of the mouse αA-crystallin promoter have been investigated in numerous studies. Genomic footprinting in αTN4-1 cells and *in vitro* footprinting with lens nuclear extracts showed corresponding patterns of protected regions (Kantorow *et al.*, 1993). Similar regions were also footprinted with nuclear extracts from L929 fibroblasts. Although the same sites were occupied, electrophoretic mobility shift assays (EMSA) demonstrated that different proteins were bound in the non-lens environment. The same promoter region may serve as an activating element in the lens and an inhibitory element in non-lens cells through the selective binding of different factors.

The activity of the mouse promoter was originally shown to depend on two

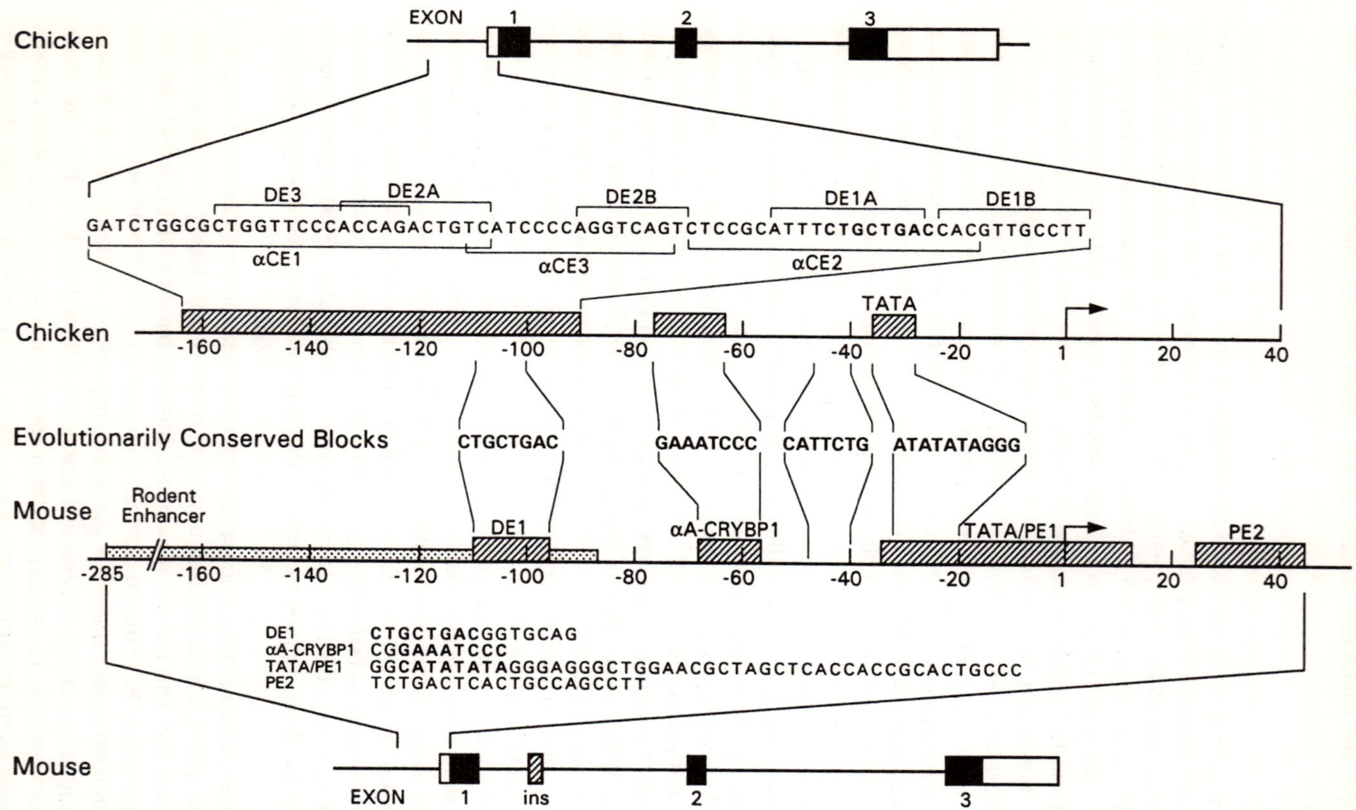

Figure 3. The complex promoters of αA-crystallin genes from chicken and mouse. Identified functional elements are shown together with evolutionarily conserved blocks which occupy similar location in both genes. Hatched boxes indicate proposed binding sites for *trans*-acting factors. Sequences of identified regions are shown above and below the promoter schematics. The gene structures for bird and rodent αA-crystallins are also shown. Coding sequences of exons are shown in black. Some mammals have an alternatively spliced insert exon (ins) (see de Jong, Leunissen and Voorler, 1993).

interacting domains, a distal element (–111/–88), and a proximal element (–88/–60) (Chepelinsky, Sommer and Piatigorsky, 1987). Further definition of the proximal element led to the expression cloning of the transcription factor αA-CRYBP1 and αTN4-1 cells (Nakamura *et al.*, 1990). αA-CRYBP1 binds to the sequence 5'-GGGAAATCCC-3' at positions –66/–57 in the mouse promoter, overlapping one of the evolutionarily conserved sequences. The αA-CRYBP1 site is similar to enhancers in that it functions independently of orientation; but, unlike enhancers, spatial location is important to αA-CRYBP1, and the extent of activation is low (threefold) (Sax, Kelement and Piatigorsky, 1990). Enhancer activity of the αA-CRYBP1 site in transfection studies appears to operate only in αTN4-1, the mouse lens-derived cell line from which it was cloned, but multimerization of the site allows activity in non-murine and non-lens cells (Sax, Klement and Piatigorsky, 1990). Among the mutations of the αA-CRYBP1 site which abolish its activity is the change of GGG to GAG (Sax, Klement and Piatigorsky, 1990), which introduces the variant of the αA-CRYBP1 site found in humans, chickens and mole rats (Hendriks *et al.*, 1987; Jaworski, Chepelinsky and Piatigorsky, 1991). Thus, αA-CRYBP1 recognizes a site present in the mouse but not in the chicken or human promoters; nevertheless, the homologue of αA-CRYBP1 has been identified in chicken lens (Donovan *et al.*, 1992).

Other regions important for transcriptional activity include DE1 (a portion of the distal element spanning –111/–97), TATA/PE- 1 (–32/+12) and PE-2 (+24/+43). Like the αA-CRYBP1 site, the DE1 site overlaps a sequence block completely conserved in mouse, chicken and human αA-crystallins; PE-1 and PE-2 correspond to regions well conserved among mammals (Jaworski, Chepelinsky and Piatigorsky, 1991; Sax and Piatigorsky, 1994). DE1 lies within a rodent αA enhancer element, defined in the hamster as –180/–85 (Chepelinsky, Sommer and Piatigorsky, 1987; Yu, Tsui and Breitman, 1990), that can activate the minimal promoter for γF-crystallin. Interestingly, DE1 and αA-CRYBP1 appear to be functionally redundant sites in transgenic mouse experiments (Sax, Ilagan and Piatigorsky, 1993), since promoter activity is abolished only if both sites are simultaneously mutated.

As noted above, more upstream sequences, extending to –162, are required for activity of the chicken αA-crystallin promoter. This segment can act as an enhancer in a heterologous system using the minimal promoter for β-actin (Matsuo and Yasuda, 1992). Nearly the entire length of the enhancer region is covered by a series of neighbouring and overlapping sequences that have been implicated as protein-binding sites or functional *cis*-acting elements; varying nomenclature has arisen for these elements, depending on the laboratories and techniques used to identify them (*Figure 3*).

The linked elements DE1A and DE1B of the chicken gene (Klement, Cvekl and Piatigorsky, 1993), which coincide with the element αCE2 (Jahn, Verkamp and Söll, 1992), contain the same conserved sequence block covered by DE1 of the mouse αA-crystallin promoter. Both the mouse DE1 and chicken DE1A/B bind nuclear extracts in EMSA, but the mouse element fails to compete with that of the chicken for binding of chicken nuclear extracts (Klement, Cvekl and Piatigorsky, 1993). Like the analogous mouse sequence, the chicken DE1A/B element appears to co-operate with other promoter regions for activity, since a requirement for interaction between the regions defined as αCE2 and αCE1 has been demonstrated (Matsuo and Yasuda, 1992). A protein, αCE1, has been identified that binds specifically to the αCE1 element

(Matsuo *et al.*, 1991; Matsuo, Takeuchi and Yasuda, 1992). Recent work has suggested that several other factors, including USF, CREB and Pax6, bind to the chicken αA-crystallin promoter (Cvekl *et al.*, 1994). Pax6 binding may have particular significance for lens-specific expression (see below).

There are many differences between the αA-crystallin promoters of mouse and chicken. Substantially different minimal sequence lengths are required, and the extent of sequence similarity is meagre. Furthermore, analyses of sequences that are conserved between the mouse and chicken (the αA-CRYBP1 and DE1 regions) show little functional correspondence. The lack of concordance between the mouse and chicken αA-crystallin promoters is intriguing in view of the fact that each apparently functions appropriately in a heterologous system. Currently, the predominant similarity between the mouse and chicken promoters is that each operates through a multiplicity of factors whose interactions co-ordinately regulate transcription.

αB-CRYSTALLIN

Unlike αA-crystallin, αB-crystallin is expressed at significant levels in non-lens tissues such as kidney, lung, cardiac and skeletal muscle and in trace amounts in brain and spleen, in addition to high levels in lens (Duguid, Rohwer and Seed, 1988; Dubin, Wawrousek and Piatigorsky; 1989; Iwaki *et al.*, 1989; Iwaki, Kume-Iwaki and Goldman, 1990). There have not been extensive studies of the human αB-crystallin gene promoter, but it has been shown that a fragment from the promoter –537/+21 linked to the CAT (chloramphenicol acetyltransferase) reporter gene expressed well in lens cells but failed to express in an astrocytoma cell line, suggesting that nerve-specific elements may lie elsewhere (Dubin *et al.*, 1990).

Most functional studies of the αB-crystallin promoter have focused on the mouse gene (*Figure 4*). Primer extension of mouse lens αB-crystallin mRNA identifies a transcription start site 45 bp upstream of the ATG codon and 25 bp down from a TATA box (Dubin, Wawrousek and Piatigorsky, 1989). The same start site is used in heart, skeletal muscle and kidney. A minor transcription start site at –40 to –50 is also used in heart and skeletal muscle. αB-crystallin is induced by heat shock in NIH 3T3 cells and a perfect heat-shock consensus sequence is found at position –53 to –39 in the mouse and human genes (Klemenz *et al.*, 1991b).

A fragment of the mouse gene promoter (–661 to +44) linked to CAT was shown to express in lens and skeletal muscle in transgenic mice when present as one copy, while in mice with multiple copies significant CAT activity was also detected in heart and other tissues (Dubin *et al.*, 1991). Transfection experiments in mouse muscle and lens cells with deletion mutants of this construct showed that the region –426 to –257 is necessary for expression in myotubes while the sequence downstream of –115 is essential for lens expression, albeit at lower levels than for longer promoter fragments (Dubin *et al.*, 1991). Further study of this –426 to –257 fragment linked to the heterologous TK (thymidine kinase) promoter expressing human growth hormone (hGH) as reporter, defined this region as a transcriptional enhancer which stimulated hGH production in differentiated myotubes, but showed less stimulation in lens cells (Dubin *et al.*, 1991). DNase I footprint analysis showed protection of three areas of the αB promoter: αBE-1 (–407 to –397), αBE-2 (–360 to –327) and αBE-3 (–317 to

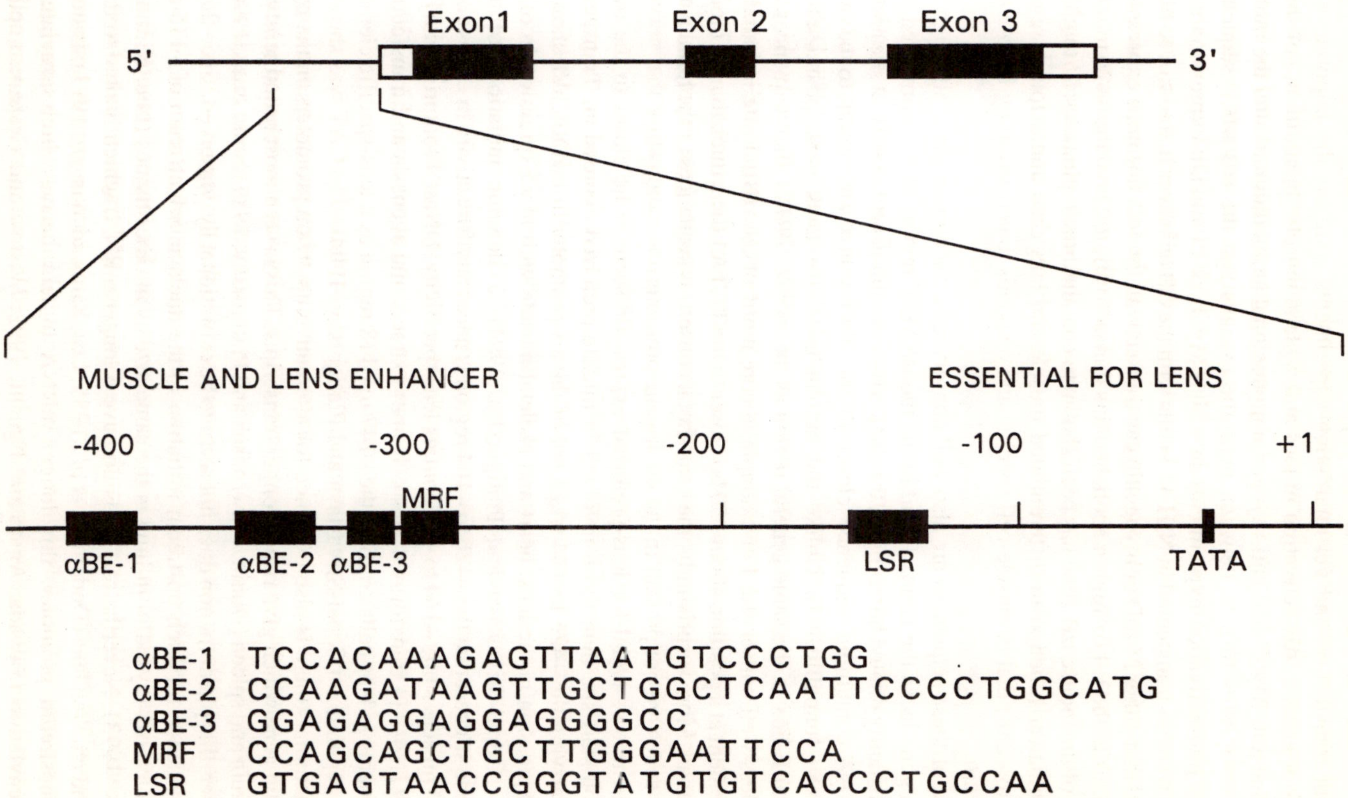

Figure 4. Functional elements of the mouse αB-crystallin gene promoter. The gene structure common to bird and mammal αB-crystallins is shown with coding sequences of exons in black. LSR, lens specific region.

–306) with myotube and αTN4 extracts, and a fourth region, MRF (–300 to –288), with myotube extract only (Gopal-Srivastava and Piatigorsky, 1993). This latter region contains a consensus E-box sequence, CANNTG. Site mutagenesis with transient transfection and gel retardation experiments confirms the importance of these four areas in αB activation in lens and skeletal muscle. In addition, gel shifts show that anti-MyoD and anti-myogenin monoclonal antibodies can alter the mobility of complexes formed between myotube extracts and the αB MRF sequence. Finally, transactivation experiments in which MyoD or myogenin expression constructs were co-transfected in NIH 3T3 cells with the αB enhancer linked to TK-hGH showed that either MyoD or myogenin can transactivate growth hormone expression. Mutations in the E-box region abolished this transactivation. Thus, the –426 to –257 region of the mouse αB gene has been shown to contain three regions (–407 to –306) with enhancer function in differentiated muscle and lens cells and a fourth region (–300 to –288) that is muscle-cell specific and responds to activation by MyoD and myogenin.

Most of these sequences are also well conserved in the human gene, although the human gene lacks the E-box found in the mouse MRF region. When compared to a more distantly related homologue, the duck αB-crystallin gene, the αBE-2 site and an AATAAA-containing sequence closer to the TATA box are found to be well conserved (unpublished). Unlike the human gene, the duck gene possesses an identical E-box in the same general region as the mouse MRF. Other sequences are much less well conserved. For example, a core motif of the αBE-1 site is found in complement at a position about 150 bp closer to the TATA box in duck than in mouse and human. More surprisingly, the heat-shock consensus sequences of the human and mouse genes are not present in the duck gene promoter.

Sequences essential for lens-preferred expression seem to lie closer to the transcription start site than those involved in muscle-preferred expression. Transgenic mice carrying the –426 to +44 fragment of the aB promoter linked to CAT showed high expression in the lens, heart and skeletal muscle with low expression in other tissues (Gopal-Srivastava and Piatigorsky, 1994). 5' deletion mutations of this transgene suggest that the –426 to –164 region is necessary for activity in cardiac and skeletal muscle while –164 to +44 confers lens specificity. DNase I footprint analysis using αTN4-1 (a transformed lens cell line) extracts, mutagenesis and transfection experiments in lens cells confirm the –147 to –118 region as a lens-specific element for αB regulation (Gopal-Srivastava and Piatigorsky, 1994).

In mammals there is also evidence for alternative upstream promoters in the αB-crystallin gene which give rise to longer transcripts. This is seen most prominently in lung, brain and spleen (Dubin, Wawrousek and Piatigorsky, 1989; Iwaki *et al.*, 1990). However, this phenomenon does not seem to be evolutionarily conserved, since there is no evidence for such upstream initiation in the duck, in which the predominant mRNA for αB-crystallin in lung is the same size as in other tissues (Lee, Kim and Wistow, 1993). Nevertheless, birds do have a longer mRNA, which is observed in duck and chicken lens (Dodemont *et al.*, 1985; Lee, Kim and Wistow, 1993). Instead of an upstream promoter, this longer mRNA in lens derives from alternative polyadenylation (Sawada, Agata and Eguchi, 1992; Wistow and Graham, unpublished).

β-CRYSTALLINS

β-crystallins have not been studied as extensively as some other classes. However, there has been some promoter analysis for two β-crystallin genes, both from chickens. Several deletion fragments of the chicken βB1-crystallin gene from –434/+30 to –126/+30 were able to drive expression of the CAT reporter gene more efficiently in primary explants of embryonic chicken lens epithelial cells than in muscle fibroblasts or HeLa cells (Roth, Das and Piatigorsky, 1991). Four functional elements, PL-1, PL-2, OL-1 and OL-2, defined by transfection studies, footprinting and EMSA, were located between positions –126 and –53 of the promoter.

Two β-crystallin polypeptides, βA3- and βA1-crystallins, are encoded by a single gene in both mammals and birds (Hogg *et al.*, 1986; McDermott, Peterson and Piatigorsky, 1992). The βA3/A1-crystallin gene has a similar structure in both human and chicken. A promoter construct containing the fragment –382/+22 of the chicken gene promotes expression of the CAT reporter gene in chicken primary lens epithelial cells but not in dermal fibroblasts (McDermott, Peterson and Piatigorsky, 1992). Deletion of sequences between –382 and –143 greatly reduces promoter activity.

γ-CRYSTALLINS

In mammals, there are six tightly clustered γ-crystallin genes with standardized names γA–γF (Breitman *et al.*, 1984; Lok *et al.*, 1984; Moormann *et al.*, 1985; Graw *et al.*, 1993). These are the classic γ-crystallins which are an important part of the embryonic lens and which, as described above, are absent from bird lenses. The proximal promoters of these genes are generally well conserved and have strongly lens-preferred expression in various systems. In rodents all six γ-crystallins are induced as fibre cells form in the embryonic lens, and their expression decreases differentially after birth until only γB-crystallin transcripts are detectable in the adult rat lens (van Leen *et al.*, 1987; Goring, Breitman and Tsui, 1992). Since γ-crystallins appear to be highly specific for the terminally differentiated lens fibre cells, and these cells cannot be grown in culture, a variety of approaches have been adopted for promoter analysis. For example, the expression of the mouse γF-crystallin gene (formerly designated γ2) has been examined in detail in primary explants of chicken lens epithelial cells (Lok *et al.*, 1989) (*Figure 5*). Surprisingly, this has been successful even though chickens do not possess homologous γ-crystallin genes. In this system the γF gene has been shown to have a lens-specific promoter consisting of two upstream enhancer-like elements (which can be successfully substituted by similar regions of the hamster αA-crystallin gene) and a proximal promoter (Yu, Tsui and Breitman, 1990). At early stages in lens development either the enhancers or the proximal promoter can direct gene expression, while later in development, co-operation between these elements is required for expression in fibre cells (Goring *et al.*, 1993). Within the proximal promoter, close to the TATA box are sequences, showing no homology to other lens-specific elements, which are able to bind a factor, γF-1, which can specifically direct expression to adult fibre cells (Liu *et al.*, 1991; Goring *et al.*, 1993). Expression of the γF-crystallin promoter in explanted chicken embryonic lens epithelial cells is enhanced by retinoic acid, a potent inducer of cell differentiation, and a novel type of retinoic acid response element, γF-RARE, has been identified in the mouse γF-crystallin gene enhancer region (Tini *et al.*, 1993).

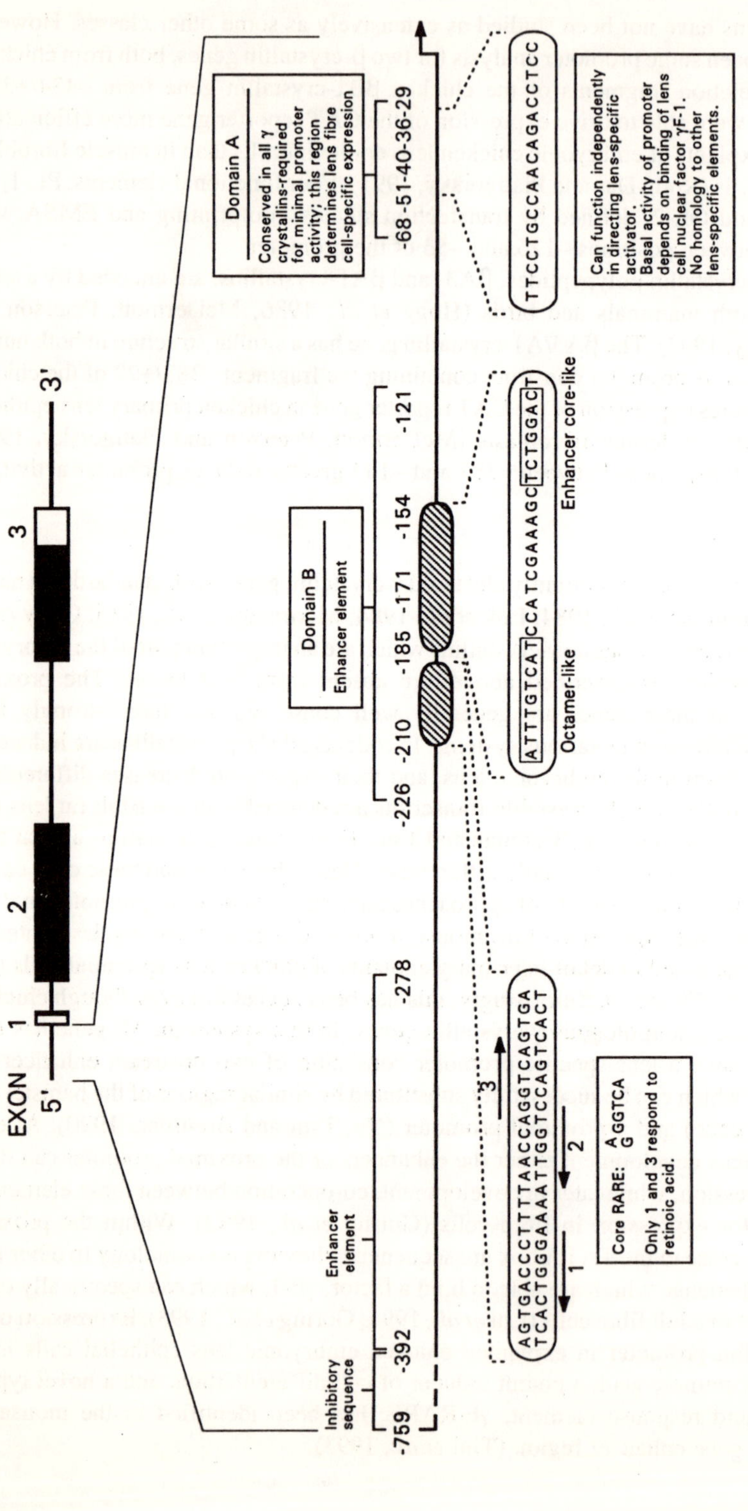

Figure 5. Functional elements of the promoter of the mouse γF-crystallin gene. The structure of a typical γ-crystallin gene is shown with coding sequences of exons in black.

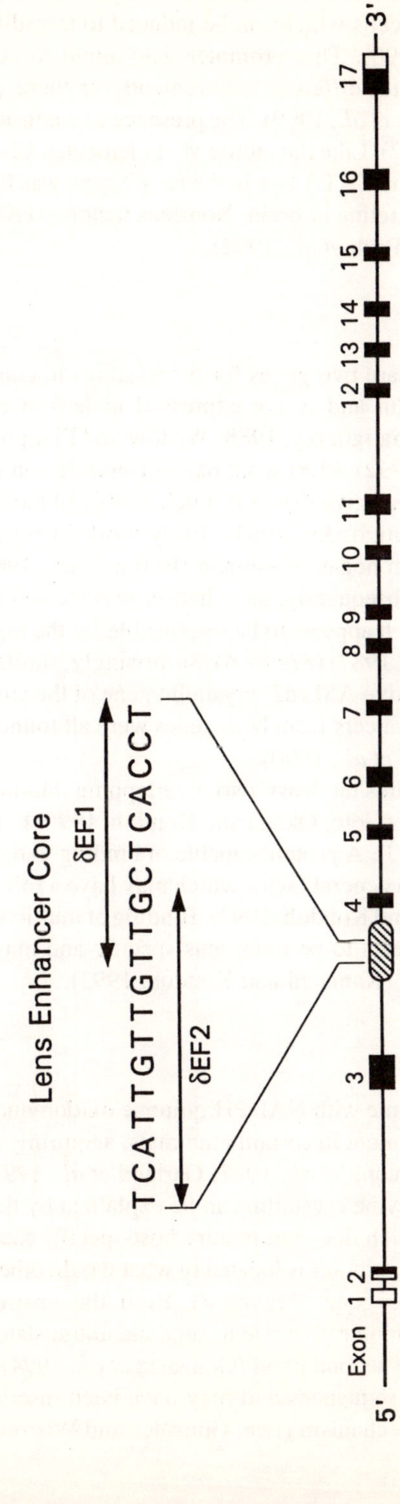

Figure 6. Schematic of the chicken δ1-crystallin gene with the location of the enhancer element. Coding sequences of exons are shown in black.

Somewhat similar results have been obtained for the rat γD-crystallin gene. In this case the promoter was transfected either into primary mouse lens epithelial cells or into chicken neural retina cells which can be induced to transdifferentiate into lens-like cells (Peek *et al.*, 1990). This promoter was found to contain at least three important elements but with different requirements for these elements in the two cultured cell systems (Peek *et al.*, 1990). The presence of a non-lens silencer was also suggested (Peek *et al.*, 1992). Like the mouse γF-1 element, a 12-nucleotide sequence immediately upstream of the TATA box in the rat γD gene was found to bind a factor found in rat lens but not in retina or brain. Non-lens factors were found to bind to the putative silencing region (Peek *et al.*, 1992).

δ-CRYSTALLINS

As described above, there are two genes for δ-crystallins in chickens, one of which codes for ASL/δ2-crystallin and is not expressed in lens at crystallin-like levels (Parker, Wawrousek and Piatigorsky, 1988; Wistow and Piatigorsky, 1988; Kondoh *et al.*, 1991; Piatigorsky, 1992). Most work has concentrated on the more specialized δ1-crystallin gene which may give rise to as much as 90% of the soluble protein in the embryonic chick lens (Piatigorsky, 1984). Early work identified a lens-preferred promoter, with an upstream negative element (Borras *et al.*, 1985; Borras, Peterson and Piatigorsky, 1988). Subsequently, an enhancer was located in the third intron of the δ1-crystallin gene, which appears to be responsible for the high expression of this gene in lens (Hayashi *et al.*, 1987) (*Figure 6*). Surprisingly, similar sequences are also present in the much less active ASL/δ2- crystallin gene of the chicken, and combinations of promoters and enhancers from both genes were all found to be preferentially expressed in lens (Thomas *et al.*, 1990).

The δ1 enhancer contains at least two overlapping binding sites for factors designated δEF1 and δEF2 (Goto, Okada and Kondoh, 1990; Funahashi *et al.*, 1991; Kamachi and Kondoh, 1993). A protein capable of binding to the δEF1 site has been cloned and turns out to be a general factor which may have a role in suppressing non-lens expression (Kamachi and Kondoh, 1993). Binding of multicomponent complexes through the δEF2 site appears to be more lens-specific and may be responsible for high expression in the lens (Kamachi and Kondoh, 1993).

ζ-CRYSTALLIN

ζ-Crystallin is a novel enzyme with NADPH:quinone oxidoreductase activity, which has undergone gene recruitment in certain mammals, acquiring a second function as a lens structural protein (Huang *et al.*, 1987; Garland *et al.*, 1991; Lee *et al.*, 1993). The recruitment of this enzyme crystallin can be explained by the lens specificity of an alternative promoter which does not require host-specific factors (Lee, Gonzalez and Winstow, 1994). This promoter is located in what would otherwise have been the first intron of the enzyme gene (*Figure 7*). Both the upstream 'housekeeping promoter' and the lens promoter give rise to separate untranslated first exons which splice to a common, coding second exon (Gonzalez *et al.*, 1994). The lens promoter is flanked by direct repeat sequences and may have been inserted into the gene by some kind of transposon mechanism (Lee, Gonzalez and Wistow, 1994). The strong

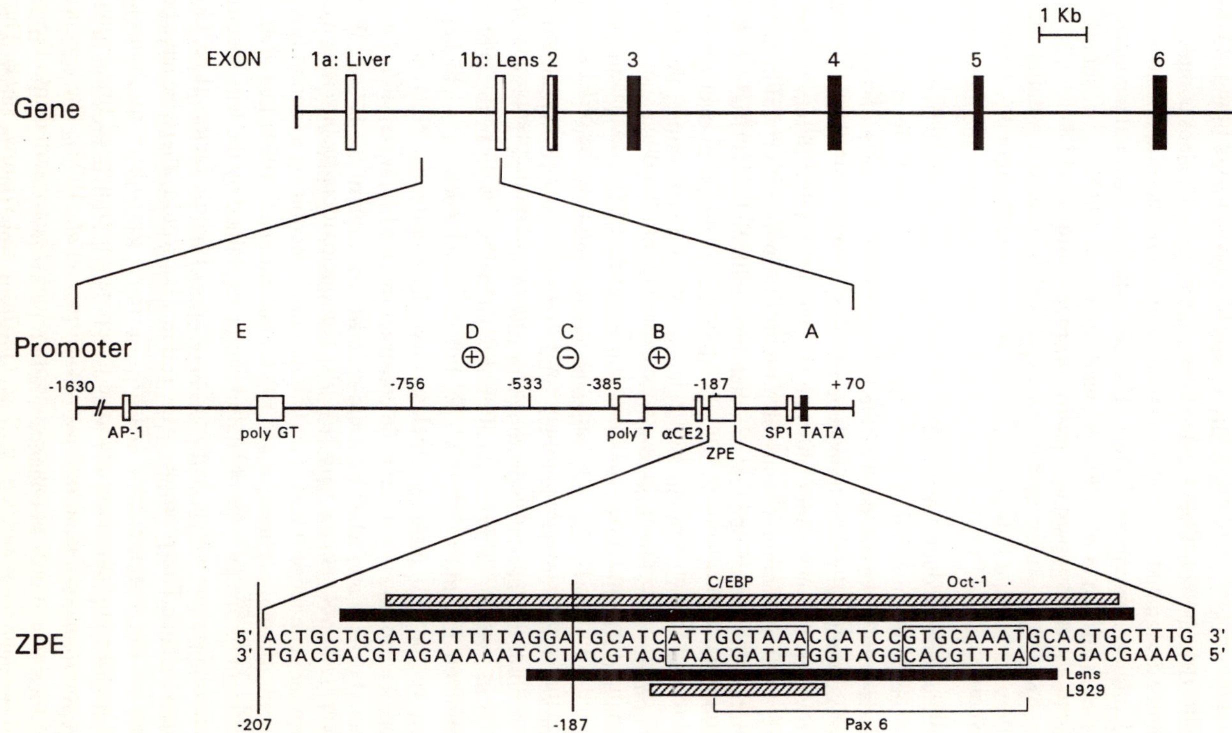

Figure 7. The guinea-pig ζ-crystallin gene has alternative promoters. Both alternative first exons are non-coding (white). Functional elements of the lens-specific promoter are shown. A–E show regions tested by transfection analysis (Lee, Gonzalez and Wistow, 1994).

tissue preference of the lens promoter is apparent in both cultured cell transfections and in transgenic mice (Lee, Gonzalez and Wistow, 1994). While proximal regions of the promoter have some activity in the brain of transgenic mice, this is abolished by the addition of more distal regions. The minimal active promoter is differentially footprinted by extracts from lens and non-lens cells. Deletion within the major region footprinted in lens, ZPE (ζ protected element), abolishes promoter function. More detailed analysis of this region by competitive EMSA has shown that this sequence gives rise to two major gel-shifted complexes, one of which is found in extracts of non-lens cells while a higher-mobility complex predominates in extracts of lens-derived cells (Richardson, Cvekl and Wistow, submitted). The lens complex depends on a sequence designated ZE-1 which is flanked by consensus binding sites for C/EBP and Oct-1 family transcription factors. The ZE-1 site is the core of a consensus Pax-6 binding site (Epstein *et al.*, 1994; Richardson, Cvekl and Wistow, submitted) and antisera to Pax-6 abolish the lens-specific EMSA complex.

α-ENOLASE/τ-CRYSTALLIN

The complete gene for duck α-enolase/τ-crystallin has been cloned (Kim *et al.*, 1991). Although the endogenous gene shows preferential expression in embryonic duck lens, no lens-preferred expression of the gene promoter has been observed in either transfected cultured cells or in transgenic mice (Kim *et al.*, 1991; Kim and Wistow, 1993). However, this gene does illustrate the way in which different regions of the lens express different complements of proteins. α-Enolase/τ-crystallin may be the most abundant protein component of avian lens epithelial cells (Rudner, Katar and Maisel, 1990). Its pattern of expression parallels that of the proto-oncogene c-*myc* in the same system (Harris, Talian and Zelenka, 1992). Furthermore, the duck gene promoter contains an E-box, a consensus c-Myc binding site (Blackwell *et al.*, 1990; Halazonetis and Kandil, 1991), which is conserved in the homologous human gene (Giallongo *et al.*, 1990), the only other α-enolase gene that has been sequenced. Nuclear proteins from chicken lens cells bind to the consensus c-Myc binding site in the duck α-enolase/τ-crystallin promoter in EMSA and DNase I protection assays (Warwar *et al.*, 1992). A monoclonal antibody detects c-Myc in the protein complexes involved in binding to this site and co-transfection of a chicken c-Myc expression vector into primary chick embryo lens epithelial cells increases the expression of a promoter/reporter gene construct fourfold. Transcriptional activation in transfected cells is abolished by a site-specific mutation of the E-box (Warwar *et al.*, 1992). This suggests that c-Myc is among the positive regulators of this promoter.

LACTATE DEHYDROGENASE B/ε-CRYSTALLIN

The gene for duck LDHB/ε-crystallin has been cloned and its expression examined in chicken lens and heart cells (Kraft *et al.*, 1993). Identical start sites for transcription were found in both tissues (Hodin and Wistow 1993; Kraft *et al.*, 1993), although a second start site 28 bp upstream was also used in heart. High expression in both cell types requires sequences from the first intron. The intron does not contain a classic enhancer since it is unable to enhance a heterologous promoter (Kraft *et al.*, 1993). Although the gene is expressed at higher levels in lens than in heart *in vivo*, no difference in promoter activity was observed in cultured cells. This suggests that the cultured chicken lens cells do not appropriately mimic the characteristics of intact lens required for ε-crystallin over-expression.

Transgenics

Crystallin gene promoters have been used to direct expression of a number of foreign genes to the lens. These have been either the bacterial reporter genes chloramphenicol acetyltransferase (CAT) or β-galactosidase (lacZ), used for straightforward functional analysis, or else genes for transforming proteins or toxins used to disrupt normal lens development.

Most constructions have involved the use of the mouse αA-crystallin gene promoter. A promoter fragment containing 5' flanking and untranslated sequences of exon 1 (–366/+45) has been used to drive the expression of CAT (Overbeek *et al.*, 1985), SV40 large T antigen (Mahon *et al.*, 1987), diphtheria toxin A (DT-A) (Kaur *et al.*, 1989; Harrington *et al.*, 1991), ricin A (Landel *et al.*, 1988), retinoic acid receptor- α-lacZ fusion (Balkan *et al.*, 1992), cellular retinoic acid-binding protein I (Perez-Castro, Tran and Nguyen-Huu, 1993), polyoma virus large T antigen (Griep *et al.*, 1989), human immunodeficiency virus TAT gene (Khillan *et al.*, 1988), dbl oncogene (Eva *et al.*, 1991), human papilloma virus type 16 E6 and E7 oncogenes (Griep *et al.*, 1993), human or murine urokinase-type plasminogen activator (uPA) (Miskin *et al.*, 1990), γ-interferon (Egwuagu *et al.*, 1994) and FGF-3/Int-2 (Chepelinsky *et al.*, 1994). In a similar manner, a fragment of the hamster αA-crystallin promoter covering the same region (–347 to +43) has been fused to CAT and DT-A (Yu, Tsui and Breitman, 1990). These constructs have shown a clear tissue specificity and the expression of the transgenes is co-regulated with the endogenous αA crystallin during lens development. It was also shown that the chicken αA-crystallin gene fused to CAT requires 162 bp of 5' flanking sequences to direct lens-specific expression in transgenic mice (Klement, Wawrousek and Piatigorsky, 1989).

The other principal system used is that of the mouse γF-crystallin promoter. γF-crystallin sequences –759 to +45, which include the core promoter and two upstream enhancer elements, contain sufficient information to direct gene expression to terminally differentiated lens fibre cells. This promoter fragment has been used to generate CAT, lacZ, DT-A and SV40 T-antigen transgenics (Breitman *et al.*, 1989; Yu, Tsui and Breitman, 1990; Bryce *et al.*, 1993; Goring *et al.*, 1993).

The chicken δ1-crystallin 5' flanking region (–2200/–51) fused to the xanthine–guanine phosphoribosyl transferase gene was found to be expressed primarily in the lens of chimeric mice (Takahashi *et al.*, 1988).

The promoter for an enzyme crystallin, guinea-pig ζ-crystallin, also shows clear lens preference, as evidenced by the expression of the CAT gene in the lens of transgenic mice (Lee, Gonzalez and Wistow, 1994). A line of transgenic mice was also constructed containing the complete duck α-enolase/τ-crystallin (αENO) gene with 5' and 3' sequences (Kim and Wistow, 1993). The duck αENO mRNA was expressed in the same pattern as the endogenous mouse isozyme, showing no apparent tissue specificity.

Is there a master lens-specific factor for crystallin expression?

The lack of an obvious common promoter sequence among crystallin genes has been striking and frustrating. Nevertheless, crystallin genes from species as divergent as chickens and mice will function in each other's lens cells (see Kondoh, Yasuda and Okada, 1983; Kondoh *et al.*, 1987; Lok *et al.*, 1985; Klement, Wawrousek and Piatigorsky, 1989; Piatigorsky, 1992). This suggests that there are lens-specific transcription factors which are essential for expression in any lens and which recognize all crystallin genes. An interesting candidate for one of these putative factors has now emerged. Pax-6 is a transcription factor of the paired-domain homeodomain family. It is essential for normal eye development (see Hill *et al.*, 1991; Glaser, Walton and Maas, 1992; Li *et al.*, 1994). As mentioned above, there are growing indications that Pax-6 may actually be playing an important role as a transcription factor for crystallin genes (Cvekl *et al.*, 1994; Richardson, Cvekl and Wistow, submitted) Since Pax-6 is also expressed in brain (Stoykova and Gruss, 1994), this could explain the overlapping expression in neural tissue which has been noted for some crystallin promoter constructs (see, for example, Lee, Gonzalez and Wistow, 1994). Since Pax-6 is very highly conserved throughout evolution (Hill and Hanson, 1992) and has such an important role in development of head tissues, it could be an ancient common factor of eyes even in vertebrates and invertebrates (Quiring *et al.*, 1994). This protein could be one of the long-sought unifying factors of lens-specific gene expression in all species.

Implications for biotechnology and gene therapy

Recent work on the lens has shed some light on basic questions of molecular evolution, development and tissue-specific gene expression. To this point there has been no prospect of applied spin-off from this basic research. However, this work does provide some interesting lessons for biotechnology and gene therapy, and more speculatively some distant possibilities for applied exploitation of the lens system.

A common objective in biotechnology is the over-expression of functional proteins. From that perspective, the eye lens is a natural biotechnological factory. Extraordinarily high concentrations of active enzymes can accumulate rapidly in developing lenses without problems such as inclusion-body formation, the results of successful evolutionary 'genetic engineering' in the lens. Future experiments which aim at over-expressing other proteins in the lens, mimicking the gene recruitment process, should tell us how robust the system is and how broad a spectrum of proteins can become crystallins.

One interesting feature of the lens which may contribute to its ability to maintain a high concentration of various soluble proteins is that the ubiquitous stress protein crystallins, α-crystallins, and maybe even β- and γ-crystallins, may act to 'chaperone' other components of the lens. If this is true, and it remains to be demonstrated, then it may be possible to achieve similar high levels of expression for a wide variety of foreign proteins targeted to the lens.

As for using the lens itself as a vehicle for recombinant protein production, this is at least distantly possible. The characterization of crystallin gene promoters has provided the tools for expression of foreign genes in transgenic animals. So far there has been little attention paid to achieving high-level expression but this goal is now being pursued. It is conceivable that important proteins could be produced at crystallin-like levels in transgenic lenses. Such a system might be advantageous for the production of high concentrations of active cytosolic proteins which have problems of processing or solubility in micro-organisms and for which other techniques, such as secretory expression in mammals (Westphal, 1989; Janne *et al.*, 1992), are not suitable. A disadvantage of lens expression is that animals must be killed and dissected to harvest expressed protein. However this problem could be minimized by the use of transgenic birds.

Although mice are the best-studied transgenic hosts, their lenses are small, particularly at birth. Chickens are a more attractive choice. Only two weeks after fertilization chicken embryos *in ovo* have large, obvious eyes and much larger lenses than mice. In the laboratory it is possible to harvest hundreds of lenses in a few hours. Extraction of soluble protein is a simple matter of homogenization (Wistow, 1993a). The technology for production of transgenic birds is less advanced than that for mice but it is progressing rapidly (Love *et al.*, 1994). Established lines of transgenic chickens could be used for continuous production of eggs carrying the desired product.

Even more speculatively, another interesting but completely unexplored strategy for high expression of a foreign protein in chicken lens would be homologous recombination into the δ1-crystallin gene. Since this gene accounts for as much as 90% of the crystallin content of embryonic chicken lenses, introduction of a foreign coding sequence could achieve very high expression. This technology is quite advanced in mammals in which it is now routinely possible to introduce genes into embryonic stem cells and then transfer those cells to embryos to create chimeric animals which can then be bred to homozygosity (for example see Fung-Leung and Mak, 1992; Koller and Smithies, 1992). At present the same kind of techniques for manipulating birds are not so advanced, although homologous recombination in chicken B cells can be achieved (Buerstedde and Takeda, 1991).

Apart from any future possibilities of biotechnological applications, crystallins provide some important, if indirect, lessons for another area of genetic manipulation, namely gene therapy. Gene therapy in the lens itself seems an unlikely prospect in the near future. There are no diseases of the lens itself which are serious enough to warrant the extreme measure of gene therapy in that tissue when cataract surgery is a simple process with a good outcome. However, the lens could possibly serve as a reservoir for expression of a secreted factor, such as an anti-inflammatory protein, which might be beneficial to other parts of a diseased eye. If the therapeutic protein in question could pass through the lens capsule, it might be preferable to introduce an expression vector by injection into the lens than into the much more sensitive retina.

However, the more immediate impact of crystallin research on gene therapy may be in the form of general lessons to be considered when targeting therapies to specific genes. The crystallins illustrate the way in which a single gene may exert multiple phenotypes through protein multifunctionality. On one hand this allows defects in one gene to have multiple effects. On the other hand, it warns that introduction of a transgene to address one problem, such as an enzyme deficiency, could have unforeseen side-effects if the introduced gene has inappropriate temporal or spatial expression. Consider the familiar glycolytic enzyme lactate dehydrogenase (LDH). As described above, it may also be a structural protein in the eye lens in some species and may have additional unexpected functions. For example, LDH isoforms have been identified as single-stranded DNA binding proteins associated with actively transcribing regions of chromatin (see Williams, Reddigari and Patel, 1985). Similarly, α-enolase, another glycolytic enzyme which may serve a structural role in lens, has been identified as a cell-surface plasminogen receptor (Miles *et al.*, 1991), while the related γ-enolase has been found associated with the centrosome of HeLa cells (Johnstone, Waisman and Rattner, 1992). In yeast, enolase serves as a heat-shock protein (Iida and Yahara, 1985). These observations and others like them suggest that it might be very important to introduce genes in gene therapy with as much of their normal flanking and intronic DNA as possible to ensure native modes of expression.

References

BAGBY, S., HARVEY, T.S., EAGLE, S.G., INOUYE, S. AND IKURA, M. (1994a). NMR-derived three-dimensional solution structure of protein S complexed with calcium. *Structure* **2**, 107–122.

BAGBY, S., HARVEY, T.S., KAY, L.E., EAGLE, S.G., INOUYE, S. AND IKURA, M. (1994b). Unusual helix-containing greek keys in development- specific Ca^{2+}-binding Protein S. ^{1}H, ^{15}N and ^{13}C assignments and secondary structure determined with the use of multidimensional double and triple resonance heteronuclear NMR spectroscopy. *Biochemistry* **33**, 2409–2421.

BALKAN, W., KLINTWORTH, G.K., BOCK, C.B., AND LINNEY, E. (1992). Transgenic mice expressing a constitutively active retinoic acid receptor in the lens exhibit ocular defects. *Developmental Biology* **151**, 622–625.

BARBOSA, P., CIALKOWSKI, M. AND O'BRIEN, W.E. (1991). Analysis of naturally occurring and site-directed mutations in the argininosuccinate lyase gene. *Journal of Biological Chemistry* **266**, 5286–5290.

BARBOSA, P., WISTOW, G.J., CIALKOWSKI, M., PIATIGORSKY, J. AND O'BRIEN, W.E. (1991). Expression of duck lens δ-crystallin cDNAs in yeast and bacterial hosts: δ2-crystallin is an active argininosuccinate lyase. *Journal of Biological Chemistry* **266**, 22319–22322.

BARNSTABLE, C.J. (1993). Glutamate and GABA in retinal circuitry. *Current Opinion in Neurobiology* **3**, 520–525.

BAX, B., LAPATTO, R., NALINI, V., DRIESSEN, H., LINDLEY, P.F., MAHADEVAN, D., BLUNDELL, T.L. AND SLINGSBY, C. (1990). X-ray analysis of βB2-crystallin and evolution of oligomeric lens proteins. *Nature* **347**, 776–780.

BERBERS, G.A., HOEKMAN, W.A., BLOEMENDAL, H., DE JONG, W.W., KLEINSCHMIDT, T. AND BRAUNITZER, G. (1984). Homology between the primary structures of the major bovine β-crystallin chains. *European Journal of Biochemistry* **139**, 467–479.

BLACKWELL, T.K., KRETZNER, L., BLACKWOOD, E.M., EISENMAN, R.N. AND WEINTRAUB, H. (1990). Sequence-specific DNA binding by the c-Myc protein. *Science* **250**, 1149–1151.

BLOEMENDAL, H., (ED.) (1981). *Molecular and Cellular Biology of the Eye Lens*. Wiley, New York.

BLUNDELL, T., LINDLEY, P., MILLER, L., MOSS, D., SLINGSBY, C., TICKLE, I., TURNELL, B. AND WISTOW, G. (1981). The molecular structure and stability of the eye lens: x-ray analysis of γ-crystallin II. *Nature* **289**, 771–777.

BOHREN, K.M., BULLOCK, B., WERMUTH, B. AND GABBAY, K.H. (1989). The aldo-keto reductase superfamily. cDNAs and deduced amino acid sequences of human aldehyde and aldose reductases. *Journal of Biological Chemistry* **264**, 9547–9551.

BONDY, C.A. AND LIGHTMAN, S.L. (1989). Developmental and physiological pattern of aldose reductase mRNA expression in lens and retina. *Molecular Endocrinology* **3**, 1417–1425.

BORRAS, T., PERSSON, B. AND JORNVALL, H. (1989). Eye lens ζ-crystallin: relationships to the family of long-chain alcohol/polyol dehydrogenases. *Biochemistry* **28**, 6133–6139.

BORRAS, T., PETERSON, C.A. AND PIATIGORSKY, J. (1988). Evidence for positive and negative regulation in the promoter of the chicken δ1-crystallin gene. *Developmental Biology* **127**, 209–219.

BORRAS, T., NICKERSON, J.M., CHEPELINSKY, A.B. AND PIATIGORSKY, J. (1985). Structural and functional evidence for differential promoter activites of the two linked δ-crystallin genes in the chicken. *EMBO Journal* **4**, 445–452.

BRAKENHOFF, R.H., ARTS, H.J., REEK, F.H., LUBSEN, N.H. AND SCHOENMAKERS, J.G. (1990). Human γ-crystallin genes. A gene family on its way to extinction. *Journal of Molecular Biology* **216**, 519–532.

BREITMAN, M.L., LOK, S., WISTOW, G., PIATIGORSKY, J., TRETON, J.A., GOLD, R.J. AND TSUI, L.C. (1984). γ-Crystallin family of the mouse lens: structural and evolution relationships. *Proceedings of the National Academy of Sciences, USA* **81**, 7762–7766.

BREITMAN, M.L., BRYCE, D.M., GIDDENS, E., CLAPOFF, S., GORING, D., TSUI, L.-C., KLINTWORTH, G.K. AND BERNSTEIN, A. (1989). Analysis of lens cell fate and eye morphogenesis in transgenic mice ablated for cells of the lens lineage. *Development* **106**, 457–463.

BRYCE, D.M., LIU, Q., KHOO, W., TSUI, L.C. AND BREITMAN, M.L. (1993). Progressive and regressive fate of lens tumors correlates with subtle differences in transgene expression in γF-crystallin-SV40 T antigen transgenic mice. *Oncogene* **8**, 1611–1620.

BUERSTEDDE, J.M. AND TAKEDA, S. (1991). Increased ratio of targeted to random integration after transfection of chicken B cell lines. *Cell* **67**, 179–188.

CARPER, D., NISHIMURA, C., SHINOHARA, T., DIETZSCHOLD, B., WISTOW, G., CRAFT, C., KADOR, P. AND KINOSHITA, J.H. (1987). Aldose reductase and ρ-crystallin belong to the same protein superfamily as aldehyde reductase. *FEBS Letters* **220**, 209–213.

CARPER, D.A., WISTOW, G., NISHIMURA, C., GRAHAM, C., WATANABE, K., FUJII, Y., HAYASHI, H. AND HAYAISHI, O. (1989). A superfamily of NADPH-dependent reductases in eukaryotes and prokaryotes. *Experimental Eye Research* **49**, 377–388.

CHEPELINSKY, A.B., SOMMER, B. AND PIATIGORSKY, J. (1987). Interaction between two different regulatory elements activates the murine αA-crystallin gene promoter in explanted lens epithelia. *Molecular and Cellular Biology* **7**, 1807–1814.

CHEPELINSKY, A.B., ROBINSON, M.L., OVERBEEK, P.A., PARKER, D.M., CHAN, C.C., JAMIESON, S. AND DICKSON, C. (1994). FGF-3 ectopic expression induces differentiation of central lens epithelia and appearance of secretory epithelia in the eyes of transgenic mice. *Investigative Ophthalmology and Visual Science* **35**, 1998.

CHIESA, R., GAWINOWICZ-KOLKS, M.A. AND SPECTOR, A. (1987). The phosphorylation of the primary gene products of α-crystallin. *Journal of Biological Chemistry* **262**, 1438–1441.

CHIESA, R., GAWINOWICZ-KOLKS, M.A., KLEIMAN, N.J. AND SPECTOR, A. (1987). The phosphorylation sites of the B2 chain of bovine α-crystallin. *Biochemical and Biophysical Research Communications* **144**, 1340–1347.

CHIESI, M., LONGONI, S. AND LIMBRUNO, U. (1990). Cardiac α-crystallin. III. Involvement during heart ischemia. *Molecular and Cellular Biochemistry* **97**, 129–136.

CHIOU, S.H. (1988). A novel crystallin from octopus lens. *FEBS Letters* **241**, 261–264.

CLARKE, F.M., MORTON, D.J., STEPHAN, P. AND WIEDEMANN, J. (1985). The functional duality of glycolytic enzymes: potential integrators of cytoplasmic structure and function. In *Cell Motility: Mechanism and Regulation (10th Yamada Conference)* (H. Ishikawa, S. Hatano and H. Sato, Eds), pp. 235–250. Tokyo University Press, Tokyo.

COOPER, R.L., CONSTABLE, I.J. AND DAVIDSON, L. (1984). Aqueous humor catecholamines. *Current Eye Research* **3**, 809–813.

COX, R.L., GLICK, D.L. AND STRUMWASSER, F. (1991). Isolation and protein sequence identification of *Aplysia californica* lens crystallins. *Biological Bulletin (Woods Hole)* **181**, 333–335.

CVEKL, A., SAX, C.M., BRESNICK, E.H. AND PIATIGORSKY, J. (1994). Complex array of positive and negative elements regulate the chicken αA-crystallin gene: involvement of Pax-6, USF, CREB and/or CREM and AP1 proteins. *Molecular and Cellular Biology* **14**, 7363–7376.

DASGUPTA, S., HOHMAN, T.C. AND CARPER, D. (1992). Hypertonic stress induces αB-crystallin expression. *Experimental Eye Research* **54**, 461–470.

DE JONG, W.W. (1981). Evolution of lens and crystallins. In *Molecular and Cellular Biology of the Eye Lens* (H. Bloemendal, Ed.), pp. 221–278. Wiley-Interscience, New York.

DE JONG, W.W., LEUNISSEN, J.A.M. AND VOORTER, C.E.M. (1993). Evolution of the α-crystallin/small heat-shock protein family. *Molecular Biology and Evolution* **10**, 103–126.

DELAYE, M. AND TARDIEU, A. (1983). Short-range order of crystallin proteins accounts for eye lens transparency. *Nature* **302**, 415–417.

DODEMONT, H., GROENEN, M., JANSEN, L., SCHOENMAKERS, J. AND BLOEMENDAL, H. (1985). Comparison of the crystallin mRNA populations from rat, calf and duck lens. Evidence for a longer αA2-mRNA and two distinct αB2 in the birds. *Biochimica Biophysica Acta* **824**, 284–294.

DONOVAN, D.M., SAX, C.M., KLEMENT, J.F., LI, X., CHEPELINSKY, A.B. AND PIATIGORSKY, J. (1992). Conservation of mouse αA-crystallin promoter activity in chicken lens epithelial cells. *Journal of Molecular Evolution* **35**, 337–345.

DRIESSEN, H.P., HERBRINK, P., BLOEMENDAL, H. AND DE JONG, W.W. (1980). The β-crystallin Bp chain is internally duplicated and homologous γ-crystallin. *Experimental Eye Research* **31**, 243–246.

DUBIN, R.A., WAWROUSEK, E.F. AND PIATIGORSKY, J. (1989). Expression of the murine αB-crystallin gene is not restricted to the lens. *Molecular and Cellular Biology* **9**, 1083–1091.

DUBIN, R.A., ALLY, A.H., CHUNG, S. AND PIATIGORSKY, J. (1990). Human αB-crystallin gene and preferential promoter function in lens. *Genomics* **7**, 594–601.

DUBIN, R.A., GOPAL-SRIVASTAVA, R., WAWROUSEK, E.F. AND PIATIGORSKY, J. (1991). Expression of the murine αB-crystallin gene in lens and skeletal muscle: identification of a muscle-preferred enhancer. *Molecular and Cellular Biology* **11**, 4340–4349.

DUGUID, J.R., ROHWER, R.G. AND SEED, B. (1988). Isolation of cDNAs of scrapie-modulated RNAs by subtractive hybridization of a cDNA library. *Proceedings of the National Academy of Sciences, USA* **85**, 5738–5742.

EGWUAGU, C.E., SZTEIN, J., CHAN, C.C., REID, W., MAHDI, R., NUSSENBLATT, R.B. AND CHEPELINSKY, A.B. (1994). Ectopic expression of γ-interferon in the eyes of transgenic mice induces ocular pathology and MHC class II gene expression. *Investigative Ophthalmology and Visual Science* **35**, 332–341.

EPSTEIN, J., CAI, J., GLASER, T., JEPEAL, L. AND MAAS, R. (1994). Identification of a Pax paired domain recognition sequence and evidence for DNA-dependent conformational changes. *Journal of Biological Chemistry* **269**, 8355–8361.

EVA, A., GRAZIANI, G., ZANNINI, M., MERIN, L.M., KHILLAN, J.S. AND OVERBEEK, P.A. (1991). Dominant dysplasia of the lens in transgenic mice expressing the dbl oncogene. *New Biologist* **3**, 158–168.

FLAHERTY, K.M., MCKAY, D.B., KABSCH, W. AND HOLMES, K.C. (1991). Similarity of the three-dimensional structures of actin and the ATPase fragment of a 70-kDa heat shock cognate protein. *Proceedings of the National Academy of Sciences, USA* **88**, 5041–5045.

FUNAHASHI, J.I., KAMACHI, Y., GOTO, K. AND KONDOH, H. (1991). Identification of nuclear factor ΔEF1 and its binding site essential for lens-specific activity of the δ1-crystallin enhancer. *Nucleic Acids Research* **19**, 3543–3547.

FUNG-LEUNG, W.P. AND MAK, T.W. (1992). Embryonic stem cells and homologous recombination. *Current Opinion in Immunology* **4**, 189–194.

GAO, C.Y., BASSNETT, S. AND ZELENKA, P.S. Cyclin B, p34cdc2 and H1-kinase activity in terminally differentiating lens fiber cells. Submitted.

GARLAND, D., RAO, P.V., DEL CORSO, A., MURA, U. AND ZIGLER, J.S., JR (1991). ζ-Crystallin is a major protein in the lens of *Camelus dromedarius*. *Archives of Biochemistry and Biophysics* **285**, 134–136.

GIALLONGO, A., OLIVA, D., CALI, L., BARBA, G., BARBIERI, G. AND FEO, S. (1990). Structure of the human gene for α-enolase. *European Journal of Biochemistry* **190**, 567–573.

GLASER, T., WALTON, D.S. AND MAAS, R.L. (1992). Genomic structure, evolutionary conservation and aniridia mutations in the human PAX6 gene. *Nature Genetics* **2**, 232–239.

GONZALEZ, P., HERNANDEZ-CALZADILLA, C., RAO, P.V., RODRIGUEZ, I.R., ZIGLER, J.S., JR AND BORRAS, T. (1994). Comparative analysis of the ζ-crystallin/quinone reductase gene in guinea pig and mouse. *Molecular Biology and Evolution* **11**, 305–315.

GOPAL-SRIVASTAVA, R. AND PIATIGORSKY, J. (1993). The murine αB-crystallin/small heat shock protein enhancer: identification of αBE-1, αBE-2, αBE- 3, and MRF control elements. *Molecular and Cellular Biology* **13**, 7144–7152.

GOPAL-SRIVASTAVA, R. AND PIATIGORSKY, J. (1994). Identification of a lens-specific regulatory region (LSR) of the murine αB-crystallin gene. *Nucleic Acids Research* **22**, 1281–1286.

GORING, D.R., BREITMAN, M.L. AND TSUI, L.C. (1992). Temporal regulation of six crystallin transcripts during mouse lens development. *Experimental Eye Research* **54**, 785–795.

GORING, D.R., BRYCE, D.M., TSUI, L.C., BREITMAN, M.L. AND LIU, Q. (1993). Developmental regulation and cell type-specific expression of the murine γF-crystallin gene is mediated through a lens-specific element containing the γF-1 binding site. *Developmental Dynamics* **196**, 143–152.

GOTO, K., OKADA, T.S. AND KONDOH, H. (1990). Functional cooperation of lens-specific and nonspecific elements in the δ1-crystallin enhancer. *Molecular and Cellular Biology* **10**, 958–964.

GRAW, J., LIEBSTEIN, A., PIETROWSKI, D., SCHMITT-JOHN, T. AND WERNER, T. (1993). Genomic sequences of murine γB- and γC-crystallin-encoding genes: promoter analysis and complete evolutionary pattern of mouse, rat and human γ-crystallins. *Gene* **136**, 145–156.

GRIEP, A.E., KUWABARA, T., LEE, E.J. AND WESTPHAL, H. (1989). Perturbed development of the mouse lens by polyomavirus large T antigen does not lead to tumor formation. *Genes and Development* **3**, 1075–1085.

GRIEP, A.E., HERBER, R., JEON, S., LOHSE, J.K., DUBIELZIG, R.R. AND LAMBERT, P.F. (1993). Tumorigenicity by human papillomavirus type 16 E6 and E7 in transgenic mice correlates with alterations in epithelial cell growth and differentiation. *Journal of Virology* **67**, 1373–1384.

HALAZONETIS, T.D. AND KANDIL, A.N. (1991). Determination of the c-MYC DNA-binding site. *Proceedings of the National Academy of Sciences, USA* **88**, 6162–6166.

HARDING, J.J. AND CRABBE, M.J.C. (1984). The lens: development, proteins, metabolism and cataract. In *The Eye*, Vol. 1B (H. Davson, Ed.), pp. 207–492. Academic Press, New York.

HARRINGTON, L., KLINTWORTH, G.K., SECOR, T.E. AND BREITMAN, M.L. (1991). Developmental analysis of ocular morphogenesis in αA-crystallin/diphtheria toxin transgenic mice undergoing ablation of the lens. *Developmental Biology* **148**, 508–516.

HARRIS, L.L., TALIAN, J.C. AND ZELENKA, P.S. (1992). Contrasting patterns of c-*myc* and N-*myc* expression in proliferating, quiescent and differentiating cells of the embryonic chicken lens. *Development* **115**, 813–820.

HARTL, F.U., HLODAN, R. AND LANGER, T. (1994). Molecular chaperones in protein folding: the art of avoiding sticky situations. *Trends in Biochemical Sciences* **19**, 20–25.

HAYASHI, S., GOTO, K., OKADA, T.S. AND KONDOH, H. (1987). Lens-specific enhancer in the third intron regulates expression of the chicken δ1-crystallin gene. *Genes and Development* **1**, 818–828.

HEAD, M.W., PETER, A. AND CLAYTON, R.M. (1991). Evidence for the extralenticular expression of members of the β-crystallin gene family in the chick and a comparison with δ-crystallin during differentiation and transdifferentiation. *Differentiation* **48**, 147–156.

HECKER, M., SESSA, W.C., HARRIS, H.J., ANGGARD, E.E. AND VANE, J.R. (1990). The metabolism of L-arginine and its significance for the biosynthesis of endothelium-derived relaxing factor: cultured endothelial cells recycle L-citrulline to L-arginine. *Proceedings of the National Academy of Sciences, USA* **87**, 8612–8616.

HENDRIKS, W., LEUNISSEN, J., NEVO, E., BLOEMENDAL, H. AND DE JONG, W.W. (1987). The lens protein αA-crystallin of the blind mole rat, *Spalax ehrenbergi*: Evolutionary change and functional constraints. *Proceedings of the National Academy of Sciences, USA* **84**, 5320–5324.

HENDRIKS, W., MULDERS, J.W.M., BIBBY, M.A., SLINGSBY, C., BLOEMENDAL, H. AND DE JONG, W.W. (1988). Duck lens ε-crystallin and lactate dehydrogenase B4 are identical: single-copy gene product with two distinct functions. *Proceedings of the National Academy of Sciences, USA* **85**, 7114–7118.

HILL, R.E. AND HANSON, I.M. (1992). Molecular genetics of the Pax gene family. *Current Opinion in Cell Biology* **4**, 967–972.

HILL, R.E., FAVOR, J., HOGAN, B.L., TON, C.C., SAUNDERS, G.E., HANSON, I.M., PROSSER, J.,

JORDAN, T., HASTIE, N.D. AND VAN HEYNINGEN, V. (1991). Mouse small eye results from mutations in a paired-like homeobox-containing gene. *Nature* **354**, 522–525.

HODIN, J. AND WISTOW, G. (1993). 5'-RACE PCR of mRNA for three taxon-specific crystallins: for each gene one promoter controls both lens and non-lens expression. *Biochemical and Biophysical Research Communications* **190**, 391–396.

HOGG, D., TSUI, L.-C., GORIN, M. AND BREITMAN, M.L. (1986). Characterization of the human β-crystallin gene Hu βA3/A1 reveals ancestral relationships among the βγ-crystallin superfamily. *Journal of Biological Chemistry* **261**, 12420–12427.

HORWITZ, J. (1992). A-Crystallin can function as a molecular chaperone. *Proceedings of the National Academy of Sciences, USA* **89**, 10449–10453.

HUANG, Q.L., RUSSELL, P., STONE, S. AND ZIGLER, J.S. (1987). ζ-Crystallin, a novel lens protein from the guinea pig. *Current Eye Research* **6**, 725–732.

IIDA, H. AND YAHARA, I. (1985). Yeast heat-shock protein of Mr 48,000 is an isoprotein of enolase. *Nature* **315**, 688–690.

INGOLIA, T.D. AND CRAIG, E.A. (1982). Four small *Drosophila* heat shock proteins are related to each other and to mammalian α-crystallin. *Proceedings of the National Academy of Sciences, USA* **79**, 2360–2364.

IWAKI, A., IWAKI, T., GOLDMAN, J.E. AND LIEM, R.K. (1990). Multiple mRNAs of rat brain α-crystallin B chain result from alternative transcriptional initiation. *Journal of Biological Chemistry* **265**, 22197–22203.

IWAKI, T., KUME-IWAKI, A. AND GOLDMAN, J.E. (1990). Cellular distribution of αB-crystallin in non-lenticular tissues. *Journal of Histochemistry and Cytochemistry* **38**, 31–39.

IWAKI, T., KUME-IWAKI, A., LIEM, R.K. AND GOLDMAN, J.E. (1989). αB-crystallin is expressed in non-lenticular tissues and accumulates in Alexander's disease brain. *Cell* **57**, 71–78.

JAHN, D., VERKAMP, E. AND SÖLL, D. (1992). Glutamyl-transfer RNA: a precursor of heme and chlorophyll biosynthesis. *Trends in Biochemical Sciences* **17**, 215–218.

JANNE, J., HYTTINEN, J.M., PEURA, T., TOLVANEN, M., ALHONEN, L. AND HALMEKYTO, M. (1992). Transgenic animals as bioproducers of therapeutic proteins. *Annals of Medicine (Helsinki)* **24**, 273–280.

JAWORSKI, C.J., CHEPELINSKY, A.B. AND PIATIGORSKY, J. (1991). The αA-crystallin gene: conserved features of the 5'-flanking regions in human, mouse, and chicken. *Journal of Molecular Evolution* **33**, 495–505.

JOHNSTONE, S.A., WAISMAN, D.M. AND RATTNER, J.B. (1992). Enolase is present at the centrosome of HeLa cells. *Experimental Cell Research* **202**, 458–463.

KAMACHI, Y. AND KONDOH, H. (1993). Overlapping positive and negative regulatory elements determine lens-specific activity of the δ1-crystallin enhancer. *Molecular and Cellular Biology* **13**, 5206–5215.

KANTOROW, M. AND PIATIGORSKY, J. (1994). α-Crystallin/small heat shock protein has autokinase activity. *Proceedings of the National Academy of Sciences, USA* **91**, 3112–3116.

KANTOROW, M., CVEKL, A., SAX, C. AND PIATIGORSKY, J. (1993). Protein–DNA interactions of the mouse αA-crystallin control regions. *Journal of Molecular Biology* **230**, 425–435.

KATO, K., SHINOHARA, H., KUROBE, N., GOTO, S., INAGUMA, Y. AND OHSHIMA, K. (1991a). Immunoreactive αA-crystallin in rat non-lenticular tissues detected with a sensitive immunoassay method. *Biochemica Biophysica Acta* **1080**, 173–180.

KATO, K., SHINOHARA, H., KUROBE, N., GOTO, S., INAGUMA, Y. AND OHSHIMA, K. (1991b). Tissue distribution and developmental profiles of immunoreactive αB-crystallin in the rat determined with a sensitive immunoassay system. *Biochimica Biophysica Acta* **1074**, 201–208.

KAUR, S., KEY, B., STOCK, J., MCNEISH, J.D., AKESON, R. AND POTTER, S.S. (1989). Targeted ablation of α-crystallin-synthesizing cells produces lens-deficient eyes in transgenic mice. *Development* **105**, 613–619.

KHILLAN, J.S., DEEN, K.C., YU, S.H., SWEET, R.W., ROSENBERG, M. AND WESTPHAL, H. (1988). Gene transactivation mediated by the TAT gene of human immunodeficiency virus in transgenic mice. *Nucleic Acids Research* **16**, 1423–1430.

KIM, R.Y. AND WISTOW, G.J. (1993). Expression of the duck α-enolase/τ-crystallin gene in transgenic mice. *FASEB Journal* **7**, 464–469.

KIM, R.Y., GASSER, R. ABD WISTOW, G.J. (1992). μ-Crystallin is a mammalian homologue of *Agrobacterium* ornithine cyclodeaminase and is expressed in human retina. *Proceedings of the National Academy of Sciences, USA* **89**, 9292–9296.

KIM, R.Y., LIETMAN, T., PIATIGORSKY, J. AND WISTOW, G.J. (1991). Structure and expression of the duck α-enolase/τ-crystallin encoding gene. *Gene* **103**, 193–200.

KLEMENT, J.F., CVEKL, A. AND PIATIGORSKY, J. (1993). Functional elements DE2A, DE2B and DE1A and the TATA box are required for activity of the chicken αA-crystallin gene in transfected lens epithelial cells. *Journal of Biological Chemistry* **268**, 6777–6784.

KLEMENT, J.R., WAWROUSEK, E.F. AND PIATIGORSKY, J. (1989). Tissue-specific expression of the chicken αA-crystallin gene in cultured lens epithelia and transgenic mice. *Journal of Biological Chemistry* **264**, 19837–19844.

KLEMENZ, R., FROHLI, E., AOYAMA, A., HOFFMANN, S., SIMPSON, R.J., MORITZ, R.L. AND SCHAFER, R. (1991a). αB-Crystallin accumulation is a specific response to Ha-*ras* and v-*mos* oncogene expression in mouse NIH 3T3 fibroblasts. *Molecular and Cellular Biology* **11**, 803–812.

KLEMENZ, R., FROHLI, E., STEIGER, R.H., SCHAFTER, R. AND AOYAMA, A. (1991b). αB-Crystallin is a small heat shock protein. *Proceedings of the National Academy of Sciences, USA* **88**, 3652–3656.

KNULL, H.R. AND WALSH, J.L. (1992). Association of glycolytic enzymes with the cytoskeleton. *Current Topics in Cellular Regulation* **33**, 15–30.

KOLLER, B.H. AND SMITHIES, O. (1992). Altering genes in animals by gene targeting. *Annual Review of Immunology* **10**, 705–730.

KOMORI, N., USUKURA, J. AND MATSUMOTO, H. (1992). Drosocrystallin, a major 52 kDa glycoprotein of the *Drosophila melanogaster* corneal lens. Purification, biochemical characterization and subcellular localization. *Journal of Cell Science* **102**, 191–201.

KONDOH, H., YASUDA, K. AND OKADA, T.S. (1983). Tissue-specific expression of a cloned chick δ-crystallin gene in mouse cells. *Nature* **301**, 440–442.

KONDOH, H., KATOH, K., TAKAHASHI, Y., FUJISAWA, H., YOKOYAMA, M., KIMURA, S., KATSUKI, M., SAITO, M., NOMURA, T., HIRAMOTO, Y. AND OKADA, T.S. (1987). Specific expression of the chicken δ-crystallin gene in the lens and the pyramidal neurons of the piriform cortex in transgenic mice. *Developmental Biology* **120**, 177–185.

KONDOH, H., ARAKI, I., YASUDA, K., MATSUBASA, T. AND MORI, M. (1991). Expression of the chicken 'δ2-crystallin' gene in mouse cells: evidence for encoding of argininosuccinate lyase. *Gene* **99**, 267–271.

KRAFT, H.J., HENDRIKS, W., DE JONG, W.W., LUBSEN, N.H. AND SCHOENMAKERS, J.G.G. (1993). Duck lactate dehydrogenase B/ε-crystallin gene: lens recruitment of a GC-promoter. *Journal of Molecular Biology* **229**, 849–859.

LANDEL, C.P., ZHAO, J., BOK, D. AND EVANS, G.A. (1988). Lens-specific expression of recombinant ricin induces developmental defects in the eyes of transgenic mice. *Genes and Development* **2**, 1168–1178.

LEE, D.C., GONZALEZ, P. AND WISTOW, G. (1994). ζ-Crystallin: A lens-specific promoter and the gene recruitment of an enzyme as a crystallin. *Journal of Molecular Biology* **236**, 669–678.

LEE, D.C., KIM, R.Y. AND WISTOW, G.J. (1993). An avian αB-crystallin. Non lens expression and sequence similarities with both small (hsp27) and large (HSP70) heat shock proteins. *Journal of Molecular Biology* **232**, 1221–1226.

LEE, D.C., GONZALEZ, P., RAO, P.V., ZIGLER, J.S., JR AND WISTOW, G.J. (1993). Carbonyl-metabolizing enzymes and their relatives recruited as structural proteins in the eye lens. In *Enzymology and Molecular Biology of Carbonyl Metabolism*, Vol. 4 (H. Weiner, Ed.), pp. 159–168. Plenum Press, New York.

LI, H.S., YANG, J.M., JACOBSON, R.D., PASKO, D. AND SUNDIN, O. (1994). Pax-6 is first expressed in a region of ectoderm anterior to the early neural plate: implications for stepwise determination of the lens. *Developmental Biology* **162**, 181–194.

LI, L.K. AND SPECTOR, A. (1967). The optical rotatory dispersion and circular dichroism of calf lens α-crystallin. *Journal of Biological Chemistry* **242**, 3234–3236.

LI, L.K. AND SPECTOR, A. (1974). Circular dichroism and optical rotatory dispersion of the aggregates of purified polypeptides of α-crystallin. *Experimental Eye Research* **19**, 49–57.

LI, X., WISTOW, G.J. AND PIATIGORSKY, J. Linkage and expression of the argininosuccinate lyase/δ-crystallin genes of the duck: insertion of a CR1 element in the intragenic spacer. *Biochim. Biophys. Acta*, in press.

LIU, Q.R., TINI, M., TSUI, L.C. AND BREITMAN, M.L. (1991). Interaction of a lens cell transcription factor with the proximal domain of the mouse γF-crystallin promoter. *Molecular and Cellular Biology* **11**, 1531–1537.

LOK, S., TSUI, L.-C., SHINOHARA, T., PIATIGORSKY, J., GOLD, R. AND BREITMAN, M. (1984). Analysis of the mouse г-crystallin gene family: Assignment of multiple cDNAs to discrete genomic sequences and characterization of representative gene. *Nucleic Acids Research* **12**, 4517–4529.

LOK, S., BREITMAN, M.L., CHEPELINSKY, A.B., PIATIGORSKY, J., GOLD, R.J.M. AND TSUI, L.-C. (1985). Lens-specific promoter activity of a mouse γ-crystallin gene. *Molecular and Cellular Biology* **5**, 2221–2230.

LOK, S., STEVENS, W., BREITMAN, M.L. AND TSUI, L.C. (1989). Multiple regulatory elements of the murine г2-crystallin promoter. *Nucleic Acids Research* **17**, 3563–3582.

LOVE, J., GRIBBIN, C., MATHER, C. AND SANG, H. (1994). Transgenic birds by DNA microinjection. *Bio/Technology* **12**, 60–63.

LOWE, J., LANDON, M., PIKE, I., SPENDLOVE, I., MCDERMOTT, H. AND MAYER, R.J. (1990). Dementia with β-amyloid deposition: involvement of αB-crystallin supports two main diseases. *Lancet* **336**, 515–516.

MCDERMOTT, J.B., PETERSON, C.A. AND PIATIGORSKY, J. (1992). Structure and lens expression of the gene encoding chicken βA3/A1-crystallin. *Gene* **117**, 193–200.

MILES, L.A., DAHLBERG, C.M., PLESCIA, J., FELEZ, J. FELEZ, J., KATO, K. AND PLOW, E.F. (1991). Role of cell-surface lysines in plasminogen binding to cells: identification of α-enolase as a candidate plasminogen receptor. *Biochemistry* **30**, 1682–1691.

MIRON, T., VANCOMPERNOLLE, K., VANDEKERCKHOVE, J., WILCHEK, M. AND GEIGER, B. (1991). A 25-kD inhibitor of actin polymerization is a low molecular mass heat shock protein. *Journal of Cell Biology* **114**, 255–261.

MISKIN, R., AXELROD, J.H., GRIEP, A.E., LEE, E., BELIN, D., VASSALLI, J.D. AND WESTPHAL, H. (1990). Human and murine urokinase cDNAs linked to the murine αA-crystallin promoter exhibit lens and non-lens expression in transgenic mice. *Eur. J. Biochem.* **190**, 31–38.

MONTGOMERY, M.K. AND MCFALL-NGAI, M.J. (1992). The muscle-derived lens of a squid bioluminescent organ is biochemically convergent with the ocular lens. Evidence for recruitment of aldehyde dehydrogenase as a predominant structural protein. *Journal of Biological Chemistry* **267**, 20999–21003.

MOORMANN, R.J., DEN DUNNEN, J.T., HEUYERJANS, J., JONGBLOED, R.J., VAN LEEN, R.W., LUBSEN, N.H. AND SCHOENMAKERS, J.G. (1985). Characterization of the rat γ-crystallin gene family and its expression in the eye lens. *Journal of Molecular Biology* **182**, 419–430.

MULDERS, J.W., HENDRIKS, W., BLANKESTEIJN, W.M., BLOEMENDAL, H. AND DE JONG, W.W. (1988). λ-Crystallin, a major rabbit lens protein, is related to hydroxyacyl-coenzyme A dehydrogenases. *Journal of Biological Chemistry* **263**, 15462–15466.

NAKAMURA, T., DONOVAN, D.M., HAMADA, K., SAX, C.M., NORMAN, B., FLANAGAN, J.R., OZATO, K., WESTPHAL, H. AND PIATIGORSKY, J. (1990). Regulation of the mouse AA-crystallin gene: isolation of a cDNA encoding a protein that binds to a *cis* sequence motif shared with the major histocompatibility complex class I gene and other genes. *Molecular and Cellular Biology* **10**, 3700–3708.

NICHOLL, I.D. AND QUINLAN, R.A. (1994). Chaperone activity of α-crystallins modulates intermediate filament assembly. *EMBO Journal* **13**, 945–953.

NICKERSON, J.M., WAWROUSEK, E.F., HAWKINS, J.W., WAKIL, A.S., WISTOW, G.J., THOMAS, G., NORMAN, B.L. AND PIATIGORSKY, J. (1985). The complete sequence of the chicken δ1-crystallin gene and its flanking region. *Journal of Biological Chemistry* **260**, 9100–9105.

NICKERSON, J.M., WAWROUSEK, E.F., BORRAS, T., HAWKINS, J.W., NORMAN, B.L., FILPULA, D.R., NAGLE, J.W., ALLY, A.H. AND PIATIGORSKY, J. (1986). Sequence of the chicken δ2-crystallin gene and its intergenic spacer. Extreme homology with the δ1 crystallin gene. *Journal of Biological Chemistry* **261**, 552–557.

NUMATA, O., TAKEMASA, T., TAKAGI, I., HIRONO, M., HIRANO, H., CHIBA, J. AND WATANABE, Y. (1991). Tetrahymena 14-nm filament-forming protein has citrate synthase activity. *Biochemical and Biophysical Research Communications* **174**, 1028–1034.

OSTRER, H. AND PIATIGORSKY, J. (1980). β-Crystallins of the adult chicken lens: relatedness of the polypeptides and their aggregates. *Experimental Eye Research* **30**, 679–689.

OVERBEEK, P.A., CHEPELINSKY, A.B., KHILLAN, J.S., PIATIGORSKY, J. AND WESTPHAL, H. (1985). Lens-specific expression and developmental regulation of the bacterial chloramphenicol acetyltransferase gene driven by the murine αA-crystallin promoter in transgenic mice. *Proceedings of the National Academy of Sciences, USA* **82**, 7815–7819.

PARKER, D.S., WAWROUSEK, E.F. AND PIATIGORSKY, J. (1988). Expression of the δ-crystallin genes in the embryonic chicken lens. *Developmental Biology* **126**, 375–381.

PEEK, R., VAN DER LOGT, P., LUBSEN, N.H. AND SCHOENMAKERS, J.G.G. (1990). Tissue- and species-specific promoter elements of rat γ-crystallin genes. *Nucleic Acids Research* **18**, 1189–1197.

PEEK, R., KRAFT, H.J., KLOK, E.J., LUBSEN, N.H. AND SCHOENMAKERS, J.G.G. (1992). Activation and repression sequences determine the lens-specific expression of the rat γD-crystallin gene. *Nucleic Acids Research* **20**, 4865–4871.

PEREZ-CASTRO, A.V., TRAN, V.T. AND NGUYEN-HUU, M.C. (1993). Defective lens fiber differentiation and pancreatic tumorigenesis caused by ectopic expression of the cellular retinoic acid-binding protein I. *Development* **119**, 363–375.

PIATIGORSKY, J. (1981). Lens differentiation in vertebrates. A review of cellular and molecular features. *Differentiation* **19**, 134–153.

PIATIGORSKY, J. (1984). δ-Crystallin and their nucleic acids. *Molecular and Cellular Biochemistry* **59**, 33–56.

PIATIGORSKY, J. (1992). Lens crystallins. Innovation associated with changes in gene regulation. *Journal of Biological Chemistry* **267**, 4277–4280.

PIATIGORSKY, J. AND ZELENKA, P.S. (1992). Transcriptional regulation of crystallin genes: *cis* elements, *trans*-factors, and signal transduction systems in the lens. *Advances in Developmental Biochemistry* **1**, 211–256.

PIATIGORSKY, J., O'BRIEN, W.E., NORMAN, B.L., KALUMUK, K., WISTOW, G.J., BORRAS, T., NICKERSON, J.M. AND WAWROUSEK, E.F. (1988). Gene sharing by δ-crystallin and argininosuccinate lyase. *Proceedings of the National Academy of Sciences, USA* **85**, 3479–3483.

PIATIGORSKY, J., HORWITZ, J., KUWABARA, T. AND CUTRESS, C.E. (1989). The cellular eye lens and crystallins of cubomedusan jellyfish. *Journal of Comparative Physiology A* **164**, 577–587.

QUAX-JEUKEN, Y., DRIESSEN, H., LEUNISSEN, J., QUAX, W., DE JONG, W. AND BLOEMENDAL, H. (1985). βs-Crystallin: structure and evolution of a distinct member of the βγ-superfamily. *EMBO Journal* **4**, 2597–2602.

QUIRING, R., WALDORF, U., KLOTER, U. AND GEHRING, W.J. (1994). Homology of the eyeless gene of Drosophila to the small eye gene in mice and Aniridia in humans. *Science* **265**, 785–789.

RAO, G.N. AND COTLIER, E. (1984). Urea cycle enzymes in retina, ciliary body-iris, lens and senile cataracts. *Experimental Eye Research* **39**, 483–495.

RAO, P.V. AND ZIGLER, J.S., JR (1990). Extremely high levels of NADPH in guinea pig lens: correlation with ζ-crystallin concentration. *Biochemical and Biophysical Research Communications* **167**, 1221–1228.

RAO, P.V., KRISHNA, C.M. AND ZIGLER, J.S. (1992). Identification and characterization of the enzymatic activity of ζ-crystallin from guinea pig lens. A novel NADPH:quinone oxidoreductase. *Journal of Biological Chemistry* **267**, 96–102.

RENKAWEK, K., DE JONG, W.W., MERCK, K.B., FRENKEN, C.W., VAN WORKUM, F.P. AND BOSMAN, G.J. (1992). αB-crystallin is present in reactive glia in Creutzfeldt-Jakob disease. *Acta Neuropathologica* **83**, 324–327.

RICHARDSON, J., CVEKL, A. AND WISTOW, G. Pax-6 is essential for lens-specific expression of ζ-crystallin. Submitted.

RODOKANAKI, A., HOLMES, R.K. AND BORRAS, T. (1989). ζ-Crystallin, a novel protein from the guinea pig lens is related to alcohol dehydrogenases. *Gene* **78**, 215–224.

ROQUEMORE, E.P., DELL, A., MORRIS, H.R., PANICO, M., REASON, A.J., SAVOY, L.-A., WISTOW, G.J., ZIGLER, J.S., JR, EARLES, B.J. AND HART, G.W. (1992). Vertebrate lens α-crystallins are modified by *O*-linked *N*-acetylglucosamine. *Journal of Biological Chemistry* **267**, 555–563.

ROTH, H.J., DAS, G.C. AND PIATIGORSKY, J. (1991). Chicken βB1-crystallin gene expression: presence of conserved functional polyomavirus enhancer-like and octamer binding-like promoter elements found in non-lens genes. *Molecular and Cellular Biology* **11**, 1488–1499.

RUDNER, G., KATAR, M. AND MAISEL, H. (1990). Enolase in the avian and turtle lens. *Current Eye Research* **9**, 139–150.

RUSSELL, P., MEAKIN, S.O., HOHMAN, T.C., TSUI, L.C. AND BREITMAN, M.L. (1987). Relationship between proteins encode by three human γ-crystallin genes and distinct polypeptides in the eye lens. *Molecular and Cellular Biology* **7**, 3320–3323.

SANS, N., SCHINDLER, U. AND SCHRÖDER, J. (1988). Ornithine cyclodeaminase from Ti plasmid C58: DNA sequence, enzyme properties and regulation of activity by arginine. *European Journal of Biochemistry* **173**, 123–130.

SAWADA, K., AGATA, K. AND EGUCHI, G. (1992). Crystallin gene expression in the process of lentoidogenesis in cultures of chicken lens epithelial cells. *Experimental Eye Research* **55**, 879–887.

SAX, C. AND PIATIGORSKY, J. (1994). Expression of the α-crystallin/small heat shock proteins/molecular chaperone genes in the lens and other tissues. In *Advances in Enzymology and Related Areas in Molecular Biology* Vol. 69, pp. 155–201. John Wiley, New York.

SAX, C.M., ILAGAN, J.G. AND PIATIGORSKY, J. (1993). Functional redundancy of the DE-1 and αA-CRYBP1 regulatory sites of the mouse αA-crystallin promoter. *Nucleic Acids Research* **21**, 2633–2640.

SAX, C.M., KLEMENT, J.F. AND PIATIGORSKY, J. (1990). Species-specific lens activation of the thymidine kinase promoter by a single copy of the mouse αA-CRYBP1 site and loss of tissue specificity by multimerization. *Molecular and Cellular Biology* **10**, 6813–6816.

SMOLICH, B.D., TARKINGTON, S.K., SAHA, M.S. AND GRAINGER, R.M. (1994). *Xenopus* γ-crystallin gene expression: Evidence that the γ-crystallin gene family is transcribed in lens and non-lens tissues. *Molecular and Cellular Biology* **14**, 1355–1363.

STAPEL, S.O. AND DE JONG, W.W. (1983). Lamprey 48-kDa lens protein represents a novel class of crystallins. *FEBS Letters* **162**, 305–309.

STAPEL, S.O., ZZWEERS, A., DODEMONT, H.J., KAN, J.H. AND DE JONG, W.W. (1985). ε-crystallin, a novel avian and reptilian eye lens protein. *European Journal of Biochemistry* **147**, 129–136.

STOYKOVA, A. AND GRUSS, P. (1994). Roles of Pax-genes in developing and adult brain as suggested by expression patterns. *Journal of Neuroscience* **14**, 1395–1412.

SUMMERS, L., WISTOW, G., NAREBOR, M., MOSS, D., LINDLEY, P., SLINGSBY, C., BLUNDELL, T. AND BARTUNIK, H. (1984). X-ray structure of the lens specific proteins: the crystallins. In *Peptide and Protein Reviews* Vol. 3 (M.T.W. Hearn, Ed.), pp. 147–168. Marcel Dekker, New York.

TAKAHASHI, Y., HANAOKA, K., HAYASAKA, M., KATOH, K., KATO, Y., OKADA, T.S. AND KONDOH, H. (1988). Embryonic stem cell-mediated transfer and correct regulation of the chicken δ-crystallin gene in developing mouse embryos. *Development* **102**, 259–269.

THOMAS, G., ZELENKA, P.S., CUTHBERTSON, R.A., NORMAN, B.L. AND PIATIGORSKY, J. (1990). Differential expression of the two δ-crystallin/argininosuccinate lyase genes in lens, heart and brain of chicken embryos. *New Biologist* **2**, 903–914.

THOMPSON, M.A., HAWKINS, J.W. AND PIATIGORSKY, J. (1987). Complete nucleotide sequence of the chicken αA-crystallin gene and its 5' flanking region. *Gene* **56**, 173–184.

TINI, M., OTULAKOWSKI, G., BREITMAN, M.L., TSUI, L.C. AND GIGUERE, V. (1993). An everted repeat mediates retinoic acid induction of the γF-crystallin gene: evidence of a direct role for retinoids in lens development. *Genes and Development* **7**, 295–307.

TOMAREV, S.I. AND ZINOVIEVA, R.D. (1988). Squid major lens polypeptides are homologous to glutathione S-transferases subunits. *Nature* **336**, 86–88.

TOMAREV, S.I., ZINOVIEVA, R.D. AND PIATIGORSKY, J. (1991). Crystallins of the octopus lens.

Recruitment from detoxification enzymes. *Journal of Biological Chemistry* **266**, 24226–24231.

TOMAREV, S.I., ZINOVIEVA, R.D., DOLGILEVICH, S.M., LUCHIN, S.V., KRAYEV, A.S., SKRYABIN, K.G. AND GAUSE, G.G., JR. (1984). A novel type of crystallin in the frog eye lens. 35-kDa polypeptide is not homologous to any of the major classes of lens crystallins. *FEBS Letters* **171**, 297–302.

TOMAREV, S.I., CHUNG, S., ZINOVIEVA, R.D. AND PIATIGORSKY, J. (1994). Structure and *in vitro* expression of S-crystallins and gluthathione S-transferases of the squid. *Investigative Ophthalmology and Visual Science* **35**, 1997.

TRETON, J.A., JONES, R.E., KING, C.R. AND PIATIGORSKY, J. (1984). Evidence against γ-crystallin DNA or RNA sequences in the chick. *Experimental Eye Research* **39**, 513–522.

VAN DEN HEUVEL, R., HENDRIKS, W., QUAX, W. AND BLOEMENDAL, H. (1985). Complete structure of the hamster αA-crystallin gene. Reflection of an evolutionary history by means of exon shuffling. *Journal of Molecular Biology*, **185**, 273–284.

VAN LEEN, R.W., VAN ROOZENDAAL, K.E.P., LUBSEN, N.H. AND SCHOENMAKERS, J.G.G. (1987). Differential expression of crystallin genes during development of the rat eye lens. *Developmental Biology* **120**, 457–464.

VAN RENS, G.L., DE JONG, W.W. AND BLOEMENDAL, H. (1991). One member of the γ-crystallin gene family, γs, is expressed in birds. *Experimental Eye Research* **53**, 125–138 [letter].

VOORTER, C.E.M., MULDERS, J.W.M., BLOEMENDAL, H. AND DE JONG, W.W. (1986). Some aspects of the phosphorylation of α-crystallin A. *European Journal of Biochemistry* **160**, 203–210.

VOORTER, C.E., DE HAARD-HOEKMAN, W.A., ROERSMA, E.S., MEYER, H.E., BLOEMENDAL, H. AND DE JONG, W.W. (1989). The *in vivo* phosphorylation sites of bovine αB-crystallin. *FEBS Letters* **259**, 50–52.

WARWAR, R.E., KIM, R.Y., WISTOW, G.J. AND ZELENKA, P.S. (1992). The τ-crystallin/α-enolase gene: a candidate for regulation by c-*myc*. *Investigative Ophthalmology and Visual Science* **33**, 794.

WATANABE, K., FUJII, Y., NAKAYAMA, K., OHKUBO, H., KURAMITSU, S., KAGAMIYAMA, H., NAKANISHI, S. AND HAYAISHI, O. (1988). Structural similarity of bovine lung prostaglandin F synthase to lens ε-crystallin [renamed ρ-crystallin] of the European common frog. *Proceedings of the National Academy of Sciences, USA* **85**, 11–15.

WAWROUSEK, E.F., CHEPELINSKY, A.B., MCDERMOTT, J.B. AND PIATIGORSKY, J. (1990). Regulation of the murine αA-crystallin promoter in transgenic mice. *Developmental Biology* **137**, 68–76.

WEINER, H. (Ed.) (1990). *Enzymology and Molecular Biology of Carbonyl Metabolism* (3rd edn). Plenum Press, New York.

WESTPHAL, H. (1989). Transgenic mammals and biotechnology. *FASEB Journal* **3**, 117–120.

WILLIAMS, K.R., REDDIGARI, S. AND PATEL, G.L. (1985). Identification of a nucleic acid helix-destabilizing protein from rat liver as lactate dehydrogenase-5. *Proceedings of the National Academy of Sciences, USA* **82**, 5260–5264.

WILLIAMS, L.A., DING, L., HORWITZ, J. AND PIATIGORSKY, J. (1985). τ-Crystallin from the turtle lens: Purification and partial characterization. *Experimental Eye Research* **40**, 741–749.

WISTOW, G. (1990). Evolution of a protein superfamily: relationships between vertebrate lens crystallins and micro-organism dormancy proteins. *Journal of Molecular Evolution* **30**, 140–145.

WISTOW, G. (1993a). Lens crystallins: a model system for gene recruitment. In *Methods in Enzymology: Molecular Evolution. Producing the Biochemical Data*, Vol. 224 (E.A. Zimmer, T.J. White, R.L. Cann and A.C. Wilson, Eds), pp. 563–575. Academic Press, San Diego, CA.

WISTOW, G. (1993b). Lens crystallins: gene recruitment and evolutionary dynamism. *Trends in Biochemical Science* **18**, 301–306.

WISTOW, G. (1993c). Possible tetramer-based quaternary structures for α-crystallins and small heat-shock proteins. *Experimental Eye Research* **56**, 729–732.

WISTOW, G. AND KIM, H. (1991). Lens protein expression in mammals: taxon-specificity and the recruitment of crystallins. *Journal of Molecular Evolution* **32**, 262–269.

WISTOW, G. AND PIATIGORSKY, J. (1987). Recruitment of enzymes as lens structural proteins. *Science* **236**, 1554–1556.

WISTOW, G. AND PIATIGORSKY, J. (1988). Lens crystallins: evolution and expression of proteins for a highly specialized tissue. *Annual Review of Biochemistry* **57**, 479–504.

WISTOW, G. AND PIATIGORSKY, J. (1990). Gene conversion and splice-site slippage in the argininosuccinate lyases/δ-crystallins of the duck lens: members of an enzyme superfamily. *Gene* **96**, 263–270.

WISTOW, G. AND SEGOVIA, L. (1994). μ-Crystallin: an unexpected marker for photoreceptors. *Investigative Ophthalmology and Visual Science* **35**, 2073.

WISTOW, G., ANDERSON, A. AND PIATIGORSKY, J. (1990). Evidence for neutral and selective processes in the recruitment of enzyme-crystallins in avian lenses. *Proceedings of the National Academy of Sciences, USA* **87**, 6277–6280.

WISTOW, G.J., MULDERS, J.W. AND DE JONG, W.W. (1987). The enzyme lactate dehydrogenase as a structural protein in avian and crocodilian lenses. *Nature* **326**, 622–624.

WISTOW, G., SUMMERS, L. AND BLUNDELL, T. (1985). *Myxococcus xanthus* spore coat protein S may have a similar structure to vertebrate lens βγ-crystallins. *Nature* **315**, 771–773.

WISTOW, G., SLINGSBY, C., BLUNDELL, T., DRIESSEN, H., DE JONG, W. AND BLOEMENDAL, H. (1981). Eye-lens proteins: the three-dimensional structure of β-crystallin predicted from monomeric γ-crystallin. *FEBS Letters* **133**, 9–16.

WISTOW, G., TURNELL, B., SUMMERS, L., SLINGSBY, C., MOSS, D., MILLER, L., LINDLEY, P. AND BLUNDELL, T. (1983). X-ray analysis of the eye lens protein γ-II crystallin at 1.9 Å resolution. *Journal of Molecular Biology* **170**, 175–202.

WISTOW, G.J., LIETMAN, T., WILLIAMS, L.A., STAPEL, S.O., DE JONG, W.W., HORWITZ, J. AND PIATIGORSKY, J. (1988). τ-crystallin/α-enolase: One gene encodes both an enzyme and a lens structural protein. *Journal of Cell Biology* **107**, 2729–2736.

YASUDA, K., KITAMURA, M., OKAZAKI, K. AND TAKEUCHI, M. (1988). Mechanism involved in regulation α-crystallin gene expression. In *UCLA Symposium on the Molecular Biology of the Eye Lens: Genes, Vision, and Ocular Disease*, Vol. 8 (J. Piatigorsky, T. Shinohara and P.S. Zelenka, Eds), pp. 205–213. Alan R. Liss, New York.

YU, C. C.-K., TSUI, L.-C. AND BREITMAN, M.L. (1990). Homologous and heterologous enhancers modulate spatial expression but not cell-type specificity of the murine γF-crystallin promoter. *Development* **110**, 131–139.

2
Natural and Synthetic Saliva: A Stimulating Subject

JOHN GIBSON[1] AND JOSIE A. BEELEY[2]

[1] Department of Oral Surgery/Oral Medicine and [2] Department of Oral Science, University of Glasgow Dental Hospital and School, 378 Sauchiehall Street, Glasgow G2 3JZ, UK

Introduction

Saliva performs many important functions in the mouth and upper aerodigestive tract. Only those unfortunate individuals who suffer from xerostomia (a dry mouth due to impaired salivary secretion) can fully understand such functions at a basic physiological level – and indeed the impaired quality of life which such xerostomia brings. The mouth and pharynx become dry, uncomfortable and overtly sore; eating and speaking are difficult and painful processes; the sense of taste is diminished and the wearing of dentures may be an ordeal. Those with natural dentition suffer from increased gingivitis and an aggressive level of dental caries. The oral musoca is vulnerable to damage and open to infection. The mouth painfully and progressively breaks down.

This chapter aims to present a review of saliva composition, function and its role in maintaining oral health. This will lead to a brief discussion of the conditions that cause xerostomia and our current knowledge of artificial saliva preparations and stimulants.

Saliva: its origin

The human salivary glands consist of three pairs of major glands (parotid, submandibular and sublingual) and multiple minor glands (buccal, palatal, labial and lingual) dispersed throughout the mouth (Mason and Chisholm, 1975). Also present in the mouth are bacteria and their products of degradation, food debris and exfoliated epithelial cells. In addition, in dentate individuals there is gingival exudate (or crevicular fluid). This is a serum-like fluid which seeps from between the teeth and gingival margins. The mixed saliva present in the oral cavity at any one point in time comes from these various sources and is termed whole saliva.

The three pairs of major glands produce around 95% of the total saliva volume – the

Abbreviations: EGF, epidermal growth factor; NGF, nerve growth factor.

Biotechnology and Genetic Engineering Reviews — Vol. 12, December 1994
0264–8725/94/12/39–61 $20.00 + $0.00 © Intercept Ltd, P.O. Box 716, Andover, Hampshire SP10 1YG, UK

parotid glands providing 60–65%, the submandibular glands 20–30% and the sublingual glands 2–5%. The minor glands produce relatively small amounts of saliva but their important role is increasingly recognized in contributing to musocal integrity by way of the hydrated mucosal surface film (Ferguson, 1989).

CONTROL OF SALIVARY SECRETION

Much work has been done on the control of salivary secretion (Emmelin, 1972) and it is now firmly established that the autonomic nervous system is fundamental and central to the control process (Garrett, 1982). The sympathetic and parasympathetic nerves are not simply opposite in nature but work together in a complex manner to stimulate, antagonize and control secretion (Garrett, 1987). The autonomic output to the glands is influenced by taste and tactile sensation from the mouth, with taste the most potent stimulus (Watanube and Dawes, 1988a) – in particular the acidic taste (Watanube and Dawes, 1988b). Visual and olfactory stimuli and inputs from higher cortical centres (e.g. fear and anxiety) also influence the control of saliva secretion via the autonomic nervous system. Mastication is also a potent stimulus to salivary flow. The composition is modulated on stimulation and also subject to circadian variation (Edgar, 1992).

Each salivary gland is made up of a number of acini (sing. acinus). These acini comprise the terminal or secretory end-piece of the gland, situated furthest away from the oral environment. Each acinus is composed of a number of acinar cells on a basement membrane and positioned around a central duct lumen. Each acinus may have cells which are exclusively serous or mucous in secretion type or contain a combination of both cell types. The acinar cells of the submandibular gland and sublingual gland and many minor glands are predominantly mucus secreting. Serous cells form the bulk of acinar cells in the parotid glands and in the minor glands found around the sulci of the circumvallate papillae on the tongue (serous glands of von Ebner). Serous cells also occur in the submandibular gland and in the minor glands anteriorly on the tongue (Waterhouse, Beeley and Mason, 1990).

Each acinus, regardless of predominant cell type, opens into a central duct lumen, which opens into an intercalated duct, lined by cuboidal epithelium with small projecting villi. The intercalated duct opens into a short, wider striated duct, lined by columnar cells with cell-surface projections. The striated duct opens into an excretory duct and on into the terminal duct, lined by stratified squamous epithelium in continuity with the oral cavity. The functions of each part of this duct system are discussed below.

A further cell population identified around the acini and ducts is the myoepithelial cell, which contains myofibrils and is contractile – compressing the acini and duct tissue to facilitate salivary flow.

MECHANISM OF SALIVARY SECRETION

Thaysen and co-workers proposed a two-stage hypothesis of saliva formation in 1954. They suggested that saliva was formed initially as an isotonic secretion in salivary gland acini and that it was subsequently modified by the addition or removal of specific ions as it passed through the ductal system. This hypothesis was confirmed by duct micropuncture techniques (Martinez, Holzgrove and Frick, 1966). Further details have emerged more recently on the two main stages of water and electrolyte secretion in saliva, i.e. acinar secretion and ductal modification (Martinez, 1987):

1. The primary acinar secretion is isotonic and similar to plasma.
2. Stimulation of salivary glands alters the rate of formation of this primary secretion, but not its composition or osmolarity.
3. The duct cells act to alter the electrolyte content, but not the water content, of the primary secretion.
4. Autonomic stimulation of the duct system modifies the ion transport function of the duct cells.

Saliva: composition

Duct saliva is an aqueous hypotonic solution consisting of 99% water and containing inorganic ions, low molecular weight organic constituents and proteins (concentra-

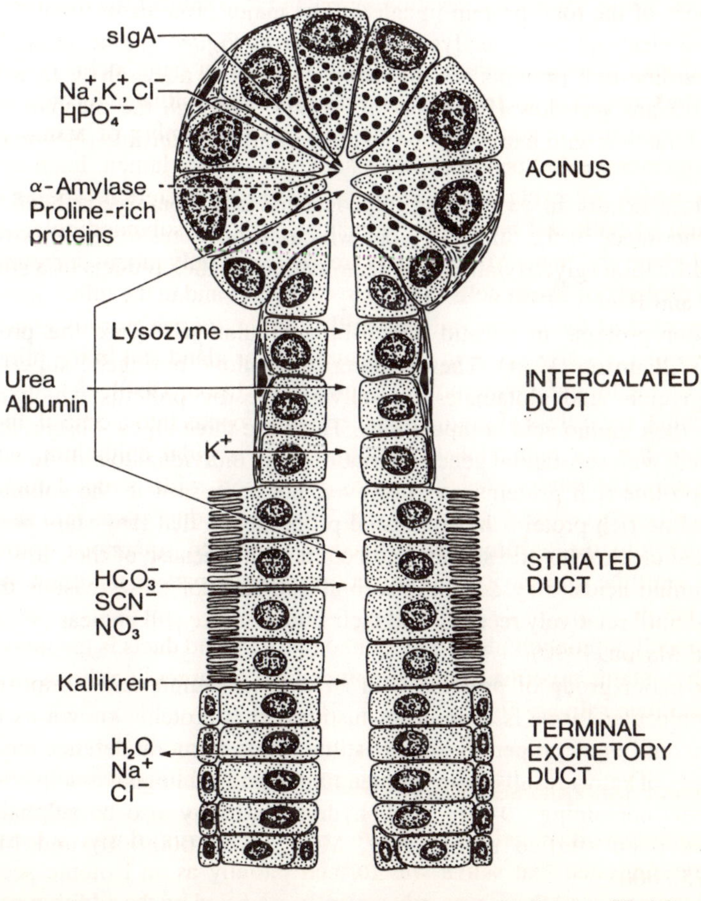

Figure 1. Sites of secretion within the gland of the constituents of saliva (modified from a diagram kindly provided by Dr D.B. Ferguson; Waterhouse, Beeley and Mason, 1990).

tion ~1–3 mg ml^{-1}). The sites of secretion within the gland of some of these components are shown in *Figure 1*. The inorganic ions include sodium, potassium, calcium, zinc, magnesium, phosphate, bicarbonate, chloride, fluoride, thiocyanate, iodide and nitrate, and their concentrations have been published elsewhere (Mason and Chisholm, 1975; Young, 1979; Waterhouse, Beeley and Mason, 1990). The major monovalent cations are sodium and potassium, with the concentration of potassium far exceeding that of sodium. The major divalent cation is calcium, which occurs both as the free ion and bound to proline-rich proteins. Phosphate also occurs in the free and bound state, with about 25% being covalently bound in phosphoproteins. Changes in the levels of the inorganic components are associated with certain pathological disorders (Ferguson, 1987).

The low molecular weight organic compounds include urea (levels of which may rise in patients on renal dialysis), uric acid, creatinine, glucose, amino acids, lactate and some steroid hormones; drugs may also be secreted in saliva.

The major proteins in duct saliva are the fascinating polymorphic families of polyfunctional proteins, α-amylases, proline-rich proteins and histatins (Beeley, 1993; Lamkin and Oppenheim, 1993). These, together with mucins, constitute about 90% or more of the total protein in saliva. The major proteins in parotid saliva, as revealed by electrophoretic analysis, are shown in *Figure 2*. Curiously, however, although proline-rich proteins are the major proteins in duct saliva, their levels in mixed saliva are very low (Beeley *et al*., 1991); although the mechanism is still unclear, interaction with bacteria is probably partially responsible (Newmann *et al*., 1993).

α-Amylase occurs in two forms, glycosylated and non-glycosylated; there is familial inheritance of the different types which probably results from post-translational modification (glycosylation and deamidation) of the products of a single gene (Eckersall and Beeley, 1981).

The major proteins in parotid and submandibular saliva are the proline-rich proteins (Williamson, 1994). These belong to a unique multigene superfamily of proline-, glycine- and glutamate-rich salivary specific proteins which are quite unusual in their amino acid compositions. There are three types – basic, acidic and glycosylated, with substantial genetic variation from individual to individual (*Figure* 2). Basic proline-rich proteins are, however, not expressed in the submandibular glands. Proline-rich proteins have unusual properties in that they stain poorly with conventional electrophoretic staining procedures and because of their low levels of aromatic amino acids, they do not absorb at 280 nm; for these reasons they were overlooked until relatively recently and their functions are still unclear (Waterhouse, Beeley and Mason, 1990).

Another major group of proteins with an unusual amino acid composition and multiple molecular forms is a group of histidine-rich proteins known as histatins, which have antifungal properties and constitute a non-humoral defence mechanism. The viscosity of saliva results from submandibular and sublingual mucins, which are glycoproteins containing 70–80% carbohydrate and may also be sulphated. Two types have been studied extensively, MG1 (M_r>1 000 000) and MG2 (M_r 200 000–250 000).

A variety of other proteins of salivary gland origin also occurs in human saliva. These include a tyrosine-rich polypeptide, statherin, a group of cysteine protease

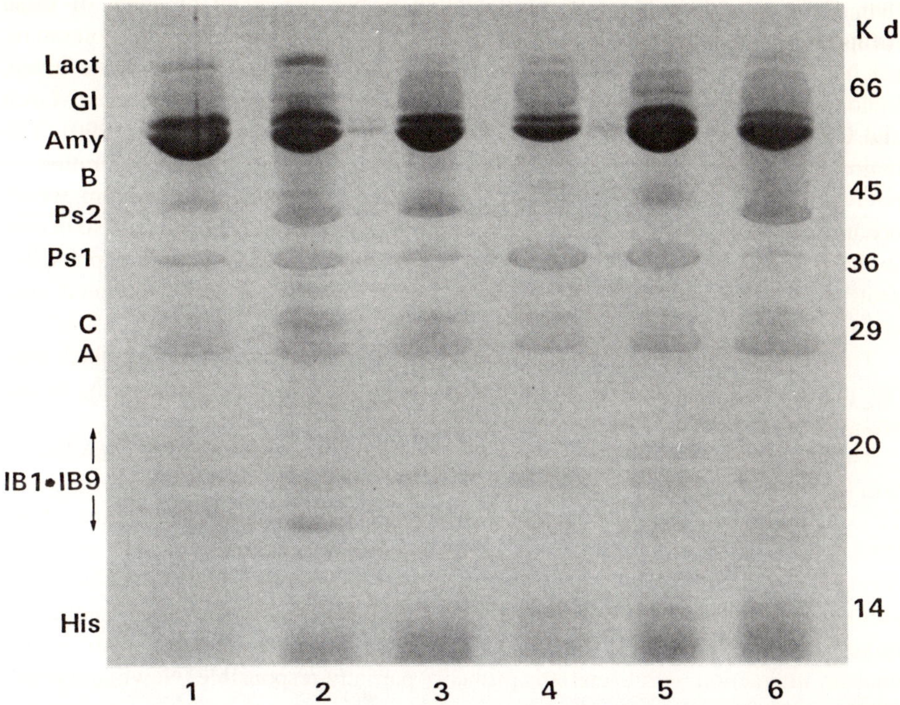

Figure 2. Sodium dodecyl sulphate polyacrylamide gel electrophoresis (Beeley *et al.*, 1991) of parotid salivary proteins from six different individuals. Lact, lactoferrin; Gl, major glycosylated proline-rich protein; Amy, α-amylase; Ps1 and Ps2, basic proline-rich proteins; B, glycosylated proline-rich proteins; A and C, acidic proline-rich proteins; IB1-IB9, basic proline-rich proteins; His, histatins. The gel was stained with Coomassie Blue R-250; in this system proline-rich proteins stain pink-violet and all other proteins stain blue.

inhibitors known as cystatins and the zinc-containing protein, gustin, which is necessary for taste sensation. The major immunoglobulin in duct saliva is IgA, 85% of it is made up of sIgA and a third of this originating from the minor glands. The main source of IgG in whole saliva is gingival exudate. Free secretory component is also present in duct saliva but levels of albumin are extremely low, except in inflammatory disorders. Lactoferrin may also be secreted by salivary glands. Blood group substances (ABO and Lewis antigens) are present in the saliva of 'secretors' and are associated with the high molecular weight mucins which originate from the submandibular and minor glands (Prakobphol, Leffler and Fisher, 1993). Saliva is also an important source of polypeptide hormones such as EGF (epidermal growth factor) and NGF (nerve growth factor). Other enzymes present include carbonic anhydrase, lysozyme, lactoperoxidase, kallikrein and lipase (from minor glands) (Castle, Arvan and Cameron, 1987; Waterhouse, Beeley and Mason, 1990). Variations in organic components may also be associated with disease states (Ferguson, 1987). A list of the proteins of gland origin present in saliva is shown in *Table 1*.

Saliva: functions

Saliva is the fluid that keeps our mouths healthy. Its complexity, both in terms of composition and formation, would suggest myriad functions and, interestingly, many of its components are polyfunctional.

Table 1. Human salivary proteins

Protein	M_r	pI	Function
[a] Proline-rich proteins			Binding to hydroxyapatite
acidic	16 300, 9000	3.5–4.5	Ca^{2a} binding
basic	6000–1200	>8.0	Inhibition of crystal growth
glycosylated	38 900	>8.0	Adhesion of micro-organisms
			Protection against dietary tannins
			Mucosal pellicle and enamel pellicle formation
[a] α-Amylase		6.5–7.5	Digestion
glycosylated	63 000		Antibacterial
non-glycosylated	59 000		Adhesion of micro-organisms
			Antibacterial
[a] Histatins	4500–7000	7.0, >9.5	Antifungal (non-immune)
			Pellicle formation
			Regulation of mineralization
			Buffering
[a] Statherin	5380	4.2	Inhibition of $CaPO_4$ precipitation
			Regulation of mineralization/inhibition of crystal growth
			Binding to hydroxyapatite
			Lubrication
[a] Cystatins	14 000		Cysteine and serine protease inhibitors
			Inhibition of $CaPO_4$ precipitation
			Binding to hydroxyapatite
			Antibacterial, antiviral
[a] Acid phosphatase	—		—
Mucins			Lubrication (speech, chewing and deglutition)
MG1	>1000 000		Bacterial agglutination/antibacterial
MG2	200 000–250 000		Pellicle formation
sulphated	15 000–300 000		Antiviral
blood group substances (ABO and Le)			Blood group antigens
Kallikrein	9600	3.8–4.5	Protease
			? Activation of polypeptide hormones
Immunoglobulins (sIgA, etc.)	405 000		Antibacterial
Lactoferrin	77 000		Antibacterial
Lysozyme	15 000		Anibacterial
Lactoperoxidase	75 000–78 000		Antibacterial
Gustin	37 000		Taste sensation
Albumin	68 000		?
Secretory component	67 000		
Carbonic anhydrase	42 000–45 000		Maintenance of bicarbonate buffering
Lipase			? Digestion
Polypeptide hormones, e.g. epidermal growth factor, nerve growth factor, etc.			Maintenance of oral and gastrointestinal mucosa
Transcalciferin	—		—

Modified from Beeley (1993).
[a] Polymorphic proteins.
Le, Lewis antigens.

FOOD-RELATED ROLES

Cleansing the mouth

This aqueous secretion both washes the oral tissues and contains components that assist in this action. α-Amylase (activated by chloride ions in saliva), generally considered to be a digestive enzyme, also has a key role in removal of residual dietary carbohydrate from the mouth. Indeed, as its pH optimum is 6.8, on passing into the acid environment of the stomach it will quickly become inactivated, although some activity may remain in the centre of a food bolus until inward seeping of gastric juice occurs.

The digestion of fat may well also begin in the oral cavity, with a lipase secreted by the minor salivary glands around circumvallate papillae (Hamosh and Burns, 1977), but a cleansing role for this enzyme may also be involved. However, despite the proteolytic activity of gastric juice, salivary lipase appears to continue acting in the acidic environment in the stomach but its activity is miniscule in comparison with pancreatic lipase.

Taste

A very common complaint of patients with xerostomia is that food has become bland and tasteless. Saliva acts to solubilize food particles to allow interaction with taste bud receptors; it also avoids competition with food substances due to its low salt and sugar concentrations. Shatzman and Henkin (1981) suggested the presence in saliva of gustin, a zinc-containing protein, as a mediator of taste sensation, and, indeed, individuals who are zinc deficient may also have impaired taste sensation.

Mastication and deglutition

The ease with which food is transferred around the oral cavity during mastication and subsequent swallowing is due in no small part to the properties of saliva. Food is wetted by the serous secretions of the parotid gland, secretion of which is stimulated on taking in food, and the lubricant properties of the viscous mucins from the submandibular, sublingual and minor glands facilitate movement during mastication and adhesion to form a bolus (Hatton *et al.*, 1985).

Although these ‘food related’ functions of saliva are essential, its ‘protective’ functions are more numerous.

MAINTENANCE OF ORAL HEALTH

Dentition

The buffering properties in saliva minimize pH fluctuations in the mouth, thereby protecting the tooth mineral from solubilization when the pH falls and deposition of calculus when the pH rises. Although both phosphate and bicarbonate, together with proteins (especially the basic histatins), act as buffers, the major salivary buffer is bicarbonate. Because the concentration of bicarbonate increases substantially when flow is stimulated, the saliva produced while chewing has a much higher buffering capacity than that produced when resting (Edgar, 1976; Shellis and Dibdin, 1988; Watanube and Dawes, 1988b; Wikner and Nedlich, 1988). Carbonic anhydrase helps to maintain this high bicarbonate level in solution. In addition to buffering the pH, saliva can also help to raise the pH of the mouth as a result of the formation of ammonia by bacterial action on urea or degradation of amino acids.

Saliva is also a potent remineralization fluid. Indeed, as the pH of the mouth fluctuates, dental enamel may loose mineral ions as the pH falls and subsequently be remineralized as the pH rises again. Only when excessive remineralization has occurred does the process become irreversible and a carious lesion is formed. Because calcium phosphate is very insoluble, the effective calcium reserve of saliva is substantially increased by the presence of acidic proline-rich proteins which bind large numbers of calcium ions. Furthermore, precipitation of calcium phosphate is inhibited by statherin, proline-rich proteins and histatins, hence saliva is supersaturated with respect to calcium phosphate. However, remineralization must be a regulated process because the topography of the tooth does not change; for example, one never gets 'spikes' forming on the teeth. Presumably salivary components are involved in this regulation process.

The protective role of saliva is seen as the tooth erupts into the mouth. At this stage, the enamel is hypomineralized and relatively porous, explaining the susceptibility of newly erupted teeth to dental caries. However, the enamel undergoes post-eruptive maturation by diffusion of calcium phosphate, magnesium and fluoride into its surface layer. Thus the surface exhibits increased hardness and decreased permeability, which are essential features in a hostile oral environment.

Similarly, the erupting tooth has an enveloping protective pellicle which is soon removed by masticatory activity. This pellicle is constantly replaced by a coating of proteins absorbed from the saliva – the so-called acquired pellicle – which acts as a diffusion barrier to acid ingress and limits dissolved mineral egress (Slomiany *et al.*, 1986). Although proline-rich proteins, and more recently α-amylase also, are frequently regarded as being involved in pellicle formation, the role of crevicular fluid, a rich source of protein, must not be overlooked. However, the protective role of pellicle is debatable, because acquired pellicle also facilitates the adhesion of plaque bacteria to the tooth surface, especially by the early colonizers of the tooth surface in plaque formation.

Some of the most important work in recent years in the sphere of caries research has revealed the significance of fluoride in remineralization of early carious lesions. Early work by McCann and Brudevold (1966) has been built upon to reveal the positive influence of fluoride ion concentrations in promoting remineralization and to encourage the formation of an enamel surface which is progressively more resistant to acid attack during the demineralization–remineralization cycles occurring during bacterial digestion of fermentable carbohydrate in plaque.

Antimicrobial activity

Saliva has an array of antimicrobial activities: antibacterial, antifungal and antiviral.

The mechanisms of antibacterial action are diverse (Waterhouse, Beeley and Mason, 1990). The major immunoglobulin is sIgA together with low concentrations of IgA, IgG and IgM. In addition to antibodies, it contains other antibacterial constituents, including lactoferrin, lactoperoxidase and lysozyme. Lactoferrin is a ferric ion-binding bifunctional glycoprotein; the apo-(iron-free) form occurs in saliva and binds ferric ions, thereby depriving micro-organisms of this nutrient and inhibiting their growth. Lactoperoxidase utilizes microbially produced hydrogen peroxide to oxidize salivary thiocyanate (SCN^-) to hypothiocyanite ($OSCN^-$) which inactivates the bacterial hexokinase (Adamson and Pruitt, 1981). Lysozyme (muramidase), which acts on the peptidoglycan of Gram-positive bacterial cell membranes, causing increased permeability or lysis, is also active against some oral bacteria. α-Amylase, together with cystatins, histatins and mucins, may also have antibacterial functions (MacKay *et al.*, 1984). Mucins, too, play an antibacterial role in causing agglutination and thereby facilitating elimination of bacteria from the mouth by swallowing.

Antifungal activity is displayed by the histidine-rich peptides, histatins, which at very low concentration are able to inhibit growth and kill off *Candida albicans* (Pollock *et al.*, 1984; Sabatini *et al.*, 1989).

Although saliva also has antiviral activity, including anti-HIV activity, the nature of this is still unclear (Malamud and Friedman, 1993). Although it may be modulated via secretory IgA (e.g. oral polio vaccine), activity has been associated with mucins (Heineman and Greenberg, 1980) and, more recently, with cystatins.

Although saliva is generally considered to have a variety of antimicrobial properties, bacteria are normal inhabitants of the mouth. Whereas some species colonize the oral surfaces, other are actively expelled. Perhaps, therefore, 'antimicrobial' is a misnomer and 'microbial modulating' would be more accurate.

Oral soft tissues

Saliva is essential for maintaining the normal properties and health of the soft tissues of the mouth. They are kept moist by water and lubricated by mucins and statherin (Douglas *et al.*, 1991). Especially important in this role are the secretions of the minor glands. These features are essential for speech as well as mastication, deglutition and the retention of dentures. A suitable osmotic environment for the cells of the soft tissues is provided largely by the relatively high levels of sodium, potassium and chloride, along with bicarbonate and phosphate.

Also important is the mucosal pellicle, which results from the formation of a surface coating of proteins, probably involving proline-rich proteins linked by the action of transglutaminase (Bradway *et al.*, 1992); presumably this also plays a protective role.

The polypeptide hormones EGF and NGF are important in maintaining the integrity of the oral mucosa, a function which may be extended to the oesophageal and gastric mucosa too. These polypeptide hormones are secreted as inactive precursors and are activated by the trypsin-like protease kallikrein.

Every day, the oral and oesophageal mucosa is exposed to potentially harmful substances in the diet. High among the list of toxins/potential carcinogens/antidigestive compounds in the human diet are tannins, which occur in tea, coffee, red wine and a variety of fruits. Recent work has shown that proline-rich proteins bind to dietary tannins to form insoluble complexes, a process which may protect us from the harmful effects of these compounds (Mehansho, Butler and Carlson, 1987).

Saliva indeed has myriad functions, but it also contains myriad polyfunctional components to enable it to perform these roles.

Xerostomia: perception and aetiology

This article is at pains to detail the complexity of saliva in terms of composition and function in an attempt to highlight the serious physiological and pathological disturbances that ensue when an individual suffers from long-standing xerostomia (FDI, 1992). Since the salivary flow rate falls to virtually nil during sleep (Schneyer *et al.*, 1956), it is likely that most of us have experienced, many on a regular basis, the discomfort of wakening with a profoundly dry mouth. For a significant number of individuals with poorly or non-functioning salivary glands for whatever reason, that discomfort continues for months, years or a lifetime.

Xerostomia is a clinical sign (the patient often refers to the symptom of 'dry mouth') which implies a decreased quantity of saliva in the mouth, often related to salivary gland dysfunction. Xerostomia itself is not a disease entity. It was first described by Bartley in 1868 and has therefore been recognized for over 100 years (Bertram, 1967). It is clearly outwith the remit of this article to consider the numberous ways of assessing human salivary flow, either as whole saliva or individual (or paired) glands, and unstimulated or stimulated. For further information on this area of clinical interest, the reader is referred to a review article by Dawes (1987) where it is suggested that the sensation of dry mouth is reported when the normal flow rate of unstimulated saliva is reduced by 40–50%.

The sensation of dry mouth is related to dehydrated areas of oral mucosa and/or the rubbing together of such adjacent layers.

Clearly, saliva can be removed from the mouth by swallowing, absorption from the oral mucosa and evaporation. Since saliva entering the mouth is thought to induce swallowing (Mansson and Sandberg, 1975), it is difficult to understand why anyone with even a minimal quantity of saliva would report a dry mouth. Some individuals who complain of a dry mouth have normal parotid salivary flow rates on investigation and it may well be that minor gland hypofunction is more important in giving a sensation of 'dryness' than is generally recognized.

There is a sizeable volume of evidence to suggest that saliva may be absorbed from oral mucosa (in particular from areas of inflammation) but, at present, there is doubt as to the rate or significance of this.

It seems likely that evaporation of saliva from the mouth may also be important in creating the sensation of dryness. Around 10–15% of the population are mouth-breathers at rest and in such individuals, about 50% of inspired air passes through the mouth (Niinimaa *et al.*, 1981). Levine (1989) has calculated that an obligate mouth-breather would lose up to 350 ml day^{-1} of water by this route.

It is therefore evident that 'physiological' xerostomia may occur even in individuals with normal salivary gland function, but there are many other causes of xerostomia and a brief overview of some of these will be given here.

PSYCHOLOGICAL INFLUENCES

It has been recognized for some years now that emotional state may influence salivary flow rates (Bates and Adams, 1968). Depression and anxiety are implicated in xerostomia and this is often complicated by drug treatment, which itself may cause further inhibition of salivary flow (see below).

SALIVARY GLAND AGENESIS

The failure of one or more major salivary glands to develop is a rare but recognized condition of unknown aetiology.

SALIVARY GLAND AND DUCT CALCULI

The formation of mucus plugs or calcified stones in the duct of a major gland may produce swelling, discomfort and reduced or absent salivary flow. However, xerostomia is unlikely to result if all other glands are functionally intact.

Similar problems may ensue following salivary gland surgery or in trauma, where the nerve supply to the gland may be damaged.

RADIOTHERAPY

Post-radiation atrophy of salivary tissue is a well recognized complication of radiotherapy to the head and neck. This atrophy is due in part to direct radiation damage to acinar tissue; and in part to decreased blood supply to the gland as a result of vascular fibrosis.

The resultant xerostomia is often permanent, but regeneration of acinar tissue may result in a partial return of salivary flow in some patients.

DRUG THERAPY

Drugs are probably the most common cause of salivary gland hypofunction (Schubert and Izutsu, 1987). Documentation of drug-induced salivary gland dysfunction goes as far back as 2000BC, with the Assyrians writing of belladonna being used to stop the flow of saliva (Michei-Pellegrini and Polayes, 1976). Modern-day surgeons and physicians may well use drugs specifically targeted at drying up secretions in the upper aerodigestive tract (e.g. anticholinergic agents) in general anaesthesia, endoscopic procedures or in sialorrhoea. However, the side-effects of many prescribed drugs result in troublesome xerostomia for some individuals. Lavelle (1988) lists over 50 such drugs and Bahn (1972) lists 250, although many of these are combination drugs.

Regardless of absolute numbers, it is clear that a significant number of drugs cause troublesome xerostomia as a side-effect. These include: antidepressants, antipsychotics, antihypertensives, diuretics, antispasmodics, antihistamines, anxiolytics, decongestants and antiparkinsonism drugs. Cancer chemotherapy agents may also cause xerostomia by direct action on the gland acinar and duct cells.

SJÖGREN'S SYNDROME

This is a relatively common chronic inflammatory autoimmune exocrinopathy characterized by decreased lacrimal and salivary gland function, leading to dry eyes and dry mouth. It is now generally agreed that two forms of the syndrome exist (Scully, 1986):

1. *primary*: where the patient has dry eyes and dry mouth and no clinical evidence of any underlying connective tissue disorder but may have positive serological markers; and
2. *secondary*: where the patient has an underlying connective tissue disorder (e.g. rheumatoid arthritis or systemic lupus erythematosus) and dry eyes and/or dry mouth.

Diagnosis of Sjögren's syndrome is mainly on clinical grounds with important laboratory confirmation. The criteria include: serum immunological markers, sialometry (the measurement of salivary flow rates), sialography (the visualization of gland structure using radiopaque dye techniques), histological examination of minor gland tissue from the lip and scintiscanning. Salivary protein abnormalities are also associated with this disorder (Beeley, Khoo and Lamey, 1991).

The complexity of immunological status in Sjögren's syndrome is outside the scope of this article, but the reader is referred to one of many standard texts on the subject (Talal, Moutsopoulos and Kassan, 1987).

In the analysis of salivary flow rates, some authors use whole saliva (Ben-Aryeh *et al.*, 1984) while others use parotid saliva (Mason and Chisholm, 1975). Reference data are available to compare normal age- and sex-matched controls for both the stimulated and unstimulated situation (Heft and Baum, 1984).

Sialography is performed on parotid or submandibular glands, using either oil- or water-based contrast media. The presence of the 'characteristic pattern' of sialectasis is reported to vary with the type of media used, and sialectasis has been found in 15–20% of normal individuals anyway (Dijkstra, 1980).

Labial gland biopsy has been in use for over 25 years (Chisholm and Mason, 1968) and assumes that changes in minor glands are reflections of the changes in major glands (Bertram and Hjorting-Hansen, 1970). The diagnosis of Sjögren's syndrome centres on the finding of a chronic focal lymphocytic sialadenitis in the minor gland tissue.

Scintiscanning studies of major salivary glands uses a radionuclide (^{99m}Tc sodium pertechnetate) injected intravenously and selectively taken up by the glands. At the time of presentation, around 50–70% of patients with Sjögren's syndrome have some gland function (Stuchell, Mandel and Baurmash, 1984) but gland uptake of the radionuclide is delayed, reduced or absent (Daniels *et al.*, 1979). It has been shown that there is good correlation between stimulated parotid flow rates and scintigraphic findings (Lamey, 1989).

The impact of newer functional scanning techniques (e.g. single photon emission computed tomography – SPECT) on salivary gland disease remains to be seen.

Clearly, evidence from scintiscanning for the presence or absence of functional salivary tissue has enormous implications for patient management, and this concept is central to the next part of our discussion.

With current knowledge, patients with no functional salivary gland tissue require provision of an artificial saliva, whereas patients with some residual functioning tissue may be treated successfully by drug manipulation (Fox *et al.*, 1986), e.g. the parasympathomimetic agent, pilocarpine (Greenspan and Daniels, 1987).

Therapy for xerostomia

The basic principles of management apply to all cases of xerostomia, regardless of cause. However, all reversible aetiological factors should be dealt with, and where the dryness is thought to be drug induced, the possibility of changing to a structurally unrelated drug should be discussed with the patient's physician.

The following basic principles of management are important in xerostomia:

EDUCATION

The patient should be given time with an experienced clinician to discuss, in clear and simple terms, the diagnosis, prognosis and proposed management. This will be followed up appropriately at regular review appointments.

ORAL HYGIENE

A good, understandable oral hygiene regime should be established. Photographic material depicting the effect of xerostomia on the teeth (caries), tooth-bearing tissues (gingivitis and periodontitis) and oral soft tissues (fungal infection and traumatic erosions) is often helpful as an educational tool.

Dentate patients should be encouraged in the regular use of fluoride toothpaste (a preparation specially formulated for dry mouth should be used, if available), together with a daily rinse of a fluoride mouthwash (e.g. 0.05% NaF). The consumption of refined carbohydrate should be discouraged. The use of a chlorhexidine mouthwash as a means of chemical plaque control should also be encouraged.

Patients with dentures should be encouraged in meticulous denture hygiene measures – particularly removing the dentures at bed-time. The use of sucrose- and glucose-containing sweets to keep the mouth moist is to be discouraged as the constant bathing of the oral mucosa with a sugar-rich solution, particularly in denture wearers, may predispose to oral fungal infections.

MONITORING

Routine clinical examination coupled with microbiological sampling is helpful to detect overt or latent oral infection (e.g. *Staphylococcus aureus* or *Candida* sp.).

SALIVARY GLAND HYPOFUNCTION

An educated decision must be made on appropriate management of the dry mucous membrane, i.e. stimulation of natural salivary flow or its replacement with an artificial saliva or combination therapy. This is discussed more fully below.

REGULAR REVIEW

Regular review of patients with xerostomia is essential, as encouragement is a constant feature of management strategy.

Stimulants for salivary gland hypofunction

Stimulants for salivary gland hypofunction may be broadly categorized as 'topical' agents and systemic agents.

'TOPICAL' AGENTS

It has been suggested elsewhere in this article that the act of chewing itself reflexly leads to increased salivary flow and this can be augmented by the use of a flavoured chewing gum. A sorbitol, or xylitol, non-cariogenic formulation is the gum of choice. There may also be another benefit here as the use of sorbitol-containing chewing gum has proven beneficial after a sugary meal or snack in normalizing plaque pH (Jensen, 1986).

Citric acid has been recognized to be one of the most effective salivary stimulants over the years (Watanube and Dawes, 1988b). The use of citric-acid flavoured boilings, sweets or the new chewing gums may well provide effective symptomatic relief from xerostomia, but consideration must always be given to their cariogenic potential in dentate patients, especially as citrate also chelates calcium ions.

SYSTEMIC AGENTS

Systemically administered agents for salivary gland hypofunction have chiefly centred upon Sjögren's syndrome. Numerous anecdotal reports can be cited to suggest that drugs affecting the underlying immune response may lead to an improvement in salivary gland function, e.g. gold (Godfrey *et al.*, 1983) and prednisolone (Talal *et al.*, 1975; Tabbara and Frayha, 1983).

Several agents have been proposed as systemic agents in the treatment of salivary gland hypofunction, often as case reports with a lack of adequate objective measurements or control subjects (Fox, 1987). These include pyridostigmine, bethanechol chloride, pilocarpine hydrochloride, bromhexine, trithio-paramethoxyphenylpropene (also known as anethole trithione), nicotinic acid and vitamin A. The last two of these agents have not been studied in sufficient depth as salivary stimulants to merit any mention.

Bromhexine

This is a mucolytic agent used in chronic bronchitis. Its mode of action is one of increasing the quantity of secretions produced but at a lower viscosity. However, most studies (e.g. Manthorpe *et al.*, 1981; Prause *et al.*, 1984) recorded no change in salivary flow but, interestingly, an increase in lacrimal flow. Avisar *et al.* (1981) reported a subjective improvement in salivary flow but it seems that bromhexine is of little benefit in the treatment of xerostomia.

Trithio-paramethoxyphenylpropene

The mechanism of action of this agent is really unknown, but it has been used, with reported improvement, in drug-induced xerostomia (De Buck, Titeca and Pelc,

1973), chemotherapy (Falkson, 1975), post-radiotherapy (Pazat, 1971) and in Sjögren's syndrome (Epstein, Decoteau and Wilkinson, 1983; Schiodt, Oxholm and Jacobsen, 1986). It has also been of significant reported benefit in combination with pilocarpine (Epstein and Schubert, 1987). This latter study used a group of nine post-radiotherapy patients on a combined regime of pilocarpine, 4 mg four times daily, and trithioparamethoxyphenylpropene, 25 mg three times daily. Significant increases in resting and stimulated whole saliva flow rates were reported. This suggests a possible synergistic effect between the two drugs.

Paramethoxyphenylpropene has no intrinsic cholinergic activity but it may increase the number of membrane muscarinic receptors or increase their sensitivity to intrinsic acetylcholine or exogenous pilocarpine (Ukai *et al.*, 1984, 1988).

Bethanechol

This agent is an analogue of acetylcholine with both muscarinic and nicotinic cholinergic activity. Given orally at a dose of 25 mg three times daily, it has been effective in counteracting xerostomia caused by tricyclic antidepressants (Everett, 1975) and phenothiazines (Schubert, 1979). Information on use of the drug in other conditions causing xerostomia is strictly limited.

Pilocarpine hydrochloride

The Indians of Paraguay and Brazil were aware for centuries that chewing leaves from the jaborandi plant caused sweating and salivation. This was investigated by missionaries and the plant was introduced to Europe in the late 19th century (Ferguson, 1993). Pilocarpine was isolated from jaborandi (Gerrard, 1875a) and its hydrochloride and nitrate salts were subsequently prepared (Gerrard, 1875b). Pilocarpine is a tertiary amine with cholinergic agonist action at the muscarinic receptors. A β-adrenergic action on the salivary glands is also known, stimulating protein secretion (Zelles, Blazsek and Keleman, 1990).

Numerous trials now point to pilocarpine as an effective agent against xerostomia in various conditions. The use of pilocarpine in this sphere of clinical practice has been suggested for many years (Mason and Glen, 1967), but it is only relatively recently that clinical trials have progressed our theoretical knowledge in any way.

Szabo has used pilocarpine to treat xerostomia in over 500 patients – post-radiotherapy, drug-induced, gland morphology defects and Sjögren's syndrome (Szabo, 1985). A high level of success with minimal side-effects is reported, accomplished by adjusting the pilocarpine dose to each patient's needs. Fox, Atkinson and Macynski (1991) reported the use of pilocarpine capsules, 5 mg three times daily, in a double-blind study for 6 months in 31 patients. Subjective improvement was reported by almost 90% of patients, but side-effects were noted by many (including sweating, flushing, micturition disturbance and gastrointestinal upset).

Ferguson *et al.* (1991) reported the use of pilocarpine solution in doses ranging from 1 to 15 mg, four times daily for up to 2 years in 100 patients with xerostomia. Side-effects were minimized by starting at low doses and increasing gradually to individually titrated quantities of the drug. They reported significant improvement in both salivation and lacrimation.

It would seem, therefore, that pilocarpine is an effective and relatively safe drug to stimulate residual functioning salivary gland tissue and that side-effects, although potentially serious, can be minimized by cautious prescribing and individual patient titration of drug dose. There is considerable variation in response between patients.

Pyridostigmine

This agent is a cholinesterase inhibitor and has a long duration of action. It exhibits both muscarinic and nicotinic activity. Teichman *et al.* (1987) reported a double-blind cross-over study with pyridostigmine at dose of 180 mg twice daily in the treatment of disopyramide-induced xerostomia. Significant symptomatic improvement in xerostomia was reported over a placebo and there was a clear increase in tear production.

OTHER CONSIDERATIONS

Numerous other therapies have been tired in xerostomia, with variable success, e.g. electrical stimulation (Steller, Chou and Daniels, 1988), reflexotheraphy (Pierminova, Goidenko and Rudenko, 1981) and acupuncture (Goidenko, Pierminova and Sitiel, 1985).

A more recent study (Blom, Dawidson and Angmar-Mansson, 1992) showed that patients with xerostomia of whatever cause who received acupuncture showed improved salivary flow rates during and after treatment, with some attaining normal flow rates. Improvement persisted for the follow-up year. Patients who received placebo acupuncture showed positive changes in flow rates during treatment but the changes reverted after placebo treatment was over.

Saliva substitutes

From previous discussion, it is apparent that natural saliva is a complex fluid with multiple functions and any currently available artificial saliva is a poor substitute. The ideal artificial saliva would be long-lasting, lubricant, protective of the oral mucosa to drying and inhibitory to undesired micro-organisms (Levine *et al.*, 1987). Many research centres are presently analysing the structural components of the salivary molecules responsible for various functions and it seems likely that genetic engineering and computer-assisted molecular design will provide a near-ideal saliva substitute in the not too distant future (Levine, 1993).

Currently available saliva substitutes fall into two main categories: those containing carboxymethylcellulose; and those containing natural mucins.

Sorbitol or xylitol is added as a sweetening agent. Electrolytes are added to mimic those in natural saliva, and calcium, phosphate and fluoride may be included to provide remineralization potential (Gelhard *et al.*, 1983). Various preservative agents (e.g. benzoate salts), flavourings (e.g. lemon or peppermint) and colourings may also be added.

Natural mucins are currently prepared from pig gastric mucosa or bovine submandibular gland – a fact that may be offensive to certain religious groups. Formulations containing natural mucins have been used in clinical studies (Vissink *et al.*, 1987; Blixt-Johansen *et al.*, 1992) and have been shown to exhibit relative lubricating values comparable with those of human saliva, while saliva substitutes with no mucin display very low lubricating properties (Hatton *et al.*, 1987; Levine *et al.*, 1987). Visch *et al.* (1986) also showed that mucin-containing artificial salivas were preferred to those containing carboxymethylcellulose by patients with xerostomia of various causes. Duxbury, Thakker and Wastell (1989) demonstrated that a mucin-containing preparation was the most preferred artificial saliva tested, although not all subjects found it totally satisfactory. The effect of both mucin-based and carboxymethylcellulose-based preparations, assessed either objectively by friction measurement or subjectively, is similar for both preparations and lasts little more than 15 minutes (Olsson and Axéll, 1991). A new thermogelling system, based on ethyl (hydroxyethyl) cellulose and an ionic surfactant, which increases in viscosity on warming, was effective for no longer time periods than carboxymethylcellulose preparations (Olsson *et al.*, 1993). Unfortunately, the electrolytes in saliva probably affect the thermogelling properties of this preparation.

Other artificial saliva agents available for treatment of xerostomia are glycerine and lemon mouthwash (Weisenfeld, Stewart and Mason, 1983) where the glycerine acts as a surface-wetting agent and 2.5% citric acid as a gland stimulant; glycerine lemon swabs may be used in nursing care. Use of polyethylene oxide solutions has also been reported (Vissink *et al.*, 1984). Preparations containing acids such as citric acid and malic acid are, however, limited in their usefulness due to their erosive potential on enamel and dentine in dentate individuals (Weisenfeld, Stewart and Mason, 1983). Whereas the rheological properties of mucins and human saliva are similar, carboxymethylcellulose and polyethylene oxide solutions are non-Newtonian liquids and therefore somewhat different. The addition of albumin to mucin solutions results in a preparation with rheological properties even more similar to human whole saliva (Mellema *et al.*, 1992).

Other ideas for the treatment of xerostomia with artificial saliva have included the construction of a reservoir in a denture which can be filled with an artificial saliva from a syringe. The patient then moistens the mouth by sucking some artificial saliva out of the reservoir (Vissink, Huisman and S'-Gravenmade, 1986; Hirvikangas, Posti and Makila, 1989).

Currently available proprietary artificial salivas which can be prescribed in the UK are (British National Formulary, 1994):

1. Luborant® (Antigen, Trafalgar House, Union Street, Southport PR9 0QS, UK) which is licensed for any condition giving rise to a dry mouth and contains carboxymethylcellulose and fluoride.
2. Glandosane® (Frensenius Ltd, 6–8 Christleton Court, Stuart Road, Manor Park, Runcorn, Cheshire WA7 1ST, UK) which is licensed only for dry mouth associated with radiotherapy or sicca syndrome and contains carboxymethylcellulose. It comes in neutral, lemon or peppermint flavours.
3. Saliva Orthana® (Nycomed (UK) Ltd, Nycomed House, 2111 Coventry Road, Sheldon, Birmingham B26 3EA, UK) which is licensed only for dry mouth associated with radiotherapy or sicca syndrome. It contains porcine gastric mucin and is available in two forms: aerosol spray and lozenges. The spray is available with and without fluoride; the lozenges have no fluoride content.
4. Salivix® (Thames Laboratories Ltd, Abbey Road, The Industrial Estate, Wrexham,

Clywd LL13 9PW, UK) which contains acacia and malic acid. It is available only as a pastille and is licensed for use in radiotherapy-induced xerostomia or sicca syndrome.

The future

The complexity of natural saliva, in terms of composition and function, makes finding the ideal substitute very difficult. Current formulations have aimed only to simulate the inorganic composition and rheological properties of saliva. This is not unreasonable because it is only within the past decade that the nature and unusual polyfunctional properties of the families of proteins in saliva have become apparent. Future preparations will therefore need to contain active components which modulate the oral microflora by mechanisms analogous to natural constituents, as well as by regulating remineralization and protecting the dentition and the soft tissues.

But the genes that code for these bioactive proteins are already being cloned and it seems reasonable to assume that the recombinant gene technology already exists which would enable them to be synthesized in the same way as human growth hormone and insulin are now being prepared. But would such a saliva need to contain the normal range of proteins and peptides? Or could it be possible that because many of the proteins are polyfunctional, a selection would be adequate? Could such proteins be of use in maintaining oral health in patients with normal salivary flow rates, perhaps in the prevention of dental caries or treatment of oral soft-tissue infections?

As the three-dimensional structures and the nature of bioactive domains of salivary proteins are also now beginning to be investigated, the possibility of computer-designed salivary molecules with specific amino acid sequences and carbohydrate side-chains which can be chemically synthesized is no longer in the realms of fantasy. Might it be possible to design molecules which are more effective than those that have evolved over the millennia?

The vehicle or mode of delivery of saliva substitutes, however, continues to present a particular problem. How can compounds be retained in a cavity, one biological function of which is the reflex removal by swallowing of substances ingested? The answer may ultimately lie with tissue transplants of biocompatible cells.

A vast cohort of individual patients suffering from the very handicapping condition of xerostomia deserve the best efforts of science, medicine and dentistry to design and develop a saliva substitute which will overcome so many of the problems encountered daily by them. Artificial saliva – a stimulating subject, indeed.

Acknowledgements

Figure 1 is reproduced by permission of Baillière Tindall. The authors thank Mrs F. Newman and Mr K. Hunter for preparation of the gel shown in *Figure 2*. Our sincere thanks are due to Miss Grace Martin for secretarial assistance.

References

ADAMSON, M. AND PRUITT, K.M. (1981). Lactoperoxidase catalysed inhibition of hexokinase. *Biochimica et Biophysica Acta* **658**, 238–247.

AVISAR, R., SAVIR, H., MACHTEY, I., OVAKIN, L., SHAKED, P., MENACHIE, R. AND ALLALOUF, D. (1981). Clinical trial of bromhexine in Sjögren's syndrome. *Annals of Ophthalmology* **13**, 971–973.

BAHN, S.L. (1972). Drug-related dental destruction. *Oral Surgery* **33**, 49–54.

BATES, J.F. AND ADAMS, D. (1968). The influence of mental states on the flow of saliva. *Archives of Oral Biology* **13**, 593–596.

BEELEY, J.A. (1993). Fascinating families of proteins: electrophoresis of human saliva. *Biochemical Society Transactions* **21**, 133–138.

BEELEY, J.A., KHOO, K.S. AND LAMEY, P.-J. (1991). Two-dimensional electrophoresis of human parotid saliva from normal and connective tissue disorder subjects using immobilised pH gradients. *Electrophoresis* **12**, 493–499.

BEELEY, J.A., SWEENEY, D., LINDSAY, J.C.B., BUCHANAN, L.M., SARNA, L. AND KHOO, K.S. (1991). Sodium dodecylsulphate polyacrylamide gel electrophoresis of human parotid salivary proteins. *Electrophoresis* **12**, 1032–1041.

BEN-ARYEH, T., MIRON, D., SZARGEL, R. AND GUTMAN, D. (1984). Whole saliva secretion rates in old and young healthy subjects. *Journal of Dental Research* **63**, 1147–1148.

BERTRAM, U. (1967). Xerostomia. *Acta Odontologica Scandinavica* **25** (Suppl. 49).

BERTRAM, U. AND HJORTING-HANSEN, E. (1970). Punch biopsy of minor salivary glands in the diagnosis of Sjögren's syndrome. *Scandinavian Journal of Dental Research* **78**, 295–300.

BLIXT-JOHANSEN, G., EK, A.C., GANOWIAK, W., GRANERUS, A.K., VON SCHENCK, H., UNOSSON, M. AND WIESEL, K. (1992). Improvement of oral mucosa with mucin containing artificial saliva in geriatric patients. *Archives of Gerontology and Geriatrics* **14**, 193–201.

BLOM, M., DAWIDSON, I. AND ANGAR-MANSSON, B. (1992). The effect of acupuncture on salivary flow rates in patients with xerostomia. *Oral Surgery, Oral Medicine, Oral Pathology* **73**, 293–298.

BRADWAY, S.D., BERGEY, E.J., SCANNAPIECO, F.A., RAMASUBBU, N., ZAWACKI, S. AND LEVINE, M.J. (1992). Formation of salivary-mucosal pellicle: the role of transglutaminase. *Biochemical Journal* **284**, 557–564.

British Medical Association and Royal Pharmaceutical Society of Great Britain (1994). British National Formulary, No. 27. BMA and RPSGB.

CASTLE, J.D., ARVAN, P. AND CAMERON, R. (1987). Protein production and secretion in exocrine cells. *Journal of Dental Research* **66**, (Special Issue), 633–637.

CHISHOLM, D.M. AND MASON, D.K. (1968). Labial salivary gland biopsy in Sjögren's syndrome. *Journal of Clinical Pathology* **21**, 656–660.

DANIELS, T.E., POWELL, M.P., SYLVESTER, R.A. AND TALAL, N. (1979). An evaluation of salivary scintigraphy in Sjögren's syndrome. *Arthritis and Rheumatism* **22**, 809–814.

DAWES, C. (1987). Physiological factors affecting salivary flow rate, oral sugar clearance, and the sensation of dry mouth in man. *Journal of Dental Research* **66** (special issue), 648–653.

DE BUCK, R., TITECA, R. AND PELC, I. (1973). Etude contrôlée de l'action de l'anetholtrithione [Sulfarlem] sur les asiates et les hyposiates provoquées par les médicaments psychotropes. *Acta Psychiatrica Belgica* **73**, 520–519.

DIJKSTRA, P.F. (1980). Classification and differential diagnosis of sialographic characteristics in Sjögren's syndrome. *Seminars in Arthritis and Rheumatism* **10**, 10–17.

DOUGLAS, W.H., REETH, E.S., RAMASUBBU, N., BHANDARY, K.K., RAJ, P.A. AND LEVINE, M.J. (1991). Statherin: a major boundary lubricant of human saliva. *Biochemical and Biophysical Research Communications* **180**, 91–97.

DUXBURY, A.J., THAKKER, N.S. AND WASTELL, D.G. (1989). A double-blind cross-over trial of a mucin-containing artificial saliva. *British Dental Journal* **166**, 115–120.

ECKERSALL, P.D. AND BEELEY, J.A. (1981). Genetic analysis of human salivary α-amylase isozymes by isoelectric focusing. *Biochemical Genetics* **19**, 1055–1062.

EDGAR, W.M. (1976). The role of saliva in the control of pH change in human dental plaque. *Caries Research* **10**, 241–254.

EDGAR, W.M. (1992). Saliva: its secretion and composition. *British Dental Journal* **172**, 305–312.

EMMELIN, N. (1972). Control of salivary glands. In *Oral Physiology* (N. Emmelin and Y. Zotterman, Eds), pp. 1–16. Pergamon Press, Oxford.

EPSTEIN, J.B. AND SCHUBERT, M.M. (1987). Synergistic effect of sialogogues in management of xerostomia after radiation therapy. *Oral Surgery, Oral Medicine, Oral Pathology* **64**, 179–182.

EPSTEIN, J.B., DECOTEAU, W.E. AND WILKINSON, A. (1983). Effect of sialor in the treatment of xerostomia in Sjögren's syndrome. *Oral Surgery, Oral Medicine, Oral Pathology* **56**, 495–499.

EVERETT, H.C. (1975). The use of bethanechol chloride with tricyclic antidepressants. *American Journal of Psychiatry* **132**, 1202–1204.

FALKSON, H.C. (1975). Relief of dry mouth. *South African Medical Journal* **49**, 690.

FDI (1992). Saliva: its role in health and disease. *International Dental Journal* **42**, 291–304.

FERGUSON, D.B. (1987). Current diagnostic uses of saliva. *Journal of Dental Research* **66**, 420–424.

FERGUSON, M.M. (1989). Diseases of the salivary glands and sensory impairment. In *World Workshop on Oral Medicine* (H.D. Millard and D.K. Mason, Eds). Year Book Medical Publishers, Chicago.

FERGUSON, M.M. (1993). Pilocarpine and other cholinergic drugs in the management of salivary gland dysfunction. *Oral Surgery, Oral Medicine, Oral Pathology* **75**, 186–191.

FERGUSON, M.M., HAYES, P., HIGHTON, J., JONES, D.S., MACFADYEN, E.E. AND PALMER, D.G. (1991). Pilocarpine oral solution. *British Dental Journal* **170**, 251.

FOX, P.C. (1987). Systemic therapy of salivary gland hypofunction. *Journal of Dental Research* **66** (Special Issue), 689–692.

FOX, P.C., ATKINSON, J.C. AND MACYNSKI, A.A. (1991). Pilocarpine treatment of salivary gland dysfunction and dry mouth (xerostomia). *Archives of Internal Medicine* **151**, 1149–1152.

FOX, P.C., VAN DER VEN, P.F., BAUM, B.J. AND MANDEL, I.D. (1986). Pilocarpine for the treatment of xerostomia associated with salivary gland dysfunction. *Oral Surgery* **61**, 243–245.

GARRETT, J.R. (1982). Adventures with autonomic nerves. Perspectives in salivary glandular innervations. *Proceedings of the Royal Microbiological Society* **17**, 242–253.

GARRETT, J.R. (1987). The proper role of nerves in salivary stimulation, a review. *Journal of Dental Research* **66**, 387–397.

GELHARD, T.B.M., FIDLER, V., S'-GRAVENMADE, E.J. AND VISSINK, A. (1983). Remineralisation of softened human enamel in mucin- or CMC- containing artificial salivas. *Journal of Oral Pathology* **12**, 336–341.

GERRARD, A.W. (1875a). The alkaloid and active principle of jaborandi. *Pharmaceutical Journal Transactions*, p. 865.

GERRARD, A.W. (1875b). Experiments upon the alkaloid of jaborandi – crystallisation of its salts. *Pharmaceutical Journal Transactions*, p. 965.

GODFREY, N., SIMON, T.M., KUNISHIMA, D.H. AND LORBER, A. (1983). Sjögren's syndrome with pseudolymphoma treated with chrysotherapy. *Journal of Rheumatology* **10**, 957–960.

GOIDENKO, V.S., PIERMINOVA, I.S. AND SITIEL, A.B. (1985). Use of auricular acupuncture reflexotherapy in treating Sjögren's disease. *Stomatologia* **64**, 47–48.

GREENSPAN, D. AND DANIELS, T.E. (1987). Effectiveness of pilocarpine in post-radiation xerostomia. *Cancer* **59**, 1123–1125.

HAMOSH, M. AND BURNS, W.A. (1977). Lipolytic activity of human lingual glands (Ebner). *Laboratory Investigation* **37**, 603–608.

HATTON, M.N., LOOMIS, R.E., LEVINE, M.J. AND TABAK, L.A. (1985). Masticatory lubrication. *Biochemical Journal* **230**, 817–820.

HATTON, M.N., LEVINE, M.J., MARGARONE, J.E. AND AGUIRRE, A. (1987). Lubrication and viscosity features of human saliva and commercially available saliva substitutes. *International Journal of Oral and Maxillofacial Surgery* **45**, 496–499.

HEFT, M.W. AND BAUM, B.J. (1984). Unstimulated and stimulated parotid salivary flow rate in individuals of different ages. *Journal of Dental Research* **63**, 1182–1185.

HEINEMAN, H.S. AND GREENBERG, M.S. (1980). Cell protective effects of human saliva specific for herpes simplex virus. *Archives of Oral Biology* **25**, 257–261.

HIRVIKANGAS, M., POSTI, J. AND MAKILA, E. (1989). Treatment of xerostomia through use of dentures containing reservoirs of saliva substitute. *Proceedings of the Finnish Dental Society* **85**, 47–50.

JENSEN, M.J. (1986). Responses of interproximal plaque pH to snack foods and effect of chewing sorbitol-containing gum. *Journal of the American Dental Association* **113**, 262–266.

LAMEY, P.-J. (1989). Diseases of the salivary glands and sensory impairment. In *World Workshop on Oral Medicine* (H.D. Millard and D.K. Mason, Eds). Year Book Medical Publishers, Chicago.

LAMKIN, M.S. AND OPPENHEIM, F.G. (1993). Structural features of salivary proteins. *Critical Reviews in Oral Biology and Medicine* **4**, 251–260.

LAVELLE, C.L.B. (1988). *Applied Oral Physiology*, (2nd edn.). Wright, London.

LEVINE, M.J. (1993). Development of artificial salivas. *Critical Reviews in Oral Biology and Medicine* **4**, 279–286.

LEVINE, M.J., AGUIRRE, A., HATTON, M.N. AND TABAK, L.A. (1987). Artificial salivas: present and future. *Journal of Dental Research* **66** (Special Issue), 693–698.

LEVINE, R.S. (1989). Saliva: 3. Xerostomia – aetiology and management. *Dental Update* (Special Suppl.), pp. 12–15.

MCCANN, H.G. AND BRUDEVOLD, F. (1966). The mechanism of caries inhibiting effect of fluoride. In *Environmental Variables in Oral Disease* (S.J. Kreshover and F.C. McClure, Eds), pp. 103–128. American Academy of Advanced Science, Washington DC.

MACKAY, B.J., DENEPITIYA, L., IACONO, V.J., KROST, S.B. AND POLLOCK, J.J. (1984). Growth inhibitory and bactericidal effects of human parotid salivary histidine-rich polypeptides on *Streptococcus mutans*. *Infection and Immunity* **44**, 695–701.

MALAMUD, D. AND FRIEDMAN, H.M. (1993). HIV in the oral cavity: virus, viral inhibitory activity and antiviral antibodies – a review. *Critical Reviews in Oral Biology and Medicine* **4**, 461–466.

MANSSON, I. AND SANDBERG, N. (1975). Salivary stimulus and swallowing reflex in man. *Acta Oto-Laryngologica* **79**, 445–450.

MANTHORPE, R., FROST-LARSEN, K., HOJ, L., ISAGER, H. AND PRAUSE, J.V. (1981). Bromhexine treatment of Sjögren's syndrome: effect on lacrimal and salivary secretions, and on proteins in tear fluid and saliva. *Scandinavian Journal of Rheumatology* **10**, 177–180.

MARTINEZ, J.R. (1987). Ion transport and water movement. *Journal of Dental Research* **66** (Special Issue), 638–647.

MARTINEZ, J.R., HOLZGROVE, H. AND FRICK, A. (1966). Micropuncture study of submaxillary gland of adult rats. *Pflügers Archiv* **290**, 124–133.

MASON, D.K. AND CHISHOLM, D.M. (1975). *Salivary Glands in Health and Disease*. W.B. Saunders, London.

MASON, D.K. AND GLEN, A.I.M. (1967). The aetiology of xerostomia (dry mouth). *Dental Magazine Oral Topics* **84**, 235–238.

MEHANSHO, H., BUTLER, L.G. AND CARLSON, D.M. (1987). Dietary tannins and salivary proline-rich proteins: interactions, induction and defence mechanisms. *Annual Review of Nutrition* **7**, 423–440.

MELLEMA, J., HOLTERMAN, H.J., WATERMAN, H.A., BLOM, C. AND S'-GRAVENMADE, E.J. (1992). Rheological properties of mucin-containing solutions and saliva substitutes. *Biorheology* **29**, 231–249.

MICHEI-PELLEGRINI, V. AND POLAYES, I.M. (1976). Historical background. In *Diseases of the Salivary Glands* (R.M. Rankow and I.M. Polayes, Eds), pp. 1–16. W.B. Saunders, Philadelphia.

NEWMAN, F., BEELEY, J.A., MACFARLANE, T.W., GALBRAITH, J. AND BUCHANAN, L.M. (1993). Salivary protein interactions with oral bacteria – an electrophoretic study. *Electrophoresis* **14**, 1322–1327.

NIINIMAA, V., COLE, P., MINTZ, S. AND SHEPHARD, R.J. (1981). Oronasal distribution of respiratory airflow. *Respiratory Physiology* **43**, 69–75.

OLSSON, H. AND AXÉLL, T. (1991). Objective and subjective efficacy of saliva substitutes containing mucin and carboxymethylcellulose. *Scandinavian Journal of Dental Research* **99**, 316–319.

OLSSON, H., AXÉLL, T., CARLSSON, A. AND BOGENTOFT, C. (1993). Objective and subjective efficacy evaluation of various polymer based saliva substitutes. *Scandinavian Journal of Dental Research* **101**, 37–39.

PAZAT, P. (1971). Traitement de la sécheresse buccale postradiothérapeutique par le trithioparaméthoxyphénylpropène. *Vie Méd* **2**, 2751–2754.

PIERMINOVA, I.S., GOIDENKO, V.C. AND RUDENKO, I.V. (1981). Experience with using reflexotherapy in treating Sjögren's syndrome. *Stomatologia* **60**, 37–38.

POLLOCK, J.J., DENEPITIYA, L., MACKAY, B.J. AND IACONO, V.J. (1984). Fungistatic and fungicidal activity of the human parotid salivary histidine-rich polypeptides on *Candida albicans*. *Infection and Immunity* **44**, 702–707.

PRAKOBPHOL, A., LEFFLER, H. AND FISHER, S.J. (1993). The high molecular weight human mucin is the carrier of ABH, Le^a and Le^b blood group antigens. *Critical Reviews in Oral Biology and Medicine* **4**, 325–333.

PRAUSE, J.V., FROST-LARSEN, K., HOJ, L., ISAGER, H. AND MANTHORPE, R. (1984). Lacrimal and salivary secretion in Sjögren's syndrome: the effect of systemic treatment with bromhexine. *Acta Ophthalmologica* **62**, 489–497.

SABATINI, L.M., WARNER, T.F., SAITOH, E. AND AZEN, E.A. (1989). Tissue distribution of RNA's for cystatins, histatins, statherin and proline-rich salivary proteins. *Journal of Dental Research* **68**, 1138–1145.

SCHIODT, M., OXHOLM, P. AND JACOBSEN, A. (1986). Treatment of xerostomia in patients with primary Sjögren's syndrome with Sulfarlem. *Scandinavian Journal of Rheumatology* **61** (Suppl.), 250–252.

SCHNEYER, L.H., PIGMAN, W., HANAHAN, L. AND GILMORE, R.W. (1956). Rate of flow of human parotid, sublingual and sub-maxillary secretions during sleep. *Journal of Dental Research* **35**, 109–114.

SCHUBERT, D.S.P. (1979). Use of bethanechol chloride with phenothiazines: a case report. *American Journal of Psychiatry* **136**, 110–111.

SCHUBERT, M.M. AND IZUTSU, K.T. (1987). Iatrogenic causes of salivary gland dysfunction. *Journal of Dental Research* **66** (Special Issue), 680–688.

SCULLY, C. (1986). Sjögren's syndrome: clinical and laboratory features, immunopathogenesis and management. *Oral Surgery* **62**, 510–523.

SHATZMAN, A.R. AND HENKIN, R.I. (1981). Gustin concentration changes relative to salivary zinc and taste in humans. *Proceedings of the National Academy of Sciences, USA* **78**, 3867–3871.

SHELLIS, R.P. AND DIBDIN, G.H. (1988). Analysis of the buffer systems in dental plaque. *Journal of Dental Research* **67**, 438–446.

SLOMIANY, B.L., MURTY, V.L.N., ZDEBSKA, E., SLOMIANY, A., GWOZDZINSKI, K. AND MANDEL, I.D. (1986). Tooth surface-pellicle lipids and their role in the protection of enamel against lactic-acid diffusion in man. *Archives of Oral Biology* **31**, 187–191.

STELLER, M., CHOU, L. AND DANIELS, T.E. (1988). Electrical stimulation of salivary flow in patients with Sjögren's syndrome. *Journal of Dental Research* **67**, 1334–1337.

STUCHELL, R.N., MANDEL, I.D. AND BAURMASH, H. (1984). Clinical utilisation of sialochemistry in Sjögren's syndrome. *Journal of Oral Pathology* **13**, 303–309.

SZABO, G. (1985). Xerostomia kezelése pilocarpinnal. *Fogorv Sz* **78**, 65–69.

TABAK, L.A., LEVINE, M.J., MANDEL, I.E. AND ELLISON, S.A. (1982). Role of salivary mucins in the protection of the oral cavity. *Journal of Oral Pathology* **11**, 1–17.

TABBARA, K.F. AND FRAYHA, R.A. (1983). Alternate-day steroid therapy for patients with primary Sjögren's syndrome. *Annals of Opthalmology* **15**, 358–361.

TALAL, N., MOUTSOPOULOS, H.M. AND KASSAN, S.S. (Eds) (1987). *Sjögren's Syndrome: Clinical and Immunological Aspects*. Springer-Verlag, New York.

TALAL, N., GREY, H.M., ZVAIFLER, N., MICHALSKI, J.P. AND DANIELS, T.E. (1975). Elevated salivary and synovial fluid B_2-microglobulin in Sjögren's syndrome and rheumatoid arthritis. *Science* **188**, 1196–1198.

TEICHMAN, S.L., FERRICK, A., KIM, S.G., MATOS, J.A., WASPE, L.E. AND FISHER, J.D. (1987). Disopyramide–pyridostigmine interaction: selective reversal of anticholinergic symptoms with preservation of antiarrhythmic effect. *Journal of the American College of Cardiology* **10**, 633–641.

THAYSEN, J.H., THORN, N.A. AND SCHWARTZ, I.L. (1954). Excretion of sodium, potassium, chloride and carbon dioxide in human parotid saliva. *American Journal of Physiology* **178**, 155–159.

UKAI, Y., TANIGUCHI, N., TAKESHITA, K., KIMURA, K. AND ENOMOTO, H. (1984). Chronic anethole trithione treatment enhances the salivary secretion and increases the muscarinic acetylcholine receptors in the rat submaxillary gland. *Archives Internationales de Pharmacodynamie et de Therapie* **271**, 206–212.

UKAI, Y., TANIGUCHI, N., TAKESHITA, K., OGASAWARA, T. AND KIMURA, K. (1988). Enhancement of salivary secretion by chronic anethole trithione treatment. *Archives Internationales de Pharmacodynamie et de Therapie* **294**, 248–258.

VISCH, L.L., S'-GRAVENMADE, E.J., SCHAUB, R.M.H., VAN PUTTEN, W.L.J. AND VISSINK, A. (1986). A double-blind crossover trial of CMC- and mucin-containing saliva substitutes. *International Journal of Oral and Maxillofacial Surgery* **15**, 395–400.

VISSINK, A., HUISMAN, M.C. AND S'-GRAVENMADE, E.J. (1986). Construction of an artificial saliva reservoir in an existing maxillary denture. *Journal of Prosthetic Dentistry* **56**, 70–74.

VISSINK, A., WATERMAN, H.A., S'-GRAVENMADE, E.J., PANDER, A.K. AND VERMEY, A. (1984). Rheological properties of saliva substitutes containing mucin, carboxymethylcellulose or polyethylene oxide. *Journal of Oral Pathology* **13**, 22–28.

VISSINK, A., SCHAUB, R.M., VAN RIJN, L.J., S'-GRAVENMADE, E.J., PANDERS, A.K. AND VERMEY, A. (1987). The efficacy of mucin-containing artificial saliva in alleviating symptoms of xerostomia. *Geriodontology* **3**, 95–101.

WATANUBE, D. AND DAWES, C. (1988a). A comparison of the effect of tasting and chewing foods on the flow rate of whole saliva. *Archives of Oral Biology* **33**, 761–764.

WATANUBE, D. AND DAWES, C. (1988b). The effects of different foods and concentrations of citric acid on the flow rate of whole saliva in man. *Archives of Oral Biology* **33**, 1–5.

WATERHOUSE, J.P., BEELEY, J.A. AND MASON, D.K. (1990). Introduction and biological basis. In *Oral Manifestations of Systemic Disease* (H.K. Jones and D.K. Mason, Eds), pp. 1–29. Ballière Tindall, London.

WEISENFELD, D., STEWART, A.M. AND MASON, D.K. (1983). A critical assessment of oral lubricants in patients with xerostomia. *British Dental Journal* **155**, 155–157.

WIKNER, S. AND NEDLICH, U. (1988). A clinical evaluation of the Dentobuff method to estimate the buffer capacity of saliva. *Swedish Dental Journal* **9**, 45–47.

WILLIAMSON, M.P. (1994). The structure and function of proline-rich regions in proteins. *Biochemical Journal* **297**, 249–260.

YOUNG, J.A. (1979). Salivary secretions of inorganic electrolytes. *International Review of Physiology* **19**, 1–58.

ZELLES, T., BLAZSEK, J. AND KELEMAN, E. (1990). Analysis of the effectiveness of pilocarpine in the management of xerostomia. *Fogorv Sz* **83**, 319–321.

3

Modification of Flower Colour using Genetic Engineering

PAULA ELOMAA[1] AND TIMOTHY HOLTON[2]

[1] *Institute of Biotechnology, University of Helsinki, Finland and* [2] *Florigene Pty Ltd, Collingwood, Victoria, Australia*

Introduction

Plant breeding, besides producing better crop plants, has also focused on species that are attractive and have aesthetically valuable characteristics. Today, a whole new area of the flower industry has evolved and rapidly expanded world-wide (Mol, Stuitje and van der Krol, 1989). Traditionally, breeding of ornamental plants has been based mainly on continuous crossing and selection. Several decades of traditional breeding have combined commercially important characteristics into élite genotypes. Developments in tissue culture and methods of molecular genetics now give an alternative approach to change single characteristics in these genotypes. Furthermore, possibilities of producing completely new characteristics are no longer limited by the natural genetic variation existing in the target species, since genes of any origin can be transferred into the pre-existing gene pool. This is not possible by using traditional methods.

Breeding of ornamental species via genetic engineering focuses on many characteristics that are the same as in traditional breeding programmes. One of the most important goals from the flower industry's perspective is to modify production characteristics such as flower productivity, timing and synchrony of flowering, disease and pest resistance and vase life. Nevertheless, flower morphology and flower colour are of the greatest importance in directing the markets in the flower industry, as characteristics that are judged by the consumer. Ever-changing preferences and fashion increase the demand for rapid production of new characteristics in ornamental plants.

Abbreviations: AMT, anthocyonin methyltransferase; An13, anthocyonin 13; ANS, anthocyanidin synthase; CaMV, Cauliflower mosaic virus; cDNA, complementary DNA; C4H, cinnamate-4-hydroxylase; CHI, chalcone-flavanone isomerase; CHS, chalcone synthase; 4CL, 4-coumaryl-CoA ligase; DFR, dihydroflavonol- 4-reductase; DHK, dihydrokaempferol; DHM, dihydromyricetin; DHQ, dihydroquercetin; EPSPS, 5-enolpyruvylshikimate-3-phosphate synthase; F3H, flavanone 3-hydroxylase; F3'H, flavonoid 3'-hydroxylase; F3'5'H, flavonoid 3', 5'-hydroxylase; FLS, flavonol synthase; FS, flavone synthase; GA, gibberellic acid; 3GT, UDPG-flavonoid-3-*O*-glucosyltransferase; GUS, β-glucuronidase; PAL, phenylalanine ammonialyase; PCR, polymerase chain reaction; 3RT, anthocyanidin-3-glucoside rhamnosyltransferase; T-DNA, transfer DNA; UV, ultraviolet.

Biotechnology and Genetic Engineering Reviews — Vol. 12, December 1994
0264–8725/94/12/63–88 $20.00 + $0.00 © Intercept Ltd, P.O. Box 716, Andover, Hampshire SP10 1YG, UK

There are still many limitations and problems to be overcome before molecular methods can be used effectively in breeding. Molecular flower-breeding requires transformation and regeneration methods for the target species. The gene to be transferred has to be isolated, characterized and suitable constructs with selectable marker genes have to be made. The gene must then be stably integrated into the genome of the target plant and expressed in a correct manner and have the desired, and stable, effect on the phenotype. The main limitation to the widespread use of molecular breeding is that there is no general transformation method suitable for all species. This is because regeneration and transformation capacities vary strongly in different genotypes, and even in various explants. Also, many target characteristics are encoded by several genes. Molecular breeding will, by no means, replace traditional breeding methods, but instead it may provide additional tools to introduce new, desirable characteristics into pre-existing varieties.

Modification of flower colour is one of the first examples of how genetic engineering can be applied in breeding new varieties. This has been possible through detailed knowledge of the genetics and biochemistry of the phenylpropanoid pathway leading to synthesis of anthocyanin pigments, and isolation of genes encoding the enzymes of the pathway. Although huge natural variation exists with respect to flower colour in several important ornamental plants, no plant species can produce varieties in the full spectrum of colours (Mol, Stuitje and van der Krol, 1989). Blue and purple varieties are still missing from important ornamental species such as rose, carnation, chrysanthemum, gerbera and tulip. There is also an absence of yellow colour in pelargonium, cyclamen and saintpaulia, as well as bright red varieties in cyclamen, iris, saintpaulia and lisianthus (Forkmann, 1991). Molecular flower breeding provides an attractive approach to change flower pigmentation in a highly controlled way, in otherwise superior genotypes achieved through traditional breeding programmes.

Besides producing new flower colours in ornamental plants, transformation with genes affecting the flower pigmentation can be used as phenotypic markers to visualize transgene regulation in general and thereby provide additional information for other transformation applications.

This chapter describes the present state in development of transformation methods in ornamental species and outlines factors influencing pigmentation. The main focus of this chapter is to introduce the information considering applications of genetic engineering in connection to flower colour. Colour modification in *Petunia hybrida*, which has been used as a model plant in this field, will be presented, together with the latest achievements in the most important ornamental plants.

Transformation of ornamental species

Development of a transformation method for any plant species or cultivar requires an effective tissue culture technique in order to regenerate the transformed tissue. Regeneration capacity is the main limitation for transformation in many species. However, regeneration media for the commercially most important ornamental species, rose, carnation, chrysanthemum and gerbera, have been developed in recent years (Hutchinson *et al*., 1992). This has enabled the use of *Agrobacterium*-mediated transformation in these species, though with variable efficiency.

Agrobacterium-mediated transformation systems have so far been the most successful for dicotyledonous plants. The mechanism and interactions underlying this method have been under extensive study and have been reviewed comprehensively by Zambryski (1988), Hooykaas (1989) and Winans (1992). Therefore, we will only briefly describe some basic details of the method.

Transformation using the *Agrobacterium* vector system is based on the capability of a naturally occurring Gram-negative soil bacterium to infect the target plant and introduce part of its DNA (T-DNA) from the Ti-plasmid into wounded plant cells. Variable copies of this DNA are integrated into random positions in the plant genome (Zambryski, 1988). In nature, infection by *Agrobacterium tumefaciens* results in crown gall tumour formation in most dicotyledonous plants. The T-DNA directs plant hormone synthesis, leading to the transformation of plant cells into tumour cells. Some of the genes in the T-DNA region encode for enzymes that synthesize opines. These compounds are excreted from the transformed plant cells and consumed by the tumour-inducing *Agrobacterium*. The T-DNA region is surrounded by 25 bp direct repeats that have been shown to define the transferred DNA. Separate from the T-DNA region, virulence (*vir*) genes encode *trans*-acting products that are associated with the DNA transfer process. The transfer mechanism is complicated and it is assumed that the T-DNA enters the plant nuclear genome as a single strand and integrates at random positions. The T-DNA can be integrated as a single copy or multiple copies, either tightly linked or sometimes dispersed throughout the genome (reviewed by Winans, 1992).

The use of the *Agrobacterium* system for gene transfer has been made possible by removing the tumour-inducing genes from the T-DNA (disarmed vectors) and inserting foreign genes to be transferred. There are two kinds of vector systems – binary and co-integrate. In co-integrate vectors the introduced DNA is in the *cis*-position in respect of *vir* genes, in place of the T-DNA (Zambryski *et al.*, 1983; Deblaere *et al.*, 1985). In binary vectors the T-DNA is in a separate plasmid from the *vir* genes needed for the transformation (An *et al.*, 1985). Both vector systems have been used to transform ornamental species.

The major limitation of this system is that it is applicable only to those species that are susceptible to infection by *Agrobacterium*. Activation of the virulence system is caused by phenolic compounds formed in the wounding of the plant tissue (Bolton, Nester and Gordon, 1986). Deficiency of these inducers in monocotyledonous plants has been proposed to be the reason that they are not susceptible to agro-infection. However, it has been shown that *Agrobactium* infection does occur in a few monocots, including species of Liliaceae and Amaryllidaceae (Hooykaas-Van Slogteren, Hooykaas and Schilperoort, 1984).

Another limitation of the *Agrobacterium*-mediated transformation system is that it requires an efficient regeneration method. Although there is a lot of literature on regeneration methods for different ornamental species, there are still only a few published transformation protocols using the *Agrobacterium* method in ornamental plants. Most of the methods rely on various modifications of the leaf-disc transformation protocol (Horsch *et al.*, 1985).

Lu *et al.* (1991) published a transformation protocol for the carnation (*Dianthus caryophyllus* L.) cultivars White, Red and Crowley Sim, using two marker genes encoding NPTII and GUS. Transgenic plants were regenerated from stem and petal explants with relatively high frequency and without any differences between the cultivars tested. It was also noticed that adding 20 μM acetosyringone to the co-cultivation medium for induction of *vir* genes increased the transformation frequency threefold.

A similar system was used to transfer the *nptII* and *uidA* marker genes into chrysanthemum (*Dendranthema indicum* L. Korean genotype) by Ledger, Deroles and Given (1991), using leaf explants. A total of 80 kanamycin-resistant plants were produced using the *nos-nptII* gene, and three transformed shoots using *nptII* and *uidA*. Van Wordragen *et al.* (1991) used an *Agrobacterium*-mediated method to produce transgenic calli from chrysanthemum (*Dendranthema grandiflora* cv. Parliament) that showed NPTII and GUS activity. The susceptibilities of various genotypes and *Agrobacterium* strains were studied and it was concluded that the transformation of chrysanthemum strongly depends on the genotype and bacterial strain used.

Transformation methods using the *Agrobacterium* system have been developed for four cultivars of the increasingly popular cut flower, lisianthus (*Eustoma grandiflorum* Grise.) and, despite a low transformation frequency (1–3% of the inoculated leaf pieces), kanamycin-resistant shoots were regenerated (Deroles *et al.*, 1993). Eleven cultivars of pelargonium were inoculated with two *Agrobacterium* strains containing plasmids with the *uidA* marker gene. Leaf discs of all cultivars showed GUS activity, with at least one strain/plasmid combination. Development of regeneration systems for three of these cultivars is in progress (Boase and Smith, 1994).

An *Agrobacterium* co-cultivation method was used to transform pieces of petioles of *Gerbera hybrida*, variety Terra Regina (Elomaa *et al.*, 1993), and also quite recently a detailed transformation protocol for various chrysanthemum varieties has been submitted by Robinson *et al.*, as reported by Courtney-Gutterson *et al.* (1994). These methods were used to alter the floral pigmentation in the ornamental species. These cases will be discussed further in connection with colour modification.

Transformation of ornamental species has, in most reported cases, relied on the *Agrobacterium* system. However, many important ornamental species, such as tulip, freesia, iris and daffodil, are monocots and therefore may not be susceptible to *Agrobacterium* infection. Development of the particle bombardment method (Klein *et al.*, 1987), based on the direct transformation of foreign genes into embryogenic tissue, has turned out to be successful in the transformation of many crop plants (Christou, 1992). This method may also be an alternative approach in ornamental plants, and has already been used to introduce the phosphinothricin (Basta) resistance gene and *uidA* marker gene into tulips. Transgenic shoots were formed from immature flower-stem segments (Wilmink, van de Ven and Dons, 1994). The expression of the *bar* and *uidA* genes was enhanced using monocot promoters.

Other alternative approaches to transform plant cells are being studied. Silicon carbide fibres have been used to deliver plasmid DNA into embryogenic maize suspension-culture cells, and fertile transgenic maize plants were recovered (Thompson *et al.*, 1994). The possibilities of using yeast artificial chromosomes in order to introduce large DNA inserts into plants are also under study (Mullen *et al.*, 1994; van Wordragen *et al.*, 1994).

The flavonoid pathway

Flavonoids are phenolic compounds common to all higher plants. They have several unrelated roles in plants (van Tunen and Mol, 1991; Koes, Quattrocchio and Mol, 1994). Coloured anthocyanins are important flower pigments, where their major role is undoubtedly to attract pollinators. Flavonoids also have a role in protection against UV light and pathogens (phytoalexins), and they function as signal molecules in symbiotic plant–microbe interactions, such as nodulation in legumes. They are important in the formation of various plant structures and in sexual reproduction.

The structure, biochemistry and synthesis of flavonoids has been studied extensively (Forkmann, 1991; Stevenson, 1991; van Tunen and Mol, 1991). Also, the genetics of the flavonoid pathway in maize, petunia and snapdragon are thoroughly known, and have been reviewed extensively (Dooner, Robbins and Jorgensen, 1991; Martin and Gerats, 1993). Therefore, only the main steps of the flavonoid biosynthesis and their role in flower pigmentation will be outlined here.

The colours of flowers are mainly formed by three chemically different types of natural pigments: flavonoids, carotenoids and betalains. These compounds can, in many cases, produce very similar colours, and very often contribute in mixtures to flower colour (Forkmann, 1991). Carotenoids are lipid-soluble pigments located in plastids, and account for the majority of yellow hues in flowers. Betalains are water-soluble nitrogenous pigments which contribute to flower colours from ivory and yellow and various shades of orange and red to violet (Forkmann, 1991).

Flavonoids, which accumulate in vacuoles of epidermal cells as water-soluble glycosides, are the most common and most important pigments found in most vascular plants. The basic structure of flavonoids consists of two aromatic rings (A and B) and a central heterocycle (C) (*Figure 1*). The degree of oxidation of the central C-ring determines to which class of flavonoids the compound belongs.

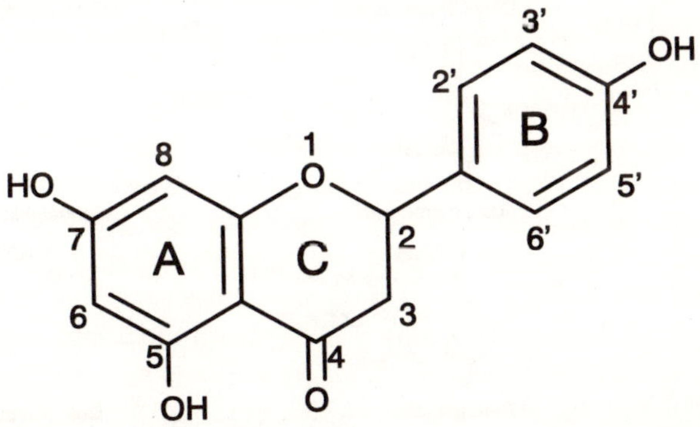

Figure 1. The basic structure of flavonoids consists of two aromatic rings (A and B) and a central heterocycle (C).

Flavonoids are synthesized via the phenylpropanoid pathway, starting from phenylalanine and leading to synthesis of coloured anthocyanins (*Figure 2*). Phenylalanine is converted through a series of enzymatic reactions catalysed by phenylalanine ammonialyase (PAL), cinnamate- 4-hydroxylase (C4H) and 4-coumaryl-CoA ligase (4CL) to 4-coumaryl-CoA, which then participates in a reaction catalysed by chalcone synthase (CHS), the first enzyme dedicated to the anthocyanin pathway.

Chalcone synthase has been extensively studied in many plant species. Reaction of 4-coumaryl-CoA with three molecules of malonyl-CoA leads to the formation of yellow chalcone. At the next step, chalcone–flavanone isomerase (CHI) catalyses the

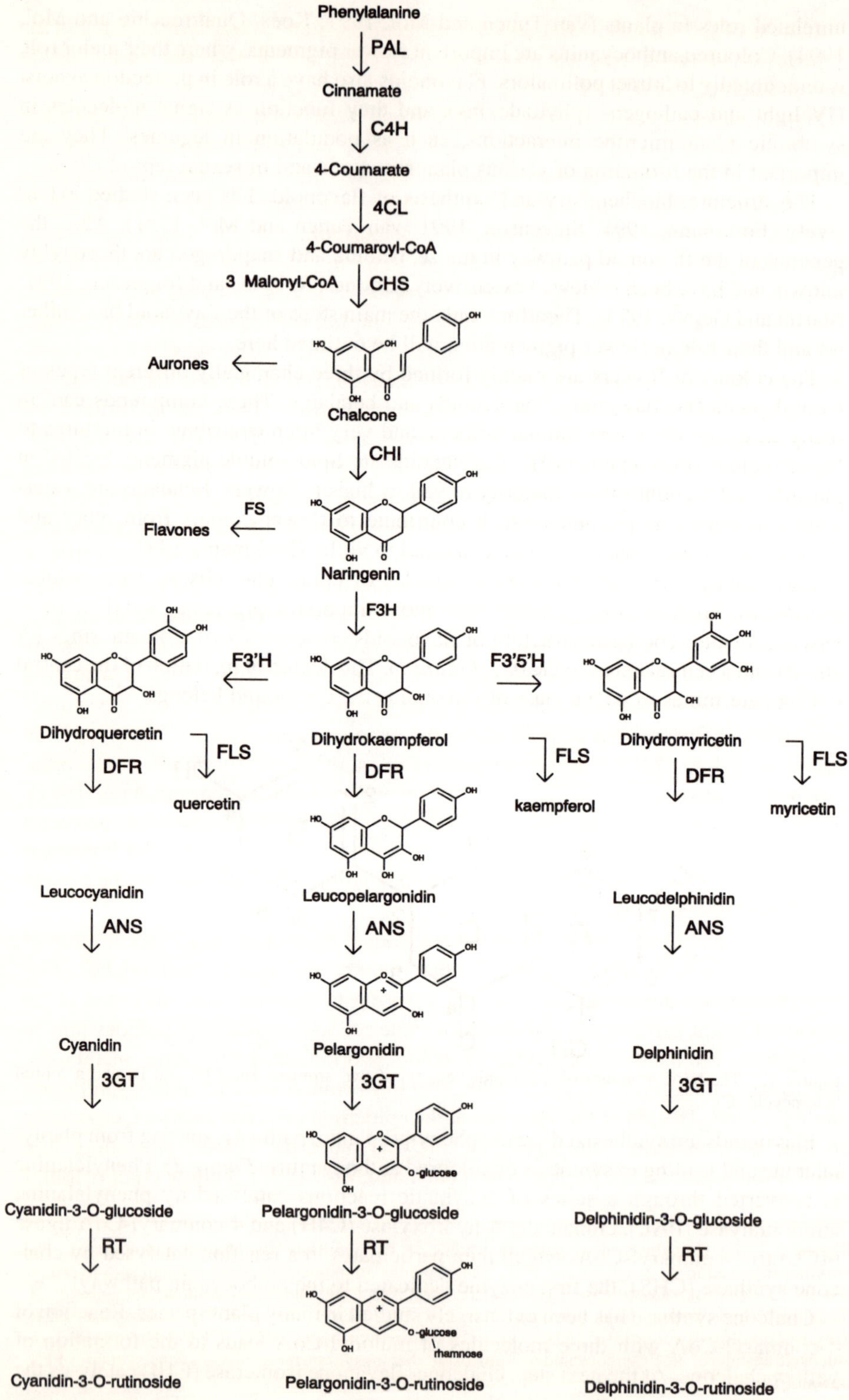
Phenylalanine
PAL
Cinnamate
C4H
4-Coumarate
4CL
4-Coumaroyl-CoA
3 Malonyl-CoA
CHS
Aurones
Chalcone
CHI
FS
Flavones
Naringenin
F3H
F3'H
F3'5'H
Dihydroquercetin
Dihydrokaempferol
Dihydromyricetin
FLS
DFR
quercetin
kaempferol
myricetin
Leucocyanidin
Leucopelargonidin
Leucodelphinidin
ANS
Cyanidin
Pelargonidin
Delphinidin
3GT
O-glucose
Cyanidin-3-O-glucoside
Pelargonidin-3-O-glucoside
Delphinidin-3-O-glucoside
RT
rhamnose
Cyanidin-3-O-rutinoside
Pelargonidin-3-O-rutinoside
Delphinidin-3-O-rutinoside

formation of the colourless flavanone, naringenin, which is then hydroxylated at the 3 position of the C-ring by flavanone 3-hydroxylase (F3H) and converted to dihydrokaempferol (DHK).

Hydroxylation of the B-ring of DHK by 3'-hydroxylase and 3'5'-hydroxylase results in formation of dihydroquercetin (DHQ) and dihydromyricetin (DHM), respectively. The synthesis of these compounds determines the type of anthocyanins produced later. Dihydromyricetin, with a 3'4'5'-hydroxylation pattern in the B-ring, is converted to blue- and purple-coloured delphinidins, while dihydroquercetin, hydroxylated at 3' and 4' positions, is the precursor of red-coloured cyanidins and 4'-hydroxylated dihydrokaempferol is the precursor of orange-coloured pelargonidins. The synthesis of the coloured anthocyanins is complex and requires several enzymes. Dihydroflavonols are first reduced by dihydroflavonol-4-reductase (DFR) to corresponding leucoanthocyanidins, which are then converted by one or two, as yet uncharacterized, enzymes to unstable anthocyanidins. They are stabilized by glycosylation at the 3 position of the C-ring by UDPG-flavonoid-3-*O*-glucosyltransferase (3GT). Finally, anthocyanidin 3-glucosides can be further modified in ordered sequence by glycosylation (e.g. glucose or rhamnose added), acylation and methylation. Methylation results in three common groups of flavonoids (peonidins, petunidins and malvidins), and has a reddening effect on colour (Forkmann, 1991). The type of modification pattern varies among different species and even varieties, determining the type of anthocyanin that is produced (van Tunen and Mol, 1991).

In addition to anthocyanins, other flavonoids are also produced via the phenylpropanoid pathway. These compounds have several distinct roles in plants and also affect floral pigmentation by acting as co-pigments. Yellow aurones are synthesized from chalcones, but little is known of the enzymatic reactions involved (van Tunen and Mol, 1991). Flavone synthase (FS) catalyses the formation of flavones, which have been shown to be part of plant waxes (van Tunen and Mol, 1991). Flavonol synthase (FLS) catalyses the formation of flavonols (kaempferol, quercetin and myricetin) from corresponding dihydroflavonols. Flavonols are also important co-pigments and have an important role in pollen germination and pollen tube growth (Mo, Nagel and Taylor, 1992).

Although the flower colour is mainly determined by the type and concentration of anthocyanin present in the vacuoles, other factors also affect the final hue. Co-pigmentation of anthocyanins with the colourless flavonols and flavones is an important factor influencing pigmentation. The relative amounts of anthocyanin to co-pigments can have a strong effect on colour (Forkmann, 1991). Also, metal ions, such as iron, aluminium and magnesium, can form complexes with anthocyanins and modify colour. The pH of the vacuoles of the epidermal cells has an effect, especially on blue pigmentation. The higher the pH, the bluer the colour, presuming that delphinidin pigments are present (van Tunen and Mol, 1991). Yellow colours found

---

Figure 2. The phenylpropanoid pathway leading to the synthesis of coloured anthocyanins. The enzymes involved in the reactions are: PAL, phenylalanine ammonialyase; C4H, cinnamate-4-hydroxylase; 4CL, 4-coumaryl-CoA ligase; CHS, chalcone synthase, CHI, chalcone–flavanone isomerase, F3H, flavanone 3-hydroxylase; F3'H, flavonoid 3'-hydroxylase; F3'5'H, flavonoid 3',5'-hydroxylase; DFR, dihydroflavonol-4-reductase; ANS, anthocyanidin synthase; 3GT, UDPG-flavonoid-3-*O*-glucosyltransferase; 3RT, anthocyanidin-3-glucoside rhamnosyltransferase; FS, flavone synthase; FLS, flavonol synthase.

in nature are the result of accumulation of chalcones, aurones or flavonols, but also of carotenoids. For a more detailed review see Forkmann (1991).

Several genes encoding the enzymes of the pathway have been cloned using various techniques (Dooner, Robbins and Jorgensen, 1991; Stevenson, 1991; van Tunen and Mol, 1991; Martin and Gerats, 1993). Many of the genes have now been isolated from *Petunia hybrida* and from the most important ornamental plants (*Table 1*).

Table 1 Flavonoid-specific cDNAs and genes cloned from petunia and other important ornamental plants

Species	Gene	Reference
Carnation	CHS	Forkmann (1991)
	CHI	Forkmann (1991)
	F3H	Britsch *et al.* (1993)
	DFR	Forkmann (1991)
	FLS	patent, PCT/AU93/00400
	ANS	Keam (unpublished)
Chrysanthemum	CHS	Courtney-Gutterson *et al.* (1994)
	DFR	Barri-Rewell (unpublished)
	FLS	patent, PCT/AU94/00400
	ANS	Barri-Rewell (unpublished)
Gerbera	CHS	Elomaa *et al.* (1993)
	DFR	Helariutta *et al.* (1993)
Lisianthus	CHS	Davies *et al.* (1993)
	CHI	Davies *et al.* (1993)
	DFR	Davies *et al.* (1993)
	F3'5'H	Nielsen (unpublished)
	FLS	Nielsen (unpublished)
Rose	CHS	Courtney-Gutterson (unpublished)
	DFR	Tanaka (unpublished)
	FLS	Tanaka (unpublished)
	ANS	Tanaka (unpublished)
Petunia	CHS	Reif *et al.* (1985)
	CHI	van Tunen *et al.* (1988)
	F3H	Britsch, Ruhnau-Britsch and Forkmann (1992)
	F3'H	patent, PCT/AU93/00127
	F3'5'H	Holton *et al.* (1993)
	FLS	Holton, Brugliera and Tanaka (1993)
	DFR	Beld *et al.* (1989)
	ANS	Weiss *et al.* (1993)
	An13	Quattrocchio *et al.* (1993)
	3RT	Brugliera *et al.* (1994), Kroon *et al.* (1994)
	AMT	Quattrocchio *et al.* (1993)

Genetic modification of the pathway

There are several possibilities of altering flower pigmentation by using genetic engineering. By suppressing the expression of certain genes of the phenylpropanoid pathway, anthocyanin synthesis can be blocked at several steps and a range of paler colours or altered colour hues can be produced. This can be done by affecting the key reactions in the pathway, but also by modifying the expression of the genes affecting the synthesis of co-pigments or later modification steps of the anthocyanin molecules.

Another strategy is to over-express a specific gene in order to intensify pigmentation. It is also possible to introduce genes of heterologous origin, encoding enzyme activities missing from the target plant. This would allow formation of novel pigmentation, not found naturally in this species. Expression of regulatory genes that affect the expression of several biosynthetic genes, can also turn on anthocyanin synthesis in normally acyanic organs, and thereby new pigmentation patterns can be achieved. These approaches and success in using them in various applications are presented next.

MODIFICATION OF THE FLOWER COLOUR INTENSITY

To date, suppression of gene expression at several different steps of the pathway has been successful in petunia, a model plant in this area, and in several important ornamental species, such as gerbera, chrysanthemum, carnation and rose. In the antisense approach, the gene to be introduced to the target plant is placed in reverse physical orientation under the promoter directing its expression. The exact mechanism of the antisense effect is still unknown, and it is possible that the suppression occurs in different ways in different cases. It has been suggested that an RNA duplex is formed between mRNA and antisense RNA. This duplex is then rapidly degraded, or processing, transport of the mRNA or its translation is prevented (van der Krol, Mol and Stuitje, 1988). The latest results from work on petunia suggest that genomic position may be important in the interaction between antisense and endogenous genes, occurring either via RNA or DNA, and that the mechanism operates post-transcriptionally (de Lange, 1994; van Blokland, 1994).

Attempts to intensify flower pigmentation by over-expressing either *chs* or *dfr* in petunia have so far been unsuccessful. Introduction of additional copies of flavonoid genes into target plants has, in fact, led to an opposite effect – reduction of pigmentation was detected. This phenomenon has been termed co-suppression because the expression of both the transgene and the endogenous gene is suppressed in these transformants.

Increased red pigmentation has been obtained in petunia and tobacco by suppressing the formation of flavonol co-pigments (Holton, Brugliera and Tanaka, 1993). It has also been possible to alter pigmentation by suppressing the expression of genes involved in further modification of flavonoid pigments by rhamnosylation (Brugliera *et al.*, 1994; Kroon *et al.*, 1994).

Petunia

The first modification of flower colour intensity using genetic engineering was done in *Petunia hybrida* (van der Krol *et al.*, 1988). Constitutive expression of a full-length antisense chalcone synthase gene (*chsA*, isolated from petunia) altered the flower pigmentation in petunia and also in a heterologous plant system, tobacco. Normal petunia (VR hybrid) flowers have uniformly coloured corollas, tubes and anthers, while, unexpectedly, some of the antisense transformants showed variable phenotypes, with sectored pigmentation patterns in corolla tissue or ring patterns starting from the tube region (van der Krol *et al.*, 1988). Also, fully white transformants were obtained from both petunia and tobacco. Northern and Western analysis of these transformants showed that steady-state *chs* mRNA and CHS protein levels correlated well with the observed phenotypes. The chalcone synthase gene family of petunia comprises 8–10 members, two of which are expressed in flower tissue (*chsA* and *chsJ*) (Koes, Spelt and Mol, 1989). These genes share 86% homology. In white tissues of antisense *chs* transformants transcripts of both these genes were absent (van der Krol *et al.*, 1990b). In addition, reduction was specific for *chs* mRNA and no effect on *chi* or *dfr* mRNA levels were detected. Steady-state antisense *chs* RNA levels were very low and did not correlate with the observed phenotypes (van der Krol *et al.*, 1990b). In fact, low levels of antisense *chs* were detected in transformants that had a strong effect on phenotype and, on the other hand, similar antisense levels detected in leaf tissue could lead to different phenotypes. Further study of the variable phenotypes indicated that pigmentation in these plants can be influenced by plant hormones and light (van der Krol *et al.*, 1990b). Spraying gibberellic acid (GA) on the plants with variable phenotypes resulted in an increase in pigmented area, while spraying with B9 (which inhibits endogenous GA synthesis) decreased the pigmented sector area. Light intensity also affected the floral pigmentation pattern. Transformants given extra light started to form fully white flowers. Similar effects have been reported in the natural *chs* mutant, Red Star (van der Krol *et al.*, 1990b).

Van Blokland (1994) introduced antisense *chs* or *dfr* genes into petunia to study the mechanism of antisense inhibition, in order to get more information about why inhibition occurred in some plants but not in others. Reduced flower pigmentation was detected in 1 out of 10 antisense *chs* transformants and in 6 out of 26 *dfr* transformants, correlating with reduced *chs* or *dfr* mRNA levels. Also, the expression of antisense *chs* gene in the white tissue was repressed with the endogenous *chs* genes. To study the expression of antisense genes, transcription rates in various transformants were measured by nuclear run-on transcription assays. Interestingly, antisense *chs* or *dfr* transcription rates were equal in unpigmented and pigmented tissues, thus showing no correlation with the phenotype. Even a minimal amount of antisense RNA was sufficient to repress *chs* and *dfr* activity, while excess of antisense RNA was sometimes ineffective (van Blokland, 1994). As the transcription rates of endogenous *chs* genes were not affected in the white tissues, it was concluded that antisense inhibition must involve a post-transcriptional mechanism. The exact mechanism, possibly involving some kind of interaction between homologous sequences, is still to be determined.

At the same time as the first antisense transformations of petunia were carried out, attempts to enhance the flower pigmentation by over-expressing the genes of the flavonoid pathway were made by two groups independently of each other (Napoli, Lemieux and Jorgensen, 1990; van der Krol *et al.*, 1990a). Both groups obtained similar results: addition of extra copies of *chs* (van der Krol *et al.*, 1990a; Napoli, Lemieux and Jorgensen, 1990) or *dfr* (van der Krol *et al.*, 1990a) genes under control of the 35S promoter or using genomic *chs* clones (van der Krol *et al.*, 1990a) resulted in blockage of anthocyanin biosynthesis. White and patterned phenotypes, different from those produced in antisense experiments, were obtained. No transformants with darker flowers than the parental genotype were obtained. In white flowers, suppression of both the introduced gene and the endogenous gene were detected. Further characterization of the *dfr* transformants showed that neither the copy number of the integrated transgene nor the expression levels detected in leaf tissue correlated with the phenotype (van der Krol *et al.*, 1990a).

The basic mechanisms of suppression of gene expression are still unknown but are under active investigation. Recent results of van Blokland (1994) suggest that co-suppression of *chs* genes in petunia is post-transcriptional. Introduction of extra copies of *chs* genes into petunia led to a reduction of steady-state *chs* mRNA levels. However, as in the case of antisense inhibition, transcription of both introduced and endogenous genes (measured by nuclear run-on assays) were not affected, showing that the promoters of these genes were active and not inactivated by methylation and co-suppression resulting from increased *chs* RNA turnover. Also, elongation of transcripts of co-suppressed *chs* genes was shown to be normal. However, it was found that splicing of transcripts from the suppressed genes was, to some extent, impaired. The ratio of unspliced to spliced transcripts was dependent on the degree of suppression. As a conclusion, van Blokland suggested that direct interaction between homologous sequences can lead to changes in chromatin structure. This could then inhibit RNA processing or transport, leading to transcript destabilization.

A group in the DNA Plant Technology Corporation (USA) has studied sense suppression, focusing especially on the stability of the modified phenotype for commercial applications. They have modified the flower colour in chrysanthemum and rose using this approach but also used petunia as a model to assay the possible changes in suppression in primary transformants during prolonged growth and among progeny of these transformants (Courtney-Gutterson, personal communication). Three white-flowering lines where *chs* expression had been suppressed were analysed in the experiments. Over 12 000 progeny with very similar flower phenotypes were obtained by crossings from the primary transformants (Courtney-Gutterson, personal communication). During an extended growth period (over 1 year), no revertant branches were obtained for either line, and it was concluded that the stability of the suppression phenotypes was sufficient for commercial purposes.

Holton, Brugliera and Tanaka (1993) demonstrated the modification of flower colour intensity in petunia and tobacco by introducing a flavonol synthase (FLS) gene which affects the synthesis of flavonols. Flavonols are important co-pigments but also involved in pollen germination and pollen tube growth. Flavonol synthase belongs to the 2-oxoglutarate-dependent dioxygenase gene family and a full-length cDNA was isolated from petunia by a combination of differential screening and polymerase chain reaction (PCR) amplification using degenerate primers. Introduction of antisense FLS cDNA in petunia (VR) plants resulted in redder flowers in 4 out of 12 transformants. Red transformants showed markedly less flavonols than non-transformed controls, analysed by thin-layer chromatography. Similarly, the same construct caused a reduction in flavonol production in tobacco, and red flowers, with, in some cases, levels of anthocyanins more than three times higher than control flowers, were detected (Holton, Brugliera and Tanaka, 1993). Obviously, enzymes involved in flavonol and anthocyanin synthesis compete for common dihydroflavonol substrates and, thereby, reduction of flavonol synthesis led to enhanced production of anthocyanins. These types of interactions between the flavonoid classes can make it difficult to predict the final flower pigmentation in advance.

It has also been possible to change flower pigmentation in petunia by altering the expression of genes affecting the further modification steps of the anthocyanin molecules, such as rhamnosylation. A rhamnose group can be added to the 3-*O*-bound glucose of the anthocyanidin-3-glucoside molecule by 3RT (anthocyanidin-3-glucoside rhamnosyltransferase) to produce anthocyanidin-3-rutinosides. Brugliera *et al.* (1994) isolated a cDNA clone corresponding to the *Rt* locus of petunia using a differential screening technique. The functionality of the isolated cDNA was shown by complementation of a *Rtrt* mutant line. Antisense expression of the cDNA clone in purple petunia VR plants (*Rtrt*) plants resulted in altered flower colour in 7 out of 12 transformants. In most cases, uniformly pink flowers were detected but also flowers with purple and red sectors. The antisense transformants had distinct pigment profiles, with reduced levels of malvidin and higher levels of petunidin compared to control plants, and thereby altered phenotypes were produced. Similar results were obtained by Kroon *et al.* (1994).

Chrysanthemum

A recent article reported on flower colour modification in florist's chrysanthemum by introducing a chalcone synthase cDNA in both antisense and sense orientation, using *Agrobacterium*-mediated transformation of leaf explants (Courtney-Gutterson *et al.*, 1994). As in the case of petunia, a full-length CHS coding sequence was used under the control of the CaMV (cauliflower mosaic virus) 35S promoter. Both approaches were comparable in producing white-flowering transgenics from the cultivar Moneymaker, although the frequency of obtaining *chs* suppression was rather low (3/133 sense and 3/83 antisense individuals). In contrast to petunia, no pattern formation was detected. The white-flowering transformants were shown to accumulate CHS precursors, analysed by thin-layer chromatography of flower extracts. RNase protection assay of the white-flowering sense transformants showed reduction of endogenous *chs* message levels in leaves and flowers (Courtney-Gutterson *et al.*, 1994).

To study the stability of the phenotype, a small-scale field trial was performed in three different locations in the USA, using one vegetatively propagated white-flowering sense and one antisense line (200 cuttings from both at each site). Some variation in flower colour was detected during the growth in various sites, caused by environmental influences on the pigmentation, but this is common to white-flowering cultivars in many ornamentals. General growth characteristics were not altered. The overall conclusion of this experiment was that both approaches can be applied equally well for colour modification in chrysanthemum. Plant Variety Rights have been applied for in respect of three of the Moneymaker transformants.

Gerbera

Gerbera hybrida is an important ornamental plant and was ranked sixth in sales through Dutch auctions in 1991. Modification of flower colour has been obtained by transformation with both antisense chalcone synthase cDNA (nearly full length) and antisense dihydroflavonol-4-reductase cDNA into the red-flowering variety, Terra Regina (Elomaa *et al.*, 1993). Although the transformation frequency using the *Agrobacterium* co-cultivation method was low, a range of new phenotypes was obtained. With the antisense *chs* construct one pink- and one cream-flowering transformant were obtained, and with antisense *dfr* two different pink transformants, with slightly different shades of colour, were obtained. The reduction of flower pigmentation correlated with the reduction of either *chs* or *dfr* mRNA levels, respectively. Transformation with one antisense gene did not affect the expression of the other of the two genes of the pathway. Also, in the case of antisense chalcone synthase transformants, the enzyme activity was shown to be reduced as well. Very little antisense RNA was detected in transformants with altered phenotypes (unpublished results). The mutual absence of both antisense and sense transcripts have been detected in other cases (Smith *et al.*, 1988; van der Krol *et al.*, 1988; Hamilton, Lycett and Grierson, 1990). Some variation in the extent of the reduced pigmentation in these transformants was noticed during their growth in greenhouses, reflecting effects of environmental conditions on anthocyanin synthesis.

Rose

Rosa hybrida is the most important commercial cut-flower. At the DNA Plant Technology Corporation, USA, transformation of a dark-red flowering variety, Royalty, was performed using chalcone synthase cDNA in sense orientation under the 35S promoter (Courtney-Gutterson, personal communication). The constructs were introduced into embryogenic calli using *Agrobacterium*-mediated transformation and 15 transgenic plants (out of about 100) with reduced pigmentation were obtained. The reduction in anthocyanin amount ranged from 39% to 97%. Thin-layer chromatography showed the accumulation of CHS precursors and Northern analysis showed specific reduction of *chs* mRNA levels in transformants. However, no white-flowering transformants were obtained. This could be for several reasons. As known from petunia, some antisense fragments are more effective in altering the phenotype than others (van der Krol *et al.*, 1990c; de Lange *et al.*, 1993). In this case, only a fragment of *chs* coding sequence was used. It is also possible that the introduced *chs* has not suppressed the expression of all members of the chalcone synthase gene family in rose (Courtney-Gutterson, personal communication). The colour changes in the pink transformants were stable.

BRINGING NEW COLOURS TO THE EXISTING POOL

There are already many examples of altering flower colour intensity by using either antisense or co-suppression approaches. Thereby, it has been possible to produce a range of paler or acyanic flower colours from red-flowering varieties, as described above. In order to bring totally new hues into the colour selection of certain species, more knowledge of the genetics and chemistry of the particular species is required. Introduction of the maize *A1* or gerbera *dfr* gene into a petunia mutant resulted in production of orange pelargonidins not occurring naturally in petunia. Pelargonidin pigments are also missing from chrysanthemum and lisianthus. The possibilities of producing novel red and yellow pigmentation in ornamentals are discussed. Also, progress in bringing the missing enzymes needed for the route to blue pigments into the most important cut-flowers is presented. In fact, all of the genes that should be necessary for the production of blue flowers have now been isolated (Holton and Tanaka, 1994).

Maize and gerbera dfr genes in petunia RL01

Transformation of petunia by Meyer *et al.* (1987) using a heterologous gene from maize was the first case of using genetic engineering to alter flower colour. Orange-coloured pelargonidin pigments are not produced in naturally occurring varieties of petunia, due to the substrate specificity of the petunia dihydroflavonol-4-reductase. Petunia DFR cannot reduce dihydrokaempferol, but uses instead dihydromyricetin and dihydroquercetin as substrates, leading to the synthesis of the corresponding leucoanthocyanidins (Meyer *et al.*, 1987). The previously characterized petunia mutant, RL01, accumulates dihydrokaempferol due to lack of both flavonoid 3'-hydroxylase and 3',5'-hydroxylase activity. The flower colour of this mutant is therefore nearly white, and it accumulates only traces of cyanidin and delphinidin derivatives. By transforming the *A1* gene from maize, encoding the maize dihydroflavonol-4-reductase, under the control of the CaMV 35S promoter, Meyer *et al.* (1987) produced a new pathway in petunia leading to the accumulation of the brick-red pelargonidin.

The petal colour has also been changed using the *A1* gene under the control of the EPSPS (5-enolpyruvylshikimate-3-phosphate synthase) promoter by Benfey *et al.* (1990). They used the EPSPS/A1 construct to transform the RL01 mutant and the mutant line W80, which completely lacks visible anthocyanin pigments. In RL01 a uniform pink-orange colour was produced while in W80 a lighter pink colour was detected.

Besides producing a new flower colour in petunia, the *A1* gene has proven to be an excellent phenotypic marker to follow transgene activity and to analyse the parameters that influence expression of a transgene. Transformation of the *A1* gene led to formation of an even, brick-red colouration in some of the transformants, and flowers with brick-red sectors were also detected, as well as control-like transformants. Further analysis of the variable phenotypes revealed an inverse correlation between the number of integrated *A1* copies and uniformity of pelargonidin pigmentation. Furthermore, it was shown that the decrease in *A1* expression was correlated to the methylation status of the 35S promoter (Linn *et al.*, 1990; Meyer, Heidmann and Niedenhof, 1993). In the majority of cases, uniform, continuous expression of the *A1* gene was found in plants where the transgene was single-copy integrate and its promoter was not methylated. Most white and variegated plants had multiple copies of the *A1* gene, which may cause co-suppression. However, Pröls and Meyer (1992) reported that even the transformants with one integrated copy of the *A1* gene did not show stable expression if the gene was integrated into a region of highly repetitive DNA.

In order to trap transposable elements from petunia, one of the *A1* transformant lines (RL01–17), containing only one copy of the *A1* gene, was chosen for large field tests, performed in 1990. Integration of a transposable element into *A1* should have resulted in either a white or a variegated phenotype (Meyer *et al.*, 1992). Over 30 000 transgenic plants were grown under field conditions and monitored for changes in pigmentation. White and variegated phenotypes were detected during the summer. A single white mutant was detected that bore white flowers throughout the growing season. It was shown to have arisen through a deletion of the *A1* gene. In addition, weakly coloured, white and sectored phenotypes were detected, which correlated with hypermethylation of the 35S promoter. Environmental factors, such as high temperature and high light intensity during the growth season, reduced the floral pigmentation. Also, variation in pigmentation was dependent on endogenous factors, such as the age of the parental plant from which the seed was derived, or the time at which crosses were made (Meyer *et al.*, 1992). The possibility of instability in the transgene expression must be kept in mind in other applications.

Making use of the homology of previously isolated *dfr* genes, the corresponding gene was isolated from gerbera, and the functionality of the gene was shown by introducing it into the petunia mutant RL01 (Helariutta *et al.*, 1993). The maize *A1* gene was used as a control gene, cloned into the same *Agrobacterium* vector under the control of the 35S promoter. Using the leaf-disc transformation method (Horsch *et al.*, 1985), 15 transformants were obtained harbouring the gerbera *gdfr* construct. Most of them showed intense brick-red colouration and one transformant had flowers with sectors. The intensity of the colour was stronger compared to the *A1* transformants, which also showed brick-red phenotypes. In addition to this, it was noticed that the *A1* transformants showed variable phenotypes with sometimes totally white flowers during the summer season. The transformants with the gerbera *gdfr1* gene instead showed stable colouration, even when grown outdoors. The copy numbers of the introduced genes were studied by Southern hybridization and no differences between the *A1* and *gdfr* transformants were detected – both had clones with 1–5 gene copies in the genome (unpublished results). In gerbera *gdfr1* transformants more intense colouration was detected in transformants with several transgene copies. By analysing the expression levels of the introduced genes by Northern analysis, it was shown that *gdfr* RNA levels were higher than the *A1* transcript levels, and they correlated with the detected phenotypes. Whether this was due to stability of the *gdfr* transcripts or to differences in transcription rates is still to be determined.

Production of novel red and yellow pigmentation in ornamentals

It has been demonstrated that transformation of an appropriate petunia mutant with a *dfr* gene from maize or gerbera leads to the production of pelargonidin, a pigment normally absent from petunia. However, the production of pelargonidin relies on the absence of flavonoid 3'-hydroxylase and flavonoid 3',5'-hydroxylase enzymes. The inability of the DFR enzyme to act on dihydrokaempferol has only been reported for solanaceous plants.

The major anthocyanin produced by chrysanthemum (*Dendranthema morifolium*) flowers has been identified as cyanidin 3-malonylglucoside (Saito *et al.*, 1988). Delphinidin pigments are not produced, presumably due to the absence of a *f3'5'h* gene. Pelargonidin pigments are also missing. However, the lack of pelargonidin is not due to the DFR specificity (Schwinn, Markham and Given, 1994) but due to the absence of plants lacking F3'H activity. Therefore, suppression of *f3'h* gene expression should lead to the production of pelargonidin pigments and novel red flower colours in chrysanthemum.

The colour range of lisianthus (*Eustoma grandiflorum*) flowers is limited to purple, mauve, pink, whites and creams. Petals of purple lisianthus are characterized by delphinidin glycosides accompanied by variable, but generally low, quantities of cyanidin glycosides (Markham and Ofman, 1993). When developing flower buds were fed with tetcyclasis, a cytochrome *P*450 inhibitor which prevents both F3'H and F3'5'H activities, a flower colour change occurred due to the production of pelargonidin pigment (Nielsen *et al*., 1994). Reduction of F3'5'H activity via an antisense or co-suppression gene construct should lead to the synthesis of pelargonidin or cyanidin pigments, resulting in the production of novel red-coloured varieties of lisianthus.

Cyanic colour is particularly abundant within the Bromeliaceae family, giving in many plants purple-blue to orange-red flowers. A survey of the anthocyanins in 32 species and two cultivars indicated that cyanidin derivatives were the predominant anthocyanidins in the inflorescences and leaves, in spite of the fact that the pelargonidin-like colours dominate in these tissues (Saito and Harborne, 1983). The reason for this apparent discrepancy came with the discovery of a new glycosidic type of cyanidin (3,5,3'-triglucoside) in 60% of the plants sampled. Three other 3'-glucosides were also found in the Bromeliaceae. All four novel cyanidin glucosides are bright scarlet instead of the deep crimson of other cyanidin glycosides.

A number of different glycosyltransferase genes have been isolated from plants, animals and bacteria. The flavonoid glycosyltransferases 3GT and 3RT share regions of sequence similarity with each other and with glucuronosyltransferases from animals. It is likely that the enzyme catalysing 3'-glucosylation of cyanidin pigments in the Bromeliaceae would also share sequence similarity with other glycosyltransferases. Transformation of plants such as chrysanthemum with a 3'-glucosyltransferase gene might enable the production of flowers with novel scarlet-red colours. This could provide an interesting alternative to the F3'H gene suppression approach outlined previously.

In *Sinningia cardinalis* and *Columnea* flowers a reductase was demonstrated which catalyses the NADPH-dependent reduction of flavanones to flavan-4-ols (Stich and Forkmann, 1988a,b). The latter compounds are precursors for the rare 3-deoxyanthocyanidins. The purified *Dahlia* DFR was shown to catalyse both reactions, the reduction of dihydroflavonols and of flavanones (Fischer *et al*., 1988), indicating that possibly one and the same enzyme is involved in flavanone and dihydroflavonol reduction. The 3-deoxyanthocyanidins are redder than the corresponding anthocyanidin derivatives. Therefore, a flavanone reductase gene might be used to engineer 3-deoxycyanidin pigments in plants such as chrysanthemum, which would result in the production of novel red colours.

In carnation flowers, the gene *I* controls chalcone isomerase activity. Genotypes with recessive *ii* alleles produce yellow flowers, which contain the chalcone isosalipurposide as the major petal pigment (Forkmann and Dangelmayr, 1980). Yellow carnations are also mutated in gene *a*, which prevents the conversion of dihydroflavonols into anthocyanins. In *iiAA* plants the flowers are orange due to the production of a mixture of chalcones and anthocyanins. Isosalipurposide, the 2'-glucoside of chalcone, is also the major petal pigment of *Paeonia trolloides, Asystasia gangetica* and *Aeschynanthus parvifolius* (Harborne, 1966).

Antisense *chi* experiments have been carried out in petunia in an attempt to produce yellow flowers through the accumulation of chalcone pigments. These experiments have failed to suppress *chi* gene expression, for unknown reasons (De Lange, 1994). However, even if *chi* expression was eliminated in the petals it is unlikely that chalcone would accumulate. This is because chalcone is rapidly isomerized spontaneously to the colourless naringenin in aqueous solution (Forkman, 1991). *Chi-A* mutants of petunia produce yellow pollen due to the accumulation of chalcone (van Tunen *et al*., 1991). However, flavonoids in pollen are deposited in the exine and/or its cavities (Wierman and Vieth, 1983). This hydrophobic environment may stabilize the chalcone molecules and prevent isomerization to the colourless naringenin.

In carnation CHI mutants, chalcone is stabilized by 2'-glucosylation, which prevents isomerization to naringenin. In order to genetically engineer yellow, chalcone-accumulating flowers it would be necessary to suppress CHI activity and anthocyanin production, while also introducing a chalcone 2'-glucosyltransferase, if there is no such endogenous enzyme.

Engineering of blue pigments

The hydroxylation pattern of the anthocyanin B-ring is generally determined by the activity of two different cytochrome *P*450 enzymes – flavonoid 3'-hydroxylase and flavonoid 3'5'-hydroxylase. Flavonoid 3',5'-hydroxylase activity is essential for the production of the blue delphinidin pigments. The introduction of a F3'5'H gene into pelargonidin- or cyanidin-producing plants should divert the biosynthetic pathway towards the production of delphinidin pigments. Provided that the vacuolar pH and co-pigment levels are sufficiently high, violet or blue-violet flowers should result.

F3'5'H activity relies on the transfer of electrons from a *P*450 reductase enzyme. A potential problem may therefore arise if a heterologous F3'5'H enzyme does not interact efficiently with the endogenous *P*450 reductase in a target plant (Forkmann, 1993). However, this is an unlikely scenario, given that the petunia F3'5'H enzymes were shown to be functional in yeast, using the endogenous yeast *P*450 reductase (Holton *et al*., 1993). Recombinant petunia F3'5'H genes have also proven to be functional in flowers of petunia (Holton *et al*., 1993), tobacco, carnation and chrysanthemum (unpublished results).

The initial complementation experiments in petunia used a line (Skr4xSw63) which was mutated in each of the flavonoid hydroxylase genes. Due to the substrate specificity of the DFR in petunia, this line produces very little anthocyanin and accumulates dihydrokaempferol (one of the substrates of F3'5'H). The *hf1* and *hf2* mutations could be complemented by transformation of Skr4xSw63 with cDNA clones of *Hf1* or *Hf2* under the control of the constitutive MAC promoter (Holton *et al*., 1993). Manipulation of the 5'-leader and promoter sequences has enabled the production of transgenic petunia plants accumulating 'wild-type' levels of delphinidin derivatives in the flowers by using either a MAC promoter or a chalcone synthase promoter (unpublished results).

Efficient production of delphinidin derivatives in other, more important, ornamental species may be more difficult. To obtain high levels of delphinidin production the target species must produce either of the substrates of F3'5'H, dihydrokaempferol or dihydroquercetin. The target plant must also produce enzymes capable of converting dihydromyricetin (the product of F3'5'H) through to delphinidin glycosides and accumulate these anthocyanins in the vacuole.

If a cyanidin-producing plant is transformed with a F3'5'H gene, expression of the *f3'5'h* gene must be sufficient to effectively compete for substrate with the endogenous genes producing F3'H, FLS and DFR activity. In petunia, when all of these enzyme activities are present the only anthocyanins produced are delphinidin derivatives, as well as the flavonols kaempferol and quercetin. If F3'H activity is absent, then only delphinidin derivatives and kaempferol are produced. Pelargonidin is not produced because of the substrate specificity of DFR. Similarly, myricetin is not produced in large amounts due to the substrate specificity of FLS. If the endogenous DFR enzyme of a target species has a preference for dihydrokaempferol or dihydroquercetin over dihydromyricetin, it may be difficult to produce high levels of delphinidin.

Efficient production of delphinidin is the first requirement for obtaining blue flowers. It is essential that the delphinidin is produced in flowers that have a sufficiently high vacuolar pH and co-pigment level. Stewart, Norris and Asen (1975) determined that the pH of petal extracts from a number of different blue-violet or blue-flowered species ranged from 5.1 to 7.5. A gene (*Ph6*) that modifies flower colour by affecting the pH of the corolla has been cloned by Chuck *et al.* (1993) from petunia using maize transposable element (*Activator*). Production of sufficient levels of co-pigments is very important to create blue flowers in some species, whereas in other species polyacylation of the anthocyanins can lead to efficient intramolecular co-pigmentation (Goto and Kondo, 1991). No anthocyanin acyltransferase genes have yet been isolated.

Geissmann and Mehlquist (1947) reported the presence of four anthocyanin types in different coloured carnation flowers: pelargonidin-3-monoside and -3,5-dimonoside, and cyanidin-3-monoside and -3,5- dimonoside. Later experiments, using less harsh extraction procedures, have demonstrated that the anthocyanins are glucosides acylated with malic acid (Terahara *et al.*, 1986).

In the presence of dominant alleles of the gene *R* (cyanidin type), part of dihydrokaempferol is hydroxylated in the 3'-position to dihydroquercetin, and both quercetin and kaempferol derivatives are formed in flowers (Stich, Eidenberger and Wurst, 1992). However, at the anthocyanin level, pelargonidin derivatives are nearly completely replaced by cyanidin derivatives in flowers of genotypes with dominant *R* alleles (Stich *et al.*, 1992). This pattern clearly corresponds to the substrate specificity of FLS and DFR, which both use dihydroflavonols as substrates. While for FLS no difference in the rate of the conversion of dihydrokaempferol and dihydroquercetin to the respective flavonols was observed, the DFR enzyme, being involved in anthocyanin synthesis, was found to reduce dihydroquercetin significantly faster (four times) than dihydrokaempferol. This means that FLS and DFR compete for dihydroquercetin as a substrate, whereas dihydrokaempferol is preferentially converted to kaempferol. Thus, relatively high amounts of kaempferol derivatives are formed even in plants containing F3'H activity.

Stich *et al.* (1992) showed by feeding experiments that carnation petals contain enzymes necessary for the conversion of dihydromyricetin into delphinidin derivatives. Administration of dihydroquercetin and dihydromyricetin initiated, in the petals of the red and orange strains, the formation of cyanidin and delphinidin derivatives, respectively, in addition to the pelargonidin derivatives which are naturally present. DFR enzyme activity was assayed in petal extracts. Both dihydroquercetin and dihydromyricetin were converted to the respective flavan-3,4-diols at a rate about four times higher than that for dihydrokaempferol. Carnations, which naturally accumulate pelargonidin pigments, have been transformed with a petunia *f3'5'h* gene. Flowers from the transgenic plants accumulate a mixture of pelargonidin and delphinidin pigments, and the colour of the flowers was shifted towards blue, as expected (unpublished results). Experiments aimed at increasing the efficiency of delphinidin production in carnation are currently under way.

Rose pigments in species and cultivars have been studied extensively. Yokoi (1974) analysed the anthocyanin pigments contained in the flowers of 670 cultivars and eight species of roses. The only anthocyanin pigments found in this survey were cyanidin 3-glucoside and 3,5-diglucoside, pelargonidin 3-glucoside and 3,5-diglucoside and peonidin 3,5-diglucoside. Most of the so-called 'blue' roses, which are in reality lilac or purple, contained only cyanidin 3,5-diglucoside, together with large amounts of flavonols. The introduction of the flavonoid 3',5'-hydroxylase gene into pelargonidin- or cyanidin-producing rose cultivars should divert the anthocyanin biosynthetic pathway towards the production of delphinidin glucosides and the flower colour towards blue. Feeding experiments with rose petals have demonstrated that the rose enzymes are capable of converting dihydromyricetin, the product of flavonoid 3',5'-hydroxylase, to delphinidin glucosides (Holton and Tanaka, 1994).

The pH of the outer epidermal cells has been determined in a range of rose varieties. The range of pH was 3.56 to 5.38 (Biolley and Jay, 1993) – purple- and lilac-coloured roses had the higher pH values. If such purple or lilac cultivars could be transformed with a *f3'5'h* gene and were able to efficiently synthesize delphinidin, the pH of the flowers is probably high enough to produce violet or blue-violet colours.

MODIFICATION OF GENES REGULATING THE FLAVONOID PATHWAY

Many regulatory genes are known to affect flower pigmentation, either by regulating directly the expression of structural genes or having secondary effects on flower colour by modulation of pH or concentration of anthocyanins in the vacuole (van Tunen and Mol, 1991). The genes involved in regulation of the anthocyanin pathway have been best characterized in maize, petunia and snapdragon, as reviewed by Dooner, Robbins and Jorgensen (1991), van Tunen and Mol (1991) and Martin and Gerats (1993). The regulatory proteins, with homology to the mammalian *myc* and *myb* transcription factors, are highly conserved, both structurally and functionally. Differences in pigmentation patterns in various species have evolved through changes in the expression of the conserved regulators (Goodrich, Carpenter and Coen, 1992). The set of target genes varies in the species studied so far. In maize, the regulatory genes act on the whole anthocyanin pathway, while in snapdragon and petunia only the late part of the pathway is regulated by these factors. The division between the early and late parts varies as well, being either before F3H (*Antirrhinum*) or after F3H (petunia) (Quattrocchio *et al*., 1993). Isolation of the regulatory genes has now allowed the alteration of flower pigmentation and creation of novel pigmentation patterns at a more general level. In addition, production of high concentrations of anthocyanin pigments can lead to the appearance of black colours in flowers of pansy, cornflower and tulip. Similarly, in maize the recessive intensifier, *in*, results in such a greatly increased quantity of pigment that the kernels of both *Pr* (cyanidin-producing) and *pr* (pelargonidin-producing) strains appear black (Reddy and Peterson, 1978). Therefore, by increasing production of anthocyanins in ornamental species, novel black or very dark colours might be engineered.

By expressing the genes regulating the structural genes of the phenylpropanoid pathway, it has been possible to enhance anthocyanin production in tobacco and *Arabidopsis* in normally acyanic organs (Lloyd, Walbot and Davis, 1992). Expression of the maize *C1* (the *myb* homologue) under the control of the 35S promoter alone did not have an effect on the phenotype of transformed tobacco or *Arabidopsis*. A transgenic tobacco line expressing the maize *R* (the *myc* homologue) showed instead intense red pigmentation in the corolla, tube and collar of the flower, as well as in the anther filaments, which normally are only light pink. In *Arabidopsis*, all *R* transformants produced more anthocyanin than normally, but without change in the pigmentation pattern. By crossing an *R*-expressing *Arabidopsis* transformant with a *C1*-expressing one, some of the progeny produced anthocyanins in root, petal and stamen tissues, which are normally acyanic (Lloyd, Walbot and Davis, 1992). In this way, totally new pigmentation patterns were produced. In a similar way, regulatory genes could be placed under the control of tissue- or cell-specific promoters, and thereby anthocyanin synthesis could be turned on in normally acyanic tissues (van Tunen and Mol, 1991).

In petunia, several regulatory genes affecting the flavonoid pathway have been characterized. Quattrocchio *et al.* (1993) concluded that products of *an1*, *an2* and *an11* genes control the transcription of the *dfrA* and *rt* genes in petunia. In addition, *an4* was shown to be regulatory gene controlling the expression of flavonoid biosynthetic genes in anthers. Transformation of two petunia lines V26 (*an4*) and W115 (*an2* and *an4*) with the maize *R* gene under the control of the 35S promoter activated petunia flavonoid genes in both lines. About 80% of the transformed calli developed red or purple cells, but none of them formed shoots. Some green shoots derived from these calli turned purple at early stages of growth, but did not form roots or grow further. Transformants that could be regenerated showed low or no accumulation of *R* mRNA and, therefore, it was not possible to test whether the *R* gene could have restored flower pigmentation in these mutants. Transient expression analysis using particle bombardment into leaf or floral tissue, instead, showed that *C1* and *R* genes activated *dfrA* promoter *in trans* in leaves and also in petal limbs of the three mutants. These experiments indicate that it is possible to activate floral pigmentation genes in tissues where they are normally silent.

Expression of the *Antirrhinum* anthocyanin regulatory gene, *del*, under the 35S promoter in petunia (V26), instead, disrupted anthocyanin biosynthesis (Robbins and Harbord, 1994). The corolla pigmentation was reduced and stress-induced leaf pigmentation enhanced in primary transformants. Reduction in corolla pigmentation was even more pronounced when *del* was transformed to V26/W115 background. The phenotypes of the transformants resembled the *Ph6* mutation in petunia that affects the vacuolar pH. However, it is not clear how the DEL protein interferes with the regulation of the anthocyanin pathway (Robbins and Harbord, 1994).

Summary and future prospects

Flower colour modification using molecular methods has now become reality. To date, the most successful applications have been in suppressing flavonoid biosynthesis at certain steps of the pathway, either by using the antisense or sense approach. In this way, pale or acyanic varieties have been produced in several species.

In the cases studied so far, the degree of suppression varies in different transformants and has resulted in the formation of new phenotypes with variably reduced pigmentation. This is not necessarily bad from a breeder's point of view, since a range of new colour shades can be produced from an élite genotype. However, what is more puzzling, is that the antisense (or sense) effect is not seen in every transformant. The mere presence of the antisense RNA does not guarantee the phenotypic effect (van Blokland, 1994). Obviously, additional factors influence the suppression. In chrysanthemum, an antisense effect was found in only 3 out of 83 transformants; while in gerbera, although only a few transformants were obtained, half of them showed changed phenotype. The sense approach was comparable to the antisense approach in chrysanthemum. Also, in rose it has not been possible to obtain fully white transformants using the sense approach. This could be due to properties of the gene construct used, or the chromosomal position of the integration in the transformants studied so far. It is also possible that some members of the *chs* gene family are still expressed and cause the residual pigmentation.

Although quite a few successful examples exist of suppression of gene expression at certain steps in the flavonoid pathway, it has not been possible to suppress chalcone isomerase genes using antisense methods. However, as already mentioned, suppression at this step might not result in accumulation of yellow chalcones, as they are spontaneously isomerized to colourless flavanones (Forkmann, 1991). De Lange (1994) transformed petunia plants with various partial and full-length *chi* antisense constructs. Suppression of expression of the endogenous *chi* gene was not detected in any of the transformants, despite the presence of large amounts of antisense RNAs. By crossing a plant containing the *chi* transgene with different antisense transformants, transcriptional suppression of the *chi* transgenes was detected in some of the progeny. It was concluded that different modes of suppression of the same *chi* transgene may indicate that the genomic position of the antisense genes, or the sense gene, could be involved in the interaction. The resident *chi* could be situated in a genomic region that prevents it from being silenced (De Lange, 1994).

In order to alter the flower colour in a more predictable and efficient way, the fundamental work done in petunia in order to reveal the basic mechanisms under both sense and antisense inhibition is of greatest importance. Van der Krol *et al.* (1990c) have studied the promoter and sequence requirements for the antisense effect in petunia. In most cases of colour modification the constitutive 35S promoter has been used, but inhibition can also be obtained using homologous promoters, since an excess of antisense RNA is not necessary for efficient suppression. Several studies have been made to retrieve strong antisense fragments from weak ones (van der Krol *et al.*, 1990c; de Lange *et al.*, 1993). In the case of chalcone synthase in petunia, a quarter-length RNA complementary to the 3' end of *chs* mRNA was able to affect flower pigmentation. However, an antisense gene corresponding to the 5' end had no effect on the phenotype (van der Krol *et al.*, 1990c). De Lange *et al.* (1993) have developed a system to study antisense inhibition in protoplasts, using the *uidA* marker gene in order to improve the effectiveness of the antisense inhibition and, on the other hand, selectively suppress members of the gene family. Maraña and Gerats (1994) reported that chimeric constructions containing a *uidA* marker gene under the control of a *dfrA* or *dfrC* promoter created a block in anthocyanin biosynthesis when introduced into petunia. In half of the transformants a dramatic reduction of *dfrA* expression and GUS activity were detected, and novel pigmentation patterns were observed. Using this system it is possible to monitor visually the native gene suppression, and histochemically the suppression of the marker gene (Maraña and Gerats, 1994).

It is now possible to affect the expression of single structural genes at various steps of the pathway, also including genes affecting co-pigmentation and further modification steps of the anthocyanin molecule. Still, production of pelargonidin pigmentation in petunia is the only example of bringing totally new colours into a pre-existing gene pool. In theory, all genes that should be needed for production of blue colours into the most important commercial species have been cloned, and transformation methods for them have been developed. Testing of the possibilities of colour modification in these plants is currently underway.

Cloning of regulatory genes whose gene products are needed for the expression of the structural flavonoid genes will, without any doubt, lead to flower-colour modification at a more general level. This type of modification may create more diversity in pigmentation patterns and also in colour shades.

Besides producing new phenotypes, flower-colour modification has turned out to be an excellent system to monitor visually the variations in expression and regulation of transgenes, and also to study the fundamental processes involved in the interactions between a transgene and the endogenous genes. This information is valuable also in the modification of other important pathways in plants. Moreover, the factors influencing the stability of the altered phenotype need to be examined when commercial-scale propagation is required.

Acknowledgements

We gratefully acknowledge Dr Neal Courtney-Gutterson, Dr Karen Nielsen and Dr Kevin Davies for providing us with information on their work. We also wish to thank Dr Teemu Teeri for his help with the figures and for valuable comments on the manuscript. Yrjö Helariutta and Mika Kotilainen are also thanked for critical reading of the manuscript and for their comments.

References

AN, G., WATSON, B.D., STACHEL, S., GORDON, M.P. AND NESTER, E.W. (1985). New cloning vehicles for transformation of higher plants. *EMBO Journal* **4**, 277–284.

BELD, M., MARTIN, C., HUITS, H., STUITJE, A.R. AND GERATS, A.G.M. (1989). Flavonoid synthesis in *Petunia hybrida*: partial characterization of dihydroflavonol-4-reductase genes. *Plant Molecular Biology* **13**, 491–502.

BENFEY, P.N., TAKATSUJI, H., REN, L., SHAH, D.M. AND CHUA, N.-H. (1990). Sequence requirements of the 5-enolpyruvylshikimate-3-phosphate synthase 5'-upstream region for tissue-specific expression in flowers and seedlings. *The Plant Cell* **2**, 849–856.

BIOLLEY, J.P. AND JAY, M. (1993). Anthocyanins in modern roses: chemical and colorimetric features in relation to the colour range. *Journal of Experimental Botany* **44**, 1725–1734.

BOASE, M.R. AND SMITH, J.L. (1994). Expression of the GUS gene in 11 *Pelargonium* cultivars transformed by *Agrobacterium tumefaciens. 4th International Congress of Plant Molecular Biology, Amsterdam.* (Abstracts).

BOLTON, G.W., NESTER, E.W. AND GORDON, M.P. (1986). Plant phenolic compounds induce expression of the *Agrobacterium tumefaciens* loci needed for virulence. *Science* **232**, 983–985.

BRITSCH, L., RUHNAU-BRITSCH, B. AND FORKMANN, G. (1992). Molecular cloning, sequence analysis and *in vitro* expression of flavanone 3-β-hydroxylase from *Petunia hybrida.*

Journal of Biological Chemistry **267**, 5380–5387.

BRITSCH, L., DEDIO, J., SAEDLER, H. AND FORKMANN, G. (1993). Molecular characterization of flavanone 3-β-hydroxylases. Consensus sequence, comparison with related enzymes and the role of conserved histidine residues. *European Journal of Biochemistry* **217**, 745–754.

BRUGLIERA, F., HOLTON, T.A., STEVENSON, T.W., FARCY, E., LU, C.-Y. AND CORNISH, E.C. (1994). Isolation and characterization of a cDNA clone corresponding to the *Rt* locus of *Petunia hybrida. The Plant Journal* **5**, 81–92.

CHRISTOU, P. (1992). Genetic transformation of crop plants using microprojectile bombardment. *The Plant Journal* **2**, 275–281.

CHUCK, G., ROBBINS, T., NIJJAR, C., RALSTON, E., COURTNEY-GUTTERSON, N. AND DOONER, H.K. (1993). Tagging and cloning of a petunia color gene with the maize transposable element *Activator. The Plant Cell* **5**, 371–378.

COURTNEY-GUTTERSON, N., NAPOLI, C., LEMIEUX, C., MORGAN, A., FIROOZABADY, E. AND ROBINSON, K.E.P. (1994). Modification of flower color in florist's chrysanthemum: Production of a white-flowering variety through molecular genetics. *Bio/Technology* **12**, 268–271.

DAVIES, K.E., BRADLEY, J.M., SCHWINN, K.E., MARKHAM, K.R. AND PODIVINSKY, E. (1993). Flavonoid biosynthesis in flower petals of five lines of lisianthus (*Eustoma grandiflorum* Grise.). *Plant Science* **95**, 67–77.

DEBLAERE, R., BYTEBIER, B., DE GREVE, H., DEBOECK, R., SCHELL, J., VAN MONTAGU, M. AND LEEMANS, J. (1985). Efficient octopine Ti plasmid-derived vectors for *Agro-bacterium*-mediated gene transfer to plants. *Nucleic Acids Research* **13**, 4777–4788.

DE LANGE, P. (1994). Gene suppression in *Petunia hybrida* by antisense genes. Unpublished Thesis, Free University of Amsterdam.

DE LANGE, P., DE BOER, G.-J., MOL, J.N.M. AND KOOTER, J.M. (1993). Conditional inhibition of β-glucuronidase expression by antisense gene fragments in petunia protoplasts. *Plant Molecular Biology* **23**, 45–55.

DEROLES, S.C., LEDGER, S.E., MILLER, R.M., DAVIES, K.M. AND GIVEN, N.K. (1993). Transformation in *Eustoma grandiflorum* (Lisianthus). In *Biotechnology in Agriculture and Forestry*, Vol. 22, *Plant Protoplasts and Genetic Engineering III* (Y.P.S. Bajaj, Ed.), pp. 202–212. Springer- Verlag, Berlin.

DOONER, H.K., ROBBINS, T.P. AND JORGENSEN, R.A. (1991). Genetic and developmental control of anthocyanin biosynthesis. *Annual Review of Genetics* **25**, 173–199.

ELOMAA, P., HONKANEN, J., PUSKA, R., SEPPÄNEN, P., HELARIUTTA, Y., MEHTO, M., KOTILAINEN, M., NEVALAINEN, L. AND TEERI, T.H. (1993). *Agrobacterium*-mediated transfer of antisense chalcone synthase cDNA to *Gerbera hybrida* inhibits flower pigmentation. *Bio/ Technology* **11**, 508–511.

FISCHER, D., STICH, K., BRITSCH, L. AND GRISEBACH, H. (1988). Purification and characterization of (+)dihydroflavonol (3-hydroxyflavanone) 4-reductase from flowers of *Dahlia variabilis. Archives of Biochemistry and Biophysics* **264**, 40–47.

FORKMANN, G. (1991). Flavonoids as flower pigments: the formation of the natural spectrum and its extension by genetic engineering. *Plant Breeding* **106**, 1–26.

FORKMANN, G. (1993). Control of pigmentation in natural and transgenic plants. *Current Opinion in Biotechnology* **4**, 159–165.

FORKMANN, G. AND DANGELMAYR, B. (1980). Genetic control of chalcone isomerase activity in flowers of *Dianthus caryophyllus. Biochemical Genetics* **18**, 519–527.

GEISSMAN, T.A. AND MEHLQUIST, G.A.L. (1947). Inheritance in the carnation, *Dianthus caryophyllus*. IV. The chemistry of flower colour variation, I. *Genetics* **32**, 410–433.

GOODRICH, J., CARPENTER, R. AND COEN, E.S. (1992). A common gene regulates pigmentation pattern in diverse plant species. *Cell* **68**, 955–964.

GOTO, T. AND KONDO, T. (1991). Structure and molecular stacking of anthocyanins – flower color variation. *Angewandte Chemie International* Edition in English **30**, 17–33.

HAMILTON, A.J., LYCETT, G.W. AND GRIERSON, D. (1990). Antisense gene that inhibits synthesis of the hormone ethylene in transgenic plants. *Nature* **346**, 284–287.

HARBORNE, J.B. (1966). Comparative biochemistry of flavonoids – I. Distribution of chalcone and aurone pigments in plants. *Phytochemistry* **5**, 111–115.

HELARIUTTA, Y., ELOMAA, P., KOTILAINEN, M., SEPPÄNEN, P. AND TEERI, T.H. (1993). Cloning of cDNA coding for dihydroflavonol-4-reductase (DFR) and characterization of *dfr* expression in the corollas of *Gerbera hybrida* var. Regina (Compositae). *Plant Molecular Biology* **22**, 183–193.

HOLTON, T.A. AND TANAKA, Y. (1994). Blue roses – a pigment of our imagination? *Trends in Biotechnology* **12**, 40–42.

HOLTON, T.A., BRUGLIERA, F. AND TANAKA, Y. (1993). Cloning and expression of flavonol synthase from *Petunia hybrida. The Plant Journal* **4**, 1003–1010.

HOLTON, T.A., BRUGLIERA, F., LESTER, D.R., TANAKA, Y., HYLAND, C.D., MENTING, J.G.T., LU, C.-Y., FARCY, E., STEVENSON, T.W. AND CORNISH, E.C. (1993). Cloning and expression of cytochrome P450 genes controlling flower colour. *Nature* **366**, 276–279.

HOOYKAAS, P.J.J. (1989). Transformation of plant cells via *Agrobacterium. Plant Molecular Biology* **13**, 327–336.

HOOYKAAS-VAN SLOGTEREN, G.M.S., HOOYKAAS, P.J.J. AND SCHILPEROORT, R.A. (1984). Expression of Ti plasmid genes in monocotyledonous plants infected with *Agrobacterium tumefaciens. Nature* **311**, 763–764.

HORSCH, R.B., FRY, J.E., HOFFMANN, N.L., EICHHOLTZ, D., ROGERS, S.G. AND FRALEY, R.T. (1985). A simple and general method for transferring genes into plants. *Science* **227**, 1229–1231.

HUTCHINSON, J.F., KAUL, V., MAHESWARAN, G., MORAN, J.R., GRAHAN, M.W. AND RICHARDS, D. (1992). Genetic improvement of floricultural crops using biotechnology. *Australian Journal of Botany* **40**, 765–87.

KLEIN, T.M., WOLF, E.D., WU, R. AND SANFORD, J.C. (1987). High-velocity microprojectiles for delivering nucleic acids into living cells. *Nature* **327**, 70–73.

KOES, R.E., QUATTROCCHIO, F. AND MOL, J.N.M. (1994). The flavonoid biosynthetic pathway in plants: function and evolution. *BioEssays* **16**, 123–132.

KOES, R.E., SPELT, C.E. AND MOL, J.N.M. (1989). The chalcone synthase multigene family of *Petunia hybrida* (V30): differential, light-regulated expression during flower development and UV light induction. *Plant Molecular Biology* **12**, 213–225.

KROON, J., SOUER, E., DE GRAAFF, A., XUE, Y., MOL, J. AND KOES, R. (1994). Cloning and structural analysis of the anthocyanin pigmentation locus *Rt* of *Petunia hybrida*: characterization of insertion sequences in two mutant alleles. *The Plant Journal* **5**, 69–80.

LEDGER, S.E., DEROLES, S.C. AND GIVEN, N.K. (1991). Regeneration and agrobacterium-mediated transformation of chrysanthemum. *Plant Cell Reports* **10**, 195–199.

LINN, F., HEIDMANN, I., SAEDLER, H. AND MEYER, P. (1990). Epigenetic changes in the expression of the maize A1 gene in *Petunia hybrida*: Role of numbers of integrated gene copies and state of methylation. *Molecular and General Genetics* **222**, 329–336.

LLOYD, A.M., WALBOT, V. AND DAVIS, R.W. (1992). *Arabidopsis* and *Nicotiana* anthocyanin production activated by maize regulators *R* and *C1. Science* **258**, 1773–1775.

LU, C.-Y., NUGENT, G., WARDLEY-RICHARDSON, T., CHANDLER, S.F., YOUNG, R. AND DALLING, M.J. (1991). *Agrobacterium*-mediated transformation of carnation (*Dianthus caryophyllus* L.). *Bio/Technology* **9**, 864–868.

MARAÑA, C. AND GERATS, T. (1994). Inhibition of flower pigmentation in petunia by GUS gene expression under the control of the *dfrA* promoter. *4th International Congress of Plant Molecular Biology, Amsterdam.* (Abstracts).

MARKHAM, K. AND OFMAN, D.J. (1993). Lisianthus flavonoid pigments and factors influencing their expression in flower colour. *Phytochemistry* **34**, 679–685.

MARTIN, C. AND GERATS, T. (1993). The control of flower coloration. *The Molecular Biology of Flowering* (B.R. Jordan, Ed.), pp. 219–255. CAB International, Wallingford, UK.

MEYER, P., HEIDMANN, I. AND NIEDENHOF, I. (1993). Differences in DNA-methylation are associated with a paramutation phenomenon in transgenic petunia. *The Plant Journal* **4**, 89–100.

MEYER, P., HEIDMANN, I., FORKMANN, G. AND SAEDLER, H. (1987). A new petunia flower colour generated by transformation of a mutant with a maize gene. *Nature* **330**, 677–678.

MEYER, P., LINN, F., HEIDMANN, I., MEYER, Z.A.H., NIEDENHOF, I. AND SAEDLER, H. (1992). Endogenous and environmental factors influence 35S promoter methylation of a maize A1 gene construct in transgenic petunia and its colour phenotype. *Molecular and General Genetics* **231**, 345–352.

MO, Y., NAGEL, C. AND TAYLOR, L.P. (1992). Biochemical complementation of chalcone synthase mutants defines a role for flavonols in functional pollen. *Proceedings of the National Academy of Sciences, USA* **89**, 7213–7217.

MOL, J.N.M., STUITJE, A.R. AND VAN DER KROL, A. (1989). Genetic manipulation of floral pigmentation genes. *Plant Molecular Biology* **13**, 287–294.

MULLEN, J., BLOWERS, A., ADAM, G., VAN ECK, J., FRARY, A. AND EARLE, E. (1994). Transformation of plant cells with yeast artificial chromosomes. *4th International Congress of Plant Molecular Biology, Amsterdam.* (Abstracts).

NAPOLI, C., LEMIEUX, C. AND JORGENSEN, R. (1990). Introduction of a chimeric chalcone synthase gene into petunia results in reversible co-suppression of homologous genes *in trans. The Plant Cell* **2**, 279–289.

NIELSEN, K., SCHWINN, K., DAVIES, K., DEROLES, S., MANSON, D., MILLER, R., PODIVINSKY, E. AND MARKHAM, K. (1994). Flower colour modification in lisianthus (*Eustoma grandiflorum*). *4th International Congress of Plant Molecular Biology, Amsterdam.* (Abstracts).

PRÖLS, F. AND MEYER, P. (1992). The methylation patterns of chromosomal integration regions influence gene activity of transferred DNA in *Petunia hybrida. The Plant Journal* **2**, 465–475.

QUATTROCCHIO, F., WING, J.F., LEPPEN, H.T.C., MOL, J.N.M. AND KOES, R.E. (1993). Regulatory genes controlling anthocyanin pigmentation are functionally conserved among plant species and have distinct set of target genes. *The Plant Cell* **5**, 1497–1512.

REDDY, A.R. AND PETERSON, P.A. (1978). The action of the intensifier (*in*) gene in flavonoid production in aleurone tissue of maize. *Canadian Journal of Genetics and Cytology* **20**, 337–347.

REIF, H.J., NIESBACH, U., DEUMLING, B. AND SAEDLER, H. (1985). Cloning and analysis of two genes for chalcone synthase from *Petunia hybrida. Molecular and General Genetics* **199**, 208–215.

ROBBINS, T. AND HARBORD, R. (1994). A regulatory gene from *Antirrhinum* disrupts anthocyanin biosynthesis in *Petunia hybrida. 4th International Congress of Plant Molecular Biology, Amsterdam.* (Abstracts).

SAITO, N. AND HARBORNE, J.B. (1983). A cyanidin glycoside giving scarlet coloration in plants of the *Bromeliaceae. Phytochemistry* **22**(8), 1735–1740.

SAITO, N., TOKI, K., HONDA, T. AND KAWASE, K. (1988). Cyanidin 3-malonylglucuronylglucoside in *Bellis* and cyanidin 3-malonylglucoside in *Dendranthema. Phytochemistry* **27**, 2963–2966.

SCHWINN, K.E., MARKHAM, K.R. AND GIVEN, N.K. (1994). Floral flavonoids and the potential for pelargonidin biosynthesis in commercial chrysanthemum cultivars. *Phytochemistry* **35**,. 145–150.

SMITH, C.J.S., WATSON, C.F., RAY, J., BIRD, C.R., MORRIS, P.C., SCHUCH, W. AND GRIERSON, D. (1988). Antisense RNA inhibition of polygalacturonase gene expression in transgenic tomatoes. *Nature* **334**, 724–726.

STEVENSON, T.W. (1991). Molecular biology of flavonoid pigment biosynthesis in flowers. In *Molecular Approaches to Crop Improvement* (E.S. Dennis and D.J. Llewellyn, Eds), pp. 127–148. Springer Verlag, Wien.

STEWART, R.N., NORRIS, K.H. AND ASEN, S. (1975). Microspectrophotometric measurment of pH and pH effect on color of petal epidermal cells. *Phytochemistry* **14**, 937–942.

STICH, K. AND FORKMANN, G. (1988a). Biosynthesis of 3-deoxyanthocyanins with flower extracts from *Sinningia cardinalis. Phytochemistry* **27**, 785–789.

STICH, K. AND FORKMANN, G. (1988b). Studies on columnidin biosynthesis with flower extracts from *Columnea hybrida. Zeitschrift für Naturforschung* **43c**, 311–314.

STICH, K., EIDENBERGER, T. AND WURST, F. (1992). Flavonol synthase activity and the regulation of flavonol and anthocyanin biosynthesis during flower development in *Dianthus caryophyllus* L. (carnation). *Zeitschrift für Naturforschung* **47c**, 553–560.

STICH, K., EIDENBERGER, T., WURST, F. AND FORKMANN, G. (1992). Enzymatic conversion of dihydroflavonols to flavan-3,4-diols using flower extracts of *Dianthus caryophyllus* L. (carnation). *Planta* **187**, 103–108.

TERAHARA, N., YAMAGUCHI, M.-A., TAKEDA, K., HARBORNE, J.B. AND SELF, R. (1986).

Anthocyanins acylated with malic acid in *Dianthus caryophyllus* and *D. deltoides. Phytochemistry* **25**, 1715–1717.

THOMPSON, J.A., FRAME, B.R., DRAYTON, P.R., BAGNALL, S.V., JIAO, S., LEWNAU, C.J., BULLOCK, W.P., WILSON, H.M., DUNWELL, J.M. AND WANG, K. (1994). Production of fertile transgenic maize plants by silicon carbide fibre-mediated transformation. *4th International Congress of Plant Molecular Biology, Amsterday.* (Abstracts).

VAN BLOKLAND, R. (1994). Trans-inactivation of flower pigmentation genes in *Petunia hybrida.* Unpublished Thesis, Free University of Amsterdam.

VAN DER KROL, A.R., MOL, J.N.M. AND STUITJE, A.R. (1988). Antisense genes in plants: an overview. *Gene* **72**, 45–50.

VAN DER KROL, A.R., LENTING, P.E., VEENSTRA, J., VAN DER MEER, I.M., KOES, R.E., GERATS, A.G.M., MOL, J.N.M. AND STUITJE, A.R. (1988). An anti-sense chalcone synthase gene in transgenic plants inhibits flower pigmentation. *Nature* **333**, 866–869.

VAN DER KROL, A.R., MUR, L.A., BELD, M., MOL, J.N.M. AND STUITJE, A.R. (1990a). Flavonoid genes in petunia: addition of a limited number of gene copies may lead to a suppression of gene expression. *The Plant Cell* **2**, 291–299.

VAN DER KROL, A.R., MUR, L.A., DE LANGE, P., GERATS, A.G.M., MOL, J.N.M. AND STUITJE, A.R. (1990b). Antisense chalcone synthase genes in petunia: visualization of variable transgene expression. *Molecular and General Genetics* **220**, 204–212.

VAN DER KROL, A.R., MUR, L.A., DE LANGE, P., MOL, J.N.M. AND STUITJE, A.R. (1990c). Inhibition of flower pigmentation by antisense CHS genes: promoter and minimal sequence requirements for the antisense effect. *Plant Molecular Biology* **14**, 457–466.

VAN TUNEN, A.J. AND MOL, J.N.M. (1991). Control of flavonoid synthesis and manipulation of flower colour. In *Developmental Regulation of Plant Gene Expression* (D. Grierson, Ed.), pp. 94–130. Blackie & Son Ltd. Glasgow and London.

VAN TUNEN, A.J., KOES, R.E., SPELT, C.E., VAN DER KROL, A.R., STUITJE, A.R. AND MOL, J.N.M. (1988). Cloning of the two chalcone flavanone isomerase genes from *Petunia hybrida*; coordinate, light-regulated and differential expression of flavonoid genes. *EMBO Journal* **7**, 1257–1263.

VAN TUNEN, A.J., MUR, L.A., RECOURT, K., GERATS, A.G.M. AND MOL, J.N.M. (1991). Regulation and manipulation of flavonoid gene expression in anthers of petunia: the molecular basis of the *Po* mutation. *The Plant Cell* **3**, 39–48.

VAN WORDRAGEN, M.F., DE JONG, J., HUITEMA, H.B.M. AND DONS, H.J.M. (1991). Genetic transformation of chrysanthemum using wild type *Agrobacterium* strains; strain and cultivar specificity. *Plant Cell Reports* **9**, 505–508.

VAN WORDRAGEN, M.F., BONNEMA, G., VERKERK, R., HONTELEZ, J., PEYTAVI, R., GRIMSLEY, N. AND ZABEL, P. (1994). YAC-transfer to plant cell. *4th International Congress of Plant Molecular Biology, Amsterdam.* (Abstracts).

WEISS, D., VAN DER LUIT, A.H., KROON, J.T.M., MOL, J.N.M. AND KOOTER, J.M. (1993). The petunia homologue of the *Antirrhinum majus candi* and *Zea mays A2* flavonoid genes; homology to flavanone 3-hydroxylase and ethylene-forming enzyme. *Plant Molecular Biology* **22**, 893–897.

WIERMAN, R. AND VIETH, K. (1983). Outer pollen wall, an important accumulation site for flavonoids. *Protoplasma* **118**, 230–233.

WILMINK, A., VAN DE VEN, B.C.E. AND DONS, H.J.M. (1994). A transformation procedure for tulip. *4th International Congress of Plant Molecular Biology, Amsterdam.* (Abstracts).

WINANS, S.C. (1992). Two-way chemical signaling in *Agrobacterium*–plant interactions. *Microbiological Reviews* **56**, 12–13.

YOKOI, M. (1974). Colour and pigment distribution in ornamental plants. V. Anthocyanin distribution in rose cultivars. *Technical Bulletin of the Faculty of Horticulture, Chiba University* **22**, 13–24.

ZAMBRYSKI, P. (1988). Basic processes underlying *Agrobacterium*-mediated DNA transfer to plant cells. *Annual Review of Genetics* **22**, 1–30.

ZAMBRYSKI, P., JOOS, H., GENETELLO, C., LEEMANS, J., VAN MONTAGU, M. AND SCHELL, J. (1983). Ti plasmid vector for the introduction of DNA into plant cells without alteration of their normal regeneration capacity. *EMBO Journal* **2**, 2143–2150.

4

The Genes Encoding the Major Milk-Specific Proteins and Their Use in Transgenic Studies and Protein Engineering

WAYNE S. BAWDEN, ROBERT J. PASSEY AND ANTONY G. MACKINLAY

School of Biochemistry and Molecular Genetics, University of New South Wales, Sydney, Australia

Introduction

This chapter begins with the characteristics of the genes encoding the following milk proteins: the caseins, the whey proteins α-lactalbumin, β-lactoglobulin (BLG), whey acidic protein (WAP), late lactation protein, lactoferrin and lysozyme. It then goes on to cover milk protein transgenic studies with specific reference to using the mammary gland as a bioreactor for the production of biologically active proteins and the genetic engineering of milk composition. However, this chapter does not cover aspects of the hormonal regulation of milk protein gene expression, the biosynthesis of milk proteins and the role of the cell–cell and cell–extracellular matrix interactions which have been treated by Rosen *et al.* (1986), Mepham *et al.* (1992) and Lin and Bissell (1993). The milk proteins and their biological function have recently been discussed by Sawyer and Holt (1993), while a comprehensive overview of milk protein gene organization is given by Mercier and Vilotte (1993).

Milk proteins

The primary role of milk is to provide the suckling newborn with all its nutritional requirements for normal growth and development. Milk is a rich source of vitamins,

---

Abbreviations: α_1AT, α_1-antitrypsin; bFSH, bovine follicle-stimulating hormone; BLG, β-lactoglobulin; BST, bovine somatotropin; CAT, chloramphenicol acetyltransferase; CHO, Chinese hamster ovary; DCRs, dominant control regions; ELISA, enzyme-linked immunosorbent assay; ES, embryonic stem; FIX, blood clotting factor IX; hCD4, human CD4; hCFTR, human cystic fibrosis transmembrane conductance regulator; hIL-2, human interleukin-2; hSA, human serum albumin; IGF-I, insulin-like growth factor-I; LLP, late lactation protein; MAR, matrix attachment regions; oBLG, ovine β-lactoglobulin; oTP, ovine trophoblast interferon; PCR, polymerase chain reaction; SOD, extracellular superoxide dismutase; t-PA, tissue plasminogen activator; UTR, untranslated region; WAP, whey acidic protein

Biotechnology and Genetic Engineering Reviews — Vol. 12, December 1994
0264–8725/94/12/89–137 $20.00 + $0.00 © Intercept Ltd, P.O. Box 716, Andover, Hampshire SP10 1YG, UK

minerals, lipids, carbohydrates and amino acids, constituting a complete food source for the neonate. It has been estimated that in the United States, cow's milk still provides approximately 25% of the population's total protein, calcium and riboflavin intake. Therefore, milk remains an important source of nutrition throughout life (Swaisgood, 1973).

The bovine milk proteins are probably the best characterised, chemically, physically and genetically, of all food proteins (Swaisgood, 1993). Milk proteins are divided into two distinct groups. First, the caseins, which are a group of acidic, proline-rich, phosphoproteins that precipitate out of skim milk at pH 4.6 and 20°C (Jollès, 1975). The proteins that remain in solution under these conditions are termed the whey proteins, and these constitute a large class of diverse globular polypeptides. The function of the casein fraction is to chelate colloidal calcium phosphate and so serves as a major source of amino acids, calcium and phosphate for the suckling infant.

The protein composition of the milk of different species varies greatly, both in the presence or absence of particular proteins and in their relative abundance. One or more of the α-caseins, β- and κ-caseins, and α-lactalbumin have been found in the milks of most mammals studied so far (although human milk appears to lack an α-casein).

BLG is the major whey protein of ruminants such as the cow (Braunitzer *et al.*, 1973), sheep (Ali and Clarke, 1988) and red deer (McDougall and Stewart, 1976). It is also present in the milk of many non-ruminant species, such as the pig and horse (Godovac-Zimmerman and Braunitzer, 1987), dog, dolphin and manatee (Pervaiz and Brew, 1986), kangaroo (Godavac-Zimmerman and Shaw, 1987), whale (Ullrey *et al.*, 1984), tammar wallaby (Collet, Joseph and Nicholas, 1991), monkey (Azuma and Yamauchi, 1991), cat (Halliday *et al.*, 1990) and possibly the echidna and platypus (Teahan, McKenzie and Griffiths, 1991) but is absent from the milk of mice, rats and humans (Pervaiz and Brew, 1985). BLG has been implicated in a variety of activities, including the transport of retinol in milk (Papiz *et al.*, 1986), to aiding the digestion of milk lipids through binding and so removing the fatty acids which inhibit pregastric lipase (Perez *et al.*, 1992).

WAP is the major whey protein in the mouse (Hennighausen and Sippel, 1982a), rat (Hobbs *et al.*, 1982) and rabbit (Devinoy *et al.*, 1988). WAP may interact with an unknown receptor in either the digestive system or the mammary gland, may carry trace metals or may simply have a nutritional role as a source of sulphur (Hennighausen and Sippel, 1982b).

Lactoferrin is an iron-binding glycoprotein and has a variety of functions, including inhibiting microbial growth (Arnold, Cole and McGhee, 1977), stimulating the immune system (Sawatzki and Rich, 1989) and aiding iron absorption in infants (Nemet and Simonovitz, 1985). Lactoferrin has been found in the milk of a number of species, including the guinea-pig, cow, goat, pig, mare, mouse (Masson and Heremans, 1971), and the human (Johansson, 1960).

Lysozyme is believed to be present in the milk of most mammals (Steinhoff, Senft and Seyfert, 1994). The low concentration in cow's milk (0.5 μg ml^{-1}) when compared to human milk (500 μg ml^{-1}) (Meyer, Erhard and Senft, 1981), prompted the suggestion that these low levels may derive from blood cells in the mammary gland (Steinhoff, Senft and Seyfert, 1994). There appear to be at least 10 copies of the lysozyme gene in the cow (Irwin *et al.*, 1989) but only one in the human (Peters *et al.*, 1989).

Milk protein cDNA and genomic sequences

The available cDNA sequences for the various milk proteins are listed in *Table 1* and *Table 2*, while available gene sequences are listed in *Table 3*.

Table 1. The currently available casein cDNAs. Since all the casein genes sequenced so far can be classified as homologues of the four archetypal bovine caseins, their cDNA sequences are presented accordingly

Casein type	Species	Reference
α_{S1}-like Caseins	Cow	Stewart, Willis and Mackinlay (1984); Nagao *et al.* (1984)
	Rat α	Hobbs and Rosen (1982)
	Guinea-pig B	Hall, Laird and Craig (1984)
	Sheep	Mercier *et al.* (1985)
	Rabbit	Devinoy *et al.* (1988)
	Goat	Brignon *et al.* (1990)
	Pig	Alexander and Beattie (1992a)
	Tammar wallaby	Collet, Joseph and Nicholas (1992)
α_{S2}-like Caseins	Cow	Stewart *et al.* (1987)
	Mouse ε	Hennighausen, Steudle and Sipple (1982)
	Rat γ	Hobbs and Rosen (1982)
	Guinea-pig A	Hall *et al.* (1984)
	Sheep	Boisnard and Petrissant (1985)
	Pig	Alexander, Das Gupta and Beattie (1992)
	Rabbit	Dawson *et al.* (1993)
β-Caseins	Cow	Stewart *et al.* (1987)
	Rat	Blackburn, Hobbs and Rosen (1982)
	Mouse	Yoshimura, Banerjee and Oka (1986)
	Rabbit	Schaerer *et al.* (1988)
	Sheep	Provot, Persuy and Mercier (1989)
	Pig	Alexander and Beattie (1992b)
	Human	Lonnerdal *et al.* (1990)
	Tammar wallaby	Collet, Joseph and Nicholas (1992)
κ-Caseins	Cow	Stewart, Willis and Mackinlay (1984)
	Rat	Nakhasi, Grantham and Gullino (1984)
	Mouse	Thompson, Dave and Nakhasi (1985)
	Pig	Levine *et al.* (1992)
	Guinea-pig	Hall (1990)
	Sheep	Furet *et al.* (1990)
	Rabbit	Bosze *et al.* (1993)

Chromosomal locations

The four bovine casein genes have been mapped to chromosome 6 at q31–33 (Womack *et al.*, 1989). From inheritance studies of the bovine caseins it was initially suggested that their genes occur as a cluster in the order α_{S1}–α_{S2}–β–κ (Grosclaude, Joudrier and Mahe, 1979). More recently, long-range restriction mapping has shown them to be in the order α_{S1}–β–α_{S2}–κ within a 200 kb region, although there is disagreement as to the distances separating them (Ferretti, Leone and Sgaramella, 1990; Threadgill and Womack, 1990).

Table 2. The currently available whey protein cDNAs for some milk protein genes

Protein type	Species	Reference
α-Lactalbumin	Human Guinea-pig	Hall *et al.* (1992)
	Rat	Dandekar and Qasba (1981)
	Sheep	Gaye *et al.* (1987)
	Goat	Kumagai *et al.* (1987)
	Tammar wallaby	Collet, Joseph and Nicholas (1990)
	Mouse	Vilotte, Soulier and Mercier (1992)
	Pig	Das Gupta, Alexander and Beattie (1992)
β-Lactoglobulin	Cow	Alexander *et al.* (1989)
	Sheep	Gaye *et al.* (1986)
	Pig	Alexander and Beattie (1992c)
	Tammar wallaby	Collet, Joseph and Nicholas (1991)
Whey acidic protein	Rat and mouse	Hennighausen *et al.* (1982)
Late lactation protein	Tammar wallaby	Collet, Joseph and Nicholas (1989)
Lactoferrin	Mouse	Rey *et al.* (1990)
Lysozyme	Cow	Steinhoff, Senft and Seyfert (1994)

Table 3. The currently available sequences for milk protein genes

Protein type	Species	Reference
α_{s1}-like Caseins	Cow	Koczan, Hobom and Seyfert (1991)
	Goat	Leroux, Mazure and Martin (1992)
	Rabbit	Jolivet *et al.* (1992)
	Rat 5' region	Yu-Lee *et al.* (1986)
α_{s2}-like Caseins	Cow	Groenen *et al.* (1993)
	Rat 5' region	Yu-Lee *et al.* (1986)
β-Casein	Cow	Bonsing *et al.* (1988); Gorodetskii, Tkach and Kapelinskaya (1988)
	Mouse	Yoshimura and Oka (1989)
	Rat	Jones *et al.* (1985)
	Goat	Roberts *et al.* (1992)
	Rabbit	Thepot *et al.* (1991)
	Human	Hansson *et al.* (1994a)
κ-Casein	Cow	Alexander *et al.* (1988)
α-Lactalbumin	Cow	Vilotte *et al.* (1987)
	Rat	Qasba and Safaya (1984)
	Human	Hall *et al.* (1987)
	Guinea-pig	Laird *et al.* (1988)
	Goat	Vilotte *et al.* (1991)
	Mouse	Vilotte and Soulier (1992)
β-Lactoglobulin	Cow	Alexander *et al.* (1993)
	Sheep	Ali and Clarke (1988)
Whey acidic protein	Rat and mouse	Campbell *et al.* (1984)
	Rabbit	Thepot *et al.* (1990)
Lactoferrin	Mouse partial	Cunningham, Headon and Conneely (1992)
	Human 5' region	Teng *et al.* (1992)
Lysozyme	Human	Peters *et al.* (1989)

The human β-casein gene has been mapped to position 4pter → q21 (Menon *et al.*, 1992). The casein genes have been mapped to chromosome 5 in the mouse (Gupta *et al.*, 1982), to position q32 of chromosome 4 in the sheep (Hayes *et al.*, 1992), and to position q24 of chromosome 12 in the rabbit (Gellin *et al.*, 1985). Hayes *et al.* (1992) also suggested that the assignment of the bovine caseins to chromosome 6 may be an error caused by the very similar banding patterns of chromosomes 4 and 6. However, according to Toldo *et al.* (1994), this confusion is caused by differences in the Reading and ISCNDA (International System for Cytogenetic Nomenclature of Domestic Animals) standards, where the ISCNDA chromosome 4 is actually the Reading chromosome 6. They therefore recommend that the bovine caseins be assigned to the Reading chromosome 6 and that the ISCNDA designation be changed accordingly.

The α-lactalbumin gene has been mapped to chromosome 5 in the cow and goat, to chromosome 3 in the sheep (Hayes, Popescu and Dutrillaux, 1993), and to position q13 of chromosome 12 in the human (Davies *et al.*, 1987).

The BLG gene has recently been mapped to position q28 on chromosome 11 of the cow, and to chromosomes 11 and 3 of the goat and sheep, respectively (Hayes and Petit, 1993).

The late lactation protein has been mapped to the proximal region of the long arm of chromosome 3 in the tammar wallaby (Westerman, Spencer and Collet, 1991). The lactoferrin gene has been mapped to chromosome 9 in the mouse and to chromosome 3 in the human (Teng *et al.*, 1987), while the human lysozyme gene locus has been mapped to chromosome 12 (Peters *et al.*, 1989).

Evolution and gene structure

THE CALCIUM-SENSITIVE CASEINS

It has been proposed that the α_{S1}- α_{S2}- and β-caseins evolved from a common ancestor which appeared about 220–400 million years ago (Hobbs and Rosen, 1982), by a series of intra- and intergenic duplications and exon shuffling events (Bonsing and Mackinlay, 1987). Alignments of the casein cDNAs from several different species show that there is little sequence conservation between different caseins within a species. Such conservation is confined to the sequences encoding the 5' untranslated regions, leader peptides and casein kinase phosphorylation sites of the calcium-sensitive caseins. The high degree of conservation of the leader peptide may be a consequence of it either being unusually short and so having a more limited capacity for variation, or of its nucleotide sequence being involved in gene expression. The lack of conservation of the majority of the coding region is consistent with the perceived loose structure–function relationship of the caseins in forming stable micelles (Bonsing and Mackinlay, 1987).

It has been proposed that the ancestral gene from which the caseins are derived would have required four essential elements: (1) the 5' non-coding region, (2) the leader peptide, (3) a hydrophilic sequence containing a major phosphorylation site, and (4) a hydrophobic sequence (Jones *et al.*, 1985). It is most probable that while the ancestral gene would retain these elements, duplicated copies of it would be free to evolve positively selectable functions.

Sequence comparisons of the β-casein genes from different species show that there are no major insertions, deletions or rearrangements, and substitutions tend to be conservative. Sequence comparisons of the α-like caseins show that they have undergone many insertion/deletion events. These comparisons imply that the β-casein gene is most closely related to the ancestral gene, while the α-like caseins have been free to diverge from this ancestral sequence. The divergence of the α_{S1}- and then α_{S2}-casein genes may have been in response to selective pressure for higher levels of calcium phosphate in the milk, necessary for faster growth of the infant (Bonsing and Mackinlay, 1987).

It seems that the α_{S1}-casein gene is the product of the first duplication event as the 5' non-coding regions of the β- and α_{S2}-like caseins are more similar than those of the β- and α_{S1}- like caseins (Bonsing and Mackinlay, 1987). This proposal is in agreement with the work of Groenen *et al.* (1993) who, from similarities between the first three exons of the α_{S2}- and β-casein genes, and also between the α_{S2}-casein exon 14 and the β-casein exon 5, suggested that not only are these two genes most closely related, but that they could have arisen from a common ancestral gene by as little as two duplication events. Interestingly, two α_{S2}-casein cDNAs which are sufficiently different to indicate the presence of two active α_{S2}-casein genes, have recently been isolated from the rabbit. Presumably these arose from a gene duplication event which, from their degree of sequence similarity, appears to have occurred relatively recently (Dawson *et al.*, 1993).

The genes encoding the α_{S1}- and α_{S2}-caseins are large and have many small exons, some of which are thought to be the result of internal duplications. This is especially apparent in the bovine α_{S1}-casein gene, where six exons all conform to the criterion GAX–six codons–XAG (Stewart, Willis and Mackinlay, 1984; Koczan, Hobom and Seyfert, 1991). Two of the exons (10 and 13) display 23/24 identity and are contained within duplicated 154 bp regions which may be the product of a gene conversion event. In the α_{S2}-casein gene it is apparent that exons 7–11 and 12–16 are the product of an internal duplication event (Stewart *et al.*, 1987), while the region containing exon 3 displays 75% identity to the region containing exon 12, and so may be a consequence of a similar event (Groenen *et al.*, 1993).

The organization of all three calcium-sensitive casein-encoding genes retains an arrangement whereby the first intron is near the end of the 5' untranslated region, with the second exon containing the remainder of the 5' untranslated region, the leader peptide and the first two amino acids of the mature protein. At the other end of these genes, the second to last exon is short and contributes either some or all of the stop codon, with the final exon contributing the remaining portion of the 3' untranslated region.

κ-CASEIN

The κ-casein gene consists of several small exons and a single large exon that encodes most of the mature protein. Of the various caseins, κ-casein displays the greatest conservation between species (Nakhasi *et al.*, 1984; Thompson, Dave and Nakhasi, 1985), which is in keeping with its structural role in the stabilization of casein micelles. There is no significant similarity to any of the α_{S1}-, α_{S2}- or β-casein genes, which has led to the proposal that the κ-casein gene was recruited into the casein cluster from an ancestral gene unrelated to the other caseins, possibly γ-fibrinogen (Alexander *et al.*, 1988). This proposal may need revision in the light of the suggestion by Weissbach and Greininger (1990) that the fibrinogens themselves are derived from two ancestral genes.

α-LACTALBUMIN

α-Lactalbumin is a calcium metalloprotein which associates with the enzyme galactosyltransferase in the Golgi bodies to form the lactose synthetase complex (Ebner and Schanbacher, 1974). It is thought that the α-lactalbumin gene and the members of the lysozyme gene family share a common ancestor. Not only do the cDNA sequences of human α-lactalbumin and chicken lysozyme have significant identity (Hall *et al.*, 1982), but the rat (Qasba and Safaya, 1984), human (Hall *et al.*, 1987), bovine (Vilotte *et al.*, 1987) and guinea-pig (Laird *et al.*, 1988) α-lactalbumin, and the chicken (Jung *et al.*, 1980) and human (Peters *et al.*, 1989) lysozyme genes contain three introns at similar positions, and the first three of the four exons are comparable in both sequence and size. In the mouse (Vilotte and Soulier, 1992), and the rat, human and guinea-pig (Soulier *et al.*, 1989), the α-lactalbumin gene has been reported to occur as a single copy. In both the cow and the sheep more than one α-lactalbumin-like gene has been reported (Soulier *et al.*, 1989). In each species an α-lactalbumin pseudogene was discovered which at its 3' end shares identity with the 3' half of the α-lactalbumin gene. The 5' end of these pseudogenes share identity only with each other and it appears that they arose before the two species themselves diverged (Soulier *et al.*, 1989). A second bovine α-lactalbumin pseudogene with identity to the 3' half of the α-lactalbumin gene has recently been reported. It is flanked by repeated LINE sequences which have also been found associated with the lysozyme gene family, and which may be involved in the evolution of these genes (Vilotte, Soulier and Mercier, 1993).

β-LACTOGLOBULIN

In most species only one active BLG gene is known to exist. However, horse (Conti *et al.*, 1984), dog and dolphin (Pervaiz and Brew, 1986), and donkey (Godovac-Zimmerman *et al.*, 1988) are thought to have two distinct active genes, while the cat appears to have three (Halliday *et al.*, 1993). In these latter cases the BLG proteins are monomeric and, with the exception of the cat, are termed BLGI and BLG II for the major and minor components, respectively. In the cat, BLG I, II and III are all deemed to be major components. Recently, a BLG pseudogene has been isolated from the cow which, from the predicted amino acid sequence of the ancestral translation product, displays greater identity to the BLG IIs than to bovine BLG. It may therefore be derived from a BLG II gene in the cow (Passey and Mackinlay, in preparation).

The BLG protein is structurally similar to the members of a genetically related protein family known as the lipocalins. This diverse family is predominantly composed of low molecular weight secretory proteins that have a strong and specific affinity for hydrophobic molecules (reviewed by Flower, North and Attwood, 1993). A conserved folding pattern and three-dimensional structure, rather than strong sequence similarity, are the defining characteristics for members of this family. Despite the lack of sequence identity, the gene structure of a number of the lipocalins has been shown to be similar to that of the BLG gene, suggesting that the members of this family are derived from a common ancestral gene (Ali and Clark, 1988). The lipocalin gene which is most similar to the BLGs, in terms of size, exon arrangement and sequence identity, is the human

PP14 gene (Vaisse *et al.*, 1990) which encodes a 43 kDa glycoprotein secreted in the endometrium during pregnancy (Bell *et al.*, 1985).

WHEY ACIDIC PROTEIN

No other genes related to the rat, mouse or rabbit whey acidic protein genes have been reported either in these or other species. The rat and mouse genes extend over 2.8 kb and 3.3 kb, respectively, and are both composed of four exons. The third and fourth exons each encode a cysteine domain and are thought to have arisen by intragenic duplication of a primordial exon (Hennighausen *et al.*, 1982).

Comparison of the rat and mouse cDNA sequences shows a greater degree of conservation between the 3' non-coding regions than between the coding regions (Hennighausen *et al.*, 1982). Comparison of the rat and mouse genes shows that the introns are only slightly less conserved than the exons (Campbell *et al.*, 1984). These observations not only indicate a loose structure–function relationship for the mature protein, but also suggest a role for the 3' untranslated region in the production and maintenance of a stable transcript.

LATE LACTATION PROTEIN

Although no function of the late lactation protein (LLP) is known for certain (apart from acting as a source of amino acids), its similarity to BLG has prompted the suggestion that it may transport some small ligand released during the hydrolysis of milk fat (Collet and Joseph, 1993). The late lactation protein is found in the tammar wallaby (*Macropus eugenii*), and also possibly in the red (*Macropus rufus*) and eastern grey *Macropus giganteus*) kangaroos, and in the quokka (*Setonix brachyurus*) (Nicholas *et al.*, 1987, and references therein). No LLP gene has yet been sequenced.

It has recently been suggested that the LLP is a member of the lipocalin superfamily. A search of the GenPeptide and NBRF-PIR protein databases revealed homologies to a number of the members of this family, including BLG (Collet and Joseph, 1993). Late lactation protein and BLG share 42% and 20% similarity at the cDNA and protein levels, respectively, and it has been suggested that they share a common ancestor from which they diverged 150 million years ago (Woodlee *et al.*, 1993). The members of the lipocalin family to which LLP has the greatest similarity are the non-core members – rat and human von Ebner's gland protein and rat odorant-binding protein – implying that LLP is also a non-core member (Collet and Joseph, 1993).

LACTOFERRIN

Comparison of the amino acid sequences of the lactoferrin, serum transferrin, ovotransferrin and melanotransferrin proteins showed that, in addition to their overall homology, the N-terminal and C-terminal halves are approximately 40% identical to each other (Metz-Boutigue *et al.*, 1984; Bowman, Yang and Gwendolyn, 1988). The similarity of these proteins has since been confirmed at the level of gene structure in that they have a similar exon–intron distribution pattern (Cunningham, Headon and Conneely, 1992). It has therefore been proposed that they form a gene family which originated approximately 200–500 million years ago from a common ancestral gene which was itself a product of an internal duplication event. This event was apparently favoured by selective pressure as it resulted in the duplication of an iron-binding site in the protein (Williams *et al.*, 1982).

Milk protein transgenics

The characterization and cloning of the milk protein genes, together with the advent of transgenic technology, presented the potential to express biologically active proteins in the milk of transgenic animals (Palmiter *et al.*, 1982; Lovell-Badge, 1985). One advantage of harvesting foreign gene products from the body fluids of transgenic animals is that they are a renewable source, while harvesting recombinant proteins from tissue would almost certainly necessitate slaughtering valuable animals (Clark *et al.*, 1987). By targeting gene expression to the liver, for example, foreign proteins have been secreted into the blood of transgenic animals (reviewed by Ebert and Schindler, 1993). A possible problem with this approach is that biologically active foreign proteins circulating in the bloodstream could have an adverse effect on the health of the animal. Another restriction is the volume of blood able to be harvested. Recombinant proteins have been successfully targeted via the liver to the serum of pigs and sheep, but so far expression levels have been no greater than 1 $\mu g\ ml^{-1}$ for proteins other than serum proteins (Ebert and Schindler, 1993). In contrast, the mammary gland of mammals is able to synthesize and secrete large amounts of foreign protein during each lactation. Sheep, for example, are able to produce approximately 100 g of protein/day, while cows can produce up to 1 kg/day (Clark *et al.*, 1989a). An additional advantage of utilizing the mammary gland as a bioreactor is that the technology to harvest large quantitites of milk is already well established and non-invasive, thereby reducing stress on thc animal (Logan, 1993).

Unlike bacterial expression systems, (see below), a number of transgenic experiments have demonstrated that the mammary gland is able to perform a variety of post-translational modifications on heterologous protein products. These include β-hydroxylation (Clark *et al.*, 1989b), glycosylation (Ebert *et al.*, 1991; Wright *et al.*, 1991) and vitamin K-dependent γ-carboxylation (Velander *et al.*, 1992). In addition to performing these post-translational modifications, the mammary gland is also able to synthesize cysteine-rich foreign proteins which are able to fold into a functional conformation (Meade *et al.*, 1990), as well as heterodimeric protein complexes (Greenberg *et al.*, 1991), transmembrane proteins (DiTullio *et al.*, 1992) and protease-sensitive metalloproteins (Hansson *et al.*, 1994b).

To date, a number of high-value biologically active human proteins have been synthesized and secreted by the mammary tissue of mice, rats, rabbits, pigs, sheep and goats. These include such complex proteins as blood clotting factor IX (FIX; Clark *et al.*, 1989b), tissue plasminogen activator (t-PA; Gordon *et al.*, 1987; Ebert *et al.*, 1991), α_1-antitrypsin (α_1AT; Simons *et al.*, 1988; Archibald *et al.*, 1990; Wright *et al.*, 1991; Bischoff *et al.*, 1992), interleukin-2 (hIL-2; Bühler *et al.*, 1990), urokinase (Meade *et al.*, 1990), growth hormone (Reddy *et al.*, 1991; Ninomiya *et al.*, 1994), protein C (Velander *et al.*, 1991, 1992), cystic fibrosis transmembrane conductance regulator (hCFTR; DiTullio *et al.*, 1992), serum albumin (hSA; Shani *et al.*, 1992), extracellular superoxide dismutase (Hansson *et al.*, 1994b), lactoferrin (Platenburg *et al.*, 1994) and lysozyme (Maga *et al.*, 1994).

Before reviewing the progress of milk transgenics in detail, it will be useful to consider briefly the two main alternatives for the production of pharmaceutical proteins, namely microbial fermentation and mammalian cell culture.

Microbial fermentation and mammalian cell culture

Bacterial fermentation has been the system of choice for the production of relatively small and unmodified recombinant proteins, such as insulin, and hIL-2 (Reisman, 1993). However, many human proteins require post-translational modifications such as phosphorylation, glycosylation, amidation and proteolytic cleavage in order to acquire or maintain biological activity. Generally, bacterial systems do not possess such post-translational machinery and are therefore unable to synthesize these complex compounds (Bebbington and Hentschel, 1985). An additional problem is the relatively high reducing environment within bacteria, which can result in failure to form the correct disulphide bonds and therefore the formation of a non-functional protein product (Bebbington and Hentschel, 1985). Many mammalian proteins expressed within *Escherichia coli*, for example, fail to fold correctly and are deposited within inclusion bodies (Marston, 1986). For example, some proteins, such as t-PA and prochymosin, are completely insoluble within *E. coli* (Yarranton, 1990). Considerable downstream processing is necessary to solubilize and refold these proteins to restore biological activity. This process may dramatically reduce product yield and is very expensive to perform on an industrial scale (Bebbington and Hentschel, 1985; Datar, Cartwright and Rosen, 1993).

Although yeasts possess the ability to glycosylate heterologous proteins, they do so with very low efficiency (Wood *et al.*, 1985; Cabezón *et al.*, 1984). For example, human α_1AT has been expressed in *E. coli* and yeast, however the recombinant proteins are not glycosylated, drastically reducing their serum half-life (Casolaro *et al.*, 1987). The recombinant protein produced by lactating transgenic sheep, however, seems to be identical to that isolated from human plasma (Wright *et al.*, 1991). Aberrant glycosylation has also been documented in yeast. Such is the case for human insulin-like growth factor-I (IGF-I) synthesized by the yeast *Saccharomyces cerevisiae*, some of which was found to be *O*-glycosylated in contrast to the natural human product (Gellerfors *et al.*, 1989). This hyperglycosylation also raises concerns of allergenicity and reduced serum half-life (Tekamp-Olson and Valenzuela, 1990; Hodgson, 1993).

An alternate route for pharmaceutical protein production is large-scale mammalian cell culture. Such is the case for recombinant t-PA which is synthesized by transformed Chinese hamster ovary (CHO) cells and marketed under the name of Activase by Genentech. However, large-scale mammalian cell culture is a very expensive process. A large proportion of this expense results from the high cost of culture media, especially fetal calf serum, and, in some cases, the low concentration of product secreted into the media, which results in greater downstream processing costs (Werner *et al.*, 1992; Datar, Cartwright and Rosen, 1993). Mammalian cell culture also suffers from technical difficulties due to slow doubling time, the low shear resistance of mammalian cells and problems with microbial contamination (Reisman, 1993).

Mammary gland-specific transgenic studies

This section provides a brief description of the approaches that are used in transgenesis, i.e. the generation and study of transgenic animals. It then proceeds to summarize work done using mammary-specific promoters, derived from milk protein genes, to direct expression of both milk proteins and heterologous proteins in the mammary gland.

TRANSGENESIS

A transgenic organism is defined as one containing a foreign gene integrated into its own genome (Smith, Meuwissen and Gibson, 1987). Methods of introducing exogenous DNA into the germline include pronuclear microinjection, homologous recombination in embryonic stem (ES) cells and the use of retroviral vectors (reviewed by Thompson *et al.*, 1993). Recently, methods have been developed to transform various somatic tissues of mature animals via jet injection (Furth *et al.*, 1992) and particle bombardment (Klein and Fitzpatrick-McElligott, 1993). These techniques have been termed 'somatic transgenesis' as they transform existing tissues. These methods have already proven useful in assessing transgenic constructs without the need to first introduce them into the germline (Furth *et al.*, 1992; Cheng, Ziegelhoffer and Yang, 1993).

The most common method of introducing exogenous DNA into the germline of animals is via microinjection into the pronucleus of a single-cell embryo (Gordon *et al.*, 1980). This process involves piercing the pronucleus with an extremely fine needle and directly injecting a DNA solution (approximately 2 pl), containing the transgene construct, at a concentration of 1–2 $\mu g\ ml^{-1}$. The pronuclei of mice, sheep, goats and rabbits are easily resolved and targeted using differential interference contrast microscopy; however, the opacity of porcine and bovine eggs necessitates centrifugation for visualization. This process stratifies the cytoplasm, resulting in the deposition of the pronuclei in a transparent equatorial layer of the egg, allowing them to be resolved and injected (Hammer *et al.*, 1985; Wall *et al.*, 1985). There appears to be no limit to the size of DNA molecules that can be microinjected and integrated into the host genome. Recently, fragments as large as 230 kb have been introduced into mice (Schedl *et al.*, 1993). After microinjection, surviving embryos are reimplanted into pseudopregnant foster mothers where normal development may continue. After birth, transgenic individuals may be identified using Southern hybridization with DNA probes specific for the transgene or, alternatively, by the polymerase chain reaction (PCR; Saiki *et al.*, 1988).

In the case of mice, approximately 25% of the pups that are born will contain one or more copies of the transgene (Brinster *et al.*, 1985). Usually, multiple copies of the transgene integrate randomly at a single site within the genome, therefore each transgenic individual produced via microinjection is unique (Smith, Meuwissen and Gibson, 1987; Pursel *et al.*, 1989). About 10–20% of transgenic individuals contain the transgene integrated at multiple chromosomal locations (Brinster *et al.*, 1985). When multiple copies of the injected DNA integrate into the genome at a single site they are usually arranged in a tandem head-to-tail array (Brinster *et al.*, 1981; Palmiter *et al.*, 1982). Generally, provided it is present in the germline, the integrated DNA is stably transmitted from generation to generation. The precise mechanism by which the microinjected DNA integrates into the genome remains unknown (Palmiter and Brinster, 1986).

At present, pronuclear microinjection is the only method available to alter the germline of higher mammals, and has been used successfully to transform mice, rabbits, goats, pigs, sheep and cows (Gordon *et al.*, 1980; Hammer *et al.*, 1985; Ebert *et al.*, 1991; Hill *et al.*, 1992). The use of homologous recombination in pluripotent ES cells allows better control over the site of transgene integration and copy number, as well as the ability to remove a specific gene and replace it with another (Capecchi, 1989; Valancius and Smithies, 1991; Stacey *et al.*, 1994). However, at present this technology exists only for mice.

TRANSGENIC EXPRESSION

Transgenic studies have demonstrated that genes from one mammalian species are able to be expressed within another. In addition, genes normally expressed in one particular tissue, such as milk protein genes, usually maintain tissue-specific gene expression when introduced into another mammalian species (Brinster *et al.*, 1983; Storb *et al.*, 1984; Swift *et al.*, 1984; Palmiter and Brinster, 1986; Simons, McClenaghan and Clark, 1987).

Typically, the expression of a particular transgene varies among the different transgenic lines and exhibits no correspondence between expression levels and transgene copy number (Lee *et al.*, 1988; Bonnerot *et al.*, 1990). It is thought that this is caused by the transcriptional state of the surrounding host chromatin influencing transgene expression levels (Allen, Norris and Surani, 1990; Al-Shawi *et al.*, 1990).

Recently, genetic elements, termed dominant control regions (DCRs), have been identified upstream of the human β-globin locus. These elements confer position-independent, copy number-dependent expression of both the cognate gene as well as those under the control of heterologous promoters (Grosveld *et al.*, 1987; van Assendelft *et al.*, 1989). Similar elements have been found downstream of the human CD2 gene (Greaves *et al.*, 1989). These elements are thought to delineate active chromatin regions and are thus able to insulate transgenes from the environmental effects stemming from the surrounding host chromatin. As a consequence, the level of transgene expression becomes proportional to the number of gene copies integrated (Patient, 1990; Felsenfeld, 1992).

TRANSGENIC STUDIES UTILIZING MAMMARY-SPECIFIC PROMOTERS

At present promoter elements from ovine BLG (oBLG; Archibald *et al.*, 1990; Shani *et al.*, 1992), murine WAP (Gordon *et al.*, 1987; Pittius *et al.*, 1988; Yu *et al.*, 1989; Reddy *et al.*, 1991; Hansson *et al.*, 1994b), rabbit WAP (Bischoff *et al.*, 1992), bovine α-lactalbumin (Stinnakre *et al.*, 1991), rat β-casein (Greenberg *et al.*, 1991), goat β-casein (DiTullio *et al.*, 1992) and bovine α_{s1}-casein (Meade *et al.*, 1990; Hyttinen *et al.*, 1994; Maga *et al.*, 1994; Platenburg *et al.*, 1994) have been utilized to express foreign proteins in the mammary gland of transgenic mice. Endogenous promoters have been utilized in sheep (Simons *et al.*, 1988; Clark *et al.*, 1989b; Wright *et al.*, 1991), rabbits (Bühler *et al.*, 1990) and goats (Ebert *et al.*, 1991), while murine WAP

Table 4. Mammary-specific transgene expression studies in mice; listed according to each mammary-specific promoter

Transgene construct	Flanking sequence[†]	Expression level/ml[§]	Tissue specificity[¶]	Reference
β-lactoglobulin (BLG)				
oBLG	4.0; 7.3 kb	3–23 mg	MG specific	Simons, McClenaghan and Clark (1987)
oBLG–hα_1AT	4.0 kb; α_1AT	0.5–7 mg	MG, SG	Archibald *et al.* (1990)
oBLG–hSA (minigene)	3.0 kb	1–10 mg	Not reported	M. Shani (personal communication)
Whey acidic protein (WAP)				
mWAP–ht-PA (cDNA)	2.6 kb; SV40	<20 ng–50 μg	MG, T, K, SuG	Gordon *et al.* (1987)
mWAP–hCD4 (cDNA)	1.85 kb; bGH	10–200 ng	Not reported	Yu *et al.* (1989)
mWAP	2.6; 1.6 kb	3–54% endog[‡]	Not reported	Burdon *et al.* (1991)
mWAP–hGH	2.6 kb; hGH	65 ng–410 μg; 3.5 mg	MG, serum	Reddy *et al.* (1991)
rWAP	949; 70 bp	0–500% endog[‡]	Not reported	Dale *et al.* (1992)
RWAP–h α_1AT	17.6 kb; α_1AT	6–10.5 mg	Not reported	Bischoff *et al.* (1992)
mWAP-hSOD (cDNA)	2.3; 4 kb	8 ng–700 μg	MG, brain	Hansson *et al.* (1994b)
α-lactalbumin				
b α-lactalbumin	750; 336 bp	0.0025–0.45 mg	MG specific	Vilotte *et al.* (1989)
b α-lactalbumin–oTP (cDNA)	750; 450 bp	1 μg	MG specific	Stinnakre *et al.* (1991)
gp α-lactalbumin	1195; 398 bp	Not reported	Leaky	Maschio *et al.* (1991)
g α-lactalbumin	8.5; 9.5 kb	1.2–3.7 mg	Leaky	Soulier *et al.* (1992)
β-casein				
r β-casein	3.5; 3 kb	0.01–1% endog[‡]	MG, brain	Lee *et al.* (1988)
r β-casein–bFSHα (cDNA)	524 bp; SV40	15 μg[f]	Not reported	Greenberg *et al.* (1991)
r β-casein–bFSHβ (cDNA)	524 bp; SV40		Not reported	Greenberg *et al.* (1991)
g β-casein (18 kb)	3.0; 6.0 kb	21–24 mg; 40 mg	MG, skin	Persuy *et al.* (1992)
g β-casein (18.5 kb)	4.2; 5.3 kb	1 mg	MG, skin	Roberts *et al.* (1992)
g β-casein–hCFTR (cDNA)	4.2; 5.3 kb	Not reported	MG specific	DiTullio *et al.* (1992)
α_{S1}-casein				
b α_{S1}-casein–h urokinase	21; 2 kb	1–2 mg	MG specific	Meade *et al.* (1990)
b α_{S1}-casein–hLF (cDNA)	6.2; 8 kb	0.1–36 μg	MG specific	Platenburg *et al.* (1994)
b α_{S1}-casein-hLZ (cDNA)	20; 2 kb	to 0.78 mg	MG specific	Maga *et al.* (1994)

[†] Flanking sequence around the designated transcriptional unit (size of 5' flanking; 3' flanking sequence).

[§] The range of transgene expression levels in milk. First set of values gives range of founder expression levels, i.e. hemizygote expression. Values after; give highest homozygote expression level if reported.

[‡] mRNA expression levels determined only.

f Bigenic expression level.

[¶] Transgene expression highly mammary gland (MG) specific or leaky expression into other tissues.

Abbreviations: α_1AT, α_1-antitrypsin; b, bovine; CFTR, cystic fibrosis transmembrane conductance regulator; FSH, follicle-stimulating hormone; g, goat; GH, growth hormone; gp, guinea-pig; h, human; K, kidney; LF, lactoferrin; LZ, lysozyme; m, mouse; MG, mammary gland; o, ovine; r, rat; R, rabbit; SA, serum albumin; SG, salivary gland; SOD, extracellular superoxide dismutase; SuG, sublingual gland; T, tongue; TP, trophoblast interferon; t-PA, tissue plasminogen activator.

has been utilized successfully in pigs (Ebert *et al.*, 1991; Velander *et al.*, 1992). Transgenic cattle have also been generated (Krimpenfort *et al.*, 1991; Hill *et al.*, 1992; Hyttinen *et al.*, 1994).

We will now discuss many of the transgenic experiments published using various mammary-specific promoters. Summaries of these studies in mice and other mammals are presented in *Table 4* and *Table 5*, respectively.

β-*Lactoglobulin*

β-Lactoglobulin (BLG) is the major whey protein of ruminants and is present at a concentration of 4.6 mg ml^{-1} in the milk of cows (Caffin, Poutrel and Rainard, 1985). Ovine β-lactoglobulin (oBLG) became the first transgenic milk protein to be expressed in another organism when Simons, McClenaghan and Clark (1987) introduced the genomic sequence into mice. The transgene contained 4 kb of 5' sequence, the entire oBLG transcriptional unit and 7.3 kb of 3' flanking sequence. Lactating transgenic mice were found to produce a new protein in their milk, the identity of which was confirmed as oBLG by Western blotting. Expression levels between different transgenic mice varied from 3 to 23 mg ml^{-1}, the latter being more than four times the endogenous expression level in sheep. The presence of the foreign gene product did not appear to affect the lactating mothers or the suckling pups adversely. Analysis of RNA isolated from a variety of tissues detected oBLG mRNA specifically in the lactating mammary gland, thus the oBLG gene was expressed in a tissue- and stage-specific manner. Whitelaw *et al.* (1992) have subsequently demonstrated that only the first 406 bp of 5' flanking sequence and 1.9 kb of 3' sequence are necessary to confer position-independent, copy-number-related expression of the oBLG transgene.

The ability of the oBLG promoter to drive the expression of foreign genes in sheep was examined with constructs containing both the cDNA sequence of α_1AT and FIX (Simons *et al.*, 1988). FIX is used in replacement therapy to treat sufferers of haemophilia B, or Christmas disease, a genetic deficiency of blood clotting factor IX which affects approximately 1 in 30 000 males (Brownlee, 1987). FIX is usually isolated from human serum; however, supply is limited and the blood-derived product has the added risk of viral contamination. Active protein cannot be produced by bacterial systems as it requires extensive post-translational modifications for activity. Active α_1AT, on the other hand, can be synthesized by micro-organisms, but due to the lack of carbohydrate side-chains the protein has a greatly reduced serum half-life (Garver *et al.*, 1987).

The oBLG–FIX construct contained the FIX cDNA sequence cloned into the *Pvu* II site in the 5' untranslated portion of the first exon of the oBLG gene and possessed 4.3 kb of upstream oBLG sequence (Simons *et al.*, 1988). FIX cDNA was used as the genomic sequence of human FIX is 33.5 kb in length and therefore difficult to manipulate (Yoshitake *et al.*, 1985). Lactating ewes, containing the above, construct expressed biologically active FIX in their milk; however, the expression level was very low. Maximal expression was just 25 ng ml^{-1}, 1/250th the concentration in human plasma and 1/100 000th the concentration of oBLG in milk (Clark *et al.*, 1989b). The maximal expression level obtained with the oBLG-α_1AT construct was 5 μg ml^{-1}, which is higher than the oBLG–FIX construct, but is still 1000 times lower than endogenous oBLG expression in sheep. As will be discussed below, there tends

Table 5. Mammary-specific transgene expression studies in rats, sheep, rabbits, goats, pigs and cattle, arranged according to species

Transgene construct	Flanking sequence†	Expression level/ml§	Tissue specificity¶	Reference
Rat				
b α-lactalbumin	0.8; 0.5 kb	0.2 μg–2.4 mg	Not reported	Hochi *et al.* (1992)
b α-lactalbumin–hGH	738 bp; hGH	1 μg–4.4 mg	MG, serum	Ninomiya *et al.* (1994)
b α_{S1}-casein–hGH	671 bp; hGH	87 μg–6.5 mg	MG, serum	Ninomiya *et al.* (1994)
b β-casein–hGH	1.7 kb; hGH	1 μg–10.9 mg	MG, serum	Ninomiya *et al.* (1994)
m WAP–hGH	2.6 kb; hGH	7–56 μg	MG, serum	Ninomiya *et al.* (1994)
Sheep				
oBLG–h α_1AT (cDNA)	4.9; 7.3 kb	5 μg	Not reported	Simons *et al.* (1988)
oBLG–h Factor IX (cDNA)	4.9; 7.3 kb	25 ng	Not reported	Clark *et al.* (1989b)
oBLG–h α_1AT	4.0 kb; α_1AT	1–35, 63 mg*	MG specific	Wright *et al.* (1991)
Rabbits				
R β-casein–h IL-2	2.0 kb; IL-2	50–430 ng	Not reported	Bühler *et al.* (1990)
Goats				
mWAP–ht-PA (cDNA)	2.6 kb; SV40	3–6 μg	Not reported	Ebert *et al.* (1991)
g β-casein–ht-PA (cDNA)	6.2; 7.1 kb	1–3 mg	MG specific	Ebert *et al.* (1994)
Pigs				
mWAP	2.6; 1.6 kb	1–2 mg	MG, SG	Wall *et al.* (1991)
mWAP–h protein C (cDNA)	2.6; 1.6 kb	0.001–1 mg	Not reported	Velander *et al.* (1992)
Cattle				
b α_{S1}-casein–hLF (cDNA)	15; 6 kb	Too early to assess		Krimpenfort *et al.* (1991)
b α_{S1}-casein–hEPO	1.5; 1.7 kb	Too early to assess		Hyttinen *et al.* (1994)
MMTV–hIGF-I	2.0 kb	Too early to assess		Hill *et al.* (1992)

† Flanking sequence around the designated transcriptional unit (size of 5' flanking; 3' flanking sequence).

§ The transgene expression levels in milk.

¶ Transgene expression highly mammary gland (MG) specific or leaky expression into other tissues.

* This sheep expressed the transgene at 63 mg ml^{-1} at the onset of lactation, then stabilized at 35 mg ml^{-1}.

Abbreviations: α_1AT, α_1-antitrypsin; b, bovine; EPO, erythropoietin; g, goat; GH, growth hormone; h, human;, IGF-I, insulin-like growth factor-I; IL-2, interleukin-2; LF, lactoferrin; m, mouse; MG, mammary gland; MMTV, murine mammary tumour virus; o, ovine; R, rabbit; SG, salivary gland; t-PA, tissue plasminogen activator.

to be a correlation between the use of cDNA based protein coding sequences and low transgene expression levels when compared to the use of whole gene sequences.

When the genomic sequence of α_1AT was linked to approximately 4.0 kb of the oBLG promoter, expression levels in transgenic mice increased dramatically to 0.5–7 mg ml^{-1} (Archibald *et al.*, 1990). The recombinant α_1AT was shown to be biologically active using a trypsin-inhibition assay. However, unexpectedly, some lines also expressed α_1AT in their salivary glands. This may be due to the absence from the construct of a salivary gland-specific negative regulatory element (Archibald *et al.*, 1990).

The genomic oBLG–α_1AT construct, described above, has also been introduced into sheep (Wright *et al.*, 1991). Two transgenic sheep produced α_1AT at levels of 1–5 mg ml^{-1}, while a third produced α_1AT at approximately 35 mg ml^{-1}, making the transgenic product the most abundant protein secreted into the milk. This same individual initially produced α_1AT at 63 mg ml^{-1} which then stabilized to about 35 mg ml^{-1}. Another sheep demonstrated the opposite effect, starting out at 0.9 mg ml^{-1}, then, at week 7, increased production to 3.5 mg ml^{-1}. The reasons for these variations are unknown. The identity of α_1AT present in the milk was confirmed by Western blotting, and it appeared to be completely *N*-glycosylated. These findings confirm the ability of the ovine mammary gland to perform *N*-glycosylation of foreign proteins and to secrete them in large quantities. The α_1AT was found to be bioactive and able to be isolated at greater than 95% purity by standard chromatographic procedures.

Shani *et al.* (1992) have performed extensive investigations into the requirements for the expression of foreign proteins in the milk of transgenic mice, using oBLG-based minigenes. Various constructs, comprising 3.0 kb oBLG and 0.8 kb of oBLG 5' and 3' flanking sequences respectively, hSA cDNA and hSA intron 1 or introns 1 and 2, were introduced into mice. A minigene containing the oBLG promoter, the first intron and the cDNA sequence of hSA expressed the encoded protein in the milk of transgenic mice at a level of 2.5 mg ml^{-1}. Transgenic mice possessing oBLG–hSA minigenes containing hSA introns 1+2+12–14 or 2+7–14 have been shown to express hSA at 1 mg ml^{-1} or greater (up to 10 mg ml^{-1}) in 20% of the mice generated (M. Shani, personal communication). These expression levels are higher than those reported using *E. coli* (Latta *et al.*, 1987) and yeast (Fleer *et al.*, 1991).

The bovine BLG gene has been cloned and sequenced in our laboratory (Alexander *et al.*, 1993). Recently, transgenic mice containing this gene have also been generated. All transgenic lines examined express the transgene, one in particular at a level several times that of the endogenous expression level in cow's milk (Bawden *et al.*, in preparation).

Whey acidic protein

Rat WAP has been expressed in transgenic mice (Bayna and Rosen, 1990; Dale *et al.*, 1992) while the murine gene has been expressed in both mice and pigs (Burdon *et al.*, 1991; Wall *et al.*, 1991). The murine promoter has also been utilized to express a number of heterologous proteins in the milk of transgenic mice. These include the oncogenes Ha-*ras* and c-*myc* (Andres *et al.*, 1987, 1988), human CD4 (Yu *et al.*, 1989), human growth hormone (Reddy *et al.*, 1991; Tojo *et al.*, 1993) and extracellular superoxide dismutase (SOD) (Hansson *et al.*, 1994b), while rabbit WAP has been used to express a variant of human α_1AT (Bischoff *et al.*, 1992). The murine promoter has also been utilized to express human t-PA in both mice and goats (Gordon *et al.*, 1987; Pittius *et al.*, 1988; Ebert *et al.*, 1991), human protein C in pigs (Velander *et al.*, 1992) and human growth hormone in rats (Ninomiya *et al.*, 1994).

Studies of the gene elements responsible for the tissue-specific regulation of WAP expression have been done largely with transgenic mice. Intragenic sequences, as well as the 3' untranslated region of the WAP gene, appear to be crucial for mammary-specific and high-level expression of WAP in transgenic mice (Bayna and Rosen, 1990; Burdon *et al.*, 1991; Dale *et al.*, 1992). The murine WAP promoter has been utilized to express human t-PA in the milk of transgenic mice (Gordon *et al.*, 1987; Pittius *et al.*, 1988). This protein has clinical potential to dissolve fibrin clots in victims of myocardial infarction. The transgene consisted of 2.6 kb of the upstream WAP sequence, including the endogenous cap site, with t-PA cDNA cloned at nucleotide 24 so that this construct utilized the start point of translation and secretion signal sequence within the t-PA cDNA. The polyadenylation and termination signals downstream of the cDNA sequence were supplied from SV40. Expression ranged from <20 to 50 000 ng ml^{-1}, approximately 5% of the endogenous WAP level. The production of functional t-PA was confirmed by fibrin clot lysis and enzyme-linked immunosorbent assay (ELISA). The transgene was predominantly expressed in the mammary gland, but low levels of WAP–t-PA mRNA were detected in the tongue, kidney and sublingual gland (Pittius *et al.*, 1988). When this construct was introduced into goats, one transgenic line expressed enzymatically active t-PA at a concentration of 3–6 μg ml^{-1} (Ebert *et al.*, 1991).

A soluble derivative of human CD4 (hCD4), the human immunodeficiency virus receptor and a potential therapeutic agent for the treatment of AIDS, has also been expressed in the mammary gland of transgenic mice (Yu *et al.*, 1989). Expression levels ranged from 10 to 200 ng ml^{-1} as determined by immunoassay techniques. The WAP–hCD4 transgene contained the hCD4 cDNA plus 1850 bp of 5' upstream WAP sequence, and the bovine growth hormone polyadenylation signal and downstream region cloned 3' to the hCD4 cDNA.

Animals transgenic for human growth hormone are known to develop a number of abnormalities (Pursel *et al.*, 1989). However, human growth hormone has been expressed in the milk of transgenic mice with no adverse effects detected (Reddy *et al.*, 1991). The human growth hormone gene was expressed specifically in the mammary gland using the 2.6 kb murine WAP promoter. Expression levels varied greatly, from 65 ng ml^{-1} to 410 μg ml^{-1}. The recombinant protein was also detected in the serum during lactation; however, no adverse effects were detected. When homozygous mice were generated from the highest-producing line, levels of greater than 3.5 mg ml^{-1} were achieved. This compared favourably with the 1 mg ml^{-1} produced with mammalian tissue culture (Bebbington and Hentschel, 1985).

Generally, the expression levels of the WAP constructs have been very low. Except for the above case, the highest expression of a heterologous protein was only 50 μg ml^{-1} (Gordon *et al.*, 1987). These findings suggest that sequences within the WAP coding region or the 3' untranslated region (3'UTR) contribute to high-level expression. Comparisons between the mouse, rat and rabbit WAP genes have shown that the 3'UTR is more highly conserved than the protein-coding region (Campbell *et al.*, 1984; Thepot *et al.*, 1990). Dale *et al.* (1992) have determined by deletion studies that the 3'UTR is indeed necessary for high-level expression of this transgene in mice. The site-independent, high-level expression of rat WAP has been achieved using only 949 bp of 5' and 70 bp of 3' flanking sequences. Expression levels were also found to be proportional to the transgene copy number. Replacement of the WAP 3'UTR with the corresponding SV40 downstream sequence resulted in reduced expression, while further deletion of 91 bp of the 3'UTR led to site-dependent expression. The importance of the 3'UTR for high-level expression may explain why the above WAP-based constructs (Gordon *et al.*, 1987; Pittius *et al.*, 1988; Yu *et al.*, 1989), lacking the WAP 3'UTR expressed so poorly.

Better WAP-driven expression of heterologous proteins has been obtained in pigs. The complete 7.2 kb murine WAP gene has been expressed in the milk of transgenic swine at a level of approximately 1–2 mg ml^{-1} in three different transgenic lines, 2 mg ml^{-1} being the endogenous expression level in mice (Wall *et al.*, 1991). This construct seemed to give consistent expression levels regardless of the site of integration, contrasting with the same transgene expressed in mice (Burdon *et al.*, 1991). These data indicate that the murine WAP promoter could be used to express valuable pharmaceutical proteins in the milk of transgenic pigs. To investigate this possibility, the cDNA sequence encoding human protein C, a potent anticoagulant, was cloned into the first exon of the above murine WAP gene. When introduced into swine, the recombinant product was produced at levels up to 1 mg ml^{-1} (Velander *et al.*, 1992). The recombinant human protein C demonstrated anticoagulant activity equivalent to that derived from human plasma, despite exhibiting variable levels of γ-carboxylation. The recombinant protein may be clinically useful in the treatment of victims of stroke, septic shock and those suffering from a congenital deficiency in protein C (Esmon, 1987). The projected US market value of protein C, according to the American Red Cross, is estimated at $US960 million (Pursel and Rexroad, 1993).

α-*Lactalbumin*

Guinea-pig and goat α-lactalbumin have been expressed in transgenic mice (Maschio *et al.*, 1991; Soulier *et al.*, 1992) while the bovine gene has been expressed in both mice and rats (Vilotte *et al.*, 1989; Hochi *et al.*, 1992). The bovine α-lactalbumin promoter has also been utilized to express ovine trophoblast interferon (oTP) in mice (Stinnakre *et al.*, 1991) as well as human growth hormone in rats (Ninomiya *et al.*, 1994).

The bovine transgene comprising the α-lactalbumin transcriptional unit (3.2 kb) with 750 bp upstream and 336 bp 3' flanking sequence was expressed in a tissue-specific manner in five out of six transgenic lines generated (Vilotte *et al.*, 1989). The best expressing line produced bovine α-lactalbumin at 0.45 mg ml^{-1}, which is equivalent to apprxoimately half the endogenous level expressed in the milk of both cows and mice. The same promoter has also been utilized to express oTP in the milk of transgenic mice (Stinnakre *et al.*, 1991). The transgene consisted of oTP cDNA cloned between the bovine α-lactalbumin 5' and 3' flanking regions. Out of the eight transgenic mice generated, only one transgenic line was found to express active oTP, and at a greatly reduced level of approximately 1 μg ml^{-1}.

Further transgenic expression studies have determined that as little as 477 bp of upstream sequence confers tissue-specific expression of the bovine α-lactalbumin gene in transgenic mice, with important *cis*-acting elements located within the region –477/–220 (Soulier *et al*., 1992).

Secretion of bovine α-lactalbumin into the milk of transgenic rats at levels higher than the endogenous bovine expression level has been obtained with a gene very similar to that expressed in mice (Hochi *et al*., 1992). This construct contained approximately 800 bp of 5' sequence and 500 bp of downstream sequence, and was maximally expressed at 2.4 mg ml^{-1}, five times that reported in mice (Hochi *et al*., 1992). Reasons for this dramatic increase in expression may be due to the different transcriptional machinery between the two host animals or to reported differences in the gene sequence (13 nucleotides different) or to a combination of these factors. The bovine α-lactalbumin promoter (738 bp) has also been utilized to drive the expression of human growth hormone at levels as high as 4.4 mg ml^{-1} in rats (Ninomiya *et al*., 1994).

The goat α-lactalbumin gene (21 kb in length, 8.5 kb of which is 5' upstream sequence) has also been expressed in the milk of transgenic mice up to 3.7 mg ml^{-1}, nearly twice the endogenous level in goats and five times that in mice (Soulier *et al*., 1992). However, the expression levels varied greatly between transgenic lines and were copy number-independent, indicating that some *cis*-acting regulatory elements were missing.

Mice deficient in α-lactalbumin (Stinnakre *et al*., 1994), as well as mice that have had their endogenous gene replaced with that of human α-lactalbumin gene, have recently been generated (Stacy *et al*., 1994). These experiments have special significance for the genetic engineering of milk composition, which will be discussed later.

β-*Casein*

Both rat and goat β-casein have been expressed in the milk of transgenic mice (Lee *et al*., 1988; Persuy *et al*., 1992; Roberts *et al*., 1992). β-Casein promoters from various species have been utilized to drive the expression of a number of heterologous proteins, such as bovine follicle-stimulating hormone (bFSH) and hCFTR in mice (Greenberg *et al*., 1991; DiTullio *et al*., 1992), human growth hormone in rats (Ninomiya *et al*., 1994), hIL-2 in rabbits (Bühler *et al*., 1990) and ht-PA in goats (Ebert *et al*., 1991).

The 14 kb genomic sequence containing the rat β-casein gene, as well as 3.5 kb of 5' and 3.0 kb of 3' flanking sequence, has been expressed in transgenic mice (Lee *et al*., 1988). Eight transgenic lines were generated; however, mRNA expression studies determined that rat β-casein expression was extremely low, being only 0.01–1% of the endogenous mouse β-casein expression levels. These data indicate that important regulatory elements necessary for high-level expression of this transgene were missing from this construct. However, all the transgenic lines expressed in a tissue-specific manner except for one line, which also expressed β-casein mRNA in the brain, albeit at a much lower level.

Further studies with various rat β-casein–chloramphenicol acetyltransferase fusion genes have determined that a –524/+490 minimal promoter can direct heterologous expression to the mammary gland (Lee, Atiee and Rosen, 1989). This same promoter was utilized to produce two separate constructs containing cDNA sequences for the bFSH subunits α and β respectively. Bigenic mice, generated by co-injection of both constructs, or by crossing individual lines expressing separate subunits, produced biologically active bFSH in their milk at levels up to 15 μg ml^{-1}. The presence of a biologically active product demonstrates that the mammary gland is able to synthesize and assemble bioactive heterodimeric proteins (Greenberg *et al.*, 1991).

The rabbit β-casein promoter has been used to express biologically active hIL-2 in rabbits at a maximal level of only 430 ng ml^{-1} (Bühler *et al.*, 1990). The transgene consisted of the 2 kb rabbit β-casein promoter cloned 5' to the genomic sequence of hIL-2. This construct contained vector sequences between the junction of these two genetic elements which may have contributed to its poor expression. Alternatively, it seems that β-casein transgenes require greater upstream sequence for high-level expression. For a review of the use of hIL-2 in immunotherapy, see Kaplan, Cohn and Smith (1992).

A dramatic improvement in expression levels, compared to the rat gene, was found with mice transgenic for the 18 kb caprine β-casein gene, containing 3 kb of 5' and 6 kb of 3' sequence (Persuy *et al.*, 1992). All of the nine transgenic lines expressed caprine β-casein in their milk, with two lines expressing as much as 21–24 mg ml^{-1}, twice the endogenous level in goats. A homozygous mouse expressed as much as 40 mg ml^{-1} (higher than the endogenous mouse β-casein expression level of 20–30 mg ml^{-1}). All the founder mice exhibited tissue-specific expression, while in two cases caprine β-casein mRNA was detected in the skin. This extremely high expression level indicates that the goat β-casein may be a useful promoter for the production of high-value bioactive proteins in the milk of transgenic animals. The goat β-casein promoter has been used to drive the expression of t-PA in the milk of transgenic goats at a level of 2–3 mg ml^{-1} (Ebert *et al.*, 1991, 1994).

The reason for the extremely low level of expression of the rat β-casein transgene with respect to the highly expressing goat equivalent is probably due to the divergence between the rat and goat β-casein genes (Persuy *et al.*, 1992). The rat and goat genes exhibit 75% homology over the first 200 bp of the 5' flanking region; however, beyond this point there is little homology between their upstream sequences (Roberts *et al.*, 1992).

The goat β-casein gene isolated by Roberts *et al.* (1992) has been used to drive the expression of heterologous proteins in the milk of both transgenic mice and goats (DiTullio *et al.*, 1992; Ebert *et al.*, 1991). A construct containing the cDNA sequence for hCFTR cloned between exons 2 and 7 of the goat β-casein gene (i.e. the cDNA sequence is flanked on either side by at least one intron) was microinjected into mice (DiTullio *et al.*, 1992). Transgenic expression was found to be specific to the mammary gland, recombinant hCFTR being associated with the membrane surrounding the milk fat globule. The authenticity of the protein product was confirmed by immunoprecipitation, but expression levels could not be accurately quantified. This construct has also been introduced into goats (H. Meade, personal communication). It is hoped that the production of recombinant hCFTR may help in the development of therapies for cystic fibrosis. The feasibility of utilizing hCFTR in protein replacement therapy has been demonstrated *in vitro* by Marshall *et al.* (1994).

α_{S1}-CASEIN

The bovine α_{S1}-casein promoter has been utilized to express a variety of recombinant human proteins in the milk of transgenic mice. These proteins include human urokinase (Meade *et al.*, 1990), lysozyme (Maga *et al.*, 1994) and lactoferrin (Platenburg *et al.*, 1994). Of the handful of transgenic cattle established, to date, two lines contain portions of the bovine α_{S1}-casein gene (Krimpenfort *et al.*, 1991; Hyttinen *et al.*, 1994).

Meade *et al.* (1990) constructed a mammary-specific expression vector consisting of 21 kb of 5' and 2 kb of 3' flanking sequence, including the native polyadenylation signal, of the bovine α_{S1}-casein gene joined by a synthetic linker containing sites for the rare cutting restriction enzymes, *Xho*I and *Not*I. The sequence encoding the genomic human urokinase gene was cloned into the *Xho*I site and the construct microinjected into the pronuclei of fertilized mice oocytes. Lactating transgenic mice secreted functional human urokinase in their milk at levels of 1–2 mg ml^{-1}. Analysis of RNA from various tissues demonstrated that human urokinase was expressed only in the mammary gland. The cysteine-rich human urokinase protein, which contains nine disulphide bonds, was specifically selected to investigate whether the mammary gland could synthesize proteins possessing a high proportion of sulphur-containing amino acids. The recombinant urokinase was demonstrated to be enzymatically active, as determined by casein/plasminogen gel assay (Nagamine, Sudol and Reich, 1983). It has been estimated that if the above construct functioned as well in transgenic cows, a single cow could produce approximately 20 g of this recombinant protein per day.

Maga *et al.* (1994) have used the same α_{S1}-casein promoter as above to express human lysozyme (cDNA sequence) in the milk of transgenic mice. Out of six transgenic lines established, two expressed human lysozyme RNA in the mammary gland, one line expressing more than the endogenous WAP RNA level. Protein expression studies have determined that as much as 0.78 mg ml^{-1} of active human lysozyme is expressed in the milk of these mice (J. Murray, personal communication).

Krimpenfort *et al.* (1991) have succeeded in producing a transgenic calf containing a transgene consisting of the human lactoferrin cDNA cloned between the second and second to last exons of the bovine α_{S1}-casein gene. Injection of over 1000 embryos resulted in the production of one female and one male transgenic calf. The female proved to be mosaic, and restriction enzyme analysis indicated that a portion of the transgene had been deleted. The male calf, 'Herman', contains 5–10 copies of the intact transgene in a head-to-tail arrangement. The production of lactating transgenic daughters from Herman will be necessary before expression studies can be performed. This may be as late as 1998. Expression studies in transgenic mice generated using a construct similar to the one described above have resulted in human lactoferrin expression levels of 0.1–36 μg ml^{-1} (Platenburg *et al.*, 1994). Higher expression levels were observed when the lactoferrin cDNA was replaced with the genomic sequence (De Wit and Pieper, unpublished).

Both lysozyme and lactoferrin have antimicrobial activity (Arnold, Cole and McGhee, 1977; Hughey and Johnson, 1987; Chang, 1990). Thus, the presence of these proteins within the milk of cattle should have the additional benefit of reducing intramammary infections, or mastitis, a major problem in the dairy industry which is estimated to cost over $US100 million annually in lost production and treatment costs in New York State alone (Miles, Lesser and Sears, 1992).

A transgenic calf containing the human erythropoietin sequence under the control of the bovine α_{S1}-casein promoter has recently been generated (Hyttinen *et al.*, 1994). The transgene construct consists of bases –610 to +1483 of the bovine α_{S1}-casein gene with the genomic erythropoietin sequence cloned downstream. This construct also contains the 1.3 kb chicken lysozyme A element and the mouse mammary tumour virus hormone response element (bases 7112–7295) cloned upstream of the α_{S1}-casein promoter.

General comments on mammary-specific promoters

CROSSING SPECIES BOUNDARIES

The fact that many milk protein genes from one mammal can be expressed efficiently in the mammary gland of another indicates that the genetic elements that determine mammary-specific expression are often conserved between species (Palmiter and Brinster, 1986). In some cases transgene expression levels may even exceed those in the species of origin, oBLG expression in mice being an example (Simons, McClenaghan and Clark, 1987). Similar experiences have been reported for mouse WAP in swine (Wall *et al.*, 1991; Velander *et al.*, 1992), goat α-lactalbumin (Soulier *et al.*, 1992) and β-casein (Persuy *et al.*, 1992) in mice, as well as bovine α-lactalbumin in rats (Hochi *et al.*, 1992). Therefore, to drive the expression of foreign proteins in the milk of transgenic animals it may be advantageous to utilize promoters from a different mammalian species. This has been demonstrated with the high-level expression of human protein C in transgenic swine utilizing the mouse WAP promoter (Velander *et al.*, 1992), while the same construct was expressed at a reduced level in transgenic mice (Velander *et al.*, 1991).

WHEY VERSUS CASEIN PROMOTERS

As the caseins are expressed at higher levels than the whey proteins in the milk of ruminants, it might be expected that casein transgenes would express at higher levels than whey transgenes. This, however, is not generally the case. With the exception of goat β-casein (Persuy *et al.*, 1992), whey transgenes tend to express at a higher level than their casein counterparts in mice.

Transgenic experiments have demonstrated that whey protein genes possess more compact promoters and give better expression than the caseins. High-level expression has been reported in transgenic mice for bovine α-lactalbumin (Vilotte *et al.*, 1989), rat WAP (Dale *et al.*, 1992) and oBLG (Whitelaw *et al.*, 1992) with as little as 750, 949 and 406 bp of upstream sequence, respectively. In the case of the rat WAP and oBLG transgenes, expression is also copy number-dependent (Dale *et al.*, 1992; Whitelaw *et al.*, 1992). Thus, expression of these whey protein transgenes occurs in a copy number-dependent fashion and is impervious to positional effects, an effect normally only associated with the presence of DCR sequences. In contrast, position-independent, copy number-dependent expression for casein-based transgenic constructs has not so far been observed. These results indicate that the *cis*-acting elements necessary to confer independent expression without interference from the transcriptional state of the surrounding chromatin are missing from these transgenes. An explanation for these findings may be that casein genes are naturally organized as a gene cluster and may be under the influence of a DCR similar to that discovered in the β-globin locus. In contrast, the whey protein genes are scattered throughout the genome and so their expression is independently regulated.

Ninomiya *et al.* (1994) have performed comparative studies of bovine casein transgenes in transgenic rats. This group has managed to achieve high-level expression of human growth hormone in transgenic rats using as litle as 671 bp and 1.7 kb of bovine α_{s1}-and β-casein upstream sequence, respectively. Although high-level expression was achieved, expression levels were still highly variable and subject to positional effects (87 μg ml^{-1} to 6.5 mg ml^{-1} and 1 μg ml^{-1} to 10.9 mg ml^{-1}, respectively). The presence of a DCR in the casein locus has not been identified.

THE EFFECT OF INTRONS

When designing constructs for the production of foreign proteins in the milk of transgenic animals, the protein coding region may be present as either the cDNA or genomic sequence. Generally higher expression levels, of the order of mg ml^{-1}, have been achieved with genomic sequences (Archibald *et al.*, 1990; Meade *et al.*, 1990; Wright *et al.*, 1991), than constructs utilizing cDNA coding sequences (Pittius *et al.*, 1988; Simons *et al.*, 1988; Clark *et al.*, 1989b; Yu *et al.*, 1989; Ebert *et al.*, 1991; Greenberg *et al.*, 1991; Stinnakre *et al.*, 1991). A study by Brinster *et al.* (1988) using four different pairs of constructs, each pair differing solely in the presence or absence of introns in their protein coding region, determined that transgenes containing introns generally produced 10- to 100-fold more mRNA transcripts than those that lacked introns. These findings may be due to the presence of enhancers or other *cis*-acting sequences necessary for high-level expression being located within introns (Brinster *et al.*, 1988). Indeed, regulatory elements have been identified within the intronic sequences of the human adenosine deaminase gene and the human hypoxanthine phosphoribosyltransferase gene (Aronow *et al.*, 1989; Reid *et al.*, 1990). This may also be the case with milk protein genes (Archibald *et al.*, 1990; Whitelaw *et al.*, 1991). Findings similar to those of Brinster *et al.* (1988) have been reported using the oBLG transgene in mice (Whitelaw *et al.*, 1991). The question arises as to whether this augmentation effect is due to intron sequences required for efficient splicing of the mRNA transcript or whether the absence of introns removes important genetic regulatory elements. Experiments reported by Whitelaw *et al.* (1991) suggest that it is, in fact, sequences within the introns themselves that regulate gene expression in concert with the 5' oBLG flanking sequence.

Complementary DNA sequences have frequently been used in transgenic constructs as they are usually the first form of the gene available and are easier to manipulate and clone compared to genomic sequences, which may be many kilobases long. In cases where only the cDNA of the gene of interest is available, or the genomic sequence is difficult to manipulate, studies have demonstrated that the addition of heterologous introns may improve their expression (Palmiter *et al.*, 1991; Whitelaw *et al.*, 1991; Shani *et al.*, 1992). Shani *et al.* (in preparation) have demonstrated that this augmentation effect of introns on gene expression can function synergistically with constructs containing different combinations of introns expressing at different levels. A variety of oBLG–hSA minigenes containing combinations of hSA introns 1–14 have been constructed and expression studies performed in tissue culture using COS cells. Various combinations of introns were found to give efficient expression. Similar constructs used to generate transgenic mice have also been observed to give good expression, up to 10 mg ml^{-1} (M. Shani, personal communication).

IMPROVING TRANSGENIC EXPRESSION

Two methods have been demonstrated to improve transgene expression. Both methods rely on creating a permissive region of transcription around the site of integration. This permissive region shields the transgene from the effects of the transcriptional state of the surrounding host chromatin which may repress transgene expression.

'Rescuing' transgene expression

The expression of poorly expressed transgenes can be elevated via co-integration with a highly expressing construct. This effect has been demonstrated by co-injecting the low expressing cDNA-based constructs, oBLG–FIX and oBLG–α_1AT (described above), with the high expressing oBLG transgene in mice (Clark *et al.*, 1992). The oBLG–FIX construct did not express at all in 10 different transgenic mouse lines established, but when co-integrated with the oBLG transgene, FIX expression levels were elevated to a maximum of 1 μg ml^{-1}. Similar results were found with the oBLG–α_1AT construct (up to 600 μg ml^{-1}). Clark *et al.* (1992) proposed that the highly expressing oBLG gene, which has co-integrated with the cDNA-based construct, is able to establish an active transcriptional locus, thereby overcoming the position effects responsible for the low expression of the cDNA-based constructs. This 'rescue' technique may be applicable to other low-expressing cDNA-based constructs.

Utilizing MAR sequences

The co-integration of matrix attachment regions (MAR) has been demonstrated to confer position-independent expression of murine WAP in transgenic mice (McKnight *et al.*, 1992). These MAR sequences from the chicken lysozyme locus restored normal hormonal and lactational regulation and are thought to buffer the transgene from positional effects which tend to interfere with transgene expression. In this study all WAP transgenes which had integrated next to a MAR sequence were actively transcribed. However, only 50% of the transgenes that integrated without MAR sequences were actively transcribed.

In studies performed with pigs and sheep, generally about 60% of the transgenic animals generated actively express foreign transgenes (Pursel *et al.*, 1990; Rexroad *et al.*, 1990; McKnight *et al.*, 1992). It may be possible to use MAR sequences to increase this proportion of transgenic expressors, and therefore reduce production costs. Such a sequence has been utilized in the α_{S1}-casein–human erythropoietin construct introduced into cattle by Hyttinen *et al.* (1994).

Genetic engineering of bovine milk composition

Another application of transgenic technology of interest to the dairy and related industries is the alteration of milk composition in dairy animals. Using a combination of recombinant DNA techniques and transgenics it should be possible to modify milk composition, thereby allowing the production of novel milk products.

Possible ways in which the composition of bovine milk might be altered include:

1. '*humanizing*' the milk of cows to make it more suitable to the nutritional needs of human infants;
2. the production of '*designer*' milks for those individuals unable to tolerate certain milk components;
3. the alteration of the casein content of cow's milk for improved nutrition and greater cheese yields; and, finally,
4. engineering milk proteins with altered physical and chemical properties for specific manufacturing processes.

The above aims could be realized in some cases by adding extra milk protein genes, via micro-injection, to the germline of dairy animals. For example, introducing a transgene encoding a human milk protein to the germline in order to make cow's milk closer in composition to human milk. In other cases it would be desirable to down-regulate the expression of certain endogenous milk protein genes which have been implicated in allergic reactions in some individuals. It has been proposed that milk with little or no lactose could be produced by down-regulating the α-lactalbumin gene or by the introduction into dairy cattle of the β-galactosidase transgene (Bremel, Heng-Cherl and Bleck, 1989). A study conducted by the US National Dairy Council has shown that approximately 70% of the world's population is deficient in the intestinal lactase necessary to digest the lactose present in milk (NDC, 1985). Adults exhibiting severe lactose intolerance may suffer from cramps, flatulence and diarrhoea, while children may suffer malnutrition or even death in extreme cases (Kretchmer, 1972; Saavedra and Perman, 1989). It is possible to selectively disrupt or entirely remove specific genes by utilizing 'knock out' technology via homologous recombination in ES cells (Cappecchi, 1989; Hasty *et al.*, 1991; Valancius and Smithies, 1991). Stinnakre *et al.* (1994) have succeeded in producing α-lactalbumin-deficient mice via gene disruption using homologous recombination in ES cells. Lactating transgenic females homozygous for this null mutation synthesize a milk deficient in α-lactalbumin and lactose. This milk is higher in fat and protein; however, the milk is so viscous that it cannot be secreted. This is probably due to the fact that lactose is the major osmole of bovine milk, and altering its levels in the mammary gland may have interfered with osmoregulation (Kuhn, 1983). Hemizygous females exhibited a 40% decrease in α-lactalbumin levels and only a 10–20% reduction in lactose levels compared to wild type.

An alternative approach to the specific down-regulation of milk protein genes may be the use of antisense and ribozyme sequences which are able to suppress native gene expression by blocking the flow of information from DNA to RNA to protein (reviewed by Inouye, 1988; Cazenave and Hélène, 1991; Symons, 1992; Bratty *et al.*, 1993).

Antisense technology offers the ability to down-regulate selectively a specific gene by utilizing RNA molecules complementary to the protein-coding, i.e. sense, mRNA sequence. Antisense molecules base-pair with the target RNA forming a double-stranded structure which hinders the binding of intracellular factors essential for translation (Cazenave and Hélène, 1991).

Antisense transgenes can be constructed by cloning the target sequence in reverse orietnation, under the regulation of an appropriate promoter. The successful application of antisense technology to down-regulate gene expression has been demonstrated dramatically by the production of transgenic dwarf mice by introducing a transgene containing the antisense cDNA sequence of rat growth hormone (Matsumoto *et al.*, 1993). A model system involving chloramphenicol acetyltransferase (CAT) expression driven by bovine α_{S1}-casein promoter has also given encouraging results in that the expression of the CAT gene in the mammary gland was reduced by 92% in bigenic mice containing the antisense CAT sequence under the same α_{S1}-casein promoter (Sokol *et al.*, submitted).

Ribozymes are RNA species that are able to cleave specific RNA sequences catalytically by virtue of their endoribonuclease activity. Target specificity is dictated by the antisense RNA sequence flanking the catalytic hammerhead domain via Watson–Crick base-pairing with the complementary substrate sequence (Haseloff and Gerlach, 1988). Since ribozymes should be able to act on many target RNAs, they should, theoretically, achieve greater down-regulation of gene expression than antisense RNA, although it is not clear that this occurs *in vivo*. Synthetic ribozymes have been used to down-regulate gene expression *in vivo* in bacteria (Sioud and Drlica, 1991), mammalian cells (Cameron and Jenings, 1989; Sarver *et al.*, 1990; Scanlon *et al.*, 1991), *Xenopus* oocytes (Cotten and Birnsteil, 1989; Saxena and Ackerman, 1990), plant cell protoplasts (Steinecke, Hergert and Schreier, 1992), *Drosophila* (Zhao and Pick, 1993) and the murine mammary gland (Sokol *et al.*, submitted).

In some cases, it may be advantageous to modify the amino acid sequence of the milk protein genes to engineer altered properties, such as proteins with increased nutritional value (Oh and Richardson, 1991a) or enhanced gelation characteristics (Lee, Cho and Batt, 1993). The technique of *in vitro* site-directed mutagenesis (Zoller and Smith, 1983; Botstein and Shortle, 1985; Jimenez-Flores and Richardson, 1988) allows the systematic modification of the amino acid content of any cloned protein-encoding sequence, thus permitting the rational design of food proteins with more desirable physical and chemical properties.

'HUMANIZING' COW'S MILK

Despite the superiority of the breast milk of healthy mothers over that of cow's milk-based substitutes, many mothers are either unable or choose not to breast feed their children. As a result, many infants throughout the world are raised on supplemented formulae based on cow's milk (Packard, 1982).

There are some important compositional differences between bovine and human milk. Most notable is the greater whey to casein ratio, a higher concentration of both lactoferrin and lysozyme, and the absence of BLG in human milk (Blanc, 1981).

Bovine milk alone is insufficient for the nutritional needs of growing infants, thus requiring milk formula manufacturers to develop fortified 'humanized' cow's milk-based formulae to simulate human milk (Packard, 1982). Transgenic technology may be able to achieve this via a combination of down-regulating certain bovine milk genes and/or by directing the expression of human milk protein transgenes to the bovine mammary gland (Mercier, 1986). An example is the attempt to introduce human lactoferrin into the milk of transgenic cattle (Krimpenfort *et al.*, 1991). Ward *et al.* (1992) and Liang and Richardson (1993) have succeeded in producing recombinant human lactoferrin in *Aspergillus oryzae* and *S. cerevisiae* at a level of 25 $\mu g\ ml^{-1}$ and 2 $\mu g\ ml^{-1}$ of culture media, respectively. Lactoferrin from these sources could be used in the production of supplemented infant formula. The infant formula market is valued at over $US4 billion per year (Spalding, 1992).

As mentioned above, BLG is the major whey protein of cow's milk, yet it is absent from that of humans. Since bovine milk is a major component of the Western diet, the proteins of cow's milk are usually the first foreign proteins to be encountered by human infants (Swaisgood, 1973). Therefore cow's milk can elicit an allergic response, resulting in gastrointestinal, cutaneous and respiratory hypersensitivity conditions in as many as 7.5% of newborns (Packard, 1982). BLG in particular has been reported to be one of the main agonists in cow's milk allergy in infants, and increased levels of BLG antibodies have been reported in sufferers of cow's milk protein intolerance (James and Sampson, 1992). Recently, BLG has also been implicated in interfering with the development of the newborn's gut mucosal barrier (Arvola *et al.*, 1993). For these reasons methods have been developed to remove BLG from bovine milk destined for consumption by infants (Kuwata *et al.*, 1985). Transgenic cattle deficient in BLG expression are another alternative. This end could be achieved by introducing transgenic constructs containing antisense and/or ribozyme sequences directed against BLG to down-regulate its native expression levels.

Human milk has a lower casein:whey ratio than cow's milk. Its protein fraction contains only 30% casein, mostly β-casein, compared to 80% casein in the cow, where α_{S1}-casein is the predominant type (Jenness and Sloan, 1970). Down-regulation of α_{S1}-casein expression in cows, using methods described above, would lower the casein:whey ratio in cow's milk. β-Casein would become the dominant casein species in cow's milk, resulting in a milk that more closely resembles that of humans.

ENGINEERING OF MILK PROTEINS

Altering casein levels for improved milk properties

We will now consider aspects of modifying the casein content of cow's milk for improved functionality with respect to the dairy processing industry. This may be achieved by the addition of extra copies of either native or modified bovine casein genes to the genome of dairy cattle. Increasing the level of a single native casein species can alter milk functionality and thus enhance its utility in various dairy processes (Yun, Ohmiya and Shimizu, 1982; Pearse *et al.*, 1986; Jimenez-Flores and Richardson, 1988; Hennighausen, Ruiz and Wall, 1990).

Adding extra casein genes to the bovine genome may increase the protein content of milk, thereby increasing its nutritional value. There could also be considerable advantages for the cheese industry. Cheese yield, which is dependent on casein content, would also increase (Richardson, 1985; Kang, Jimenez-Flores and Richardson, 1986). It has been estimated that an increase of just 20% in the α_{S1}-casein content of milk would result in an extra $US190 million per year for the US dairy industry (Hennighausen, Ruiz and Wall, 1990).

The casein proteins in milk are predominantly present as a colloidal aggregate complexed with calcium phosphate to form spherical structures that are referred to as casein micelles (McMahon and Brown, 1984). On the removal of calcium, the casein micelles dissociate to form smaller non-colloidal protein complexes termed submicelles. The casein micellar structure is stabilized by an outer layer of κ-casein (Holt and Dalgleish, 1986). Waugh and Talbot (1971) have demonstrated that there is an approximately linear relationship between the surface area of artificial casein micelles and their κ-casein content. It has been proposed that cattle expressing higher concentrations of κ-casein in their milk would therefore contain smaller micelles. Indirect evidence to this effect has been obtained in our laboratory (Pearse *et al.*, 1986). Such milk should possess altered physical properties, such as enhanced thermal stability, which would be advantageous in reducing the gelation and coagulation properties of milk during heat-mediated sterilization (Jimenez-Flores and Richardson, 1988).

Cleavage of κ-casein by chymosin leads to the destabilization of the micelle and the precipitation of the caseins, the initial step in cheese-making (Delfour *et al.*, 1965). Milk with reduced levels of κ-casein would be expected to contain a greater proportion of larger micelles with reduced stability, and therefore would be more amenable to cheese production (Jimenez-Flores and Richardson, 1988).

Experiments to determine the effects of varying the concentration of various casein species, within artificial micelle milk, on rennet coagulation time (rate of curd formation) and syneresis (expulsion of whey from the curd) have been performed (Pearse *et al.*, 1986). It was found that increasing the concentration of both κ-casein and β-casein individually had the effect of reducing rennet clotting time. Higher levels of β-casein had the additional effect of an increased rate and extent of syneresis. The qualification here is the extent to which artificial micelles accurately mirror the *in vivo* situation. Yun, Ohmiya and Shimizu (1982) have also demonstrated that the addition of β-casein to milk can increase the rigidity of the curd, which increases the yield of milk solid that can be used in the cheese manufacturing process.

Engineering the caseins for improved cheese production

In recent years *in vitro* mutagenesis has been applied to the bovine casein genes to engineer proteins with altered amino acid sequences which may possess more properties beneficial for the cheese manufacturing industry. This research has centred chiefly on modifying the various peptide bonds cleaved by chymosin, either to increase the rate of hydrolysis for more economical cheese production (Oh and Richardson, 1991b), or to eliminate the production of undesirable peptides that affect the quality of the final product (Simons *et al.*, 1993).

Chymosin-mediated cleavage of bovine κ-casein occurs at the Phe–Met peptide bond (Delfour *et al.*, 1965). Site-directed mutagenesis of this Phe–Met bond to Phe–Phe has resulted in a κ-casein species which is cleaved at a higher rate by chymosin (Oh and Richardson, 1991b). Substantial economic benefits could be obtained if this modified casein was introduced into milk destined for cheese production.

Storage accounts for approximately 6–8% of the cost of cheese production in the UK. If methods could be developed to shorten the ripening time of cheese, without adversely affecting flavour and texture, this would constitute quite a saving for the cheese industry (Law, 1987). The process of cheese maturation depends on the degradation of α_{s1}-casein via residual chymosin in the curd, which hydrolyses different peptide linkages at different rate (Creamer and Olson, 1982). The Phe23–Phe24 or Phe24–Val25 bonds in α_{s1}-casein are hydrolysed relatively quickly, while others, such as Leu149–Phe150, are hydrolysed more slowly, thus prolonging the cheese maturation process. It may be possible to engineer caseins with properties more amenable to cheese production using *in vitro* site-directed mutagenesis.

The proteolytic cleavage of β-casein by chymosin is usually slower than that of α_{s1}-casein (Grappin, Rank and Olson, 1985). The cleavage of the Leu192–Tyr193 peptide bond in β-casein results in the production of the C-terminal peptide (amino acids 193–209) responsible for the bitter taste defect in Gouda and other cheese varieties (Creamer, Mills and Richards, 1971; Visser and Slangen, 1977; Carles and Ribadeau-Dumas, 1984). Simons *et al.* (1993) have used site-directed mutagenesis to alter this Leu192–Tyr193 peptide bond to Pro192–Pro193 and Leu192–stop, resulting in an altered β-casein lacking this chymosin cleavage site.

Engineering the caseins for greater nutritional value

Although milk is almost a complete food in itself, the bovine caseins are relatively low in the sulphur-containing amino acids methionine and cysteine. The percentage of sulphur-containing amino acids present in the caseins could be increased via site-directed mutagenesis and then introduced into the bovine genome. A sulphur-enriched κ-casein has already been engineered to contain three extra methionine residues between Ala167 and Thr168 of the bovine κ-casein gene and has been expressed in *E. coli* at a level of 2 mg ml^{-1} (Oh and Richardson, 1991a).

As mentioned above, the down-regulation of BLG gene expression would aid in making the milk of cows more like that of humans. It has been postulated that down-regulation of the whey proteins would also result in increased expression of the casein genes. This rationale is based on the observation that cows homozygous for the B allele of BLG produce less BLG than the AA genotype and also produce 7% more casein (McLean, Graham and Ponzoni, 1984; Hill, 1993). This implies that the higher level of BLG production in the AA genotype is at the expense of casein production. It has been proposed that by eliminating the competition from BLG, via its genetic down-regulation, casein production may increase, making cow's milk more nutritious and of greater use to the cheese industry (Hill, 1993).

Protein engineering of the whey proteins

Recombinant bovine BLG has been expressed in *E. coli* as a preliminary to performing site-directed mutagenesis to design and produce modified BLG molecules for specific applications within the food industry (Batt *et al.*, 1990).

BLG possesses a free thiol group buried within the interior of its globular structure (Cys121). Upon heating, BLG unfolds exposing this cysteine residue, and so is then able to form aggregates with other BLG molecules and form gels (Sawyer, 1968). Upon heat treatment of milk, this free thiol group is also able to interact with the single disulphide group of κ-casein. This may interfere with the chymosin-mediated hydrolysis of this protein critical for curd formation and cheese production (Morr, 1985). The heat-induced denaturation of BLG may also lead to the fouling of dairy processing equipment (Lalande, Tissier and Corrieu, 1985). The production of milk deficient in BLG may overcome this problem (Jimenez-Flores and Richardson, 1988). Alternatively, the substitution of extra cysteine residues proximal to the native free thiol at Cys121 (Leu104 to Cys104 or Ala132 to Cys132) has resulted in recombinant BLGs that are more resistant to thermal aggregation, presumably due to the formation of an additional disulphide linkage with Cys121 (Cho *et al.*, 1994).

Recombinant BLGs engineered to contain an extra free thiol group within the hydrophobic region of the molecule (Arg40 to Cys40, or Phe82 to Cys82) have demonstrated enhanced gelation characteristics (Lee, Cho and Batt, 1993). Additionally, a thermolabile recombinant BLG variant R40C/F82C (Arg40 to Cys40 and Phe82 to Cys82) has been demonstrated to reduce whey syneresis and time for curd formation when utilized in yoghurt production (Lee *et al.*, 1993).

The process of pasteurization destroys much of the lactoferrin present in milk. Recombinant human lactoferrin has been expressed in yeast with a view to increasing its heat stability via *in vitro* site-directed mutagenesis similar to that described above for BLG (Liang and Richardson, 1993).

Outlook/future prospects

TOWARDS THE MARKETPLACE

At present, a number of pharmaceutically active human proteins have been successfully produced in the milk of transgenic animals at commercially viable levels (*Table 6*). In five cases (human α_1AT, t-PA, growth hormone, serum albumin and protein C), recombinant protein production exceeds those reported for rival systems. In less than a decade, gene 'pharming' has established itself as a competititve alternative for the production of recombinant proteins. In the case of α_1AT, t-PA and protein C, transgenic expression has been established within the mammary gland of large farm animals (sheep, goats and pigs, respectively) allowing the harvesting of large quantities of recombinant protein from each animal. Once recombinant proteins from transgenic animals are approved for clinical use, it is expected that these products will become serious competitors for the world pharmaceutical market which is valued at many hundreds of millions of dollars (see *Table 6*).

The production of α_1AT in transgenic sheep provides a good case study. α_1AT deficiency affects approximately 100 000 people world-wide and leads to the onset of the degenerative lung disease, pulmonary emphysema (Crystal, 1989). Currently, the α_1AT needed to treat deficiency disorders is derived from human plasma; however, this source cannot provide the large quantities needed to treat these patients who require approximately 200 g α_1AT/patient/year (Casolaro *et al.*, 1987). A similar situation exists for protein C which, due to its very low serum concentration, is extremely difficult to isolate from human plasma (Velander *et al.*, 1991). Hence an alternate source of these clinically important proteins is needed. Recombinant α_1AT has been expressed in yeast (Cabezón *et al.*, 1984), though this non-glycosylated product has only 2% of the *in vivo* half-life of the plasma-derived product (Casolaro *et al.*, 1987). As described above, Wright *et al.* (1991) have succeeded in producing

Table 6. A comparison of pharmaceutically important human proteins expressed in the mammary gland of transgenic animals compared with expression levels in alternate systems

Protein	Biological function/ clinical utility	Expression levels[§]	World-wide sales ($US)	Reference
FIX	Blood clotting factor	25 ng ml^{-1}	$25 000/g*	Clark *et al.* (1989b)
	haemophilia B treatment	(100 μg ml^{-1}, only 2% active)		Brownlee (1987)
α_1-antitrypsin	Neutrophil elastase	7 mg ml^{-1}	$100 m[‡]	Archibald *et al.* (1990)
	inhibitor/emphysema	(60 ng ml^{-6} cells day^{-1})		Pavirani *et al.* (1989)
Interleukin-2	Cancer, AIDS and	430 ng ml^{-1}	$20 m[†]	Bühler *et al.* (1990)
	leprosy therapy	(10 μg ml^{-1} day^{-1})		Ryll *et al.* (1990)
t-PA	Thrombolytic agent	3 mg ml^{-1}	$230 m[†]	Ebert *et al.* (1991)
	myocardial infarction	(460 μg ml^{-1}.)		Datar, Cartwright and Rosen (1993)
Growth hormone	Hypopituitary dwarfism	11 mg ml^{-1}	$575 m[†]	Ninomiya *et al.* (1994)
	chronic renal insufficiency	(200 μg ml^{-1}.)		Friedman *et al.* (1989)
Protein C	Haemostasis regulator	1.0 mg ml-1	$960 m[ʃ]	Velander *et al.* (1992)
	stroke, septic shock	(<0.4 μg ml^{-1} h^{-1})		Velander *et al.* (1991)

[§] Expression levels obtained in the mammary gland of transgenic animals. Values in brackets are production levels achieved by microbial fermentation or mammalian cell culture.
* Clark *et al.* (1987).
[‡] Spalding (1992).
[†] World-wide sales values stated in $US millions for 1992 as given in Klauser (1993).
[ʃ] American Red Cross projection of US sales as quoted in Pursel and Rexroad (1993).

a transgenic sheep that secretes fully glycosylated recombinant α_1AT in its milk at a level of approximately 35 g l^{-1}, making α_1AT the most abundant milk protein. It has been estimated that such sheep producing α_1AT could produce more than 10 kg of recombinant protein per lactation (Wright *et al.*, 1991). The world-wide market for α_1AT has been estimated to be approximately $US100 million (Spalding, 1992).

THE COST EFFECTIVENESS OF GENE PHARMING

Not only do some transgenic animals express pharmaceutically active proteins at a higher level than competing systems, but gene pharming techniques are also more cost effective when compared to conventional fermentation technology. As an example of the expense involved in producing these products using established technology, Datar, Cartwright and Rosen (1993) have critically evaluated the process economics involved in the animal cell (CHO) and bacterial fermentation processes required for the production of recombinant t-PA. Their study estimates that the fermentation tanks alone needed to culture the CHO cells cost over $US6 million as an initial investment, with an annual expenditure of an additional $US23 million simply for fermentation materials. Recombinant t-PA produced by transformed *E. coli* is synthesized in a denatured form, necessitating an additional capital investment of over $US53 million to purchase the refolding tanks required to renature the recombinant protein and restore it to a functional state (Datar, Cartwright and Rosen, 1993).

In the past a major criticism of gene pharming has been the expense of producing transgenic animals. However, the cost of producing a founder transgenic pig, goat or cow has been estimated to be as low as $US25 000, $US60 000 and $US500 000, respectively (Hennighausen, 1990). Once the capital investment necessary to build and maintain a commercial fermentation plant is compared to the costs involved in transgenic animal production, gene pharming starts to look very competitive. The higher expression levels and lower production costs of transgenic bioreactors should translate into greater product availability and lower prices for the consumer, that is the patient.

A major factor contributing to the cost of transgenic animal production is the low efficiency of transgenesis, necessitating a large number of foster mothers to bear offspring, only a small percentage of which will be transgenic. For example fewer than 1% of micro-injected sheep, pig or cow embryos develop into transgenic offspring (Murray *et al.*, 1989). The number of animals and the effort required to produce transgenic animals could be reduced by screening for integration of the transgene early in embryo development via PCR, before implantation into a foster mother (Ninomiya *et al.*, 1989; Horvat *et al.*, 1993; Bowen *et al.*, 1994). This approach would decrease the number of non-transgenic pregnancies and therefore production costs. It has already been utilized by Hyttinen *et al.* (1994) in the generation of transgenic cattle. An additional practical advantage of utilizing animals as bioreactors is that their maintenance costs are much lower than those for a fermentation plant (Datar, Cartwright and Rosen, 1993).

A major disadvantage with fermentation processes are 'scale-up' problems, which may be experienced when progressing from pilot research reactors to large-scale commercial plants (Rhodes and Birch, 1988; Reisman, 1993). Scale-up often results in unpredictable alterations in productivity, usually with very uneconomical consequences (Reisman, 1993). However, scale-up problems can be avoided in gene pharming as transgenic bioreactors are able to reproduce themselves while still retaining similar production characteristics from generation to generation (Carver *et al.*, 1993; Ebert *et al.*, 1994).

REGULATORY ISSUES AND MARKET ACCEPTANCE

As we have seen in this review, the cloning and characterization of the milk protein encoding genes has led to an avalanche of research within the past 10 years in utilizing these sequences to direct the expression of high-value pharmaceutical proteins to the mammary gland of various mammals. At present none of these recombinant proteins has appeared on the market, although α_1AT is expected to enter clinical trials in mid-1995 (Gordon Wright, personal communication).

One of the largest regulatory issues facing the use of recombinant proteins for human therapeutic use is that of product purity. Pharmaceutical proteins produced by recombinant DNA technology must be purified to remove all non-human proteins and other contaminants that may cause adverse reactions in patients (Anicetti, Keyt and Hancock, 1989). In cases where the recombinant protein constitutes a high proportion of the milk protein content, such as α_1AT expression in sheep, problems with product purification should be minimized. In addition, high expression levels in transgenic animals should allow more cost-effective protein purification than isolation from lower expressing fermentation systems, as the cost of purification is generally related to the product concentration (Wilkins and Velander, 1992). Indeed, recombinant t-PA has been obtained at greater than 98% purity from the milk of transgenic goats using a combination of acid fractionation, hydrophobic interaction and immunoaffinity chromatography (Denman *et al.*, 1991). Recombinant α_1AT derived from transgenic sheep has now been purified to greater than 99%, and will be the first protein derived from transgenic technology to be scrutinized by regulatory authorities (Gordon Wright, personal communication).

Altered milk products may also be just around the corner. For example, milk enriched with human lactoferrin destined for the production of infant formula (Krimpenfort *et al.*, 1991). As recombinant milk proteins produced in cattle are classified as food ingredients, they do not have to be exhaustively purified as do pharmaceutical proteins. Once the modified product is demonstrated to be safe for human consumption (through dietary-toxicity trials), approval by the relevant regulatory authorities is expected to be less costly and lengthy than that for recombinant proteins destined for pharmaceutical use (Spalding, 1992).

However, demonstrated product efficacy, safety and approval by the relevant regulatory authorities does not guarantee total market acceptance. Such is the case with recombinant bovine somatotropin (BST), a hormone used to increase milk production in dairy cattle, which has recently been approved for use in the USA. Although milk produced from cows treated with BST has been demonstrated to be completely safe and indistinguishable from that derived from other cows, the use of this recombinant product is facing opposition by various lobby groups in the USA (Hoyle, 1994). A study performed by Fox *et al.* (1994) indicates that the presentation of product information can greatly affect the way in which a recombinant product is perceived. For example, they found that students in urban California, who had not heard of BST before the study, initially showed a markedly negative bias against the product. However, after being presented with a balanced appraisal of the product, approximately 70% demonstrated acceptance. This bias, however, was difficult to overcome in areas that had been subjected to campaigns against the use of recombinant BST. Therefore, education may be the key to successful marketing of these genetically altered food products. It may reasonably be expected that any public opposition to the use of therapeutic proteins derived from transgenic technology will be neutralized by the benefits that these drugs will provide for patients.

Acknowledgements

The authors would like to thank Jean-Claude Mercier, Jim Murray, Moshe Shani, Gordon Wright and Atsushi Yuki for making prepublication data available to us for this review. Thanks are also due to Wendy Glenn, Judy Kaye, Steve McMaugh and André Ng-Kwai-Hang for their helpful comments. The research performed in our laboratory is supported by the Australian Dairy Research and Development Corporation.

References

ALEXANDER, L.J. AND BEATTIE, C.W. (1992a). The sequence of porcine alpha S1-casein cDNA: evidence for protein variants generated by altered RNA splicing. *Animal Genetics* **23**, 283–288.

ALEXANDER, L.J. AND BEATTIE, C.W. (1992b). The sequence of porcine β-casein cDNA. *Animal Genetics* **23**, 369–371.

ALEXANDER, L.J. AND BEATTIE, C.W. (1992c). Sequence of porcine β-lactoglobulin cDNA. *Animal Genetics* **23**, 263–265.

ALEXANDER, L.J., DAS GUPTA, N.A. AND BEATTIE, C.W. (1992). The sequence of porcine alpha S2-casein cDNA. *Animal Genetics* **23**, 365–367.

ALEXANDER, L.J., STEWART, A.F., MACKINLAY, A.G., KAPELINSKAYA, T.V., TKACH, T.M. AND GORODETSKII, S.I. (1988). Isolation and characterisation of the bovine κ-casein gene. *European Journal of Biochemistry* **178**, 395–401.

ALEXANDER, L.J., HAYES, G., PEARSE, M.J., BEATTIE, C.W., STEWART, A.F., WILLIS, I.M. AND MACKINLAY, A.G. (1989). Complete sequence of the bovine β-lactoglobulin cDNA. *Nucelic Acids Research* **17**, 6739.

ALEXANDER, L.J., HAYES, G., BAWDEN, W., STEWART, A.F. AND MACKINLAY, A.G. (1993). Complete nucleotide sequence of the bovine β-lactoglobulin gene. *Animal Biotechnology* **4**, 1–10.

ALI, S. AND CLARKE, A.J. (1988). Characterisation of the gene encoding ovine β-lactoglobulin: similarity to the genes for retinol-binding protein and other secretory proteins. *Journal of Molecular Biology* **199**, 415–426.

ALLEN, N.D., NORRIS, M.L. AND SURANI, M.A. (1990). Epigenetic control of transgene expression and imprinting by genotype-specific modifiers. *Cell* **61**, 853–861.

AL-SHAWI, R., KINNAIRD, J., BURKE, J. AND BISHOP, J.O. (1990). Expression of a foreign gene in a line of transgenic mice modulated by a chromosomal position effect. *Molecular and Cellular Biology* **10**, 1192–1198.

ANDRES, A.C., SCHÖNENBERGER, C.A., GRONER, B., HENNIGHAUSEN, L., LEMEUR, M. AND GERLINGER, P. (1987). Ha-*ras* oncogene expression directed by a milk protein promoter, tissue specificity, hormonal regulation, and tumor induction in transgenic mice. *Proceedings of the National Academy of Sciences, USA* **84**, 1299–1303.

ANDRES, A.C., VAN DER VALK, M.A., SCHÖNENBERGER, C.A., KLÜCKIGER, F., LEMEUR, M., GERLINGER, P. AND GRONER, B. (1988). Ha-*ras* and c-*myc* oncogene expression interfere with morphological and functional differentiation of the mammary epithelial cells in single and double transgenic mice. *Genes and Development* **2**, 1486–1495.

ANICETTI, V.R., KEYT, B.A. AND HANCOCK, W.S. (1989). Purity analysis of protein pharmaceuticals produced by recombinant DNA technology. *TIBTECH* **7**, 342–349.

ARCHIBALD, A.L., MCCLENAGHAN, M., HORNSEY, V., SIMONS, J.P. AND CLARK, A.J. (1990). High-level expression of biologically active human α_1-antitrypsin in the milk of transgenic mice. *Proceedings of the National Academy of Sciences, USA* **87**, 5178–5182.

ARNOLD, R.R., COLE, M.F. AND MCGHEE, J.R. (1977). A bacteriocidal effect for human lactoferrin. *Science* **197**, 263–265.

ARONOW, B., LATTIER, D., SILBIGER, R., DUSING, M., HUTTON, J., JONES, G., STOCK, J., MCNEISH, J., POTTER, S., WITTE, D. AND WIGINTON, D. (1989). Evidence for a complex regulatory array in the first intron of the human adenosine deaminase gene. *Genes and Development* **3**, 1384–1400.

ARVOLA, T., ISOLAUR, E., RANTALA, I., KIALA, M., MAJAMAA, H., VIRTANEN, E. AND ARVILOMMI, H. (1993). Increased *in vitro* intestinal permeability in sucking rats exposed to cow milk during lactation. *Journal of Pediatric Gastroenterology and Nutrition* **16**, 294–300.

AZUMA, N. AND YAMAUCHI, K. (1991). Identification of alpha-lactalbumin and beta-lactoglobulin in cynomolgus monkey (*Macaca fascicularis*) milk. *Comparative Biochemistry and Physiology* **99B**, 917–921.

BATT, C.A., RABSON, L.D., WONG, D.W.S. AND KINSELLA, J.E. (1990). Expression of recombinant bovine β-lactoglobulin in *Escherichia coli. Agricultural and Biological Chemistry* **54**, 949–955.

BAYNA, E.M. AND ROSEN, J.M. (1990). Tissue-specific, high level expression of the rat whey acidic protein gene in transgenic mice. *Nucleic Acids Research* **18**, 2977–2985.

BEBBINGTON, C. AND HENTSCHEL, C. (1985). The expression of recombinant DNA products in mammalian cells. *TIBTECH* **3**, 314–317.

BELL, S.C., HALES, M.W., PATEL, S., KIRMAN, P.H. AND DRIFE, J.O. (1985). Protein synthesis and secretion by the human endometrium and decidua during early pregnancy. *British Journal of Obsterics and Gynaecology* **92**, 793–803.

BISCHOFF, R., DEGRYSE, E., PERRAUD, F., DALEMANS, W., ALI-HADJI, D., THÉPOT, D., DEVINOY, E., HOUDEBINE, L.M. AND PAVIRANI, A. (1992). A 17.6 kbp region located upstream of the rabbit WAP gene directs high level expression of a functional human protein variant in transgenic mouse milk. *FEBS Letters* **3**, 265–268.

BLACKBURN, D.E., HOBBS, A.A. AND ROSEN, J.M. (1982). Rat β-casein cDNA: sequence analysis and evolutionary comparisons. *Nucleic Acids Research* **10**, 2295–2307.

BLANC, B. (1981). Biochemical aspects of human milk – comparison with bovine milk. *World Review of Nutrition and Diet* **36**, 1–89.

BOISNARD, M. AND PETRISSANT, G. (1985). Complete sequence of ovine α_{S2}-casein mRNA. *Biochimie* **67**, 1043–1051.

BONNEROT, C., GRIMBER, G., BRIAND, P. AND NICOLAS, J.F. (1990). Patterns of expression of position-dependent integrated transgenes in mouse embryo. *Proceedings of the National Academy of Sciences, USA* **87**, 6331–6335.

BONSING, J. AND MACKINLAY, A.G. (1987). Recent studies on nucleotide sequences encoding the caseins: A review. *Journal of Dairy Research* **54**, 447–461.

BONSING, J., RING, J.M., STEWART, A.F. AND MACKINLAY, A.G. (1988). Complete nucleotide sequence of the bovine β-casein gene. *Australian Journal of Biological Sciences* **41**, 527–537.

BOSZE, Z., DEVINOY, E., PUISSANT, C., FONTAINE, M.L. AND HOUDEBINE, L.M. (1993). Characterisation of rabbit kappa-casein cDNA: control of kappa-casein gene expression *in vivo* and *in vitro. Journal of Molecular Endocrinology* **11**, 9–17.

BOTSTEIN, D. AND SHORTLE, D. (1985). Strategies and applications of *in vitro* mutagenesis. *Science* **229**, 1193–1201.

BOWEN, R.A., REED, M.L., SCHNIEKE, A., SEIDEL, G.E., STACEY, A., THOMAS, W.K. AND KAJIKAWA, O. (1994). Transgenic cattle resulting from biopsied embryos: Expression of c-*ski* in a transgenic calf. *Biology of Reproduction* **50**, 664–668.

BOWMAN, B.H., YANG, F. AND GWENDOLYN, S.A. (1988). Transferrin: evolution and genetic regulation of expression. *Advances in Genetics* **25**, 1–38.

BRATTY, J., CHARTRAND, P., FERBEYRE, G. AND CEDERGREN, R. (1993). The hammerhead RNA domain, a model ribozyme. *Biochimica et Biophysica Acta* **1216**, 345–359.

BRAUNITZER, G., CHEN, R., SCHRANK, B. AND STANGL, A. (1973). The sequence of β-lactoglobulin. *Hopp-Seyler's Zeitschrift für Physiologische Chemie* **354**, 867–878.

BREMEL, R.D., HENG-CHERL, Y. AND BLECK, T. (1989). Alteration of milk composition using molecular genetics. *Journal of Dairy Science* **72**, 2826–2833.

BRIGNON, G., MAYE, M.F., RIBADEAU-DUMAS, B., MERCIER, J.C. AND GROSCLAUDE, F. (1990). Two of the three genetic variants of goat α_{S1}-casein which are synthesised at a reduced level have an internal deletion possibly due to altered RNA splicing. *European Journal of Biochemistry* **193**, 237–241.

BRINSTER, R.L., CHEN, H.Y., TRUMBAUER, M., SENEAR, A.W., WARREN, R. AND PALMITER, R.D. (1981). Somatic expression of herpes thymidine kinase in mice following injection of a fusion gene into eggs. *Cell* **27**, 223–231.

BRINSTER, R.L., RITCHIE, K.A., HAMMER, R.E., O'BRIEN, R.L., ARP, B. AND STORB, U. (1983). Expression of a microinjected immunoglobulin gene in the spleen of transgenic mice. *Nature* **306**, 332–336.

BRINSTER, R.L., CHEN, H.Y., TRUMBAUER, M.E., YAGLE, M.K. AND PALMITER, R.D. (1985). Factors affecting the efficiency of introducing foreign DNA into mice by microinjecting eggs. *Proceedings of the National Academy of Sciences, USA* **82**, 4438–4442.

BRINSTER, R.L., ALLEN, J.M., BEHRINGER, R.R., GELINAS, R.E. AND PALMITER, R.D. (1988). Introns increase transcriptional efficiency in transgenic mice. *Proceedings of the National Academy of Sciences, USA* **85**, 836–840.

BROWNLEE, G.G. (1987). The molecular pathology of haemophilia B. *Biochemical Society Transactions* **15**, 1–8.

BÜHLER, T., BRUYER, T., WENT, D.F., STRANZINGER, G. AND BÜRKI, K. (1990). Rabbit β-casein promoter directs secretion of human interleukin-2 into the milk of transgenic rabbits. *Bio/Technology* **8**, 140–143.

BURDON, T., SANKARAN, L., WALL, R.J., SPENCER, M. AND HENNIGHAUSEN, L. (1991). Expression of a whey acidic protein transgene during mammary development, evidence for different mechanisms of regulation during pregnancy and lactation. *Journal of Biological Chemistry* **266**, 6909–6914.

CABEZON, T., DE WILDE, M., HERION, P., LORIAU, R. AND BOLLEN, A. (1984). Expression of human α-antitrypsin cDNA in the yeast *S. cerevisiae*. *Proceedings of the National Academy of Sciences, USA* **81**, 6594–6598.

CAFFIN, J.P., POUTREL, B. AND RAINARD, P. (1985). Physiological and pathological factors influencing bovine α-lactalbumin and β-lactoglobulin concentrations in milk. *Journal of Dairy Science* **68**, 1087–1094.

CAMERON, F.H. AND JENNINGS, P.A. (1989). Specific gene suppression by engineered ribozymes in monkey cells. *Proceedings of the National Academy of Sciences, USA* **86**, 9139–9143.

CAMPBELL, S.M., ROSEN, J.M., HENNIGHAUSEN, L., STRECH, J.U. AND SIPPEL, A.E. (1984). Comparison of the whey acidic protein genes of the rat and mouse. *Nucleic Acids Research* **12**, 8685–8697.

CAPECCHI, M.R. (1989). Altering the genome by homologous recombination. *Science* **244**, 1288–1292.

CARLES, C. AND RIBADEAU-DUMAS, B. (1984). Kinetics of action of chymosin (rennin) on some peptide bonds of bovine β-casein. *Biochemistry* **23**, 6839–6843.

CARVER, A.S., DALRYMPLE, M.A., WRIGHT, G., COTTOM, D.S., REEVES, D.B., GIBSON, Y.H., KEENAN, J.L., BARRASS, J.D., SCOTT, A.R., COLMAN, A. AND GARNER, I. (1993). Transgenic livestock as bioreactors, stable expression of human alpha-1-antitrypsin by a flock of sheep. *Bio/Technology* **11**, 1263–1270.

CASOLARO, M.A., FELLS, G., WEWERS, M., PIERCE, J.E., OGUSHI, F., HUBBARD, R., SELLERS, S., FORSTROM, J., LYONS, D., KAWASAKI, G. AND CRYSTAL, R.G. (1987). Augmentation of lung antineutrophil elastase capacity with recombinant human α-1-antitrypsin. *Journal of Applied Physiology* **63**, 2015–2023.

CAZENAVE, C. AND HÉLENE, C. (1991). Antisense oligonucleotides. In *Antisense Nucleic Acids and Proteins: Fundamentals and Applications* (J.N. Mol and A.R. van der Krol, Eds), pp. 47–93. Marcel Dekker, New York.

CHANG, S.S. (1990). Antimicrobial proteins of maternal and cord sera and human milk in relation to maternal nutritional studies. *American Journal of Clinical Nutrition* **51**, 183–187.

CHENG, L., ZIEGELHOFFER, P.R. AND YANG, N.S. (1993). *In vivo* promoter activity and transgene expression in mammalian somatic tissues evaluated by using particle bombardment. *Proceedings of the National Academy of Sciences, USA* **90**, 4455–4459.

CHO, Y., GU, W., WATKINS, S., LEE, S.P., KIM, T.R., BRADY, J.W. AND BATT, C.A. (1994). Thermostable variants of bovine β-lactoglobulin. *Protein Engineering* **7**, 263–270.

CLARK, A.J., SIMONS, P., WILMUT, I. AND LATHE, R. (1987). Pharmaceuticals from transgenic livestock. *TIBTECH* **5**, 20–24.

CLARK, A.J., ALI, S., ARCHIBALD, A.L., BESSOS, H., BROWN, P., HARRIS, S., MCCLENAGHAN, M., PROWSE, C., SIMONS, J.P., WHITELAW, C.B.A. AND WILMUT, I. (1989a). The molecular manipulation of milk composition. *Genome* **31**, 950–955.

CLARK, A.J., BESSOS, H., BISHOP, J.O., BROWN, P., HARRIS, S., LATHE, E., MCCLENAGHAN, M., PROWSE, C., SIMONS, J.P., WHITEHALL, C.B.A. AND WILMUT, I. (1989b). Expression of human anti-hemophilic Factor IX in the milk of transgenic sheep. *Bio/Technology* **7**, 487–492.

CLARK, A.J., COWPER, A., WALLACE, R., WRIGHT, G. AND SIMONS, J.P. (1992). Rescuing transgene expression by co-integration. *Bio/Technology* **10**, 1450–1454.

COLLET, C. AND JOSEPH, R. (1993). A novel member of the lipocalin superfamily: tammar wallaby late-lactation protein. *Biochimica et Biophysica Acta* **1167**, 219–222.

COLLET, C., JOSEPH, R. AND NICHOLAS, K. (1989). Molecular cloning and characterization of a novel marsupial milk protein gene. *Biochemical and Biophysical Research Communications* **164**, 1380–1383.

COLLET, C., JOSEPH, R. AND NICHOLAS, K. (1990). Cloning, cDNA analysis and prolactin-dependent expression of a marsupial α-lactalbumin. *Reproduction, Fertility and Development* **2**, 693–701.

COLLET, C., JOSEPH, R. AND NICHOLAS, K. (1991). A marsupial β-lactoglobulin gene: characterisation and prolactin dependent expression. *Journal of Molecular Endocrinology* **6**, 9–16.

COLLET, C., JOSEPH, R. AND NICHOLAS, K. (1992). Molecular characterisation and *in-vitro* hormonal requirements for expression of two casein genes from a marsupial. *Journal of Molecular Endocrinology* **8**, 13–20.

CONTI, A., GODOVAC-ZIMMERMAN, J., LIBERATORI, J. AND BRAUNITZER, G. (1984). The primary structure of monomeric beta-lactoglobulin I from horse colostrum (*Equus caballus, Perissodactyla*). *Hoppe-Seyler's Zeitschrift für Physiologische Chemie* **365**, 1393–1401.

COTTON, M. AND BIRNSTIEL, M.L. (1989). Ribozyme mediated destruction of RNA *in vivo*. *EMBO Journal* **8**, 3861–3866.

CREAMER, L.K. AND OLSON, N.F. (1982). Rheological evaluation of cheese yield. *Journal of Food Science* **47**, 631–636, 646.

CREAMER, L.K., MILLS, O.E. AND RICHARDS, E.L. (1971). The action of rennets on the caseins. I Rennin action on β-casein-B in solution. *Journal of Dairy Research* **38**, 269–280.

CRYSTAL, R.G. (1989). The α_1-antitrypsin gene and its deficiency states. *Trends in Genetics* **5**, 411–417.

CUNNINGHAM, G.A., HEADON, D.R. AND CONNEELY, O.M. (1992). Structural organization of the mouse lactoferrin gene. *Biochemical and Biophysical Research Communications* **189**, 1725–1731.

DALE, T.C., KRNACIK, M.J., SCHMIDHAUSER, C., YANG, C.L.-Q., BISSELL, M.J. AND ROSEN, J.M. (1992). High-level expression of the rat whey acidic protein gene is mediated by elements in the promoter and 3' untranslated region. *Molecular and Cellular Biology* **12**, 905–914.

DANDEKAR, A.M. AND QASBA, P.K. (1981). Rat α-lactalbumin has a 17-residue-long COOH-terminal hydrophobic extension as judged by sequence analysis of the cDNA clones. *Proceedings of the National Academy of Sciences, USA* **78**, 4853–4857.

DAS GUPTA, N.A., ALEXANDER, L.J. AND BEATTIE, C.W. (1992). The sequence of a porcine cDNA encoding alpha-lactalbumin. *Gene* **110**, 265–266.

DATAR, R.V., CARTWRIGHT, T. AND ROSEN, C.G. (1993). Process economics of animal cell and bacterial fermentations: a case study analysis of tissue plasminogen activator. *Bio/Technology* **11**, 349–357.

DAVIES, M.S., WEST, L.F., DAVIS, M.B., POVEY, S. AND CRAIG, R.K. (1987). The gene for human α-lactalbumin is assigned to chromosome 12q13. *Annals of Human Genetics* **51**, 183.

DAWSON, S.P., WILDE, C.J., TIGHE, P.J. AND MAYER, J. (1993). Characterization of two novel casein transcripts in rabbit mammary gland. *Biochemical Journal* **296**, 777–784.

DELFOUR, A., JOLLES, J., ALAIS, C. AND JOLLES, P. (1965). Caseino-glycopeptides, characterisation of a methionine residue and of the N-terminal peptide. *Biochemical and Biophysical Research Communications* **19**, 425–455.

DENMAN, J., HAYES, M., O'DAY, C., EDMUNDS, T., BARTLETT, C., HIRANI, S., EBERT, K.M., GORDON, K. AND MCPHERSON, J.M. (1991). Transgenic expression of a variant of human tissue-type plasminogen activator in goat milk, purification and characterization of the recombinant enzyme. *Bio/Technology* **9**, 839–843.

DEVINOY, E., HUBERT, C., JOLIVET, G., THEPOT, D., CLERGUE, N., DESALEUX, M., DION, M., SERVELY, J.L. AND HOUDEBINE, L.M. (1988). Recent data on the structure of rabbit milk protein genes and on the mechanism of the hormonal control of their expression. *Reproduction Nutrition Development* **28**, 1145–1164.

DITULLIO, P., CHENG, S.H., MARSHALL, J., GREGORY, R.J., EBERT, K.M., MEADE, H. AND SMITH, A.E. (1992). Production of cystic fibrosis transmembrane conductance regulator in the milk of transgenic mice. *Bio/Technology* **10**, 74–77.

EBERT, K.M. AND SCHINDLER, J.E.S. (1993). Transgenic farm animals: progress report. *Theriogenology* **39**, 121–135.

EBERT, K.M., SELGRATH, J.P., DITULLIO, P., DENMAN, J., SMITH, T.E., MEMON, A., SCHINDLER, J.E., MONASTERSKY, G.M., VITALE, J.A. AND GORDON, K. (1991). Transgenic production of a variant of human tissue-type plasminogen activator in goat milk: Generation of transgenic goats and analysis of expression. *Bio/Technology* **9**, 835–838.

EBERT, K.M., DITULLIO, P., BARRY, C.A., SCHINDLER, J.E., AYRES, S.L., SMITH, T.E., PELLERIN, L.J., MEADE, H.M., DENMAN, J. AND ROBERTS, B. (1994). Induction of human tissue plasminogen activator in the mammary gland of transgenic goats. *Bio/Technology* **12**, 699–702.

EBNER, K.E. AND SCHAMBACHER, F.L. (1974). In *Lactation* (B.L. Larson and V.R. Smith, Eds), Vol. 2, pp. 77–113. Academic Press, New York.

ESMON, C.T. (1987). The regulation of natural anticoagulant pathways. *Science* **235**, 1348–1352.

FELSENFELD, G. (1992). Chromatin as an essential part of the transcriptional mechanism. *Nature* **355**, 219–223.

FERRETTI, L., LEONE, P. AND SGARAMELLA, V. (1990). Long range restriction analysis of the bovine casein genes. *Nucleic Acids Research* **18**, 6829–6833.

FLEER, R., YEH, P., AMELLAL, N., MAURY, I., FOURNIER, A., BACCHETTA, F., BADUEL, P., JUNG, G., L'HOTE, H., BECQUART, J., FUKUHARA, H. AND MAYAUX, J.F. (1991). Stable multicopy vectors for high-level secretion of recombinant human serum albumin by *Kluyveromyces* yeasts. *Bio/Technology* **9**, 968–975.

FLOWER, D.R., NORTH, A.C.T. AND ATTWOOD, T.K. (1993). Structure and sequence relationships in the lipocalins and related proteins. *Protein Science* **2**, 753–761.

FOX, J.A., HAYES, D.J., KLIEBENSTEIN, J.B. AND SHOGREN, J.F. (1994). Consumer acceptability of milk from cows treated with bovine somatotropin. *Journal of Dairy Science* **77**, 703–707.

FRIEDMAN, J.S., COFER, C.L., ANDERSON, C.L., KUSHNER, J.A., GRAY, P.P., CHAPMAN, G.E., STUART, M.C., LAZARUS, L., SHINE, J. AND KUSHNER, P.J. (1989). High expression in mammalian cells without amplification. *Bio/Technology* **7**, 359–362.

FURET, J.P., MERCIER, J.C., SOULIER, S., GAYE, P., HUE-DELAHAIE, D. AND VILOTTE, J.-L. (1990). Nucleotide sequence of ovine kappa-casein cDNA. *Nucleic Acids Research* **18**, 5286.

FURTH, P.A., SHAMAY, A., WALL, R.J. AND HENNIGHAUSEN, L. (1992). Gene transfer into somatic tissues by jet injection. *Analytical Biochemistry* **205**, 365–368.

GARVER, R.I., CHYTIL, I., KARLSSON, S., FELLS, G.A., BRANTLY, M.L., COURTNEY, M., KANTOFF, P.W., NIENHUIS, A.W., ANDERSON, W.F. AND CRYSTAL, R.G. (1987). Production of glycosylated physiologically 'normal' human α_1-antitrypsin by mouse fibroblasts modified by insertion of a human α_1-antitrypsin cDNA using a retroviral vector. *Proceedings of the National Academy of Sciences, USA* **84**, 1050–1054.

GAYE, P., HUE-DELAHAIE, D., MERCIER, J.C., SOULIER, S., VILOTTE, J.L. AND FURET, J.P. (1986). Ovine β-lactoglobulin messenger RNA: nucleotide sequence and mRNA levels during functional differentiation of the mammary gland. *Biochimie* **68**, 1097–1107.

GAYE, P., HUE-DELAHAIE, D., MERCIER, J.C., SOULIER, S., VILOTTE, J.L. AND FURET, J.P. (1987). Complete nucleotide sequence of ovine alpha-lactalbumin mRNA. *Biochimie* **69**, 601–608.

GELLERFORS, P., AXELSSON, K., HELANDER, A., JOHANSSON, S., KENNE, L., LINDQVIST, S., PAVLU, B., SKOTTNER, A. AND FRYKLUND, L. (1989). Isolation and characterisation of a glycosylated form of human insulin-like growth factor I produced in *Saccharomyces cerevisiae*. *Journal of Biological Chemistry* **264**, 11444–11449.

GELLIN, J., ECHARD, G., YERLE, M., DALENS, M., CHEVALET, C. AND GILLOIS, M. (1985). Localization of the α- and β-casein genes to the q24 region of chromosome 12 in the rabbit (*Oryctolagus cuniculus* L.) by *in situ* hybridization. *Cytogenetics and Cell Genetics* **39**, 220.

GODAVAC-ZIMMERMAN, J. AND BRAUNITZER, G. (1987). Modern aspects of the primary structure and function of β-lactoglobulins. *Milchwissenschaft* **42**, 294–297.

GODAVAC-ZIMMERMAN, J. AND SHAW, D. (1987). The primary structure, binding site and possible function of β-lactoglobulin from Eastern Grey kangaroo (*Macropus giganteus*). *Biological Chemistry Hoppe-Seyler* **368**, 879–886.

GODAVAC-ZIMMERMAN, J., CONTI, A., JAMES, L. AND NAPOLITANO, L. (1988). Microanalysis of the amino-acid sequence of monomeric beta-lactoglobulin Id from donkey (*Equus asinus*) milk. The primary structure and its homology with a superfamily of hydrophobic molecule transporters. *Biological Chemistry Hoppe-Seyler* **369**, 171–179.

GORDON, J.W., SCANGOS, G.A., PLOTKIN, D.J., BARBOSA, J.A. AND RUDDLE, F.H. (1980). Genetic transformation of mouse embryos by microinjection of purified DNA. *Proceedings of the National Academy of Sciences, USA* **77**, 7380–7384.

GORDON, K., LEE, E., VITALE, J.A., SMITH, A.E., WESTPHAL, H. AND HENNIGHAUSEN, L. (1987). Production of human plasminogen activator in transgenic mouse milk. *Bio/Technology* **5**, 1183–1187.

GORODETSKII, S.I., TKACH, T.M. AND KAPELINSKAYA, T.V. (1988). Isolation and characterisation of the *Bos taurus* β-casein gene. *Gene* **66**, 87–96.

GRAPPIN, R., RANK, T.C. AND OLSON, N.F. (1985). Primary proteolysis of cheese proteins during ripening. A review. *Journal of Dairy Science* **68**, 531–540.

GREAVES, D.R., WILSON, F.D., LANG, G. AND KIOUSSIS, D. (1989). Human CD2 3'-flanking sequences confer high level, T cell-specific, position-independent gene expression in transgenic mice. *Cell* **56**, 979–986.

GREENBERG, N.M., ANDERSON, J.W., HSUEH, A.J.W., NISHIMORI, K., REEVES, J.J., DEAVILA, D.M., WARD, D.N. AND ROSEN, J.M. (1991). Expression of biologically active heterodimeric bovine follicle-stimulating hormone in the milk of transgenic mice. *Proceedings of the National Academy of Sciences, USA* **88**, 8327–8331.

GROENEN, M.A., DIJKHOF, R.J., VERSTEGE, A.J. AND VAN DER POEL, J.J. (1993). The complete sequence of the gene encoding bovine alpha S2-casein. *Gene* **123**, 187–193.

GROSCLAUDE, F., JOUDRIER, P. AND MAHE, M.F. (1979). A genetic and biochemical analysis of a polymorphism of bovine alpha-S2-casein. *Journal of Dairy Research* **46**, 211–213.

GROSVELD, F., VAN ASSENDELFT, G.B., GREAVES, D.R. AND KOLLIAS, G. (1987). Position-independent, high-level expression of human β-globin gene in transgenic mice. *Cell* **51**, 975–985.

GUPTA, P., ROSEN, J.M., D'EUSTACHIO, P. AND RUDDLE, F.H. (1982). Localisation of the casein gene family to a single mouse chromosome. *Journal of Cell Biology* **93**, 199–204.

HALL, L. (1990). Nucleotide sequence of guinea-pig kappa-casein cDNA. *Nucleic Acids Research* **18**, 6129.

HALL, L., LAIRD, J.E. AND CRAIG, R.K. (1984). Nucleotide sequence determination of guinea pig casein B mRNA reveals homology with rat α_{S1}-caseins and conservation of the non-coding regions of the mRNA. *Biochemical Journal* **222**, 561–570.

HALL, L., CRAIG, R.K., EDBROOKE, M.R. AND CAMPBELL, P.N. (1982). Comparison of the nucleotide sequence of cloned human and guinea-pig pre α-lactalbumin cDNA with that of chick pre-lysozyme cDNA sugests evolution from a common ancestral gene. *Nucleic Acids Research* **10**, 3503–3515.

HALL, L., LAIRD, J.E., PASCALL, J.C. AND CRAIG, R.K. (1984). Guinea pig casein A cDNA nucleotide sequence analysis and comparison of the deduced protein sequence with that of bovine α_{S2}-casein. *European Journal of Biochemistry* **138**, 585–589.

HALL, L., EMERY, D.C., DAVIES, M.S., PARKER, D. AND CRAIG, R.K. (1987). Organisation and Sequence of the human α-lactalbumin gene. *Biochemical Journal* **242**, 735–742.

HALLIDAY, J.A., BELL, K., MCKENZIE, H.A. AND SHAW, D.C. (1990). Feline whey proteins: Identification, isolation and initial characterisation of α-lactalbumin, β-lactoglobulin and lysozyme. *Comparative Biochemistry and Physiology* **95B**, 773–779.

HALLIDAY, J.A., BELL, K., MCANDREW, K. AND SHAW, D.C. (1993). Feline β-lactoglobulins I, II and III, and canine β-lactoglobulins I and II: amino acid sequences provide evidence for the existence of more than one gene for β-lactoglobulin in the cat and the dog. *Protein Sequence Data Analysis* **5**, 201–205.

HAMMER, R.E., PURSEL, V.G., REXROAD, C.E., WALL, R.J., BOLT, D.J., EBERT, K.M., PALMITER, R.D. AND BRINSTER, R.L. (1985). Production of transgenic rabbits, sheep and pigs by microinjection. *Nature* **315**, 680–683.

HANSSON, L., EDLUND, A., JOHANSSON, T., HERNELL, O., STROMQVIST, M., LINDQUIST, S., LONNERDAL, B. AND BERGSTROM, S. (1994a). Structure of the human β-casein encoding gene. *Gene* **13**, 193–199.

HANSSON, L., EDLUND, M., EDLUND, A., JOHANSSON, T., MARKLUND, S.L., FROMM, S., STRÖMQVIST, M. AND TÖRNELL, J. (1994b). Expression and characterization of biologically active human extracellular superoxide dismutase in milk of transgenic mice. *Journal of Biological Chemistry* **269**, 5358–5363.

HASELOFF, J. AND GERLACH, W.L. (1988). Simple RNA enzymes with new and highly specific endoribonuclease activities. *Nature* **334**, 585–591.

HASTY, P., RAMIREZ-SOLOIS, R., KRUMLAUF, R. AND BRADLEY, A. (1991). Subtle mutation in the nox2.6 locus in ES cells. *Nature* **350**, 243–246.

HAYES, H.C. AND PETIT, E.J. (1993). Mapping of the β-lactoglobulin gene and of an immunoglobulin M heavy chain-like sequence to homologous cattle, sheep and goat chromosomes. *Mammalian Genome* **4**, 207–210.

HAYES, H.C., POPESCU, P. AND DUTRILLAUX, B. (1993). Comparative gene mapping of lactoperoxidase, retinoblastoma, and alpha-lactalbumin genes in cattle, sheep, and goats. *Mammalian Genome* **4**, 593–597.

HAYES, H.C., PETIT, E.J., LEMIEU, N. AND DUTRILLAUX, B. (1992). Chromosomal localization of the ovine beta-casein gene by non-isotopic *in situ* hybridisation and R-banding. *Cytogenetics and Cell Genetics* **61**, 286–288.

HENNIGHAUSEN, L. (1990). The mammary gland as a bioreactor: production of foreign proteins in milk. *Protein Expression and Purification* **1**, 3–8.

HENNIGHAUSEN, L., RUIZ, L. AND WALL, R. (1990). Transgenic animals – production of foreign proteins in milk. *Current Opinion in Biotechnology* **1**, 74–78.

HENNIGHAUSEN, L.G. AND SIPPEL, A.E. (1982a). Characterisation and cloning of the mRNAs specific for the lactating mouse mammary gland. *European Journal of Biochemistry* **125**, 131–141.

HENNIGHAUSEN, L.G. AND SIPPEL, A.E. (1982b). Mouse whey acidic protein is a novel member of the family of 'four disulphide' core proteins. *Nucleic Acids Research* **10**, 2677–2684.

HENNIGHAUSEN, L.G., STEUDLE, A. AND SIPPEL, A.E. (1982). Nucleotide sequence of cloned cDNA coding for mouse ε-casein. *European Journal of Biochemistry* **126**, 569–572.

HENNIGHAUSEN, L.G., SIPPEL, A.E., HOBBS, A.A. AND ROSEN, J.M. (1982). Comparative sequence analysis of the mRNAs coding for mouse and rat whey acidic protein. *Nucleic Acids Research* **10**, 3733–3744.

HILL, J.P. (1993). The relationship between β-lactoglobulin phenotypes and milk composition in New Zealand dairy cattle. *Journal of Dairy Science* **76**, 281–286.

HILL, K.G., CURRY, J., DEMAYO, F.J., JONES-DILLER, K., SLAPAK, J.R. AND BONDIOLI, K.R. (1992). Production of transgenic cattle by pronuclear injection. *Theriogenology* **37**, 222 (abstract).

HOBBS, A.A. AND ROSEN, J.M. (1982). Sequence of the rat α- and γ-casein mRNAs: evolutionary comparison of the calcium dependent rat casein multigene family. *Nucleic Acids Research* **10**, 8079–8098.

HOBBS, A.A., RICHARDS, D.A., KESSLER, D.J. AND ROSEN, J.M. (1982). Complex hormonal regulation of rat casein gene expression. *Journal of Biological Chemistry* **257**, 3598–3605.

HOCHI, S.I., NINOMIYA, T., WAGA-HOMMA, M., SAGARA, J. AND YUKI, A. (1992). Secretion of bovine α-lactalbumin into the milk of transgenic rats. *Molecular Reproduction and Development* **33**, 160–164.

HODGSON, J. (1993). Expression systems: a user's guide. *Bio/Technology* **11**, 887–893.

HOLT, C. AND DALGLEISH, D.G. (1986). Electrophoretic and hydrodynamic properties of bovine casein micelles interpreted in terms of particles with an outer hairy layer. *Journal of Colloid Interface Science* **114**, 513–524.

HORVAT, S., MEDRANO, J.F., BEHBOODI, E., ANDERSON, G.B. AND MURRAY, J.D. (1993). Sexing and detection of gene construct in microinjected bovine blastocysts using the polymerase chain reaction. *Transgenic Research* **2**, 134–140.

HOYLE, R. (1994). FDA's BST policy wreaks havoc in the market. *Bio/Technology* **12**, 570–571.

HUGHEY, V.L. AND JOHNSON, E.A. (1987). Antimicrobial activity of lysozyme against bacteria involved in food spoilage and food-borne disease. *Applied Environmental Microbiology* **53**, 2165–2170.

HYTTINEN, J.M., PEURA, T., TOLVANEN, M., AALTO, J., ALHONEN, L., SINERVITA, R., HALMEKYTÖ, M., MYÖHANEN, S. AND JÄNNE, J. (1994). Generation of transgenic dairy cattle from transgene-analysed and sexed embryos produced *in vitro*. *Bio/Technology* **12**, 606–608.

INOUYE, M. (1988). Antisense RNA: its functions and applications in gene regulation – a review. *Gene* **72**, 25–34.

IRWIN, D.M., SIDOW, A., WHITE, R.T. AND WILSON, A.T. (1989). Multiple genes for ruminant lysozymes. In *The Immune Response to Structurally Defined Proteins: The Lysozyme Model* (S.J. Smith-Gill and E.E. Sercarz, Eds), pp. 73–85. Adenine Press, Schenectady, N.Y.

JAMES, J.M. AND SAMPSON, H.A. (1992). Immunologic changes associated with the development of tolerance in children with cow milk allergy. *Journal of Pediatrics* **121**, 371–377.

JENNESS, R. AND SLOAN, R.E. (1970). The composition of milks of various species. *Dairy Science Abstracts* **32**, 599–612.

JIMENEZ-FLORES, R. AND RICHARDSON, T. (1988). Genetic engineering of the caseins to modify the behaviour of milk during processing: a review. *Journal of Dairy Science* **71**, 2640–2654.

JOHANSSON, B.G. (1960). Isolation of an iron-containing red protein from human milk. *Acta Chimica Scandinavica* **14**, 510.

JOLIVET, G., DEVINOY, E., FONTAIN, M.L. AND HOUDEBINE, L.M. (1992). Structure of the gene encoding rabbit alpha S1-casein. *Gene* **113**, 257–262.

JOLLES, P. (1975). Structural aspects of the milk clotting process. Comparative features with the blood clotting process. *Molecular and Cellular Biochemistry* **7**, 73–85.

JONES, W.K., YU-LEE, L.-Y., CLIFT, S.M., BROWN, T.L. AND ROSEN, J.M. (1985). The rat casein multigene family. Fine structure and evolution of the β-casein gene. *Journal of Biological Chemistry* **260**, 7042–7050.

JUNG, A., SIPPEL, A.E., GREZ, M. AND SCHUTZ, G. (1980). Exons encode functional and structural units of chicken lysozyme. *Proceedings of the National Academy of Sciences, USA* **77**, 5759–5763.

KANG, T., JIMENEZ-FLORES, R. AND RICHARDSON, T. (1986). Casein genes and genetic engineering of the caseins. In *Genetic Engineering of Animals; An Agricultural Perspective* (J.W. Evens and A. Hollaender, Eds), pp. 95–111. Plenum Press, New York.

KAPLAN, G., COHN, Z.A. AND SMITH, K.A. (1992). Rational immunotherapy with interleukin 2. *Bio/Technology* **10**, 157–162.

KLAUSER, A. (1993). Back to the future: biotech product sales 1983–1993. *Bio/Technology* **11**, S35–S37.

KLEIN, T.M. AND FITZPATRICK-MCELLIGOTT, S. (1993). Particle bombardment: a universal approach for gene transfer to cells and tissues. *Current Opinion in Biotechnology* **4**, 583–590.

KOCZAN, G., HOBOM, G. AND SEYFERT, H.-M. (1991). Genomic organisation of the bovine α_{S1}-casein gene. *Nucleic Acids Research* **19**, 5591–5596.

KRETCHMER, N. (1972). Lactose and lactase. *Scientific American* **227**, 70–78.

KRIMPENFORT, P., RADEMAKERS, A., EYESTONE, W., VAN DE SCHANS, A., VAN DEN BROEK, S., KOOIMAN, P., KOOTWIJK, E., PLATENBURG, G., PIEPER, F., STRIJKER, R. AND DE BOER, H. (1991). Generation of transgenic dairy cattle using '*in vitro*' embryo production. *Bio/ Technology* **9**, 844–847.

KUHN, N.J. (1983). The biosynthesis of lactose. In *Biochemistry of Lactation* (T.B. Mepham, Ed.), pp. 351–378. Elsevier Science Publishers, Amsterdam.

KUMAGAI, I., TAMAKI, E., KAKINUMA, S. AND MIURA, K.I. (1987). Molecular cloning and sequencing of cDNA encoding goat pre alpha-lactalbumin. *Journal of Biochemistry* **101**, 511–517.

KUWATA, T., PHAM, A.M., MA, C.Y. AND NAKAI, S. (1985). Elimination of β-lactoglobulin from whey to simulate human milk protein. *Journal of Food Science* **50**, 605–609.

LAIRD, J.E., JACK, L., HALL, L., BOULTON, A.P., PARKER, D. AND CRAIG, R.K. (1988). Structure and expression of the guinea-pig α-lactalbumin gene. *Biochemical Journal* **254**, 85–94.

LALANDE, M., TISSIER, J.P. AND CORRIEU, G. (1985). Fouling of heat transfer surfaces related to β-lactoglobulin denaturation during the heat processing of milk. *Biotechnology Progress* **1**, 131–139.

LATTA, M., KNAPP, M., SARMIENTOS, P., BREFORT, G., BECQUART, B., GUERRIER, L., JUNGM, G. AND MAUYAUX, J.F. (1987). Synthesis and purification of mature serum albumin from *E. coli*. *Bio/Technology* **5**, 1309–1314.

LAW, B. (1987). Proteolysis in relation to normal and accelerated cheese ripening. In *Cheese: Chemistry, Physics and Microbiology* (P.F. Fox, Ed.), pp. 365–392. Elsevier Applied Science, Amsterdam.

LEE, K.F., ATIEE, S.H. AND ROSEN, J.M. (1989). Differential regulation of rat β-casein–chloramphenicol acetyltransferase fusion gene expression in transgenic mice. *Molecular and Cellular Biology* **9**, 560–565.

LEE, K.F., DEMAYO, F.J., ATIEE, S.H. AND ROSEN, J.M. (1988). Tissue-specific expression of the rat β-casein gene in transgenic mice. *Nucleic Acids Research* **16**, 1027–1041.

LEE, S.P., CHO, Y. AND BATT, C.A. (1993). Enhancing the gelation of β-lactoglobulin. *Journal of Agricultural and Food Chemistry* **41**, 1343–1348.

LEE, S.P., KIM, D.S., WATKINS, S. AND BATT, C.A. (1994). Reducing whey syneresis in yogurt by the addition of a thermolabile variant of β-lactoglobulin. *Bioscience, Biotechnology and Biochemistry* **58**, 309–313.

LEROUX, C., MAZURE, N. AND MARTIN, P. (1992). Mutations away from splice site recognition sequences might *cis*-modulate alternative splicing of goat α_{S1}-casein transcripts. Structural organisation of the relevant gene. *Journal of Biological Chemistry* **267**, 6147–6157.

LEVINE, W.B., ALEXANDER, L.J., HOGANSON, G.E. AND BEATTIE, C.W. (1992). Cloning and sequencing of the porcine kappa-casein cDNA. *Animal Genetics* **23**, 361–363.

LIANG, Q. AND RICHARDSON, T. (1993). Expression and characterisation of human lactoferrin in yeast *Saccharomyces cerevisiae*. *Journal of Agricultural and Food Chemistry* **41**, 1800–1807.

LIN, C.Q. AND BISSELL, M.J. (1993). Multi-faceted regulation of cell differentiation by extracellular matrix. *FASEB Journal* **7**, 737–743.

LOGAN, J.S. (1993). Transgenic animals: beyond 'funny milk'. *Current Opinion in Biotechnology* **4**, 591–595.

LONNERDAL, B., BERGSTROM, S., ANDERSSON, Y., HJALMARSSON, K., SUNDQVIST, A.K. AND

HERNELL, O. (1990). Cloning and sequencing of a cDNA encoding human milk beta-casein. *FEBS Letters* **269**, 153–156.

LOVELL-BADGE, R.H. (1985). New advances in the field. *Nature* **300**, 611–615.

MCDOUGALL, E.I. AND STEWART, J.C. (1976). the whey proteins of the milk of red deer (*Cervus elaphus* L.). A homologue of bovine beta-lactoglobulin. *Journal of Biochemistry* **153**, 647–655.

MCKNIGHT, R.A., SHAMAY, A., SANKARAN, L., WALL, R.J. AND HENNIGHAUSEN, L. (1992). Matrix-attachment regions can impart position-independent regulation of a tissue-specific gene in transgenic mice. *Proceedings of the National Academy of Sciences, USA* **89**, 6943–6947.

MCLEAN, D.M., GRAHAM, E.R.B. AND PONZONI, R.W. (1984). Effects of milk protein genetic variants on milk yield and composition. *Journal of Dairy Research* **51**, 531–546.

MCMAHON, D.J. AND BROWN, R.J. (1984). Composition, structure and integrity of casein micelles: a review. *Journal of Dairy Science* **67**, 499–512.

MAGA, E.A., ANDERSON, G.B., HUANG, M.C. AND MURRAY, J.D. (1994). Expression of human lysozyme mRNA in the mammary gland of transgenic mice. *Transgenic Research* **3**, 36–42.

MARSHALL, J., FANG, S., OSTEDGAARD, L.S., O'RIORDAN, C.R., FERRARA, D., AMARA, J.F., HOPPE, IV, H., SCHEULE, R.K., WELSH, M.J., SMITH, A.E. AND CHENG, S.H. (1994). Stoichiometry of recombinant cystic fibrosis transmembrane conductance regulator in epithelial cells and its functional reconstitution into cells *in vitro*. *Journal of Biological Chemistry* **269**, 2987–2995.

MARSTON, F.A.O. (1986). The purification of eukaryotic polypeptides synthesised in *Escherichia coli*. *Biochemical Journal* **240**, 1–12.

MASCHIO, A., BRICKEL, P.M., KIOUSSIS, D., MELLOR, A.L., KATZ, D. AND CRAIG, R.K. (1991). Transgenic mice carrying the guinea-pig α-lactalbumin gene transcribe milk protein genes in their sebaceous glands during lactation. *Biochemical Journal* **275**, 459–472.

MASSON, P.L. AND HEREMANS, J.F. (1971). Lactoferrin in milk from different species. *Comparative Biochemistry and Physiology* **39**, 119–129.

MATSUMOTO, K., KAKIDANI, H., TAKAHASHI, A., NAKAGATA, N., ANZAI, M., MATSUZAKI, Y., TAKAHASHI, Y., MIYATA, K., UTSUMI, K. AND IRITANI, A. (1993). Growth retardation in rats whose growth hormone gene expression was suppressed by antisense RNA transgene. *Molecular Reproduction and Development* **36**, 53–54.

MEADE, H., GATES, L., LACY, E. AND LONBERG, N. (1990). Bovine α_{S1}-casein gene sequences direct high level expression of active human urokinase in mouse milk. *Bio/Technology* **8**, 443–446.

MENON, R.S., CHANG, Y.-F., JEFFERS, K.F., JONES, C. AND HAM, R.G. (1992). Regional localization of human β-casein gene (CSN2) to 4pter–q21. *Genomics* **13**, 225–226.

MEPHAM, T.B., GAYE, P., MARTIN, P. AND MERCIER, J.C. (1992). Biosynthesis of milk protein. In *Advanced Dairy Chemistry* Volume 1: *Proteins* (P.F. Fox, Ed.), pp. 491–543. Elsevier Applied Science, London.

MERCIER, J.C. (1986). Genetic engineering applied to milk producing animals: some expectations. In *Exploiting New Technologies in Animal Breeding* (C. Smith, J. King and J. McKay, Eds), pp. 122–131. Oxford University Press, London.

MERCIER, J.C. AND VILOTTE, J.L. (1993). Structure and function of milk protein genes. *Journal of Dairy Science* **76**, 3079–3098.

MERCIER, J.C., GAYE, P., SOULIER, S., HUE-DELAHAIE, D. AND VILOTTE, J.L. (1985). Construction and identification of recombinant plasmids carrying cDNAs coding for ovine α_{S1}-, α_{S2}-, β-, κ-casein and β-lactoglobulin. Nucleotide sequence of α_{S1}-casein cDNA. *Biochimie* **67**, 959–971.

METZ-BOUTIGUE, M.H., JOLLES, J., MAZURIER, J., SCHOENTGEN, F., LEGRAND, D., SPIK, G., MONTREUIL, J. AND JOLLES, P. (1984). Human lactotransferrin: amino acid sequence and structural comparisons with other transferrins. *European Journal of Biochemistry* **145**, 659–676.

MEYER, F., ERHARD, G. AND SENFT, B. (1981). Umweltbedingte und genetische aspekte des lysozyms in der kuhmilch. *Zuchtungskunde* **53**, 17–21.

MILES, H., LESSER, W. AND SEARS, P. (1992). The economic implications of bioengineered mastitis control. *Journal of Dairy Science* **75**, 596–605.

MORR, C.V. (1985). Functionality of heated milk proteins in dairy and related foods. *Journal of Dairy Science* **68**, 2773–2781.

MURRAY, J.D., NANCARROW, C.D., MARSHALL, J.T., HAZELTON, I.G. AND WARD, K.A. (1989). Production of transgenic merino sheep by microinjection of ovine metallothionein–ovine growth hormone fusion genes. *Reproduction, Fertility and Development* **1**, 147–155.

NAGAMINE, Y., SUDOL, M. AND REICH, E. (1983). Hormonal regulation of plasminogen activator mRNA production in porcine kidney cells. *Cell* **32**, 1181–1190.

NAGAO, M., MAKI, M., SASAKI, R. AND CHIBA, H. (1984). Isolation and sequence analysis of bovine α_{S1}-casein cDNA clone. *Agricultural and Biological Chemistry* **48**, 1663–1667.

NAKHASI, H.L., GRANTGAM, F.H. AND GULLINO, P.M. (1984). Expression of kappa-casein in normal and neoplastic rat mammary gland is under the control of prolactin. *Journal of Biological Chemistry* **259**, 14894–14898.

NDC (1985). Nutritional implications of lactose and lactase activity. *Dairy Council Digest* **56**, 25.

NEMET, K. AND SIMONOVITZ, I. (1985). The biological role of lactoferrin (review). *Haematologia* **18**, 3–12.

NICHOLAS, K.R., MESSER, M., ELLIOTT, C., MAHER, F. AND SHAW, D.C. (1987). A novel whey protein synthesized only in late lactation by the mammary gland from the tammar (*Macropus eugenii*). *Biochemical Journal* **241**, 899–904.

NINOMIYA, T., HOSHI, M., MIZUNO, A., NAGAO, M. AND YUKI, A. (1989). Selection of mouse preimplantation embryos carrying exogenous DNA by polymerase chain reaction. *Molecular Reproduction and Development* **1**, 242–248.

NINOMIYA, T., HIRABAYASHI, M., SAGARA, J. AND YUKI A. (1994). Functions of milk protein gene 5' flanking regions on human growth hormone gene. *Molecular Reproduction and Development* **37**, 276–283.

OH, S. AND RICHARDSON, T. (1991a). Genetic engineering of bovine κ-casein to improve its nutritional quality. *Journal of Agricultural and Food Chemistry* **39**, 422–427.

OH, S. AND RICHARDSON, T. (1991b). Genetic engineering of bovine κ-casein to enhance proteolysis by chymosin. In *Interactions of Food Proteins* (N. Parris and R. Barford, Eds), pp. 195–211. American Chemical Society, Washington, DC.

PACKARD, V.S. (1982). Infant food: use and misuse. In *Human Milk and Infant Formula* (G.F. Stewart, B.S. Schweigert and J. Hawthorn, Eds), pp. 187–201. Academic Press, New York.

PALMITER, R.D. AND BRINSTER, R.L. (1986). Germline transformation of mice. *Annual Review of Genetics* **20**, 465–499.

PALMITER, R.D., BRINSTER, R.L., HAMMER, R.E., TRUMBAUER, M.E., ROSENFELD, M.G., BIRNBERG, N.C. AND EVANS, R.M. (1982). Dramatic growth of mice that develop from eggs microinjected with metallothionein-growth hormone fusion genes. *Nature* **300**, 611–615.

PALMITER, R., SANDREN, E., AVARBOCK, M., ALLEN, D. AND BRINSTER, R. (1991). Heterologous introns can enhance expression of transgenes in mice. *Proceedings of the National Academy of Sciences, USA* **88**, 478–482.

PAPIZ, M.Z., SAWYER, L., ELIOPOULOS, E.E., NORTH, A.C.T., FINDLAY, J.B.C., SIVAPRASADARAO, R., JONES, T.A., NEWCOMER, M.E. AND KRAULIC, P.J. (1986). The structure of β-lactoglobulin and its similarity to plasma retinol binding protein. *Nature* **324**, 383–385.

PATIENT, R.K. (1990). Control of gene expression: tissue-specific expression. *Current Opinion in Biotechnology* **1**, 151–158.

PAVIRANI, A., SKERN, T., LE MEUR, M., LUTZ, Y., LATHE, R., CRYSTAL, R.G., FUCHS, J.P., GERLINGER, P. AND COURTNEY, M. (1989). Recombinant proteins of therapeutic interest expressed by lymphoid cells derived from transgenic mice. *Bio/Technology* **7**, 1049–1054.

PEARSE, M.J., LINKLATER,. P.M., HALL, R.J. AND MACKINLAY, A.G. (1986). The effect of casein composition and casein dephosphorylation on the coagulation and syneresis of artificial micelle milk. *Journal of Dairy Science* **53**, 381–390.

PEREZ, M.D., SANCHEZ, L., ARANDA, P., ENA, J.M., ORIA, R. AND CALVO, M. (1992). Effect of β-lactoglobulin on the activity of pregastric lipase. A possible role for this protein in ruminant milk. *Biochimica et Biophysica Acta* **1123**, 151–155.

PERSUY, M.A., STINNAKRE, M.G., PRINTZ, C., MAHA, M.F. AND MERCIER, J.C. (1992). High expression of the caprine β-casein gene in transgenic mice. *European Journal of Biochemistry* **205**, 887–893.

PERVAIZ, S. AND BREW, K. (1985). Homology of β-lactoglobulin, serum retinol-binding protein, and protein HC. *Science* **228**, 335–337.

PERVAIZ, S. AND BREW, K. (1986). Purification and characterization of the major whey proteins from the milks of the bottlenose dolphin (*Tursiops truncatus*), the Florida manatee (*Trichechus manatus latirostris*), and the beagle (*Canis familiaris*). *Archives of Biochemistry and Biophysics* **246**, 846–854.

PETERS, C.W., KRUSE, U., POLLWEIN, R., GRZESCHIK, K.-H. AND SIPPEL, A. (1989). The human lysozyme gene: sequence organization and chromosomal localization. *European Journal of Biochemistry* **182**, 507–516.

PITTIUS, C.W., HENNIGHAUSEN, L., LEE, E., WESTPHAL, H., NICOLS, E., VITALE, J. AND GORDON, K. (1988). A milk protein gene promoter directs the expression of human tissue plasminogen activator cDNA to the mammary gland in transgenic mice. *Proceedings of the National Academy of Sciences, USA* **85**, 5874–5878.

PLATENBURG, G.J., KOOTWIJK, E.P.A., KOOIMAN, P.M., WOLOSHUK, S.L., NUIJENS, J.H., KRIMPENFORT, P.J.A., PIEPER, F.R., DE BOER, H.A. AND STRIJKER, R. (1994). Expression of human lactoferrin in milk of transgenic mice. *Transgenic Research* **3**, 99–108.

PROVOT, C., PERSUY, M.A. AND MERCIER, J.C. (1989). Complete nucleotide sequence of ovine β-casein cDNA: interspecies comparison. *Biochimie* **71**, 827–832.

PURSEL, V.G. AND REXROAD, C.E. (1993). Status of research with transgenic farm animals. *Journal of Animal Science* **71**, (Suppl. 3), 10–19.

PURSEL, V.G., PINKERT, C.A., MILLER, K.F., BOLT, D.J., CAMPBELL, R.G., PALMITER, R.D., BRINSTER, R.L. AND HAMMER, R.E. (1989). Genetic engineering of livestock. *Science* **244**, 1281–1288.

PURSEL, V.G., HAMMER, R.E., BOLT, D.J., PALMITER, R.D. AND BRINSTER, R.L. (1990). Integration, expression and germline transmission of growth-related genes in pigs. *Journal of Reproduction and Fertility* **41**, (Suppl.), 77–87.

QASBA, P.K. AND SAFAYA, S.K. (1984). Similarity of the nucleotide sequences of rat α-lactalbumin and chicken lysozyme genes. *Nature* **308**, 377–380.

REDDY, V.B., VITALE, J.A., WEI, C., MONTOYA-ZAVALA, M., STICE, S.L., BALISE, J. AND ROBL, J.M. (1991). Expression of human growth hormone in the milk of transgenic mice. *Animal Biotechnology* **2**, 15–29.

REID, L.H., GREGG, R.G., SMITHIES, O. AND KOLLER, B.H. (1990). Regulatory elements in the introns of the human HPRT gene are necessary for its expression in embryonic stem cells. *Proceedings of the National Academy of Sciences, USA* **87**, 4299–4303.

REISMAN, H.B. (1993). Problems in scale-up of biotechnology production processes. *Critical Reviews in Biotechnology* **13**, 195–253.

REXROAD, C.E., HAMMER, R.E., BEHRINGER, R.R., PALMITER, R.D. AND BRINSTER, R.L. (1990). Insertion, expression and physiology of growth-regulating genes in ruminants. *Journal of Reproduction and Fertility* **41** (Suppl.), 119–124.

REY, M.W., WOLOSHUK, S.L., DE BOER, H.A. AND PIEPER, F.R. (1990). Complete nucleotide sequence of human mammary gland lactoferrin. *Nucleic Acids Research* **18**, 5288.

RHODES, M. AND BIRCH, J. (1988). Large-scale production of proteins from mammalian cells. *Bio/Technology* **6**, 518–523.

RICHARDSON, T. (1985). Chemical modification and genetic engineering of food proteins. *Journal of Dairy Science* **68**, 2753–2762.

ROBERTS, B., DITULLIO, P., VITALE, J., HEHIR, K. AND GORDON, K. (1992). Cloning of the goat β-casein-encoding gene and expression in transgenic mice. *Gene* **121**, 255–262.

ROSEN, J.M., RODGERS, J.R., COUCH, C.H., BISBEE, C.A., DAVID-INOUYE, Y., CAMPBELL, S.M. AND YU-LEE, L.Y. (1986). Multihormonal regulation of milk protein gene expression (Review). *Annals of the New York Academy of Sciences* **478**, 63–76.

RYLL,. T., LUCKI-LANGE, M., JÄGER, V. AND WAGNER, R. (1990). Production of recombinant interleukin-2 with BHK cells in a hollow fibre and a stirred tank reactor with protein-free medium. *Journal of Biotechnology* **14**, 377–392.

SAAVEDRA, J.M. AND PERMAN, J.A. (1989). Current concepts in lactose malabsorption and intolerance. *Annual Review of Nutrition* **9**, 475–502.

SAIKI, R.K., GELFAND, D.H., STOFFEL, S., SCHARF, S.J., HIGUCHI, R., HORN, G.T., MULLIS, K.B. AND ERLICH, H.A. (1988). Primer-directed enzymatic amplification of DNA with a thermostable DNA polymerase. *Science* **239**, 487–491.

SARVER, N., CANTIN, E.M., CHANG, P.S., ZAIA, J.A., LADNE, P.A., STEPHENS, D.A. AND ROSSI, J.J. (1990). Ribozymes as potential anti-HIV-1 therapeutic agents. *Science* **247**, 1222–1225.

SAWATZKI, G. AND RICH, I.N. (1989). Lactoferrin stimulates colony stimulating factor production *in vitro* and *in vivo*. *Blood Cells* **15**, 371–385.

SAWYER, L. AND HOLT, C. (1993). The secondary structure of milk proteins and their biological function. *Journal of Dairy Science* **76**, 3062–3078.

SAWYER, W.H. (1968). Heat denaturation of bovine β-lactoglobulins and relevance of disulfide aggregation. *Journal of Dairy Science* **51**, 323–329.

SAXENA, S.K. AND ACKERMAN, E.J. (1990). Ribozymes correctly cleave a model substrate and endogenous RNA *in vivo*. *Journal of Biological Chemistry* **265**, 17106–17109.

SCANLON, K.J., JIAO, L., FUNATO, T., WANG, W., TONE, T., ROSSI, J.J. AND KASHANI-SHABET, M. (1991). Ribozyme-mediated cleavage of c-*fos* mRNA reduces gene expression of DNA synthesis enzymes and metallothionein. *Proceedings of the National Academy of Sciences, USA* **88**, 10591–10595.

SCHAERER, E., DEVINOY, E., KRAEHENBUHL, J.-P. AND HOUDEBINE, L.-M. (1988). Sequence of the rabbit β-casein cDNA: comparison with other casein cDNA sequences. *Nucleic Acids Research* **16**, 11814.

SCHEDL, A., LARIN, Z., MONTOLIU, L., THIES, E., KELSEY, G., LEHRACH, H. AND SCHÜTZ, G. (1993). A method for the generation of YAC transgenic mice by pronuclear microinjection. *Nucleic Acids Research* **21**, 4783–4787.

SHANI, M., BARASH, I., NATHAN, M., RICCA, G., SEARFOSS, G.H., DEKEL, I., FAERMAN, A., GIVOL, D. AND HURWITZ, D.R. (1992). Expression of human serum albumin in the milk of transgenic mice. *Transgenic Research* **1**, 195–208.

SIMONS, G., VAN DEN HEUVEL, W., REYNEN, T., FRIJTERS, A., RUTTEN, G., SLANGEN, C.J., GROENEN, M., DE VOS, W.M. AND SIEZEN, R.J. (1993). Overproduction of bovine β-casein in *Escherichia coli* and engineering of its main chymosin cleavage site. *Protein Engineering* **6**, 763–770.

SIMONS, J.P., MCCLENAGHAN, M. AND CLARK, A.J. (1987). Alteration of the quality of milk by expression of sheep β-lactoglobulin in transgenic mice. *Nature* **328**, 530–532.

SIMONS, J.P., WILMUT, I., CLARK, A.J., ARCHIBALD, A.L., BISHOP, J.O. AND LATHE, R. (1988). Gene transfer into sheep. *Bio/Technology* **6**, 179–183.

SIOUD, M. AND DRLICA, K. (1991). Prevention of human immunodeficiency virus type I integrase expression in *Escherichia coli* by a ribozyme. *Proceedings of the National Academy of Sciences, USA* **88**, 7303–7307.

SMITH, C., MEUWISSEN, T.H.E. AND GIBSON, J.P. (1987). On the use of transgenes in livestock improvement. *Animal Breeding Abstracts* **55**, 1.

SOKOL, D.L., PASSEY, R.L., MACKINLAY, A.G. AND MURRAY, J.D. Regulation of CAT protein levels by an antisense ribozyme transgene in mice. *Transgenic Research*, submitted.

SOULIER, S., MERCIER, J.C., VILOTTE, J.L., ANDERSON, J., CLARK, A.J. AND PROVOT, C. (1989). The bovine and ovine genomes contain multiple sequences homologous to the alpha-lactalbumin-encoding gene. *Gene* **83**, 331–338.

SOULIER, S., VILOTTE, J.L., STINNAKRE, M.G. AND MERCIER, J.C. (1992). Expression analysis of ruminant α-lactalbumin in transgenic mice: developmental regulation and general location of important *cis*-regulatory elements. *FEBS Letters* **297**, 13–18.

SPALDING, B.J. (1992). Transgenic pharming advances. *Bio/Technology* **10**, 498–499.

STACEY, A., SCHNIEKE, A., MCWHIR, J., COOPER, J., COLMAN, A. AND MELTON, D.W. (1994). Use of double-replacement gene targeting to replace the murine α-lactalbumin gene with its human counterpart in embryonic stem cells and mice. *Molecular and Cellular Biology* **14**, 1009–1016.

STEINECKE, P., HERGERT, T. AND SCHREIER, P.H. (1992). Expression of chimeric ribozyme gene results in endocucleolytic cleavage of target mRNA and a concomitant reduction of gene expression *in vivo. EMBO Journal* **11**, 1525–1530.

STEINHOFF, U.M., SENFT, B. AND SEYFERT, H.-M. (1994). Lysozyme-encoding bovine cDNAs from neutrophile granulocytes and mammary gland are derived from a different gene than stomach lysozymes. *Gene* **143**, 271–276.

STEWART, A.F., WILLIS, I.M. AND MACKINLAY, A.G. (1984). Nucleotide sequences of bovine α_{S1}- and κ-casein cDNAs. *Nucleic Acids Research* **12**, 3895–3907.

STEWART, A.F., BONSING, J., BEATTIE, C.W., SHAH, F., WILLIS, I.M. AND MACKINLAY, A.G. (1987). Complete nucleotide sequences of bovine α_{S2}- and β-casein cDNAs: comparisons with related sequences in other species. *Molecular Biology and Evolution* **4**, 231–241.

STINNAKRE, M.G., VILOTTE, J.L., SOULIER, S., L'HARIDON, R., CHARLIER, M., GAYE, P. AND MERCIER, J.C. (1991). The bovine α-lactalbumin promoter directs expression of ovine trophoblast interferon in the mammary gland of transgenic mice. *FEBS Letters* **284**, 19–22.

STINNAKRE, M.G., VILOTTE, J.L., SOULIER, S. AND MERCIER, J.C. (1994). Creation and phenotypic analysis of α-lactalbumin-deficient mice. *Proceedings of the National Academy of Sciences, USA* **91**, 6544–6548.

STORB, U., O'BRIEN, R.L., MCMULLEN, M.D., GOLLAHON, K.A. AND BRINSTER, R.L. (1984). High expression of cloned immunoglobulin κ gene in transgenic mice is restricted to B lymphocytes. *Nature* **310**, 238–241.

SWAISGOOD, H.E. (1973). The caseins. *CRC Critical Reviews in Food Technology* **3**, 375–414.

SWAISGOOD, H.E. (1993). Genetic perspectives on milk proteins; comparative studies and nomenclature. *Journal of Dairy Science* **76**, 3054–3061.

SWIFT, G.H., HAMMER, R.E., MACDONALD, R.J. AND BRINSTER, R.L. (1984). Tissue specific expression of the rat pancreatic elastase I gene in transgenic mice. *Cell* **38**, 639–646.

SYMONS, R.H. (1992). Small catalytic RNAs. *Annual Review of Biochemistry* **61**, 641–671.

TEAHAN, C.G., MCKENZIE, H.A. AND GRIFFITHS, M. (1991). Some monotreme milk 'whey' and blood proteins. *Comparative Biochemistry and Physiology* **B99**, 99–118.

TEKAMP-OLSON, P. AND VALENZUELA, P. (1990). Gene expression and engineering in yeast and other fungi. *Current Opinion in Biotechnology* **1**, 28–35.

TENG, C.T., PENTECOST, B.T., MARSHALL, A., SOLOMAN, A., BROWMAN, B.H., LALLEY, P.A. AND NAYLOR, S.L. (1987). Assignment of the lactoferrin gene to human chromosome 3 and to mouse chromosome 9. *Somatic Cell and Molecular Genetics* **13**, 689–693.

TENG, C.T., LIU, Y., YANG, N., WALMER, D. AND PANELLA, T. (1992). Differential molecular mechanism of the estrogen action that regulates lactoferrin gene in human and mouse. *Molecular Endocrinology* **6**, 1969–1981.

THEPOT, D., DEVINOY, E., FONTAINE, M.L., HUBERT, C. AND HOUDEBINE, L.M. (1990). Complete sequence of the rabbit whey acidic protein gene. *Nucleic Acids Research* **18**, 3641.

THEPOT, D., DEVINOY, E., FONTAINE, M.L. AND HOUDEBINE, L.M. (1991). Structure of the gene encoding rabbit beta-casein. *Gene* **97**, 301–306.

THOMPSON, M.D., DAVE, J.R. AND NAKHASI, H.L. (1985). Molecular cloning of mouse mammary gland κ-casein: comparison with rat κ-casein and rat and human γ-fibrinogen. *DNA* **4**, 263–271.

THOMPSON, T.A., GOULD, M.N., BURKHOLDER, J.K. AND YANG, N.S. (1993). Transient promoter activity in primary rat mammary epithelial cells evaluated using particle bombardment gene transfer. *In Vitro Cellular and Developmental Biology* **29A**, 165–170.

THREADGILL, D.W. AND WOMACK, J.E. (1990). Genomic analysis of the major bovine milk protein genes. *Nucleic Acids Research* **18**, 6935–6942.

TOJO, H., TANAKA, S., MATSUZAWA, A., TAKAHASHI, M. AND TACHI, C. (1993). Production and characterization of transgenic mice expressing a hGH fusion gene driven by the promoter of mouse whey acidic protein (mWAP) putatively specific to the mammary gland. *Journal of Reproduction and Development* **39**, 145–155.

TOLDO, S.S., MEZZELANI, A., HAWKINS, G.A., BISHOP, M.D., OLSAKER, I., MACKINLAY, A.G., FERRETTI, L. AND FRIES, R. (1994). Identification of bovine chromosomes 1 to 7 based on fluorescence *in situ* hybridization, (submitted).

ULLREY, E., SCHWARTZ, C.C., WHETTER, T., RAO, T.R., EUBER, J.R., CHENG, S.G. AND BRUNNER, J.R. (1984). Blue-green color and composition of Stejnegars beaked whale (*Mesoplodon Stejnegeri*) milk. *Comparative Biochemistry and Physiology* **79B**, 349–352.

VAISSE, C., ATGER, M., POTIER, B. AND MILGROM, E. (1990). Human placental protein 14 gene: sequence and characterization of a short duplication. *DNA and Cell Biology* **9**, 401–413.

VALANCIUS, V. AND SMITHIES, O. (1991). Testing an 'in-out' targeting procedure for making subtle genomic modifications in mouse embryonic stem cells. *Molecular and Cellular Biology* **11**, 1402–1408.

VAN ASSENDELFT, G.B., HANSCOMBE, O., GROSVELD, F. AND GREAVES, D.R. (1989). The β-globin dominant control region activated homologous and heterologous promoters in a tissue-specific manner. *Cell* **56**, 969–977.

VELANDER, W.H., PAGE, R.L., MORCOL, T., RUSSEL, C.G., CANSECO, R., DROHAN, W.N., GWAZDAUSKAS, F.C., WILKINS, T.D. AND JOHNSON, J.L. (1991). Production of biologically active human protein C in the milk of transgenic mice. *Annals of the New York Academy of Science* **665**, 391–403.

VELANDER, W.H., JOHNSON, J.L., PAGE, R.L., RUSSEL, C.G., SUBRAMANIAN, U., WILKINS, T.D., GWAZDAUSKAS, F.C., PITTIUS, C. AND DROHAN, W.N. (1992). High-level expression of a heterologous protein in the milk of transgenic swine using the cDNA encoding human protein C. *Proceedings of the National Academy of Sciences, USA* **89**, 12003–12007.

VILOTTE, J.L. ABD SOULIER, S. (1992). Isolation and characterisation of the mouse α-lactalbumin-encoding gene: interspecies comparison, tissue- and stage-specific expression. *Gene* **119**, 287–292.

VILOTTE, J.L., SOULIER, S. AND MERCIER, J.C. (1992). Sequence of the murine α-lactalbumin-encoding cDNA: interspecies comparison of the coding frame and deduced preprotein. *Gene* **112**, 251–255.

VILOTTE, J.L., SOULIER, S. AND MERCIER, J.C. (1993). Complete sequence of a bovine alpha-lactalbumin pseudogene: the region homologous to the gene is flanked by two directly repeated LINE sequences. *Genomics* **16**, 529–532.

VILOTTE, J.L., SOULIER, S., MERCIER, J.C., GAYE, P., HUE-DELAHAIE, D. AND FURET, J.-P. (1987). Complete nucleotide sequence of bovine α-lactalbumin gene: comparison with its rat counterpart. *Biochimie* **69**, 609–620.

VILOTTE, J.L., SOULIER, S., STINNAKRE, M.G., MASSOUD, M., MERCIER, J.C. (1989). Efficient tissue-specific expression of bovine α-lactalbumin in transgenic mice. *European Journal of Biochemistry* **186**, 43–48.

VILOTTE, J.L., SOULIER, S., PRINTZ, C. AND MERCIER, J.C. (1991). Sequence of the goat α-lactalbumin-encoding gene: comparison with the bovine gene and evidence of related sequences in the goat genome. *Gene* **98**, 271–276.

VISSER, S. AND SLANGEN, C.J. (1977). On the specificity of chymosin (rennin) in its action on β-casein. *Netherlands Milk and Dairy Journal* **31**, 16–30.

WALL, R.J., PURSEL, V.G., HAMMER, R.E. AND BRINSTER, R.L. (1985). Development of porcine ova that were centrifuged to permit visualisation of pronuclei and nuclei. *Biology of Reproduction* **32**, 645–651.

WALL, R.J., PURSEL, V.G., SHAMAY, A., MCKNIGHT, R., PITTIUS, C. AND HENNIGHAUSEN, L. (1991). High-level synthesis of a heterologous milk protein in the mammary glands of transgenic swine. *Proceedings of the National Academy of Sciences, USA* **88**, 1696–1700.

WARD, P.P., LO, J.Y., DUKE, M., MAY, G.S., HEADON, D.R. AND CONNEELY, O.M. (1992). Production of biologically active recombinant human lactoferrin in *Aspergillus oryzae*. *Bio/Technology* **10**, 784–789.

WAUGH, D. AND TALBOT, B. (1971). Equilibrium casein micelle systems. *Biochemistry* **10**, 4153–4163.

WEISSBACH, K. AND GRUEBUBGER, M.G. (1990). Bipartite mRNA for chicken alpha-fibrinogen potentially encodes an amino acid sequence homologous to beta- and gamma-fibrinogens. *Proceedings of the National Academy of Sicences, USA* **87**, 5198–5202.

WERNER, R.G., WALZ, F., NOÉ, W. AND KONRAD, A. (1992). Safety and economic aspects of continuous mammalian cell culture. *Journal of Biotechnology* **22**, 51–68.

WESTERMAN, M., SPENCER, J. AND COLLETT, C. (1991). Chromosomal localization of the gene for the late lactation (LLP) in the tammar wallaby (*Macropus eugenii*). *Cytogenetics and Cell Genetics* **56**, 182–184.

WHITELAW, C.B.A., ARCHIBALD, A.L., HARRIS, S., MCCLENAGHAN, M., SIMONS, J.P. AND CLARK, A.J. (1991). Targeting expression to the mammary gland: intronic sequences can enhance the efficiency of gene expression in mice. *Transgenic Research* **1**, 3–13.

WHITELAW, C.B.A., HARRIS, S., MCCLENAGHAN, M., SIMONS, J.P. AND CLARK, A.J. (1992). Position-independent expression of the ovine β-lactoglobulin gene in transgenic mice. *Biochemical Journal* **286**, 31–39.

WILKINS, T.D. AND VELANDER, W. (1992). Isolation of recombinant proteins from milk. *Journal of Cellular Biochemistry* **49**, 333–338.

WILLIAMS, J., ELLEMAN, T.C., KINGSTON, I.B., WILKINS, A.G. AND KUHN, K.A. (1982). The primary structure of hen ovotransferrin. *European Journal of Biochemistry* **122**, 297–303.

WOMACK, J.E., THREADGILL, D.W., MOLL, Y.D., FABER, L.K., FOREMAN, M.L., DIETZ, A.B., TOBIN, T.C., SKOW, L.C., ZNEIMER, S.M., GALLAGHER, L.S. AND ROGERS, D.S. (1989). Syntenic mapping of 37 loci in cattle. Chromosomal conservation with mouse and man. *Cytogenetics and Cell Genetics* **51**, 1109.

WOOD, C.R., BOSS, M.A., KENTON, J.M., CALVERT, J.E., ROBERTS, N.A. AND EMTAGE, J.S. (1985). The synthesis and *in vivo* assembly of functional antibodies in yeast. *Nature* **314**, 446–449.

WOODLEE, G.L., GOOLEY, A.A., COLLET, C. AND COOPER, D.W. (1993). Origin of late lactation protein from beta-lactoglobulin in the tammar wallaby. *Journal of Heredity* **84**, 460–465.

WRIGHT, G., CARVER, A., COTTOM, D., REEVES, D., SCOTT, A., SIMONS, P., WILMUT, I., GARNER, I. AND COLMAN, A. (1991). High level expression of active human alpha-1-antitrypsin in the milk of transgenic sheep. *Bio/Technology* **9**, 830–834.

YARRANTON, G.F. (1990). Mammalian recombinant proteins: vectors and expression systems. *Current Opinion in Biotechnology* **1**, 133–140.

YOSHIMURA, M. AND OKA, T. (1989). Isolation and structural analysis of the mouse β-casein gene. *Gene* **78**, 267–272.

YOSHIMURA, M., BANERJEE, M.R. AND OKA, T. (1986). Nucleotide sequence of a cDNA encoding mouse beta casein. *Nucleic Acids Research* **14**, 8224.

YOSHITAKE, S., SHACH, B.G., FOSTER, D.C., DAVIE, E.W. AND KURACHI, K. (1985). Nucleotide sequence of the gene for human factor IX. *Biochemistry* **24**, 3736–3750.

YU, S.H., DEEN, K.C., LEE, E., HENNIGHAUSEN, L., SWEET, R.W., ROSENBERG, M. AND WESTPHAL, H. (1989). Functional CD4 protein produced in milk of transgenic mice. *Molecular Biology and Medicine* **6**, 255–261.

YU-LEE, L.-Y., RICHTER-MANN, L., COUCH, C.H., STEWART, A.F., MACKINLAY, A.G. AND ROSEN, J.M. (1986). Evolution of the casein multigene family: conserved sequences in the 5' flanking and exon regions. *Nucleic Acids Research* **14**, 1883–1902.

YUN, S.E., OHMIYA, K. AND SHIMIZU, S. (1982). Role of β-casein in milk curdling. *Agricultural and Biological Chemistry* **46**, 443–449.

ZHAO, J.J. AND PICK, L. (1993). Generating loss-of-function phenotypes of the *fushi tarazu* gene with a targeted ribozyme in *Drosophila*. *Nature* **365**, 448–451.

ZOLLER, M.J. AND SMITH, M. (1983). Oligonucleotide-directed mutagenesis of DNA fragments cloned into M13 vectors. *Methods in Enzymology* **100**, 468–500.

5
The Bioremediation of Polychlorinated Biphenyls (PCBs): Problems and Perspectives

G.K. ROBINSON AND M.J. LENN*

Research School of Biosciences, University of Kent, Canterbury, Kent CT2 7NJ, UK

Introduction

MANUFACTURE AND HISTORY

Polychlorinated biphenyls (PCBs) were synthesized between the early 1920s and 1978 by a variety of manufacturers in the US, Europe and Japan. They were manufactured for their excellent thermal and electrical properties and were used in a wide variety of commerical applications. These varied from closed-system applications, such as dielectric fluids (capacitors and transformers) and industrial fluids (hydraulic systems and gas turbines), to open-system applications, such as in the manufacture of adhesives, textiles and printing. They were made as commercial mixtures of polychlorinated biphenyls and were manufactured by the direct chlorination of biphenyl using anhydrous chlorine. This was done at temperatures above 150°C, keeping the biphenyl molten to prevent clogging, in the presence of a catalyst such as iron filings or ferric chloride. The crude product was then purified with 0.3% alkali and distilled under vacuum. The degree of chlorination is dependent upon the contact time of the chlorine and biphenyl, which is usually between 12 and 36 hours. Theoretically, there are 209 possible congeners of polychlorinated biphenyls, but due

---

Abbreviations: APEG, alkali metal polyethylene glycolate; 2,4- D, 2,4-dichlorophenoxyacetic acid; DRE, destruction and removal efficiency; ECD, electron capture detector; EI, electron impact; EPA, Environmental Protection Agency; FID, flame ionization detector; GC, gas chromatography; GCMS, gas chromatography–mass spectrometry; HECD, Hall electrolytic conductivity detector; HOPDA,2-hydroxy-6-oxo-6-phenylhexa-2,4-dienoic acid; HPLC, high-pressure liquid chromatography; ICES, International Council for Exploration of the Sea; ICRCL, Interdepartmental Committee on the Reclamation of Contaminated Land; LC, liquid chromatography; MS, mass spectrometry; OECD, Organization for Economic Co-operation and Development; PCBs, polychlorinated biphenyls; RRT, relative retention time; SEC, steric effect coefficient; TLC, thin-layer chromatography; TOC, total organic carbon; TSCA, Toxic Substances Control Act; UV, ultraviolet.

---

* *Current Address: Viridian Bioprocessing Ltd, 114–116 John Wilson Business Park, Thanet Way, Chestfield, Whitstable, Kent CT5 3QT, UK.*

Biotechnology and Genetic Engineering Reviews — Vol. 12, December 1994
0264–8725/94/12/139–188 $20.00 + $0.00 © Intercept Ltd, P.O. Box 716, Andover, Hampshire SP10 1YG, UK

to mechanistic and statistical constraints 20 PCBs are absent in commercial preparations (Hutzinger, Safe and Zitko, 1974).

Commercial production of PCBs started in 1929. Since that time 57×10^7 kg have been made in the US alone. About an equal amount was manufactured in Europe, Russia and Japan. Major manufacturers over this time period were Monsanto (US), Mitsubishi-Monsanto (Japan), Prodelec (France), Bayer (Germany) and Caffaro (Italy). The primary reason for cessation of manufacture was the persistence and recalcitrance of PCBs in the environment. Jensen had shown this as early as 1966 (Jensen, 1972). Under the Toxic Substances Control Act (1976) the US Congress banned the manufacture of PCBs although Monsanto had restricted PCB use to closed-system applications in 1971. Finally, in 1978 PCB manufacture in the US was halted and dioctyl phthalate was introduced as a replacement. In the UK PCB usage has been progressively restricted from the early 1970s until their use in new plant and equipment was banned in 1986. After the US ban, Prodelec (France) continued to manufacture Chloralkylene 12, a product containing 20% dichlorinated biphenyls. In addition, Caffaro (Italy) manufactured Apirollio 1531C and 14812, which contained 60% PCBs and 40% trichlorobenzenes, albeit for closed-system application.

In 1990, at the Third International Ministerial Conference on the Protection of the North Sea, the UK agreed, along with the other participating states, to phase out and destroy the remaining PCBs in use by 1999.

NOMENCLATURE

Polychlorinated biphenyls contain two benzene rings and, using IUPAC convention, are labelled according to ring and position. One ring is labelled clockwise and the other counterclockwise with prime numbers as shown in *Figure 1*. Additionally, each of the possible congeners has been ordered by Ballschmiter and Zell (1980) to give a further system of nomenclature. Of the commercial mixtures, the Aroclor series were the most widely used and are named according to the number of carbon atoms in the biphenyl nucleus and the percentage by weight of chlorine (for example, Aroclor 1242 contains 42% chlorine). They also differ in physical appearance, with Aroclor 1242–1248 being clear liquids, 1254–1262 light-coloured oils and 1268 an off-white powder.

SOLUBILITY

The problems associated with PCBs are largely due to the diversity of congeners

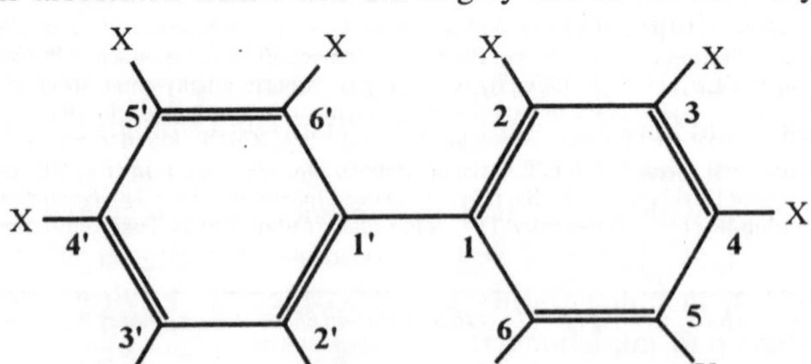

Figure 1. Conventional numbering of substituent positions in the PCB molecule X = H or Cl.

present in commercial mixtures and their associated chemical inertness. The Aroclor series of PCBs is perhaps the most widely cited preparation, largely due to the scale of its manufacture. The preparations manufactured vary in composition as outlined elsewhere. Of those available, 1242 was the most widely used Aroclor and contained a range of chlorobiphenyls containing up to five chlorine molecules. Approximately 60–80 congeners have been identified in the mixture and the properties of the Aroclor are dependent upon the congener composition. They are remarkably hydrophobic compounds and mixtures, solubility in water being governed by the percentage of higher chlorinated congeners present and the position of the chlorine atoms. Aroclor 1242 is soluble at 200 p.p.b. whereas 1260 is much less soluble, at 25 p.p.b. Of the individual congeners, 2-chlorobiphenyl is the most soluble (6 p.p.m.) and decachlorobiphenyl (0.015 p.p.m.) the least soluble in water (Hutzinger, Safe and Zitko, 1974). From the additional data shown in *Table 1*, it can be seen that even within the dichlorobiphenyls, the predicted solubility varies between 1.1 p.p.m. (2,4-dichlorobiphenyl) and 0.64 p.p.m. (2,4'-dichlorobiphenyl). Solubility is known to be a function of the total organic carbon (TOC) present in the aqueous phase. It is well known that they adsorb well to soils and sediments, with specific adsorption capacity governed by the particulate surface area and the total organic carbon content of the soil or sediment (Griffin and Chian, 1979; Lee *et al.*, 1979)

Table 1. Predicted solubilities of individual PCB congeners in water (data adapted from Dunnivant *et al.*, 1992)

PCB Congener	Solubility (p.p.m.)
2-MCB	5.8
3-MCB	4.8
4-MCB	0.76
2,2'-DCB	1.3
2,3-DCB	1.4
2,4-DCB	1.1
2,4'-DCB	0.64
2,5-DCB	1.5
2,6-DCB	2.6
3,3'-DCB	0.71
3,4-DCB	0.40
3,5-DCB	0.65
4,4'-DCB	0.02

ENVIRONMENTAL ACCUMULATION

It is estimated that 10×10^7 kg of PCBs still reside in the biosphere in the soil, air, rivers and waste streams, and lipoidal components of plants and animals (Boyle *et al.*, 1992). The average concentration of PCBs throughout US soils is 100 p.p.b., although this hides the true picture, with only a few p.p.b. present in rural areas, and up to 100 p.p.m. present in urban areas (Pal, Weber and Overcash, 1980). Of PCB-contaminated sludges disposed of in Indiana, only 15% exceeded the Environmental Protection Agency (EPA) 50 p.p.m. guidelines (Bergh and People, 1977). While this may seem to undermine the problem of PCB contamination, it must be realized that hot-spots of PCB contamination are a central concern and, due to the estimated capacitor life span of

20 years, incidents are going to increase. The PCB Spill Policy has data relating to frequency and magnitude of PCB transformer leaks; it has determined that 3.3% of such transformers will leak or spill every year and each spill comprises 66 lb of PCB (EPA Federal Register, Vol. 52). Furthermore, the EPA has recently determined that 'of the 1.6 million PCB capacitors in use in electrical substations (in US), EPA expects that 12 000 leak each year . . . Unlike PCB transformer spills, the majority of PCB capacitor spills involve the violent rupture of the capacitor and the spraying of PCBs . . . Available data indicates for over 80 percent of capacitor spills, PCBs are distributed as far as 11 feet from the centre of the spill.' (EPA Federal Register, Vol. 52).

The Organization for Economic Co-operation and Development (OECD, 1987) estimated that there were 12 000 tonnes of PCB waste in the UK. This figure is agreed in the revised Waste Management Paper No. 6 (Department of the Environment, 1994) and it is suggested that 4000–4500 tonnes have already been destroyed. Moreover, the most recent study into the deposition of PCBs, using archived vegetation samples covering the period 1965–89, showed a significant reduction of air contamination of PCBs. However, it was also noted that, in soils, the rate of deposition exceeds that of removal (Jones *et al.*, 1992).

TOXICITY

The toxicity of PCBs is a subject of much debate and research, and is related to many of the aforementioned factors which ultimately govern bioavailability. It is well known that PCBs bind covalently to DNA, both *in vivo* and *in vitro*. Discernible mutagenic activity is shown in rodents when PCBs containing 57% chlorine are studied. However, the acute toxicity may not reside in the PCBs *per se* but may be due to either the manufacture or transformation of by-products from use (Stone, 1992). Such supertoxic species have largely evaded detection due to the difficulties in analysing them (Storr-Hansen, 1991). However, it is well known that co-planarity of the chlorinated biphenyl is considered very important in acute toxicity (Huang and Gibson, 1992). This occurs in congeners which have no substitution at either of the *ortho* carbons (2,6 and 2',6'), allowing the molecule to assume a co-planar configuration. Chronic human toxicity has been well chronicled, with chloracne, pruritus, liver problems and reduced vital capacity of the respiratory system amongst the common symptoms (Kimbrough, 1980). Irrespective of the symptoms and whatever the modes of toxicity, the hydrophobicity of the congeners facilitate their accumulation in biological systems and potentiate the concern about their ultimate fate.

Bioaccumulation has been studied in bacteria (Hamdy and Lin, 1983), yeast (Cole, Reichart and Button, 1979) and fungi, but perhaps the most dramatic evidence of bioaccumulation comes from studies using oysters and fish. Aroclor 1254 (1 mg^{-1l}) has been shown to be concentrated 101 000-fold in oysters (Lowe *et al.*, 1972). Fish from the Hudson River, where GEC manufactured transformers and purchased 15% of total US PCB output between 1966 and 1974, were able to concentrate Aroclor 1016 15 000-fold in only 14 days (Skea *et al.*, 1979). Bioaccumulation, like adsorption, is probably dependent upon the log P of the congeners (a measure of their hydrophobicity) and the steric effect coefficient (SEC), which is an indicator of degree and position of chlorine substitution. Obviously, this has more pronounced deleterious effects as you ascend the food chain, and the ultimate fate of the PCB is largely dependent upon the potential of the organism to detoxify the compounds. This has been dealt with in relation to human toxicity in the submission made to the Welsh Affairs Committee (Johnston and Simmonds, 1989). Of the information currently available, detoxification requires two adjacent unsubstituted carbon atoms for the formation of the intermediate arene oxide (Matthews and Anderson, 1975).

ENVIRONMENTAL LEGISLATION REGARDING CONTAMINATED LAND AND WATER

PCBs are, along with dioxins, some of the worst of all contaminants. This is not necessarily directly related to human toxicity, although it may be regarded as such for dioxins. The main problems stem from the course of remedial action which must be taken when such pollutants are present. PCBs are extremely hydrophobic molecules and it may be agued that their health risk is not significant in terms of the possibility of getting into the water course.

The use of PCBs in the UK is restricted through the Control of Pollution (Supply and Use of Injurious Substances) Regulations 1986 SI No. 902, which defined PCBs at the 50 p.p.m. concentration level (as amended by the Environmental Protection [Controls on Injurious Substances] Regulations 1992 SI No. 31). PCBs are dealt with under the legislation on waste and special waste contained within the Control of Pollution Act (1974), Control of Pollution (Special Waste) Regulations (1980) and the Control of Pollution (Licensing of Wastes Disposal Regulations) (1976) (OECD, 1987). Additionally, a technical memorandum and Code of Practice for the disposal of PCBs and PCB-contaminated wastes has been published and is due for revision and update in 1994 (Department of the Environment, 1994). This forms part of the measures necessary to implement the European Community Directive for the disposal of PCBs and PCTs (76/403/EEC).

Legislation regarding contaminated land in the UK is extremely patchy for several reasons. Remediation is inextricably linked with end-use, such that final use as a children's playground demands more clean-up than if it were to be used as a car park. Therefore, it might be expected that trigger values have been assigned to such environmental pollutants, but this is not the case. Unlike The Netherlands (Moen, 1988), which favours multifunctionality (the remediation of land permitting *any* end-use), it is not clear what values must be attained in the UK. A value of 10 p.p.m. is suggested in Waste Management Paper No. 6 (1994) although the Department of Environment's Interdepartmental Committee on the Reclamation of Contaminated Land (ICRCL) does not have clear trigger values for PCBs related to end-use (ICRCL, 1987). Although this may seem an oversight, it must be realized that the setting of trigger values is somewhat arbitrary at the present time. Ideally, one would hope to have as full a toxicological profile as possible, but this is itself dependent upon many variables, including soil type, analytical systems employed and the presence of co-contaminants.

Treatment methods

INCINERATION

The incineration of PCBs and other classes of chemicals is an established disposal technique. They are destroyed by high-temperature thermal oxidation. In the US the Toxic Substances Control Act (TSCA) requires a DRE (destruction and removal efficiency) of 99.9999% for thermal treatment systems. In the UK, the guidelines are set by the Department of the Environment through their Waste Management Paper No. 6. High temperatures and sufficient residence time are required to conform to the guidelines and destroy all organic components. Strict adherence to the guidelines is required to ensure that thermal oxidation to the even more problematic dioxins does not result.

Apart from liquid PCB waste, most decontaminations are probably point-source depositions in the soil. If these are large spillages, the cost of removal and disposal will be increased by the transportation costs to the incinerator. Unlike in the US, no transportable incinerators able to remove PCBs are operational in the UK. Presently there are three incinerators able to deal with PCBs in the UK. The largest of these is operated by Rechem in Panteg, South Wales. Disposal costs are estimated at approximately £1000 per tonne with the incinerator able to destroy 100–400 kg PCB per hour. The operating temperature of the kiln is 950°C, with an afterburner temperature of 1100°C. In the US, satisfactory destruction of PCBs is suggested to be achieved by incineration for 2 seconds at 1100°C with 3% excess oxygen. The incinerator (8 m × 4 m) operated by Rechem fulfils these requirements, operates with 6% excess oxygen and gives a final concentration of PCB in the stack of 2 $\mu g\ m^{-3}$, well below the occupational exposure limit of 1 $mg\ m^{-3}$ (Health and Safety Executive, 1993). It was estimated that manufacturing waste arisings were 400 tonnes per year when the Department of the Environment produced Waste Management Paper No. 6 (1976). Within the past few years there has been considerable increase in the UK capacity to deal with chlorinated wastes and the total available capacity is now approximately 7000 tonnes per year (Waste Management Paper No. 6; Department of the Environment, 1994).

CHEMICAL METHODS

Since 1978 a family of alkali metal polyethylene glycolate (APEG) reagents have been developed which are reported to dechlorinate PCBs effectively and rapidly in soils and liquids. Reagents may be prepared as either the potassium or sodium derivatives. The potassium derivative is reportedly more effective for PCBs. The reagent is prepared by reacting a polyethylene glycol with potassium hydroxide. The resulting alkoxide reacts with a chlorine on the aryl ring, most efficiently at 60–100°C, to produce an ether and potassium chloride. Substitution of the chlorine with the ether-linked polyethylene glycol detoxifies the PCB (Amend and Lederman, 1992).

This would seem to oversimplify the problems of PCB degradation as commercial mixtures may be heavily substituted and not all chlorine atoms would be expected to be removed using this methodology. Moreover, addition of the polyethylene glycol will simply increase the water solubility of the derivatized PCB, aiding its removal from the contaminated media but still leaving an ether-derivatized chloroaromatic. Despite this, the method has been implemented in the US to treat soils contaminated with up to 3500 p.p.m. PCBs. It is not known whether the soils are destroyed in the process, but the removal of PCBs is of the order of 99%.

The problems of bioavailability of the PCBs have been alluded to and it is possible that this sort of methodology could be a useful initial step for other processes. Several vendors offer this service (Kokozka and Flood, 1985) but, to the authors' knowledge, it is not available for use in the UK.

BIOLOGICAL METHODS

Historically, biological treatment has been favoured by industry for dilute aqueous solutions. Although incineration is cheaper for particular waste streams, and this is largely based upon the heating value, no environmental considerations are presently made. If a waste has an organic content of over 10%, it is usual for it to be incinerated. However, waste treatment costs are based on weight or volume and not on the actual waste content. Therefore, dilution of a waste may permit its treatment by biological systems, but this is not usually favoured because of the increased cost implications. Despite this, it has been estimated that if there is a choice between incineration and biological treatment, biodegradation costs are perhaps one-half to one-tenth those of incineration (League, 1990). Moreover, it was recognized in the First Report on Contaminated Land in the UK that bioremediation is 'the only true clean-up technique in use in the UK' (House of Commons, 1990).

The methodology used in biodegradation has developed rapidly since micro-organisms were first used for wastewater treatment approximately 100 years ago (Visscher and Brinkman, 1989). Although bioengineering has improved dramatically over the years, the number of reactor subgroups that exist are still relatively small. Up until 1980 the aerobic reactor option was much favoured but more recent research has switched focus to the better potential offered by anaerobic systems.

Water is the medium routinely used in biological treatment methodologies, whether it be in activated-sludge systems, anaerobic digesters, trickling filters or oxygenated lagoons. Such systems have been investigated for their potential to degrade xenobiotics. Irrespective of the system that is ultimately applied, it is the biodegradative capacity of the micro-organism that is the starting point for the development of a process.

Traditional biodegradation relied on the use of complex uncharacterized communities of organisms. This bioenrichment approach is still applied in most processes but it has sometimes been supplemented with a policy of bioaugmentation. This process differs from the former by the addition of one or more microbial strains which are specifically able to degrade a number of chemical components. Despite these additions, there is still overlap between enrichment and augmentation because the ultimate goal is mineralization, a process requiring a diversity of phenotypes. With this in mind, several processes have been formulated which use different units in series, or recycle the waste (Ng, Yap and Sivadas, 1989; Shivaraman and Parhad, 1990).

The immobilization of micro-organisms has facilitated the development of many processes. Such biofilm reactors include trickling filters (Alkins, 1981), submerged filters, rotating discs (Howard, 1990) and fluidized beds (Dillon, 1981). All of these are well suited to aerobic application, whereas downflow and upflow filters are more amenable to anaerobic operation (Krumme and Boyd, 1988). Much work has also concentrated on the support materials used in these operations. Activated sludge has little support and consists of acclimated organisms in a non-defined matrix. Use of a trickling filter allows the adhesion of the micro-organism to a variety of supports. A very thin film of waste is sprayed on to the filter bed and this facilitates good aeration and aerobic degradation.

The problems with bioremediation of contaminated soils are different in terms of the matrix but the problem of chemical recalcitrance is still pre-eminent. Many laboratory experiments have shown the degradation of target pollutants in soil microcosms and in field lysimeters (Mikesell and Boyd, 1988; Hill *et al.*, 1989; Havel and Reineke, 1993b). Such experiments have explored both bioaugmentation and bioenrichment as feasible options for field application. Additionally, freeze-dried bacterial cultures which are designed to degrade specific waste types are commercially available (Haley *et al.*, 1990). On the larger scale, one of the EPA site programme demonstrations includes a liquid/solid contact digestion system in which either sludges or soils containing between 2000 and 800 000 p.p.m. total organic carbon, from a wood-preserving facility, are degraded biologically in several stages. In this process, an acclimated bacterial seed is used along with the option to incorporate emulsifiers (EPA, 1989).

In the degradation of PCBs, the carbon–chlorine bond is the target. Although halocarbons exist in nature, they are not common (Suida and DeBarnardis, 1973; Lovelock, 1975; King, 1986). Despite this, it has been shown repeatedly that chlorinated xenobiotics may be broken down by microbial populations. The advancement of carbon–chlorine degradation has not been facilitated by the usual approach of indexing degradative pathways simply against the disappearance of the parent compound. This is best illustrated by studies of chlorinated solvents in ground water. Although such solvents are degraded by micro-organisms (Galli and McCarty, 1989a,b; Oldenhuis *et al.*, 1989) degradation may result in the formation of recalcitrant chloro-organics such as trichloroethylene and vinyl chloride.

The biodegradation of many haloaromatic species has been investigated and reviewed by Hardman (1991). More specifically, halogenated aliphatics (Vogel, Criddle and McCarty, 1987), substituted benzenes (Goulding, Gullin and Bolton, 1988), pentachlorophenol (Topp and Hanson, 1990), PCBs (Abramowicz, 1990; Boyle *et al.*, 1992), 2,4-dichlorophenoxyacetic acid (2,4-D) (Kelly, Hallberg and Tuovinen, 1989) and chloroguaiacols, chloroveratroles and chlorocatechols (Neilson *et al.*, 1987) have all been studied and reviewed. Despite the vast array of studies undertaken, it must be stated that the problems encountered in the degradation of low-level waste/chloroaromatic compounds are related to questions of bioavailability and the build-up of intermediates.

Obviously, metabolic versatility is a key attribute when isolating organisms able to remediate a particular waste stream. To this end, pseudomonads are often cited as remarkably catholic organisms. This has been further embraced by the work of Reineke and colleagues (Hartmann *et al.*, 1989; Brunsbach and Reineke, 1993; Havel and Reineke, 1993b). Other organisms attracting a great deal of attention are the white rot fungi (most notaby *Phanerochaete chrysosporium*). These organisms have been reviewed extensively by Bumpus *et al.* (1990) and their use in appropriate bioreactors for hazardous waste remediation has been chronicled (Lewandowski, Armenants and Pak, 1990).

Analysis of PCBs

As has already been stated, PCBs were sold as mixtures of congeners. The largest manufacturer was Monsanto with their Aroclor range. The latter were sold and characterized according to the percentage of chlorine (by weight). The most popular was Aroclor 1242 which contains 42% chlorine. Other products contained as much as 60% or as little as 21% chlorine. As many as 70–80 congeners may exist within these mixtures and these differ only in the number and position of the chlorine substituents. These factors hinder the complete elucidation of metabolic events that may occur in aerobic or anaerobic systems. This is further compounded by the heterogeneous nature of the matrices from which many of these PCBs are extracted. New developments in thermal desorption chromatography permit the PCBs to be concentrated on to selective adsorbents using steam distillation and then to be thermally desorbed on to the analytical column (Brown *et al.*, 1991). This facilitates clean-up and reproducibility but is not commonly used at the present time.

When analysing the metabolic events occurring in the degradation of PCBs by micro-organisms, two main parameters are usually quantified. Both the PCBs and the released chloride are usually measured. Despite this, the problems inherent in the analysis of PCBs have yet to be completely overcome. Two systems, gas chromatography (GC) and high-pressure liquid chromatography (HPLC), are commonly employed for the routine analysis of PCBs and these have recently been supplemented with the wider use of gas chromatography–mass spectroscopy (GCMS) for both routine and in depth analysis.

Gas chromatography is by far the most common method of analysis used at the present time. The development of capillary columns has facilitated the resolution of complex chemical mixtures including the PCBs. However, the major benefit in the use of GC is the ability to couple the increased resolving power with the tried and tested detectors that are available. Two main detectors, excluding mass spectrometry (MS), are routinely used and each has advantages and disadvantages.

By far the most common general-purpose detector is that employing flame ionization (FID). The response from this detector is related to the thermally excited ions produced in a hydrogen/air flame and is directly related to the number of burnable carbon atoms. A fundamental problem with the use of the FID is the necessity to undertake extensive clean-up when sampling directly from the environment. This is necessary as soils and sediments contain many organic species that will give a response in the FID and are likely to give erroneous results when analysing the full spectrum of PCB congeners. Although the FID gives a response related to the number of carbon atoms in the nucleus, and hence gives a theoretically equal response per biphenyl molecule, it has been shown that the sensitivity is a thousand-fold less than either the electron capture (EC) or Hall electrolytic conductivity (HEC) methods of detection.

The most widely used method of detection of PCBs, or halogenated compounds in general, is the electron capture detector (ECD). It is specific for sample constituents which exhibit an affinity for electrons and a low response to other types of sample material. It contains a radioactive source (^{63}Ni) which ionizes the carrier gas and the electrons are collected at the detector electrode. Obviously, if the chemical species under investigation has an affinity for electrons it will reduce those being detected at the collector electrode. It differs from the Hall electrolytic conductivity detector (HECD) which measures the HCl formed from the parent molecule. The HECD is not widely used but does give a response directly related to the number of chlorine atoms present on the parent molecule. This is not true of the ECD which gives a disproportionate response that is not related to the degree of chlorination. Moreover, the ECD response is also a function of the position of the chlorine atoms, which is not predictable.

Despite the availability of detectors which are sensitive to PCBs there are several fundamental problems:

1. What to choose as the internal standard when using ECD.
2. How to resolve co-eluting peaks.
3. Distinguishing contaminants from true PCBs.
4. Identifying the non-*ortho* substituted, and hence the most toxic, congeners.

The choice of reference standards is of crucial importance. The disproportionality in ECD response is quite marked over the congener range, as illustrated in *Figure 2*. Furthermore, the choice of internal standard is restricted to those 'like-compounds' which are not metabolized by the organism or consortia under investigation. Octachloronaphthalene has been used extensively but gives a marked difference in response compared to the mono- and dichlorinated biphenyls. Currently, the pentachlorinated biphenyl 246-24 is used widely because it is not metabolized, is absent from the commercial mixtures, and does not co-elute with any other congener. This allows it to be used without interference from constituents, in the knowledge it will not be degraded, but does not circumvent the problems associated with disproportionality of response.

In reality, the problems of variable congener response and co-elution can only be tackled by establishing the linearity of response for each congener or by approximation using a multi-point calibration curve and internal standards. Fava, Cinti and Marchetti (1993) have used this approach with a four-point calibration plus the use of octachloronaphthalene and lindane as internal standards. Additionally, the efficiency of soil extraction was assessed using PCB 75 (246-4). The standard methodology employed in clean-up and analysis is that adopted by Quensen, Boyd and Tiedje (1990). Soils or sediments are extracted in the presence of octachloronaphthalene and are analysed relative to an autoclaved control. In this case it was an anaerobic dechlorination that was studied and was assessed by the removal of *meta* and *para* chlorines. The assumption was made that co-eluting isomers occur in equal proportions, and all co-eluting congeners increase or decrease to the same extent. The effect of this assumption was evaluated using two worst case scenarios and was shown to result in a maximum 10% error when studying heavily chlorinated mixtures (e.g. Aroclor 1260).

Although quantification of individual congeners relies on certain assumptions being made, it is well established that the order of congener elution and composition of the mixtures is fairly reproducible. *Table 2* outlines a selection of the congener profiles and peak assignments for Aroclor 1242 that have recently appeared in the literature. It can be seen that, apart from that outlined by May *et al.* (1992), they are fairly consistent, with co-eluting peaks largely unchanged. The surprising omission of the trichlorinated biphenyl 25-2, which forms 9% (w/w) of 1242, is unexplained. Despite this, the literature available on PCB traces permits a predictive determination of the qualitative presence of individual congeners without a requirement for the use of individual standards in the first instance. Obviously, complete quantification of the events occurring when PCBs are degraded does require individual standards, most of

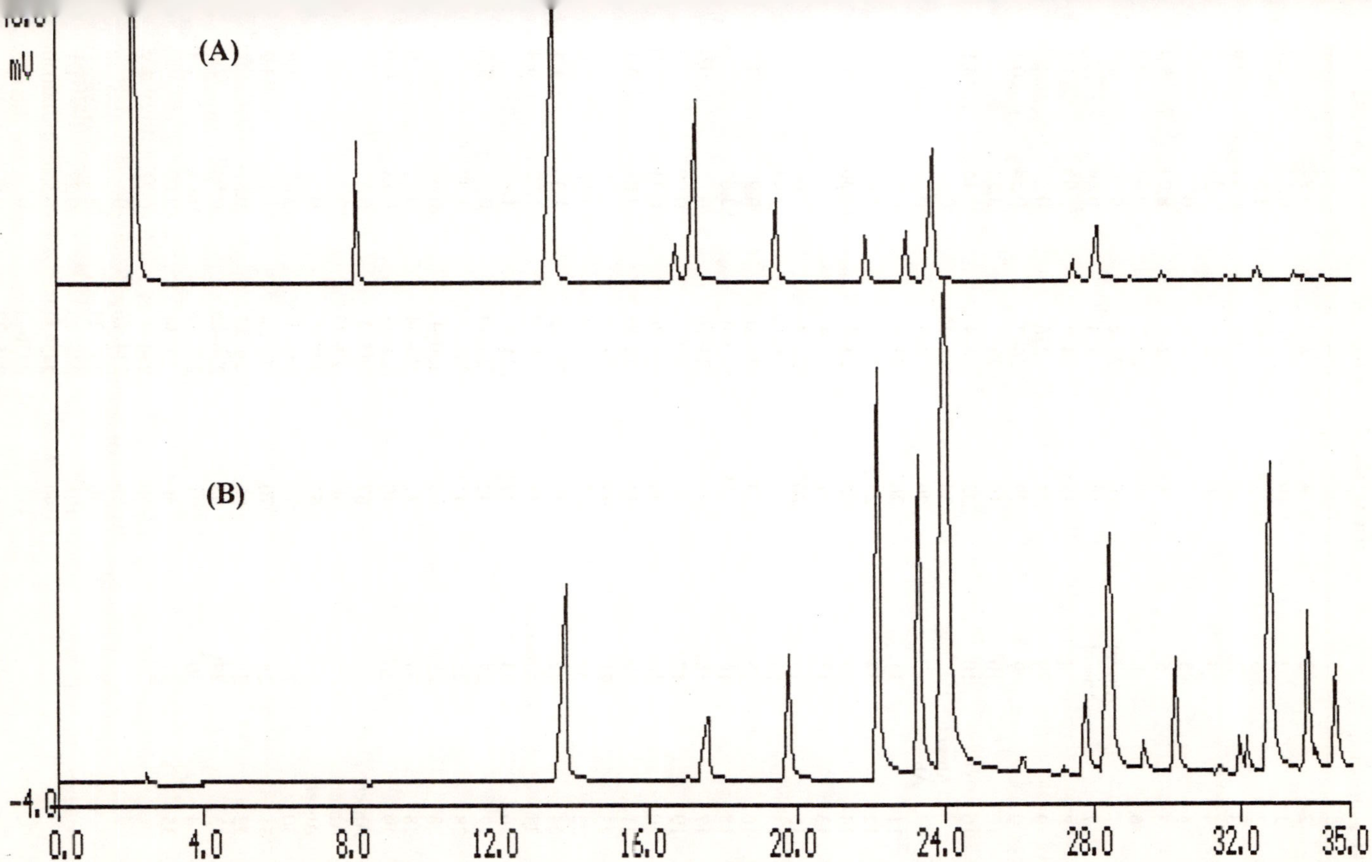

Figure 2. Gas chromatographic analysis of Aroclor 1221 using (A) flame ionization detector (FID) and (B) an electron capture detector (ECD). Using FID analysis, the relative peak areas for the different congeners reflect the relative proportions of the congeners in the mixture. The ECD does not respond to biphenyl, and generally the response increases with increasing chlorination. Note that the trace contaminants in the mixture, eluting between 29 and 35 minutes, produce a very strong signal with the ECD.

Table 2. Summary of peak assignments given in the literature

A Pk assignation			B IUPAC No.	C Congener	D % in 1242	
1	2	2	4*	2–2	3.99	
1	2	2	10	26	0.13	2.7
2	3	3	7*	24	1.04	
2	3	3	9	25	0.31	0.9
3	4	4	6	2–3	1.24	1.2
4	5	5	5	23	0	
4	5	5	8*	2–4	8.97	5.8
5	6	6	19	26–2	0.97	0.9
6	7	Absent	18	25–2	9.36	8.7
7	7	Absent	15	4–4	0.99	
7	8	8	18*	24–2	2.92	4.3
8	9	9	24	236	0	
8	9	9	27	26–3	0.54	0.7
9	10	10	16	23–2	3.25	
9	10	10	32	26–4	2.15	4.8
10	Absent		34	2–35	0	
11	Absent		29	245	0	
12	11	11	26	25–3	0.55	1.1
13	12	12	25	24–3	1.68	7.0
14	14	13	28*	24–4	13.30	6.5
14	13	13	31	25–4	4.53	7.0
15	15	15	20*	23–3	3.64	
15	15	15	33*	2–3.4	2.83	
15	16	15	53	25–26	0.97	3.3
16	18	16	22*	23–4	2.64	
16	17	16	51	24–26	0	0.9
17	19	17	45	236–2	0.90	0.5
18	20	18	46	23–26	0.31	0.31
19	21	19	52	25–25	4.08	3.1
20	23	21	47	24–24	1.65	1.1
20	22	20	49	24–25	3.28	2.6
21	24	22	48	245–2	1.33	1.4
22	25	Absent	44	23–25	1.08	3.4
23	26	24	37*	34–4	1.62	
23	27	24	42	23–24	0	3.7
24	28	25	41*	234–2	1.67	
24	28	25	64	236–4	0	
24	28	25	71	26–34	0	3.0
25	29	26	40	23–23	0.15	0.9
26	Absent		67	245–3	0	
27	Absent		58	23–35	0	
27	Absent		63	235–4	0	
28	30	27	74	245–4	2.02	1.9
29	31	28	70	25–34	1.11	3.9
30	32	29	66	24–34	0.81	
30	33	29	95	236–25	0.53	4.8
31	34	30	91	236–24	Trace	1.3
32	35	31	56*	23–34	0.60	
32			60	234–4	0.21	2.9
32			89	234–26	0	
33			90	235–24	0	
33			101	245–25	0.27	1.1
34			99	245–24	0.55	0.55
35			112	2356–3	0	
36			83	235–23	0	
37			97	245–23	0	
38			87	234–25	0.09	1.0

39	85	235–24	0	0.7
40	110	236–34	0	1.1
41	82	234–23	0	0.7
41	151	2356–25	0	
42	106	2345–3	0	
42	123	345–24	0.36	
42	149	236–245	0	0.7
43	105	234–34	0.25	1.0
44	138	234–245	0	
44	160	23456–3	0	
44	163	2356–34	0	

Column A refers to the elution order. Each of the digits relates to the peak assigned in each of three references. The first figure is the order assigned by Barriault and Sylvestre (1993), the second is the order assigned by Nies and Vogel (1991), the third is the order assigned by May *et al.* (1992). Column B refers to the IUPAC no. assigned to each congener with the position of the chlorines outlined in column C (* = major peak). Column D indicates the % (wt) of each congener in Aroclor 1242. The first figures are those of Kimbrough (1980) and the second figures are those of Lajoie *et al.* (1993). Shading relates to co-elution of congeners according to Lajoie *et al.* (1993).

which are commercially available. Of the intermediates expected to be formed from polychlorinated biphenyls, chlorobenzoates have been detected using GC after derivatization (Gerritse, van der Wande and Gottschal, 1992).

Liquid chromatography (particularly HPLC) has been used extensively for the detection of PCBs, chlorobenzoates and chlorocatechols. Resolution of the congeners is achieved on the basis of column selection and solvent system. The most common detection system employs an ultraviolet (UV) detector to analyse the signal from the aromatic nucleus. Although sensitive, the samples must be clean and it does not offer the same predictable resolution outlined for gas chromatography. The main benefit is in the detection of chlorobenzoates and chlorocatechols, but problems of co-elution persist with these groups of compounds.

Chloride release is an excellent indicator of PCB degradation but requires a high degree of sensitivity and little background interference. Many methods are available, ranging from simple colorimetric systems to complex liquid chromatography systems. The methods are surveyed in *Table 3* and each has advantages and disadvantages. None of the methods outlined should be used in isolation, but when coupled with GC or liquid chromatography (LC) they are invaluable in the elucidation of metabolic pathways involving chloro-organics.

The definitive identification of individual PCB congeners within mixtures of congeners would aid considerably the quest for methods of eradicating PCBs. As

Table 3. Summary of some of the methods available for the detection of chloride

Basis of assay	Sensitivity
End-point titration with silver ions[1]	0.1–100 p.p.m.
Atomic adsorption[1]	50 p.p.b.
Modified Mohr titration[2]	1.5–15 p.p.m.
X-ray fluorescence spectrophotometry[2]	40 p.p.b.
Ion chromatography[3]	40 p.p.b.
Ion chromatography with electrical conductivity detector[3]	1 p.p.b.
Colorimetric[4]	350 p.p.m.

Sources: 1 *Analytical Chemistry* **47**, 343 (1975); 2 *Analytical Chemistry* **63**, 318 (1991); 3 *Analytical Chemistry* **61**, 284 (1989); 4 *Analytical Chemistry* **29**, 241 (1957).

outlined above, all of the analytical methods are deficient in some way. To this end, mass spectrometry, when coupled to gas chromatography (GCMS), has aided PCB research. Full use is often made of the isotopic distribution of chlorine (75.53% ^{35}Cl and 24.47% ^{37}Cl) in identifying ion clusters corresponding to the degree of chlorination. Despite this, GCMS should not be thought of as an analytical tool that has no drawbacks. Irrespective of the detector that is used, the resolution of the individual congeners determines the success of the methodology used thereafter. Therefore, although GCMS may definively identify isolated congeners that do not co-elute, it is very difficult to identify those that do. Electron impact (EI) spectra of PCBs have been studied extensively by Safe and Hutzinger (1971, 1972) and it has been shown that co-eluting peaks have fragmentation patterns related to the expulsion of chlorine atoms. Therefore, if a peak consisted of three co-eluting congeners and these differed in the number of chlorine atoms they possessed, it would be difficult to distinguish them apart from the most chlorinated congener. Cairns and Siegmund (1981) further developed mass spectrometry with PCBs, using methane to protonate the molecular species. Only a very small percentage of ions (<2%) corresponded to the loss of HCl from the molecular species, causing a problem of identification when investigating the lower homologues. This technique permits instant analysis of each PCB mixture on the basis of the distribution of molecular weights. Therefore, it is possible to specify a particular *m/z* value and get a profile within any eluting peak. This permits the degradation of each congener to be assessed so long as the peak compositions are unequivocal. As with other methods, this again suffers from a disproportionality of response across the congener range. Highly chlorinated congeners have less sites available for protonation and sensitivity is therefore reduced.

In conclusion, the quantitative analysis of PCBs is a fundamental obstacle which must be overcome if their metabolism is to be fully understood. This is best illustrated by an American study which investigated the accuracy and precision of PCB analysis by accredited laboratories. For a prepared sample at 100 p.p.m. PCB, relatively high by environmental standards, 30% of laboratories out of a total of 120 would have classified the waste as not special, i.e. < 50 p.p.m. (Milby, Miller and Forrester, 1985).

The choice of technique, reference standard and clean-up method all have a bearing on the level of degradation which is finally reported. The advent of capillary columns has facilitated the resolution of many PCB congeners, but there are still many problems left to solve. Mullin *et al.* (1984) published an excellent paper which reported the separation of 187 PCB congeners using a 50 m capillary column. Despite this, it was reported that 11 pairs of compounds were not fully resolved and, although general trends existed within the congeners, it was not possible to predict separations. In general, relative retention time (RRT) increased with increasing chlorine content, but within each isomer group there were large variations in the magnitude of the GC response. Only when all congeners are resolved will it be possible to standardize conditions for the identification and quantification of any of the PCBs that may be present in environmental samples.

Bioavailability

The bioavailability of a compound may be defined as that fraction of the compound which may be acted upon by a biological system. For the purpose of this chapter we will restrict the discussion to that fraction which may be available to micro-organisms. Ordinarily, microbial systems are investigated in the aqueous phase with water-soluble, readily available substrates. The limitations in such a system are the transport of the substrates across the biological membrane and their utilization by either inducible or constitutive enzyme systems.

PCBs are extremely hydrophobic molecules which adsorb strongly to virtually any type of matrix. When studied in the laboratory, either as single congeners or as mixtures, they are either supplied directly to the aqueous phase and allowed to equilibrate, or are added as a concentrated stock solution in an appropriate water-miscible solvent (acetone or methanol). Although both of these techniques are accepted as standard, each gives rise to different concentrations and compositions in the aqueous phase. Murphy, Mullin and Meyer (1987) conducted equilibration studies with PCBs after allowing them to equilibrate in the water phase, but they showed that preparation by addition of solvent results in an altered composition. Each of the components of a PCB mixture will become soluble in the aqueous phase until they are saturated. Therefore, if added as a solvent solution, the least soluble components will become saturated first and give rise to a second phase composed of insoluble PCBs. This will retard the partitioning into the aqueous phase and give rise to aqueous solubilities well below those reported by equilibration. Furthermore, the mixtures in the environment are present as sub-cooled liquids and are extremely temperature dependent. Therefore, even in the simplest laboratory study into PCB degradation, toxicity and solubility of mixtures may be problematic. As an indicator, data are available on solubility, vapour pressure and Henry's Law constants for the whole congener range (Murphy, Mullin and Meyer, 1987; Dunnivant *et al.*, 1992).

The main impetus behind the study of PCBs is their recalcitrance and toxicity in the environment. In natural water systems, toxic effects arise because of their accumulation by many biological systems. Movement into the atmosphere has not received significant attention, but data from Diskey and Andren (1981) and Rice, Eadie and Ertsfeld (1983) have suggested that in Lake Michigan there appears to be transfer from the water to the atmosphere. This has given rise to the conclusion that the atmosphere, rather than the sediments, is the major sink for PCBs in Lake Michigan and similar bodies of water (Murphy, Mullin and Meyer, 1987).

Irrespective of the above, it is generally accepted that the greatest problem associated with PCB contamination is due to soil and sediment contamination. Sorption of organics has attracted much attention but little definitive experimentation. It is widely accepted that sorption limits the availability of compounds for degradation, but studies in the field have not been unanimous in their conclusions (Gordon and Millero, 1985; Ogram *et al.*, 1985; Rijnaarts *et al.*, 1990). This is entirely understandable when one considers the multiplicity of variables that govern bioavailability. In the broadest context, and without alluding to the factors already mentioned, bioavailability must be governed by the physico-chemical structure of the sorbate, the nature of the sorbent, the mechanism of sorption, the factors governing equilibration and the properties of the degradative organisms.

PCBs have a high affinity for suspended solids, which is increased significantly in the presence of increased amounts of organic carbon (Hamelink, Waybrant and Ball, 1971). PCBs are usually associated with soils or sediments in a soil–water system and the sorption of PCBs by various earth materials is shown in *Table 4* (Griffin and

Table 4. Effect of surface area (SA), total organic carbon (TOC) and type of earth material on the sorption of PCBs

Sorbent	Sorption constant (K)	TOC (%)	SA (m^2 g^{-1})
Ottawa silica sand	22	<0.1	0.4
Montmorillonite clay	172	0.93	21.1
Montmorillonite clay (LTA)	145	0.13	20.2
Catlin silt loam	532	4.73	26.5
Catlin (6 h) LTA	472	4.37	25.4
Catlin (12 h) LTA	310	3.64	24.5
Catlin (336 h) LTA	239	1.84	23.8
Medium temp. coal char 650°C	1938	74.04	253
Medium temp. coal char 650°C (LTA)	1432	64.00	214
High temp. coal char 980°C	1220	76.62	44
High temp. coal char 980°C (LTA)	1174	32.14	120

LTA denotes low-temperature ashed samples. Data from Griffin and Chian (1979).

Chian, 1979). The amount of PCBs sorbed by the earth materials could be related to the equilibrium concentration of PCBs and could be described by the simple linear relation:

$$x/m = KC$$

where x = micrograms of compound sorbed, m = weight of adsorbent (g), C = the equilibrium concentration of PCBs in solution (μg ml^{-1}) and K = sorption constant (ml g^{-1}). It can be clearly seen that the sorption of PCBs is directly related to the total organic carbon and that low-temperature ashing significantly reduces sorbed PCBs. As might be expected, the more hydrophobic, heavily chlorinated congeners adsorb much more than the lighter, chlorinated congeners (Haque and Schmedding, 1976; Lee *et al.*, 1979). Furthermore, the work of Lee *et al.* (1979) established a direct relationship between TOC and the sorption constant (K) as follows:

$$K = 255 + 18.5 \text{ TOC}.$$

This was further developed to include surface area (SA) as a variable, giving rise to the linear regression relation of:

$$K = 188 + 3.36 \text{ SA} + 11.4 \text{ TOC}.$$

Aroclor 1242 was also studied and the sorption constant (K_{OC}) normalized for the organic carbon content, had a value of 10 725. Moreover, it was possible to express the linear regression relationship as:

$$K = 12.43 + 4.06 \text{ SA} + 73.45 \text{ TOC}.$$

It can be seen that, of the two coefficients (SA and TOC), it is the total organic carbon content of the soil that largely governs sorption.

Chlorobenzoates and chlorocatechols, which are products of the biodegradation of PCBs, are far more soluble in water than the parent compounds. Solubility figures produced by Dolfing and Harrison (1992) suggest that, apart from 4-chlorobenzoate, all of the mono- and dichlorinated benzoates are soluble to levels > 2.5 mM. It might therefore be expected that sorption of these chemical species is insufficient to limit their bioavailability. Further data relating to the movement of these chemical species in different soils would be beneficial when investigating the mineralization of congener mixtures.

The movement of Aroclors in soils has been studied with respect to both the soil characteristics and the mobile phase. Using soil thin-layer chromatography (TLC), Griffin and Chian (1979) have shown that Aroclors 1242 and 1254 are virtually immobile when either water or landfill leachate is the solvent. However, if an organic solvent such as acetone, methanol, benzene or carbon tetrachloride is used, the whole congener range is eluted. Several studies have established how far PCBs move within soils of different types. It can be concluded from these studies that PCBs are not significantly leached by water. Furthermore, the distance migrated may be related to the clay and/or organic carbon content (Suzuki, 1975; Tucker, Litschgi and Mees, 1975; Moza, Weisberger and Klein, 1976).

All of the factors that have been outlined above pertain to either sorption or aqueous solublity. This permits the use of predictive techniques to ascertain the bioavailability of organic compounds. Despite this, soils are very heterogeneous environments and consist of aggregates which vary in size. Most organisms exist on the outside of aggregates and in the small pore spaces between them. Electron microscopy has facilitated the study of these microenvironments and it has been estimated that microorganisms occupy less than 1% of the total available pore space (Paul and Clark, 1989). Therefore, pore neck size and water content of the pore are also determinants of bioavailability. Furthermore, a recent study by Guerin and Boyd (1992) concluded that, all physical and physico-chemical factors aside, there are important organism-specific properties which make generalizations regarding the bioavailability of sorbed substrates inappropriate. It is therefore necessary to plan strategically with end-use being a major factor.

Microbial degradation of PCBs and their intermediates

The degradation of aromatic organochlorines occurs via four well-characterized mechanisms, as follows:

1. Oxygenolytic dechlorination

This occurs via dioxygenative attack on the aromatic nucleus and is dependent upon the position of the chlorine substituents. The oxygenolytic cleavage of the chlorine atom occurs fortuitously when both atoms of molecular oxygen are incorporated into the aromatic nucleus. PCBs are principally hydroxylated on the unsubstituted carbons at positions 2 and 3 and at unsubstituted 3,4 positions. In either case, the mechanism seems to require the presence of adjacent unchlorinated atoms.

2. Hydrolytic dechlorination

In this reaction the incorporation of a hydroxyl group leads to the concomitant release of a chlorine substituent. Unlike the above, the hydroxyl group is derived from water and the reaction is catalysed by a halidohydrolase-type dehalogenase. The requirement for water rather than molecular oxygen also permits this dechlorination to occur in anaerobic, denitrifying conditions.

3. Reductive dechlorination

Reductive dechlorination results in the removal of the chloride substituent with the concurrent addition of electrons to the molecule. Although it has been reported to occur in both aerobic and anaerobic environments, it has been shown to occur primarily in the latter.

4. Chloride release after cleavage of the aromatic ring

The present chapter will deal with all the mechanisms outlined above. Starting with anaerobic degradation, both anaerobic and aerobic degradation will be discussed. This will be related to the metabolism of the three primary classes of compounds involved in the degradation of polychlorinated biphenyls, namely PCBs, chlorobenzoates and chlorocatechols.

ANAEROBIC DEGRADATION

Polychlorinated biphenyls (PCBs)

Historically, it was anticipated that anaerobic environments were a sink for the more heavily chlorinated PCB congeners. Using both marine muds (Carey and Harvey, 1978) and silage (Fries, 1971) it was concluded that anaerobic environments did not contribute significantly to the degradation of PCBs. Since that time, many other workers have studied the anaerobic degradation of PCBs and have shown that this supposition is not valid. It must be noted that evidence of the anaerobic degradation of the biphenyl nucleus has not been demonstrated. Chen *et al.* (1988) indicated the production of $^{14}CO_2$ when Hudson River sediments were incubated anaerobically with ^{14}C-labelled monochlorobiphenyls. To the authors' knowledge, this has not been shown subsequently and has not been reported by other workers.

The role that anaerobes play in the degradation of PCBs was first shown by Brown *et al.* (1987). Again using Hudson River sediments, they showed that the congener distribution of a known PCB mixture altered radically over a long period of time. The primary finding was that highly chlorinated PCB congeners were extensively removed, giving rise to a much higher proportion of the lower-chlorinated congeners. In all sediments, the levels of 26–2, 26–3 and all dichlorinated biphenyls increased two- to six-fold, and the level of 2-chlorobiphenyl increased seven- to seventy-fold. These findings should be considered in the light of the disproportionality of congener response of the electron capture detector outlined earlier. Despite this, it is clear that the transformations observed specifically involved the removal of *meta* and *para* chlorines. Moreover, the central finding of this and subsequent work was that congener dechlorination is specific for the particular sediment from which it is derived. Three different groups have made advances in the study of anaerobic dechlorination of PCBs (Quensen *et al.*, Abramowicz *et al.* and Vogel *et al.*). Moreover, Quensen and co-workers have shown that sediment derived from the Hudson River dechlorinates a range of congeners in Aroclors 1242, 1248, 1254 and 1260. However, a consortium that is completely devoid of sediment and able to anaerobically dechlorinate PCBs has, to our knowledge, never been shown.

Four patterns of reductive dechlorination have previously been shown to occur in anaerobic sediments:

1. *meta* removal – shown in Hudson River, Sheboygan River, Lagoons and Silver Lake;
2. *para* removal – shown in Hudson River, Saginaw River;
3. *meta* and *para* removal – shown in Hudson River;
4. *ortho* removal – shown in unacclimated pond sediment.

It must be noted that the incidence of the first three classes far outweighs that of the fourth, which was only recently shown using 2356-tetrachlorobiphenyl (Van Dort and Bedard, 1991). The frequency of dechlorination (*meta* > *para* > *ortho*) follows that suggested by Tiedje *et al.* (1993) in their review of reductive dechlorination. The ability to distinguish between populations able to dechlorinate PCBs was first shown by Ye *et al.* (1992). Using Hudson River sediments, they demonstrated that heat- or ethanol-treated samples, for enrichment of spore formers, did not produce methanol and preferentially removed *meta* chlorines. Untreated samples removed both *meta* and *para* chlorines, showing that sediment treatment eliminated the population responsible for *para* dechlorination.

The range of congeners that may be degraded by the anaerobic populations cannot easily be predicted, although there appears to be some dispute between workers in the area. Quensen, Boyd and Tiedje (1990) suggested that dechlorination occurs only with up to heptachlorinated substitutions, whereas Abramowicz (1990) favoured the dechlorination of octachlorinated biphenyls. In our view, a threshold is unlikely to exist considering the phenotypic diversity of microbial populations outlined above. It is clear, however, that as the degree of chlorination increases, so does the probability of reductive dehalogenation. Moreover, as the proportion of mono- and dichlorinated biphenyls increase, the role of anaerobes is reduced.

From a toxicology standpoint, the dechlorination of *meta* and *para* chlorines is very advantageous. It will be recalled that the most toxic PCB congeners (34–34, 345–34 and 345–345) are those which are co-planar, possessing no *ortho* chlorines. These congeners are structurally similar to 2378-tetrachlorodibenzo-*p*-dioxin and exhibit dioxin-like toxicity. They are a central target for the anaerobic degradative mechanisms and consequently their removal leads to risk reduction. Those possessing *ortho* chlorines have not been extensively tested, but it is suggested in the review by Tiedje *et al.* (1993) that overall toxicity has been demonstrably reduced using both toxicity bioassays (Giesy, Jones and Mora, 1992, unpublished, quoted by Tiedje *et al.*, 1993) and by congener-specific tandem quadrapole mass spectrometry (Quensen *et al.*, 1992). However, despite the conclusion drawn by Tiedje *et al.* (1993), we believe more extensive toxicological testing is required in this area because it is quite clear that *ortho*-enriched congeners are a major component of commercial Aroclors, with Aroclor 1242 containing > 85% *ortho*-substituted congeners.

More than 50 organochlorine compounds have been shown to be reductively dechlorinated (Tiedje *et al.*, 1993). It is widely believed that anaerobes, from whichever environment, utilize chlorine as the terminal electron acceptor, with the addition of the electron to the aryl carbon–chlorine bond. Subsequently, the chloride is lost and hydrogen is abstracted. Of the potential primary electron donors (water, hydrogen or unspecified organic compound), Nies and Vogel (1991) provided evidence that the proton source was water. Using a deuterium label, the authors showed that the dechlorination of 23456-pentachlorinated biphenyl was concomitant with the incorporation of the label. However, other sources of reducing equivalents should not be discounted. Krone *et al.* (1989) have shown that vitamin B_{12}, a known hydride-transfer agent, can catalyse the reductive dechlorination of certain chloromethanes. Mohn and Tiedje (1992) offered several possible routes to reductive dechlorination and concluded that direct electron addition and nucleophilic attack is involved. This could involve transition metal-containing coenzymes, hydride ions or NADH.

Chlorobenzoates

Chlorobenzoates are the most noted intermediates in the aerobic degradation of polychlorinated biphenyls. In many ways they also mirror the degradation of the parent compounds, with a diversity of mono-, di- and trichlorinated congeners possible. However, unlike the parent compounds (PCBs), it may be argued that they are of more importance in that the problems presented by individual congeners are magnified as they are channelled towards fewer central chlorinated intermediates.

The anaerobic degradation of chlorobenzoates was first shown by Horowitz, Suflita and Tiedje (1983) and Suflita, Robinson and Tiedje (1983). The latter studied the reductive dehalogenation of halobenzoates and demonstrated that a methanogenic consortium from sludge was able to dehalogenate iodo-, bromo- and chlorobenzoates, but not fluorobenzoates, and showed a specificity towards the *meta* position of chlorobenzoate. Shelton and Tiedje (1984) isolated, and DeWeerd *et al.* (1990) characterized an organism, *Desulfomonile tiedje* strain DCB1, from this consortium, which has become a benchmark in the field of reductive dehalogenation of aromatic compounds. With this bacterium as the key dechlorinating organism, a defined 3-chlorobenzoate-degrading methanogenic consortium was constructed and shown to be able to grow with 3-chlorobenzoate as the sole carbon and energy source (Dolfing and Tiedje, 1986). This finding established the fundamental precedent and underlying biochemical mechanism, which showed that micro-organisms can grow using chlorinated substrates as their sole electron acceptor. In the consortium, H_2 was the source of reducing equivalents and the sole function of *D. tiedje* was to furnish benzoate from 3-chlorobenzoate. It must be stressed that no energy-yielding reactions other than reductive dechlorination are known for this organism, leading to the conclusion that *D. tiedje* DCB1 is able to conserve energy from the dechlorination reaction (Dolfing and Tiedje, 1987). More recently, the isolate was used in axenic culture to show that (1) the cell yield was higher with 3-chlorobenzoate than with benzoate, and (2) the level of ATP was higher in 3-chlorobenzoate-metabolizing cells compared with benzoate-metabolizing cells (Dolfing, 1990; Mohn and Tiedje, 1990). Having established the precedent, much research has focused on the ubiquity of micro-organisms able to utilize chlorinated organics as electron acceptors in anaerobic environments. Despite theoretical studies (Dolfing and Harrison, 1992) which have established the potential of chloroaromatics to serve as electron acceptors, we are unaware of any other pure isolates that have been proven to derive energy in a manner analogous to *D. tiedje* DCB1.

As with polychlorinated biphenyls, the characterization of anaerobic chlorobenzoate-degrading microbial consortia has lagged behind the establishment of reproducible precedents for site and congener-specific dechlorination. Gerritse, van der Wande and Gottschal (1992) have shown *ortho* dechlorination, and have outlined some of the arbiters of dechlorination. Although the *ortho* and *para* positions might be expected to be most susceptible to dechlorination, the presence of the carboxyl group favours the following order of dechlorination (*meta* > *ortho* > *para*). Despite this generality, *ortho* dechlorination is by no means commonplace. Gerritse, van der Wande and Gottschal (1992) have shown *ortho* dechlorination, but this was achieved at the expense of long incubation times (50% dechlorination of 246- to 24-dichlorobenzoate took 400 days). Moreover, this consortium was unable to further dechlorinate monochlorinated benzoates. These findings further illustrate the specificity of individual consortia and potentiate the belief that predictive ground-rules will be very difficult or impossible to formulate.

Chlorocatechols

Surprisingly, very little research appears to have been done on the anaerobic degradation of chlorocatechols. This may, in part, be due to the reliance on aerobic mechanisms of degradation for this class of compounds. Alternatively, in the context of PCB biodegradation, it may be expected that reductive, or other methods of dechlorination, will precede the incorporation of dioxygen into the aromatic nucleus. This would yield mono- or dichlorocatechols, thus restricting the opportunity to utilize anaerobes for their degradation. As with PCBs and chlorobenzoates, metabolically stable enrichment cultures are isomer specific. Allard *et al.* (1991) have studied microbial consortia from the Gulf of Bothnia and the Baltic Sea (Neilson *et al.*, 1987, 1988). A systematic examination of all possible chlorocatechols with a range of anaerobic consortia revealed that certain chlorocatechols (3,4; 3,6 and 3,4,6) were resistant to dechlorination under the test conditions. Moreover, even for the chlorocatechols which were dechlorinated, differences in both the rate of reaction and yield were observed. In the longer term, two further pitfalls in the use of these consortia were suggested: (1) over several years the rate of dechlorination was substantially reduced; and (2) the products of certain dechlorinations were altered.

In conclusion, it can be seen that the anaerobic degradation of chloroaromatics has received substantial attention. However, it must be concluded that the fate of PCBs, chlorobenzoates and chlorocatechols, under anaerobic conditions, cannot at present be resolved on a general and rational basis. Apart from the obvious environmental factors (pH, temperature, nutrient availability, bioavailability and presence of co-contaminants) several determinants are common to each class of compound:

1. The rate of anaerobic dechlorination is often, but not always, a function of the number and position of the chlorine atoms.
2. The concentration of the aryl chloride may determine the acclimation period (PCBs are very slowly degraded at < 50 p.p.m. compared with 100–1000 p.p.m.).
3. Consortia are often undefined and uncharacterized and inextricably associated with the matrix from which they originate.
4. Consortia are isomer/congener specific.

It is difficult to attribute the level of importance of each of the above and to understand their interactions. However, it is clear that each contributes to the acclimation period. A specific study of this process, using chlorobenzoates as target compounds, has shown that the acclimation period is inversely related to the subsequent rate of dehalogenation (Linkfield, Suflita and Tiedje, 1989). These authors identified induction, genetic change, diauxy and the selection of a rare population as possible explanations. Ultimately, the authors identified induction as the most likely arbiter, in the system they studied, but it seems equally feasible that the other factors have a role to play in other cases.

AEROBIC BIODEGRADATION OF PCBS AND THEIR INTERMEDIATES

There have been many reports of aerobic microbial degradation of PCBs. However, it is hard to find evidence of single species that utilize individual, or mixtures of, PCB congener(s) as the sole source of carbon for growth. It is essential when studying PCB biodegradation to distinguish between *degradation*, which can refer to any single or multi-step bioconversion of the PCB molecule, and *mineralization*, which refers to the complete utilization of the PCB substrates to give biomass, carbon dioxide and chloride.

Aerobic PCB metabolism, whether or not it leads to mineralization of the substrates, is often based on the well-documented pathway for biphenyl catabolism. Before discussing the various routes of microbial PCB metabolism in any detail, it is useful to describe the overall route by which biphenyl is assimilated by bacteria. The biphenyl degradation pathway, shown in *Figure 3*, can be considered to consist of an *upper* pathway, for the oxidation of biphenyl to benzoic acid, and a *lower* pathway, by which benzoic acid is mineralized via oxidation to catechol.

The upper pathway has been demonstrated in a number of bacterial species (Lunt and Evans, 1970; Catelani, Sorlini and Trecanni, 1971; Catelani *et al*., 1973; Gibson *et al*., 1973) and is known to be the result of the action of four enzymes; biphenyl 2,3-dioxygenase, an NAD^+-dependent dihydrodiol dehydrogenase, 2,3-dihydroxybiphenyl dioxygenase and 2-hydroxy-6-oxo-6-phenylhexa-2,4-dienoic acid (HOPDA) hydrolase. The action of this last enzyme produces benzoic acid and a five-carbon hydroxy-acid which is thought to be mineralized via the corresponding keto-acid.

In the lower pathway, benzoic acid is mineralized via oxidation to form catechol. Catechol is a central intermediate in the microbial degradation of aromatic compounds, and is readily catabolized via the *meta*- or *ortho*-cleavage pathways (Dagley, 1978).

Co-metabolism of PCBs by biphenyl degrading bacteria

In a number of bacterial strains, the enzymes of the upper pathway for biphenyl degradation have a sufficiently broad substrate specificity to accommodate the oxidation of many PCB congeners. The great majority of reports concerning aerobic bacterial PCB degradation involve co-metabolism of certain PCB congeners by cells growing or pre-grown on biphenyl. This process was first recognized by Ahmed and Focht (1973) who described the degradation of mono- and dichlorobiphenyls to the corresponding chlorobenzoates by two species of *Achromobacter*. The co-metabolism of PCBs by bacteria expressing the upper biphenyl degradation pathway has

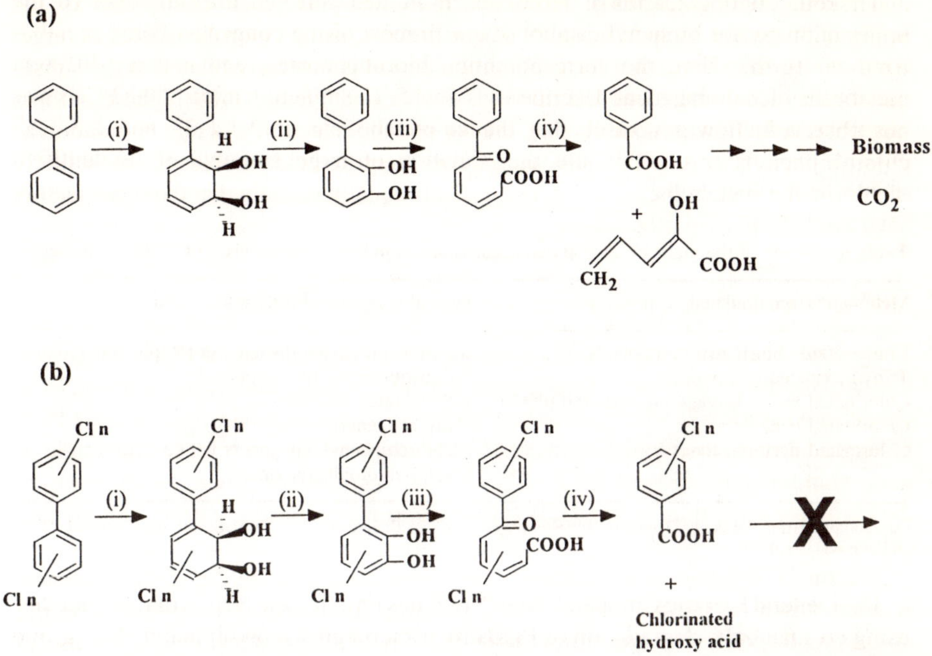

Figure 3. The upper pathway for (a) bacterial biphenyl catabolism and (b) co-metabolism of PCBs by bacteria grown with biphenyl as the sole source of carbon. (i) biphenyl 2,3-dioxygenase; (ii) dihydrodiol dehydrogense; (iii) 2,3-dihydroxybiphenyl dioxygenase; (iv) 2-hydroxy-6-oxo-6-phenylhexa-2,4-dienoic acid hydrolase. Co-metabolism of PCBs rarely proceeds beyond the formation of chlorobenzoates.

since been extensively investigated, particularly by Furukawa, and by the General Electric Company, New York.

Furukawa and co-workers have published a number of studies into the co-metabolism of PCBs by *Alcaligenes* sp. strain Y42 and *Acinetobacter* sp. strain P6, which were originally isolated by enrichment culture using biphenyl and 4-chlorobiphenyl, respectively, as the sole source of carbon for growth and energy (Furukawa, Matsumura and Tonomura, 1978; Furukawa, Tonomura and Kamibayashi, 1978; Furukawa, Tomizuka and Kamibayashi, 1979). In these experiments, the bacteria were pre-grown using biphenyl or 4-chlorobiphenyl as the sole source of carbon before being challenged with a range of 36 PCB congeners. Of these, a total of 23 compounds were metabolized by *Alcaligenes* sp. strain Y42 and 33 by *Acinetobacter* sp. strain P6. A number of correlations were obtained between PCB structure (number and position of chlorine substitutions) and biodegradability, which can be summarized as follows:

1. Degradation decreased as the degree of chlorination increased.
2. 2,2'- and 2,6-chlorinated congeners exhibited a striking resistance to degradation.
3. Congeners possessing chlorine substitutions on both rings were less susceptible to degradation than those having the same number of chlorines on a single ring.
4. Ring fission occurred preferentially on the less- or non-chlorinated ring.
5. The formation and stable accumulation of a yellow *meta*-cleavage product was always observed in the degradation of 4'-chloro substituted congeners.

The routes of degradation of the congeners studied were generally analogous to the upper pathway for biphenyl catabolism, with most of the congeners being metabolized no further than the corresponding chlorobenzoates, and several different metabolites accumulated, as described in *Table 5*. The mineralization of the PCBs was not observed. It was notable that the co-metabolism of 2,3,2',3'- and 2,3,2',5'-chlorobiphenyls resulted in the accumulation of large amounts of unidentified dichlorinated metabolites.

Table 5. Some of the metabolites that can accumulate from the co-metabolism of PCBs by bacteria

Metabolites accumulated	Typical congeners forming the metabolites
Unidentified chlorinated compounds	2,2'-Substituted tetra/penta-chlorobiphenyls
Trihydroxylated compounds	2'-Substituted trichlorobiphenyls
Chlorinated *meta*-cleavage products (HOPDA)	4'-Substituted congeners
Chlorinated benzoic acids	Many congeners
Chlorinated acetophenones	4-Chlorobiphenyl, congeners with 3-' 2,5- or 2,4,5-chlorophenyl rings

Data compiled from Furukawa, Tomizuka and Kamibayashi (1979), Bedard *et al.* (1987a,b), Barton and Crawford (1988) and Bedard (1990).

The General Electric Company, New York, has reported the results of many studies using co-metabolic degradation of PCBs by bacteria grown on biphenyl. This group has isolated two strains, *Alcaligenes eutrophus* strain H850 and *Pseudomonas* sp. strain LB400, both of which have the ability to co-metabolize a particularly broad spectrum of PCB congeners (Bedard *et al.*, 1986, 1987a, b; Bopp, 1986). In comparison with the organisms studied by Furukawa, both of these bacteria have a superior ability to degrade *ortho*-substituted chlorophenyl rings. This is particularly noteworthy, as 2-; 2,4-; 2,5-; 2,3,6- and 2,4,5-chlorophenyl rings are found on many of the congeners present in commercial PCB mixtures (Bedard, 1990). In addition, these strains have only a very limited ability to degrade 4'-substituted congeners. On the basis of these results, Bedard *et al.* (1987b) and subsequently Nadim *et al.* (1988) proposed the presence of an additional 3,4-dioxygenation activity in these strains. 3,4-Dioxygenation results in the formation of the corresponding 3,4-dihydrodiol which cannot be dehydrogenated by these organisms. This compound undergoes a second 3,4-dioxygenase attack on the other chlorophenyl ring to generate a *bis*-3,4-diol which accumulates (Nadim *et al.*, 1988).

An additional property of *A. eutrophus* H850 and *Pseudomonas* sp. strain LB400 is that in both organisms the biphenyl 2,3-dioxygenase can, and may even show a preference for, attack at an *ortho*-chlorinated carbon (Bedard, 1990). It is proposed that in this reaction, instead of forming a dihydrodiol, a chloride ion is spontaneously lost from an unstable intermediate to form a dihydroxychlorobiphenyl, which is a normal intermediate in the upper chlorobiphenyl pathway.

It is not yet certain whether the ability of bacteria such as *A. eutrophus* H850 and *Pseudomonas* sp. strain LB400 to degrade a wide range of chlorinated biphenyls is due to the action of a single broad-specificity dioxygenase or two different enzymes, conferring a 2,3- or 3,4-dioxygenation, respectively. It does appear, however, that only one enzyme may be responsible. Gibson *et al.* (1993) have reported that homogeneous preparations of the oxygenase component of the *Pseudomonas* sp.

strain LB400 biphenyl dioxygenase system oxidize 2,5,2',5'-tetrachlorobiphenyl at the 3,4-position. Further, attempts by Mondello (1989) to demonstrate the presence of more than one biphenyl dioxygenase in *Pseudomonas* sp. strain LB400 were not successful. Whether or not one enzyme can be responsible for the different types of dioxygenase attack, it is clear that different classes of dioxygenase activity do exist. Bedard and Haberl (1990) proposed that the bacterial strains studied by the group at General Electric represent four distinct classes of PCB dioxygenase activity.

The genes encoding for enzymes of the upper PCB degradative pathway have been isolated from a number of bacteria and cloned (Furukawa and Miyazaki, 1986; Khan and Walia, 1989; Mondello, 1989; Hayase, Taira and Furukawa, 1990). Yates and Mondello (1989) conducted DNA–DNA hybridization experiments, comparing genes encoding PCB-degradative enzymes from several PCB-degrading strains. These genes were found to be closely related in *A. eutrophus* strain H850 and *Pseudomonas* sp. strain LB400, but genetically distinct from the six other strains tested. Strong conservation of restriction sites in these organisms was only found in the region of DNA encoding PCB degradation; no other sequence similarities were detected between the two genomes. Comparison of PCB-degradative genes (designated *bph A, B, C* and *D*, encoding biphenyl 2,3-dioxygenase, biphenyl dihydrodiol dehydrogenase, 2,3-dihydroxybiphenyl dioxygenase and 2-hydroxy-6-oxo-6-phenylhexa-2,4-dienoic acid (HOPDA) hydrolase, respectively) at the nucleotide-sequence level has revealed a high degree of conservation between different species (Furukawa, Arimura and Miyazaki, 1987; Furukawa *et al.*, 1989; Kimbara *et al.*, 1989; Peloquin and Greer, 1993). These genes are chromosomally encoded, and it is not clear whether sequence homologies have arisen by DNA transfer or whether the organisms share a common phylogenetic origin.

Metabolites formed by PCB co-metabolism

The degree of PCB removal that can be obtained by co-metabolizing bacteria growing with biphenyl supplied as a source of carbon and energy is undeniably impressive. Indeed, Unterman *et al.* (1988) reported that treatment of Aroclor 1242 with a combination of *Corynebacterium* sp. strain MB1 and *Pseudomonas* sp. strain LB400, representing two different classes of PCB dioxygenase, resulted in some degradation of almost all congeners in the Aroclor mixture.

The scheme for PCB degradation outlined above, in which PCBs are co-metabolized by a pathway analogous to the upper pathway for biphenyl degradation, is very much a simplified overview. In fact, resting-cell experiments by many workers have demonstrated that a large number of different chlorinated metabolites are produced using this approach. The types of products that can accumulate are largely independent of the bacterial strain used, and are determined primarily by the chlorine substitution pattern of the reacting ring (Bedard and Haberl, 1990). One way around this problem may be to adopt the use of mixed cultures of PCB-degrading bacteria. However, Bedard and Haberl (1990) concluded that despite the presence of different classes of PCB dioxygenase, the specificities of the enzymes required for further degradation were similar in eight different strains studied. Hence it is not surprising that, in addition to chlorinated benzoic acids, chlorinated dihydroxybiphenyls and chlorinated HOPDA are major products of PCB co-metabolism.

In addition to the chlorinated intermediates from the upper chlorobiphenyl pathway, a number of other products accumulate from PCB co-metabolism. As reported above, accumulation of the corresponding *bis* diol results from 3,4-dioxygenation of PCBs. Barton and Crawford (1988) reported the generation of large quantities of 4-chloroacetophenone by *Pseudomonas* sp. MB86 when grown with 4-chlorobiphenyl as the sole source of carbon and energy. This compound was believed to be a dead-end metabolite, and growth of the organism was inhibited by the build-up of this metabolite. Following a detailed study on the degradation of 4-chlorobiphenyl by *Achromobacter* sp. strain B-218 and *Bacillus brevis* strain B-257, Massé *et al.* (1984) reported the accumulation of a large number of different chlorinated aromatic acids in the growth medium. Some of these products were postulated to have been generated via a 3,4-dihydroxylation of 4-chlorobiphenyl. Other differences in these minor degradation products were thought to be due to reduction of one or both double bonds in HOPDA.

Ring-chlorinated acetophenones can also result from 2,3-dioxygenase attack on a 3-chlorophenyl ring (Bedard, 1990), or from attack on a 2,5- or 2,4,5-chlorophenyl ring of a PCB by a route that has not yet been elucidated (Bedard *et al.*, 1987b). The degradation of 2,4,6-trichlorobiphenyl by *Acinetobacter* sp. strain P6, grown with 4-chlorobiphenyl as the sole carbon source, has been shown to result in the stable accumulation of a trihydroxy compound (Furukawa, Tomizuka and Kamibayashi, 1979). In the same paper it was reported that several 2,3-chlorinated PCBs were readily metabolized by both *Acinetobacter* sp. strain P6 and *Alcaligenes* sp. strain Y42 to give unidentified products.

It can be seen that the use of bacteria able to co-metabolize PCBs, although often providing a high degree of PCB removal, is likely to result in the production of a number of different, sometimes uncharacterized metabolites. One consequence may be that accumulation of these chlorinated aromatic compounds may inhibit the degradation of PCBs, as has been shown to be the case with chlorobenzoates (Sondossi, Sylvestre and Ahmed, 1992). When considering the use of this approach for the clean-up of contaminated sites, the consequences of the potential release of these metabolites must be a consideration. This can be seen to be even more important when one considers the large amount of information that is available about the toxicology of PCBs, whereas the toxicity of many of the potential products is largely unknown.

Microbial degradation of chlorobenzoates

Degradation of chlorobenzoates represents the lower pathway in the aerobic mineralization of PCBs. Chlorinated benzoates can be degraded aerobically by a number of routes, involving dechlorination via ring cleavage and lactonization, or by hydrolytic or oxidative dehalogenation (Reineke and Knackmuss, 1988).

The principle route of aerobic chlorobenzoate degradation involves oxygenation by benzoate dioxygenase, giving chlorinated 1,2-dihydro-1,2-dihydroxybenzoates. These products undergo an enzymic dehydrogenation to give chlorocatechols (Reineke and Knackmuss, 1978a, b). Subsequent degradation of the chlorocatechols by *ortho* cleavage produces chloromuconates. Lactonization by chloromuconate cycloisomerases subsequently results in the spontaneous elimination of hydrogen chloride (see *Figure 4*).

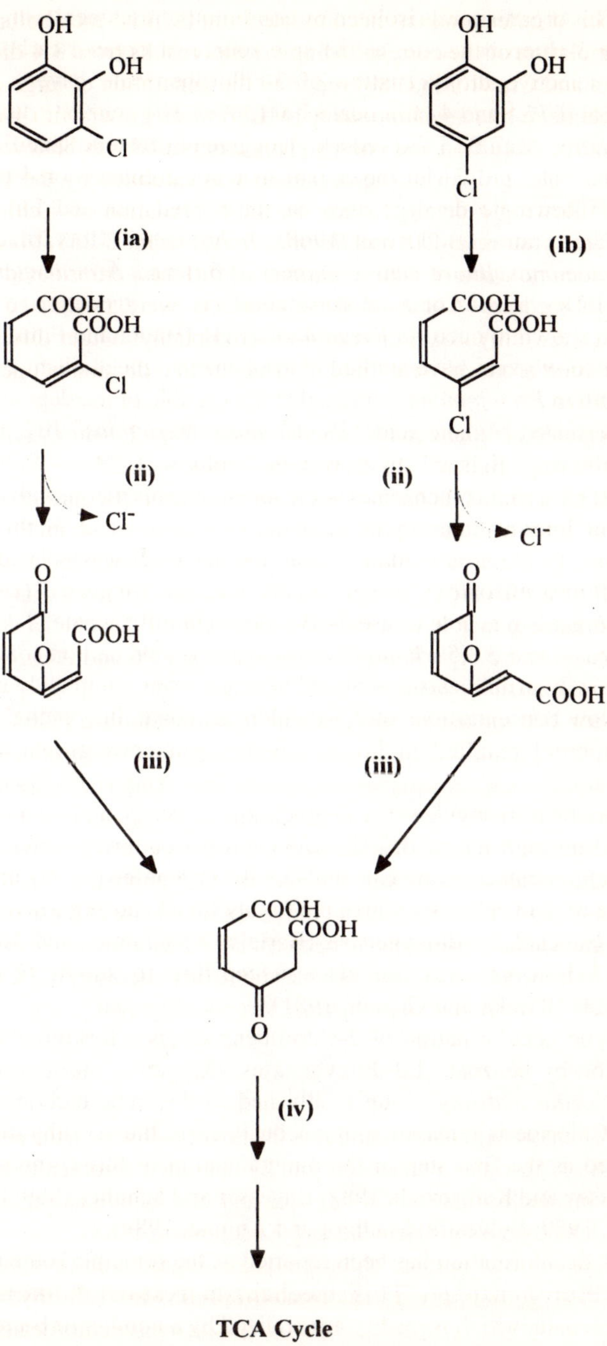

Figure 4. Mineralization of chlorocatechols by the modified *ortho*-cleavage pathway in *Pseudomonas* sp. strain B13. (ia) Catechol-1,2-dioxygenase (pyrocatechase type I); (ib) chlorocatechol-1,2-dioxygenase (pyrocatechase type II); (ii) *cis, cis*-chloromuconate cycloisomerase; (iii) 4-carboxymethylenebut-2-en-4-olide hydrolase; (iv) maleyl acetate reductase.

Reineke and Knackmuss have studied this route extensively in *Pseudomonas* sp. strain B13. This organism was isolated by continuous enrichment using benzoate, and subsequently 3-chlorobenzoate, as the sole source of organic carbon (Dorn *et al.*, 1974). Non-selective dioxygenation of 3-chlorobenzoate gave a mixture of 3-chlorocatechol (67%) and 4-chlorocatechol (33%). The benzoate dioxygenase from *Pseudomonas* sp. strain B13 showed a high degree of substrate specificity, being able to oxidze benzoate and 3-chlorobenzoate but not 2-chloro- or 4-chlorobenzoate. Specificity of benzoate dioxygenases has also been reported for enzymes from *Nocardia* (Cain, Tranter and Darrah, 1968), *Arthrobacter* (Horvarth and Alexander, 1970), *Pseudomonas fluorescens* (Hughes, 1965) and *Azotobacter* (Walker and Harris, 1970). Degradation of 3-chlorobenzoate via dioxygenation to chlorocatechol has also been shown to occur in *Pseudomonas* H1 (Haller and Finn, 1979).

Focht and co-workers have studied chlorobenzoate metabolism extensively, and have isolated two *Pseudomonas* strains that are capable of the degradation of a wide range of chlorinated benzoic acids. *Pseudomonas aeruginosa* JB2, isolated from a PCB-contaminated soil, is able to grow with 2-chloro-, 3-chloro-, 2,3-dichloro-, 2,5-dichloro- or 2,3,5-trichloro-benzoates as the sole source of carbon, with chlorocatechols as the central intermediates in mineralization (Hickey and Focht, 1990). *Pseudomonas putida* P111 is a particularly versatile strain which was isolated by enrichment using 2,5-dichlorobenzoate as the sole source of carbon for growth (Hernandez *et al.*, 1991). This organism is able to use 2-, 3- and 4-chlorobenzoate, 2,3-; 2,4- and 2,5-dichlorobenzoate and 2,3,5-trichlorobenzoate as the sole carbon source for growth. However, growth on all *ortho*-substituted benzoates was completely inhibited by the presence of low concentrations of 3,5-dichlorobenzoate. Interestingly, the presence of this compound resulted in higher yields from cells grown with 3-, or 4-chlorobenzoate.

It is important to recognize that degradation of compounds via chlorocatechols must proceed through the modified *ortho*-cleavage pathway outlined above. *Meta* cleavage of chlorocatechols by catechol-2,3-dioxygenases can occur, but results in the formation of acid chlorides which are highly toxic, causing irreversible inactivation of the catechol-2,3-dioxygenase (Bartels, Knackmuss and Reineke, 1984). Catechol-2,3-dioxygenases are also susceptible to substrate inhibition by chlorocatechols (Klecka and Gibson, 1981).

Oxygenolytic dechlorination of 2-chlorobenzoate is a fortuitous event resulting from oxidation by benzoate-1,2-dioxygenases. Oxygen is introduced into the ring such that an *ortho* hydroxy group is attached to the same carbon as the chlorine substituent. Chloride is released spontaneously from the *cis* dihydrodiol. This has been observed as the first step in the mineralization of this compound by several bacteria (Zaitsev and Karasevich, 1982; Engesser and Schulte, 1989; Fetzner, Muller and Lingens, 1989; Sylvestre, Mailhiot and Ahmad, 1989).

Hydrolytic dechlorination has been reported as the principle route in the degradation of 4-chlorobenzoate. This mechanism results in the formation of 4-hydroxybenzoate, which is readily mineralized by a number of bacteria. This route has been shown to occur in *Acinetobacter* sp. strain 4CB1, which was isolated from a PCB-contaminated soil by Adriens *et al.* (1989). It was established that the dehalogenase enzyme uses water as the hydroxyl donor. 4-Hydroxybenzoate was oxidized by an NADH-dependent mono-oxygenase to form protocatechuate, which was subsequently degraded by both *ortho* and *meta* cleavage. The 4-chlorobenzoate dehalogenase from *Pseudomonas* sp. strain CBS3 was purified by Elsner *et al.* (1991) and demonstrated to be a three-component complex. Ring dehalogenation to give 3-hydroxybenzoate has also been reported as the first step in the degradation of 3-chlorobenzoate by a *Pseudomonas* sp. (Johnson, Briggs and Alexander, 1972).

Mineralization of chlorobenzoates by co-metabolism and by constructed strains

Although the use of chlorobenzoates as the sole source of carbon for growth by naturally occurring bacteria appears to be fairly limited, the scope for chlorobenzoate mineralization can be broadened considerably by exploiting co-metabolic effects and by the construction of specialized hybrid strains.

The degradation of 3,4-dichlorobenzoate by *Acinetobacter* sp. strain 4-CB1 was elucidated by Adriens and Focht (1991). This organism was unable to use 3,4-dichlorobenzoate as a sole carbon source for growth. However, cells grown on 4-chlorobenzoate were able to co-metabolize 3,4-dichlorobenzoate via hydrolytic dechlorination to give 3-chloro-4-hydroxybenzoate. Subsequent degradation of 3-chloro-4-hydroxybenzoate was accomplished without the requirement for a co-substrate. *Pseudomonas putida* strain P111 is able to use 3,5-dichlorobenzoate as a growth substrate only in the presence of 3- or 4-chlorobenzoate. In this instance, the monochlorobenzoates induce the formation of a functional dihydrodiol dehydrogenase, which is not induced by 3,5-dichlorobenzoate (Hernandez *et al.*, 1991).

Pseudomonas sp. strain JB1, isolated by enrichment using biphenyl as a carbon source, is able to use benzoic acid as the sole carbon source, but cannot grow with chlorobenzoates (Parsons *et al.*, 1988). Benzoate-grown cultures of this organism were able to degrade 3-chlorobenzoate, and to a lesser extent 4-chlorobenzoate, 3,5- and 3,4-dichlorobenzoates. 2-Chlorobenzoate, 2,5- and 2,6-dichlorobenzoate were not co-metabolized. Analysis of intermediates, although not conclusive, suggested that co-metabolism of the chlorobenzoates proceeded through *meta* cleavage of chlorocatechols. As discussed above, such transformations are likely to be counter-productive, resulting in production of toxic chloro-2-hydroxymuconate semialdehydes.

Pseudomonas aeruginosa strain JB2, isolated by Hickey and Focht (1990), was detailed above as a bacterium with a broad chlorobenzoate-degrading capacity. Although this organism could not use 2,4-dichlorobenzoate as a growth substrate, cells grown on benzoate could effect the mineralization of this compound.

The growth spectrum of *Pseudomonas* sp. strain B13 was extended considerably by introduction of the TOL plasmid from *Pseudomonas putida* strain mt-2, which encodes a broad specificity benzoate-1,2-dioxygenase (Reineke and Knackmuss, 1979). Exconjugants were prepared by combining donor and recipient cultures on a filter and subsequent selection on 4-chlorobenzoate agar plates. The resultant strains were able to utilize 4-chloro- and 3,5-dichlorobenzoates. In *Pseudomonas* sp. strain B13 and its derivative strains, chlorocatechols are assimilated by an *ortho*-cleavage pathway requiring the induction of one of two sets of enzymes, depending on the position of chlorine substitution. Hence 3-chlorobenzoate-grown cells possess both pyrocatechase type I (the ring fission enzyme normally induced by catechol) and the iso-functional pyrocatechase type II, which has a high activity with chlorosubstituted catechols. In cells grown with 4-chlorobenzoate as the sole source of carbon, induction of the type II pyrocatechase is proportionally much higher (Reineke and Knackmuss, 1980).

The degradative capacity of *Pseudomonas* sp. strain B13 was extended to include 2-chlorobenzoate by Hartmann *et al.* (1989). The 3-methylsalicylate-degrading bacterium, *Pseudomonas* sp. strain WR401 (Engelberts, Schmidt and Reineke, 1989) was mated with *Pseudomonas* sp. strain B13 to obtain transconjugants able to grow with 5-chlorosalicylate as the sole source of carbon and energy. Subsequent selection and adaption led to the isolation of strains able to mineralize 2-chlorobenzoate.

Microbial degradation of chlorinated acetophenones

The build-up of chlorinated acetophenones poses a potential problem in PCB bioremediation, as these compounds are toxic, and can inhibit the growth of PCB-degrading bacteria (Barton and Crawford, 1988; Havel and Reineke, 1993a).

Acetophenone is readily degraded by bacteria. The mineralization of acetophenone by an *Arthrobacter* sp. was investigated by Cripps (1975), and shown to occur through a reaction equivalent to the Baeyer–Villiger oxidation, giving phenyl acetate which was further degraded via phenol and catechol. This organism was unable to use chlorinated acetophenones as a carbon source for growth. Degradation of acetophenone by a *Nocardia* sp. was shown to occur by the same route (Cripps, Trudgill and Whateley, 1978). Little is known, however, about the degradation of chlorinated acetophenones. Higson and Focht (1990) showed that all monochlorinated acetophenones and 2,4- and 2,5-dichloroacetophenones can be degraded by an *Alcaligenes* sp., grown with acetophenone as the sole source of carbon. Degradation proceeded via an analogous Baeyer–Villiger type oxidation, giving chlorinated phenyl acetates which were further oxidized to the corresponding chlorophenols. Some degree of chloride release was observed with each substrate, suggesting that mineralization did occur. *Ortho*-chlorinated acetophenones were dehalogenated to a far lesser extent that other isomers. None of the chloroacetophenones supported growth of this organism, due to either feedback inhibition by chlorophenol or a lack of induction of the Baeyer–Villiger oxygenase.

In the only available report of microbial growth on, and mineralization of, a chlorinated acetohenone, Havel and Reineke (1993a) described the isolation and characterization of a mixed culture which was able to mineralize 4-chloroacetophenone. The culture consisted of an *Arthrobacter* species and a *Micrococcus* species. *Arthrobacter* sp. strain M5 was capable of the mineralization of 4-chloroacetophenone only under very carefully controlled conditions of substrate supply. This was due to the toxic effects of 4-chlorophenol which rapidly accumulated in the medium. The function of *Micrococcus* sp. strain B1 was to rapidly remove 4-chlorophenol formed by the *Arthrobacter* sp. Resting-cell experiments suggested that *Arthrobacter* sp. strain M5 might be able to grow with other chlorinated acetophenones if grown in co-culture with a bacterium capable of mineralization of the corresponding chlorophenols. These experiments suggest that the mineralization of these PCB metabolites may require a delicate control of conditions that could be difficult to achieve in environmental samples. Further complications may arise from the stable accumulation of chloro-1-phenylethanols which were reported to result from the reduction of the ketone.

Mineralization of PCBs by micro-organisms

Athough bacteria able to use biphenyl as the sole source of carbon for growth are ubiquitous, no polychlorinated biphenyl has been shown to support the growth of a naturally occurring micro-organism in pure culture. Indeed, Focht (1993) suggests that bacteria that use PCBs as sole carbon sources probably do not exist in nature. The only chlorobiphenyl that has been reported to support the growth of naturally occurring micro-organisms in pure culture is 4-chlorobiphenyl. Growth on 4-chlorobiphenyl appears to be usually limited to breakdown of the non-chlorinated ring, with the corresponding accumulation of 4-chlorobenzoate. This has been shown in *Achromobacter* (Ahmed and Focht, 1973; Massé *et al.*, 1984), *Pseudomonas* sp. (Barton and Crawford, 1988), *Acinetobacter* sp. (Furukawa, Tonomura and Kamibayashi, 1978) and *Bacillus brevis* (Massé *et al.*, 1984). It is also worth noting that the five-carbon hydroxy-acid resulting from fission of the non-chlorinated ring has not been isolated from cultures, and its degradation has not been unequivocally established.

Shields, Hooper and Sayler (1985) isolated strains of *Alcaligenes* and *Acinetobacter* species from a mixed culture degrading monochlorinated biphenyls. Both strains harboured a plasmid encoding the complete pathway for mineralization of 4-chlorobiphenyl, which was degraded via 4-chlorobenzoate. Mineralization was demonstrated by measurement of $^{14}CO_2$ production from *U*-^{14}C-4-chlorobiphenyl.

Some mineralization of lower-chlorinated biphenyls by microbial consortia certainly occurs in the environment. However, only a few PCB-degrading mixed cultures have been successfully demonstrated in the laboratory. Even with microbial consortia, mineralization is generally limited to *para*-chlorinated biphenyls. Chlorinated benzoic acids have a role in regulating bacterial PCB degradation; hence for significant degradation of PCBs to occur in the environment there is the requirement for the presence of organisms possessing genes encoding both the upper pathway (PCB degradation) and the lower pathway (chlorobenzoate degradation). In an attempted simulation of a natural river environment, Kong and Sayler (1983) showed that all monochlorinated and monobrominated biphenyls were degraded at low concentrations. Degradation occurred through the corresponding halogenated benzoates. However, only in the case of 4-chlorobiphenyl was mineralization demonstrated. Further work by this group identified a freshwater bacterial consortium which was able to mineralize both 4-chlorobiphenyl and 4,4'-dichlorobiphenyl, when either compound was provided as the sole source of carbon (Pettigrew *et al.*, 1990). This consortium consisted of *Pseudomonas testosteroni*, which facilitated the degradation of 4-chlorobiphenyl and 4,4'-dichlorobiphenyl to 4-chlorobenzoate, and an *Arthrobacter* sp., which was able to mineralize 4-chlorobenzoate. The successful mineralization of these compounds by this consortium depended on the presence of a third organism, a strain of *Pseudomonas putida*, for which a role in the consortium was not established.

Furukawa and Chakrabarty (1982) achieved a 98% release of chloride from 4-chloro- and 3,5-dichlorobiphenyl by co-culturing *Acinetobacter* sp. strain P6 with pseudomonads able to use chlorobenzoates. The mineralization of 3-chlorobiphenyl has also been achieved by an aerobic bacterial consortium (Fava and Marchetti, 1991). This consortium consisted of two pseudomonads capable of degrading the substrate to 3-chlorobenzoate, and a strain of *Pseudomonas fluorescens* which was able to co-mineralize 3-chlorobenzoate when supplied with benzoic acid.

Fava *et al.* (1994) obtained a three-membered bacterial consortium capable of mineralizing all three monochlorobiphenyls as well as 2,5-, and 3,4-dichlorobiphenyl when cultured in shake-flasks. Significant degradation and dechlorination of several other di- and trichlorinated biphenyls was also observed. *Pseudomonas* sp. strain CPE1 was isolated from this co-culture and its ability to mineralize 4-chlorobiphenyl was demonstrated when supplied as the sole carbon source in pure culture. Degradation was routed through 4-chlorobenzoic acid. Mineralization was demonstrated by the measurement of near-stoichiometric amounts of chloride released into the medium.

With the development of specialized strains able to effect the mineralization of several chlorinated benzoic acids, it is theoretically possible to devise co-cultures able to mineralize lowly chlorinated PCBs. However, Havel and Reineke (1991) demonstrated that the stability of co-cultures of chlorobiphenyl- and chlorobenzoate-degrading strains can be difficult to achieve. *Pseudomonas* sp. strain B13 was cultured with biphenyl-degrading isolates and incubated in the presence of 3-chlorobiphenyl. Chloride release was never greater than 50% due to the inhibitory effect of black-coloured products that accumulated in the medium. The build-up of black-coloured products, which has been investigated more fully by Haller and Finn (1979) was due to an accumulation of chlorocatechols, which undergo auto-oxidation to give chloroquinones which readily polymerize. For the same reason, no greater than 20% chloride release was attained with co-cultures theoretically able to completely mineralize 2-chlorobiphenyl.

PCB biodegradation by fungi

Dmochewitz and Ballschmiter (1988) demonstrated that *Aspergillus niger* is able to degrade the lower-chlorinated PCBs present in Clophen A30 (a commercial PCB preparation equivalent to Aroclor 1242). Aroclors 1254 and 1260 were not degraded. *Para*-chlorination appeared to inhibit attack; 4,4'-dichlorination essentially preventing attack by the fungus.

Phanerochaete chrysosporium has been shown to degrade many PCBs. It is believed to utilize the same enzymes involved in lignin degradation. Dechlorination of PCBs is thought to occur via the production of hydroxy radicals. *Phanerochaete chrysosporium* mineralizes highly chlorinated congeners and Aroclor 1254 (Bumpus *et al.*, 1985, 1988); however, only very low PCB concentrations can be tolerated (e.g. 250 p.p.b. Aroclor 1254; 5.5 p.p.b. 3,4,3',4'-chlorobiphenyl). Eaton (1985) reported 7% mineralization of Aroclor 1254 (0.3 p.p.m.) after 22 days' incubation with nitrogen-limited cultures of *P. chrysosporium*. Sustained rates of degradation require the addition of complex carbohydrate carbon sources (Bumpus *et al.*, 1988). Another limitation to the use of this organism is that growth is unlikely to occur at temperatures below 15°C (Lamar *et al.*, 1988; Viney and Bewley, 1988). The successful application of this white-rot fungus to PCB degradation will require demonstration of activity at much higher concentrations than those reported to date.

CONSTRUCTED STRAINS

The development of individual strains possessing both the upper (chlorobiphenyl) and lower (chlorobenzoate) pathways has been achieved by a number of workers, and may offer the greatest opportunities for PCB degradation. This approach has a number of potential advantages over the use of mixed cultures. In a co-culture system, the organism responsible for the oxidation of PCBs to chlorobenzoates derives no carbon or energy from the process and requires either another carbon source or must successfully compete with other bacteria for the water-soluble five-carbon acid released from ring cleavage. Furthermore, the requirement for the addition of biphenyl as a substrate to initiate and maintain activity can be obviated if the organism can also assimilate carbon from chlorobenzoate degradation.

A number of workers have used the principles of *in vivo* genetic exchange to construct strains able to mineralize monochlorinated biphenyls when grown on these compounds as the sole source of organic carbon. Much of this work can be regarded as an extension of the development of chlorobenzoate-degrading organisms detailed earlier in this report. Mokross, Schmidt and Reineke (1990) obtained a number of 3-chlorobiphenyl-degrading bacteria by mating *P. putida* strain BN10 and *Pseudomonas* sp. strain B13. *Pseudomonas putida* strain BN10 was isolated for its ability to mineralize biphenyl and to convert 3-chlorobiphenyl to 3-chlorocatechol. Mating with strain B13, which is able to transfer 3-chlorobenzoate-degradative ability to other pseudomonads (Reineke *et al.*, 1982), produced *P. putida* BN210, which mineralized 3-chlorobiphenyl when grown with this compound as the sole source of organic carbon. Growth of this organism on 3-chlorobiphenyl induced the following enzyme activities; phenyl catechol-2,3-dioxygenase, benzoate dioxygenase, catechol -1,2-dioxygenase, chloromuconate cycloisomerase and 4-carboxymethylenbut-2-en -4-olide hydrolase. This organism grew on 2- or 4-chlorobiphenyl, with accumulation of the corresponding chlorobenzoate.

Using similar techniques, this group has also obtained a hybrid strain of *P. cepacia* (JHR22) that mineralized 2-, 3- or 4-chlorobiphenyl and 2,4- or 3,5-dichlorobiphenyl when grown with any of these compounds as the sole source of carbon (Havel and Reineke, 1991). To a large extent, use of this hybrid strain overcame the problem of black-coloured products that were found in co-cultures. Growth with 4-chlorobiphenyl resulted in only 50% chloride release, and the formation of 50% 4-chloro-dihydrodihydroxybenzoate. The organism was able to tolerate high concentrations of 4-chlorobiphenyl. Stoichiometric chloride release from 2-chlorobiphenyl was observed when this substrate was supplied in low concentrations, but growth was inhibited by the formation of black-coloured products when concentrations greater than 4 mM were provided in the medium. *Pseudomonas cepacia* JHR22 was also able to grow with the commercial PCB mixture Aroclor 1221 as the sole source of carbon, with corresponding mineralization of several congeners.

Another approach to obtaining hybrid organisms able to mineralize chlorinated biphenyls is the use of multi-chemostat enrichment. By this process, Adams *et al.* (1992) constructed a 3-chlorobiphenyl- mineralizing recombinant (*Pseudomonas* sp. strain CB15) by intergenic mating of a biphenyl utilizer and a chlorobenzoate utilizer. The accumulation of coloured products from polymerization of chloroquinones was noted in these cultures. Combination of the upper and lower pathways in this single strain allowed some interesting studies on inhibition of the enzymes responsible for ring cleavage in both pathways. 3- Chlorocatechol was found to be a potent inhibitor of 2,3-dihyroxybiphenyl dioxygenase, causing 98% inhibition when present at a concentration of 0.33 mM. Similarly, 2,3-dihydroxybiphenyl was found to competitively inhibit the *ortho* fission of chlorocatechol. 2,3-Dihyroxybiphenyl dioxygenase was also subject to substrate inhibition.

This group also obtained an organism that mineralized 2-chloro- and 2,5-dichlorobiphenyl by mating *P. aeruginosa* strain JB2, which mineralizes chlorobenzoates, with *Arthrobacter* sp. strain B1Barc, which degrades 2-chlorobiphenyl to 2-chlorobenzoate (Hickey, Brenner and Focht, 1992). In contrast to the hybrid strains obtained by Reineke and co-workers, this strain, *P. aeruginosa* strain UCR2, spontaneously lost the ability to mineralize 2-chlorobiphenyl when cultured in the absence of 2-chlorobiphenyl or 2-chlorobenzoate, instead accumulating near-stoichiometric amounts of 2-chlorobenzoate from 2-chlorobiphenyl. Genetic analysis was not successful in demonstrating the genetic events involved during strain construction, and suggested that the pathway by which strain UCR2 was derived was more complex than a single unilateral genetic exchange.

APPLICATION OF BIODEGRADATION TO THE CLEAN-UP OF PCB-CONTAMINATED SOILS AND SEDIMENTS

Given the range of microbial processes leading to PCB degradation (anaerobic reductive dechlorination, co-metabolism by cells grown on biphenyl, mineralization of 4-chlorobiphenyl, the development of constructed strains and fungal dechlorination), one might expect that bioremediation of PCB-contaminated sites was a feasible option. This may prove to be the case, but to date no truly successful PCB bioremediation process has been demonstrated. In the preceding discussion we have shown that anaerobic reductive dechlorination of higher-chlorinated PCBs results in the accumulation of predominantly mono-, di-, and tri-chlorinated congeners, and that these lowly chlorinated congeners are more amenable to aerobic biodegradation. Recently, the application of a sequential anaerobic–aerobic process for complete mineralization of PCBs in environmental samples has been considered.

Any claims to effect the bioremediation of hydrophobic pollutants such as PCBs must be carefully examined, as problems arising from sorption to soils or bioreactor materials, or by leaching, can produce false-positive results. Such difficulties have been discussed in detail by Shannon and Unterman (1993). Another important factor in the assessment of bioremediation processes, which we believe tends to be overlooked in the literature, is the formation of metabolites which may be unidentified, could be far more mobile (more hydrophilic) than the target compounds, and which have less-defined toxicological characteristics.

Brunner, Sutherland and Focht (1985) undertook a study to determine likely methods for the stimulation of PCB biodegradation in soil. Importantly, mineralization of PCBs was determined by measuring release of $^{14}CO_2$ from [^{14}C]Aroclor 1242 in addition to GC analyses. [^{14}C]Aroclor 1242 was added to soil to a concentration of 100 p.p.m. Over a period of 210 days, improving the nutrient status of the soil by sludge or straw amendment showed little significant mineralization of Aroclor 1242. Similarly, incubation under anaerobic conditions had no effect. Inoculation with the PCB co-metabolizing *Acinetobacter* sp. strain P6 (Furukawa, Tomizuka and Kamibayashi, 1979) resulted in significant depletion of the lower-chlorinated congeners from Aroclor 1242, but little mineralization was observed. Application of biphenyl to the soil (3.66 g biphenyl per kg soil) was far more successful and effected the mineralization of many of the congeners present. This process, whereby a structurally similar non-chlorinated analogue is used as a growth substrate and to stimulate co-metabolism of the target molecule(s) has been termed 'analogue enrichment'. Clearly, the indigenous microbial population was able to co-metabolize PCBs when the required enzyme activities were induced by growth on biphenyl. This effect could be stimulated further by addition of both *Acinetobacter* sp. strain P6 and biphenyl. That biphenyl amendment was by far the most important factor was illustrated by the fact that repeated biphenyl additions immediately increased the rate of $^{14}CO_2$ evolution.

Researchers at the General Electric Company have performed degradation trials with PCBs bound to soils using their PCB co-metabolizing strains. Unterman *et al.* (1986) reported initial experiments using soil spiked with Aroclor 1242. Aqueous buffer and washed cells pre-grown on biphenyl were added to give a soil slurry culture (10% v/v soil/water). The slurries were incubated with agitation. Using *Pseudomonas* sp. strain LB400 and *Corynebacterium* sp. strain MB1, 95% of 50 p.p.m. Aroclor 1242 was degraded in 24 hours. When the study was repeated by adding solid biphenyl and inoculating soil slurries with LB400 (1% inoculum) the organism grew well, but significantly less degradation of the PCBs occurred. This contrasts with studies by Kohler, Kohler-Staub and Focht (1988), who showed that growing cultures of *Arthrobacter* sp. strain B1B and *Acinetobacter* sp. strain P6 were far superior to resting-cell suspensions in terms of the extent of PCB degradation and the range of congeners that were transformed. Bioremediation was also investigated using soil from an actual PCB-contaminated site (Unterman *et al.*, 1988). In this case, cells were pre-grown using biphenyl but were not washed prior to addition to the soil. By this means, biphenyl and bacteria were simultaneously added to the soil. The soil was inoculated three times per week for 15 weeks with *Pseudomonas* sp. strain LB400 applied to the surface. Significant depletion of PCB congeners only occurred in the top 1 cm of the soil. When the soil system was also regularly mixed, 35% depletion of PCB congeners was achieved after 23 weeks. Improtantly, it was noted that the addition of nutrients and biphenyl alone was sufficient to stimulate some PCB degradation by the indigenous microflora.

Barriault and Sylvestre (1993) studied the degradation of Aroclor 1242 in soils using *Pseudomonas testosteroni* strain B-356. Again, repeated additions of biphenyl were found to be the most effective way to maintain co-metabolic activity. The viable cell count of the introduced bacteria remained fairly constant throughout the 3 week experiment, whereas addition of biphenyl resulted in a large increase in the native soil population. This demonstrates the need to account for the contribution of the indigenous population to any effects observed in bioremediation studies. Also noted was the importance of the bioavailability of the PCBs in determining their degradation in soil. Inoculation with a biosurfactant-producing strain in addition to *P. testosteroni* B-356 and repeated biphenyl additions was the most successful treatment, providing for degradation of approximately 30% of added Aroclor 1242. One important conclusion from this work was that the accumulation of toxic metabolites from biphenyl and PCBs inhibit the biodegradation process. Such effects have also been noted by Sondossi, Sylvestre and Ahmad (1992), with regard to the accumulation of chlorobenzoates and their degradation products. Chlorobenzoates were found to inhibit biphenyl and chlorobiphenyl oxidation by *P. testosteroni* B-356; effects were notable at concentrations of 3-chlorobenzoate of only 160 μM. Chlorocatechols and muconic semi-aldehydes resulting from chlorobenzoate degradation were also found to inhibit the HOPDA dehydrogenase in this organism; only 0.5 μM 3-chlorocatechol was required to inhibit 99% of this enzyme activity. The practical implications with regard to PCB bioremediation are illustrated in work by Hickey, Searles and Focht (1993) who demonstrated that PCB mineralization in soils was enhanced by inoculation with chlorobenzoate-degrading bacteria. Inoculation with only pseudomonad strains able to mineralize chlorobenzoates was more effective at stimulating PCB mineralization than inoculation with PCB co-metabolizers or chlorobiphenyl-mineralizing strains in combination with chlorobenzoate-degrading strains.

Difficulties with multiple inoculation of soils were also found by Havel and Reineke (1992). Soil microcosms were inoculated with three bacterial strains which were theoretically able to mineralize several congeners present in Aroclor 1221. In pre-sterilized soil, about 70% of the congeners were degraded. However, in non-sterilized soil 4-chlorobenzoate was metabolized by the indigenous bacteria, giving toxic metabolites which rapidly inhibited the growth of the introduced strains. Use of the constructed hybrid strain *Pseudomonas cepacia* strain JHR22, which was able to mineralize components of Aroclor 1221 in axenic culture, was more successful (Havel and Reineke, 1993b). The organism survived in non-sterile soil for several weeks and mineralized all congeners in Aroclor 1221 which were chlorinated on only one phenyl ring. The hybrid pathway for PCB mineralization, which originated from three different strains was more successful than consortia because 4-chlorobenzoate was degraded before it accumulated sufficiently to be metabolized to toxic products by the indigenous soil population.

A range of different approaches to the bioremediation of PCB-contaminated soil was investigated by Viney and Bewley (1988, 1990). The use of surfactants applied to the soil to increase the bioavailability of the PCBs was investigated, but all surfactants examined were shown to have some inhibitory effect on PCB metabolism by the selected bacteria and fungi. The most effective treatment in large-scale soil systems (10 kg horticultural sand) involved inoculation with a mixture of bacterial strains (including *A. eutrophus* H850, *Pseudomonas* sp. strain LB400 and *Corynebacterium* sp. strain MB1 from GEC) and *P. chrysosporium*. Soil systems were incubated at 25°C, turned frequently, and received repeated additions of biphenyl and phosphate-buffered mineral salts. Under these conditions 42% of the total PCB from Aroclor 1242 was depleted. The more highly chlorinated congeners from this mixture were not degraded.

Apart from the limitations in the degree of PCB degradation that can be achieved by exploiting co-metabolism by biphenyl utilizers, it must be recognized that biphenyl is itself fairly insoluble, toxic and costly; the addition of biphenyl to large-scale soil environments may itself not prove to be feasible. Viney and Bewley (1988) calculated a biphenyl requirement of 1% (w/w) of the total soil to be treated.

Harkness *et al.* (1993) conducted a study of the *in situ* aerobic biodegradation of PCBs in sediment in the Hudson River. The PCBs present in the Hudson River sediment at the time of the experiment had previously undergone considerable transformation, presumably by microbial reductive dechlorination, since their release into the river (the date of PCB release is not given). Although the original PCB mixture was predominantly Aroclor 1242, PCBs in the sediment samples were mainly mono- and di-chlorobiphenyls (these constitute approximately 9% of the congeners in Aroclor 1242). Addition of inorganic nutrients, biphenyl and oxygen resulted in the loss of 35–55% of the PCBs present, with the resultant formation of chlorobenzoates, over a period of 73 days. Repeated inoculation with a PCB-degrading bacterium (*A. eutrophus* strain H850) failed to improve biodegradative ability of the system. Approximately 50% of the PCBs in the sediment were resistant to degradation due to their sorption to the organic components of the sediment matrix. In small-scale laboratory experiments, Anid, Ravest-Webster and Vogel (1993) investigated the stimulation of aerobic PCB degradation in anaerobic Hudson River sediments by addition of hydrogen peroxide as an oxygenation agent. Again, the PCB congener distribution of the sediment indicated that considerable reductive dechlorination had occurred previously. In this study, Aroclor 1242 was added to the sediment to give a final concentration of 300 p.p.m. Sediments were incubated under anaerobic conditions for 76 weeks before oxygenation by addition of hydrogen peroxide. Analysis demonstrated that the anaerobic treatment reduced the levels of di-, tri-, tetra-, penta- and hexachlorinated congeners by 11%, 73%, 66%, 73% and 94%, respectively: correspondingly the amount of monochlorobiphenyls increased by 76%. At the highest dose of hydrogen peroxide (1.7 g H_2O_2 per litre sediment/water), subsequent oxygenation resulted in a 95% reduction in the total mass of PCBs present in the sediment. As no chloride measurements were reported during these experiments it is not possible to determine the extent to which mineralization of the PCBs occurred. This group also noted that a certain proportion of the PCBs were resistant to degradation, and this was attributed to sorption of PCBs on the sediment matrix.

Although it is recognized that the extreme hydrophobicity of PCBs limits their bioavailability both in soils and aqueous systems, few efforts have been made to overcome this problem. The hydrophobicity of these compounds may also influence materials that can be used to construct bioreactors, as the PCBs adhere strongly to a wide range of surfaces.

The use of surfactants is well established in oil recovery and hydrocarbon degradation; however, the implications for PCB degradation are not clear. Aronstein and Alexander (1993) have recently reported that addition of non-ionic surfactants can enhance biodegradation of biphenyl and phenanthrene in laboratory soil columns, even at concentrations lower than those required to desorb the compounds from soil, suggesting that physiological factors may also be involved. Similarly, Guerin and Jones (1988) found that the addition of anionic surfactants to liquid cultures of *Mycobacterium* sp. strain BG1 enhanced the rate of degradation of phenanthrene, but also that the order of enhancement did not correlate with increased solubility. Surfactants have also been shown to enhance significantly ligninase production by *P. chrysosporium* (Asther *et al.*, 1987; Lestan *et al.*, 1993). The potential of surfactants to facilitate the biodegradation of PCBs merits considerable further attention. Kiu (1980) found that the rate of degradation of PCBs by *Pseudomonas* sp. strain 7509 could be enhanced by growing the cells in a stable PCB–ligninsulfonate emulsion. Biodegradation took place at the PCB–water interface.

There is clearly scope for the use of *in vitro* DNA manipulation in PCB bioremediation. Mondello (1989) achieved the expression of the *P. putida* LB400 *bphABCD* genes in *Escherichia coli*. The recombinant strains were able to effect PCB degradation in liquid culture without the requirement of growth on biphenyl. However, *E. coli* is unlikely to be a suitable host, since this is not a native soil organism and does not grow readily at ambient temperatures. Lajoie *et al.* (1993) have recently exploited genetic engineering techniques in an elegant attempt at overcoming both the insolubility of PCBs and the need for a growth substrate to allow co-metabolism of PCBs. Broad-substrate-specificity PCB-degrading genes were cloned from a naturally occurring *Pseudomonas* species into a plasmid (pRK293). The plasmids were transferred to a host (*Pseudomonas paucimobilis* 1IGP4) which uses a non-toxic, water-soluble, non-ionic surfactant as a selective growth substrate. Treatment of soil contaminated with Aroclor 1242 in the presence of the surfactant (Igepal CO-720, 1% w/w) with the recombinant strain resulted in the degradation of many PCB congeners in the absence of biphenyl. The authors proposed that the ability of the surfactant to enhance the bioavailability of the PCBs could be better exploited in a soil–slurry bioreactor.

Conclusions

In this chapter we have attempted to discuss all the factors relevant to the development of strategies for PCB bioremediation. The rational design of processes for the degradation of PCBs have been impeded by the very physico-chemical characteristics which made them attractive chemicals. The extreme hydrophobicity of PCBs and their high sorption coefficients provide the first obstacle to degradation, namely the very low bioavailability of PCBs. Typical bioremediation strategies rely on the target compounds being at least partially soluble in the aqueous phase in order to employ conventional biotreatment processes. One option is to pretreat the soil or sediment to give a PCB-rich liquor. Surfactant or solvent soil washing processes that are efficient, inexpensive and compatible with microbial PCB degradation have not yet been demonstrated. While not precluding this approach, it seems unlikely that suitable surfactants or solvents are available which would permit their use in a bioreactor. They are likely to be both inhibitory to any microbial population and to create a soil or sediment which, though PCB free, is unable to be returned to source without further treatment. The complexity of commercial PCB mixtures, which contained up to 70 congeners, has proved a major obstacle to their biodegradation. Much research has consequently focused upon single congeners or defined congener mixtures which are not representative of real contaminated sites.

The first step in any bioremediation process must be to consider the nature of the problem and the desired outcome within the framework of legislative guidelines. Legislation varies between countries and it is outside the scope of this chapter to compare global guidelines for PCBs. However, the situation in the UK serves to illustrate several factors. At present we have no system for measuring PCBs in which the individual congeners are assigned a weighted toxicity equivalent, thus enabling a standardized total figure to be reported. This has been recommended by the International Council for Exploration of the Sea (DoE, 1994). However, even the projected ICES guidelines recommend the use of only seven congeners (Ballschmiter numbers 7, 28, 52, 101, 118, 136, 153 and 180). Each of these are *ortho* substituted, and thus co-planar PCBs are not considered, even though it is known that these are the most toxic congeners (Safe, 1990). In the UK, guidelines for PCB-contaminated soils and sediments are presently related to end-use for the particular site. While appreciating that this is preferable to multifunctionality, target values must be more clearly defined and be related to potential toxicity.

Whether setting guidelines or attempting to attain them, the validity and reproducibility of the analyses is of fundamental importance. We have attempted to present a review of the methods currently employed, each of which has inherent problems. It is worth reiterating that validation of results is of primary concern and the findings of Milby, Miller and Forrester (1985) and Holden (1986) show that even accredited laboratories are unable to provide reliable determination of PCBs. When this is coupled to the multiplicity of matrices, the inability to resolve clearly all congeners, the disproportionality of detector responses and the variety of internal standards used, it is not surprising that objective criticism of such data is difficult.

When developing bioremediation strategies, and assuming targets are known and measurable, the major task is to obtain isolates and consortia that are able to degrade, and ultimately mineralize, PCBs. Mineralization has to be the ultimate target in remediation strategies as it obviates further problems caused by known and unknown breakdown products. To achieve this requires both anaerobic and aerobic microorganisms for the reasons explained above. In both cases it seems clear that the origin of the isolate(s), the congener range, the presence of co-contaminants and the matrix characteristics are primary determinants of initial attack. Moreover, many of these determinants are site specific, and this should be borne in mind during laboratory investigations. Alongside these factors, the findings of key workers in the field have provided us with invaluable information relating to the microbial catalysts.

By far the least well-understood catabolic route is that which occurs under anaerobic conditions. Reductive dechlorination of PCBs by either an axenic culture or consortia free from sediment has never been shown. This has hampered the study of anaerobic PCB biodegradation and left many questions, such as the role of the sediment, unanswered. Despite this, it is clear that different consortia can have different congener specificites and the rate of dechlorination is often a function of the number and position of the chlorine atoms. One clear trend is that *meta* and *para* dechlorination of PCBs occurs far more frequently than *ortho* dechlorination. This is also true of the *ortho* chlorobenzoates, impeding the possible degradative routes of commercial PCB mixtures.

Much greater effort has been expended on the aerobic degradation of PCBs and their intermediates. Axenic cultures, consortia and hybrid strains have all been isolated and well characterized. Common rules relating to catabolism have been formulated, but the application of such strains has been limited. This is partly due to the same site-specific constraints applied to the anaerobes. Additionally, our understanding of the biochemistry and regulation of the processes involved in aerobic haloaromatic catabolism has alerted us to the limitations of these organisms. Cometabolic degradation of PCBs has been studied extensively and shown to facilitate the degradation of many congeners. However, although PCBs have been shown to be removed, the production of a number of different, and sometimes uncharacterized, metabolites has been shown. Chlorobenzoates, a major bottleneck in PCB mineralization, have been shown to be inhibitory to PCB biodegradation. It would therefore seem likely that these and other intermediates may inhibit or retard PCB biodegradation. The requirement for addition of biphenyl to stimulate co-metabolism may prove to be unfeasible on a large scale. This requirement may be obviated by developments in strain construction and genetic engineering, but fully competent strains have not yet been developed. In addition, legislation concerning the release of genetically engineered organisms into the environment does not currently favour this approach.

In conclusion, many factors influence the biodegradation of PCBs. Once clear on the legislative guidelines and the analytical methodology, a number of hurdles remain to be overcome. Despite the availability of extensive literature, clear ground-rules on the isolation and employment of micro-organisms for PCB mineralization have yet to be formulated. It seems unlikely that this will ever be possible, site investigation always being a prerequisite to bioremediation. This aside, current research will hopefully provide a basis for the sequential anaerobic and aerobic biotreatment of PCB-contaminated soils and sediments. Clearly, for such a process the challenge is as great in terms of engineering as it is for development of competent microbial consortia. Unlike many studies to date, success must not be determined solely by PCB disappearance, a yardstick too often used to promote incomplete bioremediation strategies. The problems associated with PCBs are by no means exclusive to these compounds and any rational remediation strategy is likely to be a blueprint for bioremediation of other hydrophobic pollutants.

Acknowledgements

The authors gratefully acknowledge the European Community who are funding the studies into PCB biodegradation in the laboratories of Prof. C.J. Knowles at the University of Kent. This is being done in collaboration with the laboratories of Prof. W. Reineke (Wuppertal, Germany), Dr G. Schraa (Wageningen, Holland) and Dr D. van der Lelie (Mol, Belgium). The authors acknowledge Prof. C.J. Knowles for his advice in the preparation of this chapter.

References

ABRAMOWICZ, D.A. (1990). Aerobic and anaerobic biodegradation of PCBs: a review. *Critical Reviews in Biotechnology* **10**, 241–251.

ADAMS, R.H., HUANG, C.M., HIGSON, F.K., BRENNER, V. AND FOCHT, D.D. (1992). Construction of a 3-chlorobiphenyl utilising recombinant from an intergenic mating. *Applied and Environmental Microbiology* **58**, 647–654.

ADRIENS, P. AND FOCHT, D.D. (1991). Cometabolism of 3,4-dichlorobenzoate by *Acinetobacter* sp. strain 4-CB1. *Applied and Environmental Microbiology* **57**, 173–179.

ADRIENS, P., KOHLER, H.-P.E., KOHLER-STAUB, D. AND FOCHT, D.D. (1989). Bacterial dehalogenation of chlorobenzoates and coculture biodegradation of 4,4'-dichlorobiphenyl. *Applied and Environmental Microbiology* **55**, 887–892.

AHMED, M. AND FOCHT, D.D. (1973). Degradation of polychlorinated biphenyls by two species of *Achromobacter*. *Canadian Journal of Microbiology* **19**, 47–52.

ALKINS, P.R. (1981). The pesticide manufacturing industry – Current waste treatment and disposal practices. EPA Project No. 12020-FYE NTIS Cat No. PB-211129. Jan. 1972. In *Pesticide Destruction and Detoxification Processes and Techniques* (A.P. Dillon, Ed.). Noyes Data Corp., Park Ridge, NJ.

ALLARD, A.S., HYNNING, P.A., LINDGREN, C., REMBERGER, M. AND NIELSON, A.H. (1991). Dechlorination of chlorocatechols by stable enrichment cultures of anaerobic bacteria. *Applied and Environmental Microbiology* **57**, 77–84.

AMEND, L.J. AND LEDERMAN, P.B. (1992). Critical evaluation of PCB remediation technologies. *Environmental Progress* **11**, 173–177.

ANID, P.J., RAVEST-WEBSTER, B.P. AND VOGEL, T.M. (1993). Effect of hydrogen peroxide on the biodegradation of PCBs in anaerobically dechlorinated river sediments. *Biodegradation* **4**, 241–248.

ARONSTEIN, B.N. AND ALEXANDER, M. (1993). Effect of a non-ionic surfactant added to the soil surface on the biodegradation of aromatic hydrocarbons within the soil. *Applied Microbiology and Biotechnology* **39**, 386–390.

ASTHER, M., CORRIEU, G., DRAPRON, R. AND ODIER, E. (1987). Effect of Tween 80 and oleic acid on ligninase production by *Phanerochaete chrysosporium* INA-12. *Enzyme and Microbial Technology* **9**, 245–249.

BALLSCHMITER, M. AND ZELL, M. (1980). Analysis of PCBs bý glass capillary gas chromatography. *Fresenius Zeitschrift für Analytische Chemie* **302**, 20.

BARRIAULT, D. AND SYLVESTRE, M. (1993). Factors affecting PCB degradation by an implanted bacterial strain in soil microcosms. *Canadian Journal of Microbiology* **39**, 594–602.

BARTELS, I., KNACKMUSS, H.-J. AND REINEKE, W. (1984). Suicide inactivation of catechol-2,3-dioxygenase from *Pseudomonas putida* mt-2 by 3-halocatechols. *Applied and Environmental Microbiology* **47**, 500–505.

BARTON, M. AND CRAWFORD, R. (1988). Novel biotransformation of 4-chlorobiphenyl by a *Pseudomonas* sp. *Applied and Environmental Microbiology* **54**, 594–595.

BEDARD, D. (1990). Bacterial transformation of polychlorinated biphenyls. *Advances in Biotechnology* **4**, 369–387.

BEDARD, D.L. AND HABERL, M.L. (1990). Influence of chlorine substitution pattern on the degradation of PCBs by eight bacterial strains. *Microbial Ecology* **20**, 87–102.

BEDARD, D.L., UNTERMAN, R., BOPP, L.H., BRENNAN, M.J., HABERL, M.L. AND JOHNSON, C. (1986). Rapid assay for screening and characterizing microorganisms for the ability to degrade polychlorinated biphenyls. *Applied and Environmental Microbiology* **51**, 761–768.

BEDARD, D.L., WAGNER, D.L., BRENNAN, M.J., HABERL, M.L. AND BROWN, J.F. (1987a). Extensive degradation of Aroclors and environmentally transformed polychlorinated biphenyls by *Alcaligines eutrophus* H850. *Applied and Environmental Microbiology* **53**, 1094–1102.

BEDARD, D.L., HABERL, M.L., MAY, R. AND BRENNAN, M.J. (1987b). Evidence for novel mechanisms of polychlorinated biphenyl metabolism in *Alcaligines eutrophus* H850. *Applied and Environmental Microbiology* **53**, 1103–1112.

BERGH, A.K. AND PEOPLE, R.S. (1977). PCB distribution in sewage waste and their environmental and community effects p 4–6. *Proceedings of the National Conference on the Treatment and Disposal of Industrial Wastewaters and Residues, Houston.*

BOPP, L.H. (1986). Degradation of highly chlorinated PCBs by *Pseudomonas* strain LB400. *Journal of Industrial Microbiology* **1**, 23–29.

BOYLE, A.W., SILVIN, C.J., HASSETT, J.P., NAKAS, J.P. AND TANENBAUM, S.W. (1992). Bacterial PCB biodegradation. *Biodegradation* **3**, 285–298.

BROWN, J.F., JR, WAGNER, R.E., FENG, H., BEDARD, D.L., BRENNAN, M.J., CARNAHAN, J.C. AND RAY, R.J. (1987). Environmental dechlorination of PCBs. *Environmental and Toxicological Chemistry* **6**, 579–593.

BROWN, R.S., PETTIT, K., PRICE, D. AND JONES, P.W. (1991). Thermal desorption gas chromatography: a quick screening technique for polychlorinated biphenyls. *Chemosphere* **23**, 1145–1150.

BRUNNER, W., SUTHERLAND, F.H. AND FOCHT, D.D. (1985). Enhanced biodegradation of polychlorinated biphenyls in soil by analogue enrichment and bacterial inoculation. *Journal of Environmental Quality* **14**, 324–328.

BRUNSBACH, F.R. AND REINEKE, W. (1993). Degradation of chlorobenzoates in soil slurry using special organisms. *Applied Microbiology and Biotechnology* **39**, 117–122.

BUMPUS, J.A., TIEN, M., WRIGHT, D. AND AUST, S.D. (1985). Oxidation of persistent environmental pollutants by a white rot fungus. *Science* **228**, 1434–1436.

BUMPUS, J.A., FERNANDO, T., MILESKI, G.J. AND AUST, S.D. (1988). Biodegradation of organopollutants by *Phanerochaete chrysoporium*: practical considerations. *Proceedings of the 13th Annual Research Symposium on Land Disposal Remedial Action, Incineration and Treatment of Hazardous Waste*, pp. 411–418. US Environmental Protection Agency, EPA/600/2-86/040.

BUMPUS, J.A., MILESKI, G., BROCK, B., ASHBAUGH, W. AND AUST, S.D. (1990). Biological oxidations of organic compounds by enzymes from a white rot fungus. In *Innovative Hazardous Waste Treatment Technology*. Vol. 3: *Biological Processes* (H.M. Freeman and P.R. Sferra, Eds), pp. 47–54. Technomic Publishing.

CAIN, R.B., TRANTER, E.K. AND DARRAH, J. (1968). The utilisation of some halogenated aromatic acids by *Nocardia. Biochemical Journal* **106**, 211–227.

CAIRNS, T. AND SIEGMUND, E.G. (1981). Determination of polychlorinated biphenyls by chemical ionization mass-spectrometry. *Analytical Chemistry* **53**, 1599–1603.

CAREY, A.E. AND HARVEY, G.R. (1978). Metabolism of polychlorinated biphenyls by marine bacteria. *Bulletin of Environmental Contamination and Toxicology* **20**, 527–534.

CATELANI, D., SORLINI, C. AND TRECANNI, V. (1971). The metabolism of biphenyl by *Pseudomonas putida. Experientia* **27**, 1173–1174.

CATELANI, D., COLOMBI, A., SORLINI, C. AND TRECANNI, V. (1973). Metabolism of biphenyl. *Biochemical Journal* **134**, 1063–1066.

CHEN, M., HONG, C.S., BUSH, B. AND RHEE, G.Y. (1988). Anaerobic biodegradation of polychlorinated biphenyls by bacteria from Hudson River sediments. *Ecotoxicology and Environmental Safety* **16**, 95–105.

COLE, M.A., REICHART, P.B. AND BUTTON, D.K. (1979). Reversible bioconcentration of monochlorinated biphenyls by *Rhodotorula rubra*, correlations with aqueous solubility. *Bulletin of Environmental Contamination and Toxicology* **23**, 44.

CRIPPS, R.E. (1975). The metabolism of acetophenone. *Biochemical Journal* **152**, 233–241.

CRIPPS, R.E., TRUDGILL, P.W. AND WHATELEY, J.G. (1978). The metabolism of 1-phenylethanol and acetophenone by *Nocardia* T5 and an *Arthrobacter* species. *European Journal of Biochemistry* **86**, 175–186.

DAGLEY, S. (1978). Pathways for the utilisation of organic growth substrates. In *The Bacteria*, Vol. 6 (L.N. Ornston and J.R. Sokatch, Eds), pp. 305–388. Academic Press, New York.

DEPARTMENT OF THE ENVIRONMENT (1976, 1994). *Polychlorinated biphenyls.* Waste Management Paper No. 6. HMSO, London.

DEWEERD, K.A., MANDELCO, L., TANNER, R.S., WOESE, C.R. AND SUFLITA, J.M. (1990). *Desulfomonile tiedje* gen. nov. and sp. nov., a novel anaerobic, dehalogenating, sulfate-reducing bacterium. *Archives of Microbiology* **154**, 23–30.

DILLON, A.P. (1981). *Pesticide Disposal and Detoxification Processes and Techniques*. Noyes Data Corp., Park Ridge, NJ.

DMOCHEWITZ, S. AND BALLSCHMITER, K. (1988). Microbial transformation of technical mixtures of polychlorinated-biphenyls (PCB) by the fungus *Aspergillus niger. Chemosphere* **17**, 111–121.

DOLFING, J. (1990). Reductive dechlorination of 3-chlorobenzoate is coupled to ATP production and growth in an anaerobic bacterium, strain DCB-1. *Archives of Microbiology* **153**, 264–266.

DOLFING, J. AND HARRISON, B.K. (1992). Gibbs free energy of formation of halogenated aromatic compounds and their potential role as electron acceptors in anaerobic environments. *Environmental Science and Technology* **26**, 2213–2218.

DOLFING, J. AND TIEDJE, J.M. (1986). Hydrogen cycling in a three- tiered food web growing on the methanogenic conversion of 3-chlorobenzoate. *FEMS Microbial Ecology* **38**, 293–298.

DOLFING, J. AND TIEDJE, J.M. (1987). Growth yield increase linked to reductive dechlorination in a defined 3-chlorobenzoate degrading methanogenic culture. *Archives of Microbiology* **149**, 102–105.

DORN, E., HELLWIG, M., REINEKE, W. AND KNACKMUSS, H.J. (1974). Isolation and characterisation of a 3-chlorobenzoate degrading pseudomonad. *Archives of Microbiology* **99**, 61–70.

DOSKEY, P.V. AND ANDREN, A.W. (1981). Concentrations of airborne PCBs over Lake Michigan. *Journal of Great Lakes Research* **7**, 15–20.

DUNNIVANT, F.M., ELZERMAN, A.W., JURS, P.C. AND HASON, M.N. (1992). Quantitative structure–property relationships for aqueous solubilities and Henry's Law constants of polychlorinated biphenyls. *Environmental Science and Technology* **26**, 1567–1573.

EATON, D.C. (1985). Mineralisation of polychlorinated biphenyls by *Phanerochaete chrysosporium*, a lignolytic fungus. *Enzyme and Microbial Technology* **7**, 194–196.

ELSNER, A., LOFFLER, F., MIYASHITA, K., MULLER, R. AND LINGENS, F. (1991). Resolution of 4-chlorobenzoate dehalogenase from *Pseudomonas* sp. strain CBS3 into three components. *Applied and Environmental Microbiology* **57**, 324–326.

ENGELBERTS, K., SCHMIDT, E. AND REINEKE, W. (1989). Degradation of *O*-toluate by *Pseudomonas* sp. strain WR401. *FEMS Microbiology Letters* **59**, 35–38.

ENGESSER, K.H. AND SCHULTE, P. (1989). Degradation of 2-bromo, 2- chloro- and 2-fluorobenzoate by *Pseudomonas putida* CLB 250. *FEMS Microbiology Letters* **60**, 143–148.

EPA (1989). The Superfund Innovative Technology Evaluation Programme. Technology Profiles Report No. EPA/540/5-89/013. EPA Risk Reduction Engineering Laboratory, Cincinnati, Ohio.

EPA FEDERAL REGISTER Vol. 52, 63/P 10695.

FAVA, F. AND MARCHETTI, L. (1991). Degradation and mineralisation of 3-chlorobiphenyl by mixed aerobic bacterial culture. *Applied Microbiology and Biotechnology* **36**, 240–245.

FAVA, F., CINTI, S. AND MARCHETTI, L. (1993). Dechlorination of Fenchlor 54 and of a synthetic mixture of PCBs by anaerobic microorganisms. *Applied Microbiology and Biotechnology* **38**, 808–814.

FAVA, F., DI GIOIA, D., CINTI, S., MARCHETTI, L. AND QUATTRONI, G. (1994). Degradation and dechlorination of low-chlorinated biphenyls by a three-membered bacterial co-culture. *Applied Microbiology and Biotechnology* **41**, 117–123.

FETZNER, S., MULLER, R. AND LINGENS, F. (1989). A novel metabolite in the microbial degradation of 2-chlorobenzoate. *Biochemical and Biophysical Research Communications* **161**, 700–705.

FOCHT, D.D. (1993). Microbial degradation of chlorinated biphenyls. *Soil Biochemistry* **8**, 341–407.

FRIES, F.G. (1971). Degradation of chlorinated hydrocarbons under anaerobic conditions. *Advanced Chemistry Series* **111**, 256–270.

FURUKAWA, K. AND CHAKRABARTY, A.M. (1982). Involvement of plasmids in total degradation of chlorinated biphenyls. *Applied and Environmental Microbiology* **44**, 619–626.

FURUKAWA, K. AND MIYAZAKI, T. (1986). Cloning of a gene cluster encoding biphenyl and chlorobiphenyl degradation in *Pseudomonas pseudoalcaligenes. Journal of Bacteriology* **166**, 392–398.

FURUKAWA, K., ARIMURA, M. AND MIYAZAKI, T. (1987). Nucleotide sequence of the 2,3-dihydroxybiphenyl dioxygenase gene of *Pseudomonas pseudoalcaligenes. Journal of Bacteriology* **169**, 427–429.

FURUKAWA, K., MATSUMURA, F. AND TONOMURA, K. (1978). *Alcaligines* and *Acinetobacter* strains capable of degrading polychlorinated biphenyls. *Agricultural and Biological Chemistry* **42**, 543–548.

FURUKAWA, K., TOMIZUKA, N. AND KAMIBAYASHI, A. (1979). Effect of chlorine substitution on the bacterial metabolism of various polychlorinated biphenyls. *Applied and Environmental Microbiology* **38**, 301–310.

FURUKAWA, K., TONOMURA, K. AND KAMIBAYASHI, A. (1978). Effect of chlorine substitution on the biodegradability of polychlorinated biphenyls. *Applied and Environmental Microbiology* **35**, 223–227.

FURUKAWA, K., HAYASE, N., TAIRA, R. AND TOMIZUKA, N. (1989). Molecular relationship of chromosomal genes encoding biphenyl/polychlorinated biphenyl catabolism; some soil bacteria possess a highly conserved *bph* operon. *Journal of Bacteriology* **171**, 5467–5472.

GALLI, R. AND MCCARTY, P.L. (1989a). Biotransformation of 1,1,1-trichloroethane, trichloromethane and tetrachloromethane by a *Clostridium* sp. *Applied and Environmental Microbiology* **55**, 837–844.

GALLI, R. AND MCCARTY, P.L. (1989b). Kinetics of the biotransformation of 1,1,1-trichloromethane by *Clostridium* sp. strain TCA 11B. *Applied and Environmental Microbiology* **55**, 845–851.

GERRITSE, J., VAN DER WANDE, B.J. AND GOTTSCHAL, J.C. (1992). Specific removal of chlorine from the *ortho* position of halogenated benzoic acids by reductive dechlorination in anaerobic enrichment cultures. *FEMS Microbiology Letters* **100**, 273–280.

GIBSON, D.T., CRUDEN, D.L., HADDOCK, J.D., ZYLSTRA, G.J. AND BRAND, J.M. (1993). Oxidation of polychlorinated biphenyls by *Pseudomonas* sp. strain LB400 and *Pseudomonas pseudoalcaligines* KF707. *Journal of Bacteriology* **175**, 4561–4564.

GIBSON, D.T., ROBERTS, R.L., WELLS, M.C. AND KOBAL, V.M. (1973). Oxidation of biphenyl by a *Beijerinckia* species. *Biochemical and Biophysical Research Communications* **50**, 211–219.

GORDON, A.S. AND MILLERO, F.J. (1985). Adsorption mediated decrease in the biodegradation rate of organic compounds. *Microbial Ecology* **11**, 289–298.

GOULDING, C., GULLIN, C.J. AND BOLTON, E. (1988). Biodegradation of substituted benzenes. *Journal of Applied Bacteriology* **65**, 1–5.

GRIFFIN, R.A. AND CHIAN, E.S.K. (1979). Attenuation of water soluble PCBs by earth materials. *EGN 86 Illinois State Geological Survey*, Champaign, p. 98.

GUERIN, W.F. AND BOYD, S.A. (1992). Differential bioavailability of soil-sorbed naphthalene to two bacterial species. *Applied and Environmental Microbiology* **58**, 1142–1152.

GUERIN, W.F. AND JONES, G.E. (1988). Mineralisation of phenanthrene by a *Mycobacterium* sp. *Applied and Environmental Microbiology* **54**, 937–944.

HALEY, M.V., VICKERS, E.J., CHEG, T.-U., DEFRANK, J., JUSTUS, T.A. AND LANDIS, W.G. (1990). Biodegradation and aquatic toxicity reduction of the persistent riot control material, 1,4-dibenz-oxazephine. In *American Society for Testing and Materials ASTM STP 1096* (W.G. Landis and W.M. van der Scahlie, Eds).

HALLER, H.D. AND FINN, R.R. (1979). Biodegradation of 3-chlorobenzoate and the formation of black colour in the presence and absence of benzoate. *European Journal of Applied Microbiology and Biotechnology* **8**, 191–205.

HAMDY, N.K. AND LIN, Y.C. (1983). Biodegradation of chlorinated hydrocarbons. *83rd AGM of the American Society of Microbiology, New Orleans, 6 March.*

HAMELINK, J.L., WAYBRANT, R.C. AND BALL, R.C. (1971). A proposal: exchange equilibria control the degree chlorinated hydrocarbons are biologically magnified in Lentic environments. *Transactions of the American Fish Society* **100**, 207.

HAQUE, R. AND SCHMEDDING, D. (1976). Studies on the adsorption of selected polychlorinated biphenyls and isomers on several surfaces. *Journal of Environmental Science and Health* **B11**, 129.

HARDMAN, D.J. (1991). Biotransformation of halogenated compounds. *Critical Reviews in Biotechnology* **11**, 1–40.

HARKNESS, M.R., MCDERMOTT, J.B., ABRAMOWICZ, D.A., SALVO, J.J., FLANAGAN, W.P., STEPHENS, M.L., MONDELLO, F.J., MAY, R.J., LOBOS, J.H., CARROL, K.M., BRENNAN, M.J., BRACCO, A.A., FISH, K.M., WARNER, G.L., WILSON, P.R., DIETRICH, D.K., LIN, D.T., MORGAN, C.B. AND GATELY, W.L. (1993). *In situ* stimulation of aerobic PCB degradation in Hudson River sediments. *Science* **259**, 503–507.

HARTMANN, J., ENGELBERTS, K., NORDHAUS, B., SCHMIDT, E. AND REINEKE, W. (1989). Degradation of 2-chlorobenzoate by *in vivo* constructed hybrid pseudomonads. *FEMS Microbiology Letters* **61**, 17–22.

HAVEL, J. AND REINEKE, W. (1991). Total degradation of various chlorobiphenyls by cocultures and *in vivo* constructed hybrid pseudomonads. *FEMS Microbiology Letters* **78**, 163–170.

HAVEL, J. AND REINEKE, W. (1992). Degradation of Aroclor 1221 and survival of strains in soil microcosms. *Applied Microbiology and Biotechnology* **38**, 129–134.

HAVEL, J. AND REINEKE, W. (1993a). Microbial degradation of chlorinated acetophenones. *Applied and Environmental Microbiology* **59**, 2706–2712.

HAVEL, J. AND REINEKE, W. (1993b). Degradation of Aroclor 1221 in soil by a hybrid pseudomonad. *FEMS Microbiology Letters* **108**, 211–218.

HAYASE, N., TAIRA, K. AND FURUKAWA, K. (1990). *Pseudomonas putida* KF715 *bphABCD* operon encoding polychlorinated biphenyl degradation: cloning, analysis, and expression in soil bacteria. *Journal of Bacteriology* **172**, 1160–1164.

HEALTH AND SAFETY EXECUTIVE (1993). Occupational Exposure Limits 1993, EH40/93.

HERNANDEZ, B.S., HIGSON, F.K., KONDRAT, R. AND FOCHT, D.D. (1991). Metabolism of and inhibition of chlorobenzoates in *Pseudomonas putida* P111. *Applied and Environmental Microbiology* **57**, 3361–3366.

HICKEY, W.J. AND FOCHT, D.D. (1990). Degradation of mono, di, and trihalogenated benzoic acids by *Pseudomonas aeroginosa* JB2. *Applied and Environmental Microbiology* **56**, 3842–3850.

HICKEY, W.J., BRENNER, V. AND FOCHT, D.D. (1992). Mineralisation of 2-chloro and 2,5-dichlorobiphenyl by *Pseudomonas* sp. strain UCR2. *FEMS Microbiology Letters* **98**, 175–180.

HICKEY, W.J., SEARLES, D.B. AND FOCHT, D.D. (1993). Enhanced mineralisation of polychlorinated biphenyls in soil inoculated with chlorobenzoate degrading bacteria. *Applied and Environmental Microbiology* **59**, 1194–1200.

HIGSON, F.K. AND FOCHT, D.D. (1990). Bacterial degradation of ring chlorinated acetophenones. *Applied and Environmental Microbiology* **56**, 3678–3685.

HILL, D.L., PHELPS, T.J., PALUMBO, A.V., WHITE, D.C., STRANDBERG, G.W. AND DONALDSON, T.L. (1989). Bioremediation of polychlorinated biphenyls: degradation capabilities in field lysimeters. *Applied Biochemistry and Biotechnology* **20/21**, 233–243.

HOLDEN, A.V. (1986). The reliability of PCB analysis. In *PCBs and the Environment*, Vol. 1 (J.S. Waid, Ed.), pp. 66–78. CRC Press, Boca Raton.

HOROWITZ, A., SUFLITA, J.M. AND TIEDJE, J.M. (1983). Reductive dehalogenations of halobenzoates by anaerobic lake sediment microorganisms. *Applied and Environmental Microbiology* **45**, 1459–1465.

HORVARTH, R.S. AND ALEXANDER, M. (1970). Cometabolism of *m*-chlorobenzoate by an *Arthrobacter*. *Applied Microbiology* **20**, 254–258.

HOUSE OF COMMONS SELECT COMMITTEE ON THE ENVIRONMENT (1990). First Report on Contaminated Land. Vol. I. xliv (143).

HOWARD, H.K. (1990). Biodegradation of aqueous hazardous waste leachates in a pilot-scale rotating biological contractor. In *Innovative Hazardous Waste Treatment Technology*. Vol. 3: *Biological Processes* (H.M. Freeman and P.R. Sferra, Eds), pp. 1–6. Technomic Publishing.

HUANG, S. AND GIBSON, G.G. (1992). Species and congener specific induction of hepatic cytochrome P450 4a by PCBs. *Biochemical Pharmacology* **43**, 637–639.

HUGHES, D.E. (1965). The metabolism of halogen substituted benzoic acids by *Pseudomonas fluorescens*. *Biochemical Journal* **96**, 181–188.

HUTZINGER, O., SAFE, S. AND ZITKO, V. (1974). *The Chemistry of PCBs*. CRC Press, Cleveland, Ohio.

ICRCL (1987). *Guidance on the Assessment and Redevelopment of Contaminated Land*. Guidance Note ICRCL 59/83, (2nd Edn). ICRCL, DoE, Romney House, 43 Marsham St, London SW1 3PY, UK.

JENSEN, S. (1972). The PCB story. *Ambio* **1**, 123–131.

JOHNSON, H.W., BRIGGS, G.G. AND ALEXANDER, M. (1972). Metabolism of 3-chlorobenzoic acid by a pseudomonad. *Soil Biology and Biochemistry* **4**, 187–190.

JOHNSTON, P.A. AND SIMMONDS, M.P. (1989). *A brief overview of the health and environmental effects of PCBs*. Submission to the Welsh Affairs Committee (12/12/89).

JONES, K.C., SANDERS, G., WILD, R., BURNETT, V. AND JOHNSTON, A.E. (1992). Evidence for a decline of PCBs and PAHs in rural vegetation and air in the United Kingdom. *Nature* **356**, 137–140.

KELLY, M.P., HALLBERG, K.B. AND TUOVINEN, O.H. (1989). Biological degradation of 2,4-dichlorophenoxyacetic acid: chloride mass balance. *Applied and Environmental Microbiology* **55**, 2717–2719.

KHAN, A. AND WALIA, S. (1989). Cloning of bacterial genes specifying degradation of 4-chlorobiphenyl from *Pseudomonas putida* OU843. *Applied and Environmental Microbiology* **55**, 2717–2719.

KHAN, A. AND WALIA, S. (1989). Cloning of bacterial genes specifying degradation of 4-chlorobiphenyl from *Pseudomonas putida* OU843. *Applied and Environmental Microbiology* **55**, 798–805.

KIMBARA, R., HASHMOTO, T., FUKUDA, M., ROANA, T., TAKAGI, M., OISHI, M. AND YANO, R. (1989). Cloning and sequencing of two tandem genes involved in degradation of 2,3-dihydroxybiphenyl to benzoic acid in the polychlorinated biphenyl-degrading soil bacterium *Pseudomonas* sp. strain KKS102. *Journal of Bacteriology* **171**, 2740–2747.

KIMBROUGH, R.D. (1980). Occupational exposure. In *Halogenated Biphenyls, Terphenyls, Naphthalenes, Dibenzodioxins and Related Products* (R.D. Kimbrough, Ed.), Chapter 9c. Topics in Environmental Health, Vol. 4. Elsevier/North-Holland Biomedical Press, Amsterdam.

KING, G.M. (1986). Inhibition of microbial activity in marine sediments by a bromophenol from a hemichordate. *Nature* **323**, 257.

KLECKA, G.M. AND GIBSON, D.T. (1981). Inhibition of catechol-2,3- dioxygenase from *Pseudomonas putida* by 3-chlorocatechol. *Applied and Environmental Microbiology* **41**, 1159–1165.

KOHLER, H.-P.E., KOHLER-STAUB, D. AND FOCHT, D.D. (1988). Cometabolism of polychlorinated biphenyls: enhanced transformation of Aroclor 1254 by growing bacterial cells. *Applied and Environmental Microbiology* **54**, 1940–1945.

KOKOZKA, L. AND FLOOD, J. (1985). A guide to EPA-approved PCB disposal methos. *Chemical Engineering*, 8 July, 41–43.

KONG, H.-L. AND SAYLER, G.S. (1983). Degradation and total mineralisation of monohalogenated biphenyls in natural sediment and mixed bacterial culture. *Applied and Environmental Microbiology* **46**, 666–672.

KRONE, U.E., LAUFER, K., THAUER, R.K. AND HOGENKAMP, H.P.C. (1989). Reductive dehalogenation of chlorinated C^1-hydrocarbons mediated by corrinoids. *Biochemistry* **28**, 4908–4914.

KRUMME, M.L. AND BOYD, S.A. (1988). Reductive dechlorination of chlorinated phenol in anaerobic upflow reactors. *Water Research* **22**, 171–177.

LAJOIE, C.A., ZYLSTRA, G.J., DEFLAUN, M.F. AND STROM, P.F. (1993). Development of Field application vectors for bioremediation of soils contaminated with polychlorinated biphenyls. *Applied and Environmental Microbiology* **59**, 1735–1741.

LAMAR, R.T., LARSEN, M.J., KIRK, T.K. AND GLASER, J.A. (1988). Growth of the white-rot fungus *Phanerochaete chrysosporium* in soil. *Proceedings of the 13th Annual Research Symposium on Land Disposal Remedial Action, Incineration and Treatment of Hazardous Waste*, pp. 411–418. US Environmental Protection Agency, EPA/600/2-86/040.

LEAGUE, J. (1990). Liquid solids contact (LSC) or how to bug hazardous waste to death. In *Innovative Hazardous Waste Treatment Technology*. Vol. 3: *Biological Processes* (H.M. Freeman and P.R. Sferra, Eds), pp. 19–24. Technomic Publishing.

LEE, M.C., GRIFFIN, R.A., MILLER, M.L. AND CHIAN, E.S.K. (1979). Adsorption of water soluble PCB Aroclor 1242 and used capacitor fluid by soil materials and coal chars. *Journal of Environmental Science and Health* **A14**, 415.

LESTAN, D., CERNILEC, M., STRANCAR, A. AND PERDIH, A. (1993). Influence of some surfactants and related compounds on ligninolytic activity of *Phanerochaete chrysosporium*. *FEMS Microbiology Letters* **106**, 17–22.

LEWANDOWSKI, G.A., ARMENANTS, P.M. AND PAK, D. (1990). Reactor design for hazardous waste treatment using a white rot fungus. *Water Research* **24**, 75–82.

LINKFIELD, T.G., SUFLITA, J.M. AND TIEDJE, J.M. (1989). Characterisation of the acclimation period before anaerobe dehalogenation of halobenzoates. *Applied and Environmental Microbiology* **55**, 2773–2778.

LIU, D. (1980). Enhancement of PCBs biodegradation by sodium ligninsulfonate. *Water Research* **14**, 1467–1475.

LOVELOCK, J.E. (1975). Natural halocarbons in the air and in the sea. *Nature* **256**, 193.

LOWE, J.I., PARRISH, P.R., PATRICK, J.M., JR AND FORRESTER, J. (1972). Effects of the PCB Aroclor 1254 on the American oyster *Crassastrea virginica*. *Marine Biology* **17**, 209.

LUNT, D. AND EVANS, C. (1970). The microbial metabolism of biphenyl. *Biochemical Journal* **118**, 54–55.

MASSÉ, R., MESSIER, F., PELOQUIN, L., AYOTTE, C. AND SYLVESTRE, M. (1984). Microbial degradation of 4-chlorobiphenyl: a model compound of chlorinated biphenyls. *Applied and Environmental Microbiology* **47**, 947–951.

MATTHEWS, H.B. AND ANDERSON, M. (1975). PCB dechlorinated versus PCB distribution and excretion. *Proceedings of the National Conference on PCBS* (F.A. Ayer, compiler), p. 50. US EPA Office of Toxic Substances, US Dept. Commerce, Washington, DC.

MAY, H.D., BOYLE, A.W., ALLENPRICE II, W. AND BLAKE, C.K. (1992). Subculturing of a polychlorinated biphenyl – dechlorinating anaerobic enrichment on solid media. *Applied and Environmental Microbiology* **58**, 4051–4054.

MIKESELL, M.D. AND BOYD, S.A. (1988). Enhancement of pentachlorophenol degradation in soil through induced anaerobiosis and bioaugmentation with anaerobic sewage sludge. *Environmental Science and Technology* **22**, 1411–1414.

MILBY, T.H., MILLER, P.E. AND FORRESTER, T.L. (1985). PCB-containing transformer fires: decontamination guidelines based on health consideration. *Journal of Occupational Medicine* **27**, 351–356.

MOEN, J.E.T. (1988). Soil protection in the Netherlands. In *Contaminated Soil '88*, Vol. 2. (W. Wolf, W.J. van der Brink and F.J. Colon, Eds), p. 1495. Kluwer Academic Publishers, Dordrecht.

MOHN, W.W. AND TIEDJE, J.M. (1990). Strain DCB-1 conserves energy for growth from reductive dechlorination coupled to formate oxidation. *Archives of Microbiology* **153**, 267–271.

MOHN, W.W. AND TIEDJE, J.M. (1992). Microbial reductive dehalogenation. *Microbiology Reviews* Sept., 482–507.

MOKROSS, H., SCHMIDT, E. AND REINEKE, W. (1990). Degradation of 3- chlorobiphenyl by *in vivo* constructed hybrid pseudomonads. *FEMS Microbiology Letters* **71**, 179–186.

MONDELLO, F.J. (1989). Cloning and expression in *Escherichia coli* of *Pseudomonas* strain LB400 genes encoding polychlorinated biphenyl degradation. *Journal of Bacteriology* **171**, 1725–1732.

MOZA, P., WEISBERGER, I. AND KLEIN, W. (1976). Fate of 2,2'-dichlorobiphenyl-^{14}C in carrots, sugarbeets and soil under outdoor conditions. *Journal of Agricultural and Food Chemistry* **24**, 881.

MULLIN, M.D., POCHINI, C.M., MCCRINDLE, S., ROMKES, M., SAFE, S.H. AND SAFE, L.M. (1984). High-resolution PCB analysis: synthesis and chromatographic properties of all 209 PCB congeners. *Environmental Science and Technology* **18**, 468–476.

MURPHY, T.J., MULLIN, M.D. AND MEYER, J.A. (1987). Equilibration of polychlorinated biphenyls and toxaphene with air and water. *Environmental Science and Technology* **21**, 155–162.

NADIM, L.M., SCHOCKEN, M.J., HIGSON, F.K., GIBSON, D.T., BEDARD, D.L., BOBB, L.H. AND MONDELLO, F.J. (1988). Bacterial oxidation of polychlorinated biphenyls. *Proceedings of the 13th Annual Research Symposium on Land Disposal Remedial Action, Incineration and Treatment of Hazardous Waste*, pp. 395–402. US Environmental Protection Agency, EPA/600/2-86/040.

NEILSON, A.H., ALLARD, A.S., LINDGREN, C. AND REMBERGER, M. (1987). Transformations of chloroguaiacols, chloroveratroles, and chlorocatechols by stable consortia of anaerobic bacteria. *Applied and Environmental Microbiology* **53**, 2511–2519.

NEILSON, A.H., ALLARD, A.S., HYNNING, P.A. AND REMBERGER, M. (1988). Transformations of halogenated aromatic aldehydes by metabolically stable anaerobic enrichment cultures. *Applied and Environmental Microbiology* **54**, 2226–2236.

NG, W.J., YAP, M.G.S. AND SIVADAS, M. (1989). Biological treatment of a pharmaceutical wastewater. *Biological Wastes* **29**, 299–311.

NIES, L. AND VOGEL, T.M. (1991). Identification of the proton source for the microbial reductive dechlorination of 2,3,4,5,6-pentachlorobiphenyl. *Applied and Environmental Microbiology* **57**, 2771–2774.

OECD (1987). *Control of PCB Waste*. Environmental Monograph No. 12.

OGRAM, A.V., JESSUP, R.E., LOU, L.T. AND RAO, P.S.C. (1985). Effects of sorption on biological degradation rates of (2,4-dichlorophenoxy) acetic acid in soils. *Applied and Environmental Microbiology* **49**, 582–587.

OLDENHUIS, R., VINCK, R.L.J.M., JANSSEN, D.B. AND WITHOLT, B. (1989). Degradation of chlorinated aliphatic hydrocarbons by *Methylosinus trichosprorium* OB3b expressing soluble methane monooxygenase. *Applied and Environmental Microbiology* **55**, 2819–2826.

PAL, D., WEBER, J.B. AND OVERCASH, M.R. (1980). The fate of PCBs in soil plant systems. *Residue Reviews* **74**, 45–98.

PARSONS, J.R., SIJM, D.T.H., VAN LAAR, A. AND HUTZINGER, O. (1988). Biodegradation of chlorinated biphenyls and benzoic acids by a *Pseudomonas* strain. *Applied Microbiology and Biotechnology* **29**, 81–84.

PAUL, E.A. AND CLARK, F.E. (1989). *Soil Microbiology and Biochemistry*. Academic Press, San Diego.

PELOQUIN, L. AND GREER, C.W. (1993). Cloning and expression of the polychlorinated biphenyl-degradation gene cluster from *Arthrobacter* M5 and comparison to analogous genes from Gram-negative bacteria. *Gene* **125**, 35–40.

PETTIGREW, C.A., BREEN, A., CORCORAN, C. AND SAYLER, G.S. (1990). Chlorinated biphenyl mineralisation by individual populations and consortia of freshwater bacteria. *Applied and Environmental Microbiology* **56**, 2036–2045.

QUENSEN, J.F., BOYD, S. AND TIEDJE, J. (1990). Dechlorination of four commercial polychlorinated biphenyl mixtures (Aroclors) by anaerobic microorganisms from sediments. *Applied and Environmental Microbiology* **56**, 2360–2369.

QUENSEN, J.F., TIEDJE, J.M., BOYD, S.A., ENKE, C., LOPSHIRE, R., GIESY, J., MORA, M., CRAWFORD, R. AND TILLIT, D. (1992). Evaluation of the suitability of reductive dechlorination for remediating PCB- contaminated soils and sediments. In *Soil Decontamination using Biological Processes*. DECHEMA, Frankfurt, Germany.

REINEKE, W. AND KNACKMUSS, H.-J. (1978a). Chemical structure and biodegradability of halogenated aromatic compounds. Substituent effects on 1,2-dioxygenation of benzoic acid. *Biochimica et Biophysica Acta* **542**, 412–423.

REINEKE, W. AND KNACKMUSS, H.-J. (1978b). Chemical structure and biodegradability of halogenated aromatic compounds. Substituent effects on dehydrogenation of 3,5-cyclohexadiene-1,2-diol-1-carboxylic acid. *Biochimica et Biophysica Acta* **542**, 424–429.

REINEKE, W. AND KNACKMUSS, H.-J. (1979). Construction of haloaromatics utilising bacteria. *Nature* **277**, 385–386.

REINEKE, W. AND KNACKMUSS, H.-J. (1980). Hybrid pathway for chlorobenzoate metabolism in *Pseudomonas* B13 derivatives. *Journal of Bacteriology* **142**, 467–473.

REINEKE, W. AND KNACKMUSS, H.-J. (1988). Microbial degradation of Haloaromatics. *Annual Reviews of Microbiology* **42**, 263–287.

REINEKE, W., WESSELS, S.W., RUBIO, M.A., LATORRE, J., SCHWEIN, U., SCHMIDT, E., SCHLOMANN, M. AND KNACKMUSS, H.-J. (1982). Degradation of monochlorinated aromatics following transfer of genes encoding chlorocatechol catabolism. *FEMS Microbiology Letters* **14**, 291–294.

RICE, C.P., EADIE, B.J. AND ERTSFELD, K.M. (1983). Enrichment of PCBs in Lake Michigan surface films. *Journal of Great Lakes Research* **8**, 265–270.

RIJNAARTS, H.H.M., BACHMANN, A., JUMELET, J.C. AND ZEHNDER, A.J.B. (1990). Effect of desorption and intraparticle mass transfer on the aerobic biomineralization of α-hexachlorocyclohexane in a contaminated calcareous soil. *Environmental Science and Technology* **24**, 1349–1354.

SAFE, S.H. (1990). Polychlorinated biphenyls (PCBs), dibenzo-*p*-dioxins (PCDDs), dibenzofurans (PCDFs) and related compounds: environmental and mechanistic considerations which support the development of toxic equivalency factors (TEFs). *Critical Reviews in Toxicology* **21**, 51–88.

SAFE, S. AND HUTZINGER, O. (1971). Chlorine randomization between phenyl groups in the electron impact-induced fragmentation of polychlorinated biphenyls. *Journal of the Chemical Society. Chemical Communications* 446–448.

SAFE, S. AND HUTZINGER, O. (1972). Mass spectra of polychlorinated biphenyls. *Journal of the Chemical Society. Perkin Transactions* **1**(5), 686–691.

SHANNON, M.J.R. AND UNTERMAN, R. (1993). Evaluating bioremediation: distinguishing fact from fiction. *Annual Reviews in Microbiology* **47**, 715–738.

SHELTON, D.R. AND TIEDJE, J.M. (1984). Isolation and partial characterisation of an anaerobic consortium that mineralizes 3-chlorobenzoic acid. *Applied and Environmental Microbiology* **48**, 840–848.

SHIELDS, M.S., HOOPER, S.W. AND SAYLER, G.S. (1985). Plasmid-mediated mineralisation of 4-chlorobiphenyl. *Journal of Bacteriology* **163**, 882–889.

SHIVARAMAN, N. AND PARHAD, N.M. (1990). Biological treatment of cyanide wastewaters. In *Innovative Hazardous Waste Treatment Technology*. Vol. 3: *Biological Processes* (H.M. Freeman and P.R. Sferra, Eds), pp. 39–46. Technomic Publishing.

SKEA, J.C., SIMONIN, H.A., BEAN, H.J., COLQUHOUN, J.R., SPAGNOLI, J.J. AND VEITH, G.D. (1979). Bioaccumulation of Aroclor 1016 in Hudson River fish. *Bulletin of Environmental Contamination and Toxicology* **22**, 332.

SONDOSSI, M., SYLVESTRE, M. AND AHMAD, D. (1992). Effects of chlorobenzoate transformation on the *Ps. testosteroni* biphenyl and chlorobiphenyl degradation pathway. *Applied and Environmental Microbiology* **58**, 485–495.

STONE, R. (1992). Swimming against the PCB tide. *Science* **255**, 798–799.

STORR-HANSEN, E. (1991). Comparative analysis of 30 PCB congeners and two capillary columns of different polarity with non-linear multi-level calibration. *Journal of Chromatography* **558**, 375–391.

SUFLITA, J.M., ROBINSON, J.A. AND TIEDJE, J.M. (1983). Kinetics of microbial dehalogenation of haloaromatic substrates in methanogenic environments. *Applied and Environmental Microbiology* **45**, 1466–1473.

SUIDA, J.F. AND DE BERNARDIS, J.F. (1973). Naturally occurring halogenated organic compounds. *Lloydia* **36**, 107.

SUZUKI, M. (1975). Translocation of polychlorobiphenyls in soil into plants. *Archives of Environmental Contamination and Toxicology* **5**, 343.

SYLVESTRE, M.V., MAILHIOT, K. AND AHMAD, D. (1989). Isolation and preliminary characterisation of a 2-chlorobenzoate degrading *Pseudomonas. Canadian Journal of Microbiology* **35**, 439–443.

TIEDJE, J.M., QUENSEN, J.F., CHEE-SANFORD, J., SCHIMEL, J.P. AND BOYD, S.A. (1993). Microbial reductive dechlorination of PCBs. *Biodegradation* **4**, 231–240.

TOPP, E. AND HANSON, R.S. (1990). Degradation of pentachlorophenol by a *Flavobacterium* species grown in continuous culture under various nutrient limitations. *Applied and Environmental Microbiology* **56**, 541–544.

TUCKER, E.S., LITSCHGI, W.J. AND MEES, W.M. (1975). Migration of polychlorinated biphenyls in soil induced by percolating water. *Bulletin of Environmental Contamination and Toxicology* **13**, 86.

UNTERMAN, R., BRENNAN, M.J., BROOKS, R.E. AND JOHNSON, C. (1986). Biological degradation of polychlorinated biphenyls. In *Biotechnology for Degradation of Toxic Chemicals in Hazardous Wastes* (R.J. Scholze, Ed.), Noyes Data Corp, Park Ridge, NJ.

UNTERMAN, R., BEDARD, D.L., BRENNAN, M.J., BOPP, L.H., MONDELLO, F.J., BROOKS, R.E., MOBLEY, D.P., MCDERMOTT, J.B., SCHWARTZ, C.C. AND DIETRICH, D.K. (1988). Biological approaches for polychlorinated biphenyl degradation. In *Environmental Biotechnology: Reducing Risks from Environmental Chemicals Through Biotechnology* (G.S. Omenn, Ed.), pp. 253–269. Plenum Press, New York.

VAN DORT, H.M. AND BEDARD, D.L. (1991). Reductive *ortho* and *meta* dechlorination of a polychlorinated biphenyl congener by anaerobic microorganisms. *Applied and Environmental Microbiology* **57**, 1576–1578.

VINEY, I. AND BEWLEY, R.J.F. (1988). *Microbial degradation of polychlorinated biphenyls.* Welsh Office Report. 0802883.

VINEY, I. AND BEWLEY, R.J.F. (1990). Preliminary studies on the development of a microbiological treatment for polychlorinated biphenyls. *Archives of Environmental Contamination and Toxicology* **19**, 789–796.

VISSCHER, K. AND BRINKMAN, J. (1989). Biological degradation of zenobiotics in waste management. *Hazardous Waste and Hazardous Materials* **6**, 201–212.

VOGEL, T.M., CRIDDLE, C.S. AND MCCARTY, P.L. (1987). Transformation of halogenated aliphatic compounds. *Environmental Science and Technology* **21**, 722–736.

WALKER, N. AND HARRIS, D. (1970). Metabolism of 3-chlorobenzoic acid by *Azotobacter* sp. *Soil Biology and Biochemistry* **2**, 27–32.

YATES, J.R. AND MONDELLO, F.J. (1989). Sequence similarities in the genes encoding PCB degradation by *Pseudomonas* strain LB400 and *Alcaligines eutrophus* H850. *Journal of Bacteriology* **171**, 1733–1735.

YE, D.Y., QUENSEN, J.F., TIEDJE, J.M. AND BOYD, S.A. (1992). Anaerobic dechlorination of polychlorobiphenyls (Aroclor 1242) by pasteurized and ethanol-treated microorganisms from sediments. *Applied and Environmental Microbiology* **58**, 1110–1114.

ZAITSEV, G.M. AND KARASEVICH, Y.N. (1982). Utilisation of 2-chlorobenzoic acid. *Mikrobiologiya* **53**, 75–80.

6
The Structures and Catalytic Mechanisms of Active-Site Serine β-Lactamases

JOSETTE LAMOTTE-BRASSEUR[1], JAMES KNOX[2], JUDITH A. KELLY[2], PAULETTE CHARLIER[1], EVELINE FONZÉ[1], OTTO DIDEBERG[3] AND JEAN-MARIE FRÈRE[1]

[1] Centre d'Ingénierie des Protéines et Laboratoire d'Enzymologie, Université de Liège, Belgium; [2] Department of Molecular and Cell Biology, University of Connecticut, CT, USA and [3] Laboratoire de Cristallographie Macromoléculaire, Institut de Biologie Structurale, Grenoble, France

Introduction: DD-transpeptidases, β-lactamases and penicillin resistance

The success of β-lactam antibiotics, penicillins, cephalosporins and related compounds in antibacterial chemotherapy rests both on their high efficiency and on their specificity. They interfere with the last stage of peptidoglycan synthesis, a transacylation reaction, most often involving a D-alanyl–D-alanine terminated peptide (*Figure 1*). This reaction, catalysed by enzymes attached to the outer face of the cytoplasmic membrane, is unique to the bacterial world and has absolutely no equivalent in eukaryotic cells (Frère *et al.*, 1992). The DD-transpeptidases are active-site serine enzymes (Frère and Joris, 1985) which perform their catalytic cycle according to an acylation/deacylation pathway, usually represented by the three-step model shown in the left branch of *Figure 2* and similar to the well- known mechanism of active-site serine proteases such as chymotrypsin and trypsin, although many DD-transpeptidases generally behave as more efficient transacylases than these latter enzymes. It is, however, rather clear that this model represents a simplification of the real situation (Jamin, Wilkin and Frère, 1993). The right branch of *Figure 2* depicts the interaction between β-lactam antibiotics and the DD-transpeptidases. By contrast to their catalytic equivalent, ES*, the adducts formed with the β-lactams are rather stable with k_{+3} values in the 10^{-4} to 10^{-6} s^{-1} range. Since these acylenzymes are unable to fulfil their physiological role, functionally impaired peptidoglycan is formed, leading eventually to cell death, a phenomenon in which the activation of cell autolysins also sometimes appears to be involved (Tomasz, 1979).

Abbreviations:CARB, carbenicillin-hydrolysing β-lactamases; EREF, enzymic rate enhancement factor; NMR, nuclear magnetic resonance; PBPs, penicillin-binding proteins; WT, wild-type.

Biotechnology and Genetic Engineering Reviews — Vol. 12, December 1994
0264–8725/94/12/189–230 $20.00 + $0.00

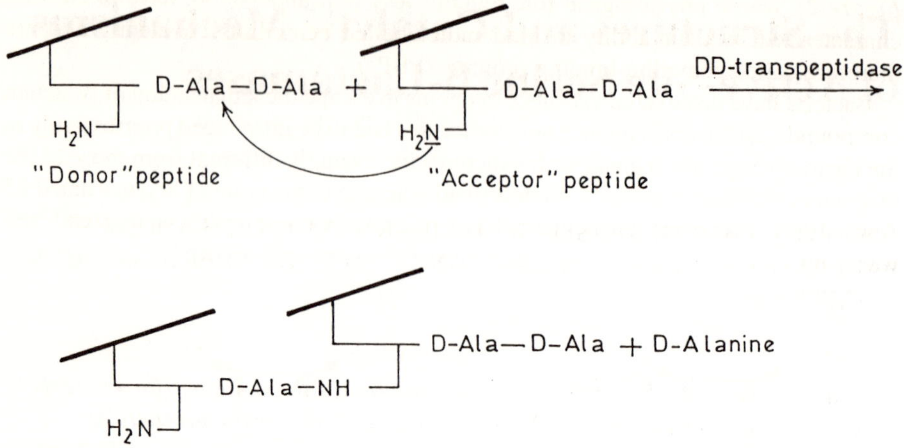

Figure 1. The transpeptidation reaction catalysed by the bacterial DD-peptidases. The heavy lines represent the glycan moiety of the nascent peptidoglycan network.

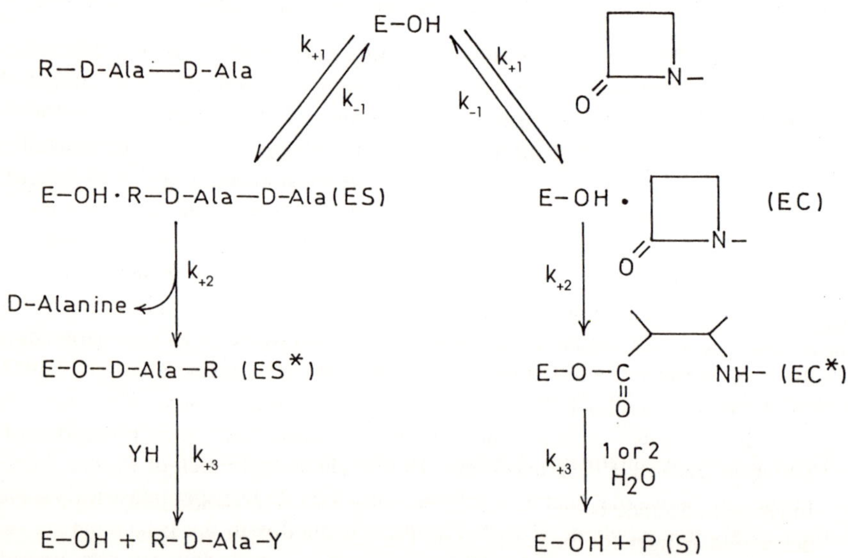

Figure 2. Interaction between the bacterial DD-transpeptidases (left branch) and their physiological substrates (R-DAla-DAla = donor substrate; YH = acceptor substrate; see *Figure 1*) and β-lactams (right branch). With the substrates, the values of k_{+2} and k_{+3} are high, thus ensuring a high catalytic turnover rate. By contrast, the efficiency of β-lactams as inactivators rests on high k_{+2}/K' values [where $K' = (k_{+2} + k_{-1})/k_{+1}$] and very low k_{+3} values, resulting in a nearly complete immobilization of the enzyme as the inactive adduct EC*. Some DD-carboxypeptidases have also been described. In this case, the acceptor is water (YH = H_2O). DD-trans- and carboxypeptidases form a family of DD-peptidases (Joris *et al.*, 1988). Deacylation of EC* can occur by the direct hydrolysis of the ester bond, or involve a rate-limiting hydrolysis of the penicillin C_5–C_6 bond, followed by rapid deacylation.

The DD-transpeptidases and the related DD-carboxypeptidases (see the legend to *Figure 2*), whose physiological role remains less well understood, have often been characterized by their ability to bind radiolabelled penicillins covalently and are therefore called penicillin-binding proteins (PBPs).

Bacteria have discovered various strategems to escape the lethal action of β-lactam compounds, and in consequence new molecules had to be introduced progressively in the chemotherapeutic arsenal, with structures increasingly different from those of the original penicillins (*Figure 3*). As described at length in the issue of *Science* dated 15 April 1994, resistance phenomena have recently become quite widespread and worrying.

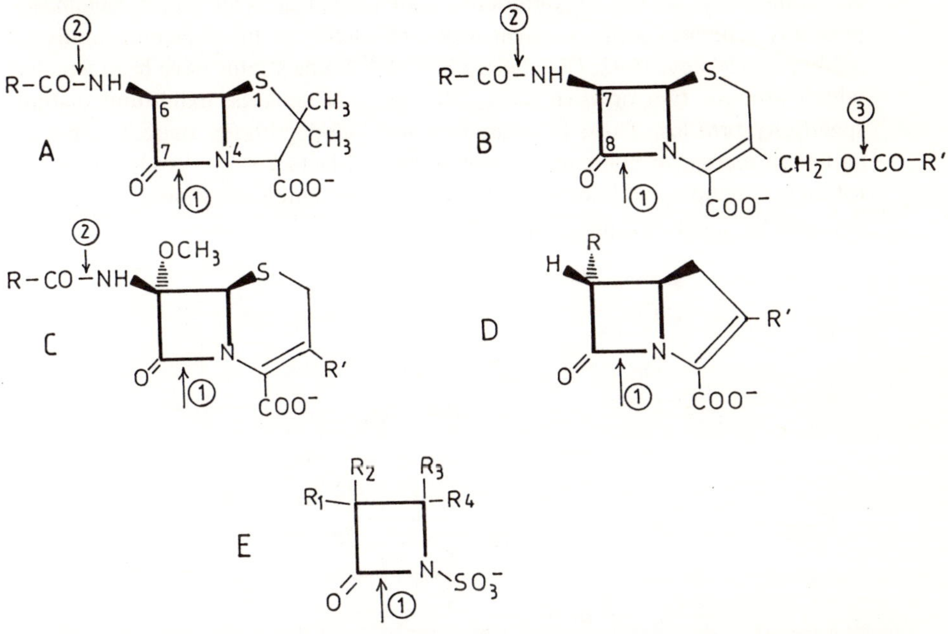

Figure 3. Structures of (A) penicillins; (B) cephalosporins; (C) cephamycins; (D) carbapenems; and (E) monobactams. The arrows indicate the sites of action of (1) β-lactamases, (2) acylases and (3) esterases. Note that some cephalosporins do not exhibit an ester side-chain on C_3. The chemical structure common to all these compounds is the four-membered β-lactam ring, highlighted by heavier lines.

Three distinct mechanisms of resistance have been identified:

1. Intrinsic resistance is due to the synthesis of DD-transpeptidases with a strongly decreased affinity for the antibiotics. The resistant PBPs are very slowly acylated by the β-lactams, the k_{+2}/K' values being as low as 10 M^{-1} s^{-1} as against up to 300 000 M^{-1} s^{-1} for penicillin-sensitive DD-peptidases (Frère and Joris, 1985; Frère *et al.*, 1992). This intrinsic resistance to β-lactams is sometimes accompanied by resistance to other families of antibiotics, which can result in the emergence of deadly strains. A frightening example is the methicillin-resistant *Staphylococcus aureus* against which the only presently available efficient compounds are those of the vancomycin family.
2. The diffusion of the antibiotic to its membrane-bound targets can be impeded by the outer layers of the cell wall itself, the 'outer membrane' in Gram-negative

bacteria (Nikaido and Normark, 1987; Waley, 1987; Frère *et al.*, 1989), and the mycolic acid layers of the Gram-positive mycobacteria. In the former, increased resistance has been attributed to modifications of the porins, proteins which form hydrophilic channels in the outer membrane and allow small, water-soluble molecules to penetrate into the periplasmic space (Nikaido and Normark, 1987).

3. Last, but not least, bacteria can produce enzymes which chemically modify the antibiotics: acylases, esterases or β-lactamases (*Figures 3* and *4*). Of these only β-lactamases appear to be seriously involved in resistance phenomena. By contrast to those formed after the action of the two other types of hydrolytic enzymes, the products obtained by the irreversible opening of the β-lactam ring are completely devoid of significant antibacterial activity and β-lactamases presently represent a major threat to the efficiency of the β-lactam family of antibiotics (Waley, 1992; Coyette *et al.*, 1994). Some strains have been isolated which produce two or even three different β-lactamases, exhibiting distinct specificity profiles. Thus, β-lactamases and DD-peptidases share a common ability to recognize compounds containing a β-lactam ring. As shown in the following sections, additional, even more striking similarities have been found between the latter enzymes and a large number of β-lactamases.

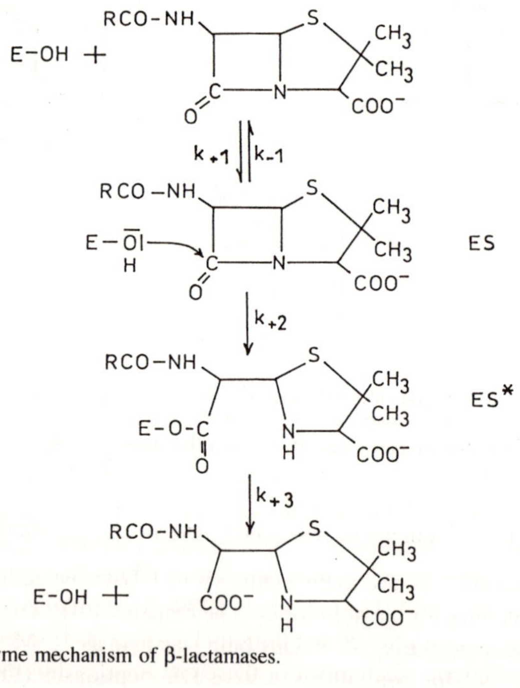

Figure 4. The acylenzyme mechanism of β-lactamases.

Metallo- and active-site serine β-lactamases

The hydrolytic properties of β-lactamases rest on two distinct catalytic mechanisms (Waley, 1992). Some enzymes are Zn^{2+}-containing metalloproteins, while the vast majority of known β-lactamases are active-site serine enzymes, just like the DD-peptidases.

For many years, only innocuous strains of *Bacillus cereus* were known to produce

Zn^{2+}-β-lactamases which were, in consequence, considered as mere biochemical curiosities. More recently, an increasing number of pathogenic strains of the *Bacteroides, Xanthomonas, Aeromonas, Enterobacter, Pseudomonas* and *Serratia* genera have been detected, characterized by a high resistance to carbapenems, compounds which generally escape the activity of the more common active-site serine β-lactamases. This resistance is due to the synthesis of Zn^{2+}-β-lactamases (Payne, 1993). It can be predicted that the increased utilization of carbapenems will result in the spreading of genes coding for the Zn^{2+}-β-lactamases to an ever larger number of strains and it would be advisable to develop the search for specific inhibitors for these enzymes. Most of them are monomeric, with *M*r values of about 23 000. The enzyme produced by *Xanthomonas maltophilia* is, however, tetrameric, but devoid of allosteric properties. The known sequences are clearly homologous (Felici *et al.*, 1993). In the *B. cereus* enzyme, three His residues are responsible for liganding the Zn^{2+}ion (Sutton *et al.*, 1987). These three His residues are found in equivalent positions in the other sequences, including that of the *Xanthomonas* monomer, but in the *Aeromonas* enzyme, one of them is replaced by Asn. A conserved Cys residue is also present, not far from the Zn^{2+} in the *B. cereus* enzyme structure, but its exact role remains mysterious, as well as the catalytic mechanism of these enzymes (for more details see Waley, 1992).

The active-site serine β-lactamases, however, represent a more immediate problem. They hydrolyse the antibiotics according to the acylation/deacylation pathway described by *Figure 4* and characterized by the following values of the steady-state kinetic parameters (Waley, 1992):

$$k_{cat} = \frac{k_{+2} \cdot k_{+3}}{k_{+2} + k_{+3}}$$

$$K_m = \frac{k_{+3} \cdot K'}{k_{+2} + k_{+3}}$$

$$\frac{k_{cat}}{K_m} = \frac{k_{+2}}{K'} \text{ where } K' = \frac{k_{+2} + k_{-1}}{k_{+1}}.$$

The efficiency of these enzymes can be really frightening. Some of them have indeed been described as 'perfect catalysts' (Christensen, Martin and Waley, 1990), with k_{+2}/K' ratios close to the diffusion-limit values (10^8 M^{-1} s^{-1}) combined with very high k_{+3} values (4000 s^{-1}). Moreover, in constitutive mutants of some Gram-negative strains, enormous amounts of enzyme are produced, yielding periplasmic β-lactamase concentrations up to 1 mM (Hechler *et al.*, 1989). Under these conditions, a sensitive antibiotic has no chance of ever reaching its DD-peptidase target. A detailed understanding of the catalytic mechanism of these enzymes is thus essential to the design of new molecules which can escape their hydrolytic action.

The interactions between β-lactams and both the active-site serine β-lactamases and DD-peptidases thus occur according to identical or very similar reaction pathways (compare *Figures 2* and *4*) with a major difference at the level of the k_{+3} value, which is usually very high with the former and very low with the latter. Practically, this quantitative difference translates into qualitatively distinct behaviours: β-lactams inactivate DD-peptidases while β-lactamases destroy the same molecules.

Genetic data and the three classes of active-site serine β-lactamases

On the basis of their primary structures these enzymes have been divided into three molecular classes A, C and D (for historical reasons, class B contains the metallo-β-lactamases). Within each class, the catalytic properties can exhibit strong variations, but the sequences are clearly homologous and newly discovered enzymes are easily distributed among these classes (Waley, 1992; Joris *et al.*, 1991).

Table 1. Molecular classification of β-lactamases and genetic characteristics

Class	Examples	Type of bacteria	Gene location	Induction
A	TEM	Gram –	Plasmid	C
	SHV	Gram –	Plasmid	C
	CARB	Gram –	Plasmid	C
	Bacillus licheniformis	Gram +	Chromosome	I[a]
	Bacillus cereus I	Gram +	Chromosome	I[a]
	Staphylococcus aureus	Gram +	Chromosome or plasmid	I[b]
	Streptomyces albus G	Gram +	Chromosome	C
	Streptomyces cacaoi ULG	Gram +	Chromosome	I
	Actinomadura R 39	Gram +	Chromosome	C
B	*Bacillus cereus* II	Gram +	Chromosome	C[a]
	Aeromonas hydrophila	Gram –	Chromosome	C
	Xanthomonas maltophilia	Gram –	Chromosome	I
	Bacteroides fragilis	Gram –	Chromosome	C[c]
	Serratia marcescens *Enterobacter cloacae* *Pseudomonas aeruginosa*	Gram –	Chromosome or plasmid	C[d]
	Pseudomonas aeruginosa	Gram –	Plasmid	C
C	*Escherichia coli* K 12	Gram –	Chromosome	C
	Citrobacter freundii	Gram –	Chromosome	I
	Serratia marcescens	Gram –	Chromosome	I[a]
	Enterobacter cloacae	Gram –	Chromosome	I[a]
	Pseudomonas aeruginosa	Gram –	Chromosome	I
	BIL-1, MIR-1	Gram –	Plasmid	C
D	OXA	Gram –	Plasmid	C
	PSE-2	Gram –	Plasmid	C

C, Constitutive; I, inducible.
[a] Constitutive mutants (*Bacillus licheniformis* 749/C, *Enterobacter cloacae* P99 and 908R) are deregulated and produce large amounts of enzyme.
[b] Constitutive production in some strains.
[c] Inducible in other *Bacteroides* strains or species.
[d] Probably the same or very similar protein.
An alternative classification, based on the catalytic profiles, has been proposed by Bush (1989a). Detailed references about the listed enzymes can be found in the same article.
CARB, carbenicillin-hydrolysing β-lactamases.

The structural β-lactamase genes are found both on the chromosome and on plasmids (*Table 1*). In this latter case, the danger of rapid spreading through an initially penicillin-sensitive population is evident. Gram-positive bacteria only produce class A and class B β-lactamases which are most often chromosome-encoded. In Gram-negative bacteria, class B and class C enzymes are usually chromosome-encoded (but some exceptions are known) and class D β-lactamases are always plasmid-encoded. The Gram-negative class A enzymes are usually plasmid-encoded, but in *Klebsiella* and *Proteus* the gene is on the chromosome, probably on a transposon. The most widespread plasmids code for the TEM and SHV β-lactamases which are consequently responsible for many clinical problems. The production of the enzymes is either constitutive or inducible. In *Staphylococci*, the β-lactamase regulation system seems to share some elements with that of the penicillin-resistant PBP 2' (or 2a) (Joris, Hardt and Ghuysen, 1994). Synthesis of the class C β-lactamase is inducible in most Enterobacteria, with the notable exception of *Escherichia coli*. The strong selective pressure exerted in the hospital environment results in the emergence, with a high frequency, of constitutive overproducers of a class C enzyme. This is due to the inactivation of the *amp*D gene, whose product, a 21 000 M_r protein, is involved in the cytoplasmic recycling of peptidoglycan fragments. In *amp*D mutants, one of these fragments accumulates and triggers a continuous production of the enzyme by binding to the AmpR protein, which in turn acts as a transcription activator (Jacobs *et al.*, 1994a, 1994b).

Interestingly, a very similar regulation system, involving an AmpD and an AmpR-like protein also prevails in *Proteus vulgaris*, but the β-lactamase is a class A enzyme (Datz *et al.*, 1994). Similarly, an *amp*R gene, controlling the production of a class A enzyme, has been identified in the Gram-positive *Streptomyces cacaoi* (Lenzini *et al.*, 1992).

Sequence and structure comparisons

A large number of class A β-lactamase sequences are known and they exhibit a high degree of diversity, reflecting a similar diversity in the catalytic profiles (Matagne *et al.*, 1990; Ambler *et al.*, 1991). Only nine residues are strictly conserved and, even if highly conservative substitutions such as Leu → Ile are neglected, this number does not increase above 25, out of a total chain length of 270–290 residues. Nevertheless, superposition of the peptide backbones of the four class A enzymes whose structures have been established by X-ray diffraction highlights a strongly conserved general architecture (Dideberg *et al.*, 1987; Herzberg, 1991; Jelsch *et al.*, 1992; Lamotte-Brasseur *et al.*, 1991; Moews *et al.*, 1990; Knox and Moews, 1991), with the exception of a short stretch (residues 80–105) where insertions occur in several enzymes. The molecules present mixed α–β structures, with an all-helical domain on the left (*Figure 5a*) and a mixed α/β domain on the right, composed of five antiparallel β-strands covered by three helices. The helices and β-sheet comprise about 45% and 25% of the polypeptide, respectively. The active-site serine is situated in a depression between the two domains, at the N-terminus of the eleven residue α_2 hydrophobic helix mostly buried in the core of the all-α domain. Despite the fact that computer-aided sequence comparisons fail to reveal significant isology between the members of class A and class C, striking similarities in the general folding of the proteins have been detected when the structures of two class C enzymes were established (Oefner *et al.*, 1990; Lobkovsky *et al.*, 1993). A similar, two-domain organization was found (*Figure 5b*), with an identical positioning of the active-site serine between the two domains at the N-terminus of the α_2 helix. In the class C enzymes, four additional β-strands form a small antiparallel sheet, which is twisted with respect to the major, five-stranded sheet common to both classes of enzymes, while helix α_{10} is antiparallel to α_{11}, rather than parallel. The most important difference occurs at the bottom of the active site, where loops are found in both structures, but running antiparallel to each other (*Figure 6*). In class A enzymes, this loop is referred to as the Ω-loop and contains one turn of helical structure, helix α_7.

(a)

(b)

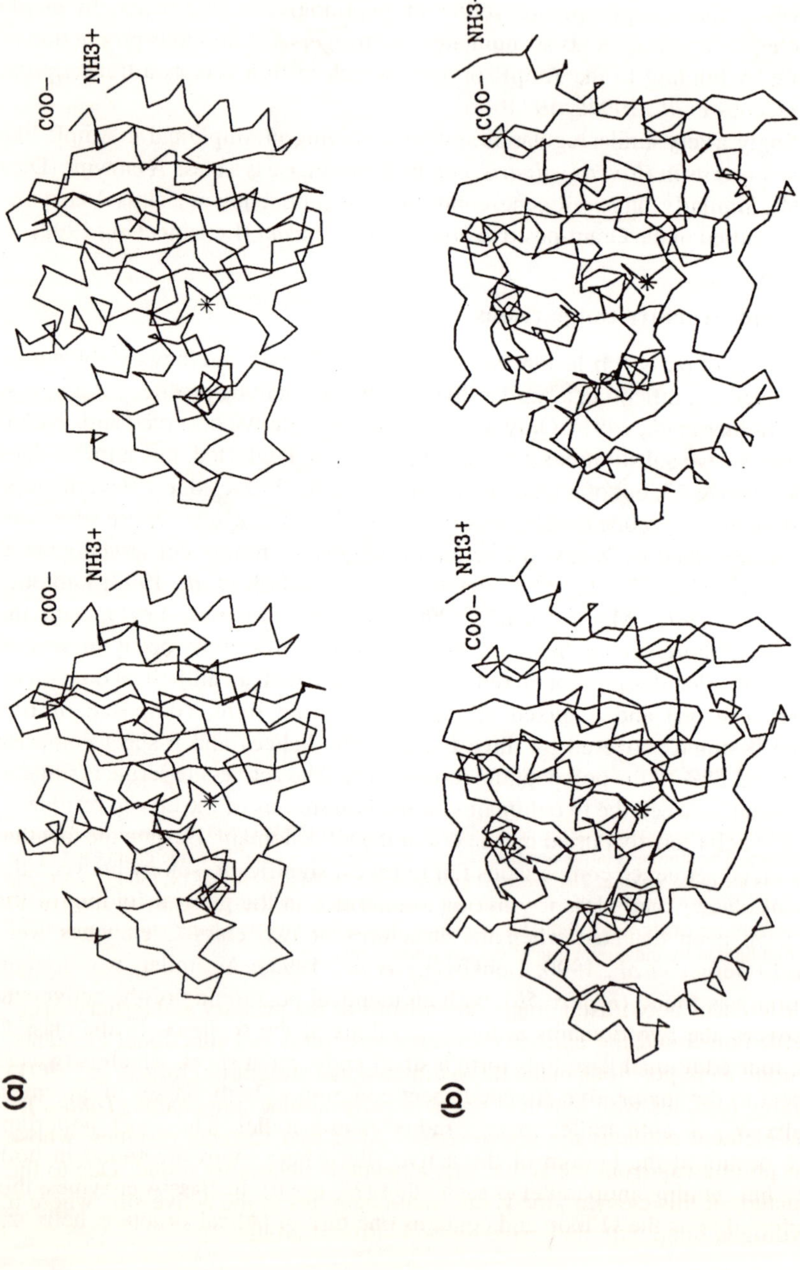

Figure 5. Tertiary structures, represented by the α-carbon traces of (a) the *Streptomyces albus* G (class A) and (b) the *Enterobacter cloacae* P99 (class C) β-lactamases. The active-site serine is labelled (*).

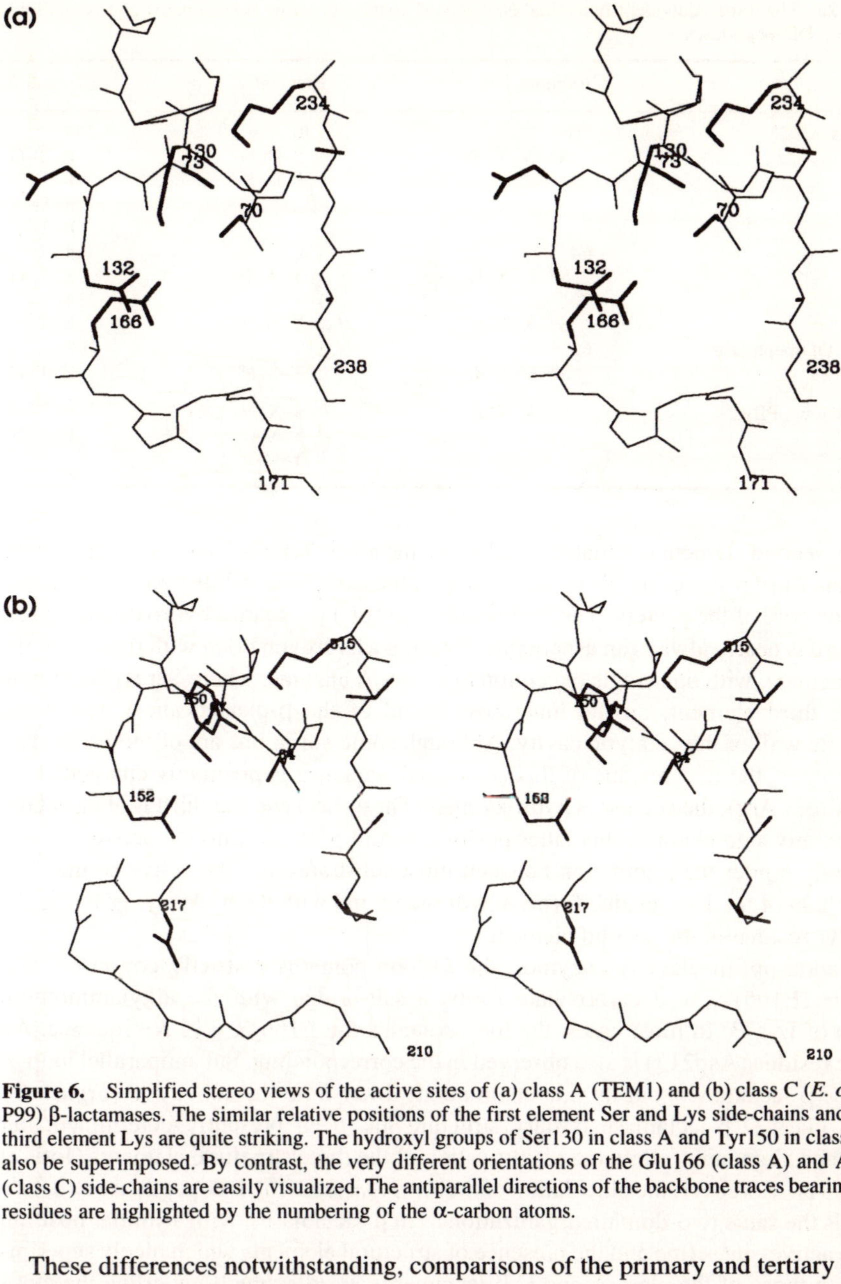

Figure 6. Simplified stereo views of the active sites of (a) class A (TEM1) and (b) class C (*E. cloacae* P99) β-lactamases. The similar relative positions of the first element Ser and Lys side-chains and of the third element Lys are quite striking. The hydroxyl groups of Ser130 in class A and Tyr150 in class C can also be superimposed. By contrast, the very different orientations of the Glu166 (class A) and Asp217 (class C) side-chains are easily visualized. The antiparallel directions of the backbone traces bearing these residues are highlighted by the numbering of the α-carbon atoms.

These differences notwithstanding, comparisons of the primary and tertiary structures have demonstrated the existence of three structural and functional elements, located in equivalent positions in the three-dimensional structures (Joris *et al.*, 1991) and containing residues with identical or chemically similar side-chains (*Table* 2).

The first element contains the active serine. After two variable residues, whose side-chains point away from the active site, a lysine is invariably found. Due to the helical structure of this element, the lysine side-chain lies in the active site, where it forms a hydrogen bond with the active serine hydroxyl group.

Table 2. The three equivalent functional elements of active-site serine β-lactamases and penicillin-sensitive DD-peptidases

	Element 1	Element 2	Element 3
Class A	70 S*–X–X–K	130 S–D–N S–D–S	234 K–T–G K–S–G R–S–G R–T–G
Class C	64 S*–X–X–K	150 Y–A–N	314 K–T–G
Class D	70 S*–X–X–K	144 Y–G–N	214 K–T–G
S. R61 DD-peptidase	62 S*–V–T–K	159 Y–S–N	298 H–T–G
Other known PBPs	S*–X–X–K	S–X–N S–X–C Y–G–N	K–T–G K–S–G

The second element is situated on a loop in the all-α domain. The side-chains of the first and third residues in this sequence border the active site, while that of the second is in the core of the protein. The first residue (Ser or Tyr) bears a hydroxyl group and the third is nearly always an asparagine, forming a hydrogen bond with the Lys of the first element, with one single exception in a class A enzyme, where Ser replaces Asn.

The third element, on the innermost strand of the β-pleated sheet, forms the opposite wall of the catalytic cavity. Although some variations are observed in class A enzymes, the first residue of this conserved sequence is positively charged (Lys, sometimes Arg), the second is hydroxylated (Thr or Ser) and the third is always Gly. In fact, any side-chain in this latter position would protrude into the active site and sterically hinder the interaction between most substrates and the active serine. The side-chain of the Lys residue forms a hydrogen bond with the hydroxyl group of the Ser/Tyr residue in the second element.

In addition, in class A enzymes, the Ω-loop contains a strictly conserved Glu residue (E166) whose carboxylate forms a salt-bridge with the alkylammonium group of Lys73. In most cases, the loop contains the E166 X E X N sequence. An acidic residue (Asp217) is also observed in the corresponding, but antiparallel loop of class C β-lactamases, but its side-chain points *away* from the active site (*Figure 6*).

No class D β-lactamase tertiary structure has been presently determined, but sequence alignments allow the identification of the three structural elements (Joris *et al.*, 1991). Moreover, the elucidation of the *Streptomyces* R61 DD-peptidase structure reveals the same two-domain organization as in β-lactamases, with a similar position of the active-site serine and the presence of structural elements which nicely superimpose on those of the class A and C β-lactamases, results which underline the close relationships between the two families of penicillin-recognizing enzymes (Kelly *et al.*, 1989). The three elements can be identified in the sequences of all penicillin-binding proteins, irrespective of the exact affinity that these enzymes exhibit for the β-lactam antibiotics (*Table 2*).

In the high-molecular-mass penicillin-binding proteins, whose M_r value can be as high as 100 000, the penicillin-binding and putative transpeptidase domain which contains these elements is located in the C-terminal portion of the protein, whereas the N-terminal part supplies the membrane-anchoring segment and, at least in some cases, a polypeptide chain catalysing the transglycosylation, the reaction which immediately precedes transpeptidation in the peptidoglycan biosynthesis (Ghuysen, 1991).

It is interesting to note that the fold of class C β-lactamases is much closer to that of the *S*.R61 DD-peptidase than is the fold of the class A enzymes (Lobovsky *et al.*, 1993). The class C and *S*.R61 enzymes also exhibit a tyrosine as the first residue of the second element, in contrast to the class A β-lactamases and most PBPs where this residue is a serine.

The similarity between β-lactamases and DD-peptidases is further underlined by the finding that various compounds behave as substrates of both types of enzymes. Indeed, even if the β-lactamases fail to recognize the peptide substrates, both types of enzymes hydrolyse linear esters, thiolesters and an aziridine (*Table 3*). The class C β-lactamases even catalyse transacylation reactions with ester and thiolester carbonyl donors and D-amino acids as acceptors (Pratt and Govardhan, 1984; Pazhanisamy, Govardhan and Pratt, 1989; Adam *et al.*, 1990; Jamin *et al.*, 1991).

Table 3. Linear substrates of DD-peptidases and β-lactamases

Substrate	k_{cat}/K_m value (M^{-1} s^{-1})		
	S.R61 DD-peptidase	Class A β-lactamase	Class C β-lactamase
Peptide I Ac_2-LLys-DAla-DAla	4000 Tp:+	0	0
Ester II[a] C_6H_5-CO-Gly-O-CH-COOH (with CH_2-C_6H_5 on CH)	5500 Tp:+	80	20 000 Tp:+
Thiolester III C_6H_5-CO-Gly-S-CH_2-COOH	100 000 Tp:+	4500	7000 Tp:+
Thiolester IV C_6H_5-CO-DAla-S-CH_2-COOH	400 000 Tp:+	<10	180
Aziridine V C_6H_5-CH_2-CO-Gly-N-CH-COOH (aziridine ring)	600	120[b] 1900[c]	130

The class A and class C β-lactamases are those of *Bacillus licheniformis* 749/C and of *Enterobacter cloacae* P99, respectively.
Tp, Transpeptidation or transacylation.
[a] The phenyllactate has a *D* configuration.
[b] and [c] Results obtained by Murphy and Pratt (1991) with other class A enzymes: [b] TEM1 and [c] *Bacillus cereus* I.

At this point, it is interesting to compare some kinetic parameters of the different enzymes. *Table 4* highlights the very low k_{+3} values of the DD-peptidase and striking differences in the rates of acylation which are generally much higher with the β-lactamases, although cefoxitin acylates the *S* R61 DD-peptidase more readily than the class A enzymes. The relative resistance of some compounds to hydrolysis by several class A enzymes is often due to low acylation rates (cefuroxime, cefotaxime and TEM-1) while in class C, this is most often due to slow deacylation (ampicillin, cefuroxime, cefotaxime and cefoxitin). The wide diversity in the catalytic properties of class A enzymes has been discussed by Matagne *et al.* (1990) and Raquet *et al.* (1994). The class C β-lactamases seem to form a more homogeneous group (Galleni,

Table 4. Some kinetic parameters characterizing the interactions between penicillin-recognizing enzymes and their substrates or inactivators

Substrate	β-lactamases								*Streptomyces* R61 DD-peptidase	
	Class A						Class C			
	Bacillus licheniformis		TEM1		*Staphylococcus aureus* PC1		*Enterobacter cloacae* P99			
	k_{cat} (s^{-1})	k_{cat}/K_m (mM^{-1} s^{-1})	k_{cat} (s^{-1})	k_{cat}/K_m (mM^{-1} s^{-1})	k_{cat} (s^{-1})	k_{cat}/K_m (mM^{-1} s^{-1})	k_{cat} (s^{-1})	k_{cat}/K_m (mM^{-1} s^{-1})	k_{+3} (s^{-1})	k_{+2}/K' (mM^{-1} s^{-1})
Benzylpenicillin	2200	29000	1600	84000	400	80000	15	35000	1.4×10^{-4}	17
Ampicillin	1500	11000	1000	33000	750	50000	1	1200	1.4×10^{-4}	0.11
Carbenicillin	400	8000	220	25000	180	70	0.003	220	1.4×10^{-4}	0.17
Cephaloridin	630	5000	1500	2200	0.07	100	700	10000	-	0.15
Cephalothin	50	2500	160	650	0.006	8	200	20000	3×10^{-6}	0.8
Cefuroxime	16	170	6	6	0.07	1.2	0.05	3000	4×10^{-6}	0.8
Cefotaxime[a]	7	30	9	1.5	Not detectable		0.015	1500	$<4 \times 10^{-6}$	0.016
Cefoxitin	-	0.02	0.004	0.006	Not detectable		0.06	2500	5×10^{-5}	1.5
References	Matagne *et al.* (1990)		Raquet *et al.* (1994)		Richard Virden, personal communication; Matagne and Frère, unpublished		Galleni, Amicosante and Frère (1988)		Frère and Joris (1985)	

a Various TEM variants hydrolyse this compound much more efficiently.
(-) Not determined.
Note that $k_{cat}/K_m = k_{+2}/K'$.

Amicosante and Frère, 1988) but these results concern a much smaller number of enzymes.

Mechanistic considerations

The catalytic mechanism of other active-site serine amidases and esterases has been studied in detail (see for instance Fersht, Blow and Fastrez, 1973; Brady *et al.*, 1990). In many of these, a histidine residue acting as a general base activates the serine during the acylation process and the hydrolytic water molecule during the deacylation. The oxyanion, which is expected to be formed at the level of the tetrahedral intermediate, is also stabilized by two hydrogen bonds, with the main chain-NH-groups of Ser195 and Gly193 in the case of chymotrypsin (*Figure 7*). Docking of the penicillin molecule in the active sites of various β-lactamases and of the *S.* R61 DD-peptidase indicates that similar 'oxyanion holes' can be located in all the structures, comprising the main chain -NH-groups of Ser70 and Ala237 in the class A enzymes, Ser64 and Ser/Ala318 in class C (*Figure 8*) and Ser62 and Thr301 in the DD-peptidase (Kelly *et al.*, 1989; Moews *et al.*, 1990; Oefner *et al.*, 1990; Herzberg, 1991; Jelsch *et al.*, 1992; Lobkovsky *et al.*, 1993). Conversely, no conserved histidine is found in β-lactamases and PBPs. The *Streptomyces albus* G class A β-lactamase does not even contain one such residue (Dehottay *et al.*, 1987). In consequence, the search

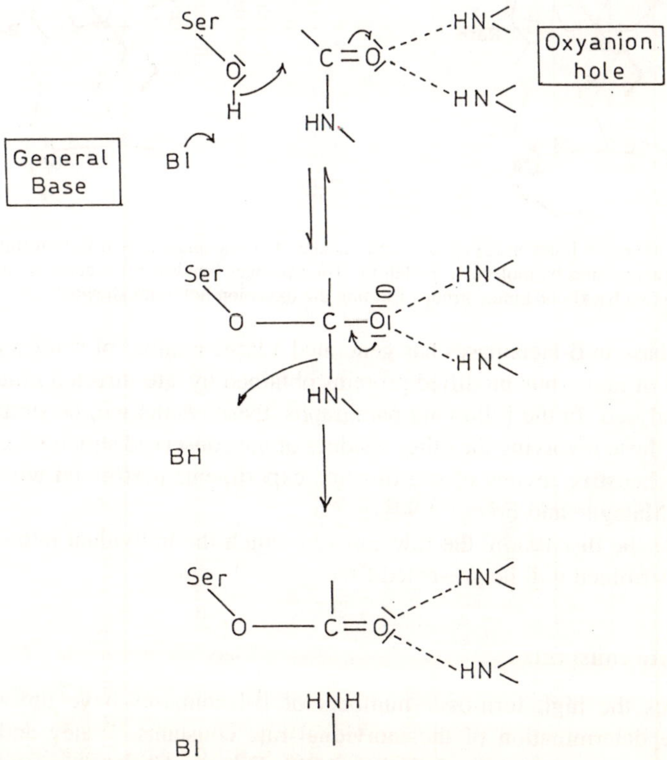

Figure 7. Formation of the tetrahedral intermediate during the hydrolysis of a peptide bond by an active-site serine enzyme, showing the combined actions of a general base (BI, His57 in chymotrypsin) and of the oxyanion hole and collapse of the tetrahedral intermediate to yield the acylenzyme.

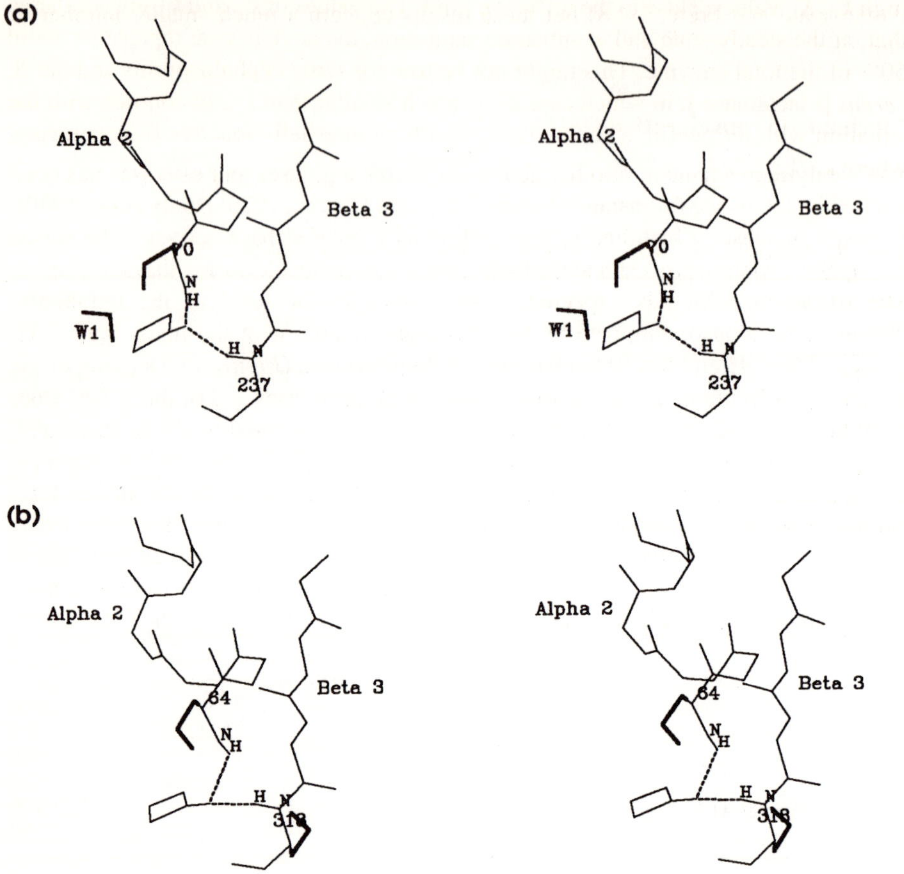

Figure 8. The oxyanion holes in (a) class A and (b) class C β-lactamases. A β-lactam ring has been positioned in the active sites by molecular modelling. The hydrogen bonds formed between the β-lactam carbonyl and the two backbone amide groups forming the oxyanion holes are shown.

for a general base in β-lactamases has generated a large number of studies in which the properties of numerous modified proteins obtained by site-directed mutagenesis have been analysed. In the following paragraphs, these results will be summarized, together with those involving the other residues of the conserved structural elements. For a more exhaustive review of site-directed experiments performed with class A enzymes see Matagne and Frère (1994).

To facilitate the discussion, the few cases in which the individual rate constants have been determined will be presented first.

Individual rate constants

In many cases the high turn-over numbers of β-lactamases have unfortunately precluded the determination of the individual rate constants. Waley and his co-workers (Christensen, Martin and Waley, 1990; Gibson, Christensen and Waley, 1990) have, however, succeeded in obtaining information with some class A enzymes and penicillin substrates (*Tables 5* and *6*). The interactions are characterized by very high k_{cat}/K_m values, close to the diffusion limit. The values of k_{+2} and k_{+3} are similar so that, at the steady-state and at substrate saturation, the acylenzyme represents about 50% of the total enzyme. This might not be true for some cephalosporins and the *B. cereus* β-lactamase I, in which case k_{+2} is much smaller than k_{+3}, in contrast with the situation with nitrocefin, a cephalosporin with an unusually reactive β-lactam ring, where $k_{+2}/k_{+3} = 0.4$.

Table 5. Individual rate constants for the hydrolysis of benzylpenicillin at 20°C

Rate constant (units)	*B. cereus* β-lactamase I	*S. aureus* PC1	TEM1
k_{+1} ($\mu M^{-1}\ s^{-1}$)	40	22	120
k_{-1} (s^{-1})	2300	200	12000
k_{+2} (s^{-1})	4100	170	2800
k_{+3} (s^{-1})	2000	100	1500
K' (μM)	160	17	120

Data from Christensen, Martin and Waley (1990).

Table 6. Values of k_{+2} and k_{+3} for some class A enzymes (at 20°C)

Enzyme	Substrate	$k_{+2}(s^{-1})$	$k_{+3}(s^{-1})$
Bacillus cereus β-lactamase I	Phenoxymethyl penicillin	5000	4500
	Phenethicillin	6500	2700
	Carbenicillin	1100	730
	Nitrocefin	30	71
Staphylococcus aureus PC1	Phenoxymethyl penicillin	130	100
	Phenethicillin	110	50

Data from Christensen, Martin and Waley (1990).

With class C β-lactamases, deacylation is generally rate-limiting, as shown by the increased reaction rates observed upon addition of methanol (Knott-Hunziker *et al.*, 1982). Values of k_{+2} could only be determined for some poor substrates, since those of good substrates and of many poor substrates were too high to allow estimation by stopped-flow methods (*Table 7*; Monnaie, Virden and Frère, 1992).

Table 7. Kinetic parameters for some substrates of the class C *Enterobacter cloacae* 908R β-lactamase (at 30°C)

Substrate	$k_{+2}(s^{-1})$	$k_{+3}(s^{-1})$	k_{+2}/k_{+3}	K'(μM)
Carbenicillin	3.3	0.003	2500	15
Ampicillin	120	1	120	100
Benzylpenicillin	>150	15	>10	>5
Cloxacillin	>150	0.006	>40000	>15
Cefotaxime	>150	0.006	>25000	>100
Imipenem	>150	0.003	>50000	>1000
Aztreonam	>150	0.0004	>700000	>800

Data from Monnaie, Virden and Frère (1992).

Site-directed mutagenesis

The major results are listed in *Tables 8–10*. A critical analysis of these data highlights a few clear conclusions but also underlines unexpected, distinct behaviours of the different enzymes, which makes it difficult to propose a 'unified' mechanism, not only for the β-lactamases and DD-peptidases, but also for the β-lactamases of different classes.

FIRST ELEMENT: THE ACTIVE-SITE SERINE

The replacement of the active Ser by Cys yields significantly poorer β-lactamases, with altered substrate profiles (Sigal *et al.*, 1984; Jacob, Joris and Frère, 1991). The acylation of the class C enzyme appears to be specifically impaired, while the k_{+3} value is barely modified (Dubus *et al.*, 1993). This indicates that the nucleophilic attack on the substrate carbonyl probably requires a more stringent geometry than the hydrolysis of the acylenzyme, which might be facilitated by the more pronounced reactivity of the thiolester. Accordingly, the equivalent mutant of the *S.* R61 peptidase completely loses both activity and penicillin-binding capacity (Hadonou *et al.*, 1992a), also suggesting defective acylation processes. In this respect, the fact that no active-cysteine β-lactamase or DD-peptidase has ever been found is worth mentioning. Surprisingly, the S70A mutant of the *S. albus* G β-lactamase retains a very low, but significant, activity, with a $kcat/K_m$ for benzyl penicillin of 200 M^{-1} s^{-1}, i.e. 0.01% of the wild-type (WT) protein. Other groups in the catalytic site somehow directly activate a hydrolytic water molecule, but the initial binding of the substrate also seems to be deficient (Jacob, Joris and Frère, 1991).

FIRST ELEMENT: THE LYSINE RESIDUE

The individual rate constants for the hydrolysis of benzylpenicillin by the K73R mutant of the *B. cereus* class A β-lactamase were determined by Gibson, Christensen and Waley (1990), who showed that k_{+2} was decreased 70-fold and k_{+3} less than 15-fold. A similarly impaired acylation step appears to prevail for the corresponding K67R and K65R mutants of the class C β-lactamase (Monnaie, Dubus and Frère, 1994) and of the DD-peptidase (Hadonou *et al.*, 1992a). With the former, this seems to be due to an increase of K' rather than to a decrease of k_{+2}, a result which underlines the fact that similar effects on the steady-state parameters can result from variations of different rate constants. The K67Q mutant of the class C enzyme is much more severly affected and this might not only suggest an electrostatic role for the positive charge on the 67 side-chain, but also highlight the central role of the dense array of hydrogen bonds in the active site, which may be partially conserved with an Arg, but not with a Gln side-chain.

With the DD-peptidase, the K65R is the only mutant where the modification of a conserved residue results in a smaller decrease of the enzymic activity than of the penicillin acylation rate, thus yielding a poor, but essentially pencillin-insensitive DD-peptidase.

Table 8. Kinetic characteristics of class A β-lactamase mutants

	Benzylpenicillin		Cephalosporin C		Cephalothin		Cephaloridine		Nitrocefin		Enzymes and references
	Relative values[a] of		Relative values[a] of		Relative values[a] of		Relative values[a] of		Relative values[a] of		
	k_{cat}	k_{cat}/K_m	k_{cat}	k_{cat}/K_m	k_{cat}	k_{cat}/K_m	k_{cat}	k_{cat}/K_m	k_{cat}	k_{cat}/K_m	
S70C	80	130	-	-	-	-	850	1500	-	-	TEM1, Sigal *et al.* (1984)
	-	6000[b]	-	>10000	-	100000	-	25000	-	-	*S. albus* G, Jacob, Joris and Frère (1991)
K73R	40	60	20	35	-	-	-	-	170	400	*B. cereus*, Gibson, Christensen and Waley (1990)
S130A	40	40	-	500	700	1000	-	3300	-	12	*S. albus* G,
S130G	>1	6	-	500	800	1000	-	800	-	4	Jacob *et al.* (1990a,b)
N132S	2	2	-	300	500	2000	430	1000	-	2000	*S. albus* G,
N132A	-	300	-	-	-	>100000	-	>10000	–	10000	Jacob *et al.* (1990b)
K234H, pH=5.5[c]	2	11	-	50	-	80	-	-	4	6	*S. albus*, G,
pH=7.0	10	63	-	-	-	440	-	-	-	48	Brannigan *et al.* (1991)
K234R	1	10	-	-	1.5	1.2	1.2	1.5	-	-	TEM1,
K234T	50	2000	-	-	>300	-	-	-	2.5	-	Lenfant, Labia and Masson (1991)
K234A	-	10000	-	-	-	-	-	-	-	7000	*B. licheniformis*,
K234E	50	3000	-	-	-	-	-	-	40	3500	Ellerby *et al.* (1990)
S235A	3	4	100	250	-	-	14	130	-	-	TEM1, Imtiaz *et al.* (1993a)
	4	3	10	260	7	270	45	300	<5	100	TEM1, Dubus *et al.* (1994a)
E166D	3300	500	200	180	-	-	-	-	170	230	*B. cereus*, Gibson, Christensen and Waley (1990)
E166A	3000	600	-	-	-	-	-	-	3000	200	*B. cereus*,
E166C	6000	1200	-	-	-	-	-	-	5000	300	Leung *et al.* (1994)
E166A	>10^7	-	-	-	-	-	-	-	>10^7	-	*B. licheniformis*, Escobar, Tan and Fink (1991)
E166Y	28000	8000	-	-	4300	80	2000	50	-	-	TEM1, Delaire *et al.* (1991)
E166N	>10^7	-	-	-	-	-	-	-	-	-	TEM1, Adachi, Ohta and Matsuzawa (1991)
R220L[d]	-	60	-	600	-	800	-	6	-	14	*S. albus* G, Jacob-Dubuisson *et al.* (1991)
R244K	0.9	1.5	-	-	-	-	1	1	1	1.3	TEM1,
R244S	1	8	-	-	-	-	1	1.6	1.5	2.5	Zafaralla *et al.* (1992)
R244T	6	60	-	-	240	300	140	1300	-	-	TEM1,
R244Q	2	20	-	-	240	300	70	700	-	-	Delaire *et al.* (1992)
M69L	1.6	2.5	-	–	5	5	4	4	-	-	TEM1, Delaire *et al.* (1992)

[a] Relative values = (parameter) wild type/(parameter) mutant; [b] Result obtained with ampicillin; [c] At pH > 8, k_{cat} is close to zero; [d] k_{cat}/K_m values increase with the methyl ester of benzylpenicillin and the lactone of cephalosporin C.

Table 9. Kinetic characteristics of class C β-lactamase mutants

	Benzylpenicillin		Cephalothin		Cephaloridine		Nitrocefin		Cefotaxime		Enzymes and references
	Relative values of		Relative values of		Relative values of		Relative values of		Relative values of		
	k_{cat}	k_{cat}/K_m	k_{cat}	k_{cat}/K_m	k_{cat}	k_{cat}/K_m	k_{cat}	k_{cat}/K_m	k_{cat}	k_{cat}/K_m	
S64C	1.5	5000	25000	100000	5000	7000	<50	2000	0.4	3000	*E. cloacae* 908R, Dubus *et al.* (1993)
K67R	1	1000	1.5	25	3	20	<4	500	0.1	250	*E. cloacae* 908R
K67Q	–	–	2500	45000	–	–	–	–	–	–	Monnaie, Dubus and Frère (1994)
Y150S	35	25	90	400	540	400	200	55	1	1.3	
Y150F	15	9	1500	6000	270	400	100	2	0.6	18	*E. coli* K12,Dubus *et al.* (1994b)
Y150E	60	150	120	6000	1800	860	800	1000	2.5	12	
Y150S	9	150	20000	7000	3800	15000	300	10	14	10000	*E. coli* K12,
Y150F	400	90	10000	350	4400	90	1800	17	100	–	Dubus and Frère, unpublished
K315H[a]	20	1400	300	300	–	–	<8	100	–	–	*E. cloacae* 908R,
K315Q	560	5600	1400	1200	–	–	<70	800	–	–	Monnaie *et al.* (1994)
T316A[a]	0.5	1.5	1.7	37	2	20	1	1.3	2	15	*E. coli* K12,
T316V	1.2	3	6	4000	<10	120	20	40	4	160	Dubus *et al.* (1994)

a Loss of transacylation properties.
–, Not determined.

Table 10. Kinetic characteristics of the *S.* R61 DD-peptidase mutants

	Peptide I		Ester II		Thiolester III			Benzyl-penicillin	Cephalosporin C,	Cefuroxime,	References
	Relative values of		Relative values of		Relative values of		Relative	Relative	relative	relative	
	k_{cat}	k_{cat}/K_m	k_{cat}	k_{cat}/K_m	k_{cat}	k_{cat}/K_m	T/H	k_{+2}/K'	k_{+2}/K'	k_{+2}/K'	
K65R	–	200	–	–	<1	1500	–	20000	–	–	Hadonou *et al.* (1992a)
Y159S	2500	3000	250	400	16	1.6	>50	1.7	3	0.5	Wilkin *et al.* (1993a)
Y159F	7500	8000	–	>5000	600	1000	>50	300	100	>70	
N161S	30	30	15	20	15	2.5	4	1	3	0.3	Wilkin *et al.* (1993b)
N161A	350	300	40	200	2.5	30	7	70	45	1	
H298K	<50	130	–	–	5	2.5	4	60	12	–	Hadonou *et al.* (1992b)
H298Q	100	180	–	–	4	10	22	100	60	–	
T299V	550	900	–	>5000	3	12	50	25	>15000	>2000	Wilkin *et al.* (1994)

Relative T/H is the $\frac{(\text{transpeptidation/hydrolysis})_{WT}}{(\text{transpeptidation/hydrolysis})_{mutant}}$ ratio under the same conditions of acceptor and donor substrate concentrations (WT = wild type).

Replacement of the following residues by uncharged side-chains did not significantly alter the activity: E172, E182, D195, D225, E228, D246, D248 (Hadonou *et al.*, 1992a; Wilkin and Frère, unpublished).

SECOND ELEMENT: THE SERINE/TYROSINE RESIDUE

Replacement of Ser130 by Gln in the *S. albus* G β-lactamase yields a nearly inactive protein since this bulky side-chain does not allow an adequate binding of the substrate. In the S130A and S130G mutants, the loss of the important 130–234 hydrogen bond results in a considerable destabilization, but the proteins retain some activity against benzylpenicillin (2.5 and 17%), methicillin (10 and 35%) and nitrocefin (8 and 25%). With all other substrates, the activity is below 1%. The *k*cat/*K*m values are generally more affected for cephalosporins (Jacob *et al.*, 1990a,b).

The analysis of the Y150F, Y150S and Y150E mutants of the *E. coli* K12 class C enzyme yields a particularly incoherent picture (Dubus *et al.*, 1994b) and significantly different results have been obtained in different laboratories (Dubus and Frère, unpublished). However, the general trend shows large decreases (100-fold or more) of *k*cat and *k*cat/*K*m for good substrates, while some poor substrates seem to be less affected. Apparently, the presence of slowly isomerizing forms of the free enzyme seriously complicates the analysis (Page, 1993).

The activity of both the Y159S and the Y159F mutants of the *S.* R61 DD-peptidase on the peptide and ester substrates is strongly impaired but, surprisingly, acylation of the former by the thiolester and β-lactams is barely modified – it is even increased with cefuroxime (Wilkin *et al.*, 1993a). The mutants are totally unable to catalyse the transpeptidation reaction.

SECOND ELEMENT: THE ASPARAGINE RESIDUE

The N132S mutant of the *S. albus* G enzyme exhibits a spectacular modification of its substrate profile: while retaining most of the penicillinase activity of the wild-type protein, it becomes a very poor cephalosporinase (Jacob *et al.*, 1990a), properties similar to those of the *S. aureus* β-lactamase, where residue 132 is, however, the usual Asn. A serine residue in position 132 is found in the *Bacillus cereus* β-lactamase III, an enzyme which is not specifically more active on penicillins than on cephalosporins. These results underline the difficulties encountered when one attempts to generalize the behaviour of even closely related proteins. Although more impaired, the behaviour of the N132A mutant of the *S. albus* G enzyme parallels that of its N132S counterpart.

The peptidase and esterase activities of the *S.* R61 DD-peptidase are also more affected by the N161A than by the N161S mutations and, strikingly, acylation of the latter by the thiolester and β-lactams are unaffected (Wilkin *et al.*, 1993b).

Preliminary results (A. Dubus, personal communication) indicate very impaired activities for the N152D, E, L and H mutants of a class C enzyme.

THIRD ELEMENT: THE LYSINE RESIDUE

In class A, the main function of this side-chain mainly appears to be to supply a positive charge. The K234R and, at low pH, the K234H mutants are fully active. Moreover, an R234 side-chain naturally occurs in the CARB class A enzymes (Lachapelle, Dufresne and Levesque, 1991; Couture, Lachapelle and Levesque, 1992) which are especially active against carbenicillin (Labia, Guionie and Barthélémy,

1981), a compound bearing a double negative charge at neutral pH (R = C6H5-CH(COO^-)–, *Figure 3A*). The other mutants, with a neutral or negative 234 side-chain are three orders of magnitude less active. Some of the results suggest that the positive charge might enhance both initial recognition and transition state stabilization (Lenfant, Labia and Masson, 1991), but the *k*cat (and *not* the *K*m) value of the K234H mutant closely parallels the titration of the imidazole group (Brannigan *et al.*, 1991), indicating a major role for the second factor.

The pH-dependency of the class C enzyme K315H mutant is quite different and the activity remains poor even at pH 5.0 (Monnaie *et al.*, 1994). Although this mutation increased its similarity to the R61 DD-peptidase, it resulted in a loss of the transacylation properties of the *E. cloacae* enzyme. Conversely, the H298K and H298Q mutants of the peptidase were significantly affected, but less so than the class C mutants, with an additional, specific impairment of the transpeptidation properties (Hadonou *et al.*, 1992b). Optimization of the respective activities of the two enzymes has thus been attained by the utilization of different side-chains. In this respect, it is also worth remembering that the *S.* R61 enzyme is the only known DD-peptidase with a His residue in the third element, all the other ones exhibiting β-lactamase-like KTG or KSG sequences.

THIRD ELEMENT: THE THREONINE/SERINE RESIDUE

The hydroxyl group of this residue is one of the very few functionalities to be strictly conserved in all the active-site serine penicillin-recognizing proteins, and its elimination was expected to result in dramatic decreases of the enzyme activities. Surprisingly, with the β-lactamases, this is only so for the cephalosporinase properties, suggesting that the 'natural' selective pressure responsible for the conservation of the hydroxyl group was due to cephalosporins rather than penicillins (Imtiaz *et al.*, 1993a; Dubus *et al.*, 1994a). The parallel strict conservation of a Ser or Thr in DD-peptidases is more easy to rationalize, since the T299V mutant of the *S.* R61 enzyme becomes a poor peptidase, practically devoid of transpeptidation properties, and consequently unable to efficiently participate in peptidoglycan biosynthesis (Wilkin *et al.*, 1994).

THE GLUTAMATE166 RESIDUE OF CLASS A β-LACTAMASES

This is certainly the active-site residue whose role in the catalytic process has been the subject of the most heated controversies. The experimental results also appear to be quite incoherent and difficult to reconcile. In several cases, substitutions by other residues (including the conservative Glu→Asp mutation) decrease both the *k*cat and *k*cat/*K*m parameters, with rather minor effects on the *K*m values. The E166D mutant of the *B. cereus* enzyme has been particularly well studied by Gibson, Christensen and Waley (1990), who observed similar, 2000-fold decreases of k_{+2} and k_{+3}. By contrast, two other mutants, E166A of *B. licheniformis* (Escobar, Tan and Fink, 1991) and E166N of TEM (Adachi, Ohta and Matsuzawa, 1991) are nearly completely unable to catalyse the deacylation step. The most surprising discrepancy is the enormous difference in the behaviours of the identical E166A mutants of the closely related *B. cereus* and *B. licheniformis* enzymes. The most recent data suggest, however, that the apparent k_{+3} value for the latter might be lowered by an isomerization of the acylenzyme into a non-productive form (Escobar *et al.*, 1994) and, in the same contribution, the authors confirm that the acid limb of the *k*cat/*K*m pH profile indeed reflects the protonation state of Glu166.

Catalytic mechanisms

THE GENERAL BASE

The nucleophilic attack of the substrate carbonyl group by an active-site serine side-chain is expected to be facilitated by a general base to which the proton of the serine hydroxyl group is transferred during this process. As noted above, this function is often fulfilled by a histidine residue in other active-site serine enzymes. The pH-dependencies of the *k*cat/*K*m values of class A β-lactamases suggest the involvement of a more acidic group, with a pK of 4.5–5.5 (Waley, 1975; Brannigan *et al.*, 1991). The properties of the E166D mutant of the *B. cereus* enzyme underline the central role of the Glu166 side-chain in both the acylation and deacylation steps.

GLU166 IN CLASS A ENZYMES

The side-chain of this residue takes part in a dense hydrogen-bond network within the active site. As seen above, no corresponding residue is found in that position in the class C enzymes or in the *S.* R61 DD-peptidase structure (*Figure 6*). Its importance in the catalytic process was suggested by the study of a natural variant of the *Staphylococcus aureus* enzyme (Herzberg *et al.*, 1991). The structures show, however, that the distance between the Glu166 carboxylate group and the active serine hydroxyl is too long to allow a direct proton transfer between the two side-chains. On the other hand, the crystallographic and modelling data (Lamotte-Brasseur *et al.*, 1991) clearly show that a conserved water molecule might serve as a relay in this proton transfer (*Figure 9*). This hypothesis is strengthened by modelling results obtained by 'docking' cefoxitin in the enzyme active site. This molecule, which contains a methoxy group on C7 (*Figure 3*) acylates the active serine of class A enzymes with an exceedingly poor efficiency (*Table 4*). It does, nevertheless, easily fit into the active site, but the 7-methoxy group displaces the water molecule (Matagne *et al.*, 1993), which can explain why no further reaction occurs after the formation of the non-covalent ES complex (*Figure 10*). Studies performed with the class A enzyme from *Staphylococcus aureus* PC1 and a covalently bound phosphonate acting as a transient state analogue seem to confirm this hypothesis (Rahil and Pratt, 1994).

By contrast, with the D166N mutant of the TEM β-lactamase, acylation appears to occur readily, while deacylation is extremely slow (Adachi, Ohta and Matsuzawa, 1991; Strynadka *et al.*, 1992). To explain these results, it has been proposed that the deprotonated side-chain of Lys73 might accept the serine proton in acylation, while the role of Glu166 would be restricted to the deacylation step. This is only possible if the enzyme active-site environment decreases the pK of the alkylammonium group of Lys73 by 5–6 pH units.

According to the authors, such an exceptionally large effect could be due to a very positive electric field created by the α2-helix dipole, the alkylammonium group of Lys234 (element 3) and, possibly, some other residues, such as Arg244 and Arg164

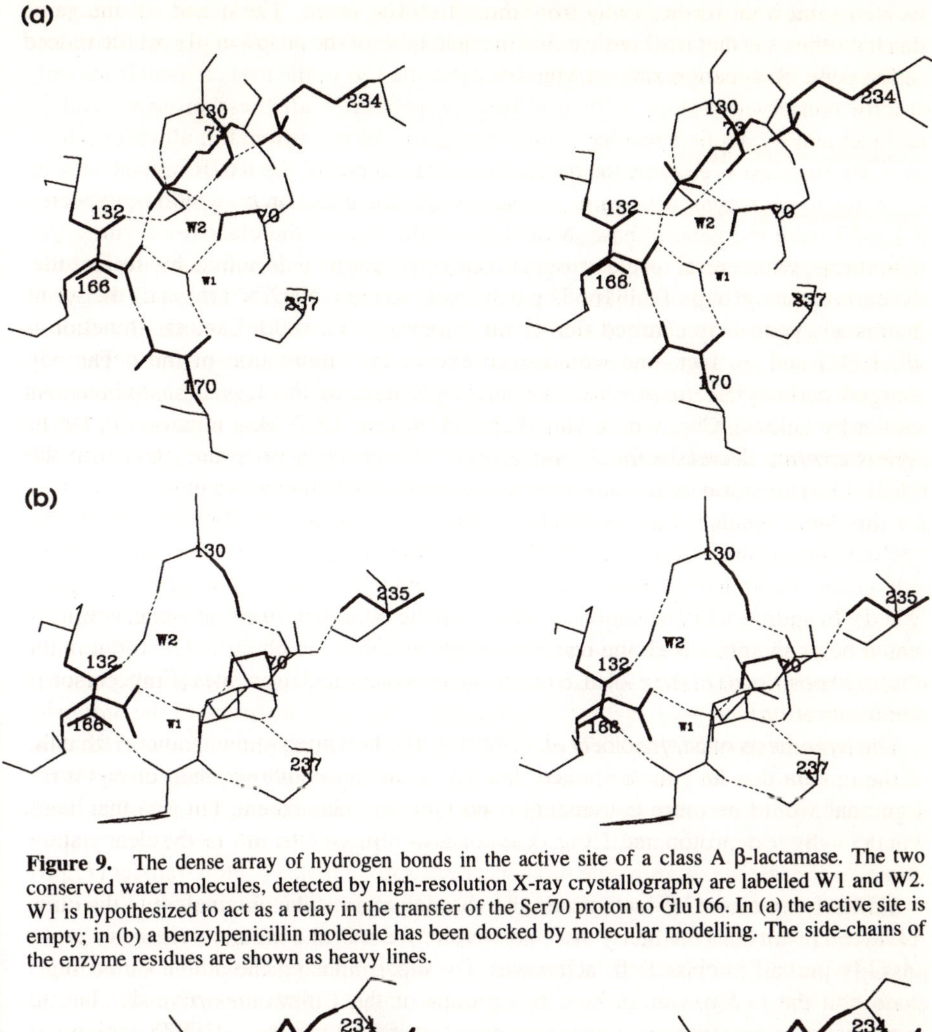

Figure 9. The dense array of hydrogen bonds in the active site of a class A β-lactamase. The two conserved water molecules, detected by high-resolution X-ray crystallography are labelled W1 and W2. W1 is hypothesized to act as a relay in the transfer of the Ser70 proton to Glu166. In (a) the active site is empty; in (b) a benzylpenicillin molecule has been docked by molecular modelling. The side-chains of the enzyme residues are shown as heavy lines.

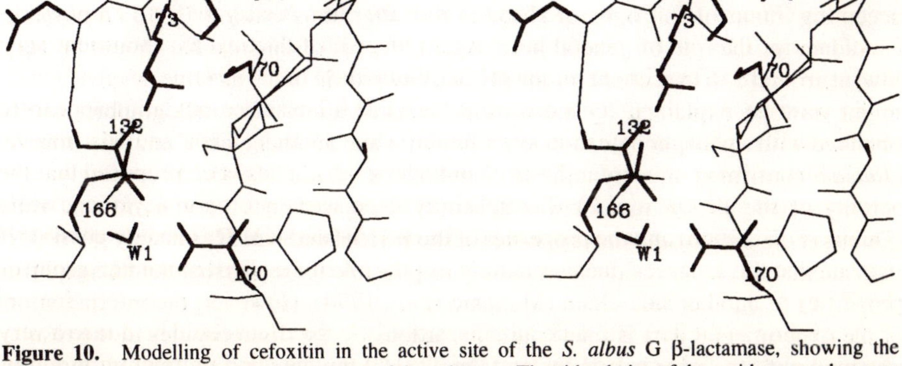

Figure 10. Modelling of cefoxitin in the active site of the *S. albus* G β-lactamase, showing the displacement of W1 by the methoxy group of the substrate. The side-chains of the residues are shown as heavy lines.

located somewhat further away from the active-site serine. The major advantage of this hypothesis is that it allows a uniform mechanism to be proposed for all active-site serine penicillin-recognizing enzymes. Indeed, the Lys of the first element is the only strictly conserved residue with acid/base properties in all these proteins, and its replacement by another residue results in significant impairment in all cases. However, for the class C enzyme the decreased acylation rate of the K67R mutant seemed to be due to an increased *K*' value, whereas a specific effect on k_{+2} would be expected if Lys67 were the general base. Moreover, in the case of the class A enzymes, this hypothesis, which rests on electrostatic factors, is clearly undermined by the fact that the carboxylate group of Glu166 is much closer to the ε-NH2/NH3⁺ group of Lys73 than is any positively charged side-chain (Moews *et al.*, 1990; Lamotte-Brasseur *et al.*, 1991) and, in fact, one would even expect the elimination of the negatively charged carboxylate to *increase* the nucleophilicity of the Lys73 side-chain. As shown by Gibson, Christensen and Waley (1990) the Lys73Arg mutation in the *B. cereus* enzyme decreases the k_{+2} value only 70-fold, i.e. significantly less than the Glu166Asp mutation in the same enzyme, a result which indicates a more crucial role for this latter residue. Finally, nuclear magnetic resonance (NMR) titration of the 13C-labelled lysine residues in the TEM enzyme failed to reveal that one of these side-chains exhibited an abnormally low pK (Raquet, Damblon and Frère, unpublished). In addition, modification of Asn170 in the *B. licheniformis* enzyme, a residue which helps maintain both the water molecule and the side-chain of Glu166 in their observed positions (*Figure 9*), also results in decreased acylation rates (Fink, personal communication).

The hypothesis of Strynadka *et al.* (1992) might, however, remain valid in the case of the mutant that they have studied. Indeed, as suggested above, their electrostatic argument would be more convincing if no Glu166 was present, but one may then wonder why a deprotonated Lys73 cannot also replace Glu166 in the deacylation step.

Similarly, in class C β-lactamases, the absence of a residue equivalent to the class A Glu166 results in a distinctly more positive active site, and the same situation might possibly prevail in class D β-lactamases. By superimposing the active serine side-chain and the two oxyanion hole NH groups of the *Citrobacter freundii* class C enzyme on the corresponding groups of chymotrypsin, Oefner *et al.* (1990) observed that the Tyr150 oxygen atom of the former was in a position equivalent to the proton-accepting imidazole nitrogen of His57 in the latter, thus making Tyr150 a possible candidate for the role of general base. Again, the pK of this residue should be very low, at around 5–6 to account for the pH-activity profile of the enzyme, a value which might now be explained by electrostatic considerations. Crystallographic results obtained with a phosphonate monoester inhibitor and another class C enzyme from *E. cloacae* confirmed this hypothesis (Lobkovsky *et al.*, 1994). Accordingly, the activity of the *Y150F* mutant was generally decreased in a very significant way (Dubus *et al.*, 1994b) and the properties of the K315H and K315Q mutants seemed to indicate that this latter residue was mainly responsible for increasing the nucleophilic properties of another side-chain (Monnaie *et al.*, 1994). However, the interpretation of the experimental data is made quite hazardous by the discrepancies in the results obtained with the poor substrates and the Tyr150 mutants and by the fact that, for technical reasons, it is not possible to determine the individual values of k_{+2} and *K*' for good substrates. In consequence, the effects of the various mutations become very difficult to assess.

PROTONATION OF THE LEAVING GROUP

In chymotrypsin catalysis, this is thought to occur via the back-delivery of the proton first accepted by the general base (His57, *Figure 7*). In class C, if Tyr150 is indeed the general base (but see above), a similar mechanism could prevail. However, in class A enzymes, this is certainly not so. Crystallographic and modelling studies of the structure of the acylenzyme indicate that the hydroxyl group of Ser130 might be ideally located to fulfil this function (Lamotte-Brasseur *et al.*, 1991; Strynadka *et al.*, 1992), which is only possible if a proton is simultaneously transferred on to the same side-chain, since the instability of the alcoholate ion precludes its formation as an end product. The dense array of hydrogen bonds within the active site would make this exchange possible, with the proton originating from Glu166 and travelling to the β-lactam nitrogen via Lys73, a second conserved water molecule (W2, *Figure 9*) and, finally, Ser130.

Despite the striking similarities found in their tertiary structures and the equivalent positions occupied by chemically identical or similar groups, it thus seems difficult to propose a unique mechanism for the hydrolysis of β-lactams by class A and class C enzymes. Since with the latter deacylation is often the rate-limiting step in the reaction pathway, it is tempting to hypothesize that Glu166 was 'invented' by class A enzymes to accelerate this final step. It would subsequently have been 'recruited' for playing a similar role in acylation. But slow deacylation is not an absolute property of class C enzymes: with some substrates, this occurs at very respectable rates (Galleni, Amicosante and Frère, 1988), an observation which is also valid for the less well-studied class D enzymes (Ledent *et al.*, 1993) which also lack an equivalent of Glu166, but, like their class C counterparts, contain a Tyr residue in the second element.

On the basis of the data that are presently available, it seems reasonable to assume that two distinct mechanisms prevail in the hydrolytic pathways of class A and class C β-lactamases. *Figures 11* and *12* illustrate the various possible intermediates in these pathways. It should be emphasized that these schemes are probably too simplistic and that other side-chains in or near the active sites might also play secondary, but yet important, roles in these phenomena (see below the examples of the Lys residue of the third element and of Arg244/Arg220 in the class A enzymes).

AND THE DD-PEPTIDASES?

When the comparison is extended to the DD-peptidases, further difficulties arise. If the intrinsic reactivity of the β-lactam amide can be considered as partly responsible for the relative efficiency of the active serine acylation by penicillins, it is not so with the physiological peptide substrates. In the *S.* R61 DD-peptidase the Tyr residue of the second element might participate in the activation of the serine acting as the general base, and back-deliver the proton to the leaving group, just as in class C enzymes. The barely affected k_{cat}/K_m value of the thiolester (which has a much better leaving group than the peptide) with the Y159S mutant would support this hypothesis. Similarly, the

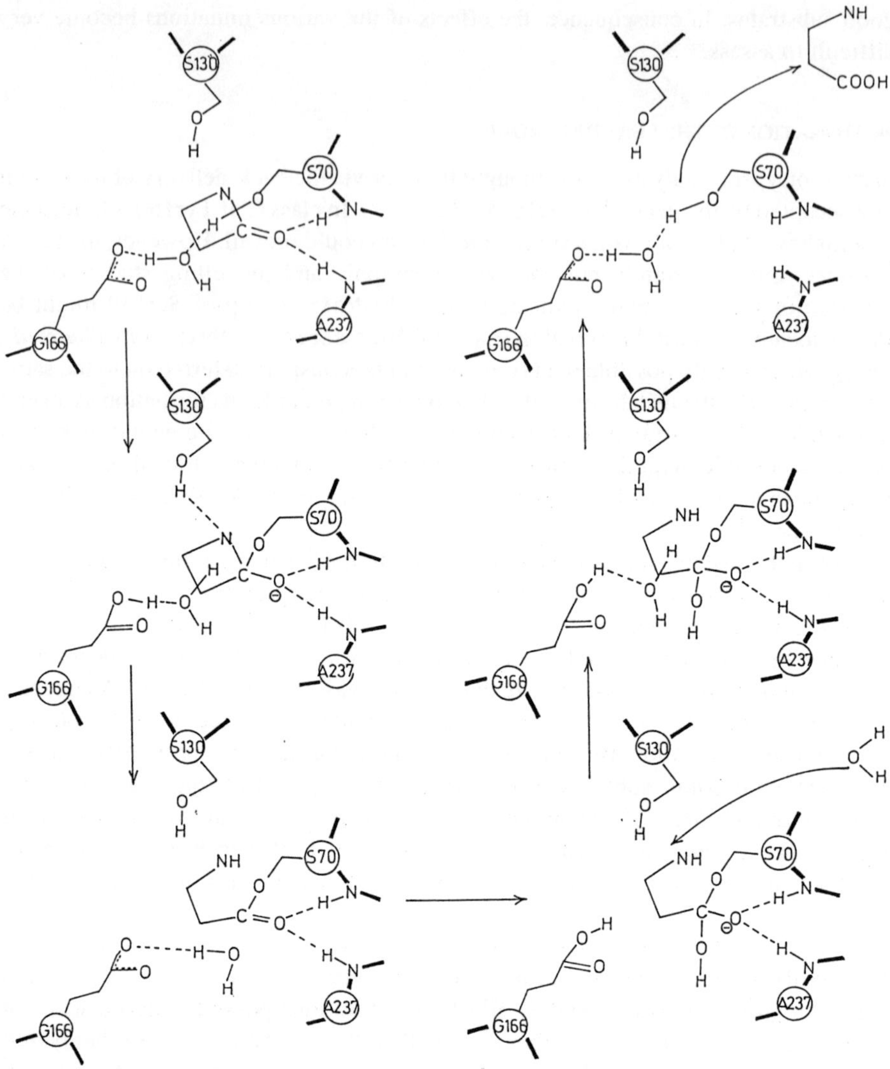

Figure 11. Hypothetical intermediates in the hydrolysis of a β-lactam by a class A enzyme.

acylation of the peptidase by β-lactams is very slow when compared to that of the β-lactamases and this relatively inefficient process is not impaired by the same mutation. A water molecule, loosely bound to the Ser159 side-chain might be sufficient to protonate the leaving group, an alternative mechanism that would not be possible with the bulky side-chain of the Y159F mutant.

One must finally remember that most DD-peptidases and PBPs contain a Ser residue in the second element, in contrast to the Tyr of the *S.* R61 enzyme. Despite this apparent handicap, the *Actinomadura* R39 DD-peptidase is more efficiently acylated by the peptide (k_{cat}/K_m = 50 000 M^{-1} s^{-1}) and β-lactams (k_2/K' = 300 000 M^{-1} s^{-1} with benzylpenicillin) than is the R61 enzyme. It seems unlikely that such high rates might be attained without the participation of additional side-chains which remain to be identified.

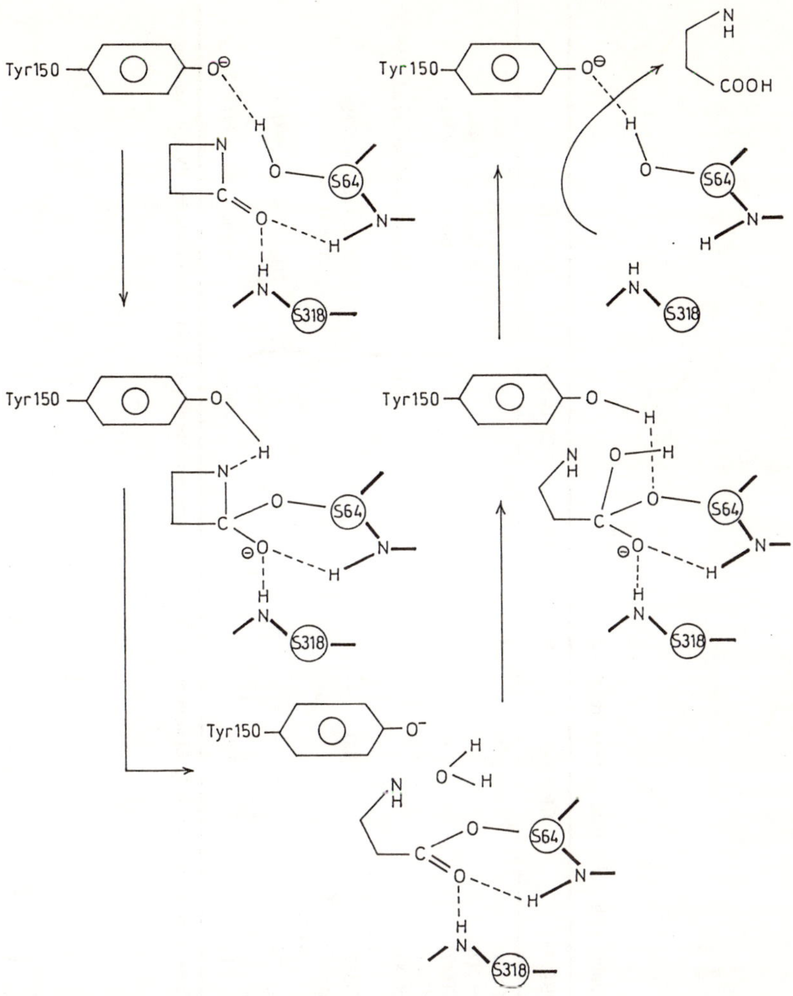

Figure 12. Hypothetical intermediates in the hydrolysis of a β-lactam by a class C enzyme.

The free carboxylate of β-lactam antibiotics

The presence of a negatively charged carboxylate on C3 of penicillins or C4 of cephalosporins is a prerequisite for antibiotic activity. In monolactams, this is replaced by a sulphonate group, which, thanks to its larger size, occupies a similar position relatively to the scissile amide bond.

Accordingly, the methylester and the amide of benzylpenicillin acylate both the DD-peptidase and the class A β-lactamases with a very poor efficiency (*Table 11*). Surprisingly, it is not so for the class C β-lactamases, which readily hydrolyse the methylester. To avoid the complications due to the different intrinsic reactivities of the various compounds, Laws and Page (1989) have introduced the enzymic rate enhancement factor (EREF), the ratio of the enzyme k_{cat}/K_m to the second-order rate constant of hydrolysis by OH^- ions, and *Table 11* compares these EREF values for pairs of compounds containing, or devoid of, a free carboxylate.

Table 11. Ratios of the EREF values for pairs of compounds with and without a free carboxylate

Enzyme			BP/BPMe	BP/BPamide	PV/PV-CH_2OH	CP/lactone	References
Peptidase	*S.* R61	WT	12700	>100000	–	28	Varetto *et al.* (1991)
		T299V	9200	–	–	–	Wilkin *et al.* (1994)
Class A	TEM1	WT	170000	–	860	–	Dubus *et al.* (1994a)
		S225A	33000	–	425	–	
	S. albus G	WT	50000	–	–	6.0	Jacob-Dubuisson *et al.* (1991)
		R220L	110	–	–	0.5	
	B. licheniformis	WT	500000	5000	–	3	Varetto *et al.* (1991)
	B. cereus	WT	100000	–	170	2.5	Laws *et al.* (1993)
Class C	*E. coli* K12	WT	6.6	–	2.5	260	Dubus *et al.* (1994a)
		T315A	4.1	–	5	80	
		T315V	1.8	–	1.8	110	
	E. cloacae 908R	WT	110	500	–	800	Varetto *et al.* (1991)

WT, Wild type; EREF, enzyme rate enhancement factor (see text); BP, benzylpenicillin; BPMe, BPmethylester; PV, phenoxymethylpenicillin; PV-CH_2OH, 3-hydroxymethyl-PV; CP, cephalosporin; lactone, corresponding lactone.

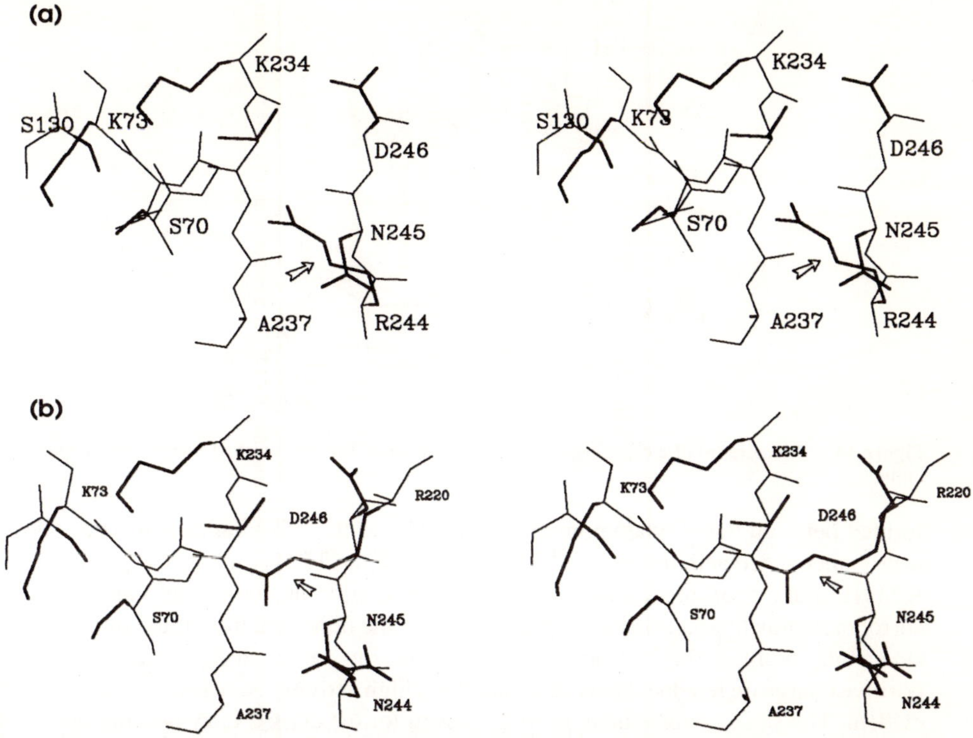

Figure 13. Similar positions of the guanidinium groups (arrows) of (a) Arg244 (*B. licheniformis* β-lactamase) and (b) Arg220 (*Streptomyces albus* G β-lactamase) near the backbone of the β3-strands.

Docking the β-lactam antibiotics in the active sites of the various enzymes always orients the negatively charged carboxylate (or sulphonate) to the side-chains of the first, hydroxylated residue of the second and the two first residues of the third elements. In addition, in the *B. licheniformis* class A β-lactamase, the guanidinium group of Arg244, whose C_α is in the β4-strand, is positioned rather close to the active site, and it was hypothesized that it could form a long ionic bond with the penicillin carboxylate. Arg244 is not conserved in all class A enzymes (Ambler *et al.*, 1991) but when it is absent, the guanidinium group of Arg220 is found lying in the same position, although the C_α of this residue is located on a different part of the protein backbone (*Figure 13*). The residues which might, directly or indirectly, interact with the negative charge of the substrate are thus Ser130, Lys234, Thr/Ser235 and Arg220/Arg244 in class A β-lactamases; Tyr150, Lys315 and Thr316 in class C; and Tyr159, His298 and Thr299 in the DD-peptidase. Unfortunately, the mutants have not been systematically studied with the methylester or with the cephalosporin lactones, other derivatives devoid of a net negative charge (*Figure 14*).

The high-pH limbs of the pH-activity profiles of class A enzymes reflect the titration of a group of pK 9.0–10.0. The results obtained with the K234H mutant of the *S. albus* G β-lactamase indicate that the titrated side-chain might be that of Lys234 (Brannigan *et al.*, 1991), although other explanations have been proposed (Laws *et al.*, 1993). Computer-aided docking of β-lactams in the active site of class A β-lactamases has often been performed with the assumption that a salt-bridge was

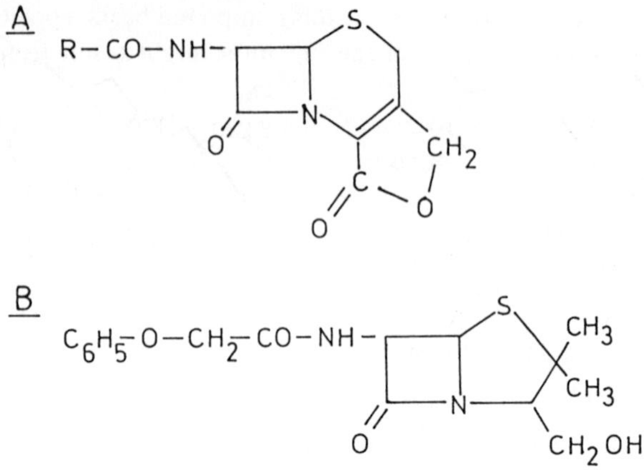

Figure 14. Structure (a) of a cephalosporin lactone, and (b) of 3-hydroxymethyl-phenoxymethylpenicillin.

formed between the alkylammonium of Lys234 and the substrate carboxylate. The nearly unchanged activities of the K234R mutant of the TEM1 and, at low pH, of the K234H mutant of the *S. albus* G enzymes might, at first sight, appear to corroborate this hypothesis but, in the latter case, the deprotonation of the imidazole side-chain modifies the *k*cat much more than the *K*m values. It is thus clear that a positive charge on residue 234 is necessary for a high activity, but its exact role is less evident. The same conclusion appears to prevail for the class C enzymes, but here a positive histidine is unable to replace the Lys315 alkylammonium, which further underlines fundamental differences in the mechanisms of both classes of enzymes. In other docking attempts, hydrogen bonds, rather than salt-bridges, are formed between the substrate carboxylate and the S130/Y150 and T(S)235/T316 hydroxyl groups, the first hydrogen bond being significantly weaker with cephalosporins (Lamotte-Brasseur *et al.*, 1991; Juteau *et al.*, 1992). This seems to be confirmed by the distinctly more pronounced effect of the disappearance of the T(S)235/T316 hydroxyl on the rates of acylation by cephalosporins when compared to those of penicillins, and this conclusion can be extended to the peptidase.

Analysis of *Table 11* also shows that, for class C enzymes, the interactions between the penicillin carboxylate and the enzymes are much weaker than for class A and for the peptidase. Conversely, lactonization of the cephalosporins has a greater effect in class C, although much less spectacular than the decreased activity of class A enzymes on the methyl ester. One may wonder if steric problems do not arise with this compound, since the differences are not so large with the hydroxymethyl derivative of penicillin (*Figure 14*). It has also been noted (Laws and Page, 1989) that the lactone oxygens bear partial negative charges, but the electrostatic potential maps of the penicillin ester and amide also show deep negative wells centred on the C3' carbonyl oxygen(s) (Varetto *et al.*, 1991). The elimination of the hydroxyl group of the third element's second residue never significantly modifies the EREF ratios, which is somewhat surprising. In fact, these ratios are only significantly modified in the R220L mutant of the *S. albus* G enzyme, indicating a possible, direct or indirect interaction between Arg220 and the substrate carboxylate. Accordingly, this mutant exhibits decreased acylation rates by all the 'normal' substrates, with the exception of cephaloridine. Since this compound bears a positively charged side-chain on C3, the lesser effect of the mutation might result from a 'substrate assisted catalysis' (Jacob-Dubuisson *et al.*, 1991).

In agreement with the structural data illustrated by *Figure 12*, the R244T and R244Q mutants of the TEM1 β-lactamase are impaired by factors quite similar to those of the R220L mutant of *S. albus* G, but here, the very different behaviour of the R244S mutant remains unexplained.

Evolution under our eyes

RESISTANCE TO MECHANISM-BASED INHIBITORS

The discovery of clavulanic acid (*Figure 15*) supplied a new chemotherapeutic tool in fighting the β-lactamase-producing bacteria. This compound, acting as a suicide substrate, efficiently acylates most class A β-lactamases. The acylenzyme can then undergo both hydrolysis and rearrangement(s) leading to more stable adducts. In consequence, and although clavulanic acid itself fails to inactivate the DD-peptidases, it has been widely utilized to vastly increase the efficiency of classical, β-lactamase-sensitive compounds, such as amoxycillin, against the TEM- and SHV-producing strains. This selective pressure has recently resulted in the spontaneous appearance of TEM mutants exhibiting a decreased sensitivity to this mechanism-based inactivator. The mutations were identified as R244S or R244C (Belaaouaj, 1992; Vedel *et al.*, 1992) and site-directed mutagenesis experiments confirmed the involvement of Arg244 in the inactivation phenomenon (Delaire *et al.*, 1992; Imtiaz *et al.*, 1993b). Its guanidinium group anchors a structurally conserved water molecule which serves as the source of a critical proton in the stepwise rearrangement of the clavulanate moiety. The M69L mutant of the same enzyme is also rather resistant to the same inactivator and remains quite efficient against the 'normal' substrates, observations which have been explained by a slight modification of the active-site structure (Delaire *et al.*, 1992). Class C enzymes are not sensitive to clavulanic acid and this seems to be mainly due to a poor recognition, resulting in an inefficient acylation (Monnaie and Frère, 1992), although an additional impairment of the rearrangement process could also be relevant (Lobkovsky *et al.*, 1993).

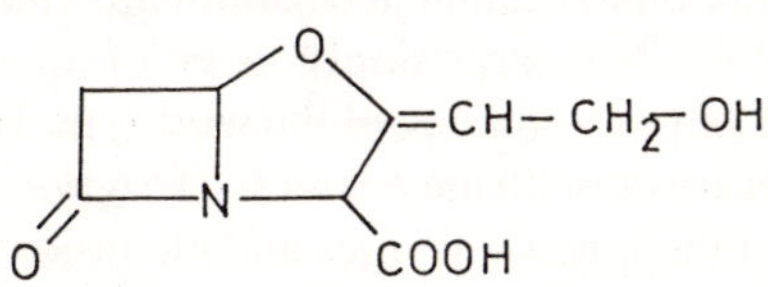

Figure 15. Structure of clavulanic acid.

A detailed review of the highly diversified mechanisms of β-lactamase inactivation has recently been presented (Pratt, 1992).

HYDROLYSIS OF THIRD-GENERATION CEPHALOSPORINS

Another strategy devised for fighting β-lactamase-producing pathogens was to try to avoid the activity of the enzymes by synthesizing 'β-lactamase stable' molecules. The utilization of methicillin (*Figure 16*) against the β-lactamase-producing *Staphylococci* was the first successful example of this strategy.

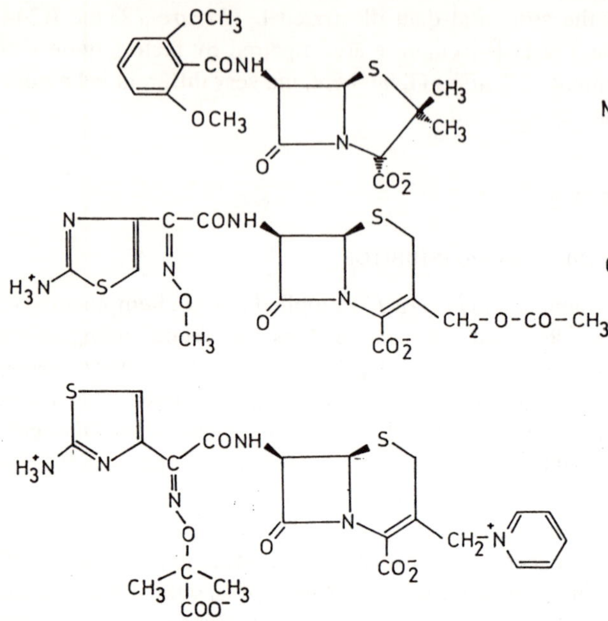

Figure 16. Structures of methicillin and of the oxyimino cephalosporins, cefotaxime and ceftazidime.

More recently, cefotaxime and ceftazidime (*Figure 16*), third-generation cephalosporins which are hydrolysed orders of magnitude more slowly than many β-lactams by the prevalent TEM and SHV enzymes, became very popular in the hospital environment. Maybe too popular. Indeed, their widespread, and sometimes abusive, utilization was responsible for the emergence of strains producing variants of these enzymes which exhibited increased hydrolytic activities against the two compounds. At least five SHV and 25 TEM variants have now been identified and were found to differ from the parent enzymes by a very limited number of amino acid substitutions (Sougakoff *et al.*, 1988; Collatz, Labia and Gutmann, 1990; Jacoby and Medeiros, 1991; Coyette *et al.*, 1994). Not surprisingly, none of the residues in the conserved elements were involved, but examination of the wild-type TEM1 structure shows that all the modified residues are close to the active site (*Figure 17*). *Table 12* summarizes the properties of some of these new enzymes and identifies the mutations.

Several authors (Lee *et al.*, 1991; Sowek *et al.*, 1991; Huletsky, Knox and Levesque, 1993) have tried to rationalize the observed alterations of substrate profiles on the basis of the structural variations, relying on the three-dimensional structures of related enzymes, generally that of the *B. licheniformis* β-lactamase. More recently, Raquet *et al.* (1994) have performed a detailed kinetic study of six TEM variants and analysed their properties on the basis of the newly determined structure of the TEM1, 'wild-type' enzyme, since no data are presently available about the structures of the variants (Jelsch *et al.*, 1992; Strynadka *et al.*, 1992; Fonzé *et al.*, in press). The data shown in *Table 12* are mostly taken from this work and although a detailed

Table 12. Properties of some TEM mutants and of other enzymes which hydrolyse third-generation cephalosporins (Raquet *et al.*, 1994)

	Sequences				Kinetic parameters					
					Absolute values[a]			Relative values		
	Residue 104	Omega loop	β3-strand		BP	CT	CR	CTX/CT	CAZ/CR	CTX/CAZ
		171	234 240							
TEM1	E	RWEPELNE	KSGAGE	k_{cat}/K_m	84	0.65	2.2	0.002	30×10^{-6}	20
				k_{cat}	1600	160	1500	0.06	0.0002	30
TEM11	E	*H*WEPELNE	KSGAGE	k_{cat}/K_m	14	0.36	1.3	0.064	0.011	1.7
								(32)	(350)	
				k_{cat}	85	110	160	0.025	0.11	0.15
								(.4)	(550)	
TEM10	E	*S*WEPELNE	KSGAG*K*	k_{cat}/K_m	25	0.16	0.12	0.062	0.23	0.35
								(31)	(8000)	
				k_{cat}	51	7	6	0.04	2.8	0.02
								(.7)	(14000)	
TEM19	E	RWEPELNE	KSGA*S*E	k_{cat}/K_m	10	2.3	2.2	0.1	0.002	46
								(50)	(70)	
				k_{cat}	62	34	105	1.9	0.25	2.5
								(30)	(1250)	
TEM3	*K*	RWEPELNE	KSGA*S*E	k_{cat}/K_m	9.5	1.9	1.1	0.53	0.036	25
								(250)	(1200)	
				k_{cat}	38	14	33	3.6	0.2	73
								(60)	(1000)	
M. fort	P	RWEVEL NS	KTGAGD	k_{cat}/K_m	1	1.9	0.7	0.05	–	–
PER1	T	ANEAQMHA	KTGT SG	k_{cat}/K_m	40[b]	68[b]	22[b]	0.85	0.77	3.4

BP, Benzylpenicillin; CT, cephalothin; CR, cephaloridine; CTX, cefotaxime; CAZ, ceftazidime.
The values between parentheses give the factor of increase when compared to TEM1.
[a] Absolute values: k_{cat}/K_m in μM^{-1} s-1, k_{cat} in s^{-1}.
M. fort = *Mycobacterium fortuitum* (Amicosante *et al.*, 1990).
PER1: [b] = relative values (Nordmann *et al.*, 1993).

Figure 17. Location of the α-carbons of Arg244 and of the residues modified in the TEM variants which hydrolyse third-generation cephalosporins (o). The conserved elements are also labelled.

discussion of these data is beyond the scope of the present review, the following points deserve attention.

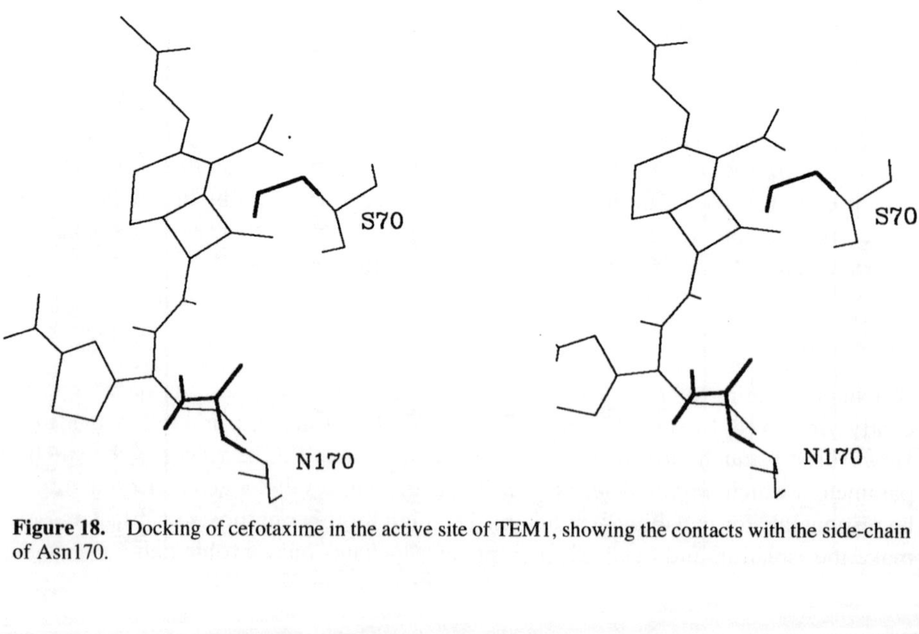

Figure 18. Docking of cefotaxime in the active site of TEM1, showing the contacts with the side-chain of Asn170.

1. Docking the optimized structures of cefotaxime and ceftazidime in the TEM1 active site leads to short contacts between the oxime substituents and the side-chain of Asn170 (*Figure 18*), resulting in a displacement of the catalytic Glu166. Moreover, with ceftazidime, the position of the catalytic water molecule W1 is also modified, which explains the even poorer activity of the enzyme against this substrate.
2. In TEM1, residue Arg164 forms a salt-bridge with Asp179, thus stabilizing the Ω loop. Substitution of Arg164 by uncharged residues in TEM10 and TEM11 makes the loop more mobile, which somewhat decreases the activity against the best substrates. By contrast, this allows more freedom for the oxime carboxylate of ceftazidime, which no longer disturbs the water molecule. This explains the increased efficiency with this substrate. The effects of the mutation on the rate of cefotaxime hydrolysis result from a negative factor, the perturbation of the Ω loop (hence, the *decreased* $k$cat value) and a larger positive effect, the disappearance of the short contacts with Asn170, mainly reflecting on the K' value. These considerations account for the larger increase in the activity of these mutants against ceftazidime, reflected in the lowered CTX/CAZ ratios in *Table 12*.
3. In the absence of three-dimensional structures for these proteins, modelling of the mutants with new Lys side-chains (TEM10 and TEM3) is made difficult by the large number of possible conformations of the long alkyl moiety of this residue.
4. The Ser238 side-chains of TEM19 and TEM3 might form a new H-bond with the oxime oxygen of cefotaxime. To avoid short contacts with the hydroxymethyl group, the side-chain of cefazidime must be oriented away from the active site, allowing the return of W1 to its efficient position.
5. The high activity of TEM3 against ceftazidime might be explained by the preferential binding of a minor conformer of the antibiotic.
6. Other enzymes have been described (*Table 12*), which exhibit relatively high activities against third-generation cephalosporins, often with significantly reduced activities against penicillins. A striking example is the enzyme produced by *Mycobacterium fortiutum* (*Table 12*). Mutations of the residues defined as strategic on the basis of the sequences of the TEM variants are sometimes found in these enzymes, but different modifications also appear to yield similar specificity profiles. The catalytic properties of class A β-lactamases thus present examples of converging evolution within the generally divergent sequence differentiations. The utilization of new compounds can, in consequence, result in the appearance of enzymes which destroy them, by the selection of strains producing either 'original' enzymes, quite different from those which are already well recognized, or variants derived from the latter by a few point mutations.

Conclusions

Our understanding of the catalytic mechanisms of β-lactamases has been significantly improved by the determination of several structures at high resolution. However, it is much more difficult to account accurately for the values of the kinetic parameters which, *in fine* determine the exact difficulties that a given enzyme might be responsible for in a clinical situation. The high turnover numbers of β-lactamases make the isolation and detailed study of intermediates quite problematic, although some low-temperature studies have been performed with success (Cartwright and Waley, 1987). Alternatively, computer modelling has also succeeded in rationalizing the behaviour of some enzyme–substrate pairs (Lamotte-Brasseur *et al.*, 1992; Matagne, Lamotte-Brasseur and Frère, 1993). However, even when a high-resolution structure is available for the parent enzyme, the quantitative interpretation of the catalytic properties of mutants and of related proteins remains difficult when the reaction pathway involves several steps and includes the formation of covalent intermediates. Catalysis is indeed a complex process and these considerations underline the difficulties which can be encountered when attempting to deduce the properties of the protein from the gene sequence.

The past successes of antibacterial chemotherapy are currently challenged by many pathogenic strains which have efficiently countered the various strategies developed to keep them under control. A few years ago, the medical community relaxed in the security afforded by 'β-lactamase-stable' compounds (Bush, 1989b) and inactivators like clavulanic acid and sulbactam. Bacteria have exhibited extraordinary adaptation capacities and, in this respect, it is quite surprising that several pharmaceutical companies appear to have lost interest in the search for new antibacterial agents.

When a new compound is introduced in the clinical arsenal, the only prediction which can safely be made is that, given enough time, bacteria will find a way to escape its effects. Only a deeper understanding of the various physiological processes which govern bacterial multiplication can help us continue to be 'one compound ahead' of the pathogens.

References

ADACHI, H., OHTA, T. AND MATSUZAWA. H. (1991). Site-directed mutants, at position 166, of RTEM-1 β-lactamase that form a stable acyl-enzyme intermediate with penicillin. *Journal of Biological Chemistry* **266**, 3186–3191.

ADAM, M., DAMBLON, C., PLAITIN, B., CHRISTIAENS, L. AND FRÈRE, J.M. (1990). Chromogenic depsipeptide substrates for β-lactamases and penicillin-sensitive DD-peptidases. *Biochemical Journal* **270**, 525–529.

AMBLER, R.P., COULSON, A.F., FRÈRE, J.M., GHUYSEN, J.M., JAURIN, B., JORIS, B., LEVESQUE, R., TIRABY, G. AND WALEY, S.G. (1991). A standard numbering scheme for the class A beta-lactamases. *Biochemical Journal* **276**, 269–272.

AMICOSANTE, G., FRANCESCHINI, N., SEGATORE, B., ORATORE, A., FATTORINI, L., OREFICI, G., VAN BEEUMEN, J. AND FRÈRE, J.M. (1990). Characterization of a β-lactamase produced in *Mycobacterium fortuitum* D316. *Biochemical Journal* **271**, 729–734.

BELAAOUAJ, A. (1992). Beta-lactamases TEM-1: emergence clinique de mutants résistants aux inhibiteurs de beta-lactamases, PhD Thesis, Université de Paris VII.

BRADY, L., BRZOZOWSKI, A., DEREWENDA, Z., DODSON, E., DODSON, G., TOLLEY, S., TURKENBURG, J., CHRISTIANSEN, L., HUGE-JENSEN, B., NORSKOV, L. AND MENGE, V. (1990). A serine protease triad forms the catalytic centre of a triacylglycerol lipase. *Nature* **343**, 767–770.

BRANNIGAN, J., MATAGNE, A., JACOB, F., DAMBLON, C., JORIS, B., KLEIN, D., SPRATT, B.G. AND FRÈRE, J.M. (1991). The mutation Lys234His yields a class A β-lactamase with a novel pH-dependence. *Biochemical Journal* **278**, 673–678.

BUSH, K. (1989a). Characterization of β-lactamases. *Antimicrobial Agents and Chemotherapy* **33**, 259–263.

BUSH, K. (1989b). Excitement in the beta-lactamase area. *Journal of Antimicrobial Chemotherapy* **24**, 831–836.

CARTWRIGHT, S.J. AND WALEY, S.G. (1987). Cryoenzymology of beta-lactamases. *Biochemistry* **26**, 5329–5337.

CHRISTENSEN, H., MARTIN, M.T. AND WALEY, S.G. (1990). β-lactamases as fully efficient enzymes. Determination of all the rate constants in the acyl-enzyme mechanism. *Biochemical Journal* **266**, 853–861.

COLLATZ, E., LABIA, R. AND GUTMANN, L. (1990). Molecular evolution of ubiquitous beta-lactamases towards extended-spectrum enzymes active against newer beta-lactam antibiotics. *Molecular Microbiology* **4**, 1615–1620.

COUTURE, F., LACHAPELLE, J. AND LEVESQUE, R.C. (1992). Phylogeny of LCR-1 and OXA-5 with class A and class D β-lactamases. *Molecular Microbiology* **6**, 1693–1705.

COYETTE, J., NGUYEN-DISTECHE, M., LAMOTTE-BRASSEUR, J.,JORIS, B., FONZÉ, E. AND FRÈRE, J.M. (1994). Molecular adaptations in resistance to penicillins and other β-lactam antibiotics. *Advances in Comparative and Environmental Physiology* **20**, 233–267.

DATZ, M., JORIS, B., AZAB, E.A.M., GALLENI, M., VAN BEEUMEN, J., FRÈRE, J.M. AND MARTIN, H.H. (1994). A common system controls the induction of very different genes: the class A β-lactamase of *Proteus vulgaris* and the enterobacterial class C β-lactamase. *European Journal of Biochemistry* **226**, 149–157.

DEHOTTAY, P., DUSART, J., DE MEESTER, F., JORIS, B., VAN BEEUMEN, J., ERPICUM, T., FRÈRE, J.M. AND GHUYSEN, J.M. (1987). Nucleotide sequence of the gene encoding the *Streptomyces albus* G β-lactamase precursor. *European Journal of Biochemistry* **166**, 345–350.

DELAIRE, M., LENFANT, F., LABIA, R. AND MASSON, J.M. (1991). Site-directed mutagenesis on TEM-1 beta-lactamase: role of Glu166 in catalysis and substrate binding. *Protein Engineering* **4**, 805–810.

DELAIRE, M., LABIA, R., SAMAMA, J.P. AND MASSON, J.M. (1992). Site-directed mutagenesis at the active site of *Escherichia coli* TEM-1 β-lactamase. Suicide inhibitor-resistant mutants reveal the role of arginine 244 and methionine 69 in catalysis. *Journal of Biological Chemistry* **267**, 20600–20606.

DIDEBERG, O., CHARLIER, P., WÉRY, J.P., DEHOTTAY, P.H., DUSART, J., ERPICUM, T., FRÈRE, J.M. AND GHUYSEN, J.M. (1987). The crystal structure of the β-lactamase of *Streptomyces albus* G at 0.3 nm resolution. *Biochemical Journal* **245**, 911–913.

DUBUS, A., MONNAIE, D., JACOBS, C., NORMARK, S. AND FRÈRE, J.M. (1993). A dramatic change in the rate-limiting step of β-lactam hydrolysis results from the substitution of the active-site serine residue by a cysteine in the class C β-lactamase of *Enterobacter cloacae* 908R. *Biochemical Journal* **292**, 537–543.

DUBUS, A., WILKIN, J.M., RAQUET, X., NORMARK, S. AND FRÈRE, J.M. (1994a). Catalytic mechanism of active-site serine β-lactamases: role of the conserved hydroxyl group of the KT(S)G triad. *Biochemical Journal* **301**, 485–494.

DUBUS, A., NORMARK, S., KANIA, M. AND PAGE, M.G.P. (1994b). The role of tyrosine 150 in catalysis of β-lactam hydrolysis by AmpC β-lactamase from *Escherichia coli* investigated by site-directed mutagenesis. *Biochemistry*, **33**, 8577–8586.

ELLERBY, L.M., ESCOBAR, W.A., FINK, A.L., MITCHINSON, C. AND WELLS, J.A. (1990). The role of lysine-234 in beta-lactamase catalysis probed by site-directed mutagenesis. *Biochemistry* **29**, 5797–5806.

ESCOBAR, W.A., TAN, A.K. AND FINK, A.L. (1991). Site-directed mutagenesis of beta-lactamase leading to accumulation of a catalytic intermediate. *Biochemistry* **30**, 10783–10787.

ESCOBAR, W.A., TAN, A.K., LEWIS, E.R. AND FINK, A.L. (1994). Site-directed mutagenesis of glutamate-166 in β-lactamase leads to a branched pathway mechanism. *Biochemistry* **33**, 7619–7626.

FELICI, A., AMICOSANTE, G., ORATORE, A., STROM, R., LEDENT, P., JORIS, B. AND FRÈRE, J.M. (1993). An overview of the kinetic parameters of class B β-lactamases. *Biochemical Journal* **291**, 151–155.

FERSHT, A.R., BLOW, D.M. AND FASTREZ, J. (1973). Leaving group specificity in the chymotrypsin catalysed hydrolysis of peptides. A stereochemical interpretation. *Biochemistry* **12**, 2035–2041.

FONZÉ, E., CHARLIER, P., TO'TH, Y., VERMEIRE, M., RAQUET, X. AND DUBUS, A. AND FRÈRE, J.M. TEM-1 beta-lactamase structure solved by moleular replacement and refined structure of the S235A mutant. *Acta Chrystallographica, Section D*, in press.

FRÈRE, J.M. AND JORIS, B. (1985). Penicillin-sensitive enzymes in peptidoglycan biosynthesis. *CRC Critical Reviews in Microbiology* **11**, 299–396.

FRÈRE, J.M., JORIS, B., CRINE, M. AND MARTIN, H.H. (1989). Quantitative relationship between sensitivity to beta-lactam antibiotics and beta-lactamase production in Gram-negative bacteria: II. Non-steady-state treatment and progress curves. *Biochemical Pharmacology* **38**, 1427–1433.

FRÈRE, J.M., NGUYEN-DISTECHE, M., COYETTE, J. AND JORIS, B. (1992). Mode of action: interaction with the penicillin binding proteins. In *The Chemistry of Beta-Lactams* (M.I. Page, Ed.), pp. 148–197. Chapman & Hall, London.

GALLENI, M., AMICOSANTE, G. AND FRÈRE, J.M. (1988). A survey of the kinetic parameters of class C β-lactamases. II Cephalosporins and other β-lactam compounds. *Biochemical Journal* **255**, 123–129.

GHUYSEN, J.M. (1991). Serine β-lactamases and penicillin-binding proteins. *Annual Review of Microbiology* **45**, 37–67.

GIBSON, R.M., CHRISTENSEN, H. AND WALEY, S.G. (1990). Site-directed mutagenesis of β-lactamase I. Single and double mutants of Glu- 166 and Lys-73. *Biochemical Journal* **272**, 613–619.

HADONOU, A.M., WILKIN, J.M., VARETTO, L., JORIS, B., LAMOTTE-BRASSEUR, J., KLEIN, D., DUEZ, C., GHUYSEN, J.M. AND FRÈRE, J.M. (1992a). Site-directed mutagenesis of the *Streptomyces* R61 DD-peptidase. Catalytic function of the conserved residues around the active site and a comparison with class-A and class-C β-lactamases. *European Journal of Biochemistry* **207**, 97–102.

HADONOU, A.M., JAMIN, M., ADAM, M., JORIS, B., DUSART, J., GHUYSEN, J.M. AND FRÈRE, J.M. (1992b). Importance of the His-298 residue in the catalytic mechanism of the *Streptomyces* R61 extracellular DD-peptidase. *Biochemical Journal* **282**, 495–500.

HECHLER, U., VAN DEN WEGHE, M., MARTIN, H.H. AND FRÈRE, J.M. (1989). Overproduced β-lactamase and the outer membrane barrier as resistance factors in *Serratia marcescens* highly resistant to β-lactamase stable β-lactam antibiotics. *Journal of General Microbiology* **135**, 1275–1290.

HERZBERG, O. (1991). Refined crystal structure of β-lactamase from *Staphylococcus aureus* PC1 at 2.0 A resolution. *Journal of Molecular Biology* **217**, 701–719.

HERZBERG, O., KAPADIA, G., BLANCO, B., SMITH, T.S. AND COULSON, A. (1991). Structural basis for the inactivation of the P54 mutant of beta-lactamase from *Staphylococcus aureus* PC1. *Biochemistry* **30**, 9503–9509.

HULETSKY, A., KNOX, J.R. AND LEVESQUE, R.C. (1993). Role of ser-238 and lys-240 in the hydrolysis of third-generation cephalosporins by SHV-type β-lactamases probed by site-directed mutagenesis and three-dimensional modeling. *Journal of Biological Chemistry* **268**, 3690–3697.

IMTIAZ, U., MANAVATHU, E.K., LERNER, S. AND MOBASHERY, S. (1993a). Critical hydrogen bonding by serine 235 for cephalosporinase activity of TEM-1 β-lactamase. *Antimicrobial Agents and Chemotherapy* **37**, 2438–2442.

IMTIAZ, U., BILLINGS, E., KNOX, J.R., MANAVATHU, E.K., LERNER, S.A. AND MOBASHERY, S. (1993b). Inactivation of class A β-lactamases by clavulanic acid: the role of Arginine-244 in a proposed nonconcerted sequence of events. *Journal of the American Chemical Society* **115**, 4435–4442.

JACOB, F., JORIS, B., DIDEBERG, O., DUSART, J., GHUYSEN, J.M. AND FRÈRE, J.M. (1990a). Engineering a novel β-lactamase by a single point mutation. *Protein Engineering* **4**, 79–86.

JACOB, F., JORIS, B., LEPAGE, S., DUSART, J. AND FRÈRE, J.M. (1990b). Role of the conserved amino acids of the SDN loop (Ser^{130}, Asp^{131} and Asn^{132}) in a class A β-lactamase studied by site-directed mutagenesis. *Biochemical Journal* **271**, 399–406.

JACOB, F., JORIS, B. AND FRÈRE, J.M. (1991). Active-site serine mutants of the *Streptomyces albus* G beta-lactamase. *Biochemical Journal* **277**, 647–652.

JACOB-DUBUISSON, F., LAMOTTE-BRASSEUR, J., DIDEBERG, O., JORIS, B. AND FRÈRE, J.M. (1991). Arginine 220 is a critical residue for the catalytic mechanism of the *Streptomyces albus* G β-lactamase. *Protein Engineering* **4**, 811–819.

JACOBS, C., HUANG, L.J., BARTOWSKY, E., NORMARK, S. AND PARK, J.T. (1994a). Bacterial cell wall recycling provides cytosolic muropeptides as effectors for β-lactamase induction. *EMBO Journal* **13**, 4684–4694.

JACOBS, C., JORIS, B., JAMIN, M., VANBEEUMEN, J., VAN HEIJENOORT, J., PARK, J.T., NORMARK, S. AND FRÈRE, J.M. (1994b). AmpD, essential for both β-lactamase regulation and cell wall recycling, is a novel cytosolic *N*-acetylmuramyl-L-alanine amidase. *Molecular Microbiology*, in press.

JACOBY, G.A. AND MEDEIROS, A.A. (1991). More extended-spectrum β-lactamases. *Antimicrobial Agents and Chemotherapy* **35**, 1697–1704.

JAMIN, M., WILKIN, J.M. AND FRÈRE, J.M. (1993). A new kinetic mechanism for the concomitant hydrolysis and transfer reactions catalysed by bacterial DD-peptidases. *Biochemistry* **32**, 7278–7285.

JELSCH, C., LENFANT, F., MASSON, J.M. AND SAMAMA, J.P. (1992). β-lactamase TEM1 of *E. coli*. Crystal structure determination at 2.5 A resolution. *FEBS Letters* **299**, 135–142.

JORIS, B., HARDT, K. AND GHUYSEN, J.M. (1994). Induction of β-lactamase and low affinity penicillin binding protein 2' synthesis in Gram-positive bacteria. *New Comprehensive Biochemistry* **27**, 505–515.

JORIS, B., GHUYSEN, J.M., DIVE, G., RENARD, A., DIDEBERG, O., CHARLIER, P., FRÈRE, J.M., KELLY, J., BOYINGTON, J., MOEWS, P. AND KNOX, J. (1988). The active-site-serine penicillin-recognizing enzymes as members of the *Streptomyces* R61 DD-peptidase family. *Biochemical Journal* **250**, 313–324.

JORIS, B., LEDENT, P., DIDEBERG, O., FONZÉ, E., LAMOTTE-BRASSEUR, J., KELLY, J.A., GHUYSEN, J.M. AND FRÈRE, J.M. (1991). Comparison of the sequences of class-A beta-lactamases and of the secondary structure elements of penicillin-recognizing proteins. *Antimicrobial Agents and Chemotherapy* **35**, 2294–2301.

JUTEAU, J.M., BILLINGS, E., KNOX, J.R. AND LEVESQUE, R.C. (1992). Site-saturation mutagenesis and three-dimensional modelling of ROB-1 define a substrate binding role of Ser130 in class A β-lactamases. *Protein Engineering* **5**, 693–701.

KELLY, J.A., KNOX, J.R., ZHAO, H., FRÈRE, J.M. AND GHUYSEN, J.M. (1989). Crystallographic mapping of β-lactams bound to a D-alanyl-D-alanine peptidase target enzyme. *Journal of Molecular Biology* **209**, 281–295.

KNOTT-HUNZIKER, V., PETURSSON, S., WALEY, S.G., JAURIN, B. AND GRUNDSTROM, T. (1982). The acyl-enzyme mechanism of β-lactamase action. *Biochemical Journal* **207**, 315–322.

KNOX, J.R. AND MOEWS, P.C. (1991). β-lactamase of *Bacillus licheniformis* 749/C-refinement at 0.2 nm resolution and analysis of hydration. *Journal of Molecular Biology* **220**, 435–455.

LABIA, R., GUIONIE, M. AND BARTHÉLÉMY, M. (1981). Properties of three carbenicillin-hydrolysing β-lactamases (CARB) from *Pseudomonas aeruginosa*: identification of a new enzyme. *Journal of Antimicrobial Chemotherapy* **7**, 49–56.

LACHAPELLE, J., DUFRESNE, J. AND LEVESQUE, R. (1991). Characterization of bla (CARB-3) encoding the carbenicillinase-3 β-lactamase of *Pseudomonas aeruginosa. Gene* **102**, 7–12.

LAMOTTE-BRASSEUR, J., DIVE, G., DIDEBERG, O., CHARLIER, P., FRÈRE, J.M. AND GHUYSEN, J.M. (1991). Mechanism of acyl transfer by the class A serine β-lactamase of *Streptomyces albus* G. *Biochemical Journal* **279**, 213–221.

LAMOTTE-BRASSEUR, J., JACOB-DUBUISSON, F., DIVE, G., FRÈRE, J.M. AND GHUYSEN, J.M. (1992). *Streptomyces albus* G. serine β-lactamase. Probing of the catalytic mechanism via molecular modelling of mutant enzymes. *Biochemical Journal* **282**, 189–195.

LAWS, A.P. AND PAGE, M.I. (1989). The effect of the carboxy group on the chemical and β-lactamase reactivity of β-lactam antibiotics. *Journal of the Chemical Society – Perkin Transactions II*, 1577–1581.

LAWS, A.P., LAYLAND, N.J., PROCTOR, D.G. AND PAGE, M.I. (1993). The roles of the carboxy group in β-lactam antibiotics and lysine 234 in β-lactamase I, *Journal of the Chemical Society – Perkin Transactions II*, 17–21.

LEDENT, P., RAQUET, X., JORIS, B., VAN BEEUMEN, J. AND FRÈRE, J.M. (1993). A comparative study of class D β-lactamases. *Biochemical Journal* **292**, 555–562.

LEE, K.Y., HOPKINS, J.D., O'BRIAN, T.F. ABD SYVANEN, M. (1991). Gly238-Ser substitution changes the substrate specificity of the SHV class A β-lactamases. *Protein Structure, Function and Genetics* **11**, 45–51.

LENFANT, F., LABIA, R. AND MASSON, J.M. (1991). Replacement of Lysine 234 affects transition state stabilization in the active site of beta-lactamase TEM1. *Journal of Biological Chemistry* **266**, 17187–17194.

LENZINI, V.M., MAGDALENA, J., FRAIPONT, C., JORIS, B., MATAGNE, A. AND DUSART, J. (1992). Induction of *Streptomyces cacaoi* β-lactamase gene cloned in *S. lividans. Molecular and General Genetics* **235**, 41–48.

LEUNG, Y.C., ROBINSON, C.V., APLIN, R.T. AND WALEY, S.G. (1994). Site-directed mutagenesis of β-lactamase I: role of Glu-166. *Biochemical Journal* **299**, 671–678.

LOBKOVSKY, E., MOEWS, P.C., HANSONG, L., HAICHING, Z., FRÈRE, J.M. AND KNOX, J.R. (1993). Evolution of an enzyme activity: Crystallographic structure of 2-A resolution of cephalosporinase from the *ampC* gene of *Enterobacter cloacae* P99 and comparison with a class A penicillinase. *Proceedings of the National Academy of Sciences, USA* **90**, 11257–11261.

LOBKOVSKY, E., BILLINGS, E.M., MOEWS, P.C., RAHIL, J., PRATT, R.F. AND KNOX, J.R. (1994). Crystallographic structure of a phosphonate derivative of the *Enterobacter cloacae* P99 cephalosporinase: mechanistic interpretation of a β-lactamase transition state analog. *Biochemistry* **33**, 6762–6772.

MATAGNE, A. AND FRÈRE, J.M. (1994). Contribution of mutant analysis to the understanding of enzyme catalysis: the case of class A β-lactamases. *Biochimica et Biophysica Acta*, in press.

MATAGNE, A., LAMOTTE-BRASSEUR, J. AND FRÈRE, J.M. (1993). Interactions between active-site serine β-lactamases and so-called β-lactamase-stable antibiotics. Kinetic and molecular modelling studies. *European Journal of Biochemistry* **217**, 61–67.

MATAGNE, A., MISSELYN-BAUDUIN, A.M., JORIS, B., ERPICUM, T., GRANIER, B. AND FRÈRE, J.M. (1990). The diversity of the catalytic properties of class A β-lactamases. *Biochemical Journal* **265**, 131–146.

MATAGNE, A., LAMOTTE-BRASSEUR, J., DIVE, G., KNOX, J.R. AND FRÈRE, J.M. (1993). Interactions between active-site serine β-lactamases and compounds bearing a methoxy side chain on the α-face of the β-lactam ring: kinetic and molecular modeling studies. *Biochemical Journal* **293**, 607–611.

MOEWS, P.C., KNOX, J.R., DIDEBERG, O., CHARLIER, P. AND FRÈRE, J.M. (1990). β-lactamase of *Bacillus licheniformis* 749/C at 2 A resolution. *Protein Structure, Function and Genetics* **7**, 156–171.

MONNAIE, D. AND FRÈRE, J.M. (1993). Interaction of clavulanate with class C β-lactamases. *FEBS Letters* **334**, 269–271.

MONNAIE, D., DUBUS, A. AND FRÈRE, J.M. (1994). The role of Lys 67 in class C β-lactamase is mainly electrostatic. *Biochemical Journal* **302**, 1–4.

MONNAIE, D., VIRDEN, R. AND FRÈRE, J.M. (1992). A rapid-kinetic study of the class C β-lactamase of *Enterobacter cloacae* 908R. *FEBS Letters* **306**, 108–112.

MONNAIE, D., DUBUS, A., COOKE, D., MARCHAND-BRYNAERT, J., NORMARK, S. AND FRÈRE, J.M. (1994). Role of residue Lys 315 in the mechanism of action of the *Enterobacter cloacae* 908R β-lactamase. *Biochemistry* **33**, 5193–5201.

MURPHY, B.P. AND PRATT, R.F. (1991). *N*-(Phenylacetyl)Glycyl-D-Aziridine-2-Carboxylate, an acyclic amide substrate of beta-lactamases – importance of the shape of the substrate in beta-lactamase evolution. *Biochemistry* **30**, 3640–3649.

NIKAIDO, H. AND NORMARK, S. (1987). Sensitivity of *Escherichia coli* to various β-lactams is determined by the interplay of outer membrane permeability and degradation by periplasmic β-lactamase: a quantitative predictive treatment. *Molecular Microbiology* **1**, 29–36.

NORDMANN, P., RONCO, E., NAAS, T., DUPORT, C., MICHEL-BRIAND, Y. AND LABIA, R. (1993). Characterization of a novel extended-spectrum β-lactamase from *Pseudomonas aeruginosa. Antimicrobial Agents and Chemotherapy* **37**, 962–969.

OEFNER, C., DARCY, A., DALY, J.J., GUBERNATOR, K., CHARNAS, R.L., HEINZE, I., HUBSCHWERLEN, C. AND WINKLER, F.K. (1990). Refined rystal structure of beta-lactamase from *Citrobacter freundii* indicates a mechanism for beta-lactam hydrolysis. *Nature* **343**, 284–288.

PAGE, M.G.P. (1993). The kinetics of non-stoichiometric bursts of β-lactam hydrolysis catalysed by class C β-lactamases. *Biochemical journal* **295**, 295–304.

PAYNE, D.J. (1993). Metallo-β-lactamases a new therapeutic challenge. *Journal of medical Microbiology* **39**, 93–99.

PAZHANISAMY, S., GOVARDHAN, C.P. AND PRATT, R.F. (1989). β-lactamase-catalysed aminolysis of depsipeptides: amine specificity and steady-state kinetics. *Biochemistry* **28**, 6863–6870.

PRATT, R.F. (1992). β-lactamase inhibition. In: *The Chemistry of β-lactams* (M.I. Page, Ed.), pp. 229–271. Chapman & Hall, London.

PRATT, R.F. AND GOVARDHAN, C.P. (1984). β-Lactamase-catalyzed hydrolysis of acyclic depsipeptides and acyl transfer to specific amino acid acceptors. *Proceedings of the National Academy of Sciences, USA* **81**, 1302–1306.

RAHIL, J. AND PRATT, R.F. (1994). Characterization of covalently bound enzyme inhibitors as transition-state analogs by protein stability measurements: phosphonate monoester inhibitors of a β-lactamase. *Biochemistry* **33**, 116–125.

RAQUET, X., LAMOTTE-BRASSEUR, J., FONZÉ, E., GOUSSARD, S., COURVALIN, P. AND FRÈRE, J.M. (1994). TEM mutants hydrolysing third generation cephalosporins: a kinetic and molecular modelling analysis. *Journal of Molecular Biology* **244**, 625–639.

SIGAL, I.S., DEGRADO, W.F., THOMAS, B.J. AND PETTEWAY, S.R. (1984). Purification and properties of Thiol β-lactamase. *journal of Biological Chemistry* **259**, 5327–5332.

SOUGAKOFF, W., GOUSSARD, S., GERBAUD, G. AND COURVALIN, P. (1988). Plasmid-mediated resistance to third-generation cephalosporins caused by TEM-type penicillinase genes. *Review of Infectious Diseases* **10**, 879–884.

SOWEK, J.A., SINGER, S.B., OHRINGER, S., MALLEY, M.F., DOUGHERTY, T.J., GOUGOUTAS, J.Z. AND BUSH, K. (1991). Substitution of lysine at position 104 or 240 of TEM-1 pTZ18R β-lactamase enhances the effect of serine-164 substitution on hydrolysis or affinity for cephalosporins and the monobactam aztreonam. *Biochemistry* **30**, 3179–3188.

STRYNADKA, N.C.J., ADACHI, H., JENSEN, S.E., JOHNS, K., SIELECKI, A., BETZEL, C., SUTOH, K. AND JAMES, M.N.J. (1992). Molecular structure of the acyl-enzyme intermediate in β-lactam hydrolysis at 1.7 A resolution. *Nature* **359**, 700–705.

SUTTON, B.J., ARTYMIUK, P.J., CORDERO-BORBOA, A.E., LITTLE, C., PHILLIPS D.C. AND WALEY, S.G. (1987). An X-ray crystallographic study of β-lactamase II from *Bacillus cereus* at 0.35 nm resolution. *Biochemical Journal* **248**, 181–188.

TOMASZ, A. (1979). The mechanism of the irreversible antimicrobial effects of penicillins: how the beta-lactam antibiotics kill and lyse bacteria. *Annual Review of Microbiology* **33**, 113–137.

VARETTO, L., DE MEESTER, F., MONNAIE, D., MARCHAND-BRYNAERT, J., DIVE, G., JACOB, F. AND FRÈRE, J.M. (1991). The importance of the negative charge of beta-lactam compounds in the interactions with active-site serine DD-peptidases and beta-lactamases. *Biochemical Journal* **278**, 801–807.

VEDEL, G., BELAAOUAJ, A., GILLY, L., LABIA, R., PHILIPPON, A., NEVOT, P. AND PAUL, G. (1991). Clinical isolates of *Escherichia coli* producing TRI β-lactamases: novel TEM-enzymes conferring resistance to β-lactamase inhibitors. *Journal of Antimicrobial Chemotherapy* **30**, 449–462.

WALEY, S.G. (1975). The pH-dependence and group modification of β-lactamase I. *Biochemical Journal* **149**, 547–551.

WALEY, S.G. (1987). An explicit model for bacterial resistance: application to beta-lactam antibiotics. *Microbiological Sciences* **4**, 143–146.

WALEY, S.G. (1992). β-lactamase: mechanism of action. In *The Chemistry of β-Lactams* (M.I. Page, Ed.), pp. 198–228. Chapman & Hall, London.

WILKIN, J.M., JAMIN, M., DAMBLON, C., ZHAO, G.H., JORIS, B., DUEZ, C. AND FRÈRE, J.M. (1993a). The mechanism of action of DD-peptidases. The role of Tyrosine 159 in the Streptomyces R61 DD-peptidase. *Biochemical Journal* **291**, 537–544.

WILKIN, J.M., JAMIN, M., JORIS, B. AND FRÈRE, J.M. (1993b). The mechanism of action of DD-peptidases. The role of Asparagine 161 in the Streptomyces R61 DD-peptidase. *Biochemical Journal* **293**, 195–201.

WILKIN, J.-M., DUBUS, A., JORIS, B. AND FRÈRE, J.M. (1994). The mechanism of action of DD-peptidases. The role of threonines 299 and 301 in the Streptomyces R61 DD-peptidase. *Biochemical Journal* **301**, 477–483.

ZAFARALLA, G., MANAVATHU, E.K., LERNER, S.A. AND MOBASHERY, S. (1992). Elucidation of the role of arginine-244 in the turnover processes of class A β-lactamases. *Biochemistry* **31**, 3847–3852.

7
Enzyme-catalysed Lipid Modification

PATRICK ADLERCREUTZ

Department of Biotechnology, Chemical Centre, Lund University, Sweden

Lipids are of great importance, for example in the food, chemical and pharmaceutical industries. For many applications, lipids isolated from biological material can be used directly or after more or less sophisticated purification procedures. However, for certain applications, synthetic lipids are more suitable, and enzymes have proven to be useful catalysts for their preparation. The synthesis can be carried out starting from building blocks like fatty acids and glycerol, but often a natural lipid material is used as starting material and enzymatic conversions are employed to modify its composition and thereby its properties. From the application point of view, key characteristics of the lipid can be its melting behaviour (Macrae, 1985), its ability to interact with other biomolecules (Millqvist *et al.*, 1994) or other physical properties. Furthermore, for lipids used in foods, the health aspects of the fatty acid content are of vital importance (Sridhar and Lakshminarayana, 1992). In this chapter the use of enzymes in the preparation of acylglycerols (tri-, di- and monoglycerides), phospholipids and fatty acid esters will be covered. A complicating factor is that these substances are poorly soluble in water, which is the solvent normally used for enzymatic conversions. Accordingly, emulsions have commonly been used as reaction media, but recently the introduction of biocatalysis in organic media has greatly widened the scope of enzymatic lipid modification (Tramper *et al.*, 1992). The enzymes used for lipid conversions are mainly lipases and phospholipases.

Lipase structure and mechanism

The lipases constitute a group of enzymes which has attracted increasing interest in recent years. One reason for this is the variety of substances that can be synthesized in lipase-catalysed reactions. Reviews concerning the use of lipases for the modification of fats and other lipids (Mukherjee, 1990) and concerning the use of lipases in organic synthesis have been published (Haraldsson, 1992). Furthermore, lipases are of fundamental interest in enzymology because of their special properties, enabling them to be active at interfaces and thereby convert water-insoluble substrates (Brockman, 1984).

Abbreviations: DHA, docosahexaenoic acid; EPA, eicosapentaenoic acid.

Biotechnology and Genetic Engineering Reviews — Vol. 12, December 1994
0264–8725/94/12/231–254 $20.00 + $0.00 © Intercept Ltd, P.O. Box 716, Andover, Hampshire SP10 1YG, UK

The first three-dimensional structures of lipases were determined only recently (Brady *et al.*, 1990; Winkler, D'Arcy and Hunziker, 1990). It was shown that both the *Rhizomucor miehei* lipase and human pancreatic lipase have active sites with a trypsin-like catalytic triad involving the amino acid residues serine, histidine and aspartic acid. Somewhat later, a *Geotrichum candidum* lipase was shown to have a slightly different catalytic triad containing glutamic acid instead of aspartic acid (Schrag *et al.*, 1991). The information concerning the three-dimensional structures, in combination with studies of the catalytic activity of lipases have shown that these enzymes work according to the same mechanism as the well-known serine proteases. The serine hydroxyl group of the lipase makes a nucleophilic attack on the carbonyl carbon of the acyl donor, and an acyl enzyme is formed as an intermediate (*Figure 1*). In aqueous solution the acyl enzyme is hydrolysed by water.

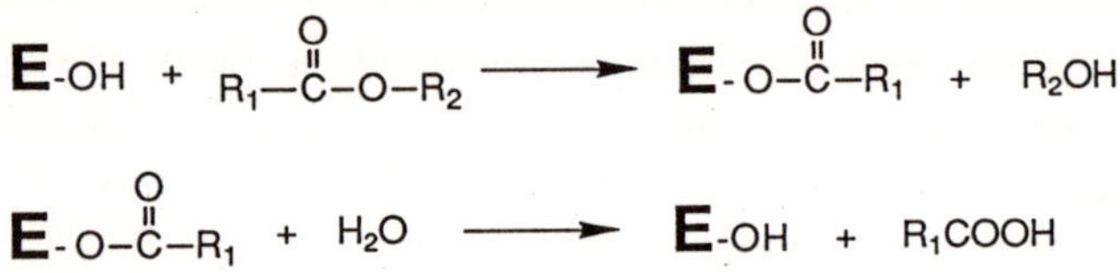

Figure 1. Mechanism of lipase-catalysed hydrolysis of an ester. E-OH: lipase molecule with the serine hydroxyl group of the active site.

Lipase-catalysed reactions

The physiological role of lipases is to hydrolyse triglycerides. Most of these substrates are not soluble in water to any appreciable extent. Therefore the reactions take place in emulsions. Lipases are adapted to this and are activated by the presence of an interface between an aqueous and an organic phase (Brockman, 1984). Consequently, with substrates of moderate solubility in water, the reaction rate of the lipase-catalysed reaction is normally very low as long as the substrate concentration is below the solubility limit of the substrate. Once this limit is exceeded and a separate substrate phase is formed, high reaction rates are obtained. The products of the hydrolysis of triglycerides are diglycerides and free fatty acids (*Figure 2*). The diglycerides can be further hydrolysed to monoglycerides and eventually to glycerol, with the liberation of the other esterified fatty acids. The substrate specificity of most lipases is quite broad, so apart from triglycerides many other esters can be hydrolysed with high reaction rates.

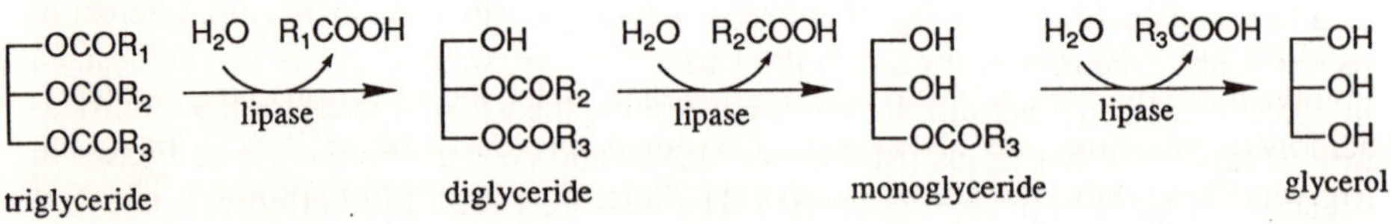

Figure 2. Lipase-catalysed hydrolysis of a triglyceride.

From the synthetic point of view, hydrolysis reactions are of limited interest. Thus the interest for lipases for synthetic purposes was modest until recently, when enzymology in organic media became an active research area (Tramper *et al.*, 1992). By using organic media with low water content, hydrolytic enzymes in general, and lipases in particular, can be used for many synthetic reactions. In organic media, other nucleophiles can compete successfully with water in the deacylation of the acyl enzyme. This is the basis of most of the recent applications of lipases in organic media. Furthermore, the equilibria of the reactions in organic media are frequently altered so that large amounts of the condensation product are present at equilibrium (Valivety *et al*., 1991).

Lipases can thus catalyse esterification, which is the reversal of the ester hydrolysis. A carboxylic acid acts as acyl donor and an alcohol carries out the deacylation of the acyl enzyme (*Figure 3*). Another way to prepare esters is to carry out an alcoholysis reaction. Alcoholysis is the reaction analogous to hydrolysis with an alcohol acting as a nucleophile instead of water (*Figure 3*). Organic media should be used to suppress the hydrolysis which occurs at high water content. Alcoholysis reactions can be used to prepare esters of fatty acids directly from triglycerides. Since triglycerides constitute an abundant raw material, alcoholysis is an attractive one-step process for the production of fatty acid esters (Shaw, Wang and Wang, 1991; Kanasawud *et al*., 1992).

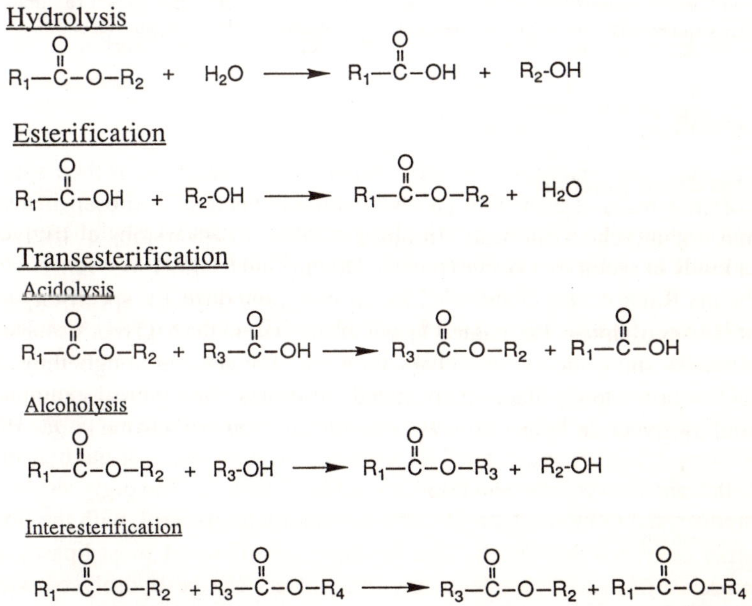

Figure 3. Lipase-catalysed reactions.

Other kinds of lipase-catalysed reactions are slightly more complex, since they involve more than one acyl-enzyme intermediate. In the modification of triglycerides, acidolysis reactions are often used (*Figures 3* and *4*) (Macrae, 1985). First, the triglyceride acylates the enzyme, leaving a diglyceride as reaction product. The acyl enzyme can then be deacylated by water or another nucleophile. The free fatty acid to be incorporated can acylate the enzyme as well, and thereby form a second acyl enzyme. If this is deacylated by a diglyceride, a triglyceride containing the new fatty acid is formed. It is clear that many different reactions occur, involving all triglycerides, partial glycerides, free fatty acids and acyl enzymes containing all the fatty acids present. A detailed kinetic analysis of these reactions is therefore quite complex (Kyotani *et al*., 1988; Reyes and Hill, 1994). If the reaction is continued long enough, an equilibrium mixture will eventually be formed. It is not necessary that the fatty acid to be incorporated into the triglyceride is present in the form of a free fatty acid. Fatty acid esters and triglycerides are other possible acyl donors, which are used in practical applications. These reactions are called interesterification reactions (*Figure 3*).

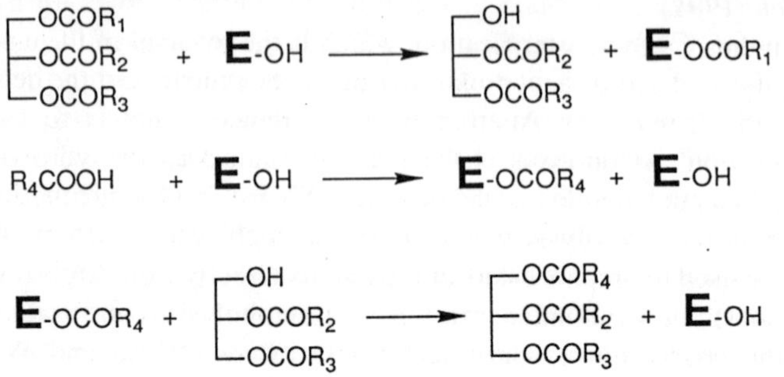

Figure 4. Mechanism of lipase-catalysed acidolysis. The lipase can form acyl enzymes, both in reactions with triglycerides (the first reaction) and with free fatty acids (the second reaction). Several nucleophiles can deacylate the acyl enzymes, for example diglycerides (the third reaction).

Lipase specificity

One of the most important features of enzyme-catalysed reactions is their specificity. This is often a main reason why an enzymatic conversion is chosen instead of a traditional organic-chemical route. In lipase-catalysed conversions of triglycerides, different kinds of specificity occur (Jensen, Dejong and Clark, 1983; Jensen, Rubano Galluzzo and Bush, 1990) (*Table 1*). Lipases can show different specificity towards different classes of lipids. The natural lipase substrates are the triglycerides. However, some enzymes show the typical characteristics of lipases although they convert triglycerides more slowly than other related substrates. One typical example is the mono- and di-glyceride lipase from *Penicillium camembertii* (Isobe *et al.*, 1992).

Table 1. Different types of lipase selectivity

Lipid-class selectivity
Regioselectivity
Fatty-acid selectivity
Nucleophile selectivity
Stereoselectivity

The most important type of lipase specificity utilized for lipid conversions is regiospecificity. A relatively large group of lipases, including pancreatic lipase and many fungal lipases, show specificity for the 1- and 3-positions of triglycerides. This has been widely used in triglyceride modification when modifications in the 1- and 3-positions are desired but the correct fatty acid is present already in the 2-position (see below). A lipase with strong selectivity for the 2-position has not yet been discovered, but a lipase from *Geotrichum candidum* showed moderate preference for this position (Sugihara, Shimada and Tominaga, 1991).

Lipases can show selectivity for certain fatty acids. The clearest example is a lipase from *Geotrichum candidum* which is selective for fatty acids having a *cis*-9 double bond (Jensen, 1974). Only some strains of *Geotrichum candidum* have a lipase with high selectivity, while others have less selective lipases or may be mixtures of lipases with different selectivity (Sonnet, Foglia and Baillargeon, 1993). A few lipases show selectivity for short fatty acids in triglycerides. Both *Syncephalastrum racemosum* lipase and *Penicillium caseicolum* lipase showed the highest rates for hydrolysis of tributyrin (Chopra, Chander and Singh, 1982; Alhir, Markakis and Chandan, 1990). The latter lipase is of importance for flavour development during the ripening of certain types of cheese. Lipases from plant seeds often show selectivity for the triglycerides present in the seeds. Examples are the selectivity of corn lipase for trilinolein and triolein, castor bean lipase for triricinolein, rapeseed lipase for trierucin and elm seed lipase for tricaprin (Lin, Yu and Huang, 1986). Furthermore, the seed lipase of *Vernonia galamensis* has recently been shown to be selective for vernolic acid (12,13-epoxy-oleic acid) (Ncube *et al.*, submitted). Most lipases except those mentioned show fairly equal activity on a wide range of fatty acids (Berger and Schneider, 1991).

The fatty-acid selectivity of lipases can be used to enrich certain fatty acids. Several lipases discriminate against fatty acids with a *cis*-5 double bond. The strong selectivity expressed in esterification reactions was used to convert the other fatty acids to esters while the $\Delta 5$ unsaturated acids were enriched in the free fatty acid fraction (Hayes and Kleiman, 1993). The same principle was used to enrich erucic acid from a rape-seed-oil hydrolysate. In this case *Geotrichum candidum* lipase catalysed the esterification, leaving erucic acid in the free fatty acid fraction (Sonnet, Foglia and Feairheller, 1993). In both cases the fatty-acid selectivity was less pronounced in hydrolysis reactions than in esterification.

It should be pointed out that the fatty-acid specificity seen in a lipase-catalysed reaction is not solely due to the specificity of the lipase. The reaction conditions, especially the temperature, may influence the specificity. Normally, fatty acids react more effectively when the temperature is above the melting temperature of that particular fatty-acid chain. By lowering the temperature it was possible to induce selectivity for linolenic acid, since at 10°C this chain was still liquid while fatty-acid chains with less unsaturation had higher melting points (Kaimal and Saroja, 1988).

In the normal hydrolytic reaction catalysed by lipases, water acts as nucleophile in the deacylation of the acyl enzyme. However, in organic media other nucleophiles can successfully compete. In fact, most lipases have quite broad specificity for the nucleophile, and many of them accept even quite bulky nucleophiles as diglycerides, as well as many alcohols of interest for synthetic purposes in organic chemistry. In most cases, primary alcohols are the best nucleophiles, while secondary alcohols are less reactive and tertiary alcohols seldom react at all (Rangheard *et al.*, 1992). One of the best-investigated lipases in this respect is the *Rhizomucor miehei* lipase, which accepts a very wide range of alcohols containing a variety of other functional groups as well (Miller *et al.*, 1988).

Still another kind of specificity is stereospecificity. Methods to study this specificity with triglyceride substrates have been developed only recently (Rogalska, Ransac and Verger, 1990). The degree of steroselectivity is normally moderate and varies for different substrates. It has even been observed that a switch in stereoselectivity occurs, so that a few lipases show preference for the sn-3 position of trioctanoin and for the sn-1 position in triolein (Rogalska *et al.*, 1993). It should be pointed out that the stereoselectivity of lipases has been widely used in the resolution of racemic mixtures of esters other than acylglycerols (Haraldsson, 1992). This is the most widespread use of lipases in organic synthesis.

Acyl migration

As mentioned above the 1,3-specific lipases have been used extensively in triglyceride conversions. A side-reaction that can occur in these transformations is the spontaneous acyl migration in the di- and mono-glycerides present in the reaction mixture (Bloomer, Adlercreutz and Mattiasson, 1991). As a result, the fatty acid composition in the 2-position might be changed, even though the enzyme itself is completely 1,3-specific. Similarly, in the preparation of pure 2-monoglycerides, 1,2-diglycerides, etc (see below), it is important to suppress acyl migration or else an isomer mixture will be formed. The rate of acyl migration is solvent dependent; it decreases with increasing solvent polarity (Sjursnes and Anthonsen, 1994).

Modes of using enzymes in organic media

Research concerning enzymatic reactions in organic media has been quite active during the past decade. In most applications an organic solvent is added to the reaction mixture. However, sometimes the substrates can constitute both substrates and solvent for the reaction. This is an attractive approach for practical applications, for example in the fat area. From the theoretical point of view, the solvent-free systems can be treated in the same way as those containing organic solvents.

There are several modes of using enzymes in organic media (Adlercreutz and Mattiasson, 1987; Tramper *et al.*, 1992) (*Figure 5*). The most straightforward way is simply to add an organic solvent to the aqueous solution of the enzyme. If the solvent is water-miscible, a homogeneous system is formed, otherwise a two-phase system. In both these cases enzyme inactivation is a key issue. In the homogeneous system, the enzyme is often active up to a certain concentration of solvent (Khmelnitsky *et al.*, 1991). In the two-phase system, partial inactivation of many enzymes occurs at the interface. However, as mentioned above, lipases normally work at interfaces and are activated by these (Brockman, 1984) and they are therefore suited for systems containing water-immiscible solvents.

In order to take full advantage of the organic reaction-medium, the water content in the reaction mixture should be kept quite low. Lyophilized enzyme powders can be directly suspended in organic media, and for some small-scale applications this is a practical way to use enzymes (*Figure 5c*). However, sometimes problems occur due to the aggregation of the enzyme powder. To decrease these problems and to facilitate the handling of the enzyme in general, it is common practice to immobilize the enzyme on a support material before using it in the organic medium (Reslow, Adlercreutz and Mattiasson, 1987) (*Figure 5d*). Because the enzyme is not soluble in the organic environment, there is no need for covalent bonds between the support and the enzyme. Therefore simple immobilization methods such as adsorption or deposition can be used. This kind of enzyme preparation is used at present in potential industrial processes employing enzymes in organic media (Macrae, 1985).

Still another mode of using enzymes in organic media is to solubilize the enzymes

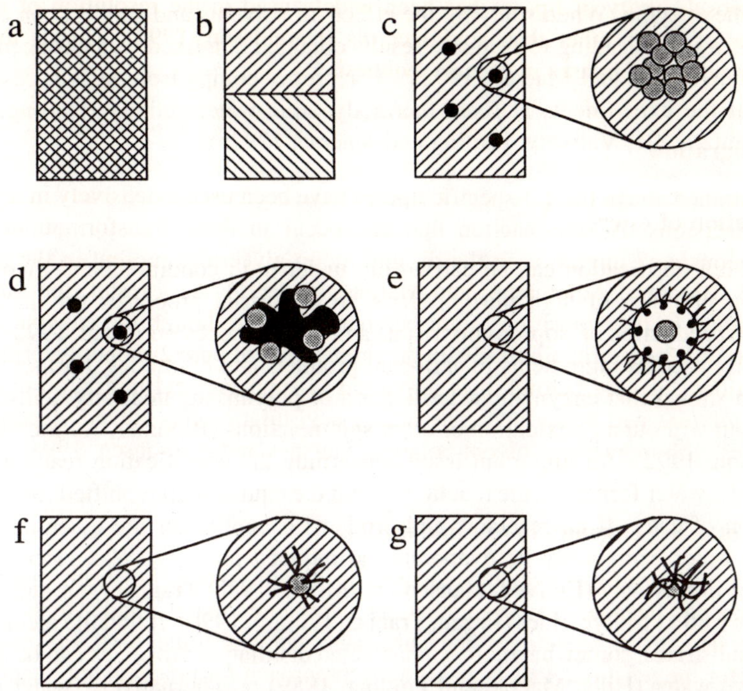

Figure 5. A schematic presentation of different ways to use enzymes in combination with organic solvents: (a) homogeneous mixture of water and water-miscible solvent; (b) two-phase system containing water and a water-immiscible solvent; (c) enzyme powder suspended in solvent, the small circles in the enlarged figure represent enzyme molecules; (d) enzyme on a support suspended in solvent; (e) enzyme solubilized in a microemulsion containing water, organic solvent and surfactant; (f) covalently modified enzyme dissolved in organic solvent; (g) non-covalent enzyme–polymer complex dissolved in organic solvent.

in microemulsions (Martinek *et al.*, 1989) (*Figure 5e*). Provided the surfactant does not inactivate the enzyme, good enzyme stability can often be achieved in microemulsions. However, product isolation is complicated by the presence of the surfactant. Another way to solubilize enzymes in organic media is to couple them covalently to polyethylene glycol (Inada *et al.*, 1986) (*Figure 5f*). The resulting modified enzyme can be dissolved in solvents like aromatic hydrocarbons and chlorinated hydrocarbons. Mass transfer is quite rapid in these systems, but the covalent modification often results in partial inactivation of the enzyme. Recently it has been shown that non-covalent surfactant–enzyme complexes and polymer–enzyme complexes can be dissolved in organic solvents (Okahata and Ijiro, 1992; Otamiri, Adlercreutz and Mattiasson, 1992; Blinkovsky, Khmelnitsky and Dordick, 1994) (*Figure 5g*). The main drawback is the presence of the surfactant or polymer during product isolation. Otherwise, this is quite an attractive method.

In all these examples small amounts of water are needed to make the enzyme catalytically active. Water present in the reaction mixture partitions between the solvent, the enzyme and the support or other components present. The amount of water bound to the enzyme governs its catalytic activity. Because of the widely different abilities of solvents to dissolve water, and of supports to bind water, the total water concentration in the reaction mixture gives little information about the hydration of the enzyme. When studying the effects of solvents and supports on enzyme catalysis, it is misleading to compare results obtained at a fixed water concentration (Valivety, Halling and Macrae, 1992a). For proper comparisons it is necessary to carry out the experiments at fixed thermodynamic water activity (Halling, 1984; Adlercreutz, 1991; Valivety, Halling and Macrae, 1992a).

Preparation of esters

Enzymatic esterification can be carried out under mild conditions and therefore it results in products of high purity. There are industrial processes for the lipase-catalysed production of isopropylmyristate and a few other esters (Staal, 1990). These are mainly used as ingredients in skin-care products and cosmetics. Furthermore, it has been shown that enzymatic esterification of polyunsaturated fatty acids can be carried out without any oxidation or other side-reactions (Bloomer, Adlercreutz and Mattiasson, 1992). An important issue concerning the esterification reactions is to remove the water formed in the reaction so that the equilibrium is shifted towards the formation of ester. If all reactants and products except water are essentially non-volatile, water can simply be evaporated at reduced pressure. Wax esters can be produced in this way (Eigtved, Hansen and Miller, 1988; Trani, Ergan and André, 1991), as well as triglycerides (Ergan, Trani and André, 1990). If volatile alcohols are used as substrates, somewhat more sophisticated distillation processes can be applied to remove water (Hills, Macrae and Poulina, 1989) or, alternatively, water can be absorbed by molecular sieves or other absorbents (Knox and Cliffe, 1984; Omar, Nishio and Nagai, 1988). Often ester yields in excess of 99% can be achieved (Bloomer, Adlercreutz and Mattiasson, 1992). Sometimes fairly good yields can be obtained just by carrying out the reaction in a two-phase system in which the ester is extracted into the organic phase (Monot *et al.*, 1991; Borzeix, Monot and Vandecasteele, 1992). Several flavour esters have been prepared using this technique (Langrand *et al.*, 1990).

Fatty acid esters can be prepared directly from triglycerides using an alcoholysis reaction. With some lipases the reaction can be carried out without extra solvent (Shaw, Wang and Wang, 1991; Kanasawud *et al.*, 1992). The yield can sometimes be improved by adsorption of the other product, glycerol, on an adsorbent such as silica gel (Stevenson, Stanley and Furneaux, 1994).

Triglyceride modification

Fats and oils mainly consisting of triglycerides are very important products, especially in the food area. The physical properties of a fat depend on its fatty acid composition. In order to obtain products with the desired properties, fractionation is used to a large extent. However, sometimes there is a need to exchange fatty acids present in the triglycerides with other fatty acids.

It was observed long ago that lipases can be used to incorporate new fatty acids into triglycerides (Borgström, 1954). Research has been intense during the past decades with the aim to prepare fats with desirable properties. An important example is the synthesis of cocoa-butter substitutes. Cocoa butter is widely used in chocolate and related products, because of its unusual melting characteristics. It has a rather narrow melting range around 30°C due to its high content of the triglycerides oleyl-distearoylglycerol and oleyl-palmitoyl-stearoylglycerol. Because of the high price of natural cocoa butter, large efforts have been spent on the preparation of substitutes with similar properties from cheaper raw materials. In the production of cocoa-butter substitutes, lipases have proven to be quite useful (Macrae, 1985). Using 1,3-specific lipases it is possible to manipulate the fatty acid composition in these positions without changing the fatty acids in the 2-position. Since vegetable oils often contain oleic acid in the 2-position, the task for the bioconversion is to introduce the proper saturated fatty acids in the 1- and 3-positions. Processes have been developed starting from, for example, a fraction of palm oil containing 2-oleyl-1,3-dipalmitoylglycerol as the main component (*Figure 6*). Transesterification with stearic acid or an ester thereof yields a cocoa-butter substitute (Macrae, 1985). Alternatively, oils containing large amounts of oleic acid-rich triglycerides, such as olive oil, can be transesterified with, for example, hydrogenated cotton seed oil which contains mainly triglycerides of palmitic and stearic acids (Chang, Abraham and John, 1990). Other suitable starting materials are oils from sunflower seeds or rape seeds. The oleic acid present in the 2-position in the substrate should remain there. However, due to acyl migration in di- or monoglycerides, the fatty acid incorporated in the 1- and 3-positions might end up in the 2- position (see above). Consequently, triglycerides with fatty acids in all three positions are formed; these are not accepted in cocoa-butter substitutes because their melting points are too high. Because of the mechanism of the reaction (*Figure 4*) diglycerides are necessary intermediates in the reactions. In order to minimize the spontaneous acyl migration, it is important to have a high catalytic activity in the reactor so that the enzymatic reaction is fast (Bloomer, Adlercreutz and Mattiasson, 1991). Then the residence time in the reactor is short and the extent of spontaneous reactions is therefore small. It is thus possible to suppress this side-reaction so that it is negligible. Several patents concerning lipase-catalysed production of cocoa-butter substitutes have been filed (Matsuo *et al.*, 1980; Macrae and How, 1983; Nakamura *et al.*, 1987) and industrial production started a couple of years ago.

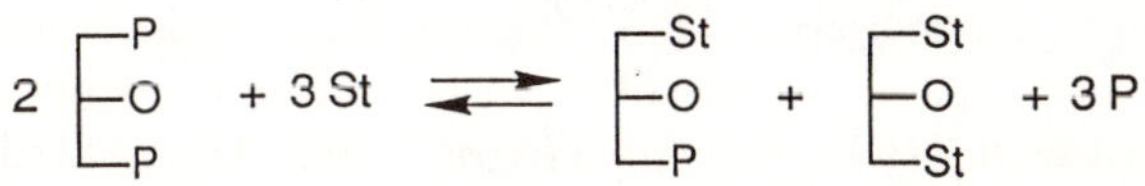

Figure 6. Schematic presentation of the conversion of 2-oleyl-1,3-dipalmitoylglycerol to 2-oleyl-3-palmitoyl-1-stearoylglycerol and 2-oleyl-1,3-distearoylglycerol in a lipase-catalysed acidolysis reaction. The reaction products are the main components of cocoa butter.

Other structured lipids containing suitable fatty acids have been evaluated for different applications. Triglycerides with medium-chain fatty acids in the 1- and 3-positions have been suggested for use in intravenous nutrition. The fatty acids in these positions are cleaved off by lipoprotein lipase in the bloodstream more effectively than longer fatty acids (Björkling, Godtfredsen and Kirk, 1991).

Since polyunsaturated fatty acids like eicosapentaenoic acid (EPA, 20:5) and docosahexaenoic acid (DHA, 22:6) have a number of health-promoting effects, large efforts have been spent on the incorporation of these fatty acids into different kinds of fats and oils. Either moderate amounts have been incorporated in fats which normally contain no polyunsaturated fatty acids or very small amounts, like groundnut oil (Sridhar and Lakshminarayana, 1992), or products highly enriched in polyunsaturated fatty acids are prepared (Li and Ward, 1992). Fish oils contain large amounts of polyunsaturated fatty acids and they are therefore suitable starting materials for further enrichment (Haraldsson *et al.*, 1989; Haraldsson and Almarsson, 1991). Phospholipids have been suggested as another type of lipid for the administration of polyunsaturated fatty acids to humans. Consequently, incorporation of these fatty acids into phospholipids has been investigated (see below; Hosokawa *et al.*, 1991, 1993).

Because of the surplus of butter fat available in some countries, research has been conducted with the aim of preparing new products from this source. Possibilities investigated are transesterification to obtain fats with modified fatty acid composition (Kalo, Huotari and Antila, 1990) and monoglyceride production (Yang and Parkin, 1994).

Hydrolysis of triglycerides to obtain free fatty acids and glycerol is not included in this chapter. This topic was covered by Mukherjee (1990).

Preparation of mono- and diglycerides

Regio-isomerically pure 1(3)-monoglycerides have been prepared from glycerol and fatty acids or fatty acid esters using 1,3-specific lipases as catalysts in organic solvents (Hayes and Gulari, 1991; Berger and Schneider, 1992) (*Figure 7*). In order to increase the interface between the polar glycerol and the apolar solvent containing the fatty acid, the glycerol was adsorbed on to a solid support. The monoglyceride yields were above 75% with diglycerides as the main by-product (Berger and Schneider, 1992). One way to increase the monoglyceride yield is to adsorb the monoglyceride continuously during the reaction (van der Padt *et al.*, 1992). 1,3-Specific lipases have been used to prepare 1,3-diglycerides as well (Berger, Laumen and Schneider, 1992).

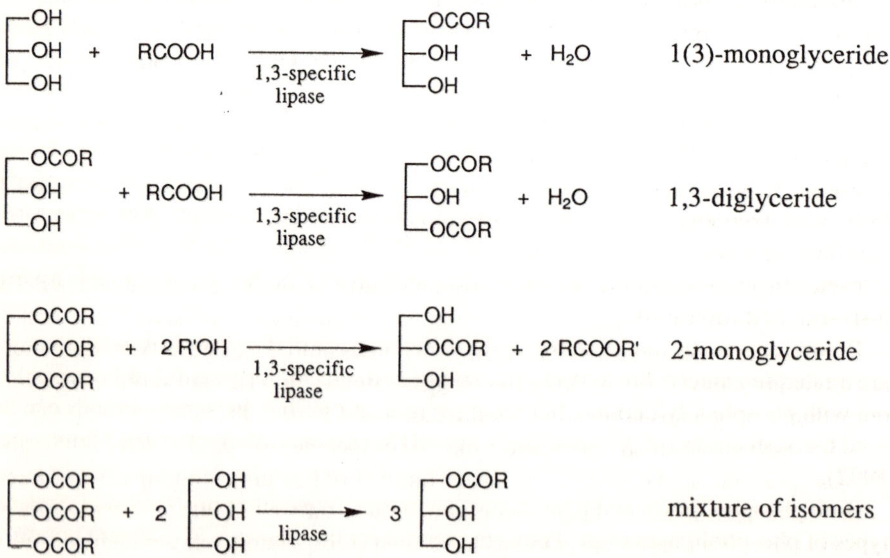

Figure 7. Lipase-catalysed preparation of mono- and diglycerides. Details concerning the different methods are given in the text.

Alternatively, monoglycerides can be prepared via lipase-catalysed glycerolysis of triglycerides (*Figure 7*). This has been achieved in a solvent-free system (McNeill, Shimizu and Yamane, 1991). High yields (67–90%) were obtained when the reactions were carried out at temperatures low enough for the precipitation of the monoglyceride. This led to solidification of the entire reaction mixture.

2-Monoglycerides have been prepared using 1,3-specific lipases for the hydrolysis (Holmberg and Osterberg, 1988) or alcoholysis (Millqvist, Adlercreutz and Mattiasson, 1994) of triglycerides (*Figure 7*). In the hydrolysis reaction carried out in a microemulsion, acyl migration was a problem, but in the alcoholysis reaction acyl migration was slow provided a suitable solvent, such as methyl-tert-butyl ether, was used.

Phospholipid modification

In nature, phospholipids play important roles as components of cell membranes, etc. Often different classes of phospholipids (having different polar groups) occur together and they constitute complex mixtures of molecular species having different fatty acid groups. Phospholipids differing in the polar part can be separated by chromatographic methods. However, separation with respect to the fatty acids often becomes too complicated due to the large number of possible combinations. For some applications these mixtures can be used as such, but for more specific applications one needs preparations which are uniform both concerning the polar part and the fatty acids. Pure phospholipids can be synthesized chemically, enzymatically or by a combination of these methods. Natural phospholipid mixtures are sometimes used as starting materials. A convenient method to prepare a new, well-defined phospholipid is to use a starting material containing the same fatty acid in both positions; these can be synthesized chemically. If a phospholipid containing two different fatty acids in the sn-1 and sn-2 positions is desired, one of the fatty acids can be exchanged using enzymatic methods (Svensson *et al.*, 1993) (*Figure 8*). The enzymatic exchange can be done either in two steps – hydrolysis and esterification – or in a one-step transesterification reaction. With the two-step procedure one can obtain a pure product provided the lysophospholipid is purified after the hydrolytic step. In the transesterification there is a need for a large excess of the new fatty acid because the fatty acid composition in the reactive positions of the product will reflect the equilibrium mixture of the fatty acids available for reaction. On the other hand, the transeterification reaction is less laborious and gives a higher yield than the hydrolysis–esterification route.

Below, enzymatic modification of the different parts of the phospholipid molecules are treated separately. Most work concerning exchange of fatty acids has been carried out with phosphatidylcholine, but there are indications that the same methods can be used for other classes of phospholipids as well (Svensson, Adlercreutz and Mattiasson, 1992).

Both phospholipases and lipases have been used to modify phospholipids. Several types of phospholipases exist. They are specific for the cleavage of different bonds in the phosolipid molecules. Phospholipases A and B cleave off the fatty acids, and phospholipase C and D hydrolyse the ester bonds on either side of the phosphate group (*Figure 9*). The use of the new techniques employing enzymes in organic

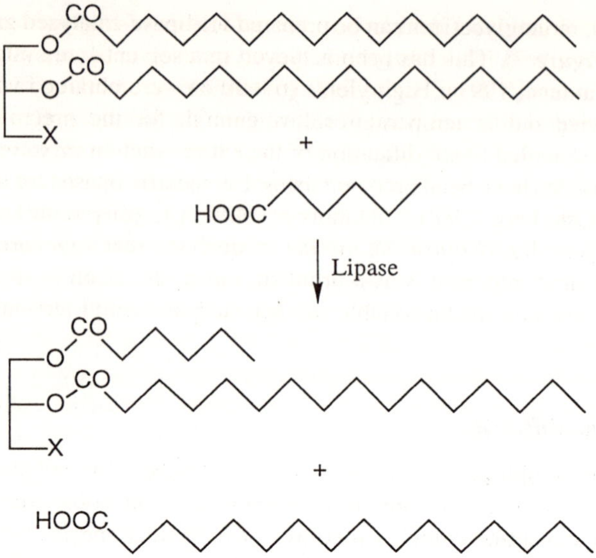

Figure 8. Conversion of dipalmitoyl-phosphatidylcholine to 1-hexyl-2-palmitoyl-phosphatidylcholine in a lipase-catalysed acidolysis reaction (Svensson *et al.*, 1993).

media, makes it possible to reverse these reactions and synthesize phospholipids with phospholipases as catalysts.

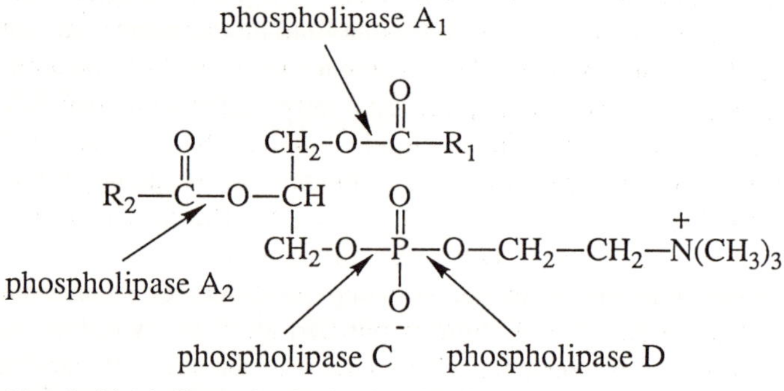

Figure 9. Phosphatidylcholine molecule showing the cleavage sites of the phospholipases.

Exchange of the fatty acid in the sn-1 position of phospholipids

Phospholipase A_1 has not been used to any large extent for lipid modification. It was long ago observed that lipases can be used to incorporate new fatty acids into phospholipids (Brockerhoff *et al.*, 1976), and since lipases are quite convenient to use they have been widely employed for the exchange of fatty acids in the sn-1 position of phospholipids. The lipases specific for the 1- and 3-positions in triglycerides have been shown to be specific for the sn-1 position of phospholipids. The reactions are often carried out as acidolysis reactions. In the first attempts, the fatty acid and the phospholipid were dispersed as micelles in an aqueous buffer, but the yields were moderate (Brockerhoff *et al.*, 1976) (*Table 2*). A main point in the synthesis is the minimization of hydrolysis. Organic media with low water contents have therefore been tried for the reactions. Compared to the acidolysis of triglycerides, the phospholipid reactions give considerably more hydrolysis even in the organic media. The transesterification results described in the literature are summarized in *Table 2*. The reports include several different ways of carrying out the reactions. Covalent modification of the lipase with polyethylene glycol made it possible to dissolve the enzyme in benzene (Yoshimoto *et al.*, 1986). A two-phase system consisting of water and hexane was used by Yagi *et al.* (1990), and resulted in slightly higher yield. Water/hexane was used by Totani and Hara (1991) as well. In order to reduce hydrolysis, water was replaced by glycerol. A microemulsion with iso-octane as organic solvent and Aerosol-OT as surfactant was used to incorporate polyunsaturated fatty acids into phosphatidylcholine (Holmberg and Eriksson, 1992). Immobilization of the lipase on a porous support and suspending this enzyme preparation in the reaction mixture have been used successfully in a number of recent reports (Svensson, Adlercreutz and Mattiasson, 1990, 1992; Pedersen, 1991; Mutua and Akoh, 1993). Toluene was a good solvent for the incorporation of saturated fatty acids into phosphatidylcholine (Svensson, Adlercreutz and Mattiasson, 1990) while hexane was the best for the incorporation of eicosapentaenoic acid (Mutua and Akoh, 1993).

Table 2. Lipase-catalysed transesterification of phospholipids (from Svensson, 1994)

Yield of modified phospholipid (%)	Incorporation of fatty acid (%)	Reference
14–23	18	Brockerhoff *et al.* (1976)
20	8.5	Yoshimoto *et al.* (1986)
25	25	Yagi *et al.* (1990)
40	25	Svensson, Adlercreutz and Mattiasson (1990)
<25	67	Pedersen (1991)
47	32	Totani and Hara (1991)
60	50	Svensson, Adlercreutz and Mattiasson (1992)
75	25	Holmberg and Eriksson (1992)
?	17.7	Mutua and Akoh (1993)

The yield of modified phospholipid and the incorporation of the new fatty acid(s) into the phospholipid are noted. Complete exchange of the fatty acid in the sn-1 position would give an incorporation of 50%.

Apart from the yield, the extent of incorporation of the new fatty acid is of importance. Since the reaction should be specific for the sn-1 position of the phospholipid, 50% incorporation is what is desired. Higher values of incorporation show that some of the new fatty acid has been incorporated in the sn-2 position, indicating poor regioselectivity in the reaction or acyl migration (Pedersen, 1991). To evaluate the importance of different reaction parameters on the reaction, it was helpful to plot the yield versus the incorporation of the new fatty acid (*Figure 10*). In order to obtain high yields of modified phospholipid, it was necessary to keep the water content low. In the system with immobilized lipase, this was achieved by adjusting the thermodynamic water activity of the reaction mixture (Svensson, Adlercreutz and Mattiasson, 1992). A low water activity resulted in a low reaction rate but the yields were considerably higher than at high water activity. The best results were obtained with *Rhizopus arrhizus* lipase immobilized on a polypropylene support (Svensson, Adlercreutz and Mattiasson, 1992). Fatty acid esters have been tried as acyl donors instead of free acids but the yields were lower (Svensson, Adlercreutz and Mattiasson, 1990).

The regioselectivity in the lipase-catalysed transesterification of phospholipids is quite high. In the sn-2 position less than 1% incorporation was found when almost complete exchange had occurred in the sn-1 position (Svensson, Adlercreutz and Mattiasson, 1992).

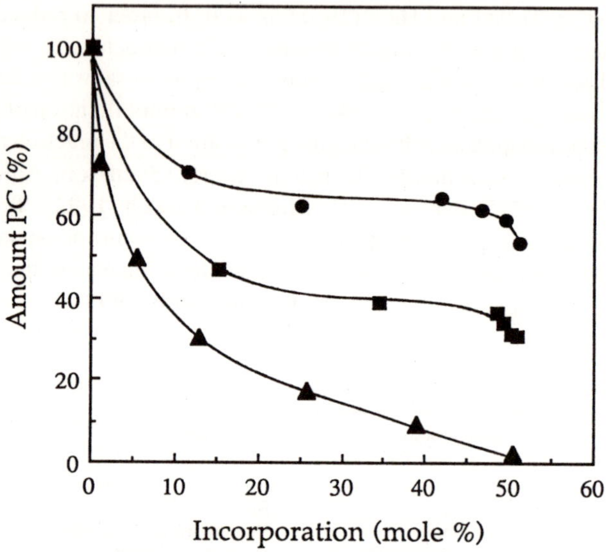

Figure 10. Lipase-catalysed acidolysis of phosphatidylcholine. The time course of the reaction is illustrated as the amount of remaining phosphatidylcholine as a function of the incorporation of the new fatty acid, heptadecanoic acid. The reactions were carried out in toluene using Lipozyme IM 20 at a water activity of 0.43 (triangles) or *Rhizopus arrhizus* lipase adsorbed on polypropylene at a water activity of 0.43 (squares) or 0.11 (circles) (Svensson, Adlercreutz and Mattiasson, 1992).

Exchange of the fatty acid in the sn-2 position of phospholipids

Only phospholipase A_2 has been successfully used to modify the fatty acid composition in the sn-2 position of phospholipids. One could hope that non-specific lipases would be useful for the exchange of fatty acids in both positions, but this has not yet been realized.

The fatty acid in the sn-2 position can easily be hydrolysed off using phospholipase A_2. The reversed reaction proceeds with rather low yields, even in organic media with low water content. In a screening of different enzymes, phospholipase A_2 from *Naja naja* provided the best yield (6.5%) (Pernas *et al.*, 1990) (*Table 3*) and toluene was used as solvent. There is a large interest in incorporating polyunsaturated fatty acids into phospholipids for use in food. The condensation of lysophospholipid and polyunsaturated fatty acids in a microemulsion provided a yield of 6% using phospholipase A_2 from porcine pancreas (Na *et al.*, 1990). By using a solvent-free system, polyunsaturated fatty acids from fish oil were incorporated into phospholipids using porcine pancreatic phospholipase A_2 (Hosokawa *et al.*, 1991). The yield reported was 15%, which was improved to around 20% (Hosokawa *et al.*, 1993). Even better yields

Table 3. Phospholipase A_2-catalysed esterification of lysophospholipid and fatty acid

Yield of phospholipid (%)	Reference
6.5	Pernas *et al.* (1990)
6	Na *et al.* (1990)
15	Hosokawa *et al.* (1991)
>20	Lilja-Hallberg and Härröd (1992)
20	Hosokawa *et al.* (1993)
14	Svensson (1994)

(>20%) were obtained in a similar study using porcine pancreatic phospholipase A_2 immobilized on a polymer carrier and with iso-octane as solvent (Lilja-Hallberg and Härröd, 1992).

In the esterification reaction, the theoretically maximal yield is determined by the equilibrium position. This in turn depends on the concentrations or activities of the reactants. If a high yield with respect to the lysophospholipid is desired, it is beneficial to use a large excess of the free fatty acid. Furthermore, the water activity in the reaction mixture should be kept low, since water is a reaction product. A systematic investigation of the influence of the water activity on the yield in the preparation of phosphatidylcholine from lysophosphatidylcholine and oleic acid has been carried out. The best yield in this study (14%) was achieved at a water activity of 0.33 (Svensson, 1994). At lower water activity the porcine pancreatic phospholipase A_2 immobilized on XAD-8 was not active. The lack of catalytic activity at low water activity is a serious drawback in this type of reaction. Lipases often show good catalytic activity even at very low water activity (Valivety, Halling and Macrae, 1992b), while phospholipases and most other enzymes require higher water activity.

It is noteworthy that transesterification reactions do not seem to work well with phospholipase A_2 (Na *et al.*, 1990; Pernas *et al.*, 1990). The reason might be due to differences in the catalytic mechanism of lipases and phospholipase A_2. The catalytic mechanism of phospholipase A_2 from bovine pancreas has been studied in detail. It shows large similarities with the mechanism of the lipases. However, phospholipase A_2 lacks the serine residue which acts as a nucleophile in the case of lipases. Instead, a water molecule is present in this position in the active site (Dijkstra *et al.*, 1981). As a result, no acyl enzyme is formed. The 'transesterification' reaction in this case would be a hydrolysis reaction followed by esterification. Since both these reactions can be carried out separately, the problem is probably that it is difficult to find conditions under which both reactions occur to an appreciable extent. There are a couple of reports that the transesterification reaction can, indeed, be carried out (Mukherjee, 1990; Mutua and Akoh, 1993).

Preparation of lysophospholipids

Lysophospholipids have been prepared from phospholipids using lipases as catalysts. Since the lipase-catalysed reactions are selective for the sn-1 position, the reaction product has the remaining fatty acid in the sn-2 position. Both hydrolysis reactions (Haas *et al.*, 1993; Morimoto *et al.*, 1993) and alcoholysis reactions (Sarney, Fregapane and Vulfson, 1994) have been used. If acyl migration occurs, the isomeric

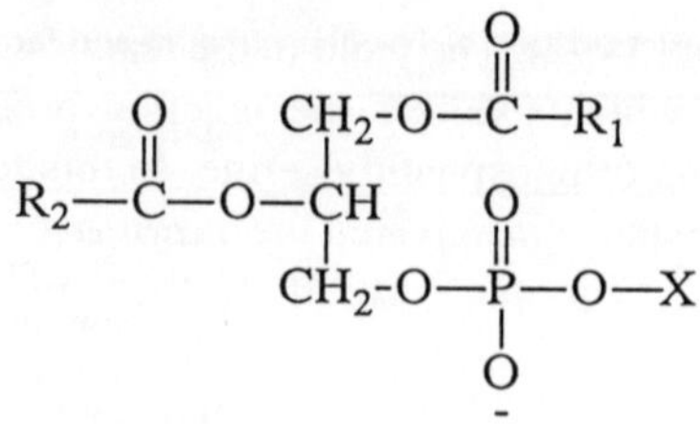

R_1,R_2: Fatty acid alkyl chains

X	
$-CH_2CH_2N^+(CH_3)_3$	Phosphatidyl choline
$-CH_2CH_2NH_3^+$	Phosphatidyl ethanolamine
$-CH_2CH(OH)CH_2(OH)$	Phosphatidyl glycerol
$-CH_2CH(NH_3^+)COOH$	Phosphatidyl serine
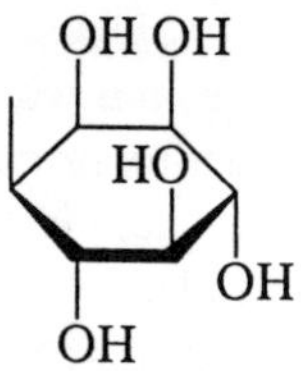	Phosphatidyl inositol
-H	Phosphatidic acid

Figure 11. Different classes of glycerophospholipids.

purity of the product might decrease. Furthermore, the yield will decrease due to further hydrolysis of the second fatty acid. Quite pure products have been obtained in boric acid–borax buffers, while other buffers provided products of lower purity (Morimoto *et al.*, 1993).

Conversions from one phospholipid class to another

Phospholipase C or D can be used to convert one phospholipid class to another (*Figure 11*). The most common starting material is phosphatidylcholine. Phospholipase D from cabbage was used to convert phosphatidylcholine to phosphatidylglycerol in a membrane reactor (Lee *et al.*, 1985). Ether was added to dissolve the phospholipid and stabilize the enzyme. Later, the same research group compared a micelle and an emulsion system as reaction media for this reaction (Juneja *et al.*, 1987a). In the micelle system, the initial reaction rate was high but the competing hydrolysis reaction resulted in large amounts of phosphatidic acid. In the emulsion of buffer/glycerol in ether containing the phosphatidylcholine, this side-reaction was effectively suppressed so almost quantitative conversion to phosphatidylglycerol was achieved. In these studies, dissolved enzyme was used. However, it was found advantageous to immobilize the enzyme by adsorption on octyl-Sepharose CL-4B (Juneja *et al.*, 1987b). With the immobilized phospholipase D, repeated batch and continuous reactions were carried out. A similar enzyme preparation, but with phospholipase D from a *Streptomyces* species was used to convert phosphatidylcholine to phosphatidylserine (Juneja *et al.*, 1992). The yield in the reaction was improved by the addition of choline oxidase which oxidizes the reaction product choline and therefore favours further formation of phosphatidylserine. In this reaction hydrogen peroxide was formed, and this substance inactivated the enzymes to some extent. The addition of catalase to decompose hydrogen peroxide further improved the process (Juneja *et al.*, 1992).

At least some phospholipase D enzymes have broad nucleophile specificity. This fact has been used to synthesize unnatural phospholipids. A phospholipase D from *Streptomyces prunicolor* accepted as nucleophiles pentoses, hexoses, glycosides, amino sugars, alcoholic sugars, di- and oligosaccharides, amino acids, nucleic acids, dinitrophenol and primary and secondary alcohols (Kudo, 1988). Other interesting examples are the transphosphatidylation of phosphatidylcholine with azasugars, nucleosides, peptides and inhibitors (Wang *et al.*, 1993). For these reactions an enzyme from *Streptomyces* was active but the cabbage enzyme and other enzymes were not. The unnatural phospholipids were intended for incorporation into liposomes for use in drug delivery and targeting. Another interesting example is the preparation of sialic acid derivatives. Here, the sialic acid was coupled to the rest of the molecule via a spacer (Ri *et al.*, 1993) (*Figure 12*).

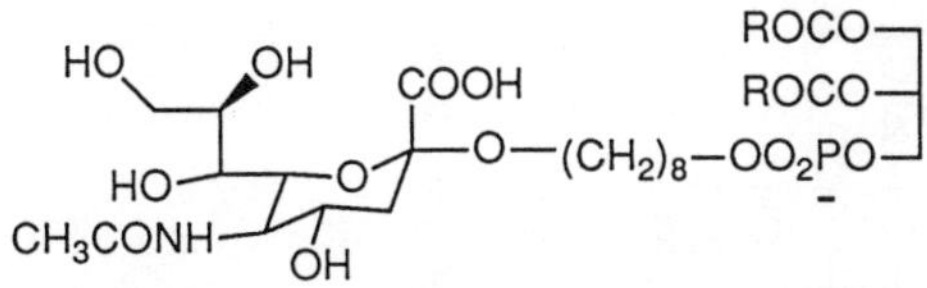

Figure 12. Sialic acid derivative prepared by phospholipase D catalysed transphosphatidylation of phosphatidylcholine (Ri *et al.*, 1993).

The natural reaction of phospholipase C is to break the bond between the glycerol backbone and the phosphorus in glycerophospholipids (*Figure 9*). However, it can hydrolyse the corresponding bond in sphingophospholipids as well, to form phosphorylcholine and a ceramide (Kanfer and Spielvogel, 1975) (*Figure 13*). Furthermore, it has been shown that phospholipase C possesses transphosphatidylation activity. The enzyme from different *Clostridia* was used to transfer the phosphorylcholine part of phosphatidylcholine to ceramide so that sphingomyelin was formed (Kanfer and Spielvogel, 1975).

A somewhat unusual reaction has been described for phospholipase D. It catalyses a transphosphatidylation reaction with phosphatidylglycerol both as donor and acceptor of the phosphatidyl group under the formation of cardiolipin (Stanacev, Stuhne-Sekalec and Domazet, 1973) (*Figure 14*). In fact, this is the route for cardiolipin biosynthesis in various species of bacteria.

Concluding remarks

Research concerning enzymatic lipid modification has developed rapidly during the past decade due to the increased knowledge about biocatalysis in organic media. In organic media, lipids can be handled effectively and the equilibrium positions are favourable for the synthesis of many interesting lipids. Compared to chemical processing, enzymatic conversions offer advantages concerning mild reaction condi-

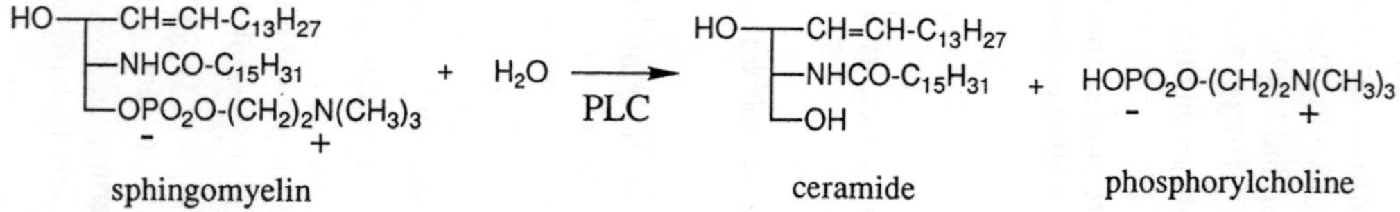

Figure 13. Preparation of ceramide by phospholipase C- catalysed hydrolysis of sphingomyelin (Kanfer and Spielvogel, 1975).

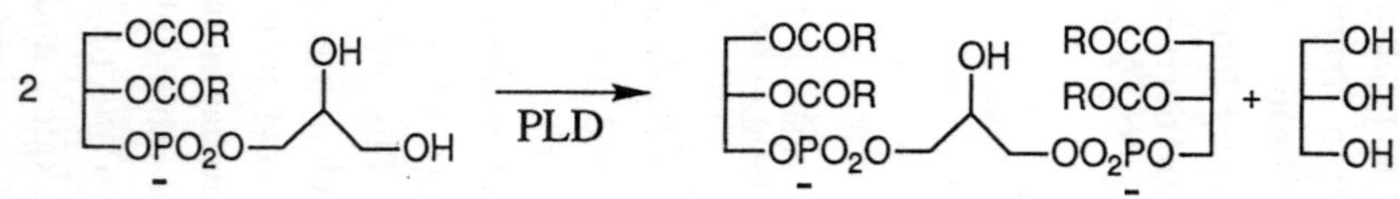

Figure 14. Phospholipase D-catalysed conversion of phosphatidylglycerol to cardiolipin (Stanacev, Stuhne-Sekalec and Domazet, 1973).

tions, which has proven to be important, for example in ester synthesis and concerning selectivity in the reactions. The selectivity of lipases has made it possible to synthesize tailor-made triglycerides, partial glycerides and phospholipids. It is likely that in the future a wide range of conversions of complex lipids will be developed. At present, our knowledge about the lipases is rapidly increasing, while much less is known about other lipid-converting enzymes. Further synthetic possibilities will certainly appear with increasing knowledge about these enzymes, and new enzymes will be discovered in screening programmes.

Acknowledgement

This work was supported by the Swedish Research Council for Engineering Sciences (TFR).

References

ADLERCREUTZ, P. (1991). On the importance of the support material for enzymatic synthesis in organic media. Support effects at controlled water activity. *European Journal of Biochemistry* **199**, 609–614.

ADLERCREUTZ, P. AND MATTIASSON, B. (1987). Aspects of biocatalyst stability in organic solvents. *Biocatalysis* **1**, 99–108.

ALHIR, S., MARKAKIS, P. AND CHANDAN, R.C. (1990). Lipase of *Penicillium caseicolum. Journal of Agricultural and Food Chemistry* **38**, 598–601.

BERGER, M. AND SCHNEIDER, M.P. (1991). Lipases in organic solvents: the fatty acid chain length profile. *Biotechnology Letters* **13**, 641–645.

BERGER, M. AND SCHNEIDER, M.P. (1992). Enzymatic esterification of glycerol II. Lipase-catalyzed synthesis of regioisomerically pure 1(3)-rac-monoacylglycerols. *Journal of the American Oil Chemist's Society* **69**, 961–965.

BERGER, M., LAUMEN, K. AND SCHNEIDER, M.P. (1992). Enzymatic esterification of glycerol I. Lipase-catalyzed synthesis of regioisomerically pure 1,3-sn-diacylglycerols. *Journal of the American Oil Chemist's Society* **69**, 955–960.

BJÖRKLING, F., GODTFREDSEN, S.E. AND KIRK, O. (1991). The future impact of industrial lipases. *Trends in Biotechnology* **9**, 360–363.

BLINKOVSKY, A.M., KHMELNITSKY, Y.L. AND DORDICK, J.S. (1994). Organosoluble enzyme-polymer complexes: a novel type of biocatalyst for nonaqueous media. *Biotechnology Techniques* **8**, 33–38.

BLOOMER, S., ADLERCREUTZ, P. AND MATTIASSON, B. (1991). Triglyceride interesterification by lipases. 2. Reaction parameters for the reduction of trisaturated impurities and diglycerides in batch reactions. *Biocatalysis* **5**, 145–162.

BLOOMER, S., ADLERCREUTZ, P. AND MATTIASSON, B. (1992). Facil synthesis of fatty acid esters in high yields. *Enzyme and Microbial Technology* **14**, 546–552.

BORGSTRÖM, B. (1954). On the mechanism of pancreatic lipolysis of glycerides. *Biochimica et Biophysica Acta* **13**, 491–504.

BORZEIX, F., MONOT, F. AND VANDECASTEELE, J.-P. (1992). Strategies for enzymatic esterification in organic solvents: comparison of microaqueous, biphasic, and micellar systems. *Enzyme and Microbial Technology* **14**, 791–797.

BRADY, L., BRZOZOWSKI, A.M., DEREWENDA, Z.S., DODSON, E., DODSON, G., TOLLEY, S., TURKENBURG, J.P., CHRISTIANSEN, L., HUGE-JENSEN, B., NORSKOV, L. AND MENGE, U. (1990). A serine protease triad forms the catalytic centre of a triacylglycerol lipase. *Nature* **343**, 767–770.

BROCKERHOFF, H., SCHMIDT, P.C., FONG, J.W. AND TIRRI, L.J. (1976). Introduction of labeled fatty acid in position 1 of phosphoglycerides. *Lipids* **11**, 421–422.

BROCKMAN, H.L. (1984). General features of lipolysis: reaction scheme, interfacial structure and experimental approaches. In *Lipases* (B. Borgström, Ed.), pp. 3–46. Elsevier, Amsterdam.

CHANG, M.-K., ABRAHAM, G. AND JOHN, V.T. (1990). Production of cocoa butter-like fat from interesterification of vegetable oils. *Journal of the American Oil Chemist's Society* **67**, 832–834.

CHOPRA, A.K., CHANDER, H. AND SINGH, J. (1982). Lipolytic activity of *Synephalastrum racemosum. Journal of Dairy Science* **65**, 1890–1894.

DIJKSTRA, B.W., DRENTH, J. AND KALK, K.H. (1981). Active site and catalytic mechanism of phospholipse A_2. *Nature* **289**, 604–606.

EIGTVED, P., HANSEN, T.T. AND MILLER, C.A. (1988). Ester synthesis with immobilized lipases. In *Proceedings of the World Conference on Biotechnology in the Fats and Oil Industry* (T.H. Applewhite, Ed.), pp. 134–137. AOCS, Champaign, Ill.

ERGAN, F., TRANI, M. AND ANDRÉ, G. (1990). Production of glycerides from glycerol and fatty acid by immobilized lipases in non-aqueous media. *Biotechnology and Bioengineering* **35**, 195–200.

HAAS, M.J., CICHOWICZ, D.J., PHILLIPS, J. AND MOREAU, R. (1993). The hydrolysis of phosphatidylcholine by immobilized lipase: optimization of hydrolysis in organic solvents. *Journal of the American Oil Chemist's Society* **70**, 111–117.

HALLING, P.J. (1984). Effects of water on equilibria catalysed by hydrolytic enzymes in biphasic reaction systems. *Enzyme and Microbial Technology* **6**, 513–516.

HARALDSSON, G.G. (1992). The application of lipases in organic synthesis. In *The Chemistry of Acid Derivatives*, Vol. 2 (S. Patai, Ed.), pp. 1395–1473. John Wiley and Sons, New York.

HARALDSSON, G.G. AND ALMARSSON, Ö. (1991). Studies on the positional specificity of lipase from *Mucor miehei* during interesterification reactions of cod liver oil with n-3 polyunsaturated fatty acids and ethyl ester concentrates. *Acta Chemica Scandinavica* **45**, 723–730.

HARALDSSON, G.G., HÖSKULDSSON, P.A., SIGURDSSON, S.T., THORSTEINSSON, F. AND GUDBJARNASON, S. (1989). The preparation of triglycerides highly enriched with w-3 polyunsaturated fatty acids via lipase catalyzed interesterification. *Tetrahedron Letters* **30**, 1671–1674.

HAYES, D. AND GULARI, E. (1991). 1-Monoglyceride production from lipase-catalyzed esterification of glycerol and fatty acid in reverse micelles. *Biotechnology and Bioengineering* **38**, 507–517.

HAYES, D.G. AND KLEIMAN, R. (1993). The isolation and recovery of fatty acids with Δ5 unsaturation from meadowfoam oil by lipase-catalyzed hydrolysis and esterification. *Journal of the American Oil Chemist's Society* **70**, 555–560.

HILLS, G.A., MACRAE, A.R. AND POULINA, R.R. (1989). Ester preparation. European Patent application 0 383 405.

HOLMBERG, K. AND ERIKSSON, C. (1992). Lipase and phospholipase catalyzed transformations in microemulsions. *Indian Journal of Chemistry* **31B**, 886–890.

HOLMBERG, K. AND OSTERBERG, E. (1988). Enzymatic preparation of monoglycerides in microemulsion. *Journal of the American Oil Chemist's Society* **65**, 1544–1548.

HOSOKAWA, M., TAKAHASHI, K., HATANO, M. AND EGI, M. (1991). Polyunsaturated fatty phosphatide synthesis by industrial phospholipase A_2. *Nippon Shokuhin Kogyo Gakkaishi* **38**, 695–698.

HOSOKAWA, M., OHSHIMA, H., KOHNO, H., TAKAHASHI, K., HATANO, M. AND ODASHIMA, S. (1993). Synthesis of phosphatidylcholine containing highly unsaturated fatty acid by phospholipase A_2 and the effect of retinoic acid induced differentiation of HL-60 cells. *Nippon Suisan Gakkaishi* **59**, 309–314.

INADA, Y., TAKAHASHI, K., YOSHIMOTO, T., AJIMA, A., MATSUSHIMA, A. AND SAITO, Y. (1986). Application of polyethylene glycol-modified enzymes in biotechnological processes: organic solvent-soluble enzymes. *Trends in Biotechnology* **4**, 190–194.

ISOBE, K., NOKIHARA, K., YAMAGUCHI, S., MASE, T. AND SCHMID, R.D. (1992). Crystallisation and characterization of monoacylglycerol and diacylglycerol lipase from *Penicillium camembertii. European Journal of Biochemistry* **203**, 233–237.

JENSEN, R.G. (1974). Characteristics of the lipase from the mold, *Geotrichum candidum*: A review. *Lipids* **9**, 149–157.

JENSEN, R.G., DEJONG, F.A. AND CLARK, R.M. (1983). Determination of lipase specificity. *Lipids* **18**, 239–252.

JENSEN, R.G., RUBANO GALLUZZO, D. AND BUSH, V.J. (1990). Selectivity is an important characteristic of lipases (acylglycerol hydrolases). *Biocatalysis* **3**, 307–316.

JUNEJA, L.R., HIBI, N., INAGAKI, M., YAMANE, T. AND SHIMIZU, S. (1987a). Comparative study on conversion of phosphatidylcholine to phosphatidylglycerol by cabbage phospholipase D in micelle and emulsion systems. *Enzyme and Microbial Technology* **9**, 350–354.

JUNEJA, L.R., HIBI, N., INAGAKI, M., YAMANE, T. AND SHIMIZU, S. (1987b). Repeated batch operations for phosphatidylglycerol synthesis from phosphatidylcholine with immobilized phospholipase D. *Applied Microbiology and Biotechnology* **27**, 146–151.

JUNEJA, L.K., TANIGUCHI, E., SHIMIZU, S. AND YAMANE, T. (1992). Increasing productivity by removing choline in conversion of phosphatidylcholine to phosphatidylserine by phospholipase D. *Journal of Fermentation and Bioengineering* **73**, 357–361.

KAIMAL, T.N.B. AND SAROJA, M. (1988). Selective removal of linolenic acid from soybean oil by lipase-catalyzed interesterification at low temperatures. *Biotechnology Letters* **10**, 337–340.

KALO, P., HUOTARI, H. AND ANTILA, M. (1990). *Pseudomonas fluorescens* lipase-catalysed interesterification of butter fat in the absence of a solvent. *Milchwissenschaft* **45**, 281–285.

KANASAWUD, P., PHUTRAKUL, S., BLOOMER, S., ADLERCREUTZ, P. AND MATTIASSON, B. (1992). Triglyceride interesterification by lipases. 3. Alcoholysis of pure triglycerides. *Enzyme and Microbial Technology* **14**, 959–965.

KANFER, J.N. AND SPIELVOGEL, C.H. (1975). Phospholipase C catalyzed formation of sphingomyelin-^{14}C from lecithin and N-(^{14}C)-oleoyl-sphingosine. *Lipids* **10**, 391–394.

KHMELNITSKY, Y.L., MOZHAEV, V.V., BELOVA, A.B., SERGEEVA, M.V. AND MARTINEK, K. (1991). Denaturation capacity: a new quantitative criterion for selection of organic solvents as reaction media in biocatalysis. *European Journal of Biochemistry* **198**, 31–41.

KNOX, T. AND CLIFFE, K.R. (1984). Synthesis of long-chain esters in a loop reactor system using a fungal cell bound enzyme. *Process Biochemistry* **1988**, 188–192.

KUDO, S. (1988). Biosurfactants as food additives. In *Proceedings of the World Conference on Biotechnology in the Fats and Oils Industry* (T.H. Applewhite, Ed.), pp. 195–201. AOCS, Champaign, Ill.

KYOTANI, S., FUKUDA, H., MORIKAWA, H. AND YAMANE, T. (1988). Kinetic studies on the interesterification of oils and fats using dried cells of fungus. *Journal of Fermentation Technology* **66**, 71–83.

LANGRAND, G., RONDOR, N., TRIANTAPHYLIDES, C. AND BARATTI, J. (1990). Short chain flavour ester synthesis by microbial lipases. *Biotechnology Letters* **12**, 581–586.

LEE, S.Y., HIBI, N., YAMANE, T. AND SHIMIZU, S. (1985). Phosphatidylglycerol synthesis by phospholipase D in a microporous membrane bioreactor. *Journal of Fermentation Technology* **63**, 37–44.

LI, Z.-Y. AND WARD, O.P. (1993). Lipase-catalyzed esterification of glycerol and n-3 polyunsaturated fatty acid concentrate in organic solvent. *Journal of the American Oil Chemist's Society* **70**, 745–748.

LILJA-HALLBERG, M. AND HÄRRÖD, M. (1992). Synthesis of phosphatidyl choline with polyunsaturated fatty acids by phospholipase A_2 in an organic solvent. In *Biocatalysis in Non-conventional Media* (J. Tramper, M. Vermüe, H.H. Beeftink and U. von Stockar, Eds), pp. 747–753. Elsevier, Amsterdam.

LIN, G., WU, F.-C. AND LIU, S.-H. (1993). Phospholipase A_2 catalysis in organic media. *Tetrahedron Letters* **34**, 1959–1962.

LIN, Y.H., YU, C. AND HUANG, A.H.C. (1986). Substrate specificities of lipases from corn and other seeds. *Archives of Biochemistry and Biophysics* **244**, 346–356.

MCNEILL, G.P., SHIMIZU, S. AND YAMANE, T. (1991). High-yield enzymatic glycerolysis of fats and oils. *Journal of the American Oil Chemist's Society* **68**, 1–5.

MACRAE, A.R. (1985). Interesterification of fats and oils. In *Biocatalysts in Organic Synthesis* (J. Tramper, H.C. van der Plas and P. Linko, Eds), pp. 195–208. Elsevier, Amsterdam.

MACRAE, A.R. AND HOW, P. (1983). Interesterification with a lipase enzyme as an interesterification catalyst. European patent application 0 093 602.

MARTINEK, K., KLYACHKO, N.L., KABANOV, A.V., KHMELNITSKY, Y.L. ABD LEVASHOV, A.V.

(1989). Micellar enzymology: its relation to membranology. *Biochimica et Biophysica Acta* **981**, 161–172.

MATSUO, T., SAWAMURA, N., HASHIMOTO, Y. AND HASHIDA, W. (1980). Producing a cacao butter substitute by transesterification of fats and oils. UK patent application 2 035 359.

MILLER, C., AUSTIN, H., POSORSKE, L. AND GONZLEZ, J. (1988). Characteristics of an immobilized lipase from the commercial synthesis of esters. *Journal of the American Oil Chemist's Society* **65**, 927–931.

MILLQVIST, A., ADLERCREUTZ, P. AND MATTIASSON, B. (1994). Lipase-catalyzed alcoholysis of triglycerides for the preparation of 2-monoglycerides. *Enzyme and Microbial Technology* **16**, 1042–1047.

MILLQVIST, A., ADLERCREUTZ, P., MATTIASSON, B., MIEZIS, Y. AND LARSSON, K. (1994). Starch complexing by enzymatically prepared 2-monoglycerides compared to effects by 1-isomers. *Starch* **46**, 347–348.

MONOT, F., BORZEIX, F., BARDIN, M. AND VANDECASTEELE, J.-P. (1991). Enzymatic esterification in organic media: role of water and organic solvent in kinetics and yield of butyl butyrate synthesis. *Applied Microbiology and Biotechnology* **35**, 759–765.

MORIMOTO, T., MURAKAMI, N., NAGATSU, A. AND SAKAKIBARA, J. (1993). Regiospecific deacylation of glycerophospholipids by use of *Mucor javanicus* lipase. *Tetrahedron Letters* **34**, 2487–2490.

MUKHERJEE, K.D. (1990). Lipase-catalyzed reactions for modification of fats and other lipids. *Biocatalysis* **3**, 277–293.

MUTUA, L.N. AND AKOH, C.C. (1993). Lipase-catalyzed modification of phospholipids: incorporation of n-3 fatty acids into biosurfactants. *Journal of the American Oil Chemist's Society* **70**, 125–128.

NA, A., ERIKSSON, C., ÖSTERBERG, E. AND HOLMBERG, K. (1990). Synthesis of phosphatidylcholine with (n-3) fatty acids by phospholipase A_2 in microemulsion. *Journal of the American Oil Chemist's Society* **67**, 766–770.

NAKAMURA, K., YOKOMICHI, H., OKISAKA, K., NISHIDE, T., KAWAHARA, Y. AND NOMURA, S. (1987). Process for transesterifying fats. European patent application 0 257 388.

NCUBE, I., GITLESEN, T., ADLERCREUTZ, P. AND MATTIASSON, B. Novel type of substrate specificity found in the *Veronia galamensis* lipase. (submitted).

OKAHATA, Y. AND IJIRO, K. (1992). Preparation of a lipid-coated lipase and catalysis of glyceride ester syntheses in homogeneous organic solvents. *Bulletin of the Chemical Society of Japan* **65**, 2411–2420.

OMAR, I.C., NISHIO, N. AND NAGAI, S. (1988). The role of water on the equilibrium of esterification by immobilized lipase packed-bed column reactor. *Biotechnology Letters* **10**, 799–804.

OTAMIRI, M., ADLERCREUTZ, P. AND MATTIASSON, B. (1992). Complex formation between chymotrypsin and ethyl cellulose as a means to solubilize the enzyme in active form in toluene. *Biocatalysis* **6**, 291–305.

PEDERSEN, K.B. (1991). Enzymic interesterification of phospholipids and fatty acids or esters. Patent application, PCT WOP 9103564.

PERNAS, P., OLIVIER, J.L., LEGOY, M.D. AND BEREZIAT, G. (1990). Phospholipid synthesis by extracellular phospholipase A_2 in organic solvents. *Biochemical and Biophysical Research Communications* **168**, 644–650.

RANGHEARD, M.-S., LANGRAND, G., TRIANTAPHYLIDES, C. AND BARATTI, J. (1992). Multicompetitive enzymatic reactions in organic media: application to the determination of lipase alcohol specificity. *Enzyme and Microbial Technology* **14**, 966–974.

RESLOW, M., ADLERCREUTZ, P. AND MATTIASSON, B. (1987). Organic solvents for bioorganic synthesis. 1. Optimisation of parameters for a chymotrypsin catalyzed process. *Applied Microbiology and Biotechnology* **26**, 1–8.

REYES, H.R. AND HILL, C.G., JR (1994). Kinetic modeling of interesterification reactions catalyzed by immibilized lipase. *Biotechnology and Bioengineering* **43**, 171–182.

RI, S., SUGINO, T., REKA, R.J., SHINHO, K., SENOO, M., NISHIMOTO, K., KIN BAND YAMAMOTO, T. (1993). Sialic acid derivatives and their preparation. Japanese patent application JP 05132496.

ROGALSKA, E., RANSAC, S. AND VERGER, R. (1990). Stereoselectivity of lipases. II. Stereoselective hydrolysis of triglycerides by gastric and pancreatic lipases. *Journal of Biological Chemistry* **265**, 20271–20276.

ROGALSKA, E., CUDREY, C., FERRATO, F. AND VERGER, R. (1993). Stereoselective hydrolysis of triglycerides by animal and microbial lipases. *Chirality* **5**, 24–30.

SARNEY, D.B., FREGAPANE, G. AND VULFSON, E.N. (1994). Lipase-catalyzed synthesis of lysophospholipids in a continuous bioreactor. *Journal of the American Oil Chemist's Society* **71**, 93–96.

SCHRAG, J.D., LI, Y., WU, S. AND CYGLER, M. (1991). Ser–His–Glu triad forms the catalytic site of the lipase from *Geotrichum candidum. Nature* **351**, 761–764.

SHAW, J.-F., WANG, D.-L. ABD WANG, Y.J. (1991). Lipase-catalysed ethanolysis and isopropanolysis of triglycerides with long-chain fatty acids. *Enzyme and Microbial Technology* **13**, 544–546.

SJURSNES, B.J. AND ANTHONSEN, T. (1994). Acyl migration in 1,2-dibutyrin. Dependence on solvent and water activity. *Biocatalysis* **9**, 285–297.

SONNET, P.E., FOGLIA, T.A. AND FEAIRHELLER, S.H. (1993). Fatty acid selectivity of lipases: erucic acid from rapeseed oil. *Journal of the American Oil Chemist's Society* **70**, 387–391.

SONNET, P.E., FOGLIA, T.A. AND BAILLARGEON, M.W. (1993). Fatty acid selectivity: the selectivity of lipases of *Geotrichum candidum. Journal of the American Oil Chemist's Society* **70**, 1043–1045.

SRIDHAR, R. AND LAKSHMINARAYANA, G. (1992). Incorporation of eicosapentaenoic and docosahexaenoic acids into groundnut oil by lipase-catalyzed ester interchange. *Journal of the American Oil Chemist's Society* **69**, 1041–1042.

STAAL, L.H. (1990). To esters via bio-technology. Presentation at the American Oil Chemist's Society World Conference, Malaysia, 1990.

STANACEV, N.Z., STUHNE-SEKALEC, L. AND DOMAZET, Z. (1973). Enzymatic formation of cardiolipin from phosphatidylglycerol by the transphosphatidylation mechanism catalyzed by phospholipase D. *Canadian Journal of Biochemistry* **51**, 747–753.

STEVENSON, D.E., STANLEY, R.A. AND FURNEAUX, R.H. (1994). Near-quantitative production of fatty acid alkyl esters by lipase-catalyzed alcoholysis of fats and oils with adsorption of glycerol by silica gel. *Enzyme and Microbial Technology* **16**, 478–484.

SUGIHARA, A., SHIMADA, Y. AND TOMINAGA, Y. (1991). A novel *Geotrichum candidum* lipase with some preference for the 2-position on a triglyceride molecule. *Applied Microbiology and Biotechnology* **35**, 738–740.

SVENSSON, I. (1994). Phospholipid modifications with enzymes in organic media. Ph.D. thesis, Lund University, Sweden.

SVENSSON, I., ADLERCREUTZ, P. AND MATTIASSON, B. (1990). Interesterification of phosphatidylcholine with lipases in organic media. *Applied Microbiology and Biotechnology* **33**, 255–258.

SVENSSON, I., ADLERCREUTZ, P. AND MATTIASSON, B. (1992). Lipase-catalyzed transesterification of phosphatidylcholine at controlled water activity. *Journal of the American Oil Chemist's Society* **69**, 986–991.

SVENSSON, I., ADLERCREUTZ, P., MATTIASSON, B., MIESIZ, Y. AND LARSSON, K. (1993). Phase behaviour of aqueous systems of enzymatically modified phosphatidylcholines with one hexadecyl and one hexyl or octyl chain. *Chemistry and Physics of Lipids* **66**, 195–197.

TOTANI, Y. AND HARA, S. (1991). Preparation of polyunsaturated phospholipids by lipase-catalyzed transesterification. *Journal of the American Oil Chemist's Society* **68**, 848–851.

TRAMPER, J., VERMÜE, M., BEEFTINK, H.H. AND VON STOCKAR, U. (Eds) (1992). *Biocatalysis in Non-conventional Media.* Elsevier, Amsterdam.

TRANI, M., ERGAN, F. AND ANDRÉ, G. (1991). Lipase-catalyzed production of wax esters. *Journal of the American Oil Chemist's Society* **68**, 20–22.

VALIVETY, R.H., HALLING, P.J. AND MACRAE, A.R. (1992a). Reaction rate with suspended lipase catalyst shows similar dependence on water activity in different organic solvents. *Biochimica et Biophysica Acta* **1118**, 218–222.

VALIVETY, R.H., HALLING, P.J. AND MACRAE, A.R. (1992b). *Rhizomucor miehei* lipase remains highly active at water activity below 0.0001. *FEBS Letters* **301**, 258–260.

VALIVETY, R.H., JOHNSTON, G.A., SUCKLING, C.J. AND HALLING, P.J. (1991). Solvent effects on biocatalysis in organic systems: equilibrium position and rates of lipase catalyzed esterification. *Biotechnology and Bioengineering* **38**, 1137–1143.

VAN DER PADT, A., KUERENTJES, J.T.F., SEWALT, J.J.W., VAN DAM, E.M., VAN DORP, L.J. AND VAN'T RIET, K. (1992). Enzymatic synthesis of monoglycerides in a membrane bioreactor with an in-line adsorption column. *Journal of the American Oil Chemist's Society* **69**, 748–754.

WANG, P., SCHUSTER, M., WANG, Y.-F. AND WONG, C.-H. (1993). Synthesis of phospholipid-inhibitor conjugates by enzymatic transphosphatidylation with phospholipase D. *Journal of the American Chemical Society* **115**, 10487–10491.

WINKLER, F.K., D'ARCY, A. AND HUNZIKER, W. (1990). Structure of human pancreatic lipase. *Nature* **343**, 771–774.

YAGI, T., NAKANISHI, T., YOSHIZAWA, Y. AND FUKUI, F. (1990). The enzymatic acyl exchange of phospholipids with lipases. *Journal of Fermentation and Bioengineering* **69**, 23–25.

YANG, B. AND PARKIN, K.L. (1994). Monoacylglycerol production from butteroil by glycerolysis with gel-entrapped microbial lipase in microaqueous media. *Journal of Food Science* **59**, 47–52.

YOSHIMOTO, T., NAKATA, M., YAMAGUCHI, S., FUNADA, T., SAITO, Y. AND INADA, Y. (1986). Synthesis of eicosapentaenoyl phosphatidylcholines by polyethylene glycol-modified lipase in benzene. *Biotechnology Letters* **8**, 771–776.

8
Enzymes in Water-in-oil Microemulsions ('Reversed Micelles'): Principles and Applications

CHRISTOPHER OLDFIELD

Energy Biosystems Corporation, 4200 Research Forest Drive, The Woodlands, Texas TX 77381, USA

Introduction

Microemulsions are thermodynamically stable, single-phase mixtures of oil, water and surfactant. They form spontaneously, usually within seconds, simply by shaking the components together in appropriate proportions. In these respects they differ fundamentally from conventional emulsions. The exact structure of the microemulsion depends on the kind of surfactant used and on the exact composition (relative amounts of water, oil and surfactant), but a common defining feature is that the surfactant is located almost exclusively at the oil–water interface, forming a close-packed monolayer such that the polar or charged headgroups are hydrated on the aqueous side of the interface and the apolar tails are likewise favourably solvated on the oil side. When the (water+surfactant) volume fraction is not too high (<0.2), the microemulsion is oil-continuous and for many, but not all, systems the water is dispersed in the form of tiny (nanometer diameter), approximately spherical, droplets which diffuse randomly and independently with the continuous oil medium. These systems are referred to as water-in-oil (w/o) microemulsions, or as reversed micellar systems (since the orientation of the surfactant molecules, with the headgroups pointing inwards, is opposite to that encountered in classical aqueous micelles). A characteristic feature of w/o

---

Abbreviations: AOT, Aerosol-OT; BTP, bis-tris-propane; CD, circular dichroic spectroscopy; CSTR, continually stirred tank reactor; α-CT, α-chymotrypsin; CTAB, cetyltrimethylammonium bromide; DDAB, didodecyldimethylammonium bromide; EPR, electron paramagnetic resonance; FFA, free fatty acid; GPNA, glutaryphenylalanyl-4-nitroanilide; GuHCl, guanidine hydrochloride; HLADH, horse-liver alcohol dehydrogenase; HLB, hydrophile-lyophile balance; HRP, horseradish peroxidase; LDH, lactate dehydrogenase; LipOH, liposamide dehydrogenase; MBGs, microemulsion-based gels; NMR, nuclear magnetic resonance; PC, phosphatidylcholine; PCS, photon correlation spectroscopy; pit, phase-inversion temperature; SAAP, *N*-succinyl-Ala-Ala-Pro-Phe-4-nitroanilide; SANS, small-angle neutron scattering; SAXS, small-angle X-ray scattering; SDS, sodium dodecyl sulphate; t-CNI, *N-trans* cinnamoyl imidazole; TG, triglyceride; TOMAC, trioctyl ammonium chloride; TP, temperature-phase; TTN, total turnover number.

Biotechnology and Genetic Engineering Reviews — Vol. 12, December 1994
0264–8725/94/12/255–327 $20.00 + $0.00 © Intercept Ltd, P.O. Box 716, Andover, Hampshire SP10 1YG, UK

microemulsions is that they are almost always optically transparent since, due to their small size, the droplets do not scatter visible light.

Water-in-oil microemulsions have attracted much interest from biotechnologists in recent years, since the discovery that enzymes can be dissolved in them with retention of activity and, in large measure, stability. The concept of the enzyme-containing water droplet as a microreactor stably dispersed within a continuous oil medium has proven to be an extremely attractive concept for those interested in biotransformations involving apolar (water-insoluble) substances. Consequently the development of these systems for such purposes is an area of great interest and active research (reviewed by Luisi and Magid, 1986; Martinek *et al.*, 1986, 1989; Luisi *et al.*, 1988).

Biotransformation is an interdisciplinary science which relies on a combination of gene-manipulation techniques to improve the activity/stability characteristics of existing biocatalysts, of fermentation techniques to meet the demand for the biocatalyst, and of medium-engineering to provide solvent systems appropriate for the biotransformation. The application of microemulsions quite clearly falls within the medium-engineering side of this equation. A well-engineered medium is expected (1) to support and stabilize biocatalyst activity, (2) to be a good solvent for reactants, products and any additional cofactors, and (3) to be accessible to post-reaction separations techniques permitting product recovery. A fundamental objective of this chapter is to give an indication of the inherent flexibility in microemulsion engineering which permits each of these requirements to be addressed in a systematic and holistic fashion. Their properties in this context will be reviewed in terms of the progress that has been made over the past few years in understanding the activity and stability characteristics of solubilized enzymes, and in the peculiarities of microemulsion structure and phase-behaviour which complement this application by facilitating cofactor regeneration, reaction monitoring and product recovery.

As a consequence of the great activity in this area, a number of new possibilities for the application of microemulsions which go beyond the basic medium-engineering concept, have become apparent. In one active area of research, microemulsions are being applied for the selective extraction of added-value products (amino acids, antibiotics and proteins) from conjugate aqueous phases such as culture-broths; in another, the biotechnological applications of microemulsion gels (that is, microemulsions with the rheological properties of highly viscous gels) are being explored. New results and trends in these fields are reported here, also.

This chapter is therefore divided into four sections, as follows: The first discusses phase-behaviour, structure and dynamics of microemulsions. This information is considered essential for a full appreciation of their applications and potential in biotransformations. The second is a discussion of the properties and behaviour (structure, activity, stability) of microemulsified enzymes. The third discusses recent advances in process engineering of microemulsions with a view to industrial-scale biotransformations and covers practical considerations such as provision of cofactors and product separation. The fourth deals with new research and topics, the final section summarizes and gives an indication of trends and future prospects.

Although warranted, there is unfortunately no room here for a detailed comparison of microemulsions with other methods of stabilizing enzyme activities in non-aqueous media (including two-phase (oil+water) systems (Lilly, 1983), enzymes rendered oil-soluble by covalent modification (Inada *et al.*, 1986) and dry (or immobilized) enzyme suspensions in organic solvents (Klibanov, 1989). Where possible, reference will be made to a few studies where direct comparisons have been attempted. See recent reviews (Dordick, 1991; Gupta, 1992) for an appreciation of current status of these other technologies.

Physical chemistry

SOME FUNDAMENTAL PROPERTIES OF MICROEMULSIONS

Definition of a microemulsion

Danielsson and Lindman (1981) give the following comprehensive definition: 'A microemulsion is defined as a system of water, oil and amphiphile which is a single phase, optically isotropic and thermodynamically stable liquid solution'. This definition reasonably excludes from consideration:

1. real emulsions (which, by definition, are unstable with respect to the corresponding two-phase systems);
2. solutions of surfactant in a single solvent (i.e. binary molecular and micellar or reversed-micellar systems);
3. liquid-crystalline systems; and
4. amphiphile-free systems of any description.

The three basic microemulsion types are shown in *Figure 1*. The oil-continuous (water-in-oil; w/o) type microemulsions, already described, consist of surfactant-coated droplets of water in an oil-continuous phase. Water-continuous (oil-in-water; o/w) microemulsions are dispersions of surfactant-coated oil droplets in a water-continuous phase. The droplets-type structure is typical for compositions which are predominantly (>80% v/v) oil or water. To a good first-approximation, the droplets in a given preparation are the same size, i.e. microemulsions are characteristically monodisperse. Bicontinuous microemulsions contain approximately equal volumes of water and oil (and a correspondingly large amount of surfactant); they are distinct from w/o and o/w systems in that the oil, water and surfactant subphases are physically continuous throughout the sample, the interfacial surfactant forming a dividing surface between the oil and water domains. It is rather more difficult to assign a structural icon for these systems, since a number of different structures may be envisaged, each consistent with the thermodynamic constraints imposed by the bicontinuous condition (Scriven, 1976). Probably, examples of both highly randomized disordered structures (*Figure 1c*) and various kinds of semi-ordered structures with elements of tetragonal (*Figure 1d*) or cubic (*Figure 1e*) topology exist, depending on the particular microemulsion system and its exact composition (Zemb *et al.*, 1987; Lindman *et al.*, 1989).

Droplets-type microemulsions are usually optically transparent, although preparations with larger droplets (radii of 10 nm or greater) and bicontinuous microemulsions often show Tyndall scattering. All microemulsions, however, are optically isotropic.

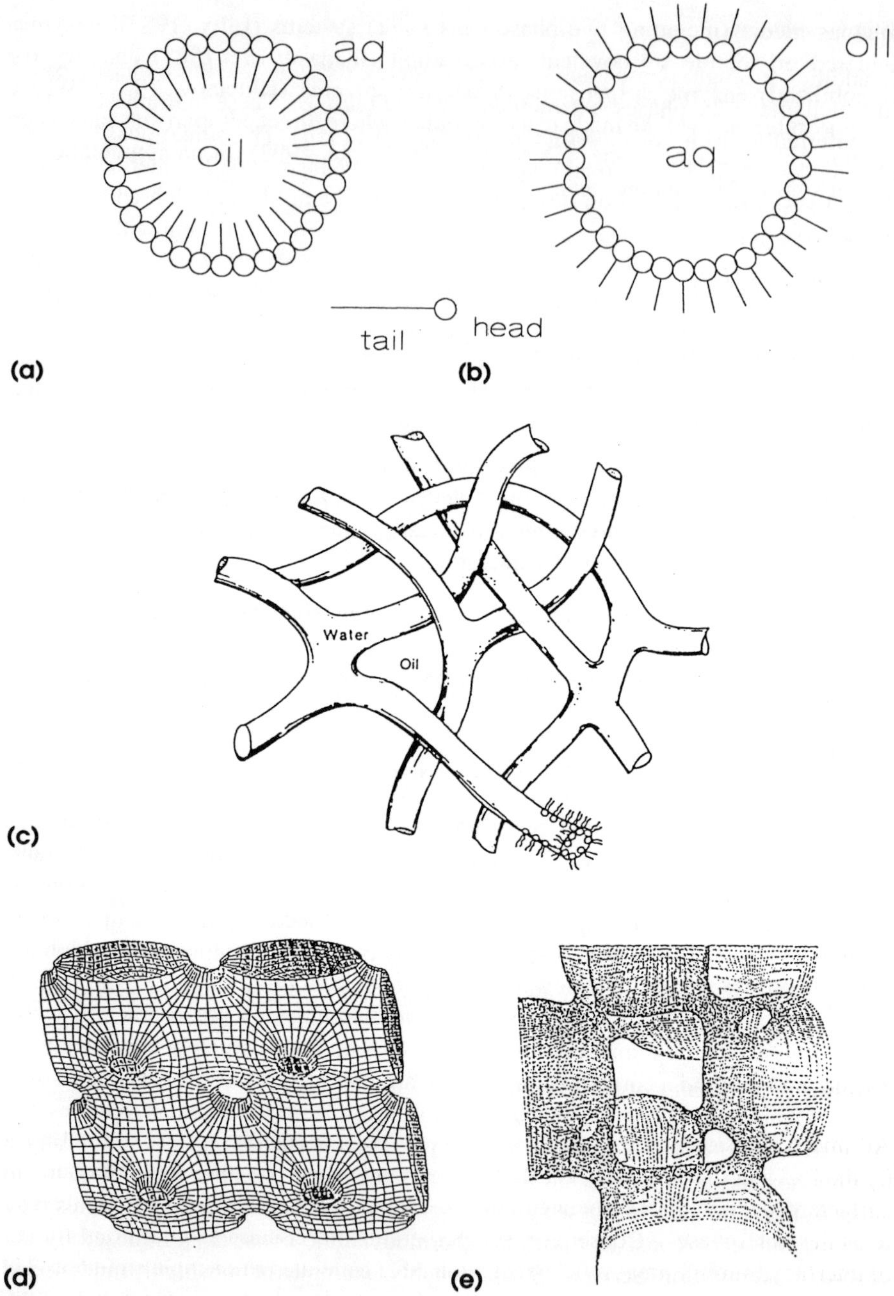

Figure 1. Examples of microemulsion structural motifs (not to scale). (a) Water droplets in an oil-continuous system; (b) oil-droplets in a water-continuous system; (c) an open, disordered bicontinuous system (Chen, S.J. Evans, D.F., Ninham, B.W., Mitchell, D.J., Blum, F.D. and Pickup, S. (1986) *Journal of Physical Chemistry* **90**, 842–847 © (1986) Americal Chemical Society); (d) one example of a system possessing low mean curvature with tetragonal symmetry; and (e) one example of a system possessing low mean curvature with body-centred cubic symmetry (Lindman, B., Shinoda, K., Olsson, U., Anderson, D., Karlström, G. and Wennerström, H. (1989) *Colloids and Surfaces* **38**, 205–224). The bicontinuous structures should be envisaged as highly dynamic, or 'melted'. Figures reproduced with permission.

Winsor systems

In addition to single-phase microemulsions, there exist two- and three-phase systems in which one of the phases is a microemulsion (*Figure 2*). These multiphase systems are referred to using the classification of Winsor (1954). The single-phase systems already described (o/w, w/o and bicontinuous) are Winsor type IV microemulsions. Type I systems consist of an o/w microemulsion in equilibrium with an excess oil phase; type II systems consist of a w/o microemulsion in equilibrium with an excess water phase; and type III systems consist of a microemulsion phase in equilibrium with both an excess water and an excess oil phase. In the last case the microemulsion is usually the middle phase and is usually bicontinuous in nature. As discussed in more detail on pp. 265–270, these systems may be interconverted by manipulating the temperature or the ionic strength and this has important implications in areas such as product recovery and selective extraction processes.

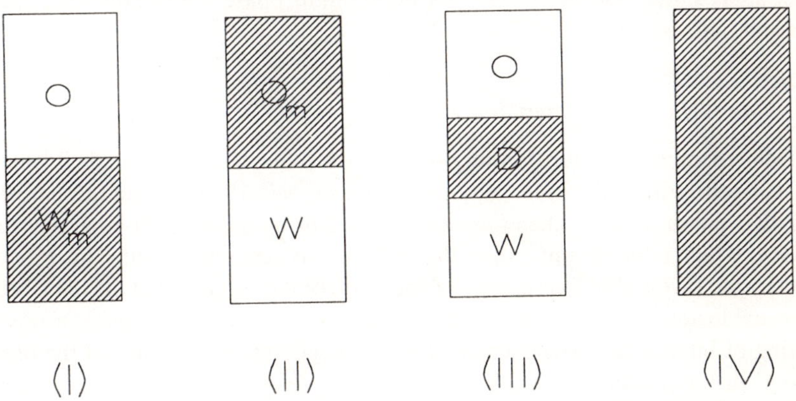

Figure 2. Winsor systems. The Winsor (I) system consists of a water-continuous microemulsion (W_m) in equilibrium with an excess oil (O) phase; The Winsor (II) system consists of an oil-continuous microemulsion (O_m) in equilibrium with an excess water phase (W). The Winsor (III) consists of a surfactant-rich phase (bicontinuous microemulsion, D), in equilibrium with excess oil and water phases. The Winsor (IV) system is a single-phase microemulsion which may be oil-, water- or bicontinuous.

Dispersion properties of microemulsions

All microemulsion phases, whether droplets-type or bicontinuous, are characterized by the fact that the surfactant is located almost entirely at the oil–water interface (the surfactant layer, with its associated water and oil of solvation, is conveniently referred to as the 'interphase'). Consequently, microemulsions contain an enormous amount of interfacial surface area , much greater than that generated in a simple emulsion. The interfacial area is dependent only on the amount of surfactant present and on the surface area occupied per surfactant molecule. For example, a typical w/o microemulsion consists of 9.5 ml oil (e.g. *n*-heptane), 0.5 ml water and 0.1 mol dm^{-3} of a surfactant. If each surfactant molecule occupies 0.5 nm^2 at the interface (a typical value), simple calculation shows that the total surface area is about 30 m^2 per cm^3 of microemulsion. The composition of a w/o microemulsion is usually related by stating the surfactant concentration and the ratio *R*, where,

$$R = \frac{[H_2O]}{[AOT]} \qquad \text{Eq. (1)}$$

(the concentration units are mol per litre of microemulsion, M^m). The droplet size is directly proportional to R, for many w/o microemulsion systems. In the preparation just described, $R = 28$ and a typical value for the radius of the water droplets is 5 nm (pp. 000–000). The surface area:volume ratio is therefore of the order of 10^8 m^{-1}. There are about 10^{17} droplets per cm^3. A droplet of radius 5 nm is roughly 11 times larger (in terms of volume) than the roughly spherical α-chymotrypsin (α-CT) molecule (M_r = 25 kDa), or about the same size as a horse-liver alcohol dehydrogenase (HLADH) molecule (M_r = 80 kDa). Since the dimensions of these droplets are very small compared with those of visible light (400–700 nm), light-scattering by the droplets is minimal and the microemulsion is optically transparent. Calculations such as these serve to show the incredible degree of dispersion achieved and the important consequence that interfacial surface area cannot be a limiting factor in determining phase-transfer rates between the water and bulk-oil phases.

Role of the surfactant in determining structure

An idealized microemulsion-forming surfactant is one which is insoluble in, and also cannot micellize in, either of the bulk phases (Overbeek, 1978). On addition to an oil-water two-phase system, therefore, the surfactant spontaneously migrates to the interface, which expands until all of the surfactant is accommodated. This process is driven by the favourable free-energy of solvation of the surfactant at the interface and by the exclusion of direct oil–water contact, and is accompanied by a dramatic lowering of interfacial surface tension (to very small positive values, of the order of 10^{-2} mN m^{-1}, compared with 50 mN m^{-1} for the tension at an oil–water interface (Aveyard *et al.*, 1986; Aveyard, Binks and Mead, 1986; Langevin, Guest and Meunier, 1986)).

As already indicated, even relatively small amounts of surfactant require an enormous amount of surface area and the exact way in which this area is accommodated within the volume of the sample (i.e. the microstructure of the microemulsion phase) is determined by the preference of the surfactant film to spontaneously adopt a certain 'natural' radius of curvature, r_N, which has both direction and magnitude, corresponding to a minimum surface free-energy (Safran and Turkevitch, 1983; Mukherjee, Miller and Fort, 1983; Lam, Falk and Schechter, 1987). Hence surfactant films which spontaneously curve such that the headgroups face 'inwards' (negative curvature) will form oil-continuous microemulsions, films which spontaneously curve with the headgroups facing 'outwards' (positive curvature) will form water-continuous microemulsions, and those which spontaneously form films with low, or zero (mean) radii of curvature, favour the open, extended surfactant films found in bicontinuous microemulsions (Auvray *et al.*, 1984; Lindman *et al.*, 1989).

The reduction in surface free-energy due to surfactant saturation, and the preference to adopt a certain curvature, are sufficient for a qualitative understanding of the stability of microemulsion phases. For example, w/o microemulsions form because the equilibrium interfacial free-energy is extremely low and because the surfactant film has negative curvature. The combined surface free-energy of the droplets, though positive (which in itself predicates phase separation), is small enough that it is counterbalanced by the entropy of dispersion of the droplets within the continuous phase (Ruckenstein and Chi, 1975). Hence there is a delicate balance between droplet surface energy and dispersion entropy, and a microemulsion will form only if the surfactant used is capable of lowering the surface energy to values which can be compensated for by the dispersion entropy. (Certain surfactants which, acting alone, cannot stabilize microemulsions may do so in the presence of a second amphipathic compound – a co-surfactant; p. 268. The action of the co-surfactant complements that of the surfactant by further lowering the surface tension, to below the threshold value for microemulsion formation.) Similarly, oil and water alone do not form microemulsions because the interfacial tension is so high that it cannot be so compensated, even for very high concentrations of droplets (high dispersion entropies). The same kind of argument applies to water-continuous microemulsions, except that the surfactant spontaneously curves positive, and for bicontinuous systems, except that zero mean curvature is preferred and the entropy factor arises from the extreme torsional flexibility of the interface.

Curvature is a powerful tool for understanding and predicting microstructure. For example, the existence of Winsor II systems is a consequence of the fact that the microemulsion phase will take up only a finite amount of water, until the radius of curvature, $r \approx r_N$ ($R = R_{max}$). Then, further water can only be incorporated at the expense of a further decrease in the radius of curvature ($r<r_N$), and an unfavourable increase in the curvature energy. The excess water therefore remains as a separate phase in (dynamic) equilibrium with the internalized phase. A similar argument applies to Winsor I systems. Winsor III systems are formed as the surfactant phase takes up (approximately) equal amounts of both oil and water to establish the preferred zero-mean curvature, leaving two excess phases.

The formation of single-phase w/o microemulsions is consistent with the amount of the water in the system being such that $R<R_{max}$. It follows that $r>r_N$, given that all the surfactant is located interfacially, and the difference, (r_N-r) must increase as R decreases. This observation is consistent with the inverse proportionality between size and concentration in droplets-type microemulsions (p. 272, equation 7b) observed when R is varied at constant surfacant concentration (constant total interfacial area): As R decreases the droplet concentration increases, so providing increased entropic compensation for the increased curvature energy of the smaller droplets. The balance between radius of curvature and droplet concentration is also consistent with the relatively low polydispersity index of droplets-type microemulsions (Kotlarchyk *et al.*, 1982; Robinson *et al.*, 1984).

The natural radius of curvature may be considered to be set, in magnitude and direction, by the balance of opposing forces within the surfactant film (Bowcott and Schulman, 1955). A very simple model considers the surfactant film as a bilayer composed of a layer of headgroups fused to layer of tails. On each side, the various kinds of intermolecular interactions (dispersion, dipole–dipole, electrostatic) determine an optimal separation between adjacent groups within a plane, and curvature arises when this optimal distance is different for the layer of heads, compared with that for the layer of tails. The curving interface permits interactions on both sides to be optimized under the bilayer constraint. Factors such as the type of oil, the electrolyte concentration and the temperature strongly influence the value of r_N for any given surfactant, and herein lies the influence of these parameters on Winsor phase equilibria as already mentioned.

Progress towards a quantitative description of structure and phase-behaviour in surfactant systems on purely thermodynamic grounds is impeded by a lack of quantitative understanding of the nature of the forces within the surfactant layer and how they interplay to determine curvature. A simpler, and very successful approach to this problem involves considering the geometrical properties – the shape – of the surfactant molecule (Oakenfull, 1980; Mitchell and Ninham, 1981). The curvature preference for a particular surfactant molecule can be determined by calculating the surfactant packing parameter, *Sp* (Mitchell and Ninham, 1981):

$$Sp = \frac{v}{a_0 \times l_c} \qquad \text{Eq. (2)}$$

where v is the volume occupied by the tail, l_c is its length (measured radially from the interface) and a_0 is the interfacial surface area occupied per molecule. *Sp* is really a measure of the ratio of hydrophobic to hydrophilic surface area, and in this sense is similar to the older hydrophile–lyophile balance (HLB) system (Becher, 1984). Microemulsion-forming surfactants are characterized by *Sp* values around unity, corresponding to the HLB concept of a balanced surfactant (i.e. one which is insoluble in bulk phases and hence preferentially migrates to the interface, the essential prerequisite, *vide supra*, for a microemulsion-forming surfactant). Functionally $Sp \approx 1$ (HLB = 8–10) surfactants are most likely to stabilize bicontinuous microemulsions, since their effective hydrophobic and hydrophilic surface areas are equal. Surfactants which are functionally $Sp > 1$ (HLB 4–6) form negatively curving structures (in this way the excess volume occupied by the tails is accommodated) and hence stabilize w/o microemulsions, and those which are functionally $Sp < 1$ (HLB 12–15) form positively curving structures in order to accommodate the excess volume of the headgroups. The condition $Sp \gg 1$ (very low HLB) and $Sp \ll 1$ (very high HLB) describes surfactants which are very oil- or water-soluble, and which therefore do not saturate interfaces (Becher, 1984). The term 'functional' is important because parameters such as temperature, electrolyte concentration and oil type determine the actual value of *Sp*, and this may be very different from a first-approximation value calculated, for example, using algorithms based on atomic volumes. Usually this difference arises as a consequence oil- or co-surfactant penetration into (solvation of) the tailgroups (increasing v and/or l_c); in the case of ionic surfactants a_0 is a function of the electrolyte concentration (a_0 decreases as the degree of electrostatic shielding increases). The corollary of this argument is that microemulsion structure can be modulated by the use of additives (co-surfactants, electrolytes) which alter *Sp* (pp. 267–270).

The 'thermodynamic' (natural curvature) and 'geometric' (packing) approaches to predicting phase behaviour and structure in surfactant systems should be considered to be complementary. The thermodynamic approach seeks to explain microstructure by elucidating the nature and balance of the forces in the surfactant layer, whereas the packing approach acknowledges the fact that molecular shape largely determines the overall shape of structures formed by non-covalent assembly (Evans, Mitchell and Ninham, 1986).

STRUCTURE AND PHASE-BEHAVIOUR OF W/O MICROEMULSIONS

Microemulsion-forming surfactants

Surfactants which approximate the *Sp* ~ 1 condition for a microemulsion-former include sodium *bis*-2-ethylhexyl sulphosuccinate (Aerosol-OT, AOT), didodecyldimethylammonium bromide (DDAB) and trioctyl ammonium chloride (TOMAC). Many other common surfactants are too hydrophilic to stabilize w/o microemulsions (*Sp* too low), although some of these will do so in the presence of a co-surfactant (p. 000). Examples include members of the Brij and Lubrol W series of polyoxyethylene ethers of aliphatic alcohols, the Triton-X and Nonidet-P series of polyoxyethylene ethers of 4-*t*-octylphenol and the Tween series of polyoxyethylene sorbitol esters (all non-ionic surfactants), the alkyl sulphates (e.g. sodium dodecyl sulphate, SDS; anionic) the alkyltrimethylammonium halides (e.g. cetyl-trimethylammonium bromide, CTAB; cationic) and phosphatidylcholines (lecithins), all of which are familiar to biologists (Helenius and Simons, 1975).

AOT as a model surfactant

The most commonly used surfactant for enzymological studies is AOT. This surfactant solubilizes large amounts of water (at least 60 mol mol^{-1}, at ambient temperatures) in a range of oil solvents which include alphatic alcohols (octanol, decanol), aromatics (benzene, xylene) and alkanes (*n*-heptane, iso-octane, cyclohexane), without the requirement for a co-surfactant.

A space-filling model of AOT (*Figure 3*) gives a visual indication of why this molecule is a good w/o microemulsion-forming surfactant. The sulphosuccinate headgroup is small in comparison with the bulk of the molecule, the alkyl tails are of medium length and are branched, so that v (equation 2) is large for a modest value of l_c. The consequence is that this surfactant preferentially forms films with negative curvature, permitting the accommodation of the bulky hydrophobic groups at the interface.

The AOT molecule possesses a further interesting structural feature, the potential for considerable rotational flexibility about the C–C bonds of the ethane skeleton of the succinate moiety. It has been shown using ^{1}H NMR spectroscopy (Maitra and Eicke, 1979; Maitre, 1984) and ^{13}C NMR and infra-red spectroscopy (Martin and Magid, 1981) that the equilibrium population of rotamers varies with *R*. Specifically, the population of the more hydrophilic (lower *Sp*) *trans* isomer increases at the expense of the more hydrophobic *gauche* isomer with increasing *R*. This isomerism permits the *Sp* to increase by a factor of two, from about 1.5 to about 3, as *R* is increased from 0 to 50. The headgroup surface area (a_0) is essentially independent of *R* and the increase in *Sp* is achieved mainly by this alteration in the relative orientations of the two hydrocarbon tails (i.e. *Sp* is explicitly a function of v). Thus it appears that AOT is capable of altering its wedge-angle (within limits) to suit the radius of curvature of the droplet formed (Kubik *et al.*, 1982; Maitra, 1984).

The presence of several medium-length alkyl tails (high v/l ratio) and a small charged headgroup is a common motif among surfactants which stabilize

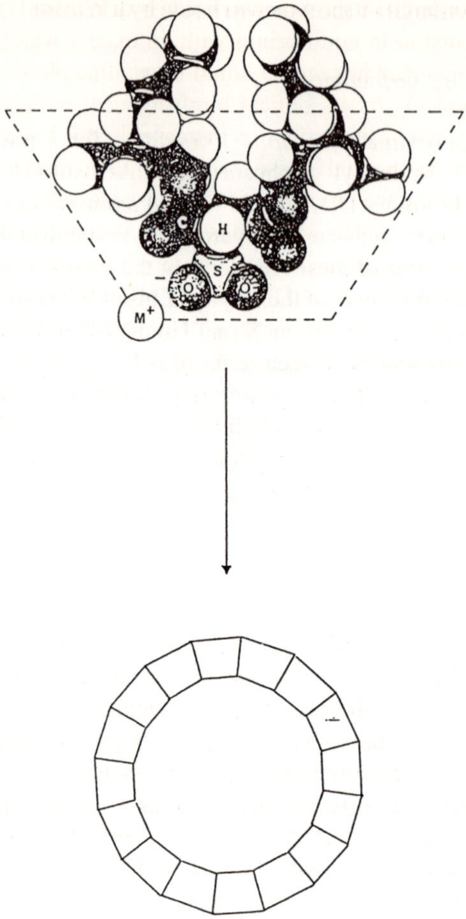

Figure 3. A space-filling model of an AOT molecule. The shape favours the formation of negatively curving structures.

microemulsions in the absence of a cosurfactant. For example, DDAB has two tails and TOMAC has three.

Phase diagram for AOT systems

Phase diagrams are used to relate structure–composition relationships in colloidal systems. Phase information for a three-component (oil, water, surfactant) system at a given temperature is presented in the form of a Gibbs phase triangle. Any point within the triangle represents a unique combination (volume, weight, or mol-fraction) of the three components. The phase-behaviour of several ternary AOT–water–oil systems has been characterized in detail (Ekwall, Mandell and Fontell, 1970; Kunieda and Shinoda, 1979, 1980; Shinoda and Kunieda, 1987).

The ternary phase diagram for the AOT–water–iso-octane system at 15°C is shown in *Figure 4* (Kunieda and Shinoda, 1979). A large area, O_m, of oil-continuous Winsor IV (single-phase) microemulsion extends from the oil-corner deep into the phase diagram. The O_m is bordered on the water-rich side by a Winsor II domain, O_m+W (oil-continuous microemulsion in equilibrium with an excess water phase) and, on the AOT-rich side, by a reversed hexagonal liquid-crystalline phase. This is distinguishable from the adjacent O_m by its optical birefringence, indicating a non-isotropic composition. As the water volume-fraction increases in this lower region of the phase triangle, the reversed hexagonal phase evolves into a lamellar liquid-crystalline phase. Extending from the water-corner of the phase triangle, in the direction of the oil-corner is a W_m phase of water-continuous microemulsion and of Winsor I W_m+O (water-continuous microemulsion in equilibrium with excess water). AOT is not a good o/w microemulsion-former, and this simply reflects the difficulty of fitting this wedge-shaped molecule on to the surface of a sphere in the 'normal' micellar configuration, as can be seen from *Figure 3*.

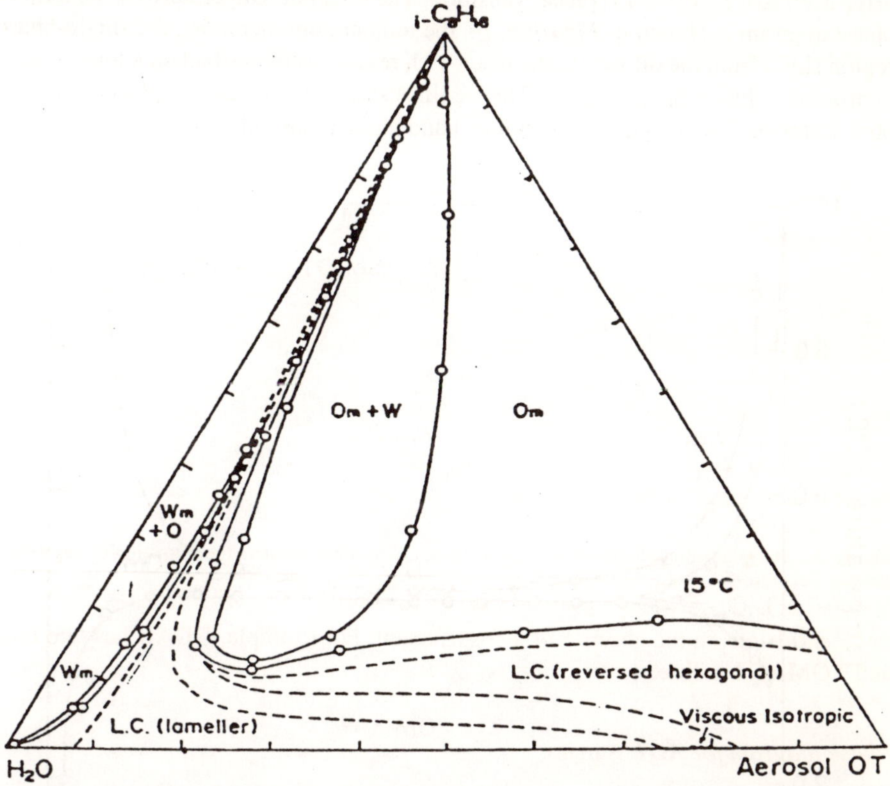

Figure 4. Ternary phase diagram for the water–AOT–iso-octane system at 15°C. Om, oil-continuous (w/o) microemulsion; Om+W, oil-continuous microemulsion plus conjugate aqueous phase; Wm, water-continuous microemulsion; Wm+O, water-continuous microemulsion plus excess oil-phase; l.c., liquid-crystalline phase. The phase diagram is pseudo-ternary since a salt (Na_2SO_4) is present at a constant molar AOT:Na_2SO_4 ratio of approximately 50:1. (Reproduced from Kunieda, H. and Shinoda, K. (1979) *Colloid and Interface Science* **70**, 577–583, with permission.)

Effects of temperature, salt on phase-behaviour

The Gibbs phase triangle gives information about the phase-behaviour at a single temperature. The temperature dependence information can be built-in by extending a fourth (temperature) axis, perpendicular to those of the Gibbs triangle, to make a phase prism. A temperature-phase (TP) plot for the AOT–iso-octane system is shown in *Figure 5* (Shinoda and Kunieda, 1987). This plot corresponds to a section through the phase prism parallel to the oil–water axis. The AOT concentration is constant at 3.5 wt% (80 mmol dm^{-3}) which falls within the range (50–200 mmol dm^{-3}) used in most published enzymological studies. In this very oil+water-rich part of the phase triangle the single-phase region occupies a relatively small area and, except at very low (<0.05) or very high (<0.9) water:oil ratios, is confined within a narrow channel located parallel to the abscissa, at about 42°C (the HLB, or phase-inversion temperature (pit)). This channel, extending as it does right across the phase plot, implies an evolution of w/o → o/w, moving from right to left. In the major part of the diagram, therefore, the effect of increasing the temperature up to, and then beyond, 42°C is to drive a Winsor I → IV → II phase transition. The effect of temperature on the ternary phase diagram is shown in *Figure 6*. As the temperature increases, the single-phase region shifts from the oil-rich to the water-rich region, as the surfactant adopts a more hydrophilic (lower *Sp*) character. The tie-lines show that the excess phase is almost pure water (below the HLB temperature) and almost pure oil (above).

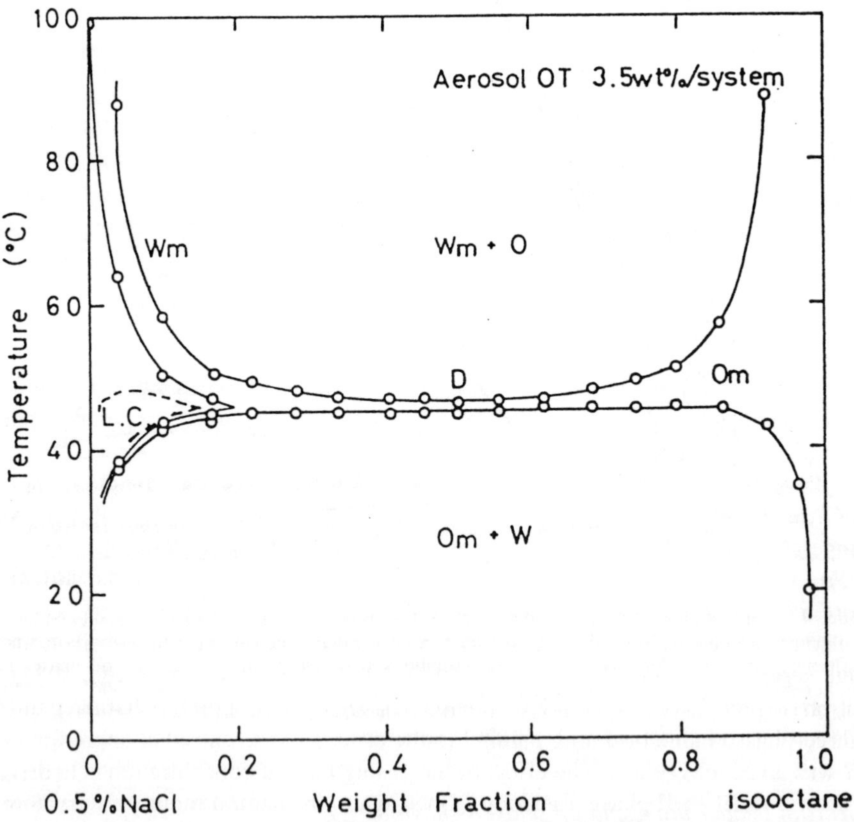

Figure 5. Temperature-phase plot for the AOT–water–isooctane system. Abbreviations are as for *Figure 4*. (Reproduced from Shinoda, K. and Kunieda, H. *Journal of Colloid and Interface Science* **118**, 586–589 (1987), with permission.)

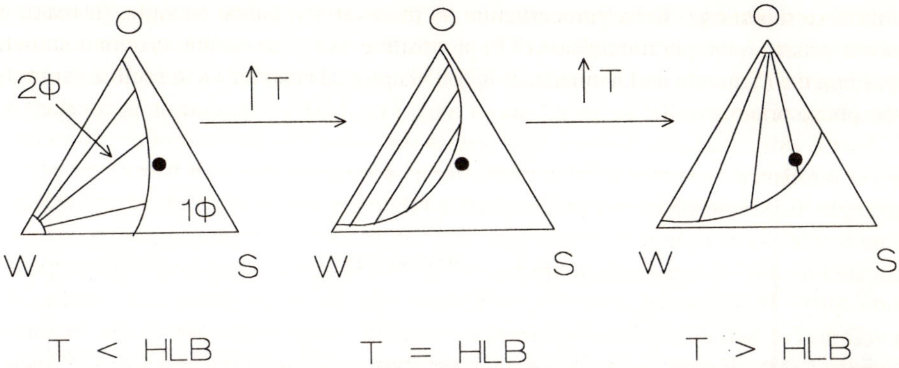

Figure 6. The effect of increasing the temperature on the ternary phase diagram for an oil–water–ionic surfactant system. The phase diagram is simplified to include only one- and two-phase regions. Tie-lines in the two-phase region indicate the compositions of the two phases. T < HLB temperature, the two-phase region consists of a w/o microemulsion phase in equilibrium with an almost pure aqueous phase. T> HLB, the two-phase region consists of an o/w microemulsion in equilibrium with an almost pure oil phase. The point (●) is in the single-phase region when T < HLB, and in the two-phase region when T > HLB.

The effect of temperature on the phase-behaviour is understandable in terms of its effect on the natural curvature of the surfactant film. Increasing the temperature is thought to increase the degree of dissociation of the counterions and surfactant headgroups, leading to decreased headgroup shielding and therefore to a decrease in *Sp* (increased a_0). As the temperature is increased from a point well below the HLB temperature, therefore (*Figure 5*), the natural radius of curvature decreases continuously, passing through zero mean-curvature at the HLB temperature in the single-phase region and becoming positive above it. In order to maintain the natural radius of curvature, r_N, at any temperature, the microemulsion phase takes up, or expels, some of the internalized phase. Thus as the temperature is increased to the HLB temperature, the oil-continuous microemulsion phase takes up water. The volume of water solubilized is maximal in the single-phase region where, in order to maintain zero mean-curvature, it is roughly equal to the volume of the oil-phase. Above the HLB temperature the microemulsion phase becomes water-continuous and expels oil.

Compared with AOT and other ionic surfactant systems, the temperature-phase diagram for non-ionic surfactant systems is inverted so that increasing the temperature drives a Winsor II → IV → I transition, as described by Lichterfield, Schmeling and Strey (1986) for $C_{12}E_5$ systems. This is probably because the effect of increasing temperature is to dehydrate the polar headgroups, so causing a decrease in a_0 (increase in *Sp*). In other words, the effect of increasing temperature is to make the surfactant more hydrophobic, instead of more hydrophilic.

The presence of electrolyte also has a profound effect on the phase-behaviour of ionic surfactant systems (Ghosh and Miller, 1987; Shinoda and Kunieda, 1987). A salt-phase plot (2 wt% AOT, 42°C) is shown in *Figure 7*. The plot is characterized by a three-phase channel running parallel to the abscissa and centred at approximately 0.7 wt% (120 mM) NaCl. The effect of increasing the salt concentration is to drive a Winsor I → III → II phase transition. Once again, the natural radius of curvature is achieved by uptake/expulsion of the internalized phase. The salt effect can also be explained in terms of its effect on the natural curvature of the surfactant film. Increasing the salt concentration causes an increase in *Sp* (decrease in a_0 due to more effective shielding). Therefore curvature increases from a negative value (favouring water-continuous microemulsions) to a positive one (favouring oil-continuous), passing through zero mean-curvature in the Winsor III channel, as the salt concentration is increased.

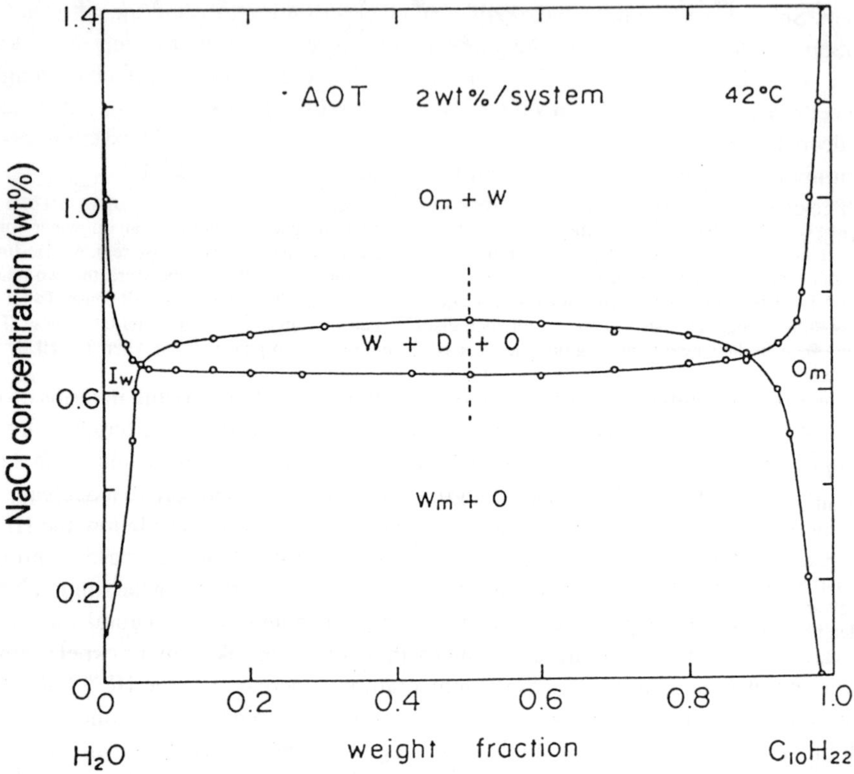

Figure 7. Salt-phase plot for the water–AOT–decane system at 42°C. Nomenclature as for *Figure 4*. (From Shinoda, K. and Kunieda, H. *Journal of Colloid and Interface Science* **118**, 586–589 (1987), with permission.)

Quaternary systems: the effect of co-surfactants

Another way of modulating surfactant curvature is by the addition of a co-surfactant. Typical co-surfactants for w/o microemulsions include medium-chain alcohols and polarizable small molecules such as chloroform and benzene. Like surfactants, co-surfactants are amphipathic but are much more oil soluble and partition extensively between the interphase and the bulk oil-phase. The interfacial surfactant:co-surfactant ratio can be measured, at least for compositions close to the O_m phase-boundary, using a titration method (Bowcott and Schulman, 1955). The interphase surfactant:co-surfactant ratio is usually relatively low. Values of around 2, are reported for both the oleate–benzene–water–hexanol system (Bowcott and Schulman, 1955) and the CTAB–octane–water–hexanol system (Hilhorst *et al.*, 1984).

Co-surfactants act at two levels. In the first place they complement the action of a surfactant in lowering the surface tension sufficiently for microemulsion formation to occur. Secondly, they dictate structure in the surfactant phase by modulating interfacial curvature. This second property is easily understood in terms of the *Sp* of the surfactant/co-surfactant combination. In the case of alcohols, for example, the hydroxyl group contributes to a_0 and there is a contribution to v and l_c from the alkyl chain. Short-chain alcohols make only a small contribution to v and l_c from the alkyl chain. Short-chain alcohols make only a small contribution to v and the net result is that *Sp* decreases. As the chain-length increases, the contribution to v increases (that to a_0 remains constant) and *Sp* increases. When the co-surfactant chain-length exceeds that of the surfactant itself there is a contribution l_c, as well as to a_0 and v, and *Sp* decreases again. Hence *Sp* passes through a maximum (most likely to stabilize oil-continuous microemulsions) as the alkanol chain-length is increased.

The effects of temperature and additives on surfactant curvature are summarized in *Table 1*.

Table 1. Factors influencing the curvature and *Sp* (equation 2) of surfactant monolayers (after Fletcher and Parrot, 1989)

Variable	Effect on curvature	Reason
Electrolyte concentration	More negative (higher *Sp*) as electrolyte concentration increases	More effective shielding reduces a_0
Oil (alkane) type	More positive (lower *Sp*) as alkane chain-length increases	Alkane penetration (solvation) of surfactant tails decreases with increasing chain-length
Co-surfactant (alkanol) chain-length	More negative (higher *Sp*) as chain-length increases until $l_{ALKANOL} = l_{SURFACTANT}$; then more positive	Co-surfactant headgroup contribution to a_0 constant; tail contribution to v increases with $l_{ALKANOL}$; *Sp* increases. l_c also increases when $l_{ALKANOL} > l_{SURFACTANT}$, then *Sp* decreases again
Temperature (ionic surfactants)	More positive with increasing temperature	Surfactant/counterion dissociation increases
Temperature (non-ionic surfactants)	More negative with increase temperature	Headgroup size reduced due to dehydration

Consideration of phase-behaviour is extremely important from an applications point-of-view. For example, in the design of a microemulsion reactor system with provision for product recovery by phase separation, a composition within the single-phase channel (*Figure 5*) is ideal, since post-reaction phase separation can be initiated by changing the temperature by just a few degrees. In this way the energy input required to effect the phase separation is minimized (obviously, much larger temperature changes are necessary to induce phase separation at compositions far from phase boundaries). In the case of AOT, and other systems prepared with ionic surfactants, an increase in temperature (triggering IV → II) is appropriate for isolating a nearly pure (water, surfactant-free) oil-phase containing apolar reaction products (*Figure 6*). In addition, the single-phase channel, in extending right across the diagram, offers the opportunity for close optimization of composition to suit the reaction under consideration (through, for example, optimization of biocatalyst activity and stability; substrate and cofactor solubility).

Further opportunities for fine-tuning phase-behaviour include changing the type of oil (in the case of AOT microemulsions, for example, the HLB temperature shifts downwards with increasing alkane chain-length) (Fletcher, Howe and Robinson, 1987). Changing an ionic for a non-ionic surfactant provides the opportunity to induce phase separation (IV $\rightarrow$ II) by decreasing, rather than increasing, the temperature. Non-ionic surfactants may therefore be used if product extraction with recovery and re-use of thermolabile enzymes is desirable. In the case of the Brij/Lubrol W type surfactants (*CiEj*), there is the opportunity for optimization of HLB temperature by systematic variation of both alkyl tail length, *i*, and headgroup length, *j*.

The effect of salt on phase-behaviour is important in the design of Winsor II systems, e.g. for protein purification by liquid–liquid extraction; one method of modulating phase-transfer characteristics is to alter the ionic strength, (p. 310). Here an understanding of phase behaviour is necessary to avoid phase inversion with loss of the w/o microemulsion phase.

Structure in the O_m domain

The presence of a transparent, isotropic, O_m phase is consistent at the macroscopic level with structural types ranging from 'pure' droplets (as originally conceptualized by Hoar and Schulman, 1943) and evolving through various types of bicontinuous structure (Lindman *et al.*, 1989), as the (surfactant+water) volume fraction increases (i.e. as one moves away from the oil-corner of the phase diagram). Detailed physical studies are required to establish the relationship between structure and composition. Scattering techniques (neutrons, light or X-rays) are usually employed to give a first-approximation analysis. However, since these techniques usually require the fitting of data to a structural model, the results are not necessarily unambiguous (Langevin, 1989; Lindman *et al.*, 1989). For this reason, the best-understood microemulsions are those for which more than one of the above techniques has been applied, and for which basic structural analyses have been complemented with additional information, the most useful being macroscopic self-diffusion coefficients measured using Fourier transform ^{13}C NMR relaxation spectroscopy (Lindman, Stilbs and Moseley, 1981; Lindman *et al.*, 1983). This technique permits the simultaneous measurement of diffusion coefficients of all the components of the system over distances (several hundred micrometres) which are large compared with the dimensions of microemulsion structural domains, hence providing information about the dynamic behaviour of the microemulsion components which is missing from the static picture obtained from scattering measurements. The principle is that molecules in the continuous phase(s) will have macroscopic self-diffusion coefficients approximately equal to that in a structureless fluid (of the order of 10^{-9} m^2 s^{-1}), whereas those contained within a domain will diffuse perhaps 10^2 times more slowly. Furthermore, in the bicontinuous case, it is possible to distinguish between alternative structures since each of these has a characteristic set of diffusion coefficients (Lindman *et al.*, 1989).

In the particular case of the AOT system, low-angle X-ray scattering (Ekwall, Mandell and Fontell, 1970), NMR self-diffusion (Lindman, Stilbs and Moseley, 1981) and SANS (Kotlarchyk *et al.*, 1984; Huang and Kotlarchyk, 1986) support a droplets-type structure deep into the ternary phase diagram, at ambient temperatures and moderate-to-high ionic strengths. This behaviour is quite remarkable when it is considered that it seems to be the case even when the (surfactant+water) volume fraction exceeds 0.74, the packing maximum for a body-centred cubic arrangement of spheres (Ekwall, Mandell and Fontell, 1970). It seems that under these conditions the droplets become distorted (non-spherical) in shape, rather than simply fusing together to form a bicontinuous structure. For the majority of systems, however, NMR self-diffusion measurements clearly indicate an evolution from droplets to bicontinuous structure as the (water+surfactant) volume-fraction increases (Lindman, Stilbs and Moseley, 1981; Lindman *et al.*, 1983).

Several other ionic and non-ionic systems consist of droplets at low (surfactant+water) volume-fractions, including the CTAB–water–(chloroform+iso-octane) system (Lang *et al.*, 1990), water–hexane microemulsions stabilized by the alkylpolyoxyethylenes $C_{10}E_4$ and $C_{12}E_5$ (Fletcher and Horsup, 1992) and the poly(oxyethylene) sorbitan trioleate (Tween 85)–water–hexane–isopropanol system (Komives, Lilley and Russell, 1994). Detailed studies on ternary DDAB microemulsions indicate droplets at low (water+DDAB) volume fractions; at higher volume fractions (precise values depend on the oil), structure is best described in terms of a bicontinuous structure consisting of interconnected water channels of the type shown in *Figure 1$_c$*. The dimensions (radius of curvature) of the channels, is are strongly dependent on the oil type (Chen *et al.*, 1986). The water–Triton-X-100–(hexane+benzene 1:1v/v)) system seems to behave similarly (Zhu, Wu and Schelly, 1992).

Droplets-type microemulsions: the aggregation model

AOT microemulsions obey a very well-known structural model which may be referred to as the 'aggregation' model (Nicholson and Clark, 1984), or as the 'equipartition' model (Zulauf and Eicke, 1979; Eicke and Kubik, 1980). It has the following features:

1. the discontinuous subphase is a monodispersion of spherical water droplets, each bounded by a monolayer of surfactant;
2. all surfactant is located at the surface of the water droplet;
3. each surfactant molecule occupies the same surface area at the droplet surface;
4. each water molecule occupies the same average volume.

Microemulsions obeying this model have structural features which can be deduced from purely geometrical considerations (Hoar and Schulman, 1943; Luisi *et al.*, 1988). The radius, r_{wp}, of the water-pool is given by:

$$r_{wp} = \frac{3v_{H_2O}}{a_0} . R \qquad \text{Eq. (3)}$$

where v_{H_2O} is the volume of a single water molecule (usually taken to be 0.03 nm^3). If v_{H_2O} and a_0 are independent of R, equation (3) predicts a linear dependence of *rwp* on the compositional parameter R. The hydrodynamic radius, *rh* of the droplet is

$$r_h = r_{wp} + l_c. \qquad \text{Eq. (4)}$$

The number of water molecules per droplet is

$$N_{H_2O} = \frac{4\pi r^3_{wp}}{3v_{H_2O}} \qquad \text{Eq. (5)}$$

The number of surfactant molecules per droplet is

$$N_{AOT} = \frac{N_{H_2O}}{R} \qquad \text{Eq. (6)}$$

and the molar droplet concentration is

$$[\bigcirc]_T = \frac{[AOT]}{N_{AOT}} = \frac{[H_2O]}{N_{H_2O}} \qquad \text{Eq. (7a)}$$

It is also worth noting that (combining equations 3 and 5 and substituting for N_{H_2O} in equation 7a),

$$[\bigcirc]_T = \frac{K[H_2O]}{R^3} \qquad \text{Eq. (7b)}$$

where constant $K = a_0^3/36\pi v_{H_2O}^2$. It follows from equations (7a) and (7b) that the droplet concentration and R may be varied independently of each other.

It is useful to relate these various manipulations to the phase diagram (*Figure 4*). If one traverses the O_m along a linear path which begins at a point on the water–surfactant axis and passes through the oil-corner, R, and therefore the droplet size, remains constant and the droplet concentration changes (decreasing in the direction of the oil-corner). The value of R is determined by the point of origin of the line on the water–surfactant axis. On the other hand, if one follows a linear path originating at the oil–surfactant axis and passing through the water-corner, R increases in direct proportion to the water volume-fraction at constant surfactant concentration. From mass-balance considerations, the droplet concentration also decreases with increasing R along this path. These are the two most common systematic variations of microemulsion composition adopted for experimental purposes. Indeed, the validity or otherwise of the aggregation model may be checked, for any given system, by measuring the droplet size along these paths. Along the oil-path, the droplets do not change in size; along the water-path, the water-pool size is directly proportional to R (equation 3).

Droplet sizes, measured using small-angle neutron scattering (SANS) (Robinson *et al*., 1984; Huang and Kotlarchyk, 1986); Huang, Sung and Wu, 1989), photon correlation spectroscopy (PCS) (Zulauf and Eicke, 1979; Nicholson and Clarke, 1984; Aveyard, Binks and Mead, 1986; Aveyard *et al*., 1986); small-angle X-ray scattering (SAXS) (North *et al*., 1986; Pileni, 1989), and nuclear magnetic resonance (NMR) (Maitra, 1984) are consistent with the aggregation model. Thus, at ambient temperatures, for $R = 5$–50, the water–AOT–*n*-heptane system obeys very well the equation $r_{wp} = 0.15\,R$ nm and the water–AOT–decane system obeys the equation $r_{wp} = 0.12\,R$ nm. From equation (3), values for a_0 of approximately 0.58 nm² and 0.75 nm² are obtained for the heptane and decane systems, respectively, assuming $v_{H_2O} = 0.03$ nm³.

Time-resolved quenching of flurorescence, or of hydrated electrons (generated by pulsed radiolysis), provides a direct measure of the droplet concentration, rather than of the droplet size (which can then be estimated from mass-balance considerations). Droplet sizes calculated for AOT microemulsions using these techniques are in very good agreement with those measured using scattering techniques (Bridge and Fletcher, 1983; Pileni *et al.*, 1984).

Use of equations (3)–(7) permits the calculation of N_{AOT}, N_{H_2O} and the droplet concentration given size data. Calculations of this kind are most illuminating. Although the droplet radius is directly proportional to R, the number of water molecules per droplet is directly proportional to $R^{1/3}$, hence the droplets 'grow' very quickly as R is increased (Nicholson and Clarke, 1984). For example, $R = 10$ droplets have a radius of 1.5 nm and therefore contain about 470 molecules (equation 5); $R = 20$ droplets have twice the radius and contain around 3800 molecules. $R = 50$ droplets each contain about 60 000 water molecules. At the same time, mass conservation requires that, assuming the surfactant concentration is held constant as R is varied, the droplet concentration decreases. Thus if [AOT] = 0.1 M (a typical working concentration), the droplet concentrations are 2100, 520 and 85 μM at R = 10, 20 and 50. Aggregation of water has important consequences for the dependence of the water activity, and hence for the rates of chemical and enzyme-catalysed reactions, on R.

The effects of temperature on phase-behaviour are reflected at the microscopic level on the structure. In AOT microemulsions, for example, phase separation at high temperatures arises because droplets grow very large and ultimately coalesce, probably because the surfactant becomes sufficiently hydrophilic (very high HLB; $Sp >> 1$) that it becomes soluble in the aqueous phase and desorbs from the interface (Zulauf and Eickc, 1979). Similarly the phase separation at low temperatures occurs because the surfactant becomes oil-soluble.

Phase-behaviour in phospholipid microemulsions

A number of amphipathic compounds produced by living organisms have been shown to be capable of stabilizing w/o microemulsions, including lecithin (Misiorowski and Wells, 1974; Poon and Wells, 1974) and triacylglucolipids (Lindblom *et al.*, 1993, reviewed by Walde *et al.*, 1990). This class of microemulsions is becoming increasingly important due to the status of phospholipids as biological surfactants since they are more likely to be approved in connection with food- and drug-related applications than are synthetic surfactants. In addition, lecithin itself is an important foodstuff and there is much interest in biotransformations involving this material (for example, in the upgrading of low-quality lecithins, (e.g. palm oil), to food-grade cocoa butter by lipase-catalysed trans- or interesterification reactions (Macrae, 1985)). Here microemulsions have a role to play in facilitating such biotransformations since the surface-active substrate molecules forming the interfacial monolayer are equally accessible to the biocatalyst in the aqueous subphase.

A number of studies of the phase-behaviour of lecithin–water–organic solvent systems have been published. Most studies utilize purified (triglyceride-free) lecithin from egg or soybean. The purest commercially available soybean lecithin is about 97% phosphatidylcholines, the remainder being other phosphatidic acid derivatives. Naturally occurring phosphatidylcholines are somewhat heterogeneous with respect to the fatty acyl content (chain-length and degree of unsaturation). Synthetic phospholipids of defined fatty acyl composition are available have been used in a few studies.

Lecithin yields an O_m phase in water–oil ternary systems with a wide range of oils of varying polarities, including diethylether, benzene and cyclohexane, and aliphatics such as hexane, octane and decane (Shervani *et al.*, 1991). With relatively polar solvents (benzene, diethylether) the O_m is a non-viscous isotropic oil-continuous droplets type microemulsion (Kumar and Raghunathan, 1986). In alkanes, the phase-behaviour is more complex, and the O_m is a liquid microemulsion only at low R. As R is increased above a critical (alkane-specific) value, the system becomes first viscous, due to the formation of non-spherical structures, and then an optically transparent, isotropic gel. The liquid–gel transition occurs over a very narrow range ($<2R$). For cyclohexane the transition occurs around $R = 4$; for octane around $R = 2$. The transition is depressed (shifted to higher R) on addition of co-surfactants; and is reversed by increasing the temperature.

Synthetic phosphatidyl cholines prepared with short- or medium-chain fatty acyl groups ($C < 8$) do not form gels and much higher degrees of water-solubilization are obtained with these 'balanced' surfactants. Lecithin carrying two pentanoate (C_5) chains can solubilize up to 80 mol water per mol lipid (in the presence of a medium-chain alcohol co-surfactant). Longer-chain (C_6, C_7) lecithins are much less effective, maximally solubilizing about 20 mol water mol^{-1} (Peng and Luisi, 1990).

The gels, like liquid microemulsions, are transparent and isotropic (non-birefringent) and therefore are not liquid-crystalline in nature, a result confirmed by SANS and SAXS (Luisi *et al.*, 1990). However, they are characterized by extremely high shear viscosities (up to 10^6 cp), even at relatively low (lecithin+water) volume fractions (Scartazzini and Luisi, 1988). A combined light- and neutron-scattering study is consistent with the existence of extended cylindrical and semiflexible rods (Capitani *et al.*, 1991; Schurtenberger *et al.*, 1992).

Waterless microemulsions, surfactantless microemulsions

In addition to water, several other polar, oil-immiscible solvents are known to form microemulsions. Microemulsion phases have been mapped for the CTAB–formamide–(cyclohexane+butan-1-ol) system (Rico and Lattes, 1984) and for the AOT–glycerol–*n*-heptane and CTAB–glycerol–(*n*-heptane+chloroform) systems (Fletcher, Galal and Robinson, 1984, 1985). The two glycerol systems apparently obey the aggregation model, with $r_h = 1.72 + 0.88\,R$ and $1.8 + 1.2\,R$ nm, respectively. Microemulsions with polar subphases consisting of glycerol–water mixtures in various ratios may also be prepared.

Surfactantless microemulsions are prepared with relatively large amounts of a co-surfactant. Typically, the amphipath is a medium chain-length aliphatic alcohol such as propan-2-ol. The ternary phase diagram for the hexane/propan-2-ol/water system is shown in *Figure 8* (Khmelnitsky *et al.*, 1986, 1989b). The phase-diagram is divided into four major domains, a three-phase region (A), a single-phase oil+surfactant-continuous microemulsion region (B), a single-phase region which consists of loosely associated H-bonded aggregates of hydrated alcohol (C) and true ternary solutions (D). The shaded region indicates an apparently droplets-like region within the microemulsion domain. These systems are not much studied. The absence of surfactant may be important in certain applications where even traces of surfactants are undesirable. Possibilities for post-reaction product recovery lie in the possibility for inducing a phase transition (B) → (A), for example, by addition of more amphipath or water, as required.

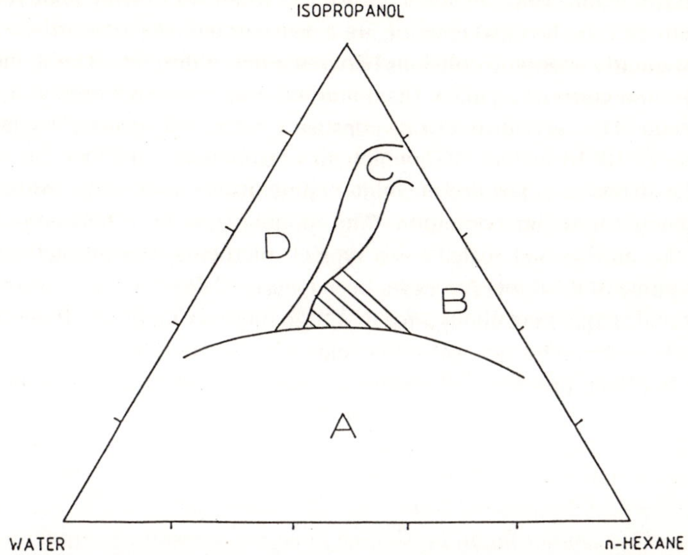

Figure 8. Ternary phase diagram for the hexane/isopropanol/water system. The apparently droplets-type microemulsion region is shaded. (Redrawn from Khmelnitsky Yu. L., van Hoek, A. Veeger, C. and Visser, A.J.W.G. (1989b) *Journal of Physical Chemistry* **93**, 872–878 © (1989) American Chemical Society, with permission.)

DYNAMIC PROPERTIES

Droplet dynamics

Individual droplets are very mobile structures. Quasi-elastic neutron scattering studies indicate that water molecules diffuse rapidly within the confines of the droplet, and the AOT molecules diffuse laterally over the droplet surface, on the sub-nanosecond time scale (Fletcher, Robinson and Tabony, 1986). ^{13}C NMR studies suggest that the AOT molecules additionally undergo restricted rotational motions about the long-axis. The droplets are in continuous random (Brownian) motion and are frequently involved in diffusion-controlled inelastic collisions with each other. The interdroplet collisions can be described essentially as hard-sphere interactions, with a small attractive potential, believed to be due to van der Waals interactions between the surfactant tails of the colliding droplets (Brouwer, Nieuwenhuis and Kops-Werkhoven, 1980; Cazabat and Langevin, 1981; Huang *et al.*, 1984).

Exchange

Perhaps the most important dynamic element of microemulsion behaviour is the exchange process which permits water-soluble solutes to transfer between droplets. Exchange can be demonstrated by choosing a chemical reaction of the type A + B →

P, where A and B are water soluble. If two microemulsions, one containing A and the other B, are mixed, formation of product, P, will be observed (Eicke, Shepherd and Steinemann, 1976). Fast-reaction techniques have been used to study the exchange mechanism for AOT microemulsions in some detail (Atik and Thomas, 1981; Fletcher and Robinson, 1981; Fletcher, Howe and Robinson, 1987). Data for several different kinds of fast chemical reaction are consistent with the formation of a fused dimer in sufficiently energetic colisions between pairs of droplets. The dimer breaks down rapidly to regenerate a pair of 'daughter' droplets compositionally identical to the original pair. The second-order rate constant for dimer formation, k_{fus} (k-fusion) is of the order of 10^7 M^{-1} s^{-1} at 25°C, which may be compared with a value of *c.* 10^{10} M^{-1} s^{-1} for the diffusion-controlled encounter rate-constant, indicating approximately 0.1% efficiency for dimer formation. This corresponds to a between-collisions lifetime on the millisecond scale, given typical microemulsion droplet concentrations. The lifetime of the dimer has never been measured directly, but probably lies in the microsecond range, long enough for thorough intermixing of the dimer contents, but too short to undergo further collision events which might lead to the formation of trimers, etc. In effect, droplets and dimers and their contents are in equilibrium,

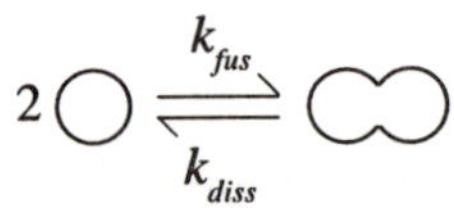

where $k_{eq} = k_{diss}/k_{fus} << 1$. Measurements of activation parameters imply a 'dumbell'-shaped transition structure, with a relatively small contact area, rather than a spherical unit (Eicke, Shepherd and Steinemenn, 1976; Fletcher, Howe and Robinson, 1987).

Some measurements of k_{fus} have been made for other microemulsions. Using fluorescence quenching of Zn-porphyrin-cytochrome *c* by ferricyanide, Vos, Lavalette and Visser (1987) reported values of 1–5 × 10^7 M^{-1} s^{-1}, depending on *R*, for CTAB microemulsions with hexanol as a co-surfactant. Using time-resolved fluorescence quenching of Tris-(2,2'bipyridyl)ruthenium(II) by methylviologen, Fletcher and Horsup (1992) reported values of the order of 10^8–10^9 M^{-1} s^{-1} for several different alkylpolyoxyether (i.e. Brij/Lubrol W) surfactants. Note that in the last case the exchange rates are close to the diffusion-limited maximum.

Assuming that exchange of solutes between droplets occurs by the mechanism just described, equations can be derived for the dependence of the rates of chemical and enzyme-catalysed reactions involving water-soluble reactants (i.e. confined within the aqueous subphase), on the droplet size and concentration and on the exchange rate. The key to the solution of the rate equation is to relate the observed rate to the rate within the droplet. Intuitively, the observed rate must be a function of the rate of the reaction within the droplet and of the number of droplets (per unit volume of microemulsion) in which the reaction is occurring (per unit time). This can be done using the steady-state approximation, i.e. by assuming that the number of droplets within which the reaction is occurring is constant during the period over which the rate is measured (Oldfield, 1991). *Figure 9* shows the scheme for the reaction, A + B → P. The reaction between A and B, initially confined within separate droplets (which is inevitably the case under steady-state conditions) requires that A and B first exchange into the same droplet. The reaction may occur either in the dimer (type (i) kinetics (Oldfield *et al.*, 1992), or later, in a daughter droplet (type (ii) kinetics), depending on the rate of dimer dissociation relative to that of the chemical reaction. The solution yields an expression for the apparent second-order rate constant, k_{ex}, for the chemical reaction, expressed in 'overall' concentration units ($M^{-1m}\,s^{-1}$, that is, litre microemulsion per mol per second), in terms of k_{+1}^{obs}, and k_{+2}^{obs}, (units, s^{-1}) the first-order rate constants for the conversion of droplet (A,B)(dimer) to (P)(dimer) and of (A,B) to (P), respectively. The equation is completed by introducing an expression for the rate within a single droplet. In the simplest case, when there is only one each of A and B in the droplet, k_{+1}^{obs} is related to k_{+1}^{d}, the second-order rate constants for the reaction within the droplet (units, $M^{-1d}\,s^{-1}$; that is, litre water (within a single droplet) per mol per second), by the expression,

$$k_{+1}^{obs} = k_{+1}^{d}\,\frac{1}{N_A V_d} \qquad \text{Eq. (8)}$$

where V_d is the volume of the droplet and N_A is Avogadro's number. The same relationship holds for k_{+2}^{d} and k_{+1}^{obs}. The complete equation is,

$$k_{ex} = \left(\frac{\dfrac{k_{+1}^{d}}{N_A 2V_d} + \dfrac{k_{+2}^{d}k_{diss}}{k_{+2}^{d} + k_{fus}\phi^{aq}}}{\dfrac{2k_{diss}}{k_{fus}} + \dfrac{k_{+1}^{d}}{k_{fus}N_A 2V_d} - \dfrac{k_{diss}\phi^{aq}}{k_{+2}^{d} + k_{fus}\phi^{aq}}} \right) \qquad \text{Eq. (9)}$$

and ϕ^{aq} replaces the product [( )] T. $N_A V_d$ (the volume fraction of the aqueous subphase is the product of the molar volume of each droplet and the total droplet concentration). The equation in this form only holds for droplets containing at most one each of A and B, and requires some modification in order to accommodate higher-order occupancies.

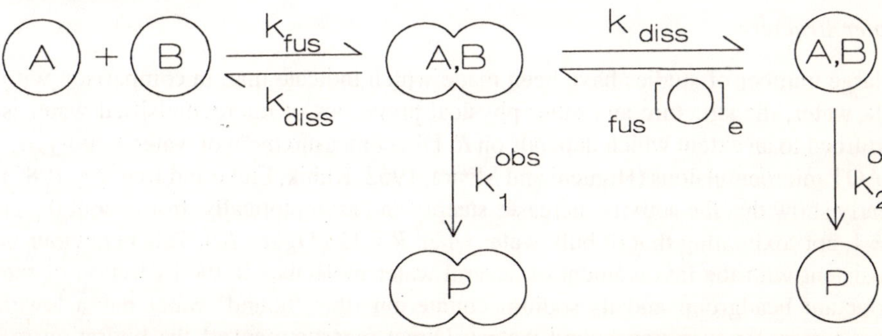

Figure 9. Scheme for the chemical reaction, A + B → P in an oil-continuous microemulsion. (Reproduced from Oldfield, C., Otero, C., Rua, M.L. and Ballesteros, A. (1992). In *Biocatalysis in Non-Conventional Media* (J. Tramper, Ed.), pp. 189–198. Elsevier, Amsterdam, with permission.

There are two interesting reductions of equation (9). When $k_{diss} << k_{+1}^{d}/(N_A V_d)$, such that the reaction can occur only within the dimer,

$$k_{ex} = k_{fus}. \qquad \text{Eq. (10)}$$

Hence this condition must be satisfied by a chemical reaction in order to make it useful for the measurement of k_{fus}. On the other hand, if $k_{diss} >> k_{+1}^{d}/(N_A V_d)$, so that the reaction can only occur after the dimer has dissociated (i.e. within one of the daughter-droplets derived from it) then it is expected that,

$$k_{ex} = \frac{k_{+2}^{d}}{\dfrac{2k_{+2}^{a}}{k_{fus}} + \phi^{aq}}. \qquad \text{Eq. (11)}$$

The factor '2' appears as a statistical factor since there are two alternative fates for (A,B) dissociation, to (A,B) + ( ) or to (A) + (B), only the former being potentially product-forming. Equation (11) predicts a linear dependence of $1/k_{ex}$ on ϕ^{aq}. Thus k_{ex} is expected to decrease as ϕ^{aq} increases, i.e. as the droplet concentration, and hence the rate of the non-productive exchange-out reaction, (A,B) → (A) + (B) (= k_{fus} [(A,B)][( )]$_T$), increases. The rate of product formation therefore decreases, reaching a minimum when $\phi^{aq} = 1$, the (hypothetical) microemulsion with the highest possible droplet concentration. Substituting a smaller number of larger droplets at constant ϕ^{aq} does not help because the rate of P-formation within the larger droplet (equation 8) is correspondingly slower. Equation (10) therefore predicts maximal reaction rates in microemulsion preparations containing a low concentration of small droplets (minimal exchange-out rate; maximal rate of reaction between A and B within a droplet). since there is no compartmentalization in aqueous solution, and hence no equivalent intermediate (A,B), second-order rate constants in microemulsions and in aqueous solution cannot be compared directly. These observations have important implications for the rates of enzyme-catalysed reactions involving water-soluble substrates (pp. 297–301).

NATURE OF THE SOLUBILIZED WATER

Water structure

A large number of studies have been made which indicate that, in comparison with bulk water, the structure and other physical properties of microemulsified water is perturbed to an extent which depends on R. Direct measurement of water activity, a_w, in AOT microemulsions (Higuchi and Mistra, 1962; Kubik, Eicke and Jönsson, 1982) clearly show that the activity increases steeply and asymptotically, from about 0.4 at $R = 1$, approximating that of bulk water when $R > 12$ (*Figure 10*). This behaviour is consistent with the involvement of several water molecules in the hydration of the surfactant headgroup and its sodium counterion (this 'bound' water has a lower vapour pressure than non-bound water). Direct measurement of the hydration requirement of the AOT and its sodium counterion in w/o microemulsions, using NMR, electron paramagnetic resonance (EPR) and differential scanning calorimetry (Wong, Thomas and Nowak, 1977; Hauser *et al.*, 1989) indicates that between two and six water molecules are relatively tightly bound to Na-AOT, with several others being more weakly bound.

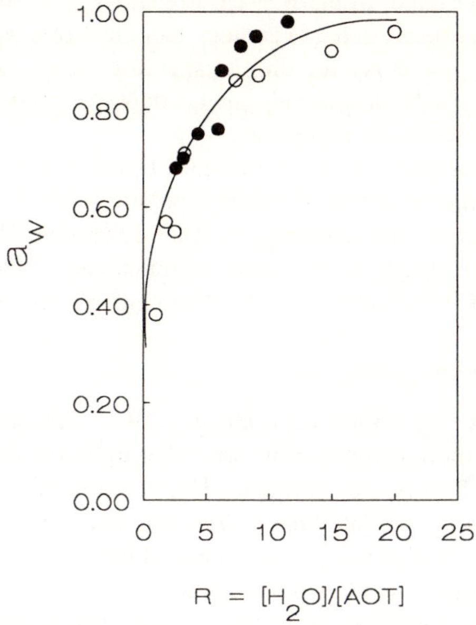

Figure 10. Dependence of water activity (a_w) on R (=$[H_2O]/[AOT]$). (O) Measured by vapour pressure osmometry (Kubik, Eicke and Jönsson (1982)); (●) Derived from proton transfer rates. Redrawn from Bardez, E., Goguillon, B.-T., Keh, E. and Valeur, B. (1984). *Journal of Physical Chemistry* **88**, 1909–1913 © (1984) American Chemical Society, with permission.

Water molecules involved in hydration of ions experience steric constraints which prevent the formation of the preferred tetrahedral clusters characteristic of bulk water, and this is clearly seen in ^{1}H NMR studies, which show chemical shifts characteristic of a non-hydrogen-bonded water structure (Wong, Thomas and Nowak, 1977; Gierasch *et al.*, 1984), in the spectral behaviour of hydrogen bond-sensitive probes, which indicate an excess of hydrogen bonding capacity (Wong, Thomas nd Nowak, 1977; Oldfield, Robinson and Freedman, 1990). Steady-state fluorescence and EPR studies of dissolved probes indicate a reduction in solvent dielectric (Wong, Thomas and Grätzel, 1976; Haering, Luisi and Hauser, 1988). As with the activity, all of these parameters relax asymptotically to near bulk-water values as R increases.

In general, it is considered that bulk-like water is absent at $R < 10$ in AOT microemulsions, but accumulates rapidly, as R increases beyond this threshold, predominating at $R > 20$. In these higher-R droplets water molecules can be considered to exchange rapidly (on the sub-nanosecond time scale) between the 'bound' (peripheral, solvating) and non-bound (bulk-water) states (Fletcher, Robinson and Tabony, 1986). The peripheral layer is considered to consist of the surfactant headgroups, those tightly bound ions forming the Helmholtz plane, and their water of hydration. The thickness of the peripheral layer is about 0.6–0.8 nm, for AOT systems (Zinsli, 1979).

The aggregation model helps to explain why the physical properties of the microdispersed water asymptote steeply to bulk-water values with increasing R. At R = 10, there can be no 'free' water, if the hydration requirement is 10 mol water per mol AOT; hence all of the 470 water molecules per droplet (equation 5) are involved in solvation. On the other hand, in the $R = 15$ droplet ($N_{AOT} = 166$; $N_{H_2O} = 2500$) the droplet may reasonably be expected to contain about 800 free water molecules. At $R = 20$ ($N_{AOT} = 190$; $N_{H_2O} = 3800$), the droplet should contain about 1800 free water molecules; and in the $R = 50$ droplet ($N_{AOT} = 1200$; $N_{H_2O} = 60\,000$) there should be a pool of about 48 000 free water molecules.

The same general properties, strong association of water with surfactant headgroups, changes in hydrogen bonding structure and micropolarity, characterize microemulsions prepared with other surfactants, including phosphatidylcholine (Wells, 1974; Boicelli, Giomini and Giuliani, 1984; Kumar and Raghunathan, 1986), oleate (Senatra, Galorielli and Guarini, 1986) and Triton X-100 (Zhu, Wu and Schelly, 1992a).

Ion distribution within the water-pool

^{23}Na NMR studies (Wong, Thomas and Grätzel, 1977) indicate that most (at least 70–80%) of the total sodium counterions are rather tightly bound to the sulphonate headgroups of the AOT molecules, forming a Helmholtz double-layer (a large degree of charge neutralization at the interface is, *a priori*, an essential prerequisite for the stability of the interface). By analogy with classical theories of charged interfaces, it is expected that the remaining, more mobile, ions are distributed radially through the droplet, along the exponential decay of the electrical field generated by the charged interface, so forming a semi-diffuse Gouy–Chapman layer. The form of this distribution is also expected to determine the radial dependence of the solvent polarity and viscosity.

A theoretical study (Karpe and Ruckenstein, 1989) indicates that the field decay, and hence the radial dependence of the sodium ion concentration, is relatively steep, such that most of the ions are confined within the peripheral water layer. The calculated pool concentrations are much lower (a few tens to a few hundred millimolar) than those calculated if the sodium ions were assumed to be evenly distributed through the water-pool (amounting to several molar). Even so, this kind of calculation ignores the rather vexed question of the activity (electrochemical potential) of the ions, and hence the ionic strength within the droplet, a knowledge of which is essential for a real understanding of the influence of the microemulsion environment on reaction rates.

pH determination and control

Another potential consequence of the disruption of water structure is a change in acid–base behaviour of the solubilized water, and hence of dissolved weak acids and bases (buffers, indicators, amino acids). Measurements of the rates of excited state pyranine deprotonation (Bardez *et al.*, 1984) are consistent with the microdispersed water behaving as a progressively weaker base (poorer proton acceptor), as R is decreased below 12. Above $R = 12$ the deprotonation rate is identical to that in bulk water. There is an excellent correlation between a_w values derived from these measurements with the vapour-pressure measurements already referred to (*Figure 10*).

Comparing the pH-dependence of the EPR order-parameter for doxylstearate (a surface-active spin-probe) in buffered and unbuffered AOT microemulsions ($R = 50$), Haering, Luisi and Hauser (1988) concluded that the pH of unbuffered water increases by approximately 1.2 units on microemulsification. This result is consistent with increases in base:acid ratio of the highly water-soluble indicators, pyranine and 4-nitrophenol-2-sulphonate (author's unpublished work), and is most likely due to ion-exchange of protons for sodium ions at the interface, with an increase in pH (decrease in a_{H^+}), which can be buffered out (El Seoud, Chinelatto and Shimizu, 1982; El Seoud and Chinelatto, 1983). It should be emphasized, however, that changes in acid:base ratio may reflect changes in the indicator pK_a, in the pH, or both, and that these alternatives cannot be distinguished without additional information.

A number of authors have discussed acid–base equilibria in terms of a pH gradient across the droplet following the decay of the interfacial field gradient (El Seoud, Chinelatto and Shimizu, 1982; El Seoud and Chinelatto, 1983; Karpe and Ruckenstein, 1989). This seems unreasonable, since although there clearly should be a concentration gradient (for H^+, Na^+ and all other ions), the thermodynamic activity (electrochemical potential) of H^+ (a_{H^+}), and Na^+ (a_{Na^+}) must be the same at all points within the microemulsion (implying an inverse relation between the field gradient and the activity coefficient), since the microemulsion is an equilibrium system. It seems more reasonable to think in terms of a single microemulsion pH (pH = $-\log 10 a_{H^+}$), with apparent pK_as for solubilized dyes determined by their (weighted) distribution between the subphases of the microemulsion (Oldfield, Robinson and Freedman, 1990). In this context the problem of pH measurement reduces from the apparently complex one of measuring the pH in a tiny water-droplet, to one of designing a probe which faithfully reports a_{H^+} of the bulk microemulsion. It has been reported that the pH, measured using the glass electrode, correlates well with the pH of the aqueous buffer solution used to prepare it (Menger and Saito, 1978). However, it is not obvious that the electrode should reliably report pH in this predominantly non-aqueous system. No further studies along this line have been reported.

van Dijk *et al.* (1992) have described a method of measuring microemulsion pH using the enzyme hydrogenase to catalyse the equilibrium between microemulsion H^+ and the standard hydrogen electrode. The technique cannot be described as routine but the thorough study carried out by these authors (covering four different microemulsion systems and five different buffers in the range pH 5–9), indicates that for the most part only small changes in pH occur when buffered aqueous solutions are solubilized. These results are consistent with ^{31}P NMR studies in AOT and in lecithin microemulsions, which show only minor changes in the base:acid ratio of solubilized phosphate buffers (Smith and Luisi, 1980; Peng and Luisi, 1990). On this basis it was recommended by Smith and Luisi (1980), and has since become convention, that the microemulsion pH should be reported as the pH of the original buffered aqueous solution used to prepare it. This convention should not be extended to unbuffered microemulsions, *vide supra*.

As a practical matter, highly water-soluble buffers are preferable over more apolar buffers since partitioning into the surfactant phase may change its pK_a. The zwitterionic Good's-type buffers are best suited for this purpose (Good *et al.*, 1966). These buffers were originally developed for the purpose of buffering biological systems containing membrane preparations since they are not soluble in phospholipid membranes (and therefore experience no pK_a shift due to partitioning). By direct analogy, therefore, they are expected also to be effective buffers for microemulsions prepared with anionic surfactants. They seem to be equally acceptable for buffering microemulsions prepared with cationic surfactants (Oldfield and Freedman, 1989).

Some early work in AOT microemulsions indicated large upward shifts in pH-profiles for several different enzymes. This phenomenon was traced to the presence of relatively large amounts of acidic contaminants (sodium hydrogen sulphate) arising from the manufacturing process, sufficient to titrate-out added buffer (reviewed by Luisi and Magid, 1986). This acidity of AOT preparations can be checked using a titration technique (Magid, Kon-no and Martin, 1981). The quality of commercially available AOT has improved considerably in recent years and no longer contains significant amounts of titratable impurities.

Enzymes in microemulsions

HISTORICAL OVERVIEW

The first example of an enzyme operating in a microemulsion environment is phospholipase A_2. An assay for this enzyme based on the hydrolysis of phosphati-

Table 2. Main examples of enzymes studied in w/o microemulsions stabilized by AOT

Enzyme	Oil	Nature of study	Reference
α-Chymotrypsin	*n*-Octane	First paper describing enzyme activity in an AOT microemulsion	Martinek *et al.* (1977)
Ribonuclease, HLADH α-chymotrypsin	Iso-octane, *n*-octane	Analytical ultracentrifugation	Bonner, Wolf and Luisi (1980) Levashov *et al.* (1982)
α-Chymotrypsin	*n*-Octane	Kinetics of N-*trans* cinnamoyl imidazole deacylation	Levashov, Klyachko and Martinek (1981)
α-Chymotrypsin	*n*-Octane	Active site spin-labelling	Belonogova *et al.* (1983)
Horseradish peroxidase	*n*-Octane	Kinetics, 'superactivity'	Klyachko, Levashov and Martinek (1984)
Chromobacterium viscosum lipase	*n*-Heptane, *n*-dodecane	Kinetics and stability	Fletcher, Galal and Robinson (1985)
HLADH	Cyclohexane	Kinetics and stability	Larsson, Adlercreutz and Mattiasson (1987)
HLADH	Isooctane	Product recovery, enzyme recycling	Larsson, Adlercreutz and Mattiasson (1990)
α-Chymotrypsin	Toluene+octane	Entrapment in nanogel particles	Khmelnitsky *et al.* (1989a)
Tyrosinase	Iso-octane	Kinetics	Bru *et al.* (1989b)
Rhizopus delemar lipase	Iso-octane	Synthesis of polyol-fatty acid esters	Hayes and Gulari (1992)
Δ^5-3-Ketosteroid isomerase	Iso-octane	Kinetics, stability	Levashov *et al.* (1992)
Penicillium simplissimum lipase	Iso-octane	Stereospecific synthesis of fatty acyl menthol esters	Stamatis, Xenakis and Kolisis (1993)
Horseradish peroxidase	Iso-octane	Synthesis of polyethyl-phenols	Rao *et al.* (1993)
Rhizopus arrhizus lipase	Iso-octane	Kinetics, spectroscopy	Brown, Yada and Marangoni (1993)

dylcholine (PC) in diethylether–PC–water microemulsion system described by Wells and co-workers (Misiorowski and Wells, 1974; Poon and Wells, 1974; Wells, 1974). Interestingly, the surfactant is also the substrate for the enzyme.

The first report of an enzyme functional in a microemulsion system prepared with a synthetic surfactant was α-chymotrypsin in an AOT-stabilized system (Martinek *et al.*, 1977), and confirmed by Menger and Yamada (1979). There followed a rapid succession of papers detailing the solubilization and properties of ribonuclease (Wolf and Luisi, 1979), cytochrome *c* (Douzou, Keh and Balny, 1979), HLADH (Meier and Luisi, 1980) and horseradish peroxidase (Levashov *et al.*, 1980), also in AOT microemulsions. Levashov *et al.* (1982) further demonstrated the activity of enzymes in microemulsions stabilized by CTAB and Brij 56, and the operation of multistep enzyme reactions, such as the conversion of phosphoenolpyruvate to lactate, via pyruvate, catalysed sequentially by pyruvate kinase and lactate dehydrogenase (see also Martinek *et al.*, 1981). These papers together served to give an indication of the potential of microemulsions as an important new class of organic solvent media for enzymes, and in doing so, opened up a new field of research.

By far, the majority of studies have been carried out using AOT microemulsions. In contrast to the several hundred studies of enzymes in these systems (see *Table 2* for some interesting examples), only a handful of studies exist for those stabilized by other kinds of surfactants (*Table 3*).

THE NATURE OF SOLUBILIZED ENZYMES

Incorporation of enzymes into microemulsions

In principle, enzymes may be introduced in three different ways (Luisi and Magid, 1986):

1. by dissolving dry enzyme powder in a pre-formed microemulsion (dry extraction);
2. by transfer from a conjugate aqueous phase (liquid extraction); and
3. by introducing the enzyme as a concentrated aqueous stock solution (injection).

The dry-extraction method is, at least in principle, the most reliable method of measuring the solubilization capacity of a single-phase microemulsion for a given protein; the liquid–liquid extraction method has found an important application in the differential extraction of proteins from aqueous phases as a method of protein purification (p. 308). The injection method is used routinely for the rapid preparation of microemulsions of defined composition.

The dry-extraction procedure simply involves stirring the pre-formed microemulsion with the enzyme powder. The progress of transfer can be followed by monitoring the A_{280} of the microemulsion phase. Excess protein is removed by low-speed centrifugation. The liquid-extraction procedure involves dissolving the protein in a Winsor II system. The system can be stirred to facilitate transfer (the microemulsion+water emulsion separates rapidly when mixing stops). For kinetic studies, the injection method is the fastest and most convenient method. Usually an aliquot of aqueous enzyme stock is injected into a pre-formed microemulsion. This additional material is rapidly incorporated into the microemulsion phase (within a

Table 3. Main examples of enzymes studied in w/o microemulsions stabilized by surfactants other than AOT

Surfactant/co-surfactant	Enzyme	Nature of study	Reference
Phosphatidylcholine	Phospholipase A_2	Routine assay of enzyme activity	Misiorowski and Wells (1974)
CTAB	20β-Hydroxysteroid dehydrogenase	Cofactor regeneration study	Hilhorst *et al.* (1984)
Brij 96	Luciferase	Kinetics	Belyaeva *et al.* (1983)
Triton X-100	Cholesterol oxidase	Kinetics and stability	Lee and Biellmann (1986)
CTAB	Bilirubin oxidase	Kinetics	Oldfield and Freedman (1989)
Brij 96	Polyphenol-oxidase	Kinetics	Sanchez-Ferrar, Bru and Garcia Carmona (1988)
Brij 56	Tryptophanase	Tryptophan synthesis: process engineering	Eggers and Blanch (1988)
TOMAC	Several	Selective protein extraction	Wolbert *et al.* (1989)
$C_{12}E_5$[a]	HLADH	Enzyme/cofactor recycling study	Larsson, Adlercreutz and Matiasson (1990)
Phosphatidylcholine	Trypsin, α-chymotrypsin	Kinetics and stability	Peng and Luisi (1990)
Phosphatidylcholine	Lipase	Entrampment in lecithin microemulsion gels	Scartazzini and Luisi (1990)
Triton X35 + Nonidet P40	Lipoxygenase	Kinetics	Piazza (1992)
Tween 85	Organophosphorus hydrolase	Kinetics	Komives, Lilley and Russell (1994); Komives, Osborne and Russell (1994)

[a] The surfactant used in this example was a single homologue of the Brij/Lubrol W class. The molecular formula is $CH_3(CH_2)_{11}(OCH_2CH_2)_5OH$, abbreviated to $C_{12}E_5$ by convention.

few seconds). This approach minimizes enzyme inactivation arising from exposure to the pre-microemulsion mixture. In some cases (e.g. in fast kinetic studies) it is desirable to prepare enzyme and substrate in separate microemulsions, which may then be mixed together (p. 276).

Size and composition of protein-containing droplets

The types of studies which have been undertaken to study the nature of the solubilized water–protein–surfactant aggregates include (1) measurements of the solubilization capacity of the microemulsion for a protein and (2) application of physical techniques including analytical centrifugation, dynamic light scattering (PCS) and small-angle neutron scattering (SANS) to determine the composition of the host structure.

Detailed studies of solubilization capacity of AOT microemulsions for α-chymotrypsin and LADH were made by Matzke *et al.* (1992). In these experiments the proteins were incorporated, using either solid-phase extraction or injection, into microemulsions buffered at pH 8. The results for both proteins show that, as R is increased, there is a sudden rise in solubilization beyond a point where the 'original' droplet size (i.e. that in a microemulsion containing no protein, but of otherwise the same composition) very roughly corresponds to the size of the enzyme (*Figure 11*). Different results were obtained for microemulsions prepared by the injection method, especially at low *R*, where considerably more protein was retained by the microemulsion than anticipated on the basis of the solid-phase extraction data. Most likely microemulsions prepared in this way are metastable (overloaded systems which are kinetically slow to lose excess protein). The differences in the microemulsion solubilization capacities for the two proteins simply reflect the differences in aqueous solubilities.

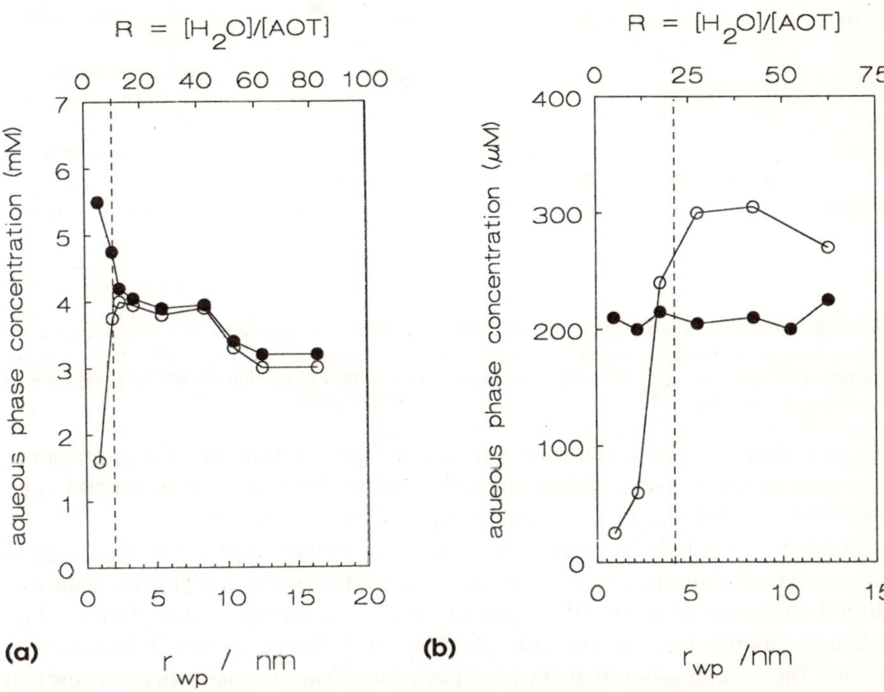

Figure 11. Solubilization of proteins as a function of original droplet size by AOT-microemulsions. (a) α-Chymotrypsin; (b) horse-liver alcohol dehydrogenase. (O) Prepared by dry-extraction; (●) prepared by injection. [AOT] = 50 mM, pH = 8.0 (200 mM^{aq} Tris-HCl). The dotted line shows the droplet radius equivalent to that of the enzyme considered as a sphere. (Redrawn from Matzke, S.F., Creagh, A.L., Haynes, C.A., Prausnitz, J.M. and Blanch, H.W. (1992). *Biotechnology and Bioengineering* **40**, 91, by permission of John Wiley & Sons, Inc.)

It is useful to compare the molar solubilization capacities (determined using the solid-extraction method) with the 'original' molar droplet concentration i.e. the concentration in microemulsions prepared with no protein.. The results obtained by Matzke *et al.* (1992) show that at low *R* the nominal α-chymotrypsin droplets ratio is <<1; rising to unity only when *R* reaches 30. Above $R = 30$, there is (nominally) more than one per original droplet. The solubilization capacity for LADH is always much lower than the droplet concentration (*Figure 12*). The assumption that the droplet concentration remains unchanged on addition of protein is unreasonable, especially in situations where the original droplet size is comparable with, or less than, that of the

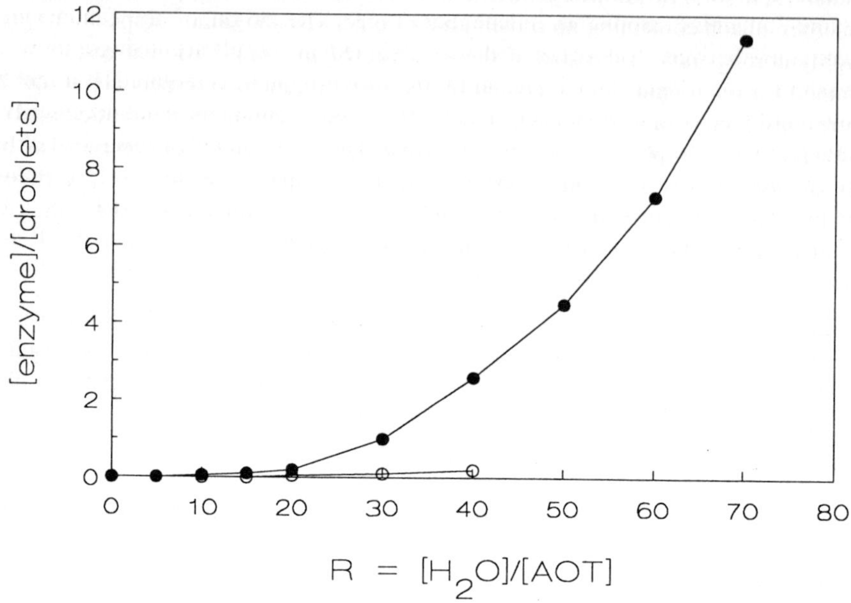

Figure 12. Apparent enzyme droplet ratio for AOT microemulsions saturated with α-chymotrypsin (●) or horse-liver alcohol dehydrogenase (O). (Redrawn from Matzke, S.F., Creagh, A.L., Haynes, C.A., Prausnitz, J.M. and Blanch, H.W. (1992). *Biotechnology and Bioengineering* **40**, 91, by permission of John Wiley & Sons, Inc.)

protein molecule, when solubilization might be expected to involve the aggregation of several droplets. Nevertheless the calculations show that, as a general rule, solubilization is not simply a function of the droplet concentration.

The precise size and composition of enzyme-containing droplets has been studied using analytical ultracentrifugation, which is a powerful tool for studying the sedimentation behaviour of a specific species in a mixed-droplet population. The protein-containing fraction, for example, may be followed by u.v. absorbance at 280 nm. These studies show that, for small proteins such as ribonuclease, lysozyme and myelin basic protein (Bonner, Wolf and Luisi, 1980; Zampieri, Jäckle and Luisi, 1986) and α-chymotrypsin (Levashov *et al.*, 1982), and at high R, the enzyme effectively sits within an 'original' droplet, and there is no massive re-distribution of water and surfactant. At lower R (the original droplet is similar in size, or smaller than, the protein), the data are consistent with the recruitment of water and surfactant from a small number (perhaps two or three) original droplets, in order to host the enzyme molecule. It seems that the enzyme-containing droplet is more water-rich than the original (i.e. the droplet value of R is larger than the nominal value, implying that R is lower for the 'unfilled' droplets (Zampieri, Jäckle and Luisi, 1986). This is reasonable given the considerable hydration requirements of the protein molecule; a quasi-elastic neutron-scattering study indicates that in AOT microemulsions at $R = 20$, a single α-chymotrypsin molecule is solvated by 250–500 molecules of water, which is roughly the same as that in bulk aqueous solution (Fletcher, Robinson and Tabony, 1986).

The size and composition of α-chymotrypsin-containing droplets has also been studied by SANS (Rahaman and Hatton, 1991). These workers find that, in AOT microemulsions prepared either by injection or by liquid extraction, the size of the enzyme-containing droplet is independent of *R*. The origin of the fundamental discrepancy between this result and that obtained in the ultracentrifuge study of Levashov *et al.* (1982) is not clear, and further investigation is required. It might be pointed out that analysis of scattering data for mixed populations is not necessarily a straightforward task, even for a two-component system, especially if the ratios of the two components are far from unity. In addition, the particular droplet population containing the enzyme cannot be identified from the scattering data alone.

The bulk of data are consistent with the 'water-shell' model (Bonner, Wolf and Luisi, 1980; Zampieri, Jäckle and Luisi, 1986) which may be described as follows:

1. Enzymes are insoluble in oils and in binary reversed micellar solutions of surfactants in oils (see, for example, Nicot *et al.*, 1985; Leser *et al.*, 1986) and therefore solubilization models in which the microdispersed water does not play a primary role as a solvent cannot be important.
2. The protein molecule is considered to be dissolved in the water-pool, which is itself contained within a close-packed surfactant monolayer.
3. No constraint is placed on the hydrodynamic radius of the droplet, and surfactant and available water are permitted to redistribute between protein-containing, and 'empty droplets'.

This model is probably adequate to describe the protein-droplet structures for globular proteins. Data for a few non-globular proteins suggest more complex solubilization behaviour, involving the formation of highly asymmetric droplet structures in order to accommodate these proteins (reviewed by Luisi and Magid, 1986). Solubilization of gelatin in AOT microemulsions leads to the formation of gels (p. 312).

The water-shell model according to Zampieri, Jäckle and Luisi (1986) might be referred to as non-conservative, since redistribution of water and surfactant, and changes in droplet dimensions, are allowed. In contrast, Martinek and co-workers (Levashov *et al.*, 1982; Martinek *et al.*, 1989) argue for a model (the 'induced fit' model) that is absolutely conservative, with no redistribution of water or surfactant between droplets, and no change in hydrodynamic radius until the original droplet size falls below that of the enzyme. Only when the corresponding pool size is less than that of the enzyme is redistribution of water and surfactant between droplets and, presumably, a change in droplet dimensions, permitted. The problem with this model, which has been criticized before (Luisi and Magid, 1986), is the fate of the water displaced by the enzyme. The water cannot leave the droplet, yet the hydrodynamic volume of the droplet cannot change, and this implies zero volume (infinite density) for this water (Levashov *et al.*, 1982). In defending the model, Martinek and his colleagues have suggested that the displaced water is expelled somewhat into the surfactant layer, and provide ^{13}C NMR data which imply greater hydration of the carbonyl carbons and of the first methylene carbons of the ethylhexyl chains of AOT in preparations containing α-chymotrypsin (Shapiro *et al.*, 1989). However, simple calculation will shows that as *R* decreases, so that the radii of the original water-pools and of the protein molecules converge, the volume of water displaced becomes so great that it would extend beyond the thickness of the surfactant layer, if reasonable values for the density of the surfactant and of the displaced water are used, thus contradicting the fixed radius element of the model.

Protein conformation; localization within the droplet

A number of spectroscopic studies have been carried out to determine how the enzyme conformation changes on solubilization, using cytochromes P_{450} and b_5 (Erjomin and Metelitza, 1983), cytochrome *c* (Vos *et al.*, 1987a), lysozyme (Steinmann, Jäcle and Luisi, 1986), Horse-liver alcohol dehydrogenase (Vos *et al.*, 1987b) and α-chymotrypsin (Dorovska-Taran, Veeger and Visser, 1993a). Most of the data pertain to studies in AOT and CTAB microemulsions.

In general, solubilization in microemulsions has a negligible effect on the UV-spectrum of proteins. Steady-state fluorescence emission spectra, on the other hand, usually show a fairly significant blue shift and increase in quantum yield, both of which are more pronounced at lower *R* (Luisi and Magid, 1986; Marzola and Gratton, 1991; Brown, Yada and Marangoni, 1993). These results suggest conformational rearrangements and interactions of the the chromophores with the surfactant layer, either or both of which result in the shifting of the chromophore to a non-polar environment. The shifts are usually relatively small (1–2 nm). Larger shifts can be expected for denatured proteins.

Secondary structure, examined using circular dichroic spectroscopy (CD), is usually only slightly perturbed, and again, the largest changes are observed at a low *R*. Major changes in CD spectra have been observed in a few cases. Lysozyme, for example, rapidly unfolds with loss of activity, in AOT microemulsions (Steinmann, Jäcle and Luisi, 1986). In contrast, solubilization of *Rhizopus arrhizus* lipase is accompanied by a large increase in secondary structure (β-sheets and turns). The enzyme remains active (Brown, Yada and Marangoni, 1993). Cytochrome *c* is solubilized in AOT microemulsions with considerable loss of conformation at high *R* ($R = 40$), whereas conformation is largely retained at lower *R* (Vos *et al.*, 1987b).

A number of studies have been carried out to determine the extent of the interactions of solubilized protein with the interphase. Petit, Brochette and Pileni (1986) studied the location of cytochrome *c*, α-chymotrypsin and ribonuclease using hydrated electron quenching and Vos *et al.* (1987a,b) studied cytochrome *c* and HLADH using fluoresence anisotropy. The interpretation of these experiments has proven not to be straightforward, but the general conclusion which may be drawn is that proteins interact with more strongly with the interphase when the pH is such that the net charge on the protein is opposite to that of the surfactant layer. Thus, at least for microemulsions prepared with ionic surfactants, electrostatic interactions are important in determining the degree of interaction.

KINETIC BEHAVIOUR OF ENZYMES IN MICROEMULSIONS

Michaelis–Menten kinetics

In microemulsions held at otherwise constant composition, enzymes usually obey Michaelis–Menten kinetics, i.e. the rate at constant substrate concentration is directly proportional to the enzyme concentration, and the rate, at constant enzyme concentration, is a rectangular-hyperbolic function of the substrate concentration. Values of k_{cat} and K_m may therefore be derived using the standard procedures. Good examples of Michaelis–Menten behaviour in AOT microemulsions include the hydrolysis of palm oil by *Rhizopus arrhizus* lipase (Kim and Cheung, 1989), of Crambe oil by several different lipases (Derksen and Cuperus, 1992) and the reduction of acetaldehyde by HLADH/NADH (Meier and Luisi, 1980).

This apparently straightforward situation is complicated by the fact that k_{cat} and K_m are found to depend rather strongly on the composition of the microemulsion and that the precise nature of this dependence is a function of the particular microemulsion system and of the enzyme and substrate combination in question (in particular, whether the substrate is oil- or water-soluble). It is reasonable to assume, *a priori*, that the actual values of k_{cat} and K_m will be complex functions of the various (qualitatively definable) elements of microemulsion structure already introduced: surfactant interactions and changes in water structure, which modulate enzyme conformation; and partitioning and exchange-related factors, which serve to influence the local substrate concentration in the microenvironment of the enzyme.

The various patterns of behaviour which have been observed are best related by considering how the kinetics vary with R and with the droplet concentration, and on whether the substrates are oil- or water-soluble, as discussed in the following sections.

Kinetics as a function of R

Enzyme activity in microemulsions is usually a strong function of R. In AOT systems the most common manifestation is the so-called bell-shaped plot, in which the activity passes through a distinct maximum. Good examples are α-chymotrypsin (Belonogova *et al.*, 1983) (*Figure 13*), *Chromobacterium viscosum* lipase (Fletcher *et al.*, 1985a) (*Figure 14*), horseradish peroxidase (Klyachko, Levashov and Martinek, 1984; Das and Maitra, 1989), *Candida rugosa* lipase (Han and Rhee, 1986), alkaline phosphatase (Gonnelli and Strambini, 1988), *Rhizopus arrhizus* lipase (Kim and Cheung, 1989) and Δ^5-3-ketosteroid isomerase (Levashov *et al.*, 1992). Bell-shaped plots have been described for enzymes in a few other microemulsion systems, including pyrophosphatase and lactate dehydrogenase in a Brij 56 system (Levashov *et al.*, 1980; Levashov, Klyachko and Martinek, 1981) and *Candida rugosa* lipase in a lecithin system (Chen and Chang, 1993) and *Rhizopus delamer* lipase, again in a lecithin system (Schmidli and Luisi, 1990).

In the case of the deacylation of *N*-transcinnamoyl α-chymotrypsin (*Figure 13*) k_{cat} rises to a maximum at $R = 10$, and then declines to a limiting high-R value which is approximately the same as that in aqueous solution. At $R = 10$, k_{cat} is about 40% higher than in aqueous solution at the same pH (the enzyme is said to be 'superactive'; p. 273). Further experiments have shown that the rotational correlation time ($1/\tau_c$) of an active-site spin label passes through a minimum at the optimal R (R_{OPT}), as shown in *Figure 13*, implying conformational changes which reduce the free volume within the active-site cleft. The kinetic and spin-label data together have been interpreted in terms of a conformational change which renders the enzyme more active than in aqueous solution. Noting that this optimal R corresponds fairly well with an 'original' droplet size which approximates that of the chymotrypsin molecule, Martinek and co-workers have concluded that the 'tight-fit' of the enzyme within this droplet is the source of the conformational restraint (the 'induced-fit' model, described above) (Belonogova *et al.*, 1983; Martinek *et al.*, 1989; Levashov *et al.*, 1990). These workers also report a linear correlation between R_{OPT} and the size of the enzyme for

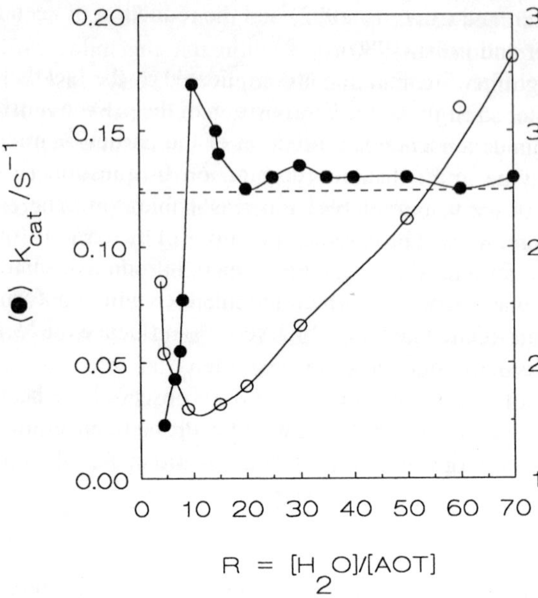

Figure 13. Dependence of k_{CAT} on R for the deacylation of *N-trans*-cinnamoylimidazole by α-chymotrypsin (●) and of the rotation frequency (O) of an active-site spin-label in an AOT microemulsion system. The broken line shows the value of k_{cat} in aqueous solution. (Reproduced from Belonogova *et al.* (1983), with permission.)

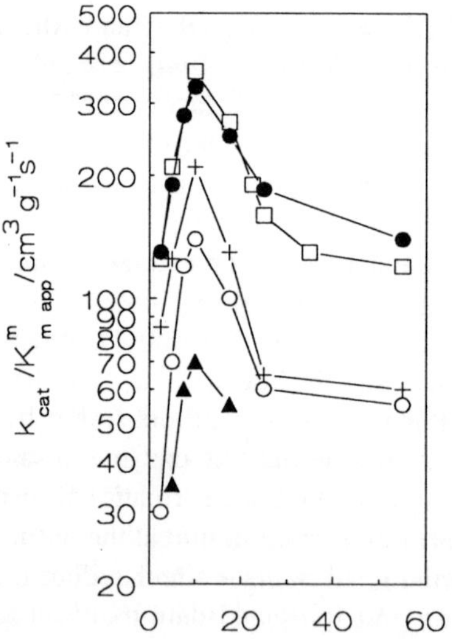

Figure 14. Dependence of k_{cat}/K for the hydrolysis of 4-nitrophenyl alkanoates by *Ch. viscosum* lipase in AOT-*n*-heptane microemulsions at pH 8. The substrates are (●) C_4, (+) C_6, (□) C_{16}. (▲) C_8 in an AOT–water–dodecane microemulsion. (Redrawn from Fletcher, P.D.I., Robinson, B.H., Freedman, R.B. and Oldfield, C. (1985a) *Journal of the Chemical Society – Faraday Transactions* **81**, 2667–2679, with permission.)

a large number of enzymes in the AOT system, implying that the induced-fit model holds for these also (Martinek *et al.*, 1989).

Elegant as this interpretation is, it is worth recalling that this induced-fit model has been criticized for being unphysical, as already discussed. In any case, it is not really necessary to invoke induced fit in order to formulate a reasonable hypothesis which explains the data. Enzymes are known to have a critical requirement for water of hydration in order to express catalytic activity, and the role of this water is generally accepted as being necessary to impart conformational mobility. Therefore in this system, as in others, it is reasonable to expect enzyme activity to increase with water content (water-activity), asymptoting to the bulk aqueous solution value as $a_w \rightarrow 1$. This underlying behaviour is recognizable in the kinetics (compare *Figures 10* and *13*). As for the superactive peak, this seems, at least in the case of α-chymotrypsin, to be a substrate-specific phenomenon. With resorufin butyrate as a substrate, k_{cat} rises sharply, to the aqueous solution value, at $R = 10$, and is thereafter constant (Dorovska-Taran *et al.*, 1993a) and with *N*-succinyl-Ala-Ala-Pro-Phe-4-nitroanilide (SAAP) as a substrate that k_{cat} simply asymptotes to the aqueous solution value at high *R* (Bru and Walde, 1991). These curves are more nearly like those expected for a straightforward 'water-of-hydration' model.

It might be suggested that the availability of water as a substrate must play a role in determining the *R*-dependence of activity for hydrolytic enzymes such as α-chymotrypsin. However, this does not seem to be the case since non-hydrolytic enzymes (a good example is Δ^5-3-ketosteroid isomerase) (Levashov *et al.*, 1992) also behave in this fashion. Furthermore, several studies involving lipases reveal the same general trend (decreasing activity with decreasing *R*) for these enzymes operating in both synthetic (condensative) and hydrolytic modes (Fletcher *et al.*, 1985a; Prazeres, Garcia and Cabral, 1992; Stamatis, Xenakis and Kolisis, 1993; Stamatis *et al.*, 1993a,b).

Several enzymes, including lactate dehydrogenase (LDH), penicillin acylase, and γ-glutamyl synthetase, have been reported by one group to display a complex *R*-dependence, characterized by multiple activity peaks (Kabanov *et al.*, 1989; Kabakov *et al.*, 1992). The authors suggest that the peaks correspond to aggregates of different sizes and subunit compositions. The results for LDH are apparently not reproducible (Khmelnitsky *et al.*, 1993). The latter authors suggest that multiple peaks in the *R*-profile reported in the earlier study are artefacts caused by the instability of the enzyme in microemulsions.

Many enzymes display *R*-dependent kinetics other than bell-shaped. For example, Fletcher *et al.* (1985a) report that the activity of *Ch. viscosum* lipase (k_{cat}/K_m for the hydrolysis of 4-nitrophenyl alkanoates) is independent of *R* in a CTAB microemulsion, in contrast to the most definite bell-shaped kinetics obtained in an AOT microemulsion, shown in *Figure 14*. A similar result was obtained by Vos *et al.*, (1987b) for LADH. Again, in a CTAB system, $k_{cat} K_m$ for the oxidation of bilirubin by bilirubin oxidase decreases smoothly by a factor of 50, as *R* is increased from 3.5 to 25 (*Figure 15a*). This is a k_{cat} effect; K_m essentially remains constant (Oldfield and Freedman, 1989). Han, Walde and Luisi (1990) report that the bell-shaped dependence of the rate of olive-oil hydrolysis by *C. rugosa* lipase in an AOT microemulsion progressively disappears as the enzyme concentration is increased tenfold until, at the highest concentrations, the rate becomes an hyperbolic function of *R* (*Figure 15b*). Perhaps most intriguing of all is the observation that the bell-shaped *R*-dependence of the rate

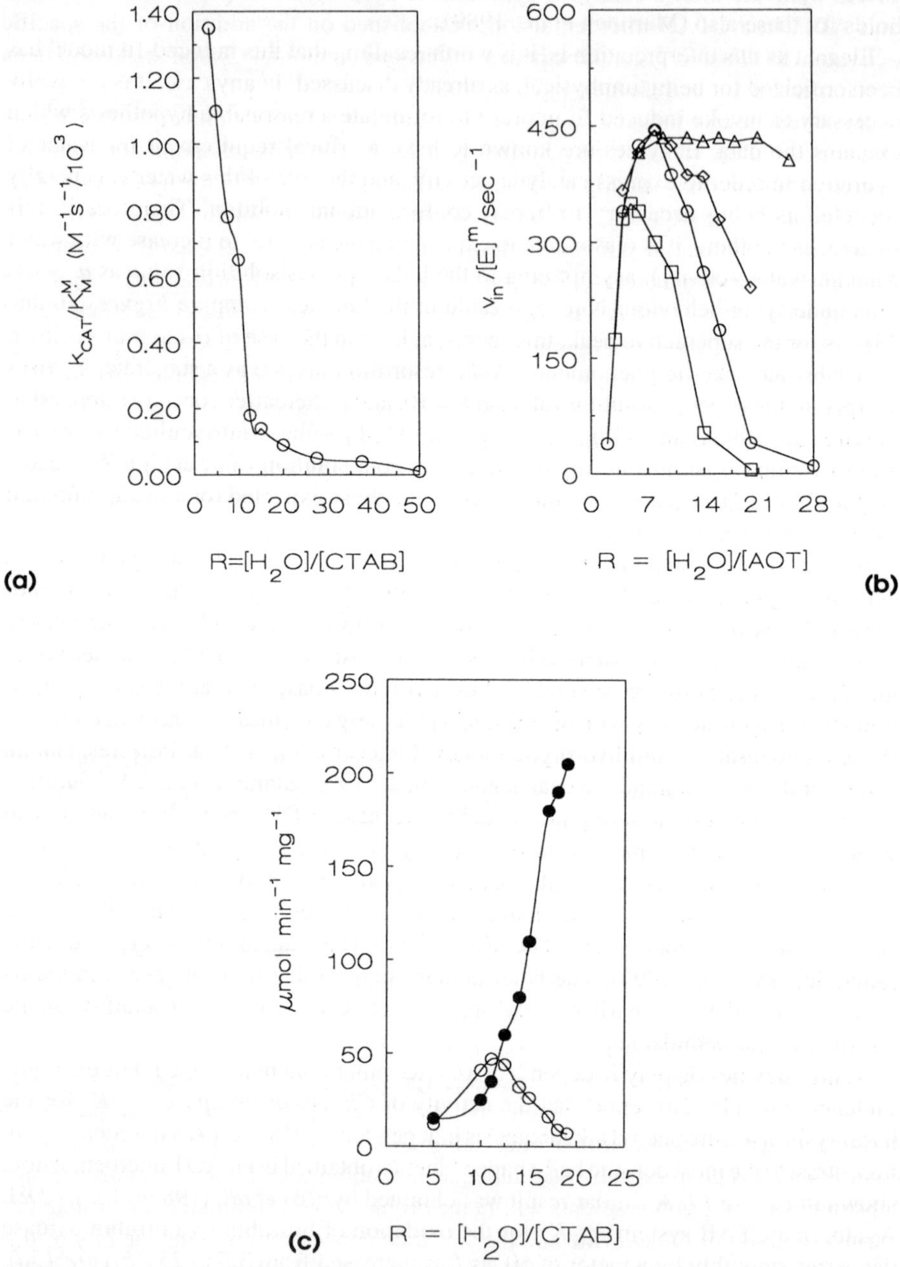

Figure 15. Dependence of rate parameters on *R* for different enzyme and microemulsion combinations. (a) Bilirubin oxidase (Oldfield, C. and Freedman, R.B. (1989). *European Journal of Biochemistry* **183**, 347–355); (b) *C. rugosa* lipase at overall concentrations of (□) 30, (O) 43, (♦), 91, and (Δ) 290 nM (Han, D., Walde, P. and Luisi, P.L. (1990). *Biocatalysis* **4**, 153–161); (c) pyruvate kinase in the presence (●) and absence (O) of the specific activator K^+ (Ramírez-Silva, L., Tuena de Gómez-Puyou, M. and Gómez-Puyou, A. (1993). *Biochemistry* **32**, 5332–5338 © (1993) American Chemical Society). Figures redrawn with permission.

of the pyruvate kinase-catalysed synthesis of pyruvate from phosphoenolpyruvate and ADP, in a CTAB microemulsion, is abolished on the addition of the specific activator, K^+ (*Figure 15c*); the kinetics are instead sigmoidal (Ramírez-Silva, Tuena de Gómez-Puyou and Gómez-Puyou, 1993).

It is often the case that enzymes have higher activity in microemulsions than in bulk aqueous solution. This 'superactivity' is a fairly common phenomenon, and the degree of activation is usually modest. As already shown, k_{cat} for the hydrolysis of N-*trans* cinnamoyl imidazole (t-CNI) is about 40% higher in the optimal $R = 10$ microemulsion than in aqueous solution at the same pH. As another example, Madamwar, Bhatt and Ray (1988) report an approximate fourfold increase in the activity of invertase (catalysing the hydrolysis of sucrose) in microemulsions stabilized by SDS or by sodium tauroglycholate.

A few enzymes appear to exhibit a much greater degree of superactivity in AOT microemulsions. The classic examples are mushroom tyrosinase (laccase), acid phosphatase and horseradish peroxidase (HRP), for which increases in k_{cat} of up to 50-, 100- and 200-fold, respectively, have been reported (Martinek *et al.*, 1986, 1989).

The origins of this kind of superactivity are not understood, but several attempts have been made to account for it and for R-dependent kinetics in general, through the construction of algorithms which are parameterized such that input of numerical values for each parameter will generate a model curve to fit the experimental data. The approach of Kabanov *et al.* (1988), extending the concept of the 'induced-fit' model, considers that droplet polydispersity accounts for the R-dependence, and the algorithm contains parameters representing droplet polydispersity and enzyme activity as a function of droplet size. On the other hand, Bru, Sánchez-Ferrar and Garcia-Carmona (1989a, 1990) considered that the enzyme activity is determined by partitioning of the enzyme between water domains characterized by three different water structures, and by partitioning of substrate between the subphases. The resulting algorithm contains a total of six adjustable parameters.

The problem with this kind of approach is that there is no independent experimental verification that the 'active' parameters actually influence enzyme activity in the way required by the model and there are no criteria for determining whether or not the values assigned to a given parameter are realistic. The algorithms have no predictive power, since they do not suggest new experiments to test the models; they are best regarded as rather elaborate curve-fitting procedures.

In another approach, Ruckenstein and Karpe (1990, 1991) considered that the (R-dependent) radial decay of the electrical field gradient generated by the fixed surfactant headgroups determines the radial distribution of solutes (hydrogen ions, counterions and charged substrate molecules), leading to pH, solvent dielectric and substrate concentration gradients within the droplet which combine to determine the intrinsic activity of the enzyme and the substrate concentration in the vicinity of the active site. Once again, there is no direct evidence to indicate the extent to which these factors are responsible for the observed R-dependencies and the question remains as to whether the concentrations, or the thermodynamic activities, of the solutes should be taken into consideration. In any case, for reactions where the enzyme operates under k_{cat} conditions (k_{cat} is a substrate concentration-independent parameter), or of enzyme-uncharged substrate combinations, the theory requires that the R-dependence of the kinetics must arise solely from pH or/and solvent dielectric effects. These kinds of considerations once again highlight the need for a better understanding of the concept and measurement of pH in order to obtain a better understanding of enzyme behaviour.

Dependence of the kinetics on the droplet concentration

As long as the droplets remain in large excess over the enzyme, so that each molecule goes into a separate, identical host-droplet (*Figure 12*), a direct proportionality between reaction rate and enzyme concentration is expected and indeed observed, since the number of identical biocatalytic units simply increases on adding more enzyme. It is also reasonable to expect kinetic parameters (k_{cat} and K_m expressed in overall concentration units) to be independent of the total droplet concentration at fixed R, since the presence of unfilled droplets should have no effect on the activity of the enzyme (the notable exception being the case of a water-soluble, as opposed to an oil-soluble or surface active substrate, in which case theory predicts, and experiment supports, a proportionality between K_m and the droplet concentration (pp. 297–301). Early studies with α-chymotrypsin (measuring k_{cat} for the hydrolysis of glutaryphenylalanyl-4-nitroanilide (GPNA) and *Ch. viscosum* lipase (k_{cat}/K_m for the hydrolysis of 4-nitrophenyl alkanoates) (Fletcher *et al.*, 1984, 1985a) indicated that this is indeed the case, although more recent data indicate numerous exceptions, and that reaction rates can often be expected to decrease with increasing droplet concentration at fixed R (i.e. increasing $\phi^{aq}(R)$. Examples are HRP (Klyachko, Levashou and Martinek, 1984; Das and Maitra, 1989), Δ^5-3-ketosteroid isomerase (Levashov *et al.*, 1992), soybean lipoxygenase-1 (Shkarina, Khün and Schewe, 1992) and *Ch. viscosum* lipase with triolein as substrate (Prazeres, Garcia and Cabral, 1992). It has been suggested that these kinetics reflect a very strong dependence of the composition of the host droplet (aggregation numbers of enzyme, water and surfactant) on $\phi^{aq}(R)$ (Klyachko, Levashov and Martinek, 1984). If this is the case, an ultracentrifugation study should reveal such changes. In the particular case of *Ch. viscosum* lipase, substrate-specific effects must once again be considered since the enzyme activity, as assayed with 4-nitrophenyl alkanoates, is independent of $\phi^{aq}(R)$ (Fletcher *et al.*, 1985a), whereas the rate decreases markedly when assayed with triolein (Prazeres, Garcia and Cabral, 1992).

Reactions involving oil-soluble substrates

Solutes will distribute between the aqueous, oil and surfactant subphases, depending whether they are primarily oil or water soluble or surface active. The kinetics of chemical and enzyme-catalysed reactions should reflect this. For apolar and surface-active substrates (e.g. steroids, lipids), which partition between the oil and surfactant subphases, the pseudophase model (Biais *et al.*, 1981) is sufficient to interpret most kinetic data (Fletcher *et al.*, 1985a; Khmelnitsky *et al.*, 1990). This model simply ignores the microdispersed nature of the aqueous subphase, treating the system instead as a three-phase (oil–surfactant–water) system of (effectively) infinite interfacial surface area. Using the steady-state approximation, the rate equation for the enzyme-catalysed transformation of an apolar substrate, in which the enzyme 'sees' only that fraction of the substrate in solution in the aqueous subphase (*Figure 16*) gives (Oldfield, 1990):

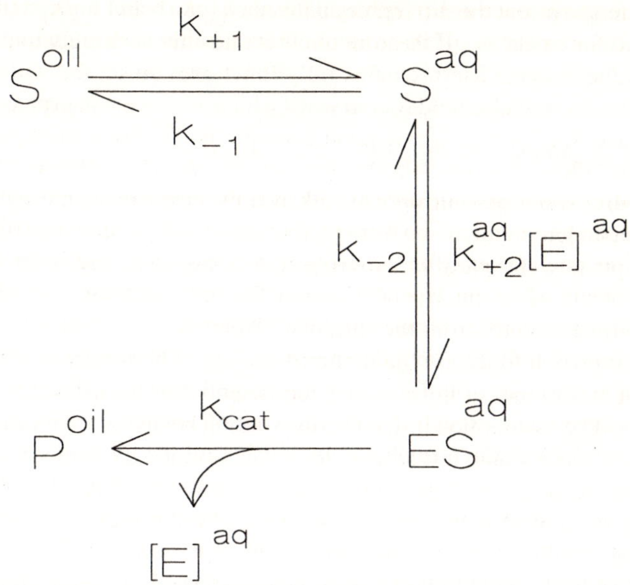

Figure 16. Kinetic scheme for the conversion of an oil-soluble substrate under pseudophase conditions. S^{aq}, substrate in the aqueous subphase; S^{oil}, substrate in the oil subphase; E^{aq}, enzyme in the aqueous subphase; ES^{aq}, enzyme–substrate complex in the aqueous subphase. The kinetics are simplified by assuming that the product, *P*, is also oil-soluble.

$$v^{aq} = \frac{k_{CAT}[E]_T^{aq}[S]^{aq}}{K_m^{aq}(K_{p\,app}\phi^{oil} + \phi^{aq}) + [E]_T^{aq}k_{cat}/k_{+1}\phi^{oil} + [S]^{aq}} \qquad \text{Eq. (12)}$$

(the volume fraction of the surfactant was ignored in this derivation). The superscript 'aq' denotes that the concentration units are mol per litre of the aqueous subphase. $K_{p\,app}$ is the apparent oil:water partition coefficient. It is extremely unlikely that phase transfer will ever be rate limiting, but the non-linear dependence of rate on the enzyme concentration predicted by this equation can be used to check for such behaviour. When phase transfer is not rate limiting ($k_{+1} >> k_{cat}$) the equation simplifies,

$$v^{aq} = \frac{k_{cat}[E]_T^{aq}[S]^{aq}}{K_{m\,app}^{aq} + [S]^{aq}} \qquad \text{Eq. (13)}$$

and Michaelis–Menten kinetics are predicted, where

$$K_{m\,app}^{aq} = K_m^{aq}(K_{p\,app}\phi^{oil} + \phi^{aq}); \qquad \text{Eq. (14)}$$

the apparent Michaelis constant, $K_{m\,app}^{aq}$ is larger than its actual value by a factor of ($K_{p\,app}\phi^{oil} + \phi^{aq}$). If the partitioning substrate is not surface active, then it is reasonable to equate $K_{p\,app}$ with the oil:water partition coefficient, $K_{OIL:AQ}$. If the substrate is surface active, then $K_{p\,app} = K_{OIL:INT} \times K_{INT:AQ}$, where $K_{OIL:INT}$ and $K_{INT:AQ}$ are respectively the oil:interphase and interphase:aqueous partition coefficients.

In a study of the catalytic behaviour of horse-liver alcohol dehydrogenase (HLADH) in AOT microemulsions, Khmelnitsky *et al.* (1990) measured $K_{OIL:INT}$ and $K_{INT:AQ}$ for the series of linear alkanols (*n*-butanol–*n*-decanol) using a calorimetric technique, and were able to show that the differences between k_{cat}/K_m^{aq} and k_{cat}/K_m^{w} (the aqueous solution values) for oxidation of these alcohols is due almost entirely to partitioning, indicating that the enzyme itself remains fully functional on solubilization.

The same kinetics will also hold for enzymes which access the interphasic substrate pool. In this case $K_{p\ app} = K_{OIL:INT}$. A good example is *Ch. viscosum* lipase in AOT microemulsions (Fletcher *et al.*, 1985b). The activity ($k_{cat}/K_{m\ app}^{aq}$) towards members of the homologous series of 4-nitrophenyl alkanoates, varies by no more than a factor of five as the chain length increases from C_4 (butyrate) to C_{16} (palmitate) (*Figure 14*), while the *n*-heptane:water partition coefficient decreases by a factor of 10^7. This is interpreted in terms of an interfacial location for the substrate, with the enzyme extracting substrate directly from the surfactant layer.

A different approach to the problem of partitioning, which circumvents the technically difficult task of measuring partition coefficients, has been described by Laane and co-workers. The assumption is that the distribution behaviour of a solute between the continuous phase and the interphase can be determined by comparing the log P value of the solute (log P_s) with those for the continuous phase, log P_{cph} and the interphase, log P_{int} (log P is the $\log_{10}$ of the octanol:water partition coefficient). A similarity in log P values for a pair of compounds implies high mutual solubility; in a multiphase system a solute will partition most favourably into the phase with the most similar log P. The log P values of each component may be calculated from tables. In the case of the microemulsion continuous and interphases, an average log P is calculated, weighted according to the mol-fraction of each component present (Laane *et al.*, 1987a,b).

In one study, the activity of 20β-hydroxysteroid dehydrogenase (catalysing the reduction of progesterone or prednisone to the corresponding alcohol) in a CTAB–water–iso-octane–hexanol microemulsion was optimized using this approach (Hilhorst *et al.*, 1984). The composition of the interphase (i.e. the surfactant:co-surfactant ratio) and of the continuous phase (i.e. the oil:co-surfactant ratio) were estimated using the titration procedure of Bowcott and Schulman (1955). The reaction rate was optimized by adjusting log P_{int} (by changing the co-surfactant (hexanol) concentration), and by adjusting log P_{cph} (by changing the type of oil) so that the (log P_{OIL} – log $P_{SUBSTRATE}$) was maximized and (log $P_{INTERPHASE}$ – log $P_{SUBSTRATE}$) was minimized (*Figure 17*). In this way the interphase concentration of substrate, hence the substrate concentration in the vicinity of the enzyme, was maximized. By the same reasoning, it should be possible to minimize product inhibition by manipulating (log $P_{INTERPHASE}$ – log $P_{PRODUCT}$) and (log P_{OIL} – log $P_{PRODUCT}$) to minimize the interphase product concentration.

There are several other examples where choice of microemulsion components strongly influences reaction rates in a fashion which can be rationalized using log P. For example, in the *C. viscosum* lipase study already mentioned (Fletcher *et al.* (1985a), a large effect on the kinetics was obtained by substituting *n*-heptane (log P = 4.0) for dodecane (log P = 6.5); the decreased $k_{cat}/K_{m\ app}^{aq}$ for the C_8 substrate is consistent with a higher oil:interphase partition coefficient for the substrate (*Figure 14*). Stamatis, Xenakis and Kolisis (1993) and Stamatis *et al.* (1993b) have shown that the initial rate of (–) menthol oleate synthesis by *Penicillium simplicissimum* lipase in ternary AOT microemulsions is quite strongly dependent on log P, with a maximum at log P = 4.5 (iso-octane).

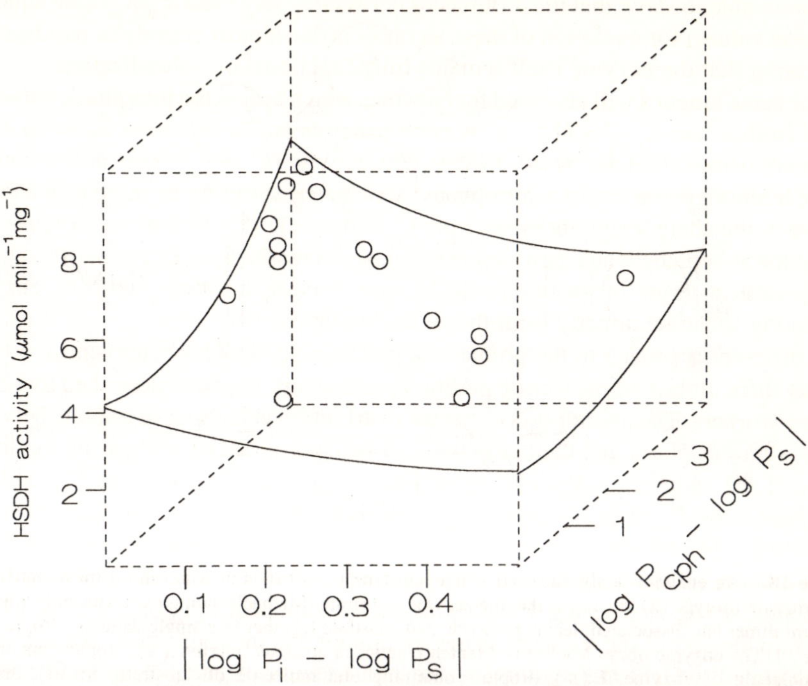

Figure 17. Dependence of the activity of 20β-hydroxysteroid dehydrogenase on log P of the substrate relative to that of the continuous subphase and interphase. The activity is maximal when the log P's are balanced such that partitioning into the interphase is favoured. (Reproduced from Hilhorst, R., Spruijt, R., Laane, C. and Veeger, C. (1984). *European Journal of Biochemistry* **144**, 459–466, with permission.)

As far as reaction rates are concerned, partitioning arguments become irrelevant for enzymes which operate under conditions of substrate saturation (k_{cat}-kinetics). However, this may not always be possible to achieve and, in any case, process design optimization must take into account factors such as substrate and/or product inhibition. In this context these studies have been instrumental in demonstrating the possibilities for optimization of microemulsion composition using a rational design approach.

Water-soluble substrates

The case of water-soluble substrates is rather different from those of oil-soluble or surface-active ones (Oldfield, 1990). In this case, under steady-state conditions, the enzyme may only receive substrate molecules by exchange, and therefore it is reasonable to expect the rate of formation of the Michaelis complex to be dependent on the droplet size and concentration and on the exchange rate, just as for other bimolecular reactions. In principle, a kinetic scheme equivalent to that shown in *Figure 9*, for a bimolecular reaction, needs to be set up. However, the solution is unwieldy, and it is useful to consider type (i) and type (ii) kinetics separately (p. 277). The scheme for type (ii) kinetics is shown in *Figure 18*. The solution for a system with droplets containing a maximum of one substrate molecule per droplet is

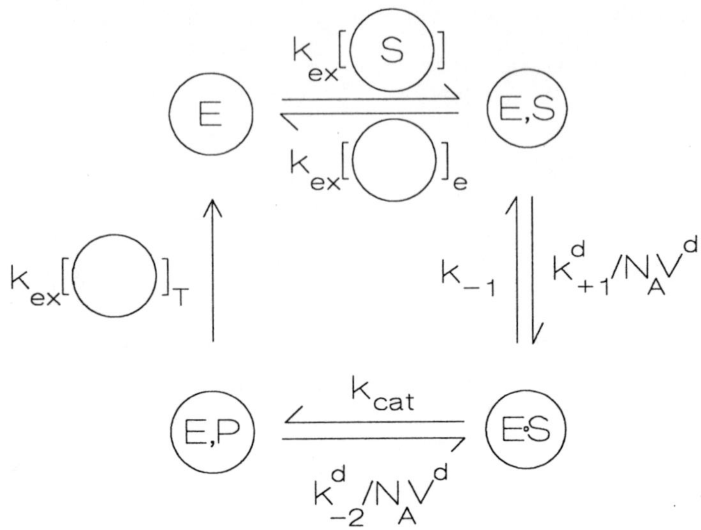

Figure 18. An enzyme-catalysed reaction obeying type (ii) kinetics in water-in-oil microemulsions. The criterion for type (ii) kinetics is that the enzyme–substrate (Michaelis complex) forms only after the transient dimer has dissociated, leaving enzyme and substrate together in a single daughter droplet (see *Figure 9*). The enzyme obeys Michaelis–Menten kinetics in aqueous solution. (E), droplet containing one molecule of enzyme, E; (S), droplet containing one molecule of substrate, S; (E,S), droplet containing one molecule each of E and S; (E•S), droplet containing a single Michaelis complex; (E•P), droplet containing one molecule each of E and product, P; ◯, empty droplet. (Reproduced from Oldfield, C., Otero, C., Rua, M.L. and Ballesteros, A. (1992). In *Biocatalysis in Non-Conventional Media* (J. Tramper, Ed.), pp. 189–198. Elsevier, Amsterdam, with permission.)

$$v3 = \frac{k_{\text{cat app}}[\text{E}]_{\text{T}}{}^{\text{m}}[\text{S}]^{\text{m}}}{K_{\text{m}}{}^{\text{m}} + [\text{S}]^{\text{m}}} \qquad \text{Eq. (15)}$$

(the superscript ‘m’ shows that the concentration units are mol per litre of microemulsion), where

$$k_{\text{cat app}} = k_{\text{cat}}{}^{\text{o}} - \frac{k_{-2}{}^{\text{d}} k_{\text{cat}}{}^{\text{o}}}{k_{-2} + k_{\text{ex}}\phi^{\text{aq}}} \qquad \text{Eq. (16)}$$

(the superscript ‘d’ indicates that the concentration units are mol per litre of microemulsion droplets and $k_{\text{cat}}{}^{\text{o}}$ is the maximum value of $k_{\text{cat app}}$, and

$$K_{\text{m}}{}^{\text{m}} = K_{\text{m}}{}^{\text{d}}\phi^{\text{aq}} + \frac{k_{\text{cat}}{}^{\text{o}}}{k_{\text{ex}}}\,; \qquad \text{Eq. (17)}$$

ϕ^{aq} replaces the product ($[\bigcirc]_{\text{T}}.N_{\text{A}}V_{\text{d}}$), as in equation (9), and

$$K_{\text{m}}{}^{\text{d}} = \frac{k_{-1} + k_{\text{cat}}{}^{\text{o}}}{k_{+1}{}^{\text{d}}} \qquad \text{Eq. (18)}$$

$K_{\text{m}}{}^{\text{d}}$ is equivalent to the Michaelis constant of aqueous solution kinetics, in the sense that it contains the same group of rate constants (equation 18). Equation (17)

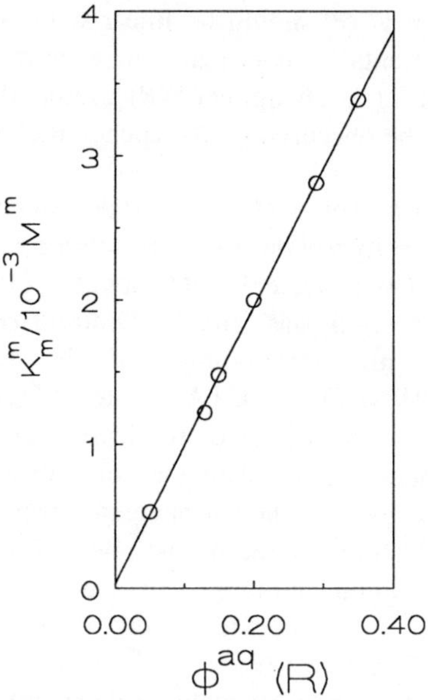

Figure 19. Dependence of K_m^m on ϕ^{aq} (R = 25) for the trypsin-catalysed hydrolysis of *N*-benzoyl DL-arginine *p*-nitroanilide in a CTAB microemulsion system. The system obeys the kinetics predicted by equation (17). For this system $k_{cat}.k_{ex}$ = 10^{-10} $M^{-1m}s^{-1}$ (Oldfield, 1990), and therefore there is direct proportionality between k_m and $\phi^{aq}(R)$ (Replotted from Martinek, K., Levashov, A.V., Klyachko, N.L., Pantin, V.I. and Berezin, I.V. (1981). *Biochimica et Biophysica Acta* **657**, 277–284, with permission.)

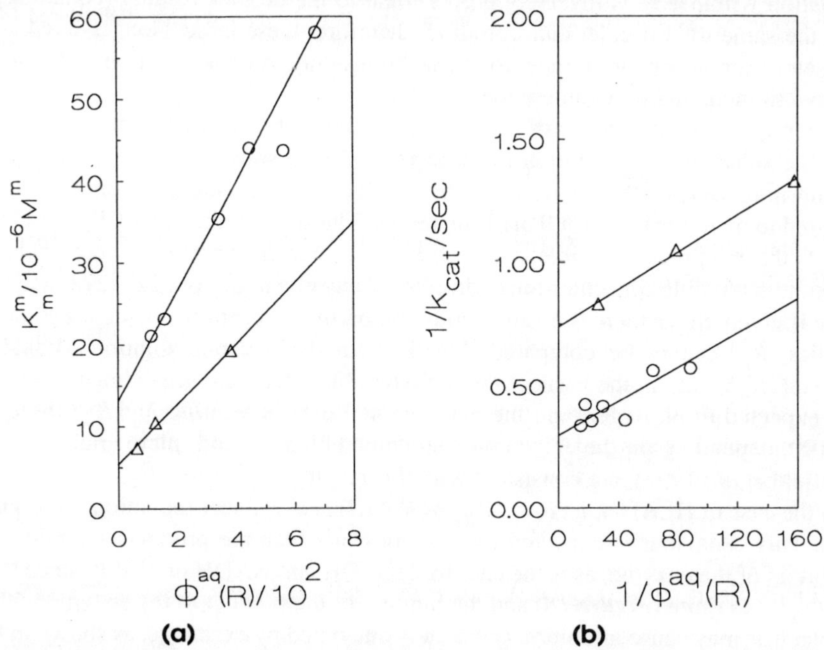

Figure 20. (a) Dependence of K_m^m on f^{aq} for the reduction of NAD$^+$ by horse-liver alcohol dehydrogenase (saturating concentrations of ethanol) in a CTAB microemulsion system; (b) dependence of $1/k_{cat}$ on $1/f^{aq}$ in the same system. (D) R = 12.5, (O) R = 7.0. (Replotted from Oldfield, C., Otero, C., Rua, M.L. and Ballesteros, A. (1992). In *Biocatalysis in Non-Conventional Media* (J. Tramper, Ed.), pp. 189–198. Elsevier, Amsterdam, with permission.)

shows that a plot of K_m^m v. ϕ^{aq} should be linear with positive slope and ordinate intercept k_{cat}/k_{ex}. Experimentally, it is better to alter ϕ^{aq} by changing the droplet concentration at constant R (i.e. changing $\phi^{aq}(R)$), rather than vice versa, because the linearity of the plot may be obscured by R-dependent effects on activity, as already discussed.

A linear dependence of K_m^m on $\phi^{aq}(R)$ was first described by Martinek *et al.* (1981), studying the kinetics of the hydrolysis of water-soluble substrates of trypsin in AOT microemulsions (*Figure 19*), and later by Fletcher *et al.* (1985b), for α-chymotrypsin in several different microemulsions, and by Oldfield *et al.* (1992), again for α-chymotrypsin in AOT microemulsions and also for the LADH-catalysed interconversion of NADH/NAD⁺ in a CTAB system (*Figure 20a*).

The dependence of K_m^m on $\phi^{aq}(R)$ (i.e. on the droplets concentration) arises because of the effect of compartmentalization of enzyme and substrate on the rate of formation of the Michaelis complex, E•S, via the intermediate species (E,S), in the same way that the reaction A + B → C is determined by the rate of formation of the intermediate species (A,B). If the droplet concentration is increased at constant substrate concentration (i.e. if the $[\bigcirc]_e$:[(S)] ratio is increased), the rate at which E,S decomposes back to separate E and S droplets (= $k_{ex}[\bigcirc]_e$[(E,S)]) increases, whilst the rate of formation (= k_{ex}[(E)]$_T$[(S)]) remains unchanged. Therefore the effect of increasing the droplet concentration is to shift the E + S → E•S pre-equilibrium to the left, which is unfavourable for product formation. Increasing the droplet volume (i.e. increasing R) at constant droplet concentration has a similar effect, since the rate of E•S formation within (E,S) is inversely proportional to the droplet volume (equation 8). For the same total droplet concentration, therefore, these larger, longer-lived (E,S) droplets are more likely to undergo exchange, leading to the separation of E and S, relative to their smaller counterparts.

The physical consequence of increasing ϕ^{aq}, therefore, is that to obtain the same rate, the substrate concentration must be increased to shift the E + S → E•S pre-equilibrium back to the right. Thus the K_m^m (the overall substrate concentration required to maintain [(E•S)] at 0.5[E]$_T$ increases. The slope of the plot of K_m^m v. $\phi^{aq}(R)$ is K_m^d, the maximum possible value of K_m^m, i.e. that in a ϕ^{aq} = 1 microemulsion (maximum possible concentration of droplets of any given size, pp. 277–278) and this value is given. Since there is no analogous compartmentalization in aqueous solution kinetics, K_m^d cannot be compared directly with the aqueous solution Michaelis constant, K_m^w, but, on the basis of the unfavourable effect of compentmentalisation, it is expected to be larger, and the experimental data, indicating K_m^d:K_m^w ratios of 40–100, depending on the enzyme–substrate combination and microemulsion type (Oldfield *et al.*, 1992), are consistent with this result.

In the case of HLADH, a plot of $1/k_{cat}$ v. $1/\phi^{aq}(R)$ is also linear (equation 16; *Figure 20b*). This behaviour is expected for reactions in which the products are potential inhibitors of the enzyme, as is the case for HLADH and NADH or NAD⁺, since then $(k_{-2}^d/N_A V_d)$ is finite (*Figure 18*) and the binding of the product to the enzyme within droplet (E,P) may cause inhibition (ultimately alleviated by exchange, as shown in the last step of *Figure 18*). Reactions such as the hydrolysis of non-specific substrates by proteases, on the other hand, give products which are poor inhibitors of the enzymes and $(k_{-2}^d/N_A V_d \approx 0)$. As an example, k_{cat} for the hydrolysis of *N*-benzoyl DL-arginine *p*-nitroanilide by trypsin is independent of $\phi^{aq}(R)$ (Martinek *et al.*, 1981).

No examples of enzymes obeying type (i) kinetics (the Michaelis complex forms within the transient dimer, p. 277) have yet been identified. This kinetic behaviour is characterized according to theory by a $K_m^{\ m}$ which is independent of the droplet concentration, but which is proportional to the droplet size (i.e. to R) but, as discussed already (pp. 289–294), multiple factors operate to determine enzyme activity as a function of R and this property may not be obvious.

Some authors have dismissed the possibility that exchange can influence the reaction rate, citing the fact that $k_{ex} >> k_{cat}$ (Bru, Sanchez-Ferrar and García-Carmona, 1989a; Ruckenstein and Karpe, 1990). Equation (17) indicates that this is the case, since k_{ex} appears only in the k_{cat}/k_{ex} term. However, this ignores the compartmentalization and its effects on the (pre-)equilibrium position for bimolecular reactions. The importance of compartmentalization was also recognized by Verhaert *et al.* (1990a), who derived a similar rate equation for a slightly different kinetic scheme. Although there are some problems with this derivation (Oldfield *et al.*, 1992), the important term '$K_m^{\ m}\phi^{aq}$' does appear in the final rate equation. However, its significance with regard to the published studies already cited, indicating a dependence of K_m on $\phi^{aq}(R)$, and the possibility of systematically varying ϕ^{aq} as a test for the validity of their kinetic scheme, was not recognized. In subsequent studies Verhaert and co-workers assumed that k_{cat} and $K_m^{\ m}$ are unchanged in the microemulsion environment (exactly the opposite conclusion to that of Oldfield, 1990) and the experimental data were fitted to the equations, primarily by manipulating the exchange rates (Tyrakowska *et al.*, 1990; Verhaert *et al.*, 1990b).

Enzyme inactivation

A number of studies have appeared which are concerned with the stability of enzymes in microemulsions relative to that in aqueous solution or in other non-aqueous solvent systems. Bearing in mind that the data set represents a fairly small number of enzymes, a number of general conclusions may be drawn, as follows:

1. Rates of loss of activity are dependent on the microemulsion type and its precise composition, and on the enzyme in question. Usually, the decay does not follow a simple kinetic (first- or second-order) pattern.
2. Loss of activity is associated with loss of active sites. This was demonstrated for LADH in an AOT microemulsion(Larsson, Oldfield and Freedman, 1989) by active-site titration and also seems to be the case for bilirubin oxidase in a CTAB microemulsion, where k_{cat} (for the oxidation of bilirubin) decreases, but $K_m^{\ m}$ remains constant (Skrika-Alexopoulis, Muir and Freedman, 1993).
3. Enzyme stability under operational conditions, or in the presence of inhibitors or other ligands, is usually greater than under storage (no substrate) conditions. Examples include bilirubin oxidase (more stable in the presence of bilirubin; (Skrika-Alexopoulis, Muir and Freedman, 1993), lactate dehydrogenase (LDH; more stable in the presence of NADH or pyruvate; (Khmelnitsky *et al.*, 1993), HLADH (more stable in the presence of AMP, NAD^+ or NADH; (Samama, Lee and Biellmann, 1987; Larsson, Oldfield and Freedman, 1989)

and tryptophanase (more stable in the presence of indole; (Eggers and Blanch, 1988).

4. Choice of microemulsion components determines stability. For example, HLADH is more stable in CTAB microemulsions than in SDS microemulsions, and more stable in CTAB microemulsions with pentan-1-ol or hexan-1-ol, rather than butan-1-ol, as co-surfactant (Samama, Lee and Biellmann, 1987).

Only a few cases have been reported in which inactivation kinetics follows a fixed, simple order with respect to time. Examples are the first-order inactivation of Δ^5-ketosteroid isomerase in an AOT microemulsion (Levashov *et al.*, 1992), of cholesterol oxidases from several microbial sources, in CTAB/butanol microemulsions (Lee and Biellmann, 1986) and of HLADH in SDS/butanol microemulsions (Samama, Lee and Biellmann, 1987). In the latter case the inactivation was shown to be caused by the hydrogen peroxide produced during the oxidation of cholesterol to cholestenone, and could be considerably improved by addition of catalase.

More usually, denaturation kinetics are more complex and cannot be assigned a simple kinetic order. By far the most common pattern of behaviour involves an (apparently) biphasic decay of activity. Examples include LADH in several different systems (Larsson, Adlercreutz and Mattiasson, 1987; Lee and Biellmann, 1987), α-chymotrypsin and *Ch. viscosum* lipase (Fletcher *et al.*, 1984, 1985a) and for bilirubin oxidase (Skrika-Alexopoulis, Muir and Freedman, 1993). In the case of bilirubin oxidase it was shown that the denaturation kinetics could be described in terms of a double exponential decay equation, consistent with commonly observed patterns of enzyme inactivation in aqueous solutions under controlled conditions. These kinetics are consistent with unimolecular inactivation mechanisms of the type N ↔ X↔ D, X ↔ N ↔ D or N ↔ D ↔ X, where N and D are the native (active) and denatured (inactive) forms, and X is an on- or off-pathway intermediate (Ikai and Tanford, 1971; Hijazi and Laidler, 1972).

Nothing is known about the molecular mechanism of inactivation. Comparative studies (Lee and Biellman, 1986; Steinmann, Jäcle and Luisi, 1986; Samama, Lee and Biellmann, 1987; Schmidli and Luisi, 1990) indicate that surfactant type (non-ionic, cationic or anionic) is a critical factor. However, it is not clear whether protein 'binding' to the interfacial surfactant layer, or the formation of discrete protein–surfactant complexes, or both, is responsible for the denaturation.

Microemulsions as media for biotransformations

OVERVIEW

At the present time, most studies of enzyme behaviour in microemulsions have focused on fundamental issues, such as the factors that determine activity and stability. In this section, the applications of microemulsions are described in more detail, beginning with a summary of the properties of microemulsions which may be useful in process design (pp. 303–304). There is not a large literature concerning the development of microemulsion-based systems for specific biotransformations, although this situation is changing. Much recent attention has focused on the use of microemulsions to facilitate lipase-mediated interconversions of lipids, and these are discussed on pp. 304–306. There are only few examples that demonstrate the potential of microemulsions as reaction media for other kinds of enzyme-catalysed reaction, and the more interesting of these are treated as casebook studies. This section concludes with a brief discussion of the application of Winsor II systems in protein purification.

GENERAL CONSIDERATIONS

From the foregoing, the unique properties of w/o microemulsions as reaction media for biotransformations can be summarized as follows:

1. compositional flexibility implies considerable potential for manipulating solvent properties to suit the reaction under consideration (a good example is the system for the production of tryptophan;
2. thermodynamic stability (spontaneous formation) implies none of the mechanical mixing problems normally associated with maintaining fine emulsion dispersions, and no scale-up problems;
3. very high surface area:volume ratios for the dispersed phase implies no phase-transfer problems, and also confers optical transparency, which permits facile monitoring of reaction progress using spectroscopic techniques;
4. manipulation of phase properties to facilitate product recovery.

Microemulsions provide an inherent flexibility in design by virtue of the fact that a large number of different oil and surfactant types, and even different polar materials, may be used. It has already been shown that microemulsion composition has a strong effect on activity and stability (pp. 288–301). In addition to those already discussed, some less obvious manipulations have turned up very interesting and useful results. For example, Yamada, Kuboi and Komasawa (1993) reported that the activity of *Ch. viscosum* lipase in AOT microemulsions can be increased by a factor of three by adding non-ionic surfactants of the Tween series, at AOT:Tween ratios of 16:1; Sebastiao, Cabral and Aires-Barros (1993) reported that the addition of hexanol (co-surfactant) to AOT microemulsions increases the stability (half-life) of a cutinase by a fctor of 45; and considerable improvements in the activity of lipoxygenase were obtained by optimizing the Nonidet P40:Triton X-35 ratio in a mixed microemulsion system (Piazza, 1992). Optimizations of this kind may well determine the economic viability of a process, and give a clear illustration of the importance of further basic research in understanding the behaviour of enzymes in these systems.

Methods for the recovery of products by taking advantage of phase-behaviour have already been outlined (pp. 263–270). The potential of this approach was clearly demonstrated by the work of Larsson, Adlercreutz and Mattiasson (1990), who showed that 4-methylcyclohexanol, produced by the horse-liver alcohol dehydrogenase (LADH)-catalysed reduction of 4-methylcyclohexanone could be recovered in the organic phase derived from AOT or by $C_{12}E_5$ w/o microemulsions by a temperature-induced Winsor IV $\rightarrow$ II shift. It was shown that multiple re-cycling of the aqueous phase, containing surfactant, enzyme and cofactors was possible with only marginal losses of enzyme activity at each step. The alternative use of ionic and non-ionic surfactants demonstrated the opportunity for recovery of the oil-phase by either increasing or decreasing the temperature. This method of product recovery certainly deserves further attention. As discussed further below, alternative methods of product recovery involve the use of membranes to stabilize the interface between w/o microemulsion and conjugate oil phases, with product recovery by partitioning.

Cofactor recycling is an important issue in determining the economics of a process. Methods for the provision of reducing power (NADH) have been reviewed by Hilhorst (1989). Four basic methods of supplying reducing equivalents are outlined in *Figure 21*. In *Figure 21a* reducing equivalents are supplied by coupling the reduction of one alcohol to the oxidation of another in an equilibrium catalysed by horse-liver alcohol dehydrogenase (Larsson, Adlercreutz and Mattiasoon, 1987). In *Figure 21b* the photosensitized oxidation of an oil-soluble electron donor in a CTAB system is linked to the methylviologen-mediated reduction of enoic acid reductase. The process is most efficient with a photosensitizer which is located in the interphase, so establishing a vectorial electron transport chain (Verhaert *et al.*, 1989) (For a brief review of photoprocesses in microemulsion systems, see Luisi *et al.*, 1988). In *Figure 21c* the oxidation of hydrogen is linked to the lipoamide dehydrogenase (LipOH)-mediated reduction of NAD^+, which is utilized (in this example) by 20β-hydroxysteroid dehydrogenase, again in a CTAB microemulsion (Hilhorst, Laane and Veeger, 1983). A methylviologen acts as a mediator in the reduction of LipOH. A total turnover number (TTN = mol product formed/mol NAD^+ consumed) of the order of 10^2 was reported, which compares well with the 10^2–10^6 range required for an economically feasible cofactor regenerating system (Chenault, Simon and Whitesides, 1988). In *Figure 21d* hydrogen is replaced by an electrode. This last approach may not be very efficient due to the low electrical conductivity of oil-continuous microemulsions.

LIPASE-CATALYSED REACTIONS

Lipases are model examples of enzymes for cell-free biotransformations since they are catalytically efficient, stable and inexpensive enzymes which catalyse reactions of recognized industrial application, including the conversion of low- to high-grade fats by transesterification. Furthermore, triglycerides and carboxylic acids are highly soluble in AOT microemulsions and phospholipids themselves stabilize microemulsions (p. 273). The massive interfacial area provided by the microemulsion environment is conducive to extremely efficient transformation rates.

Several publications deal with the application of AOT microemulsions as solvent systems for lipase-catalysed ester hydrolysis or synthesis. Han and Rhee (1985) have described an efficient procedure for the batchwise hydrolysis of olive oil using *Candida rugosa* lipase. Kim and Cheung (1989) have similarly described methods for the hydrolysis of palm oil by *Rhizopus arrhizus* lipase and Derksen and Cuperus (1992) described the application of several lipases in the hydrolysis of Crambe oil. Stamatis *et al.* (1993a) have described the *Penicillin simplicissimum* lipase-catalysed esterification of *n*-alkanols with carboxylic acids and Hedström *et al.* (1992) reported the *C. rugosa* lipase-catalysed synthesis of cholesterol oleate by direct condensation, in AOT–iso-octane–water microemulsions. Singh and Shah (1993) have made a

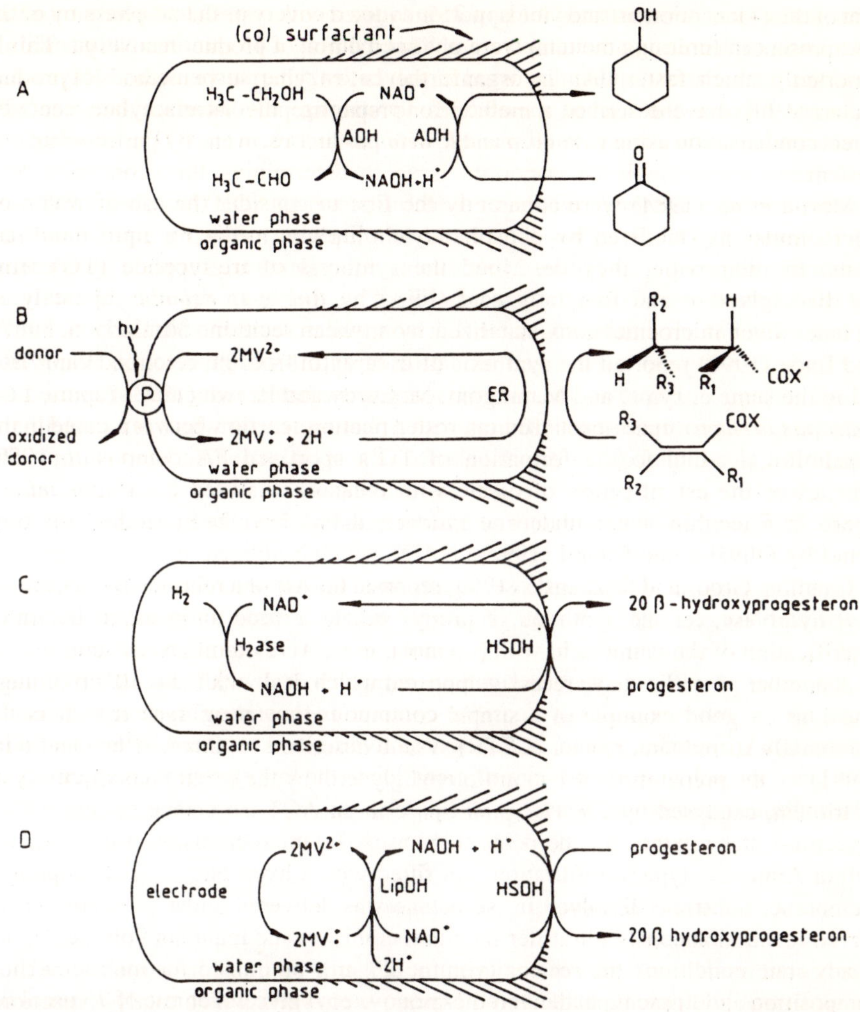

Figure 21. Enzyme-linked cofactor-regeneration in w/o-microemulsions. See text for details. (Reproduced from Hilhorst, R. (1989). In *Structure and Reactivity in Reverse Micelles* (M.P. Pileni, Ed.), pp. 323–340. Elsevier, Amsterdam, with permission.)

comparative study of the Lipozyme-catalysed esterification of glycerides in AOT-stabilized microemulsions and in lipid monolayers. Holmberg and co-workers have developed microemulsions based on AOT and other synthetic surfactants which support transesterification reactions catalysed by lipases (Holmberg and Osterberg, 1989; Osterberg, Blomstrom and Holmberg, 1989) and phospholipases A_2 (Na *et al.*, 1990).

There are also a few examples of the application of AOT-based microemulsions for the lipase-catalysed production of fine chemicals. For example Stamatis, Xenakis and Kolisis (1993) have studied the synthesis of (–)menthyl oleate from racemic menthol and medium- or long-chain carboxylic acids using *Penicillin simplissimum* lipase in AOT microemulsions. In an optimized system (iso-octane for the organic phase, R = 6), the esterification of the (–)enantiomer of menthol is made 6–8 times faster than that of the (+)enantiomer, and yields of 75% (quoted with respect to conversion of the FA, present in limiting amounts) were obtained during a 45 min incubation. This is reportedly much faster than in organic solvent/enzyme suspensions. Hayes and Gulari (1992) have described a method for preparing polyol-carboxylate esters by direct condensation using *C. rugosa* and *R. delemar* lipases, in an AOT microemulsion system.

Morita *et al.* (1984) were apparently the first to consider the use of water–oil microemulsions stabilized by phosphatidylcholines as media for lipid modifications. In their paper they described the synthesis of triglyceride (TG) from 1,2-diacylglycerol and free fatty acid (FFA) by *Rhizopus delemar* lipase in *n*-hexane–water microemulsions stabilized by soybean lecithin. Similarly Schmidli and Luisi (1990) reported the synthesis of tricaprylin from glycerol and caprylate, using the same enzyme, and Marangoni, McCurdy and Brown (1993) reported the *Rhizopus arrhizus* lipase-catalysed interesterification reaction between triolein and tripalmitin, leading to the formation of TG's of mixed FA composition. The kinetics of the esterification of oleate with ethanol, catalysed by *Mucor miehei* lipase in a lecithin–water–dodecane microemulsion, have been studied in some detail by Oliveira and Cabral (1992).

Gaathon, Groos and Rozhanski (1989) reported the use of a related enzyme, tannin acyl hydrolase, for the synthesis of propyl gallate, a food antioxidant, by trans-esterification of the tannic acid with propanol, in an AOT-stabilized system.

A number of studies have recently appeared which deal with the scale-up of these reactions. A good example of a simple continuous-flow membrane reactor is the continually stirred tank reactor (CSTR) system described by Chang, Rhee and Kim (1991) for the preparation of 1-mono-oleoyl glycerol by the continuous glycerolysis of triolein, catalysed by *Ch. viscosum* lipase in an AOT-iso-octane system with a glycerol–water mixture for the polar subphase. The microemulsion was contained within Amicon®-type ultrafiltration unit fitted with a hydrophobic (polysulphone) membrane. Substrate dissolved in iso-octane was delivered under pressure, which served to simultaneously ultrafilter the microemulsion and maintain volume. Under steady-state conditions the reactor (volume 20 ml), optimized for microemulsion composition and lipase type, delivered 1-mono-oleoyl glycerol at rate of 75 μmol h^{-1} for at least 6 days.

Prazeres *et al.* (1993a,b) have reported on the operational characteristics of a CSTR for the *Ch. viscosum* lipase-catalysed hydrolysis of triglycerides, operated either in batch mode or with circulation through a hollow-fibre system for the continuous removal of products (glycerol, free fatty acids). The hollow-fibre system is reported to be superior to the batch system due to the alleviation of product inhibition by free fatty acids. However, the microemulsion droplets may leak out of the hollow-fibre lumen into the permeate tube. In order to circumvent this generally recognized problem, Nakamura and Hakoda (1992) have proposed the use of a modified ultrafiltration in which an electrical potential is applied between the wall of the hollow fibre and a stainless-steel rod, passed through the centre of the lumen. The polarity of the electrical field is such that the droplets (which usually bear a net charge; Kim and Huang, 1986) are repelled from the wall.

TRYPTOPHAN PRODUCTION

Another example of a membrane reactor, which additionally illustrates the flexibility of design which microemulsions provide, was described by Eggers and Blanch (1988) for the synthesis of tryptophan from indole and serine, using tryptophanase (*Figure 22*). Here, the basic problem was to design a solvent system in which both indole and serine are soluble, and which also supports enzyme activity. The microemulsion selected consisted of cyclohexane, water, Brij 56 (surfactant) and Aliquat 336 (co-surfactant; commercial name for trioctylammonium bromide). The aqueous phase was buffered with potassium bis-tris-propane (BTP, pH9). This formulation was arrived at by choosing Aliquat 336 as an ion-pairing reagent for serine, so improving the solubility of the latter in the microemulsion. A non-ionic surfactant was selected on the basis that an anionic surfactant (or the counterion of a cationic surfactant) might ion-pair with the co-surfactant, leading to a complex phase-behaviour. The same considerations led to the choice of BTP as buffer; the cationic conjugate acid will not interact with the co-surfactant (the base is uncharged); an anionic buffer might ion-pair with the co-surfactant, leading to a change in pK_a, and hence pH, of unpredictable magnitude. The pH was selected to ensure stabilization of the Aliquat 336/serine ion-pair (the α-amino group, pK_a9.1, must be protonated in order to form an ion-pair), noting that the pH optimum of the enzyme lies in the range 7.2–8.8. The potassium salt of BTP was selected since this metal has a stabilizing effect on the enzyme. The oil-soluble product was recovered by partitioning into a conjugate oil phase separated from the microemulsion phase by a hydrophobic membrane. The system was reported to yield about 10 g tryptophan/g enzyme/hour, from a 1 l reactor, compared with a reported value of 1.3 $g/g/h^{-1}$ for a batch reactor utilizing an aqueous micellar Triton X-100 system as a solvent for the indole. The authors concluded with the comment that the system generated better yields of tryptophan than those obtained in a conventional two-phase system (using recombinant *E. coli* as the biocatalyst), although the high price of indole renders the system economically non-viable at the present time.

DEGRADATION OF TOXIC ORGANOPHOSPHATES

Recent work by Russell and co-workers (Komives, Lilley and Russell, 1994; Komives, Osborne and Russell, 1994; Russel and Komives, 1994) exemplifies the use of microemulsions in the area of bioremediation. These workers have described a method for continuous enzymatic detoxification, based on the use of a centrifugal reactor in conjunction with a water–*n*-hexane–Tween 85 Winsor II system, the microemulsion phase of which acts as an enzyme-containing liquid membrane. The example given describes the design for a system to remove toxic organophosphates (residual pesticides) from soil, using the enzyme organophosphorus hydrolase from *Pseudomonas diminuta*. The soil is decontaminated as it passes under centripetal force through the microemulsion phase, and is recovered from the aqueous-phase sludge. It might be noted that the chosen enzyme is extremely active, with reported k_{cat} values for dephosphorylation of para-oxon of 1100 s^{-1}. Tween 85 (sorbitan trioleate) was chosen because it is biodegradable.

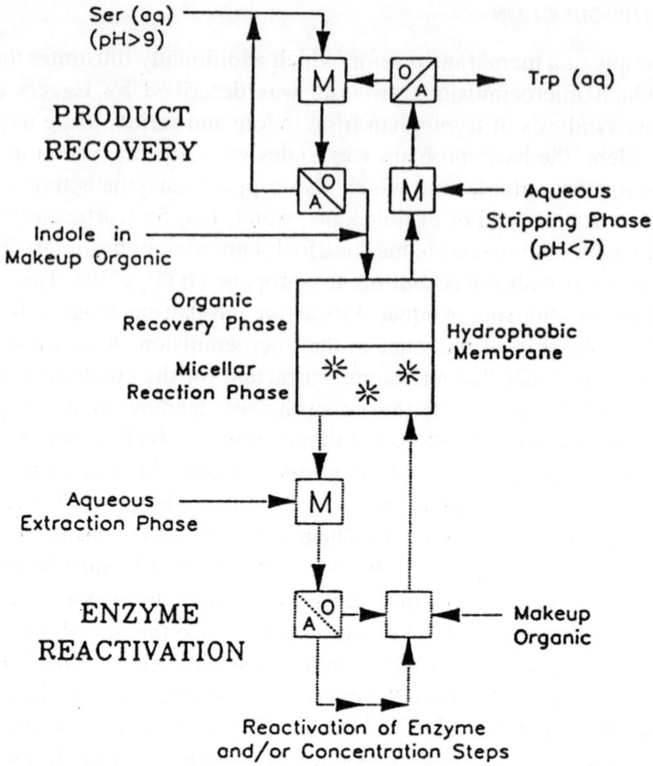

Figure 22. Reactor scheme for the production of trytophan from indole. (Reproduced from Eggers, D.K. and Blanch, H. (1988) *Bioprocess Engineering* **3**, 83–91, with permission.)

POLYMER SYNTHESIS

Rao *et al.* (1993) have described the use of w/o microemulsions as media for the synthesis of polyethylphenols from 4-ethylphenol using horseradish peroxidase. Synthesis in water–AOT–iso-octane microemulsions was found to be superior to that in iso-octane or water alone, or in iso-octane-water two-phase systems, since the microemulsion is a better solvent for both substrate (monomer) and enzyme. In practice, the polymer precipitates on reaching a critical size, which is determined by the droplet concentration. Furthermore, the polydispersity index of the material is relatively low. The microemulsion phase can be recycled (more monomer added) after filtration to recover the polymer, so suggesting the possibility of a very simple batch process. Once again, a key factor in the commercial viability of a microemulsion-based process is the high activity of the enzyme (k_{cat} = 600 s^{-1}).

PROTEIN PURIFICATION BY LIQUID–LIQUID EXTRACTION

The use of w/o microemulsions for the extraction of solutes from a conjugate aqueous phase has received considerable attention. This application follows the discovery that water-soluble solutes (salts, amino acids, proteins) partition between the microemulsion and conjugate aqueous phases of a Winsor II system, that the distribution coefficients differ for different compounds and, furthermore, that the distribution can be modulated by factors such as pH and electrolyte concentration. These observations find useful application in the purification of proteins (e.g. from culture-broths, cell-extracts) by selective extraction (Leodidis and Hatton, 1989).

At least for Winsor II systems prepared with ionic surfactants, a combination of electrostatic interactions, and the size of the protein, determine uptake. In the case of microemulsions stabilized by the cationic surfactant trioctyl ammonium chloride (TOMAC), for example, the optimal pH for phase transfer is given by the expression (Wolbert *et al.*, 1989),

$$pH_{OPT} - pI = 0.11 \times M_R - 0.97, \qquad \text{Eq. (19)}$$

where M_R is the molecular weight in kDa. This equation simply says that for good phase transfer, the pH must be about one unit above the iso-electric pH (so that the protein has a net negative charge) plus one more pH unit for each 10 kDa of molecular weight. Furthermore, there is a correlation between the charge distribution on the protein and the extent of partitioning (*Figure 23*) at the optimal pH; proteins with more asymmetric charge distributions partition more extensively into the microemulsion phase, presumably because those regions on the protein surface which bear a high degree of negative charge interact strongly with the surfactant monolayer at the microemulsion–water interface, the committing step in phase transfer.

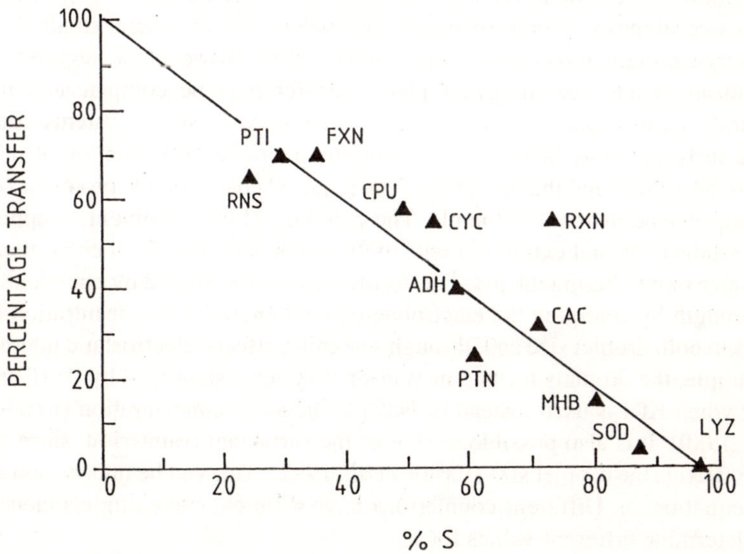

Figure 23. Dependence of degree of transfer to the microemulsion phase in a Winsor II system as a function of charge asymmetry for different proteins at their optimal extraction pH. Lower %S equates with greater charge asymmetry. Abbreviations: PTI, trypsin inhibitor; RNS, ribonuclease A; FXN, flavodoxin; CPU, parvalbumin; CYC, cytochrome *c*, ADH, liver alcohol dehydrogenase; PTN, trypsin, CAC, carbonic anhydrase; RXN, rubredoxin; MHB, haemoglobin; SOD, superoxide dismutase; LYZ, lysozyme. (Reproduced from Wolbert, R.B.G., Hilhorst, R., Voskuilen, G., Nachtegaal, H., Dekker, M., van't Riet, K. and Bijsterbosch, B.H. (1989). *European Journal of Biochemistry* **184**, 627–633, with permission.)

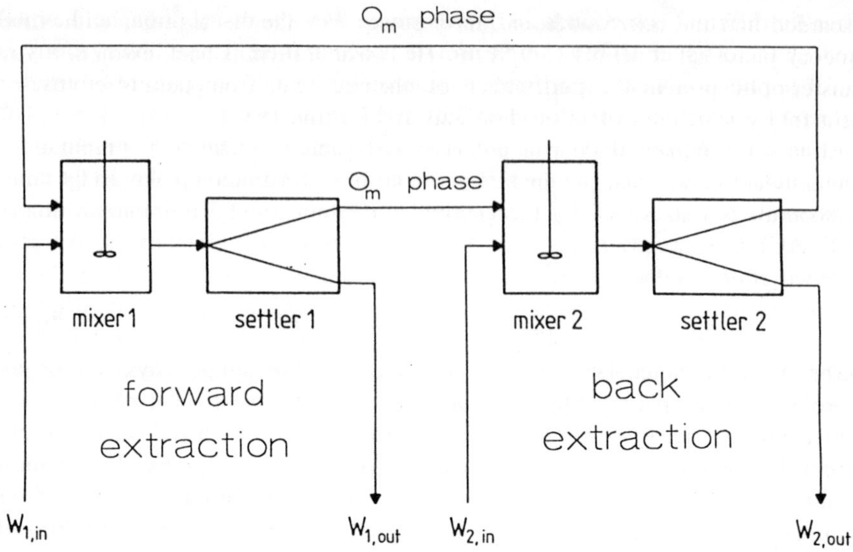

Figure 24. Reactor scheme for extraction and recovery of α-amylase. (Modified from Dekker, M., Van't Riet, K., Weijers, S.R., Baltussen, J.W.A., Laane, C. and Bijsterboch, B.H. (1986). *Chemical Engineering Journal* **33**, B27–B33, with permission.)

A fair qualitative understanding of the relation between protein charge and size and the distribution coefficient has developed (Leodidis and Hatton, 1989). The dependence on size suggests an entropy effect, the coalescence of several small droplets to house large protein molecules being unfavourable. However, a negative entropy contribution to the free-energy of phase transfer may be compensated for by a sufficiently large negative enthalpic term, arising (at least for systems containing charged surfactants) by favourable electrostatic interactions (formation of salt-links) between the protein and the interphase. The required charge on the protein is obtained by appropriate adjustment of the pH. The corollary of this argument – supported by several studies – is that extraction selectivity can be modulated either by altering the droplet size or by altering the pH. The droplet size can be altered independently of the ionic strength by changing the electrolyte type (changing the concentration leads to changes in both droplet size and, through screening effects, electrostatic interactions). For example, the droplets formed in Winsor II systems stabilized by AOT are much smaller when KCl is used instead of NaCl at the same concentration (Leodidis and Hatton, 1989). It is also possible to change the surfactant counterion, since this also strongly affects the droplet size (Eastoe *et al.*, 1992). This can be understood in terms of *Sp* (equation 2). Different counterions have different screening efficiencies and hence determine different values for a_0.

The application of Winsor II systems for large-scale protein extraction has been demonstrated for several specific proteins. *Figure 24* shows the design for a reactor system for the extraction of α-amylase from a commercially available crude enzyme preparation (Dekker *et al.*, 1986). Again, a TOMAC microemulsion system is used. The forward transfer takes place in the first CSTR (the aqueous solution contains 0.5 M Tris–HCl, pH 9.85). The equilibrated mixture then passes to the first settler unit where the microemulsion and aqueous phases separate out. The aqueous phase is discarded and the microemulsion phase passes to mixer 2, together with a fresh aqueous phase set at pH 6.9 (1.0 M tris–HCl) which permits back-extraction. After transfer of the protein to the aqueous phase the microemulsion phase is recycled. An eight-fold purification of α-amylase and 70% recovery were reported. Similarly, Barthomeuf, Régaret and Pourrat (1994) have reported a purification of tannase from fungal mycelia by partial enzymatic hydrolysis of the mycelium followed by liquid–liquid extraction using a CTAB microemulsion system. A 2.5-fold purification and 43% recovery were reported.

A variation on this theme involves disrupting cells by solubilizing them in microemulsions. After removal of cell debris e.g. by low-speed centrifugation, enzymes may be selectively back-extracted into an aqueous phase (Giovenco, Verheggen and Laane, 1987). Lee *et al.* (1989) have shown that squalene epoxide hydrolase can be extracted from rat-liver microsomes using this approach.

In a few cases, back-extraction procedures involving changing the electrolyte concentration, pH and temperature have proven to be ineffective. For example, cytochrome b_5 cannot be back-extracted from CTAB microemulsions (Pires and Cabral, 1993). A different approach to back extraction involves recovering the aqueous phase by treating the microemulsion with silica at pH 8 (Leser, Mrkoci and Luisi, 1993). This treatment afforded good recovery (yield and activity) of trypsin and α-chymotrypsin from an AOT microemulsion phase.

Apart from control of pH and electrolyte concentration, opportunities for modulation of extraction specificity include changing the surfactant type. Goto and Nakashio (1992) have reported on the extraction specificity of several different proteins with microemulsions prepared using AOT and several other surfactants. More sophisticated methods for controlling extraction specificity involve the use of affinity ligands. An elegant demonstration of this approach has been described by Paradkar and Dordick (1991, 1993), for glycoproteins such as horseradish and soybean peroxidase (pIs in the range 3.5–4.5), which do not extract into AOT microemulsions at pH>5 due to charge–repulsion. It was found that these enzymes could be extracted after complexation with concanavalin A. The lectin has a high pI (>8) and extracts readily at pH 5, as does the lectin/glycoprotein complex, presumably because the positive charge on the lectin moiety compensates for the unfavourable negative charge on the target protein. Back-extraction was achieved by equilibrating the microemulsion phase with fresh pH 5.0 buffer containing mannose (which displaces the glycoprotein ligand from the lectin). The free lectin was separated from the enzyme with a final extraction using fresh microemulsion phase (pH 5.0, 25°C), whereupon the lectin/mannose complex alone re-extracts into the microemulsion phase. Using this approach the authors reported a 30-fold purification and 30% recovery of soybean peroxidase from a commercial source of soybean hulls. In principle, the method can be extended to use antibodies for even more specific affinity-based microemulsion extraction procedures.

In an interesting elaboration on the liquid–liquid extraction theme, Hatton and co-workers have described the use of microemulsion extraction to renature proteins (Hagen, Hatton and Wang, 1990a,b). The process involves:

1. extraction of denatured proteins into the microemulsion from aqueous guanidine hydrochloride;

2. selective back-extraction of GuHCl, leaving the enzyme to refold in the microemulsion phase (reduced glutathione was addded to facitate disulphide bond formation during refolding); and
3. back-extraction to recover the enzyme from the microemulsion phase.

This procedure was successfully applied to ribonuclease, but was not successful in the case of γ-interferon, a more hydrophobic protein. It was suggested that this is because hydrophobic proteins interact much more strongly with the surfactant layer. Garza-Ramos *et al.* (1992) have described a similar approach in which GuHCl-denatured triosphosphate isomerase was renatured by injecting the protein solution into a CTAB microemulsion.

Special topics

MICROEMULSION-BASED GELS AND NANOGELS

Overview

Microemulsion-based gels (MBGs) is a generic term for systems which have the combined properties of gels (semisolid in nature; deformable with a high shear viscosity) and liquid microemulsions (usually transparent, optically isotropic, and thermodynamically stable). These gels are considered to have potential as a new generation of immobilized biocatalyst supports. Two important types are being studied at the present time: lecithin gels and gelatin gels (see below). Nanogels are prepared by polymerization of water-soluble monomers using the microemulsion as a polymerization medium and the gel may be removed from the microemulsion. Once again, nanogel-entrapped enzyme is considered to have potential as an immobilized biocatalyst. Additionally, there is some interest in the use of nanogel particles as drug-delivery systems.

Lecithin gels

First described by Scartazzini and Luisi (1988), lecithin microgels, rather than liquid microemulsions, form spontaneously in oils such as decane, hexane and cyclohexane, but not in benzene or diethylether, above a critical (water+lecithin) volume-fraction, as discussed on pp. 273–274. These gels, like droplets-type microemulsions, are optically transparent, and again the progress of chemical reactions can be followed spectroscopically. There seems to be only one report so far concerning lecithin gel-entrapped enzymes. In this example the *C. rugosa* lipase was shown to hydrolyse tricaprylin. No hydrolysis of the lecithin occurred (Scartazzini and Luisi, 1990).

Gelatin gels

Liquid microemulsions may be induced to gel by the addition of relatively large amounts of gelatin. Again the gels, prepared by mixing the microemulsion with gelatin at 55°C and cooling, are essentially transparent, isotropic and stable. Phase diagrams for the AOT–water–*n*-heptane–gelatin system are given in Atkinson *et al.*

(1991). Neutron-scattering studies suggest that gels consist of an interlaced network of surfactant-coated gelatin gel rods, co-existing with microemulsion droplets (*Figure 25*) (Atkinson *et al.*, 1989). *Chromobacterium viscosum* lipase immobilized within the gels is active, shows good stability properties, and has been shown to catalyse the condensation of a variety of alcohol/carboxylic acid combinations. The gels can be sectioned and packed in columns to make continuous-flow reactors for biotransformations (Jenta, Robinson and Thomson, 1991; Rees *et al.*, 1991). The properties and applications of gelatin-based microgels have recently been reviewed by Rees and Robinson (1993).

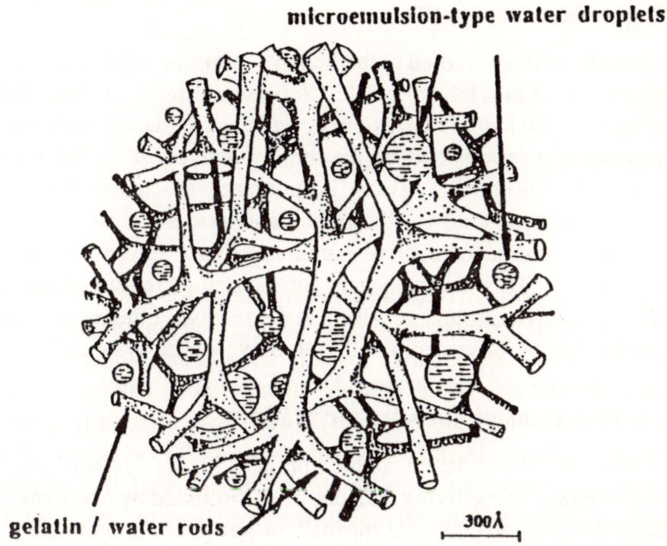

Figure 25. Proposed structure of the AOT–water–*n*-heptane–gelatine gel system. (Reproduced from Atkinson, P.J., Robinson, B.H., Howe, A.M. and Heenan, R.K. (1991). *Journal of the Chemical Society – Faraday Transactions* **87**, 3389–3397, with permission.)

Nanogels

A different kind of microemulsion gel involves the chemical or photo-polymerization of water-soluble monomers (e.g. acrylamide/bis-acrylamide, polyvinylpyrrolidone, methylcyanoacrylates), within the aqueous subphase. One can envision the creation of a range of gel structures, depending on the composition of the original microemulsion and on the rate of polymerization, from 'nanogels', generated by rapid polymerization in droplets-type systems, to extended sponge-like networks arising from slower polymerization in bicontinuous systems. The gels may be recovered from the parent microemulsion by precipitation, filtration or centrifugation.

The application sof these systems as vehicles for controlled drug-release have long been recognized (Speiser, 1984). The possibilities for gel-entrapment of enzymes were first described by Khmelnitsky *et al.* (1989b). Polymerization of acrylamide/bisacrylamide in AOT microemulsions also containing α-chymotrypsin covalently modified with a detergent, Pluronic F-108 (a co-polymer with the formula $HO(CH_2$-CH_2-$O)_{150}$-$(CH(CH_3)$-CH_2-O-$)_{55}$-$(CH_2$-CH_2-O)150-OH) generated nanodroplets of approximate radius 10–60 nm, which could be removed from the microemulsion by precipitation with acetonitrile. Covalent modification was used simply to increase the hydrodynamic volume of the enzyme in order to prevent it from diffusing out of the gel. There are also opportunities for covalent attachment of the enzyme in order to prevent it from diffusing out of the gel-bead. The enzyme was reported to retain only about 10% of its activity after modification/entrapment, but was stable for several weeks. The gels were reported to be soluble in benzene, with retention of enzyme activity as demonstrated by the esterification *N*-acetyltyrosine with ethanol (resembling, in this respect, the PEG-modified enzymes described by Inada *et al.*, 1986).

DETERGENTLESS MICROEMULSIONS

Several enzymes have been studied in the hexane–isopropanol–water system (*Figure 8*), including trypsin (Khmelnitsky *et al.*, 1986), cholesterol oxidase (Khmelnitsky, Hilhorst and Veeger, 1988) and mushroom polyphenol oxidase (Vulfson *et al.*, 1991). All three enzymes are reported to be maximally active within the microemulsion region (C; *Figure 8*). Cholesterol oxidase is apparently much less stable than trypsin. No stability data were reported for the polyphenoloxidase. *Caldariomyces fumago* chloroperoxidase is reported to be active in toluene/isopropanol/water microemulsions (Fedorak *et al.*, 1993). This model system for the demetallation of asphaltenes (the enzyme cleaves porphyrin groups, resulting in the release of the metal ion) is reported to be superior for this purpose compared with a two-phase system.

MICROEMULSIONS AS MEMBRANE-MIMETIC SYSTEMS

The application of microemulsions as membrane-mimetic systems, duplicating the lipid–aqueous interface of the living cell, has been discussed by Martinek *et al.* (1986, 1989). In the cell many enzymes are intrinsic or peripheral membrane proteins and their activity and stability characteristics may well be determined in greater or lesser degree by the nature of the membrane association and by the properties of the interfacial water. A number of recent papers have dealt with the properties of membrane-associated proteins in phosphatidylcholine-based microemulsions. Examples include enzymes of the mitochondrial respiratory electron transfer chain (Escamilla *et al.*, 1989; Escobar and Escamilla, 1992) and F_1-ATPase in the phospholipid–toluene–water system (Garza-Ramos *et al.*, 1989). These studies, which may be regarded as preliminary, indicate that the intact respiratory chain can be made to function in microemulsions.

Conclusion: trends and future prospects

The goal is to make enzymes powerful and commercially viable tools in the synthesis of useful compounds such as medicines, food additives and novel polymeric materials, and in essential processes such as decontamination (bioremediation, biodetoxification). Microemulsions have been shown to have considerable potential as solvents for many of these reactions, arising from their unique solubility characteristics for apolar, polar or charged, and surface-active compounds.

Microemulsions-based enzymology is still a relatively young field and the majority of studies have focused on the fundamentals of enzyme behaviour in these systems. Although there are still many questions to be answered, there are the beginnings of an understanding of the factors that influence reaction rates and equilibria and enzyme stability. Useful methods for product recovery, cofactor regeneration and recycling of essential components have been described. The applications detailed above show how this potential can be realized to great advantage in several different areas of bioengineering. Nearly all of these studies have been published in the past couple of years, and it is to be taken as an indication that the field has come of age that the transition from the basic to the applied is now being made.

Microemulsions-based cell-free technology, like all cell-free technologies, is nevertheless circumscribed at the present time by the relatively high cost of enzymes and cofactors, which often render them economically non-viable option compared with whole-cell biotransformations. It is reasonable to expect that, as costs come down and cell-free reactions become more competitive, more opportunities for exploiting microemulsions-based technology will appear.

The future should bring some exciting developments in the field of microemulsions prepared with biological surfactants and other novel, biodegradable surfactants. It is anticipated that microemulsions will find broader application, not only as media for the synthesis of fine chemicals (this application is heavily emphasized in the current literature), but also in bioremediation (as, for example, in the decontamination of organophosphorus-contaminated soils) and in the biopurification of petrochemicals (exemplified by use of a detergentless microemulsion as a model system for the demetallation of asphaltenes).

The further application of microemulsions for the controlled synthesis of polymers, both as added-values products and in the continued development of nanogel-encapsulated enzyme technology is anticipated. The concept of enzymes entrapped within extremely small acrylamide gel particles (good surface area:volume properties) which are active in variety of solvents and which can be stored in dry form with a long shelf-life, is one that should be actively pursued in the interests of improving the quality of off-the-shelf enzymes for use in biotransformations.

Of equal importance is the requirement for more basic research to underpin the applied effort. Good systematic characterization of the activity and stability of enzymes in different microemulsion systems would provide a useful database for applications purposes. In particular it is important to understand why certain enzymes become superactive in the microemulsion environment. Since improvements of the order described for HRP, tyrosinase and acid phosphatase can turn around the economics of a process. There is a case for the introduction and further characterization of new microemulsion systems, utilizing different surfactants (including a greater emphasis on the non-ionics); the literature bias towards AOT is understandable, but it is becoming clear that other systems have properties that make them more useful in applications. The possibility that from an holistic point of view, AOT is not the best-choice microemulsion surfactant is illustrated by the fact that only a few of the applications described above utilize this surfactant.

Acknowledgement

I wish to thank Ms Lucy Virden (Energy Biosystems Corporation) for her help in compiling the database.

References

ATIK, S.S. AND THOMAS, J.K. (1981). Transport of ions between water pools in alkanes. *Chemical Physics Letters* **79**, 351–353.

ATKINSON, P.J., GRIMSON, M.J., HEENAN, R.K., HOWE, M.H. AND ROBINSON, B.H. (1989). Structure of microemulsion-based organogels. *Journal of the Chemical Society – Chemical Communications* 1807–1811.

ATKINSON, P.J., ROBINSON, B.H., HOWE, A.M. AND HEENAN, R.K. (1991). Structure and stability of microemulsion based organogels. *Journal of the Chemical Society – Faraday Transactions* **87**, 3389–3397.

AUVRAY, L., COTTON, J.-P., OBER, R. AND TAUPIN, C. (1984). Evidence for zero mean-curvature microemulsions. *Journal of Physical Chemistry* **88**, 4586–4589.

AVEYARD, R., BINKS, B.P. AND MEAD, J. (1986). Interfacial tension minima in oil–water–surfactant systems. Effects of alkane chain-length and presence of *n*-alkanols in systems containing Aerosol-OT. *Journal of the Chemical Society – Faraday Transactions* **82**, 1755–1770.

AVEYARD, R., BINKS, B.P., CLARK, S. AND MEAD, J. (1986). Interfacial tension minima in oil–water–surfactant systems. Behaviour of alkane–aqueous NaCl systems containing Aerosol-OT. *Journal of the Chemical Society – Faraday Transactions* **82**, 125–142.

BARDEZ, E., GOGUILLON, B.-T., KEH, E. AND VALEUR, B. (1984). Dynamics of excited-state reactions in reversed micelles. 1. Proton transfer involving a hydrophilic fluorescent probe. *Journal of Physical Chemistry* **88**, 1909–1913.

BARTHOMEUF, C., RÉGARET, F. AND POURRAT, H. (1994). Improvement of tannase recovery using enzymatic disruption of mycelium in combination with reverse micellar enzyme extraction. *Biotechnology Techniques* **8**, 137–142.

BECHER, P. (1984). Hydrophile–lyophile balance: history and recent developments. *Journal of Dispersion Science and Technology* **5**, 81–96.

BELONOGOVA, O.V., LIKHTENSHTEIN, G.I., LEVASHOV, A.V., KHMELNITSKY, YU, L., KLYACHKO, N.L. AND MARTINEK, K. (1983). Use of the spin label method to study the state of the active site and microsurroundings of α-chymotrypsin, solubilized in octane, using surfactant Aerosol-OT. *Biokhimiya* **48**, 379–386. (English translation: *Soviet Journal of Biochemistry* **48**, 329–335.

BELYAEVA, E.I., BROVKO, YU., UGAROVA, M.N., KLYACHKO, N.L., LEVASHOV, A.V., MARTINEK, K. AND BEREZIN, I.V. (1983). Regulation of the catalytic activity of firefly luciferase in the system Brij 96–water–octane. *Doklady Akademii Nauk SSSR* **273**, 494–497. (English Translation: *Proceedings of the Academy of Sciences, USSR* **273**, 374–377.

BIAIS, J., BOTHOREL, P., CLIN, B. AND LALANNE, P. (1981). Theoretical behaviour of microemulsions: geometrical aspects, dilution properties and role of alcohol. Comparison with experimental results. *Journal of Dispersion Science and Technology* **2**, 67–95.

BOICELLI, C.A., GIOMINI, M. AND GIULIANI, A.M. (1984). Infrared characterization of different water types inside reverse micelles. *Applied Spectroscopy* **38**, 537–539.

BONNER, F.J., WOLF, W. AND LUISI, P.L. (1980). Micellar solubilization of biopolymers in hydrocarbon solvents. 1. A structural model for protein-containing reverse micelles. *Journal of Solid-Phase Biochemistry* **5**, 255–268.

BOWCOTT, J.E. AND SCHULMAN, J.H. (1955). Emulsions: control of droplet size and phase continuity in transparent oil–water dispersions stabilized with soap and alcohol. *Zeitschrift Electrochem.* **59**, 283–290.

BRIDGE, N.J. AND FLETCHER, P.D.I. (1983). Time-resolved studies of fluorescence quenching in a water-in-oil microemulsion. *Journal of the Chemical Society – Faraday Transactions* **79**, 2161–2169.

BROUWER, W.M., NIEUWENHUIS, E.A. AND KOPS-WERKHOVEN, M.M. (1980). Particle interactions in microemulsion systems. *Journal of Colloid and Interface Science* **92**, 57–65.

BROWN, E.D., YADA, R.Y. AND MARANGONI, A.G. (1993). The dependence of the lipolytic activity of *Rhizopus arrhizus* lipase on surfactant concentration in Aerosol-OT/isooctane reversed micelles and its relationship to enzyme structure. *Biochimica et Biophysica Acta* **1161**, 66–72.

BRU, R. AND WALDE, P. (1991). Product inhibition by α-chymotrypsin in reverse micelles. *European Journal of Biochemistry* **199**, 95–103.

BRU, R., SANCHEZ-FERRAR, A. AND GARCIA-CARMONA, F. (1989a). A theoretical study on the expression of enzymic activity in reverse micelles. *Biochemical Journal* **259**, 255–361.

BRU, R., SANCHEZ-FERRAR, A. AND GARCIA-CARMONA, F. (1989b). Characteristics of tyrosinase in reverse micelles. *Biotechnology and Bioengineering* **34**, 304–308.

BRU, R., SANCHEZ-FERRAR, A. AND GARCIA-CARMONA, F. (1990). The effect of substrate partitioning on the kinetics of enzymes acting in reverse micelles. *Biochemical Journal* **268**, 679–684.

CAPITANI, D., SEGRE, A.L., SPARAPANI, R., GIUSTINI, M., SCARTAZZINI, R. AND**4** LUISI, P.L. (1991). Lecithin microemulsion gels: an NMR study of molecular mobility based on line widths. *Langmuir* **7**, 250–253.

CAZABAT, A.M. AND LANGEVIN, D. (1981). Diffusion of interacting particles: light-scattering study of microemulsions. *Journal of Chemical Physics* **74**, 3148–3153.

CHANG, P.S., RHEE, J.S. AND KIM, J.-J. (1991). Continuous glycerolysis of olive oil by *Chromobacterium viscosum* lipase immobilized on liposome in reversed micelles. *Biotechnology and Bioengineering* **38**, 1159–1165.

CHEN, J.-P. AND CHANG, K.-C. (1993). Lipase-catalysed hydrolysis of milk-fat in lecithin reverse micelles. *Journal of Fermentation Bioengineering* **76**, 98–104.

CHEN, S.J., EVANS, D.F., NINHAM, B.W., MITCHELL, D.J., BLUM, F.D. AND PICKUP, S. (1986). Curvature as a determinant of microstructure and microemulsions. *Journal of Physical Chemistry* **90**, 842–847.

CHENAULT, H.K., SIMON, E.S. AND WHITESIDES, G.M. (1988). Cofactor regeneration for enzyme-catalysed synthesis. *Biotechnology and Genetic Engineering Reviews* **6**, 221–270.

DANIELSSON, I. AND LINDMAN, B. (1981). The definition of a microemulsion. *Colloids and Surfaces* **3**, 391–392.

DAS, S. AND MAITRA, A. (1989). Temperature-dependent catalytic behaviour of peroxidase entrapped in reverse micelles. *Colloids and Surfaces* **35**, 101–104.

DEKKER, M., VAN'T RIET, K., WEIJERS, S.R., BALTUSSEN, J.W.A., LAANE, C. AND BIJSTERBOCH, B.H. (1986). Enzyme recovery by liquid–liquid extraction using reversed micelles. *Chemical Engineering Journal* **33**, B27–B33.

DERKSEN, J.T.P. AND CUPERUS, F.P. (1992). Lipase-catalysed hydrolysis of Crambe oil in AOT-iso-octane reverse micelles. *Biotechnology Letters* **14**, 937–940.

DORDICK, J.S. (1991). Principles and applications of non-aqueous enzymology. In *Applied Biocatalysis* (H.W. Blanch and D.S. Clark, Eds), pp. 1–52. Marcel Dekker, New York.

DOROVSKA-TARAN, V., VEEGER, C. AND VISSER, A.J.W.G. (1993a). Reverse micelles as a water-property control system to investigate the hydration/activity relationship of α-chymotrypsin. *European Journal of Biochemistry* **218**, 1013–1019.

DOROVSKA-TARAN, V., VEEGER, C. AND VISSER, A.J.W.G. (1993b). Comparison of the dynamic structure of α-chymotrypsin in aqueous solution and in reversed micelles by fluorescent active site probing. *European Journal of Biochemistry* **211**, 47–55.

DOUZOU, P., KEH, E. AND BALNY, C. (1979). Cryoenzymology in aqueous media: micellar solubilized water-clusters. *Proceedings of the National Academy of Sciences, USA* **76**, 681–684.

EASTOE, J., FRAGNETO, G., ROBINSON, B.H., TOWEY, T.F., HEENAN, K. AND LANG, F.J. (1992). Variation of surfactant counterion and its effect on the structure and properties of Aerosol-OT-based water- in-oil microemulsions. *Journal of the Chemical Society – Faraday Transactions* **88**, 461–471.

EGGERS, D.K. AND BLANCH, H. (1988). Enzymatic production of L-tryptophan in a reverse micelle reactor. *Bioprocess Engineering* **3**, 83–91.

EICKE, H.-F., SHEPHERD, J.C.W. AND STEINEMANN, A. (1976). Exchange of solubilized water and aqueous electrolyte between micelles in apolar media. *Journal of Colloid and Interface Science* **56**, 168–176.

EICKE, H.-F. AND KUBIK, R. (1980). The optical matching phenomenon in water-in-oil microemulsions. *Berichte der Bunsen Gesellschaft für Physikalische Chemie* **84**, 36–41.

EKWALL, P., MANDELL, L. AND FONTELL, K. (1970). Some observations on binary and ternary Aerosol OT systems. *Journal of Colloid and Interface Science* **33**, 215–235.

EL SEOUD, O.A. AND CHINELATTO, A.M. (1983). Acid–base indicator in Aerosol-OT reversed micelles in Heptane. The use of buffers. *Journal of Colloid and Interface Science* **95**, 163–171.

EL SEOUD, O.A., CHINELATTO, A.M. AND SHIMIZU, M.R. (1982). Acid–base equilibria in the presence of aerosol-OT aggregates. Ion-exchange in reverse micelles. *Journal of Colloid and Interface Science* **88**, 420–427.

ERJOMIN, A.N. AND METELITZA, D.I. (1983). Catalysis by haemoproteins and their structural organization in reversed micelles of surfactants in octane. *Biochimica et Biophysica Acta* **732**, 377–386.

ESCAMILLA, E., AYALA, G., TUENA DE GOMEZ-PUYOU, M., GOMEZ-PUYOU, A., MILLAN, L. AND DARSZON, A. (1989). Catalytic activity of cytochrome oxidase and cytochrome *c* in apolar solvents containing phospholipids and low amounts of water. *Archives of Biochemistry and Biophysics* **272**, 332–343.

ESCOBAR, L. AND ESCAMILLA, E. (1992). Respiratory electron transfer activity in an asolectin-isooctane reverse-micellar system. *Biochemie* **74**, 161–169.

EVANS, D.F., MITCHELL, D.J. AND NINHAM, B.W. (1986). Oil, water and surfactant: Properties and conjectured structure of simple microemulsions. *Journal of Physical Chemistry* **90**, 2817–2825.

FEDORAK, P.M., SEMPLE, K.M., VAZQUEZ-DUHALT, R. AND WESTLAKE, D.W.S. (1993). Chloroperoxidase mediated modifications of asphaltenes. *Enzyme and Microbial Technology* **15**, 429–437.

FLETCHER, P.D.I. AND HORSUP, D.I. (1992). Droplet dynamics in water- in-oil microemulsions and macroemulsions stabilized by non-ionic surfactants. *Journal of the Chemical Society – Faraday Transactions* **88**, 855–864.

FLETCHER, P.D.I. AND PARROT, D. (1989). Protein-partitioning between water-in-oil microemulsions and conjugate aqueous phases. In *Structure and Reactivity in Reverse Micelles* (M.P. Pileni, Ed.), pp. 303–319. Elsevier, Amsterdam.

FLETCHER, P.D.I. AND ROBINSON, B.H. (1981). Dynamic processes in water-in-oil microemulsions. *Berichte der Bunsen Gesellschaft für Physikalische Chemie* **85**, 863–867.

FLETCHER, P.D.I., GALAL, M.F. AND ROBINSON, B.H. (1984). Structural study of Aerosol-OT stabilized microemulsions of glycerol dissolved in *n*-heptane. *Journal of the Chemical Society – Faraday Transactions* **80**, 3307–3314.

FLETCHER, P.D.I., GALAL, M.F. AND ROBINSON, B.H. (1985). Structural study of microemulsions of glycerol stabilized by cetyltrimethylammonium bromide dispersed in heptane+chloroform mixtures. *Journal of the Chemical Society – Faraday Transactions* **81**, 2053–2065.

FLETCHER, P.D.I., HOWE, A.M. AND ROBINSON, B.H. (1987). The kinetics of solubilizate exchange between water droplets of a water-in-oil microemulsion. *Journal of the Chemical Society – Faraday Transactions* **83**, 985–1005.

FLETCHER, P.D.I., ROBINSON, B.H. AND TABONY, J. (1986). A quasi-elastic neutron scattering study of water-in-oil microemulsions stabilized by Aerosol-OT. *Journal of the Chemical Society – Faraday Transactions* **82**, 2311–2321.

FLETCHER, P.D.I., FREEDMAN, R.B., MEAD, J., OLDFIELD, C. AND ROBINSON, B.H. (1984). Reactivity of α-chymotrypsin in water-in-oil microemulsions. *Colloids and Surfaces* **10**, 193–203.

FLETCHER, P.D.I., ROBINSON, B.H., FREEDMAN, R.B. AND OLDFIELD, C. (1985a). Activity of lipase in water-in-oil microemulsions. *Journal of the Chemical Society – Faraday Transactions* **81**, 2667–2679.

FLETCHER, P.D.I., REES, G.D., ROBINSON, B.H. AND FREEDMAN, R.B. (1985b). Kinetic properties of α-chymotrypsin in water-in-oil microemulsions: Studies with a variety of substrates and microemulsions. *Biochimica et Biophysica Acta* **832**, 209–214.

GAATHON, A., GROOS, Z. AND ROZHANSKI, M. (1989). Propylgallate: enzymatic synthesis in a reverse micelle system. *Enzyme and Microbial Technology* **11**, 604–608.

GARZA-RAMOS, G., DARSZON, A., TUENA DE GOMEZ-PUYOU, M. AND GOMEZ-PUYOU, A. (1989). Catalysis and thermostability of Mitochondrial F_1-ATPase in Toluene–phospholipid–low-water systems. *Biochemistry* **28**, 3177–3182.

GARZA-RAMOS, G., TUENA DE GOMEZ-PUYOU, M., GOMEZ-PUYOU, A. AND GRACY, R.W.

(1992). Dimerization and reactivation of triosephosphate isomerase in reverse micelles. *European Journal of Biochemistry* **205**, 389–393.

GHOSH, O. AND MILLER, C.A. (1987). Liquid-crystalline and microemulsion phase-behavior in alcohol-free Aerosol-OT/oil/brine systems. *Journal of Physical Chemistry* **91**, 4528–4535.

GIERASCH, L.M., THOMPSON, K.F., LACY, J.E. AND ROCKWELL, A.L. (1984). Exploring peptide interactions with interfacial water using reversed micelles. In *Reverse Micelles* (P.L. Luisi and B.E. Straub, Eds), pp. 265–277. Plenum Press, New York.

GIOVENCO, S., VERHEGGEN, F. AND LAANE, C. (1987). Purification of intracellular enzymes from whole bacterial cells using reversed micelles. *Enzyme and Microbial Technology* **9**, 470–473.

GONNELLI, M. AND STRAMBINI, G.B. (1988). Protein dynamical structure by Tryptophan phosphorescence and enzymatic activity in reverse micelles: 2. Alkaline phosphatase. *Journal of Physical Chemistry* **92**, 2854–2857.

GOOD, N.E., WINGET, G.D., WINTER, W., CONNOLLY, T.N., IZAWA, S. AND SINGH, R.M.M. (1966). Hydrogen-ion buffers for biological research. *Biochemistry* **5**, 467–477.

GOTO, M. AND NAKASHIO, F. (1992). Separation of proteins by new Reversed micelles. In *Biochemical Engineering for 2001, Proceedings of the Asia-Pacific Biochemical Engineering Conference, 1992* (S. Furusaki, I. Endo and R. Matsuno, Eds), pp. 540–543. Springer-Verlag, Heidelberg.

GUPTA, M.A. (1992). Enzyme function in organic solvents. *European Journal of Biochemistry* **203**, 25–32.

HAERING, G., LUISI, P.L. AND HAUSER, H. (1988). Characterisation by electron-spin resonance of reversed micelles consisting of the ternary system AOT–iso-octane–water. *Journal of Physical Chemistry* **92**, 3574–3581.

HAGAN, A.J., HATTON, A. AND WANG, D.I.C. (1990a). Protein-refolding in reverse micelles. *Biotechnology and Bioengineering* **35**, 955–965.

HAGAN, A.J., HATTON, A. AND WANG, D.I.C. (1990b). Protein refolding in reverse micelles: interaction of the protein with micelle components. *Biotechnology and Bioengineering* **35**, 966–975.

HAN, D. AND RHEE, J.S. (1985). Batchwise hydrolysis of olive oil by lipase in AOT–isoctane reverse micelles. *Biotechnology Letters* **7**, 651–656.

HAN, D. AND RHEE, J.S. (1986). Characteristics of lipase-catalysed hydrolysis of olive oil in AOT-iso-octane reversed micelles. *Biotechnology and Bioengineering* **28**, 1250–1255.

HAN, D., WALDE, P. AND LUISI, P.L. (1990). Dependence of lipase activity on water content and enzyme concentration in reverse micelles. *Biocatalysis* **4**, 153–161.

HAUSER, H., HAERING, G., PANDE, A. AND LUISI, P.L. (1989). Interaction of water with sodium bis(2-ethyl-1-hexyl) sulfosuccinate in reversed micelles. *Journal of Physical Chemistry* **93**, 7869–7876.

HAYES, D.G. AND GULARI, E. (1992). Formation of polyol-fatty acid esters by lipases in reversed micellar media. *Biotechnology and Bioengineering* **40**, 110–118.

HESTRÖM, G., SLOTTE, J.P., BACKLUND, M., MOLANDER, O. AND ROSENHOLM, J.B. (1992). Lipase-catalyzed synthesis and hydrolysis of cholesterol oleate in AOT/isoctane microemulsions. *Biocatalysis* **6**, 281–290.

HELENIUS, A. AND SIMONS, K. (1975). Solubilization of membranes by detergents. *Biochimica et Biophysica Acta* **415**, 29–79.

HIGUCHI, W.I. AND MISRA, J. (1962). Solubilization in nonpolar solvents. Influence of the chain-length of solvent on the solubilization of water by dioctylsulphosuccinate. *Journal of Pharmaceutical Sciences* **51**, 455–458.

HIJAZI, N.H. AND LAIDLER, K.J. (1972). Kinetic equations for consecutive reversible reactions with special reference to protein denaturation. *Journal of the Chemical Society – Faraday Transactions I* **68**, 1235–1242.

HILHORST, R. (1989). Applications of enzyme-containing reversed micelles. In *Structure and Reactivity in Reverse Micelles* (M.P. Pileni, Ed.),. pp. 323–340. Elsevier, Amsterdam.

HILHORST, R., LAANE, C. AND VEEGER, C. (1983). Enzymatic conversion of apolar compounds in organic media using an NADH regenerating system and dihydrogen as reductant. *FEBS Letters* **159**, 225–228.

HILHORST, R., SPRUIJT, R., LAANE, C. AND VEEGER, C. (1984). Rules for the regulation of enzyme activity in reserved micelles as illustrated by the conversion of apolar steroids by 20β-hydroxysteroid dehydrogenase. *European Journal of Biochemistry* **144**, 459–466.

HOAR, T.P. AND SCHULMAN, J.H. (1943). Transparent water-in-oil dispersions: the oleopathic hydro-micelle. *Nature* **152**, 102–103.

HOLMBERG, K. AND OSTERBERG, E. (1989). Process for the transesterification of triglycerides in an aqueous microemulsion reaction medium in the presence of lipase enzyme. United States Patent no. 4,838,287.

HUANG, J.S. AND KOTLARCHYK, M. (1986). Study of interfacial curvature in a three-component microemulsion with equal volumes of water and oil. *Physical Review Letters* **57**, 2587–2589.

HUANG, J.S., SUNG, J. AND WU, X.-L. (1989). The effect of H_2O and D_2O on a water-in-oil microemulsion. *Journal of Colloid and Interface Science* **132**, 34–42.

HUANG, J.S., SAFRAN, S.A., KIM, M.W., GREST, G.S., KOTLARCHYK, M. AND QUIRKE, N. (1984). Attractive interactions in micelles and microemulsions. *Physical Review Letters* **53**, 592–595.

IKAI, A. AND TANFORD, C. (1971). Kinetic evidence for incorrectly folded intermediate states in the refolding of denatured proteins. *Nature* **230**, 100–103.

INADA, Y., TAKAHASHI, K., YOSHIMOTO, T., AJIMA, A., MATSUSHIMA, A. AND SAITO, Y. (1986). Application of polyethylene glycol-modified enzymes in biotechnological processes: organic solvent-soluble enzymes. *Tibtech*. July, 190–194.

JENTA, T.R.-J., ROBINSON, B.H. AND THOMSON, A.R. (1991). Enzyme kinetic studies using lipase immobilized in microemulsion-based organo-gels. *Progress in Colloid and Polymer Science* **84**, 334–337.

JONES, J.B. (1985). An illustrative example of a synthetically useful enzyme: horse liver alcohol dehydrogenase. In *Enzymes in Organic Synthesis, CIBA Foundation Symposium 111*, pp. 3–21. Pitman, London.

KABAKOV, V.E., MERKER, S., KLYACHKO, N.L., MARTINEK, K. AND LEVASHOV, A.V. (1992). Regulation of the supramolecular structure and the catalytic activity of penicillin acylase from *Escherichia coli* in the system of reversed micelles of AOT in octane. *FEBS Letters* **311**, 209–212.

KABANOV, A.V., LEVASHOV, A.V., KLYACHKO, N.L., NAMYOTKIN, S.N., PSHEZHETSKY, A.V. AND MARTINEK, K. (1988). Enzymes entrapped in reversed micelles of surfactants in organic solvents: a theoretical treatment of catalytic activity regulation. *Journal of Theoretical Biology* **133**, 327–343.

KABANOV, A.V., NAMETKIN, S.N., EVTUSHENKO, G.N., CHERNOV, N.N., KLYACHKO, N.L., LEVASHOV, A.V. AND MARTINEK, K. (1989). A new strategy for the study of oligomeric enzymes: γ-glutamyl transferase in reversed micelles of surfactants in organic solvents. *Biochimica et Biophysica Acta* **996**, 147–152.

KARPE, P. AND RUCKENSTEIN, E. (1989). Effect of hydration on the degree of counterion binding and pH distribution in reverse micelles with aqueous core. *Journal of Colloid and Interface Science* **137**, 408–424.

KHMELNITSKY, YU. L., HILLHORST, R. AND VEEGER, C. (1988). Detergentless microemulsions as media for enzymatic reactions. Cholesterol oxidation catalysed by cholesterol oxidase. *European Journal of Biochemistry* **176**, 265–271.

KHMELNITSKY, YU. L., ZHARINOIVA, I.N., BEREZIN, I.V., LEVASHOV, A.V. AND MARTINEK, K. (1986). Detergent-free microemulsions: A new microhterogeneous medium for enzymatic reactions. *Doklady Akademii Nauk SSSR* **289**, 1178–1181. (English translation: *Proceedings of the Academy of Sciences USSR, Physical Chemistry Section* **289**, 730–733 (1987).)

KHMELNITSKY, YU. L., NEVEROVA, I.N., MOMTCHEVA, R., YAROPOLOV, A.I., BELOVA, A.B., LEVASHOV, A.V. AND MARTINEK, K. (1989a). Surface-modified polmeric nanogranules containing entrapped enzymes: a novel catalyst for use in organic media. *Biotechnology Techniques* **3**, 275–280.

KHMELNITSKY, YU. L., VAN HOEK, A., VEEGER, C. AND VISSER, A.J.W.G. (1989b). Detergentless microemulsions as media for enzymatic reactions. Spectroscopic and ultracentrifugation studies. *Journal of Physical Chemistry* **93**, 872–878.

KHMELNITSKY, YU. L., NEVEROVA, I.M., POLYAKOV, V.I., GRINBERG, V.YA., LEVASHOV, A.V. AND MARTINEK, K. (1990). Kinetic theory of enzymatic reactions in reversed micellar systems: application of the pseudophase approach for partitioning substrates. *European Journal of Biochemistry* **190**, 155–159.

KHMELNITSKY, YU. L., HILHORST, R., VISSER, A.J.W.G. AND VEEGER, C. (1993). Enzyme inactivation and protection during entrapment in reversed micelles. *European Journal of Biochemistry* **211**, 73–77.

KIM, M.W. AND HUANG, J.S. (1986). Percolation-like phenomena in oil-continuous microemulsions. *Physical Review A* **34**, 719–722.

KIM, T. AND CHEUNG, K. (1989). Some characteristics of palm kernel olein hydrolysis by *Rhizopus arrhizus* lipase in reversed micelles of AOT is iso-octane, and additive effects. *Enzyme and Microbial Technology* **11**, 528–532.

KLIBANOV, A.M. (1989). Enzymatic catalysis in anhydrous organic solvents. *Trends in Biochemical Sciences* **14**, 141–144.

KLYACHKO, N.L., LEVASHOV, A.V. AND MARTINEK, K. (1984). Catalysis by enzymes incorporated in reverse micelles of surface-active substances in organic solvents. Peroxidase in the Aerosol OT–water–octane system. *Molekulyarnaya Biologiya* **218**, 1019–1031. (English translation *Molecular Biology (Moscow)* **18**, 830–841 (1985).)

KOMIVES, C.F., LILLEY, E. AND RUSSELL, A.J. (1994). Biodegradation of pesticides in non-ionic water-in-oil microemulsions of Tween 85: relationship between micelle structure and activity. *Biotechnology and Bioengineering* **43**, 946–959.

KOMIVES, C.F., OSBORNE, D.E. AND RUSSELL, A.J. (1994). Characterization of a non-ionic surfactant reversed micellar system for enzyme catalysis. *Journal of Physical Chemistry* **98**, 369–376.

KOTLARCHYK, M., CHEN, S.H., HUANG, J.S. AND KIM, W. (1984). Structure of dense sodium di 2-ethylhexylsulfosuccinate/D_2O/decane/microemulsions. *Physical Review Letters* **53**, 941–944.

KUBIK, R., EICKE, H.-F. AND JÖNSSON, B. (1982). On the activity of water and the concept of the interfacial free-energy in W/O microemulsions. *Helvetica Chimica Acta* **65**, 170–177.

KUNIEDA, H. AND SHINODA, K. (1979). Solution behaviour of Aerosol OT/water/oil system. *Journal of Colloid and Interface Science* **70**, 577–583.

KUNIEDA, H. AND SHINODA, K. (1980). Solution behaviour and hydrophile–lyophile balance temperature in the Aerosol-OT–isooctane–brine system: correlation between microemulsions and ultralow interfacial tensions. *Journal of Colloid and Interface Science* **75**, 601–606.

KUMAR, V.V. AND RAGHUNATHAN, P. (1986). Spectroscopic investigations of the water-pool in lecithin reverse micelles. *Lipids* **21**, 764–768.

LAANE, C., BOEREN, S., VOS, K. AND VEEGER, C. (1987a). Rules for the optimization of biocatalysis in organic solvents. *Biotechnology and Bioengineering* **30**, 81–87.

LAANE, C., BOEREN, S., HILHORST, R. AND VEEGER, C. (1987b). Optimization of biocatalysis in organic media. In *Biocatalysis in Organic media* (C. Laane, J. Tramper and M.D. Lilly, Eds), pp. 65–84. Elsevier, Amsterdam.

LANG, J., MASCOLO, G., ZANA, R. AND LUISI, P.L. (1990). Structure and dynamics of cetyltrimethylammonium bromide water-in-oil microemulsions. *Journal of Physical Chemistry* **94**, 3069–3074.

LANGEVIN, D. (1989). Structure of reversed micelles. In *Structure and Reactivity in Reverse Micelles* (M.P. Pileni, Ed.), pp. 13–43. Elsevier, Amsterdam.

LANGEVIN, D., GUEST, D. AND MEUNIER, J. (1986). Correlation between interfacial tension and microstructure in Winsor equilibria. Role of the surfactant film curvature properties. *Colloids and Surfaces* **19**, 159–170.

LAM, A.C., FALK, N.A. AND SCHECHTER, R.S. (1987). The thermodynamics of microemulsions. *Journal of Colloid and Interface Science* **120**, 30–41.

LARSSON, K.M., ADLERCREUTZ, P. AND MATTIASSON, B. (1987). Activity and stability of horse-liver alcohol dehydrogenase in sodium dioctysulphosuccinate/cyclohexane reverse micelles. *European Journal of Biochemistry* **166**, 157–161.

LARSSON, K.M., ADLERCREUTZ, P. AND MATTIASSON, B. (1990). Enzymatic catalysis in microemulsions: product re-use and recovery. *Biotechnology and Bioengineering* **36**, 135–141.

LARSSON, K.M., OLDFIELD, C. AND FREEDMAN, R.B. (1989). Analysis of the inactivation of liver alcohol dehydrogenase in Aerosol-OT/iso-octane microemulsions. *European Journal of Biochemistry* **183**, 357–361.

LEE, K.M. AND BIELLMANN, J.-P. (1986). Cholesterol oxidase in microemulsion: enzymatic activity on a substrate of low water solubility and inactivation by hydrogen peroxide. *Bioorganic Chemistry* **14**, 262–273.

LEE, K.M. AND BIELLMANN, J.-P. (1987). Enzyme and organic solvents: horse liver alcohol dehydrogenase in non-ionic microemulsion: stability and activity. *FEBS Letters* **223**, 31–36.

LEE, K.M., DURIATTI, A., SCHUBER, F. AND BIELLMAN, J.-F. (1989). Squalene epoxide hydrolase and microemulsions. *FEBS Letters* **244**, 347–350.

LEODIDIS, E.B. AND HATTON, T.A. (1989). Interphase transfer for selective solubilization of ions, amino acids and proteins in reversed micelles. In *Structure and Reactivity in Reverse Micelles* (M.P. Pileni, Ed.), pp. 270–302. Elsevier, Amsterdam.

LESER, M.E., MRKOCI, K. AND LUISI, P.L. (1993). Reverse micelles in protein separation: the use of silica for the back-transfer process. *Biotechnology and Bioengineering* **41**, 489–492.

LESER, M.E., WEI, G., LUISI, P.L. AND MAESTRO, M. (1986). Application of reverse micelles for the extraction of proteins. *Biochemical and Biophysical Research Communications* **135**, 629–635.

LEVASHOV, A.V., KLYACHKO, N.L. AND MARTINEK, K. (1981). Catalysis by enzymes included in inverted micelles of surface-active agents in organic solvents. III. Influence of the degree of hydration of Aerosol-OT in octane on the rate of hydrolysis of cinnamoyl-α-chymotrypsin. *Bioorganicheskaya Khimya* **7**, 670–679. (English translation: *Soviet Journal of Bioorganic Chemistry* **7**, 364–372.)

LEVASHOV, A.V., KLACHKO, N.L., PANTIN, V.I., KHMELNITSKY, YU. L. AND MARTINEK, K. (1980). Catalysis by water-soluble enzymes included in reversed micelles of surface-active agents in non-aqueous solvents. *Bioorganicheskaya Khimya* **6**, 929–943. (English translation: *Soviet Journal of Bio-organic Chemistry* **6**, 485–498.)

LEVASHOV, A.V., KHMELNITSKY, YU. I., KLYACHKO, N.L., CHERNYAK, V.Y. AND MARTINEK, K. (1982). Enzymes entrapped into reversed micelles in organic solvents. Sedimentation analysis of the protein-Aerosol-OT–H_2O–octane system. *Journal of Colloid and Interface Science* **88**, 444–457.

LEVASHOV, A.V., KLYACHKO, N.L., BOGDANOV, N.G. AND MARTINEK, K. (1990). Fixation of a highly reactive form of α-chymotrypsin by micellar matrix. *FEBS Letters* **268**, 238–240.

LEVASHOV, A.V., KHETTAL, B., DE LAUZON, S., WAKS, M.N. AND CITTANOVA, N. (1992). Kinetics and stability of Δ^5-3-ketosteroid isomerase from *Pseudomonas testeroni* in the system of reverse micelles of AOT in iso-octane. *Biochemie* **74**, 561–564.

LICHTERFIELD, F., SCHMELING, T. AND STREY, R. (1986). Microstructure of microemulsions of the system H_2O-*n*-tetradecane-$C_{12}E_5$. *Journal of Physical Chemistry* **90**, 5762–5766.

LILLY, M.D. (1983). Two-liquid phase biocatalytic reactors. *Philosophical Transactions of the Royal Society of London* **300**, 391–398.

LINDBLOM, G., HAUKSSON, J., RILFORST, L., BERGENSTAHL, B., WIESLANDER, A. AND ERIKSSON, P.-O. (1993). Membrane lipid regulation in *Acholeoplasma laidlawwaii* grown with saturated fatty acids. Biosynthesis of a triacyglucolipid-forming reversed micelles. *Journal of Biological Chemistry* **268**, 16198–16207.

LINDMAN, B., STILBS, P. AND MOSELEY, M.E. (1981). Fourier transform NMR self-diffusion and microemulsion structure. *Journal of Colloid and Interface Science* **83**, 569–582.

LINDMAN, B., AHLNÄS, T., SÖDERMAN, O. AND WALDERHAUG, H. (1983). Fourier transform carbon-13 relaxation and self-diffusion studies of microemulsions. *Faraday Discussions of the Chemical Society* **76**, 317–329.

LINDMAN, B., SHINODA, K., OLSSON, U., ANDERSON, D., KARLSTRÖM, G. AND WENNESTRÖM, H. (1989). On the demonstration of bicontinuous structures in microemulsions. *Colloids and Surfaces* **38**, 205–224.

LUISI, P.L. AND MAGID, L.J. (1986). Solubilization of enzymes and nucleic acids in hydrocarbon micellar solutions. *CRC Critical Reviews in Biochemistry* **20**, 409–474.

LUISI, P.L., GIOMINI, M., PILENI, M.P. AND ROBINSON, B.H. (1988). Reverse micelles as hosts for proteins and small molecules. *Biochimica et Biophysica Acta* **947**, 209–246.

LUISI, P.L., SCARTAZZINI, R., HAERING, G. AND SCHURTENBERGER, P. (1990). Organogels from water-in-oil microemulsions. *Colloid and Polymer Science* **268**, 356–374.

MACRAE, A.R. (1985). Enzyme-catalysed modifications of fats and oils. *Philosophical Transactions of the Royal Society of London* **310**, 227–233.

MADAMWAR, D.B., BHATT, J.P. AND RAY, M.Y. (1988). Activation and stabilization of invertase entrapped into reversed micelles of sodium lauryl sulphate and sodium tauroglycocholate in organic solvents. *Enzyme and Microbial Technology* **10**, 302–305.

MAGID, L.J., KON-NO, K. AND MARTIN, C. (1981). Use of 1-methylcyanoformylpyridinium oximate (CPO) in evaluating the removal of acidic impurities from AOT. *Journal of Colloid and Interface Science* **83**, 307–308.

MAITRA, A. (1984). Determination of size parameters of water–aerosol-OT–oil reverse micelles from their nuclear magnetic resonance data. *Journal of Physical Chemistry* **88**, 5122–5125.

MAITRA, A. AND EICKE, H.-F. (1979). Effect of rotational isomerism on the water-solubilizing properties of Aerosol-OT as studied by ^{1}H n.m.r. spectroscopy. *Journal of Physical Chemistry* **85**, 2687–2691.

MARANGONI, A.G., MCCURDY, R.D. AND BROWN, E.D. (1993). Enzymatic esterification of triolein with tripalmitin in Canola lecithin-hexane reverse micelles. *Journal of the American Oil Chemists' Society* **70**, 737–745.

MARTIN, C.A. AND MAGID, L.J. (1981). Carbon-13 NMR investigations of Aerosol-OT water-in-oil microemulsions. *Journal of Physical Chemistry* **85**, 3936–3944.

MARTINEK, K., LEVASHOV, A.V., KLYACHKO, N.L. AND BEREZIN, I.V. (1977). Catalysis by water-soluble enzymes in organic solvents. Stabilization of enzymes against denaturation (inactivation) through their inclusion in reversed micelles of surfactant. *Doklady Akademii Nauk SSSR, English Edition* **236**, 951–953.

MARTINEK, K., LEVASHOV, A.V., KLYACHKO, N.L., PANTIN, V.I. AND BEREZIN, I.V. (1981). The principles of enzyme stabilization. VI. Catalysis by water-soluble enzymes entrapped into reversed micelles of surfactants in organic solvents. *Biochimica et Biophysica Acta* **657**, 277–281.

MARTINEK, K., LEVASHOV, A.V., KLYACHKO, N.L., KHMELNITSKY, YU. I. AND BEREZIN, I.V. (1986). Micellar enzymology. *European Journal of Biochemistry* **155**, 453–468.

MARTINEK, K., KLYACHKO, N.L., KABANOV, A.V., KHMELNITSKY, YU. L. AND LEVASHOV, A.V. (1989). Micellar enzymology: its relationship to membranology. *Biochimica et Biophysica Acta* **981**, 161–172.

MARZOLA, P. AND GRATTON, E. (1991). Hydration and protein dynamics: frequency domain fluroescence spectroscopy on proteins in reverse micelles. *Journal of Physical Chemistry* **95**, 9488–9495.

MATZKE, S.F., CREAGH, A.L., HAYNES, C.A., PRAUSNITZ, J.M. AND BLANCH, H.W. (1992). Mechanisms of protein solubilization in reverse micelles. *Biotechnology and Bioengineering* **40**, 91–102.

MEIER, P. AND LUISI, P.L. (1980). Micellar solubilization of biopolymers in hydrocarbon solvents. II. The case of horse liver alcohol dehydrogenase. *Journal of Solid Phase Biochemistry* **5**, 269–282.

MENGER, F.M. AND SAITO, G. (1978). Adsorption, displacement and ionization in water-pools. *Journal of the American Chemical Society* **100**, 4376–4379.

MENGER, F.M. AND YAMADA, K. (1979). Enzyme catalysis in water-pools. *Journal of the American Chemical Society* **101**, 6731–6734.

MISIOROWSKI, R.L. AND WELLS, M.A. (1974). The activity of phospholipase A_2 in reversed micelles of phosphatidylcholine in diethylether: effect of water and cations. *Biochemistry* **13**, 4921–4927.

MITCHELL, J. AND NINHAM, B.W. (1981). Micelles, vesicles and microemulsions. *Journal of the Chemical Society – Faraday Transactions* **77**, 601–629.

MORITA, S., NARITA, H., MATOBA, T. AND KITO, M. (1984). Synthesis of triacylglycerol by lipase in phosphatidylcholine reverse micellar system. *Journal of the American Oil Chemists' Society* **61**, 1571–1574.

MUCKHERJEE, S., MILLER, C.A. AND FORT, T. (1983). Theory of drop size and phase-continuity in microemulsions. I. Bending effects with uncharged surfactants. *Journal of Colloid and Interface Science* **91**, 223–243.

NA, A., ERIKSSON, C., ERIKSSON, S.-V., ÖSTERBERG, E. AND HOLMBERG, K. (1990). Synthesis of phosphatidylcholine with (n-3) fatty acids by phospholipase A2 in Microemulsions. *Journal of the American Oil Chemists' Society* **67**, 766–770.

NAKAMURA, K. AND HAKODA, M. (1992). Electro-ultra filtration bioreactor for enzymatic reaction in reverse micelles. In *Biochemical Engineering for 2001, Proceedings of the Asia–Pacific Biochemical Engineering Conference, 1992* (S. Furusaki, I. Endo and R. Matsuno,m Eds), pp. 540–543. Springer-Verlag, Heidelberg.

NICHOLSON, J.D. AND CLARKE, J.H.R. (1984). Photon correlation spectroscopy in the investigation of water-in-oil microemulsions. In *Surfactant in Solution* (K.L. Mittal and B. Lindmann, Eds), volume 3, pp. 1663–1674. Plenum, New York.

NICOT, C., VACHER, M., VINCENT, M., GALLAY, J. AND WAKS, M. (1985). Membrane proteins in reverse micelles: myelin basic protein in a membrane-mimetic environment. *Biochemistry* **24**, 7024–7032.

NORTH, A.M., DORE, J.C., MCDONALD, J.A., ROBINSON, B.H., HEENAN, R.K. AND HOWE, A.M. (1986). Structure and dynamics of water-in-oil microemulsions stabilized by Aerosol-OT. *Colloids and Surfaces* **19**, 21–29.

OAKENFULL, D. (1980). Constraints of molecular packing on the size and stability of microemulsion droplets. *Journal of the Chemical Society – Faraday Transactions* **76**, 1875–1886.

OLDFIELD, C. (1990). Evaluation of kinetic parameters for enzymes solubilized in water-in-oil microemulsions. *Biochemical Journal* **272**, 15–23.

OLDFIELD, C. (1991). Exchange concept in water-in-oil microemulsions: consequences for 'slow' chemical reactions. *Journal of the Chemical Society – Faraday Transactions* **87**, 2607–2612.

OLDFIELD, C. AND FREEDMAN, R.B. (1989). Kinetics of bilirubin oxidation catalysed by bilirubin oxidase in a water-in-oil microemulsion system. *European Journal of Biochemistry* **183**, 347–355.

OLDFIELD, C., ROBINSON, B.H. AND FREEDMAN, R.B. (1990). Acid–base behaviour of 4-nitrophenol and 4-nitrophenol-2-sulphonate in water-in-oil microemulsions stabilized by Aerosol-OT. *Journal of the Chemical Society – Faraday Transactions* **86**, 833–841.

OLDFIELD, C., OTERO, C., RUA, M.L. AND BALLESTEROS, A. (1992). Kinetics of enzyme-catalysed reactions in water-in-oil microemulsions. In *Biocatalysis in Non-Conventional Media* (J. Tramper, M.H. Vermüe, H.H. Beeftink and U. Van Stockar, Eds), pp. 189–198. Elsevier, Amsterdam.

OLIVIERA, A.C. AND CABRAL, J.M.S. (1993). Kinetic studies of *Mucor miehei* lipase in phosphatidylcholine microemulsions. *Journal of Chemical Technology and Biotechnology* **56**, 247–252.

OSTERBERG, E., BLOMSTROM, A.-C. AND HOLMBERG, K. (1989). Lipase-catalysed transesterification of unsaturated lipids in a microemulsion. *Journal of the American Oil Chemists' Society* **66**, 1330–1333.

OVERBEEK, J.TH.G. (1978). Microemulsions, a field at the border between lyophobic and lyophilic colloids. *Faraday Discussions of the Chemical Society* **65**, 7–19.

PARADKAR, V.M. AND DORDICK, J.S. (1991). Purification of glycoproteins by selective transport using concanavalin A-mediated reverse-micellar extraction. *Biotechnology Progress* **7**, 330–334.

PARADKAR, V.M. AND DORDICK, J.S. (1993). Affinity-based reverse-micellar extraction and separation (ARMES): a facile technique for the purification of peroxidase from soybean hulls. *Biotechnology Progress* **9**, 199–203.

PENG, Q. AND LUISI, P.L. (1990). The behaviour of proteases in reverse micelles. *European Journal of Biochemistry* **188**, 471–480.

PETIT, C., BROCHETTE, P. AND PILENI, M.P. (1986). Hydrated electrons in reverse micelles: 3. Distribution and location of probes such as ions and hydrophilic proteins. *Journal of Physical Chemistry* **90**, 6517–6521.

PIAZZA, G.J. (1992). Lipoxygenase catalysed hydroperoxide formation in microemulsions containing non-ionic surfactant. *Biotechnology Letters* **14**, 1153–1158.

PILENI, M.P. (1989). Structural changes of reverse micelles and microemulsions by adding solutes or proteins. In *Structure and Reactivity in Reverse Micelles* (M.P. Pileni, Ed.), pp. 44–53. Elsevier, Amsterdam.

PILENI, M.P., BROCHETTE, P., HICKEL, B. AND LEREBOURS, B. (1984). Hydrated electrons in reverse micelles: 2. Quenching of hydrated electron by sodium nitrate. *Journal of Colloid and Interface Science* **98**, 549–554.

PIRES, M.J. AND CABRAL, J.M.S. (1993). Liquid–liquid extraction of a recombinant protein with a reverse micelle phase. *Biotechnology Progress* **9**, 647–650.

POON, P.H. AND WELLS, M.A. (1974). Physical studies of egg phosphatidylcholine in diethylether–water solutions. *Biochemistry* **13**, 4928–4935.

PRAZERES, D.M.F., GARCIA, F.A.P. AND CABRAL, J.M.S. (1992). Kinetics and stability of a *Chromobacterium viscosum* lipase in reversed micellar and aqueous media. *Journal of Chemical Technology and Biotechnology* **53**, 159–164.

PRAZERES, D.M.F., LEMOS, F., GARCIA, F.A.P. AND CABRAL, J.M.S. (1993a). Modeling lipolysis in a reversed micellar system: Part I – conventional batch reactor. *Biotechnology and Bioengineering* **42**, 759–764.

PRAZERES, D.M.F., LEMOS, F., GARCIA, F.A.P. AND CABRAL, J.M.S. (1993b). Modeling lipolysis in a reversed micellar system: Part II – Membrane reactor. *Biotechnology and Bioengineering* **42**, 765–771.

RAHAMAN, R.S. AND HATTON, T.A. (1991). Structural characterization of α-chymotrypsin-containing AOT reversed micelles. *Journal of Physical Chemistry* **95**, 1799–1811.

RAMIREZ-SILVA, L., TUENA DE GOMEZ-PUYOU, M. AND GOMEZ-PUYOU, A. (1993). Water-induced transitions in the K^+ requirements for the activity of pyruvate kinase entrapped in reverse micelles. *Biochemistry* **32**, 5332–5338.

RAO, A.M., JOHN, V.T., GONZALEZ, R.D., AKKARA, J.A. AND KAPLAN, D. (1993). Catalytic and interfacial aspects of enzymatic polymer synthesis in reversed micellar systems. *Biotechnology and Bioengineering* **41**, 531–540.

REES, G.D. AND ROBINSON, B.H. (1993). Microemulsions and organogels: properties and novel applications. *Advanced Materials* **5**, 608–619.

REES, G.D., NASCIMENTO, M., JENTA, T.R.-J. AND ROBINSON, B.H. (1991). Reverse enzyme synthesis in microemulsion-based organogels. *Biochimica et Biophysica Acta* **1073**, 493–501.

RICO, I. AND LATTES, A. (1984). Waterless microemulsions. *Nouveau Journal de Chimie* **8**, 429–431.

ROBINSON, B.H., TOPRACIOGLU, C., DORE, J.C. AND CHIEUX, P. (1984). Small-angle neutron-scattering study of microemulsions stabilized by aerosol-OT. Part 1. Solvent and concentration variation. *Journal of the Chemical Society – Faraday Transactions 1* **80**, 13–27.

RUCKENSTEIN, E. AND CHI, J.C. (1975). Stability of microemulsions. *Journal of the Chemical Society – Faraday Transactions 2* **71**, 1690–1707.

RUCKENSTEIN, E. AND KARPE, P. (1990). On the enzymatic superactivity in ionic reverse micelles. *Journal of Colloid and Interface Science* **139**, 408–436.

RUCKENSTEIN, E. AND KARPE, P. (1991). Enzymatic super- and subactivity in non-ionic reverse micelles. *Journal of Physical Chemistry* **95**, 4869–4882.

RUSSELL, A.J. AND KOMIVES, C. (1994). May the centrifugal force be with you! A novel centrigual reactor system is used with reversed micelle enzyme catalysis to biodegrade organophosphates. *CHEMTECH*, January, 26–32.

SAFRAN, S.A. AND TURKEVITCH, L.A. (1983). Phase diagrams for microemulsions. *Physical Review Letters* **50**, 1930–1933.

SAMAMA, J.-P., LEE, K.-M. AND BIELLMANN, J.-F. (1987). Enzymes and microemulsions. Activity and kinetic properties of liver alcohol dehydrogenase in ionic water-in-oil microemulsions. *European Journal of Biochemistry* **163**, 609–617.

SANCHEZ-FERRAR, A., BRU, R. AND GARCIA CARMONA, F. (1988). Kinetic properties of polyphenol oxidase in organic solvents. A study in Brij 96 reverse micelles. *FEBS Letters* **233**, 363–366.

SCARTAZZINI, R. AND LUISI, P.L. (1988). Organogels from lecithins. *Journal of Physical Chemistry* **92**, 829–833.

SCARTAZZINI, R. AND LUISI, P.L. (1990). Reactivity of lipase in an optically transparent lecithin-gel matrix. *Biocatalysis* **3**, 377–380.

SCHMIDLI, P.K. AND LUISI, P.L. (1990). Lipase-catalysed reactions in reverse micelles formed by soybean lecithin. *Biocatalysis* **3**, 367–376.

SCHURTENBERGER, P., MAGID, L.J., LINDNER, P. AND LUISI, P.L. (1992). A sphere to flexible coil transition in lecithin reverse micellar solutions. *Progress in Colloid and Polymer Science* **89**, 274–277.

SCRIVEN, L.E. (1976). Equilibrium bicontinuous structure. *Nature* **263**, 123–125.

SEBASTIAO, M.J., CABRAL, J.M.S. AND AIRES-BARROS, M.R. (1993). Synthesis of fatty acid esters by a recombinant cutinase in reversed micelles. *Biotechnology and Bioengineering* **42**, 326–332.

SENATRA, D., GABRIELLI, G. AND GUARINI, G.G.T. (1986). Experimental evidence of interphasal water in water-in-oil microemulsions. *Europhysics Letters* **2**, 455–463.

SHAPIRO, YU. E., BUDANOV, N.A., LEVASHOV, A.V., KLYACHKO, N.L., KHMELNITSKY, YU. L. AND MARTINEK, K. (1989). ^{13}C n.m.r. study of entrapping proteins (α-chymotrypsin) and reversed micelles of surfactants (Aerosol OT) in organic solvents (n-octane). *Collection of Czechoslovak Chemical Communications* **54**, 1126–1134.

SHERVANI, Z., MAITRA, A., JAIN, T.J. AND DINESH. X. (1991). Water solubilization investigations of phosphatidylcholine reverse micelles. *Colloids and Surfaces* **60**, 161–173.

SHINODA, K. AND KUNIEDA, H. (1987). The effect of salt concentration, temperature and additives on the solvent property of Aerosol-OT solution. *Journal of Colloid and Interface Science* **118**, 586–589.

SHKARINA, T.N., KÜHN, H. AND SCHEWE, T. (1992). Specificity of soybean lipoxygenase-1 in reverse micelles of sodium *bis*(2-ethylhexyl)sulfosuccinate (AOT). *Lipids* **27**, 690–693.

SINGH, C.P. AND SHAH, D.O. (1993). Lipase catalysed esterification in monolayers and microemulsions. *Colloids and Surfaces A* **77**, 219–224.

SKRIKA-ALEXOPOULOS, E., MUIR, J. AND FREEDMAN, R.B. (1993). Stability of bilirubin oxidase in organic solvent media: a comparative study of two low-water systems. *Biotechnology and Bioengineering* **41**, 894–899.

SMITH, R.E. AND LUISI, P.L. (1980). Micellar solubilization of biopolymers in hydrocarbon solvents III. Empirical definition of an acidity scale in reverse micelles. *Helvetica Chimica Acta* **63**, 2302–2311.

SPEISER, P. (1984). Hardened reverse micelles as drug delivery systems. In *Reverse Micelles* (P.L. Luisi and B.E. Straub, Eds), pp. 339–346. Plenum, New York.

STAMATIS, H., XENAKIS, X. AND KOLISIS, F.N. (1993). Enantiomeric selectivity of a lipase from *Penicillium simplicissimum* in the esterification of menthol in microemulsions. *Biotechnology Letters* **15**, 471–476.

STAMATIS, H., XENAKIS, A., BORNSCHEUER, U., SCHEPER, T., MENGE, U. AND KOLISIS, F.N. (1993). *Pseudomonas cepecia* lipase: esterification reactions in AOT microemulsion systems. *Biotechnology Letters* **15**, 703–708.

STAMATIS, H., XENAKIS, X., PROVELEGLIOU, M. AND KOLISIS, F.N. (1993b). Esterification reactions catalysed by lipases in microemulsions: the role of enzyme localisation in relation to its selectivity. *Biotechnology and Bioengineering* **42**, 103–110.

STEINMANN, B., JÄCLE, H. AND LUISI, P.L. (1986). A comparative study of lysozyme conformation in reverse micellar systems. *Biopolymers* **25**, 1133–1156.

TYRAKOWSKA, B., VERHAERT, R.M.D., HILHORST, R. AND VEEGER, C. (1990). Enzyme kinetics in reversed micelles. 3. Behaviour of 20β-hydroxysteroid dehydrogenase. *European Journal of Biochemistry* **187**, 81–88.

VAN DIJK, C., SPRUIJT, R., LAANE, C. AND VEEGER, C. (1992). The pH of reversed micelles as imposed by the dihydrogen/proton redox couple and indicated by viologens and cytochrome c_3, using hydrogenase as redox catalyst. *European Journal of Biochemistry* **207**, 587–598.

VERHAERT, R.M.D., SCHAAFSMA, T.J., LAANE, C., HILHORST, R. AND VEEGER, C. (1989). Optimization of the photo-enzymatic reduction of the carbon–carbon double bond of α-β unsaturated carboxylates in reversed micelles. *Photochemistry and Photobiology* **49**, 209–216.

VERHAERT, R.M.D., HILHORST, R., VERMUE, M., SCHAAFSMA, T.J. AND VEEGER, C. (1990a). Description of enzyme kinetics in reversed micelles. *European Journal of Biochemistry* **187**, 59–72.

VERHAERT, R.M.D., TYRAKOWSKA, B., HILHORST, R., SCHAAFSMA, T.J. AND VEEGER, C. (1990b). Enzyme kinetics in reversed micelles. 2. Behaviour of enoate reductase. *European Journal of Biochemistry* **187**, 73–79.

VOS, K., LAVALETTE, D. AND VISSER, A.J.W.G. (1987). Triplet state kinetics of Zn-porphyrin cytochrome *c* in micellar media. Measurement of intermicellar exchange rates. *European Journal of Biochemistry* **169**, 269–273.

VOS, K., LAANE, C., WEIJERS, S. VAN HOEK, A., VEEGER, C. AND VISSER, A.J.W.G. (1987a). Time-resolved fluorescence and circular dichroism of porphyrin cytochrome *c* incorporated in reverse micelles. *European Journal of Biochemistry* **169**, 259–268.

VOS, K., LAANE, C., VAN HOEK, A., VEEGER, C. AND VISSER, A.J.W.G. (1987b). Spectroscopic properties of horse liver alcohol dehydrogenase in reversed micellar solutions. *European Journal of Biochemistry* **169**, 275–282.

VULFSON, E.N., AHMED, G., GILL, I., KOZLOV, I.A., GOODENOUGH, P.W. AND LAW, B.A. (1991). Alterations to the catalytic properties of polyphenol oxidase in detergentless microemulsions and ternary water-organic solvent mixtures. *Biotechnology Letters* **13**, 91–96.

WALDE, P., GIULANI, A.M., BOICELLI, C.A. AND LUISI, P.L. (1990). Phospholipid-based reverse micelles. *Chemistry and Physics of Lipids* **53**, 265–288.

WELLS, M.A. (1974). The nature of water inside phosphatidylcholine micelles in diethylether. *Biochemistry* **13**, 4937–4941.

WINSOR, P.A. (1954). *Solvent Properties of Amphiphilic Compounds*. Butterworths, London.

WOLBERT, R.B.G., HILHORST, R., VOSKUILEN, G., NACHTEGAAL, H., DEKKER, M., VAN'T RIET, K. AND BIJSTERBOSCH, B.H. (1989). Protein transfer from an aqueous phase into reverse micelles. The effect of protein size and charge-distribution. *European Journal of Biochemistry* **184**, 627–633.

WOLF, R. AND LUISI, P.L. (1979). Micellar solubilization of enzymes in hydrocarbon solvents. Enzymic activity and spectroscopic properties of ribonuclease in n-octane. *Biochemical and Biophysical Research Communications* **89**, 209–217.

WONG, M., THOMAS, J.K. AND GRÄTZEL, M. (1976). Fluorescence probing of inverted micelles. The state of solubilized water clusters in the alkane/diisoctyl sulfosuccinate (Aerosol OT) solution. *Journal of the American Chemical Society* **98**, 2391–2397.

WONG, M., THOMAS, J.K. AND NOWAK, T. (1977). Structure and state of H_2O in reversed micelles. 3. *Journal of the American Chemical Society* **99**, 4730–4736.

YAMADA, Y., KUBOI, R. AND KOMASAWA, I. (1993). Increased activity of *Chromobacterium viscosum* lipase in Aerosol-OT reverse micelles in the presence of non-ionic surfactants. *Biotechnology Progress* **9**, 468–472.

ZAMPIERI, G.G., JÄCKLE, H. AND LUISI, P.L. (1986). Determination of the structural parameters of reverse micelles after uptake of proteins. *Journal of Physical Chemistry* **90**, 1849–1853.

ZEMB, T.N., HYDE, S.T., DERIAN, P.-J., BARNES, I.S. AND NINHAM, B.W. (1987). Microstructure from X-ray scattering: the disordered open connected model of microemulsions. *Journal of Physical Chemistry* **91**, 3814–3820.

ZHU, D.-M., WU, X. AND SCHELLY, Z.A. (1992a). Investigation of the micropolarities in reverse-micelles of Triton X-100 in mixed solvents of benzene and *n*-hexane. *Journal of Physical Chemistry* **96**, 7121–7126.

ZHU, D.-M., WU, X. AND SCHELLY, Z.A. (1992b). Reverse micelles and water-in-oil microemulsions of Triton X-100 in mixed solvents of benzene and n-hexane. Dynamic light scattering and turbidity studies. *Langmuir* **8**, 1538–1540.

ZINSLI, P.E. (1979). Inhomogeneous interior of Aerosol OT microemulsions probed by fluorescence and polarization decay. *Journal of Physical Chemistry* **83**, 3223–3230.

ZULAUF, M. AND EICKE, H.-F. (1979). Inverted micelles and microemulsions in the ternary system H_2O/Aerosol-OT/Iso-octane as studied by photon correlation spectroscopy. *Journal of Physical Chemistry* **83**, 480–486.

9
Biomedical Applications of Enzymes and their Polyethylene Glycol Adducts

GUY FORTIER

Laboratoire d'Enzymologie Appliquée, Groupe de recherche en Biothérapeutique Moléculaire, Département de Chimie-Biochimie, Université du Québec à Montréal, Montréal (Québec), Canada

Introduction

The field involving the application of enzymes is expanding every day. We now find that enzymes are used in many branches of our economy (food, textiles, pulp and paper, electronics, medicine, chemistry and so forth), mainly as tools for specific biotransformations, for diagnostic and quality control, and as pharmaceutical drugs.

To develop and optimize the use of enzymes in the next decade, many problems have to be resolved. They concern:

1. the instability of the structure and the activity of the enzyme under various operational conditions;
2. the availability of the enzyme at a reasonable cost;
3. the obtention of a very specific biotransformation capacity; and
4. the far too rapid clearance of the enzyme in biological fluids.

Many research groups are focusing on these concerns, and approaches such as genetic and protein engineering, as well as engineering of the microenvironment of the enzyme, are presently privileged solutions. These points have been reviewed recently by many authors (Mozhaev and Martinek, 1990; Gianfreda and Scarfi, 1991; Gupta, 1991; Mattiasson and Adlercreutz, 1991; Roig and Kennedy, 1992; Janecek, 1993; Mozhaev, 1993). An enzyme should be considered as a natural 'green product' which is fully biodegradable and safe for the environment. It is easy to believe that the enzyme will become a very attractive tool in the future as the tendency of consumers to request more natural and biodegradable products and processes will be enforced.

Abbreviations: ADA, adenosine deaminase; Ado, adenosine; ALL, acute lymphoblastic leukaemia; AML, acute myeloblastic leukaemia; ASNase, asparaginase; dAdo, 2'-deoxyadenosine; dATP, deoxyadenosine triphosphate; BSA, bovine serum albumin; HRP, horseradish peroxidase; mPEG, monomethoxy polyethylene glycol; NMR, nuclear magnetic resonance; OPA, *o*-phthaldehyde; PEG, polyethylene glycol; rIL-2, recombinant interleukin-2; SDS/PAGE, sodium dodecyl sulphate polyacrylamide gel electrophoresis; SOD, superoxide dismutase; TNBS, trinitrobenzenesulphonic acid.

Biotechnology and Genetic Engineering Reviews — Vol. 12, December 1994
0264-8725/94/12/329–356 $20.00 + $0.00 © Intercept Ltd, P.O. Box 716, Andover, Hampshire SP10 1YG, UK

Table 1. Partial list of enzymes in clinical use as potential or established treatments

Disease	Enzyme deficiency	Substrate accumulation	Potential therapeutic enzyme	Specific site to be targeted	[a] Pegylated trial
Experimental treatments for storage diseases					
Aspartylglycosaminuria	Glycosylasparaginase	Aspartylglucosamine	Amidase	Lysosome	
Fabry	α-Galactosidase	Ceramide	α-Galactosidase	Lysosome	1
Gaucher	β-Glucocerebrosidase	Glucocerebroside	β-Glucocerebrosidase from placenta	Liver, spleen, bone mannose receptor, macrophage	
Mucopolysaccharidosis	β-Glucuronidase β-Glucuronidase	Mucopolysaccharide			
Pompe	α-1,4 Glucosidase	Intracellular glycogen	α-1,4 Glucosidase	Mannose cell receptors	
Wolman	Lipase, cholesteryl esterase	Cholesteryl ester	Cholesteryl esterase	Fibroblast receptor insulin or apoB	
Experimental treatments for metabolic diseases					
Acatalasaemia		Oxygen free-radical	Catalase	Systemic	2
Ascorbic acid deficiency	Gulonolactone oxidase		Gulonolactone oxidase	Systemic	3
Hyperargininaemia	Arginase	Arginine	Arginase	Systemic	4,5
Hypophosphatasia	Alkaline phosphatase		Alkaline phosphatase		
Phenylketonuria	Phenylalanine hydroxylase	Phenylalanine	Phenylalanine ammonia lyase	Systemic	6,7
Severe combined immunodeficiency	Adenosine deaminase	Deoxyadenosine homocysteine	Adenosine deaminase bovine	Equilibrium between cells and blood	8,9
Experimental treatment for organ failure					
Gout		Uric acid	Uricase	Systemic	10
Jaundice	Liver disease	Bilirubin	Bilirubin oxidase	Systemic	11
		conjugates bilirubin	UDP glucuronyl transferase	Systemic	
Liver failure		Phenol and tyrosine	Tyrosinase	Systemic	
Renal failure		Urea	Urease	Systemic	

Experimental treatments for traumatic and acute injuries					
Inflammatory	Oxygen free-radical	Superoxide dismutase	Systemic	12	
Ischaemia reperfusion	Oxygen radical	Catalase	Systemic or local		13
			Superoxide dismutase	Systemic or local	14
			Ceruloplasmin	Systemic or local	
Thromboemboly		Clot	Fibrinolysin	Fibrin	
			Streptokinase	Plasminogen activation	15
			Urokinase	Plasminogen activator	16
			Batroxobin	Defibrinogenation	17
Other enzymatic treatments					
Burn treatment		Dead tissue	Collagenase	Local	
Various cancers			Amine oxidase	}	18
			Arginase	}	19
			Asparaginase	} Systemic or local	
			Asparaginase–glutaminase	}	
			Glucose oxidase	}	
			Ribonuclease	}	
Clot formation		Heparin	Heparinase	Systemic	
Disc problem			Chymopapain	Local	
Herniated disc			Collagenase	Local	
Viral infection			Ribonuclease	Systemic	20

[a] Reference relative to the modified PEG enzyme: 1, produced by Enzon Co.; 2, Davis *et al.* (1980); 3, Hadley and Sato (1989); 4, Davis *et al.* (1980); 5, Savoca *et al.* (1984); 6, Davis *et al.* (1980); 7, Wieder *et al.* (1979); 8, Hershfield *et al.* (1987), Levy *et al.* (1988), Hershfield, Chaffee and Sorensen (1993); 9, Beauchamp, Daddona and Menapace (1984); 10, Nishimura *et al.* (1979), Davis *et al.* (1980), Nishimura, Matsushima and Inada (1981); 11, Maeda, Seymour and Miyamoto (1992); 12, Davis *et al.* (1980), Greenwald (1990); 13, Muizelaar *et al.* (1993); 14, Lehman *et al.* (1992), Tang *et al.* (1992), Thibeault, Rezaiekhaligh and Mabry (1993); 15, Fears *et al.* (1992); 16, Garman and Kalindjian (1987); 17, Nishimura *et al.* (1983); 18, Savoca *et al.* (1979, 1984); 19, see *Table 4*; 20, Laznicek *et al.* (1993).

This chapter will focus on the current biomedical applications of enzymes and their polyethylene glycol adducts, as well as potential future developments.

ENZYMES AS PHARMACEUTICAL DRUGS

For a long time in medicine, enzymes have been used as diagnostic tools (Boehringer Mannheim, 1998–89), as prognostic tools and as disease markers (Guilbault, 1983; Schwartz, 1992). More recently, enzymes have found applications as a new class of therapeutic agents (Klausner, 1983; Klein and Langer, 1986; Goldberg, 1992; Maeda *et al.*, 1992; Smith *et al.*, 1993). This new emerging field, named Enzymotherapy, should rapidly expand in the next decade when problems related to their use will be solved. Enzymes may well supply a new class of very important biological drugs. *Table 1* gives a non-exhaustive list of enzymes currently in clinical use, as well as future prospects as established treatments. For the sake of clarity, *Table 1* is subdivided on the basis of the main area in which enzymes can be useful, such as:

1. agents for replacement therapy in inherited genetic and metabolic diseases;
2. scavengers to limit the tissue damage during the course of inflammatory diseases and various ischaemia/reperfusion injuries;
3. specific detoxification agents during acute chemical poisoning;
4. temporary palliative agents during an acute organ failure;
5. agents for corrective treatments; and
6. antineoplastic agents.

Points 1 and 6 will be discussed in further detail in this chapter, using three case examples.

It is well known that several major problems limit the use and decrease the efficiency of enzymes in the clinic. One of these is the immunological reaction frequently observed following multiple injections of foreign enzymes, which can lead to antibody formation that neutralizes the enzyme or that can also lead to a severe anaphylactic shock. In addition, the half-life of the majority of enzymes in the bloodstream is impaired by rapid glomerular filtration and non-specific absorption, by the endothelial system and by the endoplasmic reticulum. The lack of specificity in targeting the enzyme to 'selected' cells or organs requires that higher doses of enzyme need to be used, directly affecting the cost of the treatment. Therefore, to overcome these limiting factors, which are acting against the expansion of the use of enzymes in humans, an approach consisting of the surface modification of clinically useful enzymes was proposed. These enzymes would be grafted with biocompatible polymer chains, with the aim of diminishing their immunogenicity and increasing their plasma half-life. During the past 20 years this has been considered as the best performing strategy, for which polyethylene glycol is the most successfully used and the most widely studied biocompatible polymer. Finally, to reduce the high cost associated with the use of pure enzymes in the treatment of patients, a solution should be found by establishing better multidisciplinary collaboration between enzymologists, clinicians and genetic engineers, that will allow the design of enzymes with both improved activity and targeting capacity.

Polyethylene glycol adducts

Many types of polymers have been tested for this purpose but the monomethoxy polyethylene glycol of molecular mass of 5000 gives the most spectacular results in terms of increasing the biological half-life of the enzyme in blood and limiting the immunological response. *Table 2* describes the changes in the characteristics of macromolecules when they are covalently modified by PEG. Some most noteworthy examples of these positive changes are:

1. the increase in water solubility, which is not a major problem with protein;
2. the increase in the resistance of the modified protein to protease action, which is of great interest for the lack of antibody response;
3. the decrease of immunogenicity due to masking of the antigenic sites of the protein by the polymer chains; and, finally,
4. the modification of its biochemical behaviour and its pharmacokinetics.

Table 2. Change in the characteristics of molecules or macromolecules modified with covalently bound polyethylene glycol

What is increased?	What is decreased?	What else?
Water solubility	Immunogenicity	Antigenicity is masked
Resistance against proteases	Toxicity	Pharmacokinetics are modified
Cellular influx	Blood clearance	
Affinity to the cellular membrane	Fouling	
Chemical and physical stability		
Biocompatibility		
Hydrodynamic volume		

CHEMISTRY OF PEG

In order to achieve the surface derivatization of an enzyme by covalent addition of polymer chains, many chemical procedures have been developed to obtain polyethylene glycol (PEG) and monomethoxy polyethylene glycol (mPEG) with reactive functional groups at their extremities. The chemistry of these functional groups, which can react with free amino, carboxylic or sulphydryl groups of the amino acids at the surface of an enzyme under mild aqueous conditions, has been reviewed recently (Harris, 1985; Harris *et al.*, 1992). The functionalization of PEG with the aim of specific cross-linking reactions with side-chains of different amino acids has been discussed in detail (Zalipsky and Lee, 1992). As an example of a reagent specific for amino acid side-chain residues, the formation of a PEG-thiol protected with a 4-thiopyridone derivative (Woghiren, Sharma and Stein, 1993) allows specific derivatization of cysteine residues of the protein.

The covalent reaction or attachment of a reactive group to the hydroxyl functions of PEG was successfully achieved with 1,1'-carbonyl diimidazole (Beauchamp *et al.*, 1983), cyanuric chloride (Nishimura *et al.*, 1983; Jackson *et al.*, 1987), 2,4,5-trichlorophenyl chloroformate or 4-nitrophenyl chloroformate (Veronese *et al.*, 1985; Fortier and Laliberté, 1993) with or without a peptide spacer arm of various lengths (Sartore *et al.*, 1991a), the Moffatt–Swern reaction (Wirth *et al.*, 1991), tresyl chloride (Delgado *et al.*, 1990), and by various *N*-hydroxysuccinimide derivatives (Anderson and Tomasi, 1988; Zalipsky, Seltzer and Nho, 1991), such as succinimidyl active ester (Knauf *et al.*, 1988) and succinimidyl carbonate (Zalipsky, Seltzer and Menon-Rudolph, 1992) or with 4-fluoro-3-nitrobenzoic acid (Ladd and Snow, 1993). The advantage of using 4-fluoro-3-nitrobenzoic acid is that the addition of PEG on to the protein surface can be followed quantitatively.

Recently, the pegylation of peptides during their solid-phase synthesis has been performed successfully (Lu and Felix, 1993). It should also be mentioned that PEG addition is not only limited to enzymes and proteins. Various small drug-PEG adducts such as doxorubicin (Caliceti *et al.*, 1993), vitamin E (Sokol *et al.*, 1993), acetominophen and other drugs (Zalipsky, Gilon and Zilkha, 1983), and various peptide-PEG adducts such as interleukin-2 (Katre, Knauf and Laird, 1987; Balemans *et al.*, 1993; Ofosu-Appiah *et al.*, 1993), hirudin (Zawilska *et al.*, 1993) and granulocyte–macrophage colony-stimulating factor (Knüsli *et al.*, 1992), have been synthesized with success.

SEPARATION AND CHARACTERIZATION OF PROTEIN–PEG ADDUCTS

One of the most difficult tasks of PEG technology is related to the determination of the number of PEG chains added on to the surface of the proteins or enzymes and to the purification of the derivatized enzymes. This should represent a crucial step for the approval of the innocuity of the product by competent authorities. For the characterization of the preparation, several techniques were developed to evaluate the molecular weight of the protein–PEG adduct, with the aim of deducing the number of PEG chains *per* protein. Gel filtration using aqueous buffer, with or without organic solvent modifiers, as well as SDS/PAGE electrophoresis failed in the majority of cases to determine the molecular weight of the preparation. Explanations such as the entanglement of PEG in the PAGE matrix (Abuchowski *et al.*, 1977; McGoff, Baziotis and Maskewicz, 1988) and insulation of the protein from the electric field by PEG (Bode, 1976), as well as a dramatic increase in the hydrodynamic volume of the modified protein or enzyme, have been considered responsible for the lack of electrophoretic mobility in SDS/PAGE electrophoresis and for the increase of the observed resident time during gel filtration of PEG-modified protein.

A denaturing buffer composed of 8 M urea was used for gel filtration chromatography of various horseradish peroxidase PEG adducts (Fortier and Laliberté, 1993). The aim was to decrease the availability of water and to reduce hydrogen interactions, thus allowing the separation of the modified proteins. During the chromatographic analysis of eight different adducts of horseradish peroxidase (HRP) prepared with PEG of molecular masses varying between 750 and 35 000, it was observed that the peak width of preparations of HRP increased with the molecular mass of PEG adducts faster than predicted by the retention time. The increase of the peak width according to the molecular mass of PEG was explained in part by the heterogeneity of the commercial preparation of HRP used and by the polydispersity of the PEG, which is more marked with high molecular mass PEGs. This was in agreement with the conclusion of the experimental work of Kunitani *et al.* (1991) who showed that protein–PEG adducts should be regarded as a homogeneous distribution of molecular weights as a function of the polydispersity of PEG. Contamination of monomethoxy polyethylene glycol by polyethylene glycol (Selisko, Delgado and Fisher, 1993) often leads to aggregation of the protein and to the misinterpretation of the molecular mass of the protein–polymer adduct.

It should be kept in mind that, due to its random coil structure, a polyethylene glycol with a molecular mass of 20 000 has a hydrodynamic volume equal to a protein of about 70 000 Da (based on its viscosity) (Tam and Tremblay, 1991). The increase in hydrodynamic volume has been well documented by a light-scattering study performed on native superoxide dismutase (SOD) and on modified SOD–$mPEG_{5000}$ (McGoff, Baziotis and Maskewicz, 1988). In this study, it was demonstrated that the apparent molecular diameter of native SOD increased from less than 3 nm to 13.2 nm when modified with mPEG with a molecular mass of 5000.

Separation of the PEG-modified protein from the unreacted PEG is a difficult task, mainly because PEG has a random coil structure which is well hydrated in aqueous solution (Bailey and Koleske, 1967). It has been demonstrated that ultrafiltration techniques can be useful for small molecular masses of PEG, but in the presence of high molecular masses of PEG (i.e. >5000) it is very difficult to separate unreacted PEG from protein-modified PEG, as was recently described in an extensive study by Lentsch and co-workers (Lentsch, Aimar and Orozco, 1993). They showed that, depending on the molecular mass of PEG in a bovine serum albumin (BSA) solution, it is difficult to increase the transmission of PEG through the ultrafiltration membrane without increasing the fouling of BSA at the membrane surface. They underlined the importance of a thick channel (thickness of the membrane) and a low cross-flow to enhance the PEG concentration polarization in order to eliminate PEG from the filtration solution.

One possible strategy which can enhance the ultrafiltration of PEG is based on the reduction of its hydrodynamic volume under conditions that are not detrimental to the activity of the modified enzyme. This can be obtained, as we have demonstrated (Fortier and Laliberté, 1993) by performing the ultrafiltration in the presence of 0.65 M potassium sulphate which is a kosmotrope salt (Chicz and Regnier, 1990), leading to a reduction in the viscosity of the PEG solution (Bailey and Koleske, 1967). Sulphate is one of the best kosmotrope salts and it helps to stabilize the enzyme or protein. Its use results in the dehydration of the PEG coil which then collapses, leading to a reduction in its hydrodynamic volume. This filtration procedure was evaluated and it was shown that the presence of this salt enhanced the efficiency of the ultrafiltration process by a factor of 2 to 3 times. Under these conditions, the pressure and the flow of the ultrafiltration device remained constant throughout the experiment, and a membrane with a smaller molecular weight cut-off was used to avoid the loss of biological material.

To address the problem of determination of the number of chains of PEG added to the surface of a protein, several techniques have been developed during the past 3 years to replace the classical spectrophotometric assay based on the reaction of the free amino groups of the protein with trinitrobenzenesulphonic acid (TNBS assay) (Fields, 1971; Snyder and Sobocinski, 1975) or the spectrofluorometric assay using fluorescamine (Stocks *et al.*, 1986). One technique consists of the addition of an amino acid spacer, such as an unnatural amino acid such as norleucine, between the protein and the PEG. The quantity of norleucine found by amino acid analysis is proportional to the number of chains added to the protein surface (Sartore *et al.*,

1991b). These authors have also shown, using seven different proteins modified by PEG to various degrees ranging from 10 to 80%, that TNBS and OPA (*o*-phthaldehyde) procedures that allow quantification of free lysine amino groups at the surface of the protein overestimated the number of amino groups that reacted with PEG compared to the amount of norleucine detected. Recently, the use of monomethoxy polyethylene glycol esters of 4-fluoro-3-nitrobenzoic acid enabled the addition of PEG on to the protein surface to be followed quantitatively by fluorescence (Ladd and Snow, 1993). More sophisticated methods were also elaborated, such as analysis of the protein adducts by nuclear magnetic resonance (NMR) (Jackson *et al.*, 1987) or analysis by mass spectrometry (Vestling *et al.*, 1993).

FACTORS AFFECTING THE CATALYTIC ACTIVITY OF AN ENZYME DURING PEG DERIVATIZATION

Table 3 summarizes the factors that can affect the catalytic activity of a biological macromolecule submitted to surface derivatization using activated PEG. *Table 4* shows a compilation of various enzymes and different derivatization procedures along with the percentage modification and the residual activity of the various enzyme–PEG adduct preparations. One of the first problems encountered during the derivatization of an enzyme is to select the PEG reagent that will minimize the loss of residual activity versus the number of polymer chains added. Strong and non-specific reagents, such as the mPEG-triazine ether derivative obtained by activation of mPEG with cyanuric chloride, often lead to a majority of enzyme derivatives with low specific activities. Many examples are given in *Table 4*, such as adenosine deaminase which has a residual activity for adenosine of 28% when cyanuric chloride is used to pegylate 40% of the lysyl residues. This is compared to a residual activity of 78% when a carbonyldiimidazole reagent is used to modify more than 80% of the lysyl residues. For other enzymes, such as catalase or superoxide dismutase, varying results have been obtained with the same reagent for the same level of modification. These variations can be related to the purity of the reagent and to the conditions of the reaction. A classical example based on the derivatization of alkaline phosphatase (Yoshinga and Harris, 1989) to a similar percentage of modification by different reagents gave varying residual activities. In this example, the succinimidyl-succinate PEG derivative acts very well, giving 61–79% modification with a small loss of specific activity of less than 10%. This is far better than the result obtained with the cyanuric chloride–PEG derivative. It was shown with some reagents, such as tresyl-chloride or cyanuric chloride-activated PEG, that the PEG adduct can be reacted with different functional groups of the protein, leading to denaturation. It became very difficult, when using the trinitrobenzene sulphonic acid assay, to extrapolate the number of PEG chains added because the assay does not account for addition of PEG to carboxyls, sugars or tyrosine due to the strong reactivity of this PEG derivative.

Another factor is related to the availability and the number of functional groups in the protein and their positioning with respect to the substrate fixation site and to the catalytic site. This can explain why some proteins such as catalase and superoxide dismutase give high residual activity after extensive modification, while others, such as phenylalanine ammonia lyase and uricase, lose most of their catalytic activity with minimal modification.

Table 3. Factors that affect the activity of an enzyme during its modification by polyethylene glycol

1. The coupling reagent bound to the PEG that reacted with the protein, they range from mild type (phenylcarbonate) to strong type (cyanuric chloride)
2. The molecular mass of the PEG, from M_r 750 to 100 000
3. The number of functional groups of PEG (monomethoxy or not, induced some polymerization)
4. The number of chains of PEG added to the surface of the enzyme
5. The functional group of the enzyme that would be derivatized, such as SH, NH_2, COOH, sugars
6. The stability of the chemical link
7. The reaction conditions, i.e. temperature, pH, co-solvent, etc.

As the importance and the usefulness of PEG adducts has increased during the past decade, the chemistry of PEG activation has attracted the attention of many organic chemists. A variety of chemical reagents specific for different functional groups of a protein are now available, and their reactions can be carried out in very mild aqueous conditions that are not detrimental to enzyme activity. The number of chains as well as the length of the PEG chain added to the surface of an enzyme can modulate the specific activity as well as the specificity of the enzyme for its substrate. For example, horseradish peroxidase modified with 3–4 PEG chains shows an increase in its specific activity towards phenylenediamine (*Table 4*). However, enzymes such as batroxobin, ribonuclease, lysozyme, metalloprotease *Vibrio* and other proteases lose their capacities to transform and to hydrolyse their natural high molecular weight substrates. This is due to steric hindrance induced by the PEG chain around their catalytic site that does not allow the macromolecular substrate to reach the catalytic centre. In the majority of cases, these modified enzymes are still very active in the presence of small synthetic substrates.

Another important factor that should be taken in consideration when performing derivatization of proteins is the durability of the chemical link between the protein and the PEG. For an *in vivo* utilization, which requires a long circulation time of the modified enzyme, a stable bond between the enzyme and the PEG is needed. To achieve this, an urethane bond is preferable to an ester bond which is more labile, having a half-life in blood of about 8 hours. Zalipsky and Lee (1992) have discussed the subject in a recent review paper.

The derivatization of an enzyme with PEG should be optimized in order to obtain a modified enzyme with high activity and with specific properties such as non-immunogenicity, high serum residence time, decrease in susceptibility to proteolysis and so forth. The user should therefore take into account the stability of the protein in the buffer at the pH at which the reaction will be carried out, to the functional group of the protein that will be modified, as well as to the number of chains that should be added to it. There are not enough detailed studies on the role of the PEG chain-length (Laliberté, Gayet and Fortier, 1994), on the nature of the functional group to use, on the ideal degree of modification to reach, versus the molecular weight of the protein and the number of available NH_2 groups, to enable the establishment of a general procedure. Thus, each protein is a unique case for which the reaction, the percentage of modifications and the choice of reagent should be optimized in a specific manner in order to take advantage of this technology.

Table 4. Effect of PEG activation reagents and the percentage modification on the residual activity of various proteins

Protein	PEG M_r	PEG activation reagent	Residual activity (substrate)	% Modification	Purpose of the study	References
Adenosine deaminase	5000	Cyanuric chloride	28% (adenosine)	40%	Pharmacokinetics of modified enzyme	Davis *et al.* (1981)
Adenosine deaminase	5000	Carbonyldiimidazole	76% (adenosine)	85%	Pharmacokinetics of modified enzyme	Beauchamp, Daddona and Menapace (1984)
Alkaline phosphatase	5000	Cyanuric chloride	67% and 33% (*p*-nitrophenylphosphate)	62% and 88%	Evaluation of the cross-linker on of the enzyme	Yoshinga and Harris (1989)
	5000	Tresyl chloride	82% and 86%	73% and 77%		
	5000	Carbonyldiimidazole	91% and 77%	56% and 78%		
	5000	Succinimidyl-succinate	98% and 93%	61% and 79%		
Arginase	5000	Trichlorophenyl-chloroformate	90% (arginine)	60%	Biochemical and pharmacokinetic behaviours	Visco *et al.* (1987)
Arginase	5000	Cyanuric chloride	65% (arginine)	53%	Pharmacokinetics	Savoca *et al.* (1979)
Asparaginase *E. coli*	5000	Not indicated	52% (asparagine)	70%	Pharmacology of the modified enzyme	Park *et al.* (1981)
Asparaginase *E. coli*	5000	Cyanuric chloride	0.9% (asparagine)	79%	Comparative study with new comb-shaped PEG	Kodera *et al.* (1992)
Bacteriorhodopsin	5000	Succinimidyl carbonate	As the native (proton pumping)	1 PEG chain	Study of the functionality of modified protein membrane	Sirokman and Fasman (1993)
Batroxobin	5000	Cyanuric chloride	2.4% (clotting), 93% (esterolytic)	29%	Evaluation of immunoreactivity, thrombin-like protease activity and blood clearance	Nishimura *et al.* (1983)
Bilirubin oxidase	5000	*p*-Nitrophenyl chloroformate	40–60% (bilirubin)	11–12 PEG added	Review on jaundice treatment	Maeda, Seymourand Miyamoto (1992)
Catalase	5000	Cyanuric chloride	100%	60–70%	Evaluation of cellular uptake of PEG-enzymes	Beckman *et al.* (1988)
Catalase	5000	Cyanuric chloride	67%, 37% and 12% (H_2O_2)	21%, 46% and 55%	Activity in organic solvents	Takahashi *et al.* (1984a)
Cholesterol esterase	5000	Cyanuric chloride	High (cholesterol linoleate)	n.d.	Cholesterol ester synthesis in benzene	Mori, Nakata and Endo (1992)
β-Galactosidase	2000, 8000, 20 000	1,1-Carbonyldiimidazole	50% (*o*-nitrophenylgalacto-pyranoside) for the different preparations	77, 60 and 54%, respectively	Study of transgalactosylation in organic solvents	Beecher *et al.* (1990)
β-Glucuronidase	5000	Cyanuric chloride			Potential treatment of mucopoly-saccharidosis	Lisi *et al.* (1982)
Gulonolactone oxidase	5000	Succinimidyl succinate	74% (ascorbate formation)	47%	Clearance and immunogenicity were not decreased	Hadley and Sato (1989)
Haemoglobin	5000	Benzene hexacarboxylic acid	Decrease in oxygen affinity	3 PEG chains	Evaluation of PEG-Hb for blood substitute	Dellacherie and Léonard (1991)
Horseradish peroxidase	750–35 000	*p*-Nitrophenyl chloroformate	110% to 75% (phenylenediamine)	65%	Effect of chain length of PEG on stability of HRP against denaturating agents (pH, T°, etc.)	Fortier and Laliberté (1993), Laliberté, Gayet and Fortier (1994)
Horseradish peroxidase	1900	Reductive alkylation	110% (phenylenediamine)	42%	Activity in organic solvents	Wirth *et al.* (1991)
	5000	Reductive alkylation	88% (phenylenediamine)	37%		
Horseradish peroxidase	5000	Cyanuric chloride	70% (phenylenediamine)	60%	Activity in organic solvents	Takahashi *et al.* (1984b)
Horseradish peroxidase	20 000	Dithioester	88% (*o*-dianisidine)	33%	Activity in organic solvents	Souppe and Urrutigoity (1989)
[illegible]	[illegible]	[illegible]	80% (esterolysis of triglyceride)	55%	Ester synthesis in organic solvents	Inada *et al.* (1984)

Lipase	5000	Cyanuric chloride	43% (hydrolytic activity)	49%	Esterification of chiral *s* alcohol in organic solvents	Kikkawa *et al.* (1989)
Metalloprotease *Vibrio*	5000	?	100% (Z–Gly–Phe), 20% (casein)	80%	Reduce the binding to δ-macroglobulin	Narukawa, Miyoshi and Shinoda (1993)
Ovalbumin	6400	Cyanuric chloride	–	0–14 PEG	Improve preparation and purification	Jackson *et al.* (1987)
Phenylalanine ammonia lyase	5000	Cyanuric chloride	40%	28%		Wieder *et al.* (1979)
Ribonuclease	5000	Hydroxysuccinimidyl ester	80% (cytidine-2,3 cyclic phosphate)	8–9 PEG chains	Evaluation of pharmacokinetics and distribution in rats	Laznicek *et al.* (1993)
Ribonuclease	1900	Trichlorophenyl-chloroformate	100% (cytidine 2,3, cyclic phosphate) 65% to 50% (ribonucleic acid)	50% to 70% 50% to 70%	Effect of modification on the activity toward different substrate sizes	Veronese *et al.* (1985)
Sphingomyelinase	5000	Hydroxysuccinimide	70% (sphingomyelin), 0% (haemolysis)	100%	Effect on membrane-attacking activity	Matsuyama *et al.* (1993)
Superoxide dismutase	5000	Cyanuric chloride	100%	60–80%	Evaluation of cellular uptake of PEG–enzymes	Beckman *et al.* (1988)
Superoxide dismutase	2100 5210	Glycol ester 4 fluoro-3-nitrobenzoic acid	100% 73%	9 chains 10 chains	Evaluation of new reagents chromophorically labelled	Ladd and Snow (1993)
Superoxide dismutase	1900	Trichlorophenyl-chloroformate	90%, 70% and 70%	30%, 50% and 75%	Evaluation of clearance, immunogenicity, toward M_r PEG and % modification	Veronese *et al.* (1985)
Superoxide dismutase	5000	Cyanuric chloride	51%	19–20 PEG chains	Pharmacokinetics of modified enzymes in rats	Pyatak, Abuchowski and Davis (1980)
Streptokinase	2000, 5000	Carbonyldiimidazole			Pharmocokinetics and immunogenicity	Fears *et al.* (1992)
Uricase	5000	Cyanuric chloride	15 and 45% (uric acid)	43 and 37%	Immunogenicity	Nishimura *et al.* (1979), Nishimura, Matsushima and Inada (1981)
Urokinase	5000	Maleic anhydride			Effect of reversible polymer on activity	Garman and Kalindjian (1987)
Proteases for peptide synthesis						
Chymotrypsin	5000	Succinimidyl carbonate	65 and 29% (P–Gly–Gly–Leu–PheNAp)	63% and 79%	Effect of modification toward MW substrates	Chiu *et al.* (1993)
Papain	5000	Cyanuric chloride	87% (Benzoil-Arg-*p*-nitroaniline)	39%	Oligopeptide synthesis in benzene	Uemura *et al.* (1990)
Papain	5000	Cyanuric chloride	70% hydrolytic activity	37%	Solid-phase peptide synthesis in organic solvents	Sakurai *et al.* (1990)
Chymotrypsin	5000	Cyanuric chloride	68% hydrolytic activity	27%		
Thermolysin	5000	Cyanuric chloride	71% hydrolytic activity	27%		
Trypsin	5000	Cyanuric chloride	71% hydrolytic activity	23%		
Pepsin	5000	Cyanuric chloride	76% hydrolytic activity	20%		
Subtilisin Carlsberg	5000	Tresyl chloride	50% (*N*-tosyl-Arg-OMe)	n.d.	Aggregates of the PEG–enzyme in organic solvents	Khan *et al.* (1992)
Thermolysin	5000	Trichlorophenyl chloroformate	>native (furylacryloyl-Gly-Leu-NH_2	20%	Peptide synthesis in organic solvents	Ferjancic, Puigserver and Gaertner (1988)
Trypsin	5000	Succinimidyl carbonate	110% to 92% (BAEE)	42 to 78% stability	Evaluation of a new reagent and its	Zalipsky, Seltzer and Menon-Rudolph (1992)

n.d., Not determined.

Behaviour of enzyme–PEG adducts

As a direct consequence of PEG modification of the surface of an enzyme, the enzyme is more resistant to various denaturing conditions (pH, elevated temperature) as was demonstrated with horseradish peroxidase (Fortier and Laliberté, 1993; Laliberté, Gayet and Fortier, 1994). The enzyme is also resistant to the action of proteases, a behaviour which is important for decreasing the immunogenicity of the protein. The plasma half-life of the enzyme is markedly increased. This will be seen below.

RESISTANCE TO PROTEASES

As was previously discussed for biomaterial covered with grafted PEG chains, repulsion of proteins to the surface results from the high segmental flexibility of the PEG that confers a strong mobility to the chain resulting in a high degree of steric exclusion (Lim and Herron, 1992). This explains in part why proteins, platelets and cells do not adhere on to PEG-treated materials (Kishida *et al.*, 1992; Amiji and Park, 1993; Llanos and Sefton, 1993). Similar reasons could explain why large substrates are not able to diffuse to the active site of the enzyme and why proteases cannot access the proteineous structure of a modified enzyme to hydrolyse peptide bonds. This proteolytic-resistant characteristic was demonstrated for several different proteins and enzymes (Kondo *et al.*, 1989; Sada *et al.*, 1991; Francis, Delgado and Fisher, 1992). The decrease in immunogenicity could be related to the lack of accessibility of the antigenic epitopes of the protein covered by PEG, which is a non-immunogenic polymer. It has been observed frequently that the protein–PEG adduct does not react with the antibody directed against the native protein (Abuchowski *et al.*, 1977; Sehon, 1992; Katre, 1993). In addition, without a complete Freund adjuvant, it is not possible to obtain antibody against PEG alone, except for those with very high molecular masses (Sehon, 1992). In the same manner, as the protein is more resistant to proteolysis, it is reasonable to think that when a modified protein is phagocytosed by macrophages, no elicitation of the humoral response occurs as there is no digestion of the modified protein. Thus, no antigen can be presented by the macrophage to the lymphocytes. There is no precise scientific explanation of why a modified protein cannot induce a humoral response, nevertheless the fact remains that the higher the number of PEG chains added at the surface, the lower are the possibilities of obtaining antibody. This phenomenon was observed for PEG of molecular mass higher than 1900.

BLOOD CLEARANCE

Modification of therapeutic proteins, enzymes or drugs with polyethylene glycol of various molecular masses offers several advantages, as mentioned in *Table 2*. One of these is related to the increase in serum residence time of the protein adducts when compared to the native proteins or enzymes. For example, *Table 5* shows the serum half-lives, in hours, of asparaginase and asparaginase–polyethylene glycol from *E. coli* in various species and as a function of the type of administration of the injection. In humans, the half-lives of asparaginase and asparaginase–PEG adducts are 7–8 hours and 384–600 hours, respectively, when injected intravenously (Park *et al.*,

1981). Numberous other examples can be found in recent review papers on the pharmacology and the pharmacokinetics of protein–PEG adducts (Francis, Delgado and Fisher, 1992; Zalipsky and Lee, 1992; Katre, 1993).

Table 5. Serum half-life (hours) of asparaginase or asparaginase–PEG adducts from *E. coli* following the administration of injection in various species

Species	Injection	ASNase	ASNase–PEG	References
Human	IM	30	144	Asselin *et al.* (1993)
	IM	33–43	–	Ho, Yap and Brown (1981)
	IV	7–8	384–600	Park *et al.* (1981)
	IV	7–29	–	Ho, Yap and Brown (1981)
Rhesus monkey	IM	146	–	Berg *et al.* (1993)
Dog	IV	24	?	Wada *et al.* (1990)
Rabbit	?	20	144	Ho *et al.* (1988)
Rat	IP	5	>72	Wada *et al.* (1990)
Mouse	IP	5	91	Abuchowski *et al.* (1984)

ASNase, asparaginase; IM, intramuscular; IP, intraperitoneal; IV, intravenous.

The increase in the half-life of a protein is of clinical importance because the therapeutic protein having a high $T_{1/2}$ in blood will maintain a therapeutic dosage during a longer time. This will reduce both the amount of enzyme–PEG adducts to be injected and the frequency of injections. Consequently, the therapeutic index will be higher for a modified protein than for the native one. This will also influence the cost of the treatment. As the modification of the surface by addition of PEG increases markedly the molecular mass and the hydrodynamic volume of the protein, the glomerular filtration of this modified protein is reduced, based on a molecular mass cut-off for the glomerular filtration of the protein of about 70 kDa.

Experiments conducted with rIL-2 (recombinant interleukin-2) have shown that the PEG adducts contribute to a decrease in renal clearance (Katre, Knauf and Laird, 1987; Knauf *et al.*, 1988) without affecting its antitumoral activity (Balemans *et al.*, 1993). The reduction of the renal clearance will favour the elimination of the protein adduct by the endothelial reticulum or by endocytosis, but at a rate of disappearance that will follow a model of second-order decay, as was demonstrated for various protein adducts. Effectively, a rapid phase is observed and this is related to a rapid endocytosis of the protein adduct, followed by the saturation of this system. A slow disappearance of the protein adduct is then observed and is probably the result of unspecific absorption. It is also expected, following surface modification, that cellular clearance of some modified proteins will be reduced due to the fact that their glycosylated parts are masked by PEG, avoiding specific interactions with cellular components. As an example, modified lactoferrin, which is normally cleared by hepatocytes by interaction between its carbohydrate components and cellular receptors, shows a dramatic decrease in its rate of clearance in the presence of hepatocytes when compared to unmodified lactoferrin (Beauchamp *et al.*, 1983).

Current biomedical applications and future developments of soluble enzymes and their pegylated derivatives

This subject has been reviewed recently (Klein and Langer, 1986; Goldberg, 1992;

Smith *et al.*, 1993). The technology of pegylation is now widespread, with the aim of increasing the therapeutic index of enzymes by limiting their loss by rapid *in vivo* degradation, by kidney clearance, by humoral response or by cellular components. *Table 1* gives an overview of diseases associated with enzyme deficiencies that can be overcome with appropriate enzyme treatments. More than 200 enzyme deficiencies related to genetic dysfunction have been reported in humans. Research in enzyme replacement therapy and gene therapy is currently in progress throughout the world to find solutions to these diseases. This is not a simple undertaking and it will take a considerable time for the establishment of an efficient gene therapy for each of these deficiencies. With this in mind, enzymotherapy is more than just an intermediate solution while awaiting the time when a successful gene therapy will be obtained. For example, in adenosine deaminase deficiency, enzymotherapy is the only palliative treatment available to limit the permanent damage during the time delay before gene therapy becomes operational in the patient. Three case applications involving the clinical use of enzymes or modified enzymes will now be presented.

CASE STUDY 1: ADENOSINE DEAMINASE DEFICIENCY

One of the most striking examples of the complementarity of the two approaches (gene therapy and enzymotherapy) is given by adenosine deaminase (ADA) deficiency. This results in defects in purine nucleotide metabolism, leading to severe immunodeficiency diseases afflicting the function of B and T lymphocytes. This enzyme catabolizes the deamination of adenosine (Ado), 2'-deoxyadenosine (dAdo) and many other 6-aminopurine nucleosides (Cory, 1992). An accumulation of deoxyadenosine triphosphate (dATP) was observed in patients with severe immunodeficiency. The dATP is an inhibitor of ribonucleoside reductase and subsequently of cellular replication. Therefore, by reducing the amount of dAdo in the blood, the accumulation of dATP is limited and the adenosine deaminase-deficient lymphocytes will be protected. The intracellular level of these metabolites is in partial equilibrium with the plasma, as shown by erythrocytes. Red cell transfusion, as well as bone marrow transplantation, has been used to palliate the ADA deficiency, despite the high risk associated with these treatments. It has been proposed (Polmar *et al.*, 1976; Hershfield *et al.*, 1987; Hershfield, Chaffee and Sorensen, 1993) to use purified ADA injected into the circulation to decrease the plasma level of the dAdo. As discussed previously, the free enzyme in circulating blood is rapidly eliminated but in the present case, no humoral response can occur since the immune system is dysfunctional. Therefore, to attain therapeutic levels of ADA, to increase the duration of the circulating enzyme in blood and to reduce the cost of treatment, the enzyme was pegylated (*Tables 1* and *4*). In a recent pharmacokinetic study in 29 patients with adenosine deaminase deficiency (Hershfield, Chaffee and Sorensen, 1993), it was shown that the pegylated ADA had a half-life in blood ranging between 3 and 6 days, when injected intramuscularly. The therapeutic level was selected to be 2–5 times greater than the normal level in total blood (red cell) and was maintained at >20 μmol h^{-1} ml^{-1}. After 6–8 months of PEG–ADA treatment, antibodies directed against ADA were found, as previously shown (Chaffee *et al.*, 1992). These did not impair treatment in the majority of patients. There was no evidence of hypersensitivity reactions. An increase of the lymphocyte responses to mitogens and restablishment of lymphocyte counts were observed in 80% of the patients. The level of dAdo decreased by 100-fold during the first 2 months of the treatment and the low level of dAdo was maintained throughout the entire study. This indicates that immune function has been restored and that the use of ADA is beneficial. The opportunistic infections were also resolved and did not recur. The patient receives a weekly injection for the rest of his or her life. Due to the high cost of this non-curative treatment, ADA deficiency is therefore a good candidate for gene therapy. Effectively, a form of gene therapy is under development and consists of exposing *in vitro* the Il-2 dependent T cells of the patient to an ADA cDNA-bearing retroviral vector. This requires that the T cells of the patient are sufficiently mature. Consequently, a PEG–ADA treatment has to be performed before and continued during the reinjection of the T cells, to ensure their normal development.

CASE STUDY 2: GAUCHER'S DISEASE

Gaucher's disease is an inherited disease of lipid catabolism resulting from an absence of glucocerebrosidase which leads to the deposition of glucocerebrosides in macrophages of the reticuloendothelial system (Brady *et al.*, 1966). It is the most common lysosomal storage disease. Numerous mutations at the acid glucosidase locus on chromosome 1q21 are responsible for the lack of glucocerebrosidase activity in macrophages (Pastores, Sibille and Grabowski, 1993). Enzyme replacement therapy during the course of this disease is effective in treating the haematologic, hepatic and splenic damage (Barton *et al.*, 1991a,b) and it is also useful for partially reconstituting the vertebral body height after 16 months of enzymotherapy, as indicated in a recent case report (Hill *et al.*, 1993). This success was primarily achieved by the use of glucocerebrosidase isolated from placenta which can selectively interact with cells bearing mannose receptors, especially the reticuloendothelial cells (Stahl *et al.*, 1978). The major disadvantage of this enzymotherapy is its cost. It was demonstrated that 1 year of therapy for a patient exceeds more than $US 200 000, depending on the protocol used, i.e. self-administration of low doses or infusion of high doses at the hospital (Barton *et al.*, 1991a,b; Beutler *et al.*, 1991; Pastores, Sibille and Grabowski, 1993; Zimran *et al.*, 1993). A major breakthrough for this treatment would be to produce the enzyme by genetic engineering and to obtain the right post-translational modifications in terms of glycosylation that would preserve the self-targeting of the native enzyme.

CASE STUDY 3: ENZYMES IN THE ARSENAL FOR CANCER TREATMENTS

An antitumoural effect of guinea-pig serum was first observed by Kidd in 1953. Broome (1961) later demonstrated that asparaginase (L-asparaginase amidohydrolase, EC 3.5.1.1.) was the active factor in this serum. Intensive studies were initiated to understand the antitumour action of asparaginase and the role of depletion of asparagine on the proliferation of lymphoblastic leukaemia cells. At the same time, asparaginase was tested against a variety of solid tumours but disappointing results were obtained.

Following these findings, the therapy of neoplasia based on deprivation of essential amino acids became an attractive approach. However, it was shown subsequently that dietary restriction of essential amino acids for control of tumour growth was inappropriate and unsuccessful (Roberts, 1981). Effectively, Marquez *et al.* (1989) and Medina, Marquez and Nunez de Castro (1992) established the dynamics of amino acid exchange between normal and neoplastic cells in mice. Three amino acids – glutamine, asparagine and serine – which are non-essential for normal cell growth, were found to be essential for the growth of tumour cells. In another study (Chuang, Yu and Wang, 1990), 17 amino acid-degrading enzymes were incubated individually with lymphocytes. The proliferation of lymphocytes was abolished only in the presence of lysine decarboxylase or asparaginase, and limited proliferations of 78% and 57%, compared to the control, were observed in the presence of arginase and tyrosinase, respectively.

Today, the use of asparaginase for the treatment of acute lymphoblastic leukaemia (ALL) and acute myeloblastic leukemia (AML) in children and adults, is one of the most widely studied and successful enzymotherapeutic applications. As the treatment of ALL or AML by asparaginase alone is ineffective, various antineoplasic drugs, such as vincristine, methotrexate, doxorubicin and VP16, are given together with the enzyme in different schedules and treatment regimens.

One of the first reviews (Capizzi and Cheng, 1981) on the use of asparaginase as a therapeutic agent against ALL discussed the toxicity and immunogenicity of asparaginase, the clinical importance of the inhibition of protein synthesis by asparagine depletion and, also, the low level of constitutive asparagine synthetase retrieved in normal cells. The explanation for tumour cell resistance to asparaginase treatment seems to be related to the fact that these cells can synthesize asparagine synthetase or that they have a high constitutive level of this enzyme. It appears difficult to evaluate the effectiveness of asparaginase treatment without evaluation of the level of constitutive asparagine synthetase in leukaemia cells. Martin *et al.* (1993) have shown by Western blot analysis of the protein content of murine leukaemia cells, both resistant (L5178Y D10/R) and non-resistant (L5178Y D10) to asparaginase treatment, that both cell types contain similar immunoreactive material with a molecular weight close to that asparagine synthetase. However, no enzymatic activity was detected in the sensitive cell preparation. It was hypothesized that sensitive cells were able to translate asparagine synthetase but with structural dysfunction. Further research was devoted to the understanding of the mechanism by which the enzyme asparaginase destroys leukaemia cells. Recently, Story *et al.* (1993) demonstrated experimentally in dogs that L-asparaginase kills lymphoma cells by apoptosis (Waring, Kos and Mullbacher, 1991), based on the analysis of cell morphology and DNA. They suggested that the enzyme is reducing or inhibiting protein synthesis to a level that prevents renewal of short-lived proteins which normally suppress activation of the mechanism of apoptosis (programmed cell death). It was also suggested that accumulation of aspartate induced a strong inhibition of the basic cellular energy mechanisms. Asparagine synthetase is responsible for the ATP-dependent synthesis of asparagine from glutamine and aspartic acid (Mehlhaff, Luehr and Schuuster, 1985).

To limit the side-effects of asparaginase treatment on normal cells, which have low levels of asparagine synthetase, and to initiate an early *de novo* synthesis of asparagine synthetase in these cells, it was proposed to deplete asparagine gradually to allow protein synthesis to continue normally. An experimental development has been presented recently based on *in vitro* procedures to evaluate the susceptibility of ALL cells of patients to asparaginase (Asselin *et al.*, 1989) and to obtain precise quantification of asparagine levels in plasma (Asselin *et al.*, 1991). This should allow the evaluation of the level of asparagine synthetase and the time required to induce synthesis of the enzyme in normal cells. Such an analysis should be helpful for limiting the undesirable side-effects (hypoproteinaemia, decrease in coagulation rate, cytotoxicity and transient ischaemic cerebral lesion) (Priest *et al.*, 1982; Cetkosky *et al.*, 1993; Pihko *et al.*, 1993) of asparaginase treatment in normal cells.

An important point to note from recent clinical studies in acute lymphoblastic or myeloblastic leukaemia in adults or in children (Clavell *et al.*, 1986; Schiller *et al.*, 1993; Weiss *et al.*, 1993; Wells *et al.*, 1993) is that the enzyme L-asparaginase or the pegylated asparaginase are part of the arsenal of chemotherapeutic drugs that can be used in multidrug resistance regimens during the phases of induction and consolidation therapy. As an example, the Children's Cancer group, in a recent survey of 1294 patients treated in their programme, concluded that high doses of cytarabine and asparaginase during the intensification phase of treatment eliminated the necessity for prolonged maintenance therapy in childhood AML and improved the overall survival rate.

Hypersensitivity to asparaginase of *E. coli* has been observed to be as high as 50% of ALL patients (Capizzi *et al.*, 1971). Intramuscular administration of the enzyme led to lower allergic side-effects. Antibodies against asparaginase of *E. coli* do not cross-react with the asparaginase obtained from *Erwinia carotovora* or *E. chysanthemi*, the other main sources of the enzyme (Wade *et al.*, 1968). The *Erwinia* enzyme therefore offers an alternative to the *E. coli* enzyme when hyper-reactivity occurs in patients. To palliate the immunogenicity and the rapid blood clearance of asparaginase, several different approaches were attempted. These consisted of immobilization of the enzyme onto or into a solid matrix, such as polyacrylamide (Mori, Tosa and Chibata, 1974; Nadler and Updike, 1974), nylon (Mori, Tosa and Chibata, 1973; Chong and Chang, 1974). cellulose nitrate membrane (Chang, 1972/73) or dacron (Cooney, Weetall and Long, 1975). These implanted devices rapidly decreased the level of asparagine in blood. Due to a limited biocompatibility, dense collagen encapsulation of these devices occurred and the diffusion of asparagine into the device was then completely restricted. Recently, liposomes (Cruz *et al.*, 1993) or erythrocyte ghosts filled with asparaginase (Naql *et al.*, 1988; DeLoach *et al.*, 1990) were successfully evaluated. The use of a biocompatible hydrogel of PEG–BSA (D'Urso and Fortier, 1994a,b) loaded with asparaginase (D'Urso *et al.*, 1994) is currently being evaluated in our laboratory.

One of the successful approaches that has been approved for clinical use is based on the surface modification of the enzyme by covalent addition of polyethylene glycol (PEG). For this purpose, the surface was modified by the addition of PEG of molecular mass of 5000 to the free amino groups of the enzyme (Abuchowski *et al.*, 1984) and recently by using comb-shaped PEG-maleic acid (Kodera *et al.*, 1992) of high molecular mass. Depending on the percentage of free NH_2 groups of lysine modified and the nature of the activated PEG, various residual activities of asparaginase were obtained, ranging from 7% to 52%. The reader is referred to the review paper of Zalipsky and Lee (1992) for more details.

Limited clinical trials for the treatment of ALL have been performed using asparaginase–PEG adducts (Park *et al.*, 1981; Ho *et al.*, 1986; Yoshimoto *et al.*, 1986;

Jurgens *et al.*, 1988). A study with ALL in human patients (Asselin *et al.*, 1993) was recently published on the pharmacokinetics of asparaginase from *E. coli* and from *Erwinia* with pegylated asparaginase provided by Enzon Inc. (South Plainfield, NJ, USA). They found that neither the age, sex and risk category of the subject nor the dose and the number of intramuscular injections received (the first one or the thirtieth one) affected the serum half-life of the protein and the enzymatic activity of asparaginase. The results obtained with the pegylated asparaginase contrast with previously published results in humans and other species in that a relatively short half-life of pegylated asparagine was obtained (*Table 5*). For patients showing signs of hypersensitivity, only the pegylated asparaginase showed persistence in blood, with a $T_{1/2}$ of about 1.8 days. No conclusions can be drawn from this study concerning the mechanism by which asparaginase is eliminated from the circulation. The results suggest, however, that neither endoplasmic reticulum nor urinary excretion mechanisms were implicated. The blood clearance probably resulted from an endocytosis mechanism.

Similar approaches based on the selective deprivation of other amino acids in plasma have also been developed with the use of the enzyme asparaginase-glutaminase isolated from *Acinetobacter glutaminsificans* (Holcenberg *et al.*, 1979) and arginase (Savoca *et al.*, 1979, 1984). Glutamine is important for the growth of many types of tumours and these tissues may use this amino acid as an energy source. As a recent example, intraperitoneal injections of glutaminase on a daily basis into mice bearing an Ehrlich ascites tumour reduced the tumour size by 72% after 15 days and increased the life span by 46% (Pal and Maity, 1992). Glutaminase is cytotoxic against various types of asparaginase-resistant leukaemia cells and against some solid lymphoid human tumours. Relatively encouraging results (prolonged survival, although no remission) were obtained with patients having tumours refractory to conventional chemotherapy (Holcenberg *et al.*, 1979). In humans, the side-effects of glutaminase are similar to those observed during asparaginase treatments, although more pronounced in the former.

The depletion of arginine was also evaluated as an antineoplasic treatment, using arginase which hydrolyses arginine into urea and ornithine. It was demonstrated that the enzyme can inhibit the proliferation of normal and tumour cells (Currie, 1978). A few years ago, Savoca showed that the use of arginase–polyethylene glycol adducts in Ehrlich ascites tumours in mice was innocuous (Savoca *et al.*, 1979, 1984). There is much controversy concerning the use of arginase as a potent antitumoral drug. It is well known that the polyamine biosynthetic pathway is very active during the growth of various cancer cells. The plasma level of polyamines can be used as an indication of growth (Jiang, 1990) and depletion of polyamines leads to inhibition of tumour growth (Janne, Poso and Raina, 1978). With this concern in mind, various inhibitors of arginase and ornithine decarboxylase have been developed and used as antitumor agents (Trujillo-Ferrara *et al.*, 1992). This seriously contrasted with the use of arginase which can lead to an increase of polyamines (spermine, putrescine, spermidine) by increasing the ornithine content of the plasma.

Oxidative products of polyamine degradation, such as aldehydes, have been implicated in programmed cell death (Parchment and Pierce, 1989). New strategies for cancer treatment are currently under development using polyamine-degrading enzymes such as bovine serum amine oxidase which is responsible for the oxidative deamination of polyamines leading to aldehydes, acrolein and hydrogen peroxide. These reaction products are highly toxic for cells (Averill-Bates *et al.*, 1993, 1994) and the toxicity is totally inhibited only in the presence of both catalase and aldehyde dehydrogenase. However, no pharmacokinetic studies of polyamine-degrading enzymes have been performed, and they are probably good candidates for derivatization by PEG, to increase their plasma half-life and to decrease their immunogenicity, for later use in humans.

Conclusion

There is no doubt that enzymotherapy is an emerging field that will provide a new class of biological drugs during the next decade. Enzymes should find wide applications as a replacement therapy in inherited genetic diseases, as scavengers to limit tissue damage during the course of inflammatory diseases and in various ischaemia/reperfusion injuries, as specific detoxification agents during acute chemical poisoning, as a temporary palliative agent during an acute organ failure and, also, as antineoplasic agents. Modification of enzymes, proteins, peptides and drugs by grafted polymers, especially polyethylene glycol, is currently the best strategy to solve problems related to the too-rapid clearance from blood and to the immunogenicity of the underivatized species. During the next decade, improvement of the purification technologies, lowering of the cost of technology and increased targeting of the enzyme–PEG adducts will be the main areas of research.

Acknowledgements

I would like to thank the Conseil de Recherches en Sciences Naturelles et en Génie and Université du Québec à Montréal for their financial support. I thank Dr D. Averill for her valuable comments and for the revision of the manuscript.

References

ABUCHOWSKI, A., VAN ES, T., PALCZUK, N.C. AND DAVIS, F.F. (1977). Alteration of immunological properties of bovine serum albumin by covalent attachment of polyethylne glycol. *Journal of Biological Chemistry* **252**, 3578–3581.

ABUCHOWSKI, A., KAZO, G.M., VERHOEST, C.R., JR, VAN ES, T., KAFKEWITZ, D., NUCCI, M.L., VIAU, A.T. AND DAVIS, F.F. (1984). Cancer therapy with chemically modified enzymes. I. Antitumor properties of polyethylene glycol–asparaginase conjugates. *Cancer Biochemistry Biophysics* **7**, 175–186.

AMIJI, M. AND PARK, K. (1993). Surface modification of polymeric biomaterials with poly(ethylene oxide), albumin, and heparin for reduced thrombogenicity. *Journal of Biomaterial Science – Polymer edition* **4**, 217–234.

ANDERSON, W.L. AND TOMASI, T.B. (1988). Polymer modification of antibody to eliminate immune complex and Fc binding. *Journal of Immunological Methods* **109**, 37–42.

ASSELIN, B.L., RYAN, D., FRANTZ, C.N., BERNAL, S.D., LEAVITT, P., SALLAN, S.E. AND COHEN, H.J. (1989). *In vitro* and *in vivo* killing of acute lymphoblastic leukemia cells by L-asparaginase. *Cancer Research* **49**, 4363–4368.

ASSELIN, B.L., LORENSON, M.Y., WHITIN, J.C., COPPOLA, D.J., KENDE, A.S., BLAKLEY, R.L. AND COHEN, H.J. (1991). Measurement of serum L-asparagine in the presence of L-asparaginase requires the presence of an L-asparaginase inhibitor. *Cancer Research* **51**, 6568–6573.

ASSELIN, B.L., WHITIN, J.C., COPPOLA, D.J., RUPP, I.P., SALLAN, S.E. AND COHEN, H.J. (1993). Comparative pharmacokinetic studies of three asparaginase preparations. *Journal of Clinical Oncology* **11**, 1780–1786.

AVERILL-BATES, D.A., AGOSTINELLI, E., PRZYBYTKOWSKI, E., MATEESCU, M.A. AND MONDOVI, B. (1993). Cytotoxicity and kinetic analysis of purified bovine serum amine oxidase in the presence of spermine in Chinese hamster ovary cells. *Archives of Biochemistry and Biophysics* **300**, 75–79.

AVERILL-BATES, D.A., AGOSTINELLI, E., PRZYBYTKOWSKI, E. AND MONDOVI, B. (1994). Aldehyde dehydrogenase and cytotoxicity of purified bovine serum amine oxidase and spermine in Chinese hamster ovary cells. *Biochemistry and Cellular Biology* **72**, 36–42.

BAILEY, F.E. AND KOLESKE, J.V. (1967). Configuration and hydrodynamic properties of the polyoxyethylene chain in solution. In *Nonionic Surfactants* (M.J. Schirk, Ed.), pp. 794–822. Marcel Dekker, New York.

BALEMANS, L.T.M., MATTIJSSEN, V., STEERENBERG, P.A., VAN DRIEL, B.E.M., DE MULDER, P.H.M. AND DEN OTTER, W. (1993). Locoregional therapy with polyethylene glycol-modified interleukin-2 of an intradermally growing hepatocellular carcinoma in the guinea pig induces T-cell-mediated antitumor activity. *Cancer Immunology Immunotherapy* **37**, 7–14.

BARTON, N.W., BRADY, R.O., DAMBROSIA, J.M., DI BISCEGLIE, A.M., DOPPELT, S.H., HILL, S.C., MANKIN, H.J., MURRAY, G.J., PARKER, R.I. AND ARGOFF, C.E. (1991a). Replacement therapy for inherited enzyme deficiency-macrophage-targeted glucocerebrosidase for Gaucher's disease. *New England Journal of Medicine* **324**, 1464–1470.

BARTON, N.W., BRADY, R.O., MURRAY, G.J., ARGOFF, C.E., GREWAL, R.P., KIAN-TI, Y., DAMBROSIA, J.M., ADRIAN, M., DI BISCEGLIE, A.M., HILL, S.C., PARKER, R.I., DOPPELT, S.H. AND MANKIN, H.J. (1991b). Enzyme-replacement therapy for Gaucher's disease. *New England Journal of Medicine* **325**, 1811.

BEAUCHAMP, C.O., DADDONA, P.E. AND MENAPACE, D.P. (1984). Properties of a novel PEG derivative of calf adenosine deaminase. *Advanced Experimental Medicine and Biology* **165A**, 47–52.

BEAUCHAMP, C.O., GONIAS, S.L., MENAPACE, D.P. AND PIZZO, S.V. (1983). A new procedure for the synthesis of polyethylene glycol–protein adducts; effects on function, receptor recognition and clearance of superoxide dismutase, lactoferrin, and α_2-macroglobulin. *Analytical Biochemistry* **131**, 25–33.

BECKMAN, J.S., MINOR, R.L., JR, WHITE, C.W., REPINE, J.E., ROSEN, G.M. AND FREEMAN, B.A. (1988). Superoxide dismutase and catalase conjugated to polyethylene glycol increases endothelial enzyme activity and oxidant resistance. *Journal of Biological Chemistry* **263**, 6884–6892.

BEECHER, J.E., ANDREWS, A.T. AND VULFSON, E.N. (1990). Glycosidase in organic solvents: II Transgalactosylation by polyethylene glycol-modified β-galactosidase. *Enzyme Microbiol Technology* **12**, 955–959.

BERG, S.L., BALIS, F.M., MCCULLY C.L., GODWIN, K.S. AND POPLACK, D.G. (1993). Pharmacokinetics of PEG–L-asparaginase and plasma and cerebrospinal fluid L-asparagine concentrations in the rhesus monkey. *Cancer Chemotherapy and Pharmacology* **32**, 310–314.

BEUTLER, W., KAY, A.C., SAVEN, A., GARVER, P., THURSTON, D.W. AND ROSENBLOOM, B.E. (1991). Enzyme-replacement therapy for Gaucher's disease. *New England Journal of Medicine* **325**, 1809–1810.

BODE, H.J. (1976). SDS-polyethyleneglycol electrophoresis, a possible alternative to SDS-polyacrylamide gel. *FEBS Letters* **65**, 56–58.

BOEHRINGER MANNHEIM (1988–1989). *Biochimica, Methods for Clinical Chemistry Research*, pp. 1–118.

BRADY, R.O., KANFER, J.N., BRADLEY, R.M. AND SHAPIRO, D. (1966). Demonstration of a deficiency of glucocerebroside-cleaving enzyme in Gaucher's disease. *Journal of Clinical Investigation* **45**, 1112–1115.

BROOME, J.D. (1961). Evidence that the L-asparaginase activity of guinea-pig serum is responsible for its antilymphoma effects. *Nature* **191**, 1114–1115.

CALICETI, P., MONFARDINI, C., SARTORE, L., SCHIAVON, O., BACCICHETTI, F., CARLASSARE, F. AND VERONESE, F.M. (1993). Preparation and properties of monomethoxy poly(ethylene glycol) Doxorubicin conjugates linked by an amino acid or a peptide as spacer. *Il Pharmaco* **48**, 919–932.

CAPIZZI, C.R.B. AND CHENG, Y.-Ċ. (1981). Therapy of neoplasia with asparaginase. In *Enzymes as Drugs* (J.S. Holcenberg and J. Roberts, Eds), pp. 1–24. Wiley Interscience, New York.

CAPIZZI, C.R.B., SKEEL, J.R., CREASEY, W., ZANES, R., OLAYON, C., PETERSON, R. AND HANDSCHUMACHER, R. (1971). L-Asparaginase: clinical, biochemical, pharmacological and immunological studies. *Annals of International Medicine* **74**, 893–901.

CETKOVSKY, P., KOZA, V., CEPELAK, V., VIT, L. AND SIGUTOVA, P. (1993). Haemostatic changes in patients with acute lymphoblastic leukaemia during and after remission induction therapy. *Fibrinolysis* **7**, 386–390.

CHAFFEE, S., MARY, A., STIEHM, E.R., GIRAULT, D., FISCHER, A. AND HERSHFIELD, M.S. (1992). IgG antibody response to polyethylene glycol modified adenosine deaminase in patients with adenosine deaminase deficiency. *Journal of Clinical Investigation* **89**, 1643–1651.

CHANG, T.M.S. (1972/73). L-asparaginase immobilized within semipermeable microcapsules: *in vitro* and *in vivo* stability. *Enzyme* **14**, 95–104.

CHICZ, R.M. AND REGNIER, F.E. (1990). High performance liquid chromatography: effective protein purification by various chromatographic modes. *Methods in Enzymology* **182**, 393–421.

CHIU, H.-C., ZALIPSKY, S., KOPECKOVA, P. AND KOPECEK, J. (1993). Enzymatic activity of chymotrypsin and its poly(ethylene glycol) conjugates toward low and high molecular weight substrates. *Bioconjugate Chemistry* **4**, 290–295.

CHONG, E.D.S. AND CHANG, T.M.S. (1974). *In vivo* effects of intraperitoneally injected L-asparaginase solution and L-asparaginase immobilized within semipermeable nylon microcapsules with emphasis on blood L-asparaginase, 'Body' L-asparaginase, and plasma L-asparagine levels. *Enzyme* **18**, 218–239.

CHUANG, J.C., YU, C.L. AND WANG, S.R. (1990). Modulation of lymphocyte proliferation by enzymes that degrade amino acids. *Clinical and Experimental Immunology* **82**, 469–492.

CLAVELL, L.A., GELBER, R.D., COHEN, H.J. ET AL. (1986). Four agent induction and intensive asparaginase therapy for treatment of childhood acute lymphoblastic leukemia. *New England Journal of Medicine* **315**, 657–663.

COONEY, D.A., WEETALL, H.H. AND LONG, E. (1975). Biochemical and pharmacologic properties of L-asparaginase bounded to dacron vascular prostheses. *Biochemical Pharmacology* **24**, 503–515.

CORY, J.G. (1992). Purine and pyrimidine nucleotide metabolism. In *Textbook of Biochemistry with Clinical Correlations* (T.M. Devlin, Ed.) (3rd Edn), pp. 529–575. Wiley-Liss, New York.

CRUZ, M.E.M., GASPAR, M.M., LOPES, F., JORGE, J.S. AND PEREZ-SOLER, R. (1993). Liposomal L-asparaginase: *in vitro* evaluation. *International Journal of Pharmaceutics* **96**, 67–77.

CURRIE, G.A. (1978). Activated macrophages kill tumor cells by releasing arginase. *Nature* **273**, 758–759.

DAVIS, F.F., ABUCHOWSKI, A., VAN ES, T., PALCZUK, N.C., SACOVA, K., CHEN, R.H.-L. AND PYATAK, P. (1980). Soluble nonantigenic polyethylene glycol-bound enzymes. In *Biomedical Polymers* (E.P. Goldberg and A. Nakajima, Eds), pp. 441–451. Academic Press, New York.

DELGADO, C., PATEL, J.N., FRANCIS, G.E. AND FISHER, D. (1990). Coupling of poly(ethylene glycol) to albumin under very mild conditions by activation by tresyl chloride: characterization of the conjugate by partitioning in aqueous two-phase systems. *Biotechnology and Applied Biochemistry* **12**, 119–128.

DELLACHERIE, E. AND LÉONARD, M. (1991). Improvement of oxygen-carrying properties of human hemoglobin by chemical modification with a benzene hexacarboxylate-monosubstituted polyoxyethylene. *Journal of Protein Chemistry* **10**, 61–67.

DELOACH, J.R., ANDREWS, K., SATTERFIELD, W. AND KEELING, M. (1990). Intraperitoneal administration of carrier erythrocytes in dogs: an improved method for delivery of L-asparaginase. *Biotechnology and Applied Biochemistry* **12**, 331–335.

D'URSO, E.M. AND FORTIER, G. (1994a). New hydrogel based on poly(ethylene glycol) cross-linked with bovine serum albumin. *Biotechnology Techniques* **8**, 71–76.

D'URSO, E.M. AND FORTIER, G. (1994b). New bioartificial polymeric material: poly(ethylene glycol) cross-linked with albumin. 1–Synthesis and swelling properties. *Journal of Bioactive and Compatible Polymers* **9**.

D'URSO, E.M., JEAN-FRANÇOIS, J., DOILLON, C.J. AND FORTIER, G. (1994). New family of hydrogels as matrix for enzyme immobilization: biomedical application. *Journal of Artificial Cells, Blood Substitutes and Immobilization Biotechnology* **22**, A14.

FEARS, R., HEARN, J., STANDRING, R., ANDERSON, J.L. AND MARDER, V.J. (1992). Lack of influence of pretreatment antistreptokinase antibody on efficacy in a multicenter patency comparison of intravenous streptokinase and anistreplase in acute myocardial infarction. *American Heart Journal* **124**, 305–314.

FERJANCIC, A., PUIGSERVER, A. AND GAERTNER, H. (1988). Unusual specificity of polyethylene glycol-modified thermolysin in peptide synthesis catalyzed in organic solvents. *Biotechnology Letters* **10**, 101–106.

FIELDS, R. (1971). The measurement of amino groups in proteins and peptides. *Biochemistry Journal* **124**, 581–590.

FLOHÉ, L. (1988). Superoxide dismutase for therapeutic use: clinical experience, dead ends and hopes. *Molecular and Cellular Biochemistry* **84**, 123–131.

FORTIER, G. AND LALIBERTÉ, M. (1993). Surface modification of horseradish peroxidase with poly(ethylene glycol) of various molecular masses. I. Preparation of the reagents and characterization of horseradish peroxidase-poly(ethylene glycol) adducts. *Biotechnology and Applied Biochemistry* **17**, 115–130.

FRANCIS, E.G., DELGADO, C. AND FISHER, D. (1992). PEG-modified proteins. In *Stability of Protein Pharmaceuticals, part B.* In Vivo *Pathways of Degradation and Strategies for Protein Stabilization* (T.J. Ahern and M.C. Manning, Eds), pp. 235–263. Plenum Press.

GARMAN, A.J. AND KALINDJIAN, S.B. (1987). The preparation and properties of novel reversible polymer–protein conjugates. 2-omega-methoxypolyethylene (5000) glycoxymethylene-3-methylmaleyl conjugates of plasminogen activator. *FEBS Letters* **223**, 361–365.

GIENFREDA, L. AND SCARFI, M.R. (1991). Enzyme stabilization: state of the art. *Molecular and Cellular Biochemistry* **100**, 97–128.

GOLDBERG, D.M. (1992). Enzymes as agents for the treatment of disease. *Clinica Chimica Acta* **206**, 45–76.

GREENWALD, R.A. (1990). Superoxide dismutase and catalase as therapeutic agents for human diseases. *Free Radical Biology and Medicine* **8**, 201–209.

GUILBAULT, G.G. (1984). *Analytical Uses of Immobilized Enzymes*. Marcel Dekker, New York.

GUPTA, M.N. (1991). Thermostabilization of proteins. *Biotechnology and Applied Biochemistry* **14**, 1–11.

HADLEY, K.B. AND SATO, P.H. (1989). Catalytic activity of administered gulonolactone oxidase polyethylene glycol conjugates. *Enzyme* **42**, 225–234.

HARRIS, J.M. (1985). Laboratory synthesis of polyethylene glycol derivatives. *JMS Review of Macromolecular Chemistry and Physics* **C25**, 325–373.

HARRIS, J.M., SEDAGHAT-HERATI, M.R., SATHER, P.J., BROOKS, D.E. AND FYLES, T.M. (1992). Synthesis of new poly(ethylene glycol) derivatives. In *Poly(ethylene glycol) Chemistry. Biotechnological and Biomedical applications* (M.J. Harris, Ed.), pp. 371–382. Plenum Press, New York.

HERSHFIELD, M.S., CHAFFEE, S. AND SORENSEN, R.U. (1993). Enzyme replacement therapy with polyethylene glycol–adenosine deaminase in adenosine deaminase deficiency: overview and case reports of three patients, including two now receiving gene therapy. *Pediatric Research* **33**, S42–S48.

HERSHFIELD, M.S., BUCKLEY, R.H., GREENBERG, M.L., MELTON, A.L., SCHIFF, R., HATEM, C., KURTZBERG, J., MARKERT, M.L., KOBAYASHI, R.H., KOBAYASHI, A.L. (1987). Treatment of adenosine deaminase deficiency with polyethylene glycol modified adenosine deaminase. *New England Journal of Medicine* **316**, 589–596.

HILL, S.C., PARKER, C.C., BRADY, R.O. AND BARTON, N.W. (1993). MRI of multiple plastyspondyly in Gaucher disease: response to enzyme replacement therapy. *Journal of Computer Assisted Tomography* **17**, 806–809.

HO, D.H., YAP, H.Y., BROWN, N. *et al.* (1981). Clinical pharmacology of intramuscularly administered L-asparaginase. *Journal of Clinical Pharmacology* **21**, 72–78.

HO, D.H., BROWN, N.S., YEN, A., HOLMES, R., KEATING, M., ABUCHOWSKI, A., NEWMAN, R.A. AND KRAKOFF, I.H. (1986). Clinical pharmacology of polyethylene glycol–L-asparaginase. *Drug Metabolism Disposition* **14**, 349–352.

HO, D.H., WANG, C.Y., LIN, J.R., BROWN, N., NEWMAN, R.A. AND KRAKOFF, I.H. (1988). Polyethylene glycol–L-asparaginase and L-asparaginase studies in rabbits. *Drug Metabolism Disposition* **16**, 27–29.

HOLCENBERG, J.S., BORELLA, L.D., CAMITTA, B.M. AND RING, B.J. (1979). Human pharmacology and toxicology of succinylated *Acinetobacter* glutaminase-asparaginase. *Cancer Research* **39**, 3145–3151.

HOVI, L., ERA, P., RAUTONEN, J. AND SIIMES, M.A. (1993). Impaired muscle strength in female adolescents and young adults surviving leukemia in childhood. *Cancer* **72**, 276–281.

INADA, Y., NISHIMURA, H., TAKAHASHI, K., YOSHIMOTO, T., SAHA, R.A. AND SAITO, Y. (1984). Ester synthesis catalyzed by polyethylene glycol-modified lipase in benzene. *Biochemical and Biophysical Research Communications* **122**, 845–850.

JACKSON, C.-J.C., CHARLTON, J.L., KUZMINSKI, K., LANG, G.M. AND SEHON, A.H. (1987). Synthesis, isolation, and characterization of conjugates of ovalbumin with monomethoxypolyethylene glycol using cyanuric chloride as the coupling agent. *Analytical Biochemistry* **165**, 114–127.

JANECEK, S. (1993). Strategies for obtaining stable enzymes. *Process Biochemistry* **28**, 435–445.

JANNE, J., POSO, H. AND RAINA, A. (1978). Polyamines in rapid growth and cancer. *Biochimica et Biophysica Acta* **473**, 241–293.

JIANG, X. (1990). Determination of polyamines in urine of normal human and cancer patients by capillary gas chromatography. *Biomedical Chromatography* **4**, 73–77.

JURGENS, H., SCHWAMBORN, D., KORHOLZ, D., WAHN, V. AND GOBEL, U. (1988). Clinical experiments with polyethylene glycol bound *E. coli* L-asparaginase in patients with multiple recurrences of acute lymphoblastic leukemia. *Klinische Paediatrie* **200**, 184–189.

KATRE, N.V. (1993). The conjugation of proteins with polyethylene glycol and other polymers. Altering properties of proteins to enhance their therapeutic potential. *Advanced Drug Delivery Reviews* **10**, 91–114.

KATRE, N.V., KNAUF, M.J. AND LAIRD, W.J. (1987). Chemical modification of recombinant interleukin 2 by polyethylene glycol increases its potency in the murine Meth A sarcoma model. *Proceedings of the National Academy of Science, USA* **84**, 1487–1491.

KHAN, S.A., HALLING, P.J., BOSLEY, J.A., CLARK, A.H., PEILOW, A.D., PELAN, E.G. AND ROWLANDS, D.W. (1992). Polyethylene glycol-modified subtilisin forms microparticulate suspensions in organic solvents. *Enzyme and Microbial Technology* **14**, 96–100.

KIKKAWA, S., TAKAHASHI, K., KATADA, T. AND INADA, Y. (1989). Esterification of chiral secondary alcohols with fatty acid in organic solvents by polyethylene glycol-modified lipase. *Biochemistry International* **19**, 1125–1131.

KISHIDA, A., MISHIMA, K., CORRETGE, E., KONISHI, H. AND IKADA, Y. (1992). Interactions of poly(ethylene glycol)-grafted cellulose membranes with proteins and platelets. *Biomaterials* **13**, 113–118.

KLAUSNER, A. (1983). New applications emerging: scientists working on enzymes for therapy. *Biotechnology* **1**, 734–735.

KLEIN, M.D. AND LANGER, R. (1986). Immobilized enzymes in clinical medicine: an emerging approach to new drug therapies. *Trends in Biotechnology* **4**, 179–186.

KNAUF, M.J., BELL, D.P., HIRTZER, P., LUO, Z.P., YOUG, J.D. AND KATRE, N.V. (1988). Relationship of effective molecular size to systemic clearance in rats of recombinant interleukin-2 chemically modified with water soluble polymers. *Journal of Biological Chemistry* **263**, 15064–15070.

KNÜSLI, C., DELGADO, C., MALIK, F., DOMINE, M., TEJEDOR, M.C., IRVINE, A.E., FISHER, D.

AND FRANCIS, G.E. (1992). Polyethylene glycol (PEG) modification of granulocyte–macrophage colony stimulating factor (GM-CSF) enhances neutrophil priming activity but not colony stimulating activity. *British Journal of Haematology* **82**, 654–663.

KODERA, Y., TANAKA, H., MATSUSHIMA, A. AND INADA, Y. (1992). Chemical modification of L-asparaginase with a comb-shaped copolymer of polyethylene glycol derivative and maleic anhydride. *Biochemical and Biophysical Research Communications* **184**, 144–148.

KONDO, A., KISHIMURA, M., KATOH, S. AND SADA, E. (1989). Improvement of proteolytic resistance of immunoadsorbents by chemical modification with polyethylene glycol. *Biotechnology and Bioengineering* **34**, 532–540.

KUNITANI, M., DOLLINGER, G., JOHNSON, D. AND KRESIN, L. (1991). On line characterization of polyethylene glycol-modified proteins. *Journal of Chromatography* **588**, 125–137.

LADD, D.L. AND SNOW, R.A. (1993). Reagents for the preparation of chromophorically labeled polyethylene glycol–protein conjugates. *Analytical Biochemistry* **210**, 258–261.

LALIBERTÉ, M., GAYET, J.-C.H. AND FORTIER, G. (1994). Surface modification of horseradish peroxidase with poly(ethylene glycol) of various molecular masses. II Relation between the molecular masses of PEG and the stability of horseradish peroxidase–PEG aducts under various denaturing conditions. *Biotechnology and Applied Biochemistry* **20**.

LAZNICEK, M., SCHIAVON, O., CALICETI, P. AND VERONESE, F.M. (1993). Pharmacokinetics and distribution of ribonuclease and its monomethoxypoly(ethylene glycol) derivatives in rats. *Pharmacological Research* **28**, 153–161.

LEHMAN, J.D., DYKE, C., ABD-ELFATTAH, A., YEH, T., JR., DING, M., EZRIN, A. AND WECHSLER, A.S. (1992). Polyethylene glycol-conjugated superoxide dismutase attenuates reperfusion injury when administered twenty-four hours before ischemia. *The Journal of Thoracic and Cardiovascular Surgery* **6**, 1597–1601.

LENTSCH, S., AIMAR, P. AND OROZCO, J.L. (1993). Separation of albumin–PEG: Transmission of PEG through ultrafiltration membranes. *Biotechnology and Bioengineering* **41**, 1039–1047.

LEVY, Y., HERSHFIELD, M.S., FERNANDEZ- MEJIA, C., POLMAR, S.H., SCUDIERY, D., BERGEN, M. AND SORENSEN, R.V. (1988). Adenosine deaminase deficiency with late onset of recurrent infections. Response to treatment with polyethylene glycol-modified adenosine deaminase. *Journal of Pediatrics* **113**, 312–317.

LIM, K. AND HERRON, J.N. (1992). Molecular simulation of protein–PEG interaction. In *Poly(ethylene glycol) Chemistry. Biotechnological and Biomedical Applications* (M.J. Harris, Ed.), pp. 29–56. Plenum Press, New York.

LISI, P.L., VAN ES, T., ABUCHOWSKI, A., PALCZUK, N.C. AND DAVIS, F.F. (1982). Enzyme therapy 1. Polyethylene glycol: β-glucuronidase conjugates as potential therapeutic agents in acid mucopolysaccaridosis. *Journal of Applied Biochemistry* **4**, 19–33.

LLANOS, G.R. AND SEFTON, M.V. (1993). Does polyethylene oxide possess a low thrombogenicity? *Journal of Biomaterial Science – Polymer Edition* **4**, 381–400.

LU, Y.-A. AND FELIX, A.M. (1993). Pegylated peptides I: solid-phase synthesis of Na-pegylated peptides using Fmoc strategy. *Peptide Research* **6**, 140–146.

MCGOFF, P., BAZIOTIS, A.C. AND MASKEWICZ, R. (1988). Analysis of polyethylene glycol modified superoxide dismutase by chromatographic, electrophoretic, light scattering, chemical and enzymatic methods. *Chemical Pharmacology Bulletin* **36**, 3079–3091.

MAEDA, H., SEYMOUR, L.W. AND MIYAMOTO, Y. (1992). Conjugates of anticancer agents and polymers: advantages of macromolecular therapeutics *in vivo*. *Bioconjugate Chemistry* **3**, 351–362.

MAEDA, H., KIMURA, M., SASAKI, I., HIROSE, Y. AND KONNO, T. (1992). Toxicity of bilirubin and detoxification by PEG–bilirubin oxidase conjugate. A new tactic for treatment of jaundice. In *Poly(ethylene glycol) Chemistry. Biotechnological and Biomedical Applications* (M.J. Harris, Ed.), pp. 153–169. Plenum Press, New York.

MARQUEZ, J., SANCHEZ-JIMÉNEZ, F., MEDINA, M.A., QUESADA, A.R. AND NUNEZ DE CASTRO, I. (1989). Nitrogen metabolism in tumor bearing mice. *Archives of Biochemistry and Biophysics* **268**, 667–675.

MARTIN, J.K., SUN, W., MORAGA-A, D., SCHUSTER, S.M. AND WYLIE, D.E. (1993). An investigation into the mechanism of L-asparaginase resistance in L5178Y murine leukemia cells. *Amino Acids* **5**, 51–69.

MATSUYAMA, H., TOMITA, M., TAGUCHI, R. AND IKEZAWA, H. (1993). Alteration in reactivity of sphingomyelinase from *Streptomyces* sp. modified with a polyethylene glycol derivative. *Biological and Pharmaceutical Bulletin* **16**, 107–111.

MATTIASSON, B. AND ADLERCREUTZ, P. (1991). Tailoring the microenvironment of enzymes in water poor systems. *Trends in Biotechnology* **9**, 394–398.

MEDINA, M.A., MARQUEZ, J. AND NUNEZ DE CASTRO, I. (1992). Interchange of amino acids between tumor and host. *Biochemical Medicine and Metabolic Biology* **48**, 1–7.

MEHLHAFF, P.M., LUEHR, C.A. AND SCHUUSTER, S.M. (1985). Studies on the ammonia-dependent reaction of beef pancreatic asparagine synthetase. *Biochemistry* **24**, 1104–1110.

MORI, S., NAKATA, Y. AND ENDO, H. (1992). Enzymatic properties of polyethylene glycol-modified cholesterol esterase in organic solvents. *Biotechnology and Applied Biochemistry* **16**, 101–105.

MORI, T., TOSA, T. AND CHIBATA, I. (1973). Enzymatic properties of microcapsules containing asparaginase. *Biochimica et Biophysica Acta* **321**, 653–661.

MORI, T., TOSA, T. AND CHIBATA, I. (1974). Preparation and properties of asparaginase entrapped in the lattice of polyacrylamide. *Cancer Research* **34**, 3066–3068.

MOZHAEV, V.V. (1993). Mechanism based strategies for protein thermostabilization. *Trends in Biotechnology* **11**, 88–95.

MOZHAEV, V.V. AND MARTINEK, K. (1990). Structure stability relationships in proteins: a guide to approaches to stabilizing enzymes. *Advanced Drug Delivery Reviews* **4**, 387–419.

MUIZELAAR, J.P., MARMAROU, A., YOUNG, H.F., CHOL, S.C., WOLF, A., SCHNEIDER, R.L. AND KONTOS, H.A. (1993). Improving the outcome of severe head injury with the oxygen radical scavenger polyethylene glycol-conjugated superoxide dismutase: a phase II trial. *Journal of Neurosurgery* **78**, 375–382.

NADLER, H.L. AND UPDIKE, S.J. (1974). Gel entrapment of enzyme. *Enzyme* **18**, 150–160.

NAQL, A., DELOACH, J.R., ANDREWS, K., SATTERFIELD, W. AND KEELING, M. (1988). Determination of parameters for enzyme therapy using L-asparaginase entrapped in canine erythrocytes. *Biotechnology and Applied Biochemistry* **10**, 365–372.

NARUKAWA, H., MIYOSHI, S.-I. AND SHINODA, S. (1993). Chemical modification of *Vibrio vulnificus* metalloprotease with activated polyethylene glycol. *FEMS Microbiology Letters* **108**, 43–46.

NISHIMURA, H., MATSUSHIMA, A. AND INADA, Y. (1981). Improved modification of yeast uricase with polyethylene glycol, accompanied with nonimmunoreactivity towards anti-uricase serum and high enzymic activity. *Enzyme* **26**, 49–53.

NISHIMURA, H., ASHIHARA, Y., MATSUSHIMA, A. AND INADA, Y. (1979). Modification of yeast uricase with polyethylene glycol. Disappearance of binding ability towards anti-uricase serum. *Enzyme* **24**, 261–264.

NISHIMURA, H., TAKAHASHI, K., SAKURAI, K., FUJINUMA, Y., IMAMURA, Y., OOBA, M. AND INADA, Y. (1983). Modification of batroxobin with activated polyethylene glycol: reduction of binding ability towards anti-batroxobin antibody and retention of defibrinogenation activity in circulation of preimmunized dogs. *Life Sciences* **33**, 1467–1473.

OFOSU-APPIAH, W., RUGGIERO, C., DZIELAK, D.J. AND ANTZELEVITCH, C. (1993). The effects of PEG–interleukin-2 and interleukin-2 on essential hypertension and cellular immune function in the spontaneously hypertensive rat. *Clinical and Experimental Hypertension* **15**, 435–457.

PAL, S. AND MAITY, P. (1992). Investigation on glutamine amidohydrolase and glutamine aminotransferase activity in liver and plasma of EAC bearing mice following glutaminase therapy. *Cancer Letters* **66**, 225–231.

PARCHMENT, R.E. AND PIERCE, G.B. (1989). Polyamine oxidation, programmed cell death and regulation of melanoma in the murine embryonic limb. *Cancer Research* **49**, 6680–6686.

PARK, Y.K., ABUCHOWSKY, A., DAVIS, S. AND DAVIS, F. (1981). Pharmacology of *Escherichia coli*–asparaginase polyethylene glycol adduct. *Anticancer Research* **1**, 373–376.

PASTORES, G.M., SIBILLE, A.R. AND GRABOWSKI, G.A. (1993). Enzyme therapy in Gaucher disease type 1: dosage efficacy and adverse effects in 33 patients treated for 6 to 24 months. *Blood* **82**, 408–416.

PIHKO, H., TYNI, T., VIRKOLA, K., VALANNE, L., SAINIO, K., HOVI, L. AND SAARINEN, U.M. (1993). Transient ischemic cerebral lesions during induction chemotherapy for acute lymphoblastic leukemia. *Journal of Pediatrics* **123**, 718–724.

POLMAR, S.H., STERN, R.C., SCHWARTZ, A.L., WETZLER, E.M., CHASE, P.A. AND HIRSCHHORN, R. (1976). Enzyme replacement therapy for adenosine deaminase deficiency and severe combined immunodeficiency. *New England Journal of Medicine* **295**, 1337–1343.

PRIEST, J.R., RAMSEY, N.K.C., STEINHERZ, P.G. *et al.* (1982). A syndrome of thrombosis and hemorrhage complicating L-asparaginase therapy for childhood acute lymphoblastic leukemia. *Journal of Pediatrics* **100**, 984–989.

PYATAK, P.S., ABUCHOWSKI, A. AND DAVIS, F.F. (1980). Preparation of a polyethylene glycol: superoxide dismutase adduct, and an examination of its blood circulating life and anti-inflammatory activity. *Research Communications in Chemical Pathology and Pharmacology* **29**, 113–127.

ROBERTS, J. (1981). Therapy of neoplasia by deprivation of essential amino acids. In *Enzymes as Drugs* (J.S. Holcenberg and J. Roberts, Eds), pp. 63–76. Wiley Interscience, New York.

ROIG, M.G. AND KENNEDY, J.F. (1992). Perspectives for chemical modifications of enzymes. *Critical Reviews in Biotechnology* **12**, 391–412.

SADA, E., KATOH, S., YAMANAKA, K. AND ITAKURA, Y. (1991). Resistance to proteolysis of antibody ligands modified with polyethylene glycol. *Journal of Fermentation and Bioengineering* **71**, 137–139.

SAKURAI, K., KASHIMOTO, K., KODERA, Y. AND INADA, Y. (1990). Solid phase synthesis of peptides with polyetylene glycol-modified protease in organic solvents. *Biotechnology Letters* **12**, 685–688.

SARTORE, L., CALICETI, P., SCHIAVON, O. AND VERONESE, F.M. (1991a). Enzyme modification by mPEG with an amino acid or peptide as spacer arms. *Applied Biochemistry and Biotechnology* **27**, 55–63.

SARTORE, L., CALICETI, P., SCHIAVON, O., MONFARDINI, C. AND VERONESE, F.M. (1991b). Accurate evaluation method of the polymer content in monomethoxy (polyethylene glycol) modified proteins based on amino acid analysis. *Applied Biochemistry and Biotechnology* **31**, 213–222.

SAVOCA, K.V., ABUCHOWSKI, A., VAN ES, T., DAVIS, F.F. AND PALCZUK, N.C. (1979). Preparation of a non-immunogenic arginase by the covalent attachment of polyethylene glycol. *Biochimica et Biophysica Acta* **578**, 47–53.

SAVOCA, K.V., DAVIS, F.F., VAN ES, T., MCCOY, J.R. AND PALCZUK, N.C. (1984). Cancer therapy with chemically modified enzymes. II. The therapeutic effectiveness of arginase and arginase modified by covalent attachment of polyethylene glycol on the Taper liver tumor and the L5178Y murine leukemia. *Cancer Biochemistry Biophysics* **7**, 261–268.

SCHILLER, G., LEE, M., TERRITO, M., GAJEWSKI, J. AND NIMER, S. (1993). Phase II study of etoposide, ifosfamide, and mitoxantrone for the treatment of resistant adult acute lymphoblastic leukemia. *American Journal of Hematology* **43**, 195–199.

SCHWARTZ, M.K. (1992). Enzymes as prognostic markers and therapeuthic indicators in patients with cancer. *Clinica Chimica Acta* **206**, 77–82.

SEHON, A.H. (1992). Suppression of antibody responses by conjugates of antigens and monmethoxypoly(ethylene glycol) in *Poly(ethylene glycol) Chemistry. Biotechnological and Biomedical Applications* (M.J. Harris, Ed.), pp. 139–152. Plenum Press, New York.

SELISKO, B., DELGADO, C. AND FISHER, D. (1993). Analysis and purification of monomethoxy-polyethylene glycol by vesicle and gel permeation chromatography. *Journal of Chromatography* **641**, 71–79.

SIROKMAN, G. AND FASMAN, G.D. (1993). Refolding and proton pumping activity of a polyethylene glycol-bacteriorhodopsin water-soluble conjugate. *Protein Science* **2**, 1161–1170.

SMITH, R.A.G., DEWDNEY, J.M., FEARS, R. AND POSTE, G. (1993). Chemical derivatization of therapeutic proteins. *Trends in Biotechnology* **11**, 397–403.

SNYDER, S.L. AND SOBOCINSKI, P.Z. (1975). An improved 2,4,6-trinitrobenzene sulfonic acid method for determination of amines. *Analytical Biochemistry* **64**, 284–288.

SOKOL, R.J., BUTLER-SIMON, N., CONNER, C., HEUBI, J.E., SINATRA, F.R., SUCHY, F.J., HEYMAN, M.B., PERRAULT, J., ROTHBAUM, R.J., LEVY, J., IANNACCONE, S.T., SHNEIDER, B.L., KOCH, T.K. AND NARKEWICZ, M.R. (1993). Multicenter trial of D-a-tocopheryl polyethylene glycol 1000 succinate for treatment of vitamin E deficiency in children with chronic cholestasis. *Gastroenterology* **104**, 1727–1735.

SOUPPE, J. AND URRUTIGOITY, M. (1989). Chemical modification of horseradish peroxidase and papain with w-diamino-poly(oxyethylene)/carboxymethyl bis(dithiocarboxylate) polycondensates. *New Journal of Chemistry* **13**, 503–506.

STAHL, P.D., RODMAN, J.S., MILLER, M.J. AND SCHLESINGER, P.H. (1978). Evidence for receptor-mediated binding of glycoproteins, glycoconjugates, and lysosomal glycosidases by alveolar macrophages. *Proceedings of the National Academy of Sciences of the USA* **75**, 1399–1403.

STOCKS, S.J., JONES, A.J.M., RAMVY, C.W. AND BROOKS, D. (1986). A fluorometric assay of the degree of modification of protein primary amines with polyethylene glycol. *Analytical Biochemistry* **54**, 232–234.

STORY, M.D., VOEHRINGER, D.W., STEPHENS, L.C. AND MEYN, R.E. (1993). L-asparaginase kills lymphoma cells by apoptosis. *Cancer Chemotherapy and Pharmacology* **32**, 129–133.

TAKAHASHI, K., AJIMA, A., YOSHIMOTO, T. AND INADA, Y. (1984a). Polyethylene glycol-modified catalase exibits unexpectedly high activity in benzene. *Biochemical and Biophysical Research Communications* **125**, 761–766.

TAKAHASHI, K., NISHIMURA, H., YOSHIMOTO, T., SAITO, Y. AND INADA, I. (1984b). A chemical modification to make horseradish peroxidase soluble and active in benzene. *Biochemical and Biophysical Research Communications* **121**, 261–265.

TAM, C.M. AND TREMBLAY, A.Y. (1991). Membrane pore characterization: comparison between single and multicomponent solute probe techniques. *Journal of Membrane Science* **57**, 271–287.

TANG, G., WHITE, J.E., GORDON, R.J., LUMB, P.D. AND TSAN, M.-F. (1992). Polyethylene glycol-conjugated superoxide dismutase protects rats against oxygen toxicity. *Journal of Applied Physiology* **74**(3), 1425–1431.

THIBEAULT, D.W., REZAIEKHALIGH, M. AND MABRY, S. (1993). Lung antioxidant enzymes and cardiopulmonary responses in young rats exposed to hyperoxia and treated intratracheally with PEG catalase and superoxide dismutase. *Experimental Lung Research* **19**, 137–151.

TRUJILLO-FERRARA, J., KOIZUMI, G., MUNOZ, O., JOSEPH-NATHAN, P. AND YANEZ, R. (1992). Antitumor effect and toxicity of two new active site directed irreversible ornithine decarboxylase and extrahepatic arginase inhibitors. *Cancer Letters* **67**, 193–197.

UEMURA, T., FUJIMORI, M., LEE, H.-H., IKEDA, S. AND ASO, K. (1990). Polyethylene glycol modified papain catalyzed oligopeptide synthesis from the esters of L-aspartic and L-glutamic acids in benzene. *Agricultural and Biological Chemistry* **54**, 2277–2281.

VERONESE, F.M., LARGAJOLLI, R., BOCCU, E., BENASSI, C.A. AND SCHIAVON, O. (1985). Surface modification of proteins. Activation of monomethoxy-polyethylene glycols by phenylchloroformates and modification of ribonuclease and superoxide dismutase. *Applied Biochemistry and Biotechnology* **11**, 141–152.

VESTLING, M.M., MURPHY, C.M., KELLER, D.A., FENSELAU, C., DEDINAS, J., LADD, D.L. AND OLSEN, M.A. (1993). A strategy for characterization of polyethylene glycol derivatized proteins. A mass spectrometric analysis of the attachment sites in polyethylene glycol-derivatized superoxide dismutase. *Drug Metabolism and Disposition* **21**, 911–917.

VISCO, C., BENASSI, C.A., VERONESE, F.M. AND MIGLIOLI, P.A. (1987). Purification, modification, physico-chemical and pharmacokinetic characterization of arginase, an enzyme of potential use in therapy. *Il Farmaco* **42**, 549–559.

WADA, H., IMAMURA, I., SAKO, M., KATAGIRI, S., TARUI, S., NISHIMURA, H. AND INADA, Y. (1990). Antitumor enzyme: polyethylene glycol-modified asparaginase. *Annals of the New York Academy of Sciences* **613**, 95–108.

WADE, H.E., ELSWORTH, R., HERBERT, D. *et al.* (1968). A new L-asparaginase with antitumor activity. *Lancet* **2**, 776–777.

WARING, P., KOS, F.J. AND MULLBACHER, A. (1991). Apoptosis or programmed cell death. *Medicine Research Review* **11**, 219–236.

WEISS, M., TELFORD, P., KEMPIN, S., KRITZ, A., SOGOLOFF, H., GEE, T., BERMAN, E., SCHEINBERG, D., LITTLE, C., GAYNOR, J., BELANGER, C., BERTINO, J., ABDREEFF, M., MCKENZIE, S. AND CLARKSON, B. (1993). Severe toxicity limits intensification of induction therapy for acute lymphoblastic leukemia. *Leukemia* **7**, 832–837.

WELLS, R.J., WOODS, W.G., LAMPKIN, B.C., NESBIT, M.E., LEE, J.W., BUCKLEY, J.D., VERSTEEG, C. AND HAMMOND, G.D. (1993). Impact of high-dose cytarabine and asparaginase intensification on childhood acute myeloid leukemia: a report from the Children's Cancer Group. *Journal of Clinical Oncology* **11**, 538–545.

WIEDER, K.J., PALCZUK, N.C., VAN ES, T. AND DAVIS, F.F. (1979). Some properties of polyethylene glycol: phenylalanine ammonia lyase adducts. *Journal of Biological Chemistry* **254**, 12579–12587.

WIRTH, P., SOUPPE, J., TRITSCH, D. AND BIELLMANN, J.-F. (1991). Chemical modification of horseradish peroxidase with ethanol-methoxypolyethylene glycol: solubility in organic solvents, activity and properties, *Bioorganic Chemistry* **19**, 133–142.

WOGHIREN, C., SHARMA, B. AND STEIN, S. (1993). Protected thiol-polyethylene glycol: a new activate polymer for reversible protein modification. *Bioconjugate Chemistry* **4**, 314–318.

YOSHIMOTO, T., NISHIMURA, H., SAITO, Y., SAKURAI, K., KAMISAKI, Y., WADA, H., SAKO, M., TSUJINO, G. AND INADA, Y. (1986). Characterization of polyethylene glycol-modified L-asparaginase from *Escherichia coli* and its application to therapy of leukemia. *Japanese Journal of Cancer Research* **77**, 1264–1270.

YOSHINGA, K. AND HARRIS, J.M. (1989). Effects of coupling chemistry on the activity of poly(ethylene glycol)-modified alkaline phosphatase. *Journal of Bioactive and Compatible Polymers* **4**, 17–24.

ZALIPSKY, S. AND LEE, C. (1992). Use of functionalized polyethylene glycols for modification of polypeptides. In *Poly(ethylene glycol) Chemistry. Biotechnological and Biomedical Applications* (M.J. Harris, Ed.), pp. 347–370. Plenum Press, New York.

ZALIPSKY, S., GILON, C. AND ZILKHA, A. (1983). Attachment of drugs to polyethylene glycols. *European Polymer Journal* **19**, 1177–1183.

ZALIPSKY, S., SELTZER, R. AND MENON-RUDOLPH, S. (1992). Evaluation of a new reagent for covalent attachment of polyethylene glycol to proteins. *Biotechnology and Applied Biochemistry* **15**, 100–114.

ZALIPSKY, S., SELTZER, R. AND NHO, K. (1991). Succinimidyl carbonates of polyethylene glycol. Useful reactive polymers for preparation of protein conjugates. In *Polymeric Drugs and Drug Delivery Systems* (R.L. Dunn and R.M. Ottenbrite, Eds), pp. 91–100. American Chemical Society Symposium Series No. 469, Washington, DC.

ZAWILSKA, K., ZOZULINSKA, M., TUROWIECKA, Z., BLAHUT, M., DROBNIK, L. AND VINAZZER, H. (1993). The effect of a long-acting recombinant hirudin (PEG–hirudin) on experimental disseminated intravascular coagulation (DIC) in rabbits. *Thrombosis Research* **69**, 315–320.

ZIMRAN, A., HOLLAK, C.E.M., ABRAHAMOV, A., VA OERS, M.H.J. KELLY, M. AND BEUTLER, E. home treatment with intravenous enzyme replacement therapy for Gaucher disease: an international collaborative study of 33 patients. *Blood* **82**, 1107–1109.

10
Conjugates of Proteins with Block Co-polymers of Ethylene and Propylene Oxides

I.N. TOPCHIEVA[1] AND N.V. EFREMOVA[2]

[1]*Department of Chemistry, Moscow State University, Moscow, Russia and* [2] *A.N. Bach Institute of Biochemistry, Russian Academy of Science, Moscow, Russia*

Introduction

Covalent attachment of non-ionic water-soluble polymers to proteins is a progressive way of imparting new useful properties to a protein molecule. Among the polymers so far studied for this purpose, polyvinylpyrrolidone (Von Spect and Brendel, 1977), dextrans (Marshall and Rabinowitz, 1975) and polyethylene glycol (PEG) (Abuchowski *et al.*, 1977a,b; Takoi *et al.*, 1989) are most often employed. PEG, with a history as a plasma expander, being non-immunogenic, amphiphatic and non-toxic, has been applied to various enzymes as a superior agent for modification (Kurganov and Topchieva, 1991). The products of the reaction, polymer–protein conjugates, are characterized by reduced immunogenicity, protracted retention in the circulation and, in some cases, by enhanced stability (Veronese *et al.*, 1989). These features are due to the effect of 'steric stabilization' of the protein globule caused by attachment of polymer chains.

An important property of PEG–protein conjugates is their solubility in organic solvents. That is why they are used as biocatalysts for peptide synthesis (Matsushima, Okada and Inada, 1984), transesterification reactions (Yoshimoto *et al.*, 1984) and the process of ester amination (Takahashi *et al.*, 1985).

The next stage in the construction of polymer–protein conjugates was the modification of the polymer chain by the insertion of foreign fragments (blocks) in the macromolecule by a co-polymerization process. Using this approach PEG was altered by using well-known, commercially available, co-polymers, namely co-polymers of

Abbreviations: BBI, soybean Bauman-Birk proteinase inhibitor; BF, benzaflavin (2', 3', 4', 5'-tetrabenzoyl-5-acetyl-1,5-dihydroriboflavin); BSA, bovine serum albumin; BTEE, *N*-benzoyl-*L*-tyrosine ethyl ester; CHT, α-chymotrypsin; Cyt-C, cytochrome *c*; DNP, 2,4-dinitrophenol; DSC, differential scanning calorimetry; HPLC, high-performance liquid chromatography; MDA, malonic dialdehyde; NMR, nuclear magnetic resonance; PEG, polyethylene glycol; PK, pyruvate kinase; PMR, proton magnetic resonance; POL, peroxidation of lipids; RPE, REP, pluronics, block co-polymers of ethylene oxide and propylene oxide; TNBS, 2,4,6-trinitrobenzene sulphonic acid; UV, ultraviolet.

Biotechnology and Genetic Engineering Reviews — Vol. 12, December 1994
0264–8725/94/12/357–382 $20.00 + $0.00 © Intercept Ltd, P.O. Box 716, Andover, Hampshire SP10 1YG, UK

ethylene oxide and propylene oxide – ('pluronics') (Lunsted and Schmolka, 1976). These polymers are non-ionic surfactants in which polyethylene oxide and polypropylene oxide form the hydrophylic and hydrophobic parts of the compounds. In contrast to PEG–protein conjugates, conjugates based on pluronics are characterized by a diversity of polymer–protein structures. The amphiphilic character of protein–pluronic conjugates provides new functions of the modified proteins, such as a capability for translocation across biological membranes, maintenance of effective solubilization, and ultimately the transport of insoluble biologically active compounds, as well as the ability to form molecular assemblies, by interaction with various amphiphilic compounds. In this chapter we report methods of synthesis, purification and characterization of pluronic–protein conjugates, as well as their functional properties.

Synthesis, isolation and characterization of conjugates

SYNTHESIS OF CONJUGATES

Conjugates of proteins with pluronics are synthesized by the interaction of protein ε-amino groups with monoaldehyde derivatives of polyalkylene oxides leading to the formation of labile amidine groups, which are subsequently reduced by sodium borohydride ($NaBH_4$) or sodium cyanoborohydride:

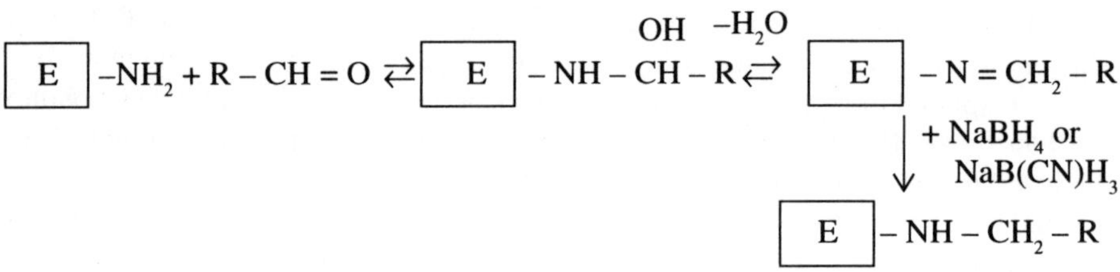

where R stands for polymer chain.

The following proteins have been used for the synthesis of protein–pluronic conjugates: bovine serum albumin (BSA) (Topchieva *et al.*, 1991), bovine α-chymotrypsin (CHT) (Efremova, Mozhaev and Topchieva, 1992), pyruvate kinase (PK) from rabbit skeletal muscles, ricin from castor-oil plant seeds and viscumin from mistletoe (Teplova *et al.*, 1993), classical soybean Bauman–Birk proteinase inhibitor (BBI) (Larionova *et al.*, 1993) and cytochrome *c* (Kirillova *et al.*, 1993).

The reactions of CHT and BBI conjugation were performed in the presence of reversible inhibitors of the active site of the enzyme. Sodium cyanoborohydride is a preferable reducing agent, for its use leads to improved preservation of the protein's native structure, while $NaBH_4$ is able to reduce S–S bonds between polypeptide chains.

The monoaldehyde derivatives of PEG and pluronics of RPE and REP types (R, starting radical; P, polypropylene oxide block; E, polyethylene oxide block) were synthesized by oxidation in the presence of MnO_2 (Boccu, Largajolli and Veronese, 1983). The following monoaldehydes have been used for protein modification:

1. monomethoxy-PEG monoaldehyde (PEG-CHO), $CH_3\text{–}(OCH_2CH_2\text{–})_{44}\text{–}OCH_2CHO$;
2. RPE block co-polymer monoaldehyde (RPE-CHO), $C_4H_9\text{–}(OCH(CH_3)CH_2)_{14}\text{–}(OCH_2CH_2)_{20}\text{–}OCH_2CHO$;
3. REP block co-polymer monoaldehyde (REP-CHO), $C_4H_9\text{–}(OCH_2CH_2)_{20}\text{–}(OCH(CH_3)CH_2)_{14}\text{–}OCH_2CHO$.

Evidently the only difference between block co-polymers (2) and (3) is in the position of polyethylene oxide and polypropylene oxide blocks towards the aldehyde group. By using these polymer reagents, two main types of conjugates with hydrophobic blocks at the periphery or inside conjugate structure were synthesized (*Figure 1*).

Molecular models of the conjugates studied in this work are presented in *Figure 1*. The conjugates have a star-shaped structure, where protein occupies the central place with polymer chains growing from it as branches. The length of polymer chain is the same order of, or even bigger than, the diameter of the protein molecule. That is why the protein in the conjugate should be sterically stabilized towards the action of macromolecular substrates, antibodies and proteolytic enzymes. It is evident that this property depends on the number and type of polymer chains distribution on the surface of protein.

Because of the amphiphilic properties of the polymeric reagents, two principal types of conjugate structure, of variable polymer chain distribution, may be obtained at the stage of synthesis:

1. polymer chains may be located in one part of the protein globule surface, forming a domain structure (IIa in *Figure 1*), or
2. polymer chains may be stochastically distributed on its surface (IIb in *Figure 1*).

When the conjugation reaction is carried out in aqueous buffering media, polymer reagents such as pluronics are present in the form of micellar microphases (Osipova, Topchieva and Kasaikin, 1990). It is highly probable that in such systems modification will lead to a significant heterogeneity of the conjugates (IIa-type structures). In contrast, in aqueous–organic mixtures pluronics are present in the form of single molecules (Osipova, 1990). The use of mixed solvents in the synthesis of conjugate would obviously lead to a stochastical type of polymer distribution on the protein globule surface (IIb-type structures in *Figure 1*). It should be noted that the ethanol concentration in the reaction medium was much lower than denaturing concentration, for the proteins under investigation (Khmelnitsky *et al.*, 1991).

It was necessary to find methods sensitive to different types of polymer distribution. One of the approaches to the investigation of the steric arrangement of the polymer chains involves the study of the ability of bound polymers to interact with the bulky cyclic reagent β-cyclodextrin. It has been shown that β-cyclodextrin forms inclusion-type complexes with pluronics (Topchieva *et al.*, 1993a, 1994b). It was natural to suppose that, in the case of the irregular type of distribution (structure IIa in *Figure 1*), polymer chains attached to protein will not be able to interact with a bulky reagent for steric reasons. It has been found that IIb-type conjugates (i.e. those

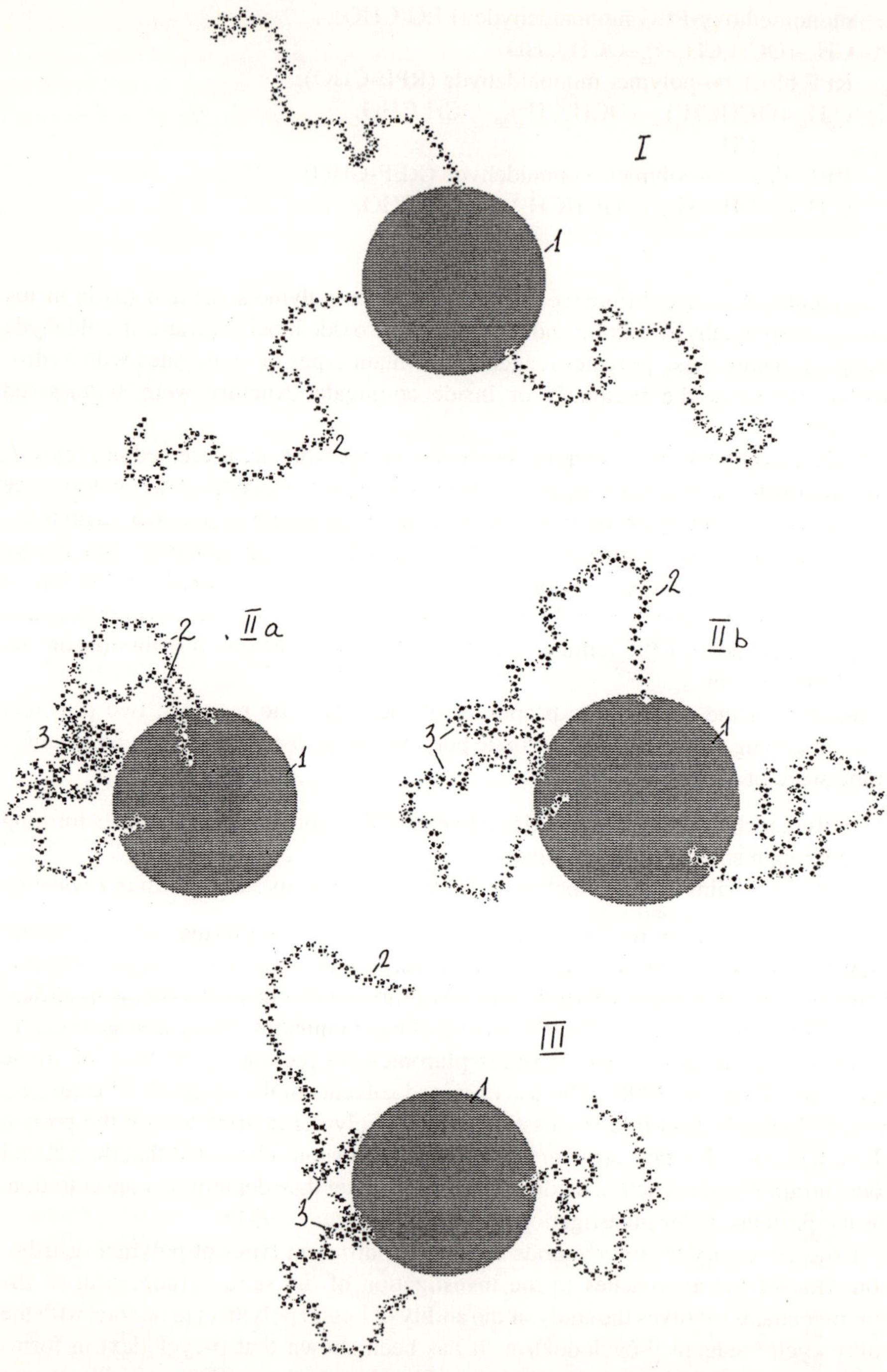

Figure 1. Molecular models of protein conjugates with PEG (I) and pluronics of various types: IIa, conjugates with irregular polymer chain distribution; IIb, with statistical distribution of polymer chains; III, with hydrophilic blocks at the periphery of the structure. 1, Protein globule; 2, polyethylene oxide block; 3, polypropylene oxide block. The radius of central globule was chosen to correspond to the average radius of a CHT molecule, 25 Å (Krigbaum and Godwin, 1968).

with a statistical distribution of polymer chains) as well as free pluronics react with β-cyclodextrin, forming water-insoluble crystalline compounds. In contrast, IIa-type conjugates (i.e. those presumed to have domain structures) cannot form complexes with β-cyclodextrin. The differences in the biological behaviour of these two structures will be reflected in the next part of this article.

PURIFICATION OF CONJUGATES

Purification of pluronic conjugates from the excess of polymer reagent offers more difficulties than purification procedures in the case of PEG–protein conjugates. This is accounted for by the amphiphilic properties of pluronics, which lead to association between unreacted reagent and polymer chains in the conjugate. This explains the low efficiency of commonly used methods of purification, such as dialysis and ultrafiltration in aqueous media. Effective purification can be achieved by using aqueous organic media, which leads to the dissociation of these complexes. Taking into account all these considerations, two methods of purification have been proposed: precipitation of conjugate by acetone and gel-permeation chromatography using 20% (v/v) ethanol as eluent. Thin-layer chromatography was used to demonstrate the absence of polymer contamination in the conjugate.

CHARACTERIZATION OF CONJUGATES

Tables 1 and *2* present characteristics of conjugates of proteins with PEG and pluronics. The protein content in the conjugates was determined either spectroscopically at 280 nm, or using a biuret reaction (Kochetov, 1980) (native proteins were used as standards). It was shown that neither PEG nor pluronics interfere with the protein assay using a spectrophotometric method (A_{280} or the biuret reaction (Efremova, Mozhaev and Topchieva, 1992). The results of both methods were in good agreement (see *Tables 1* and *2*).

Table 1. Characteristics of conjugates of CHT with PEG and pluronics

N	Polymer	Type of conjugate	Degree of modification	Wt % of CHT	Number of unmodified surface NH_2-groups (m)	Designation of conjugate
1	PEG	I[a]	3	81.0 (80.5)[b]	12	(PEG)$_3$–CHT
2	PEG	I	9	58.1 (58.8)	5	(PEG)$_9$–CHT
3	RPE	IIa	2	86.2 (86.1)	13	(RPE)$_2$–CHT-a
4	RPE	IIa	3	82.1 (79.8)	12	(RPE)$_3$–CHT-a
5	RPE	IIa	6	67.5 (66.7)	8	(RPE)$_6$–CHT-a
6	RPE	IIa	7	64.0 (64.1)	–	(RPE)$_7$ CHT-a
7	RPE	IIa	10	55.6	4	(RPE)$_{10}$–CHT-a
8	RPE	IIb	3	80.6	11	(RPE)$_3$–CHT-b
9	RPE	IIb	5	71.5 (71.0)	9	(RPE)$_5$–CHT-b
10	RPE	IIb	6	67.4 (67.5)	–	(RPE)$_6$–CHT-b
11	RPE	IIb	11	45.5 (46.0)	–	(RPE)($_{11}$–CHT-b
12	REP	III	3	80.6 (80.7)	12	(REP)$_3$–CHT
13	REP	III	7	64.2 (64.0)	–	(REP)$_7$–CHT

[a] I, IIa, IIb and III are the conjugates having structures presented in *Figure 1*.

[b] The amount of CHT present in each conjugate was determined by UV-spectroscopy and by the biuret procedure (values in parentheses).

Table 2. Characteristics of conjugates of BSA with PEG and pluronics

N	Polymer	Type of conjugate	Degree of modification	Content of BSA, wt%			Designation of conjugate
				Biuret method	280 nm	^{13}C NMR	
1	PEG	I	5	87	87	–	$(PEG)_5$–BSA
2	PEG	I	14	70	71	–	$(PEG)_{14}$–BSA
3	RPE	IIa	5	87	87	–	$(RPE)_5$–BSA-a
4	RPE	IIa	8	81	80	78	$(RPE)_8$–BSA-a
5	RPE	IIb	15	70	69	70	$(RPE)_{15}$–BSA-b
6	REP	III	6	85	84	–	$(REP)_6$–BSA
7	REP	III	11	75	75	–	$(REP)_{11}$–BSA

^{13}C NMR spectroscopy

^{13}C NMR spectroscopy presents a direct method for determination of both protein and polymer content in conjugates, because there exist regions where signals from polymer and protein do not interfere. Two conjugates of BSA with RPE co-polymer (preparations 4 and 5 in *Table 2*) were analysed by this method; samples of BSA and RPE mixtures served as standards. The results turned out to be in good agreement with data obtained by UV-spectroscopy and a biuret method (see *Table 2*). One may see that each of the methods used is suitable for the analysis of conjugate composition.

The degree of modification, *n*, which denotes the average number of polymer chains coupled with a protein molecule, was calculated from data on conjugate composition using the following formula:

$$n = \frac{\text{weight \% of polymer/MW of polymer}}{100 - \text{weight \% of polymer/MW of protein}} \qquad \text{Eq. (1)}$$

Titration of surface amino groups in conjugates

Titration of NH_2-groups with 2,4,6-trinitrobenzene sulphonic acid (TNBS) is commonly used for the determination of the degree of modification of protein in conjugates. Beforehand one should determine the number of surface primary amino groups in the parent protein. Then the degree of modification, *n*, is given by the formula:

$$n = \alpha - m, \qquad \text{Eq. (2)}$$

where α stands for the total number of surface NH_2-groups in the protein, and *m* for the number of unmodified NH_2-groups in the conjugate. In our case, however, not only unmodified NH_2-groups but also the resulting secondary amino groups of conjugate could interact with TNBS (Fields, 1971). Hence, the number of amino groups in conjugates as determined by this reaction remained the same as in the parent protein. Thus, it was necessary to use a reagent for primary amino groups only. We have chosen *o*-phthalic aldehyde and titrated NH_2-groups by the method of Roth (1971). This method was applied to a number of conjugates on the basis of CHT (column 4 in *Table 1*). The degree of modification calculated by equation (2), assuming $\alpha = 15$ for CHT (Birktoft and Blow, 1972), was in good agreement with the results obtained on the basis of protein content determination by UV-spectroscopy and the biuret method, and equation (1).

Enzymatic properties of conjugates

ENZYMATIC PROPERTIES OF CHT–PLURONIC CONJUGATES

Kinetic characteristics of the reactions of enzymatic hydrolysis of conjugates on the basis of CHT are presented in *Table 3*. These data show that:

1. Conjugates retain high enzymatic activity towards both substrates investigated, i.e. a low molecular weight substrate BTEE and casein. This presents a strong argument in favour of the retention of the protein native structure in conjugates.
2. There are insignificant changes in the values of K_m and catalytic constant for hydrolysis of low molecular weight substrate BTEE by conjugates, compared to the parent enzyme. Analogous results were obtained for hydrolysis of high molecular weight substrate, such as casein. The preservation of enzymatic properties may be explained by the fact that the modification of CHT is carried out in the presence of a reversible inhibitor (*N*-acetyl-L-tyrosine or *N*-benzoyl-L-tyrosine), whose binding to enzyme sterically screens the active site.
3. A noticeable decrease in enzymatic activity was observed for conjugates with high degrees of modification (11 and more, with maximal = 15). It is seen that the decrease in activity is due to the destruction of active sites in conjugates, but not to the lowering of the value of k_{cat} (*Table 3*, $N = 5$).

Table 3. Enzymatic properties of CHT conjugates with PEG and pluronics

N	Enzyme, conjugate	Enzymatic hydrolysis of casein	Enzymatic hydrolysis of BTEE			Content of active sites (%)
		Activity (%)	Activity (%)	k_{cat}, s^{-1}	K_m, M × 10^5	
1	CHT	100	100	31.2±0.5	6.27±0.94	80
2	$(RPE)_5$–CHT-a	102	97	42.4±2.8	13.0±2.9	61
3.	$(RPE)_3$–CHT-b	103	115	–	–	80
4.	$(RPE)_6$–CHT-b	96	98	32.5±1.2	6.18 0.82	67
5.	$(RPE)_{11}$–CHT-b	50	57	37.0±0.4	21.7 3.5	40
6.	$(PEG)_6$–CHT	97	98	40.4±0.8	9.3 1.5	80
7.	$(REP)_3$–CHT	98	100	–	–	80

THERMAL STABILITY OF CONJUGATES

It is well known that chemical modification of proteins is an effective method to change their thermal stability. In the case of PEG–protein conjugates it has been shown that attachment of polymer chains either increases or does not change thermal stability of proteins (Abuchowski *et al.*, 1977b; Veronese *et al.*, 1989). The peculiarity of conjugates with pluronics consists of the possibility of regulating thermal stability of modified protein by variation of a number of parameters, such as the degree of modification, the number and type of block arrangement, hydrophilic–lipophilic balance, and the type of polymer distribution on the surface of protein globule. For these reasons we studied the thermal inactivation of all types of conjugates of CHT. The thermal stabilities of PEG–CHT conjugate and of native CHT were studied for comparison. Kinetic curves of thermoinactivation of CHT and its conjugates with polymers are presented in *Figure 2*. The thermostability of conjugates of CHT decreases in the order: $(PEG)_9$–CHT > $(RPE)_5$–CHT-b > $(RPE)_6$–CHT-a > $(RPE)_3$–CHT-a > $(REP)_3$–CHT-a > CHT. It should be noted that all CHT–pluronic conjugates with the exception of $(RPE)_5$–CHT-b are of the IIa type.

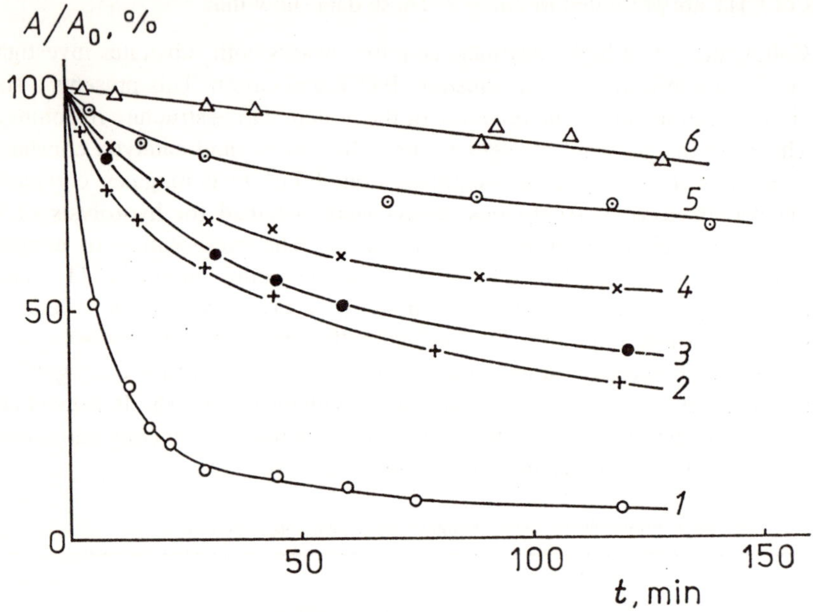

Figure 2. Kinetics of thermal inactivation of CHT and its conjugates with polyalkylene oxides: 1, CHT; 2, $(REP)_3$–CHT; 3, $(RPE)_3$–CHT-a; 4, $(RPE)_6$–CHT-a; 5, $(RPE)_5$–CHT-b; 6, $(PEG)_9$–CHT. Conditions of thermal inactivation: 0.2 M Tris–HCl, pH 8.05; 45°C. Concentration of protein in probe is 0.1 μM. The solution of CHT or its conjugate was incubated in a temperature-controlled cell at 45.0±0.05°C. After selected time intervals, the residual activities of aliquots were assayed by determining the initial rate of hydrolysis of BTEE. The enzymatic activity is expressed as a percentage of the initial activity of corresponding preparations.

It is seen that the rate of thermal inactivation of the enzyme decreases significantly as a result of the attachement of three polymer chains. The stabilizing effect becomes even more pronounced with an increase in the degree of modification of CHT. The maximal stabilizing effect is achieved for CHT–PEG conjugates with nine polymer chains. It is remarkable that the conjugate $(RPE)_5$–CHT-b, with a statistical mode of distribution, possesses higher thermostability than conjugate $(RPE)_6$–CHT-a, with a domain type of distribution. At the same time, the difference in the placement of hydrophobic blocks in conjugates with an equal number of co-polymer chains leads to change in their thermostability: a conjugate of type IIa is more stable than a conjugate of reversed type (III).

One of the unique properties of protein conjugates with pluronics is their ability to interact with both synthetic (surfactants) and natural (lipids) amphiphilic compounds

(Topchieva *et al.*, 1989; Efremova, Mozhaev and Topchieva, 1992). The functional aspect of these interactions was demonstrated by study of thermoinactivation. Thermal stability of conjugates may be enhanced by the addition of free pluronics into the incubation system. This phenomena is illustrated in *Figure 3*, which demonstrates thermal inactivation of conjugate $(RPE)_4$–CHT-a (type IIa) in the presence of added free pluronic in molar excess over the conjugate, ranging from 50 to 750. The kinetic curves in *Figure 3* show that addition of free pluronics to the solution up to a 500-fold molar excess increases the stability of enzyme in the conjugate. Further increase in the pluronic concentration does not significantly influence the thermal inactivation kinetics, perhaps due to saturation of the conjugate by associated co-polymer.

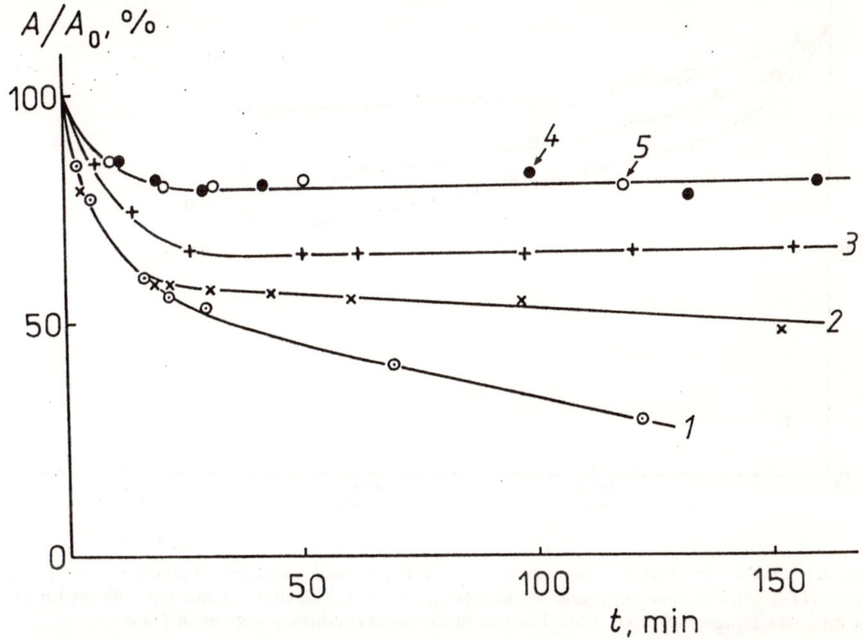

Figure 3. Kinetics of thermal inactivation of conjugate $(RPE)_4$–CHT-a alone (1) and in the presence of added free RPE co-polymer in molar excess (with respect to CHT) of 50 (2), 100 (3), 500 (4) and 750 (5). Thermal inactivation conditions were as in *Figure 2*.

One can assume that an increase in the stability of CHT–pluronic conjugates compared to the parent protein is due to the ability of polymer chains to screen hydrophobic domains at the surface of protein (Mozhaev and Martinek, 1984; Mozhaev, Martinek and Berezin, 1988). An additional increase in the stability of CHT–pluronic conjugates in the presence of free pluronics can be accounted for by two explanations. The first is complex formation between pluronics and the protein of the conjugate during incubation with heating. For steric reasons this type of interaction is mostly improbable. The second is interaction between polymer chains of conjugate and added pluronics. Probably the interaction between free and bound pluronics, leading to formation of a polymer coat, are mainly due to hydrogen bonds bridging polyether chains, involving water molecules. As a result, the contacts stabilizing the structure of the protein globule are not destroyed and encagement of the protein is realized.

Addition of egg-yolk lecithin and β-cyclodextrin produces an opposite effect on the stability of protein in conjugates (*Figure 4*). All additives themselves do not affect the behaviour of the native enzyme. Probably in the latter case these additives are stronger competitors for hydrophobic bond formation with pluronics than the corresponding domains of the protein globule. As a result, the steric arrangement of polymer chains is changed in such a way that the hydrophobic domains of protein become exposed again and destabilization of the enzyme structure as compared with conjugate takes place.

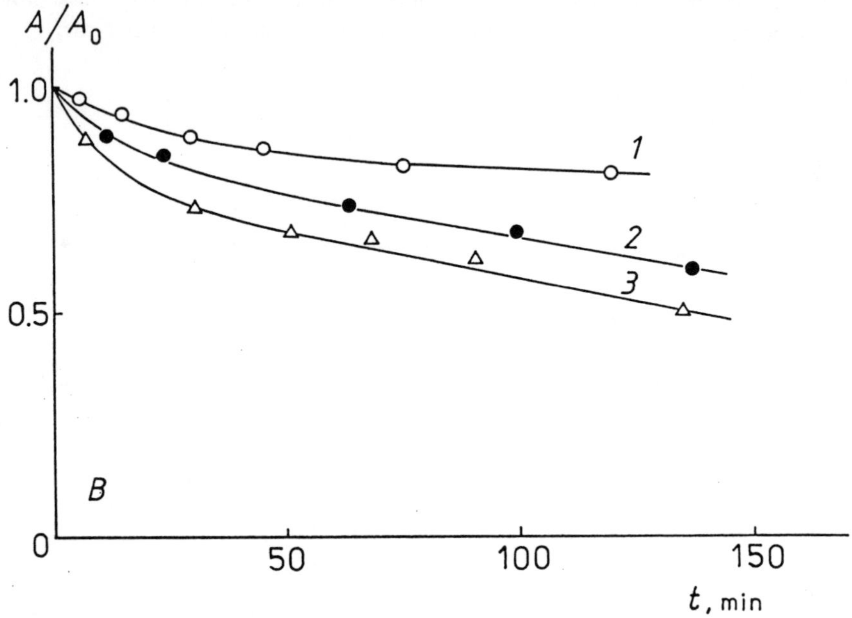

Figure 4. Kinetics of thermal inactivation of a CHT–proxanol conjugate obtained by method B $(RPE)_3$–CHT-b (1); the same conjugate in the presence of β-cyclodextrin (2) and egg-yolk lecitin (3). Molar ratio of conjugate:additive, 1:50. Thermal inactivation conditions were as in *Figure 2*.

The data presented above give examples of changes in thermal stability of conjugates due to their interaction with amphiphilic molecules. As a matter of fact, study of thermal stability may be successfully used for detection of interactions between conjugates and different partners. On the other hand, non-covalent supramolecular structures based on conjugates and additives initiate new approaches to the elucidation of the mechanisms of transport and the functioning of proteins as drugs.

Physico-chemical properties of conjugates

DIFFERENTIAL SCANNING CALORIMETRY (DSC)

Taking into consideration the results on thermal stability of conjugates, gained through investigation of their enzymatic properties (see previous section), it was interesting to obtain proof of the increase in conjugate thermal stability by the direct method of DSC.

Thus, thermally induced transitions in proteins modified with pluronics and PEG have been studied by the method of DSC. *Figure 5* shows DSC data for CHT and its conjugates with polymers in aqueous buffering media. The heat sorption peak for CHT is slightly asymmetric and possesses one maximum with T_{max}=49.6°C, which is in good agreement with the results presented by Tischenko, Tiktopulo and Privalov (1974).

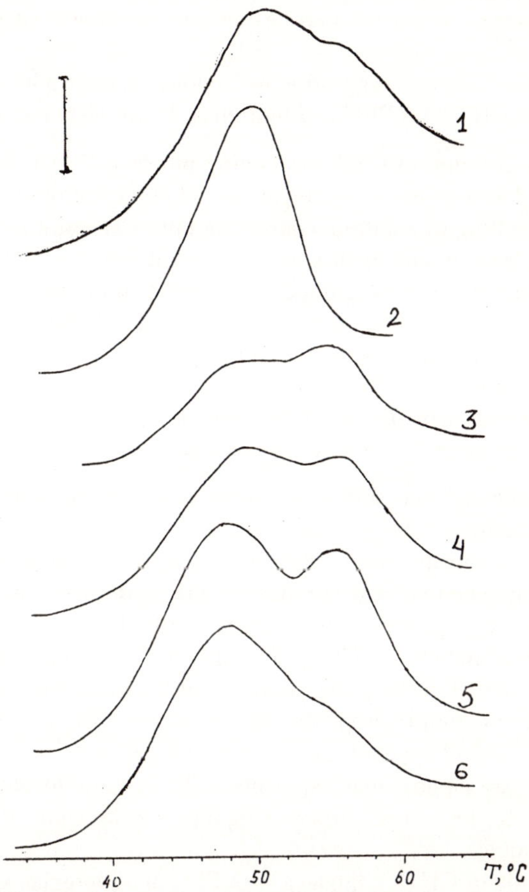

Figure 5. Temperature dependences of excess heat capacity (Δc_p) for CHT (2), and its conjugates $(PEG)_3$–CHT (3), $(RPE)_3$–CHT-b (4), $(RPE)_3$–CHT-a (5), $(REP)_3$–CHT (6), $(RPE)_7$–CHT-b (1). The concentration was 1.0 mg ml^{-1}; buffer, 20 mM phosphate, pH 7.0. The vertical bar corresponds to 1.0 kJ $K^{-1}kg^{-1}$.

Considerable changes in the form of heat sorption peaks are observed for CHT conjugates with polymers (curves 3–6 on *Figure 5*). Besides the first maximum, between the temperature limits of 47.6 and 49.6°C, there arises a second maximum on heat sorption curves of conjugates $(RPE)_3$–CHT-a, $(RPE)_3$–CHT-b and $(PEG)_3$–CHT, which is located at 54.5–55.5°C. On the heat sorption curve of $(REP)_3$–CHT there arises a shoulder in the same temperature region. A trivial explanation of this phenomenon would be that the existence of two maxima on heat sorption curves is due to the heterogeneity of conjugates, namely that the conjugate fraction with a low degree of modification gives rise to the first maximum and the fraction with higher degree of modification to the second. This suggestion is not confirmed by DSC data on conjugates with a high degree of modification (curve 1) which revealed only one maximum (T_{max} = 51.2) and a shoulder in the region 54.5–55.5°C.

One may suppose that in the course of protein globule denaturation hydrophobic parts of the protein molecule become accessible to interaction with hydrophobic blocks of co-polymer, coupled to protein. The emergence of the second maximum on heat sorption curves of conjugates may be due to the melting of such polymer–protein complexes.

In order to investigate this suggestion we conducted some additional experiments, using mixtures of CHT with PEG and pluronics. It was shown that:

1. Heating of a solution of CHT, containing pluronic (1:20), from 20°C to 60°C (1 k min^{-1}), does not lead to aggregation and precipitation of protein.
2. Analogous heating of solution, containing pure CHT or a mixture of CHT and PEG (1:20), leads to aggregation of protein followed by precipitation.
3. Protein-containing product, gained from a CHT–pluronic mixture after heating and purification from excess pluronic, contained only 50% of protein. The rest, according to PMR data, is a bound pluronic.
4. The heat sorption curve of a solution containing a mixture of CHT and pluronics has only one peak with T_{max} = 54.5°C, which coincides with the position of the second maximum on DSC curves of conjugates.

These facts demonstrate the ability of pluronics to interact with CHT in aqueous solutions under heating.

We should focus our attention on the fact that on the curve of heat sorption of $(PEG)_5$–CHT conjugate two peaks of melting were observed. An explanation of this effect relies on the concept of polyalkylene oxide solubility in water. According to the theory of Kjellander and Florin (1981), solubility of PEG is due to its ability to fit into the crystal structure of water. The coupling of hydrophobic groups or complex fragments containing both hydrophobic and polar groups to the end groups of PEG leads to a sufficient change of the conformational state of polymer chains in conjugates. One may suppose that fragments of PEG chains, located in the vicinity of the protein globule, lose their ability to fit into the structure of water and hence become hydrophobic.

Thus, DSC study of CHT conjugates with PEG and pluronics showed that:

1. Conjugates are more stable than parent protein.
2. Thermal stability of conjugates on the basis of pluronics is comparable with thermal stability of PEG–protein conjugates.
3. The emergence of the second maximum on DSC curves of conjugates is very likely due to the formation of intramolecular polymer–protein complexes.

SEDIMENTATION OF CONJUGATES

Hydrodynamic properties of conjugates were investigated using sedimentation analysis. Sedimentation coefficients obtained for the conjugates and for the initial proteins are summarized in *Table 4*. One may see that alteration of the sedimentation coefficients of proteins caused by the modification is negligible. Thus sedimentation data show that solutions of conjugates do not contain products of intermolecular interaction under the investigated conditions.

Table 4. Sedimentation coefficients for proteins and their conjugates with PEG and pluronics

N	Protein conjugate	Concentration (mg ml^{-1})	S_{20}, S
1	BSA	1	4.3±0.2
2	$(PEG)_{15}$–BSA	1	4.2±0.2
3	$(RPE)_8$–BSA-a	1	4.5±0.3
		2	4.0±0.2
		4	4.0±0.2
		6	4.2±0.2
		10	4.6±0.2
4	$(REP)_6$–BSA	1	4.1±0.2
5	CHT	1	2.0±0.15
		4	2.02±0.15
		8	2.2±0.17
		10	2.45±0.12
6	$(RPE)_3$–CHT-a	1	2.3±0.1
		3	2.42±0.15
		5	2.51±0.16
		6	2.61±0.21
		7	2.60±0.18
7	(RPE_3)–CHT-b	1	1.91±0.13
		3	1.99±0.15
		5	2.11±0.08
		10	2.25±0.13

T=20°C, 48 000 min^{-1}, 0.02 M phosphate buffer, pH 7.0.

Even in the absence of intermolecular interactions, the attachment of polymer chains might have influenced the value of sedimentation coefficients. The fact that no considerable alterations of sedimentation coefficients of proteins after modification was observed should mean that the effect of growth of molecular mass is compensated for by the flotation action of polymer chains. It should be noted that the polyalkylene oxides used in this work are characterized by a flotation factor value near 1, so they do not sediment in the ultracentrifuge, but have a tendency to flotation (Bailey and Koleske, 1966).

HYDROPHOBIC CHROMATOGRAPHY

Reverse-phase hydrophobic high-performance liquid chromatography (HPLC) was used to evaluate the relative hydrophobicity of conjugates. *Figure 6* shows the results of hydrophobic HPLC for native BBI and $(RPE)_5$–BBI conjugate (Larionova, 1993). The retention volume of the conjugate is less than that of the native protein, thus indicating its lesser affinity to the hydrophobic sorbent. The same tendency was observed for CHT and its conjugates with pluronic. At first glance, this result would seem surprising since the polymer chains entering the conjugate contain hydrophobic blocks. However, this may be accounted for by the fact that the surface of the protein globules of the conjugate is covered mainly with the hydrophilic blocks of the polymer, due to the intramolecular interactions of the hydrophobic blocks of pluronic and the hydrophobic regions of the inhibitor molecule.

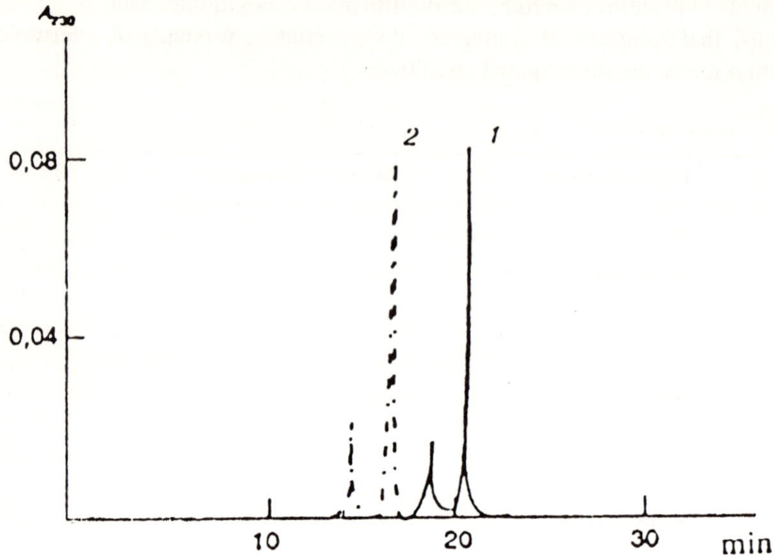

Figure 6. High-performance liquid chromatography on a Zorbax C-8 column. The eluent was 70% acetonitrile. (1) Native BBI; (2) $(RPE)_5$–BBI-a conjugate.

Conformational models of conjugates on the basis of proteins and pluronics

The whole complex of data obtained by applying hydrophobic chromatography, sedimentation and DSC, indicate the presence of general regularities in structure formation in conjugates synthesized by coupling of pluronics to proteins. These properties have been demonstrated most clearly in the case of conjugates with hydrophobic blocks at the periphery of the conjugate structure (structure II). For these conjugates excess of hydrophobicity may be compensated either by formation of intermolecular associates, or intramolecular structures. As association products of these conjugates in diluted aqueous buffers were not detected, it is natural to propose the formation of structures stabilized by intramolecular interactions. The increase in hydrophilicity of conjugates as compared to the native protein, shown by the method of hydrophobic chromatography, leads to suggestion of the formation of 'intramolecular micelles' in which polypropylene oxide blocks 'adhere' to the surface of protein.

It may be suggested that formation of compact structures of similar type presents prerequisites for complexation between polymer chains of pluronics and protein at elevated temperatures.

It should be particularly emphasized that these structures possess a certain conformational lability. This property is revealed in the presence of hydrophobic or amphiphilic 'partners', such as lipids (Topchieva, 1989), pluronics (Efremova, Mozhaev and Topchieva, 1992) and cyclodextrins (Polyakov, 1993; Topchieva, 1993a; Topchieva, 1994b), and leads to the formation ofmixed association products or complexes. Presumably, these processes may proceed by unwrapping of the intramolecular micelle of the conjugate and the interaction with hydrophobic parts of the 'partner'.

Membranotropic properties of conjugates

The ability of pluronics to cross the biological membrane (Topchieva, 1989) leads one to expect that this property could be transmitted to their conjugates with hydrophilic proteins. Translocation of conjugates through the biological membrane was investigated by using the following approach: the influence of these compounds upon the rate of oxygen consumption by liver mitochondria and thymus lymphocytes was measured. If the hydrophobic substances cross the lymphocyte plasma membrane and reach the NADH dehydrogenase site of the mitochrondrial respiratory chain, capable of binding hydrophobic substances, one should observe inhibition of oxygen consumption.

We investigated conjugates of pluronics with BSA, CHT and cytochrome *c*, possessing a variety of molecular masses, with the idea of establishing a structure–function (membranotropic properties) correlation. Detailed investigations were performed using conjugates on the basis of CHT (Topchieva, 1994b). To elucidate the influence of the position of the hydrophilic and hydrophobic blocks of pluronics in protein conjugates on their ability to suppress the respiration of lymphocytes, the effects of conjugates II and III (see *Figure 1*) were compared. As *Figure 7* demonstrates, only conjugate II, with the hydrophobic block located at the periphery of the molecule, is able to suppress the respiration of lymphocytes. At the same time, PEG–protein conjugates, conjugates with hydrophilic blocks at the periphery of the molecule (type III) and CHT itself had no influence on lymphocyte respiration. The efficiency of conjugate II can be compared with the influence of free pluronic.

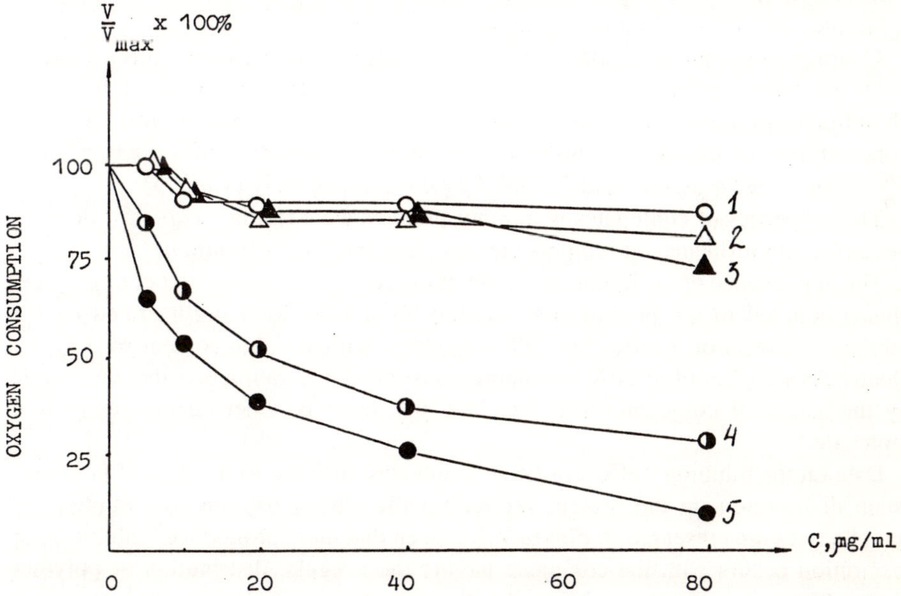

Figure 7. Influence of protein conjugates with different polymers on the rate of oxygen consumption by T lymphocytes: 1, native CHT; 2, $(PEG)_9$–CHT; 3, $(REP)_7$–CHT; 4, $(RPE)_8$–CHT-a; 5, $(RPE)_{10}$–CHT-a. Lymphocytes (10^8 cells ml^{-1}) were incubated in 8 mM MOPS buffer containing 145 mM NaCl, 3.6 mM KCl, 2 mM KH_2PO_4, and 10 mM pyruvate (pH 7.4, 37°C). Conjugates were added after oligomycin (0.2 μg ml^{-1}) and 60 μM DNP. In the control sample incubation medium was added instead of solutions of conjugates.

Table 5. The efficiency of suppression of lymphocyte and mitochondrial uncoupled respiration by block co-polymer RPE and its conjugates with α-chymotrypsin $(RPE)_6$–CHT-a), cytochrome *c* ($(RPE)_8$–Cyt-C-a) and BSA ($(RPE)_7$–BSA-a)

		Concentration of RPE and its conjugates (μM)			
		RPE	$(RPE)_6$–CHT-a	$(RPE)_8$–Cyt-C-a	$(RPE)_7$–BSA-a
a	Lymphocytes	4.9±1.2	0.66±0.1	1.1±0.1	0.75±0.1
	Mitochondria	4.7±1.2	0.48±0.06	–	–
		Concentration of RPE and its conjugates (mg ml^{-1})			
		RPE	$(RPE)_6$–CHT-a	$(RPE)_8$–Cyt-C-a	$(RPE)_7$–BSA-a
b	Lymphocytes	9.8±2.4	30.6±6.2	33.0±3.4	61.0±6.2
	Mitochondria	9.3±2.4	21.6±4.7	–	–
		Concentration of free RPE and of RPE in the conjugates (μM)			
		RPE	$(RPE)_6$–CHT-a	$(RPE)_8$–Cyt-C-a	$(RPE)_7$–BSA-a
c	Lymphocytes	4.9±1.2	6.6±1.9	3.3±0.3	5.3±0.5
	Mitochondria	4.7±1.2	4.8±1.5	–	–

Oxygen consumption was recorded in the presence of 50–60 μM DNP + 4 mM glutamate + 1 mM malate (mitochondria). Results are expressed as means±SEM for five or six experiments. The concentrations of pluronic in μM (a) and mg ml^{-1} (b) are shown which inhibit the oxygen consumption by 50%. The concentrations of free polymer and of polymer bound in a conjugate are compared for their efficiencies (c).

Concentrations of the compounds inhibiting respiration by 50% were 9.8±2.4 μg ml^{-1} (pluronic) and 30.0±6.2 μg ml^{-1} (conjugate; *Table 5*).

Conjugates produced similar effects on the respiration of isolated mitochondria. Both substances, pluronic and conjugate II, suppressed the rate of uncoupled mitochondrial respiration and lymphocyte respiration at the same concentrations. Average concentration of the same compounds reducing the respiration of mitochondria to 50% were 9.3±2.4 μg ml^{-1} and 21.6±4.7 μg ml^{-1}, respectively (*Table 5*).

The properties of conjugates with other proteins were similar (*Table 5*). Proteins *per se* had no influence on lymphocyte and mitochondrial respiration.

The dependence of the inhibitory effect of the conjugate on the number of polymer chains coupled to the protein and required for translocation is illustrated by the inhibitory effects of a series of CHT conjugates with variable content of polymer chains. As seen from *Figure 8*, the suppression of cell respiration becomes detectable by the action of conjugates with not less than three polymer chains per protein molecule.

Data on the inhibitory effect of BSA conjugates differing in the type of polymer-chain distribution on the protein surface on the rate of oxygen consumption by lymphocytes are presented in *Figure 9*. It is seen that more pronounced inhibition of respiration occurs with the conjugate having the irregular distribution of polymer chains. The same effect was observed with conjugates of CHT with similar types of structure (data not shown).

Concerning the mechanism of conjugate action on cellular membrane, some alternative explanations may be considered:

1. the conjugates interact with receptors on the external surface of the plasma

membrane, which transfer a signal into the cell;

2. the conjugates damage cells;
3. polymer chains split from proteins during lymphocyte incubation with the conjugates; or
4. the conjugates pass through the plasma membrane and reach the mitochondrial membrane.

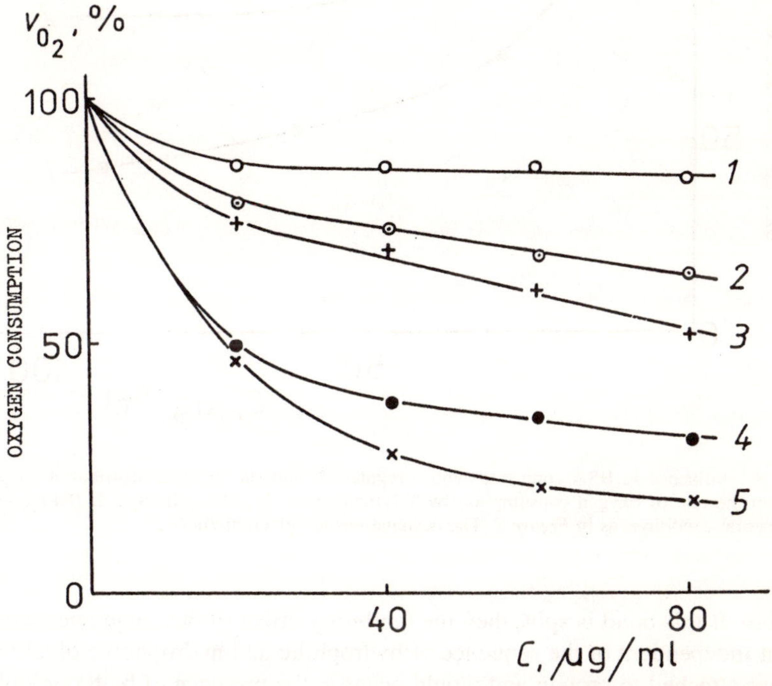

Figure 8. Influence of CHT conjugates containing different numbers of pluronic chains on the rate of oxygen consumption by T lymphocytes: 1, $(RPE)_2$–CHT-a; 2, $(RPE)_3$–CHT-a; 3, $(RPE)_6$–CHT-a; 4, $(RPE)_8$–CHT-a; 5, $(RPE)_{10}$–CHT-a. Experimental conditions were as in *Figure 7*.

The first assumption gives no explanation as to why the conjugates at virtually the same concentrations inhibit the uncoupled respiration both in whole cells and isolated mitochondria. In addition, the mitogenic lectin, concanavalin A (which is known to stimulate lymphocyte respiration by interacting with the receptors on the external surface of plasma membrane) increases their respiration rate by 20% only and has no effect on the respiration of isolated mitochondria (Lakin-Thomas and Brand, 1988). We have obtained a threefold stimulation of lymphocyte respiration by pluronic conjugates with CHT in the presence of oligomycin (Kirillova, 1993).

The experiments showing that the suppression of respiration of lymphocytes by pluronics and protein–pluronic conjugates is completely reversed by subsequent washing of the cells provide evidence that inhibitory effects were not associated with plasma-membrane injury.

The splitting of block-co-polymeric chains from the protein can hardly account for the membranotropic activity of the conjugates. The -NH-CH_2 bond, formed during conjugate formation, is resistant to hydrolysis and sterically inaccessible to enzymic

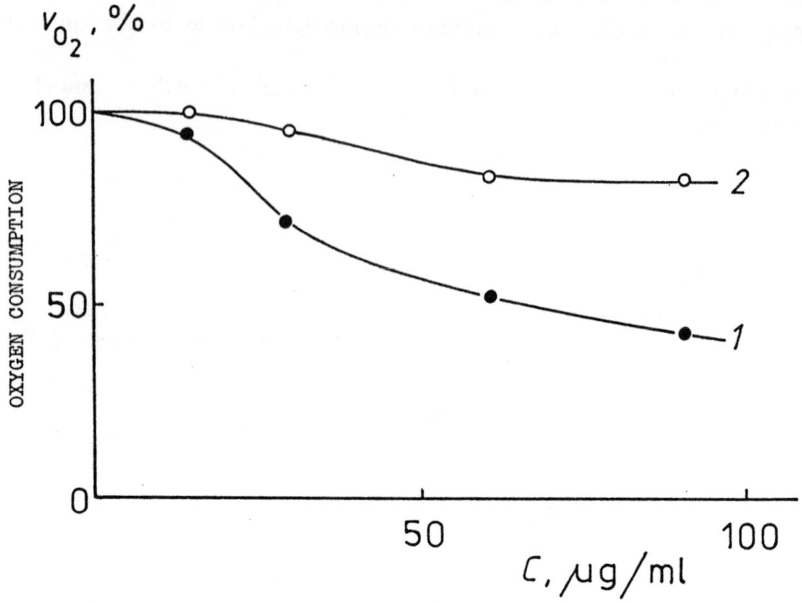

Figure 9. Influence of BSA conjugates with irregular (1) and statistical (2) distribution of pluronic chains on the rate of oxygen consumption by T lymphocytes: 1, $(RPE)_8$–BSA-a; 2, $(RPE)_{14}$–BSA-b. Experimental conditions as in *Figure 7*. The designations are given in *Table 2*.

cleavage. If this bond is split, then the inhibitory effect of the conjugates would be present independent of the sequence of hydrophilic and hydrophobic blocks of the polymer attached to protein and would occur in the presence of both conjugates II and III. However, this effect was observed in the presence of conjugate II only. We conclude that the effect of the conjugates on thymus-derived lymphocytes is a result of their transport through the plasma membrane. Hydrophilic substances, in particular proteins, are unable to enter a cell unless the plasma membrane contains corresponding carriers. However, hydrophilic substances may be translocated into a cell by means of fatty acid hydrophobization (Hu, Olson and Olson, 1986). The evidence obtained allows us to suggest that binding of synthetic polymers (pluronics) capable of crossing cellular membranes (Topchieva, 1989; Kirillova, 1993) to proteins has a similar effect to the attachement of fatty acids. It should be emphasized that hydrophobic blocks in such conjugates should be located at the periphery of the conjugate molecule and polymer chains should be distributed irregularly (structure IIa in *Figure 1*). Protein–pluronic conjugates of IIb and III types and conjugates with PEG had no influence on the respiration of these cells. In some cases PEG–protein conjugates may enter cells by endocytosis (Beckman, 1988). This process is slower by several orders of magnitude than diffusion of conjugates with membranotropic properties, and cannot occur in mitochondria.

Thus, it has been shown that covalent coupling of pluronics to hydrophilic proteins which are not transported through the cell membrane renders them membranotropic.

Solubilization efficiency of conjugates

SOLUBILIZATION OF BENZAFLAVIN BY BSA–POLYALKYLENE OXIDE CONJUGATES

On the basis of the above-described properties of pluronics, it might be expected that coupling of amphiphilic molecules to carrier proteins, e.g. BSA, would result in an enhanced ability of solubilization of water-insoluble substances, for example 2', 3', 4', 5'-tetrabenzoyl-5-acetyl-1,5-dihydroriboflavin (benzaflavin, BF) (Topchieva, 1993b).

Dihydroriboflavin esters exhibiting a wide spectrum of biological activities are important both as long-acting vitamin B preparations and as metabolic drugs (Avakumov, Kovler and Kruglikova-L'vova, 1992).

BF solubilization in aqueous solutions of BSA and its conjugates with polyalkylene oxides was estimated using a UV method at 306 nm, a wavelength corresponding to maximum absorption of BF. Absorption coefficients were measured on the basis of a plot of absorbance versus concentration of BF solubilized by BSA ($\varepsilon = 4.7 \times 10^3$ M^{-1} cm^{-1}), conjugate of BSA with PEG ($(PEG)_{14}$–BSA) ($\varepsilon = 4.7 \times 10^3$ M^{-1} cm^{-1}).

The limit of solubilization (p), being equivalent to the number of molecules bound to one molecule of the solubilizing agent, was estimated using high-speed sedimentation analysis at $\lambda = 306$ nm. The use of the sedimentation technique is based on co-sedimentation of BF bound to protein or conjugate (Topchieva, 1991).

Two methods of BF solubilization were employed, namely suspending dry material in aqueous solutions containing a solubilizing agent (method 1) and introduction of BF dissolved in ethanol into these systems (method 2). Examination of both systems using the quasi-elastic light-scattering technique showed that introduction of BF from ethanolic solutions resulted in the occurrence of small particles of the size of conjugates (less than 30 nm) along with larger ones with a radius of 200–3000 nm, presumably corresponding to BF aggregates. Solubilization using the former technique (method 1) yielded only small particles (<30 nm). To compare the two approaches, the systems containing BF and solubilizing agents of different types were subjected to centrifugation. Solubilization limit values of BF by BSA and its conjugates are presented in *Table 6*. BSA showed a marked capacity for solubilization of BF. However, direct experiments revealed an inability of pluronics to solubilize BF.

Table 6 demonstrates that all conjugate-containing systems were characterized by higher p values as compared with those containing BSA. Conjugate $(RPE)_5$–BSA-a, with hydrophobic regions located on the periphery of macromolecules, showed the highest limit of solubilization. However, method 2 allowed more BF to be introduced into a solubilization system, supposedly owing to a specific kinetic behaviour of the process. It should be emphasized that an increased number of polymer chains in the conjugate, due to a higher degree of modification (samples 4 and 5), did not appear to have any appreciable effect on the efficiency of solubilization. At the same time, a markedly enhanced density of polymer chains in the vicinity of the protein, which resulted from the interaction of the introduced free pluronics with pluronic chains covalently bound to protein, led to a decrease in the limit of solubilization, probably owing to enhanced shielding of the sites responsible for BF solubilization. Optimum incorporation of BF into the conjugate occurred at pH 7.6–9.0 (*Figure 10*). The limit of solubilization decreased with increasing ionic strength of the buffer solution.

Experiments conducted to study temperature-dependent solubilization of BF and

Table 6. Temperature-dependent solubilization of BF by BSA and its conjugates (0.02 M phosphate buffer, pH 7.2)

No.	Solubilizing agent	T(°C)	p	s_{20} (S) (solubilizing agent)	s_{20} (S) (BF complex)	Proportion of different forms (%)
1	BSA	20	17	4.2	6.3	100
		40	17	6.7	11.0	47
					6.2	53
2	$(PEG)_{14}$–BSA	20	21	4.2	6.6	100
		40	21	6.5	10.5	49
					6.8	51
3	$(REP)_{11}$–BSA	20	19	4.2	6.8	100
		40	19	6.6	13.2	47
					5.9	53
4	$(RPE)_{24}$–BSA-a	20	40	4.4	10.1	100
		40	40	6.8	14.2	21
					7.3	79
5	$(RPE)_{100}$–BSA[a]	20	18	2.9	6.2	100
		40	18	5.8	12.0	80
					6.6	20

[a] The solubilizing agent was prepared from compound 4 by addition of free pluronic. method 2 was used for introduction of BF into the system.

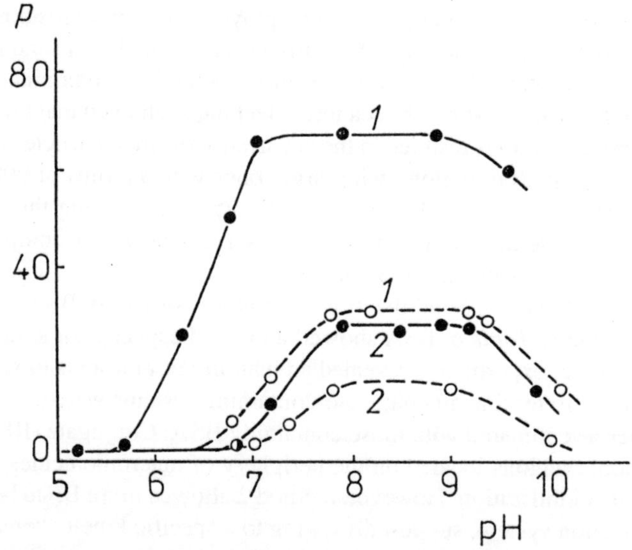

Figure 10. Relationship between BF solubilization limits (p) and pH (at 20°C). 1, Visual method; 2, sedimentation method; - - - -, BSA; ——, conjugate $(RPE)_5$–BSA-a.

BSA and its conjugates revealed marked heterogeneity of the solubilization systems at 40°C which was not normally observed in solutions containing pure protein components. Certain portions of sedimentograms indicated mean sedimentation coefficients which were approximately twice as high as those intrinsic to protein components (*Table 6*). In such systems, BF could probably facilitate the 'bridging' of protein molecules to give rise to aggregates. Comparison of the relative amounts of different particles in the experimental systems demonstrated that the proportion of aggregates in solutions containing BSA–pluronic conjugates was significantly lower than in all other BSA-containing systems. Polymer chains of these conjugates appear to be responsible for steric hindrance inherent in BF-mediated intermolecular interactions between proteins.

It is appropriate to surmise, on the basis of these findings, that BF molecules are located in hydrophobic parts of the conjugated BSA. This is confirmed by the fact that attachment of pluronics to other proteins, e.g. immunoglobulin G (IgG) or CHT, with lack of ability to bind this ligand, had no effect on their solubilizing potency.

BF TRANSFER FROM SOLUBILIZED FORMS IN RAT HEART MITOCHONDRIA

The above data suggested the possibility of using conjugates of BSA with polyalkylene oxides for the delivery of BF to target cells. Therefore, we investigated the transport of solubilized BF in biological systems, specifically BF transfer to rat iver mitochondria. This phenomenon is demonstrated by the relationship between amount of bound BF and concentration of free BF ($[BF]_{free}$) in the presence of either BSA or its conjugates. Quantitative data on the BF transfer are presented as binding isotherms in *Figure 11*. At the BF concentration range studied the plot of $[BF]_{bound}$ versus $[BF]_{free}$ is linear:

$$\frac{[BF]_{bound}}{[MT]} = K[BF]_{free}$$

where K denotes the coefficient of proportionality related to the chemical nature of the solubilizing agent, and [MT] is the concentration of mitochondrial protein. The linear correlation between the two variables indicates that for experimental results characterizing the initial portions of the binding isotherms the values of K were independent of the amount of the solubilizing agent. Also, *Figure 11* shows an identity of the ability of BSA and BSA–PEG conjugates to transfer BF to membranes. Both solubilizing agents may be characterized by the same K value: $K = 0.31 + 0.01$ (gram of mitochondrial protein per litre)$^{-1}$. Higher transfer efficiency of BSA conjugates with pluronics for BF transport, as compared to BSA and its conjugate with PEG ($K = 0.96 + 0.01$) (gram of mitochondrial protein per litre)$^{-1}$, is probably due to weaker association between the BSA–pluronic conjugate and solubilized BF. To summarize, BSA–pluronic conjugates have certain advantages over the remaining solubilization systems, both at the degree of solubilization and transport of BF. This makes them useful tools for the development of new drug formulations.

INFLUENCE OF BF ON PEROXIDATION OF LIPIDS IN MICROSOMES OF RAT LIVER

Water-soluble forms of BF may be employed to investigate the molecular mechanisms of its action in model biochemical systems. It was shown (Avakumov, Kovler and Kruglikova-L'vova, 1992) that the tetrabutyrate of riboflavin, a compound similar in structure to BF, inhibits peroxidation of lipids (POL) in rat liver mitochondria, initiated by the addition of adriamycin, an antitumour agent. Based on these observations it might be expected that BF would also be able to inhibit POL. Quantitatively the process of NADPH-dependent POL and the influence of solubi-

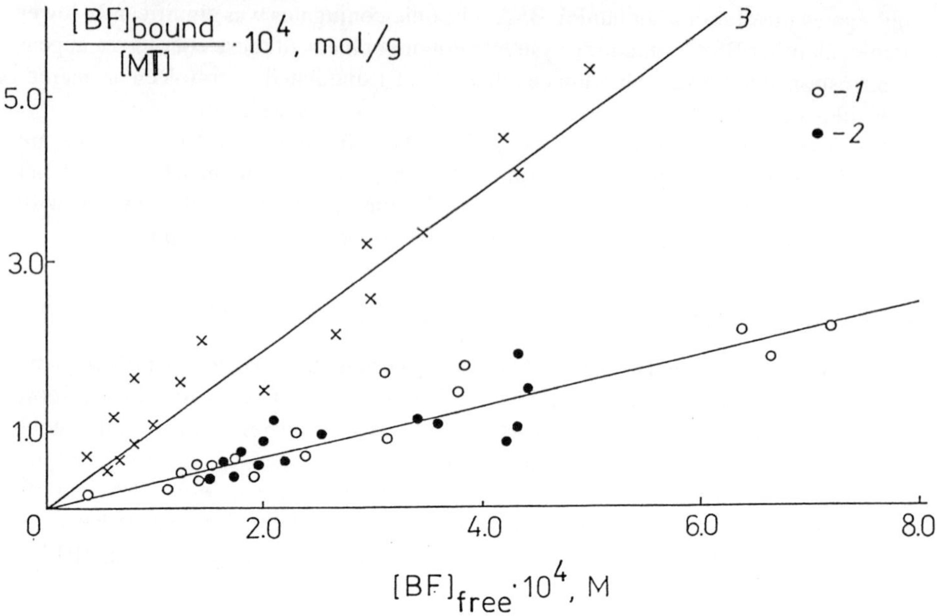

Figure 11. Relationship between concentration of bound BF (mol g^{-1} of mitochondrial protein) and concentration of free BF. Solubilizing agent: 1, BSA; 2, $(PEG)_{14}$–BSA; 3, $(RPE)_{24}$–BSA-a (structure IIa). [MT], concentration of mitochondrial protein.

lized forms of BF on its rate was studied by measuring the concentration of malonic dialdehyde (MDA), the final product of POL, in liver microsomes. In preliminary experiments the influence of BSA, PEG and pluronics on the rate of NADPH-dependent POL was studied. BSA at a concentration up to 0.7×10^{-4} M and PEG at a concentration up to 4.1×10^{-4} M had no effect on the NADPH-dependent POL. On the basis of decreasing MDA production in the presence of solubilized BF forms, we may conclude that BF, solubilized by BSA and its conjugates with polyalkylene oxides, inhibits the process of NADPH-dependent POL in rat liver microsomes. It should be noted that the inhibitory action of BF does not depend on the nature of solubilizing agent. This is depicted in *Figure 12*, which shows the relationship between the extent of POL inhibition and the concentration of BF solubilized by BSA and its conjugates with polyalkylene oxides. The concentration of BF corresponding to 50% inhibition under the conditions studied is 0.5×10^{-4} M. These investigations demonstrate the ability of BF to inhibit the POL process in biological membranes.

INFLUENCE OF BF ON THE ENZYME ACTIVITY OF D-AMINO-ACID OXIDASE FROM PIG LIVER

It has been shown in other experiments that the addition of BF dissolved in ethanol to the reaction mixture does not influence the enzymatic activity of D-amino-acid oxidase. However, solubilized forms of BF do produce an inhibitory effect. *Figure 13* shows the experimental values of the relative activity of D-amino-acid oxidase, V/V_0 (V_0 and V are the rates of enzymic reaction in the absence and in the presence of BF, respectively), plotted against the concentration of BF. The inhibitory effect of BF

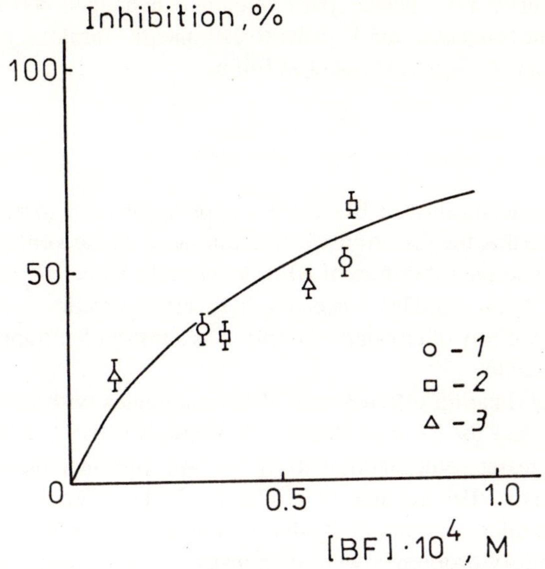

Figure 12. Relationship between degree of inhibition of MDA production in the process of NADPH-dependent POL in rat liver microsomes and concentration of BF, solubilized by BSA (1), $(PEG)_{14}$–BSA (2) and $(PRE)_{24}$–BSA-a (3). Each experimental point is an average of six measurements.

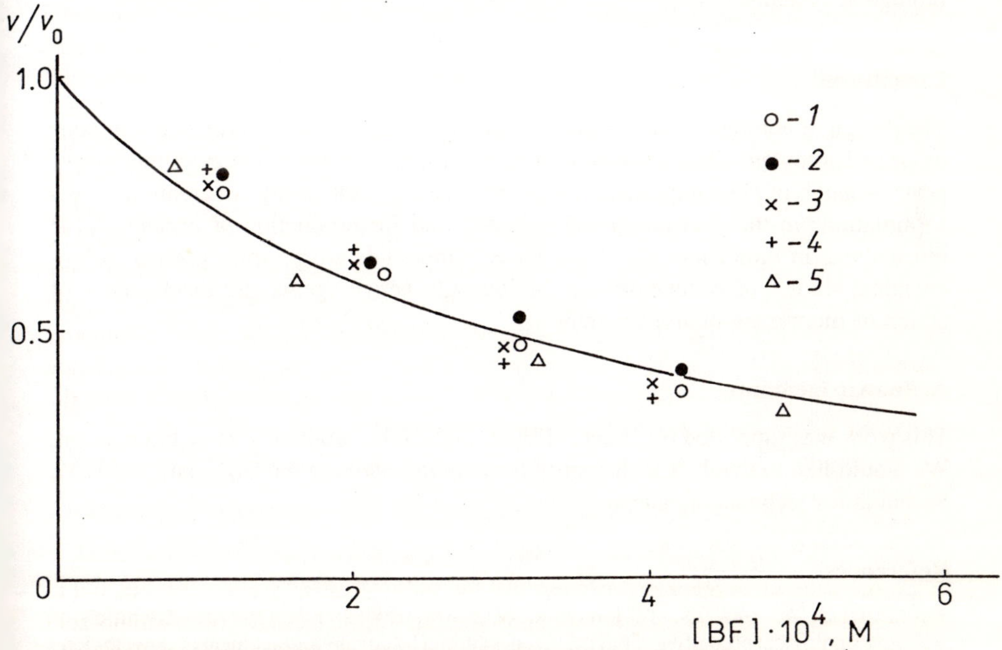

Figure 13. Inhibition of D-amino-acid oxidase by BF, solubilized by BSA (1 and 3), conjugate (PEG)14–BSA (2 and 4) and conjugate (RPE)24T–BSA (5). The enzymatic activity of D-amino-acid oxidase was registered by spectrophotometric (1,2) and polarographic (3,4,5) methods. Conditions: 0.1 M Tris–HCl, pH 8.3, 25°C; concentration of BSA in conjugates and free BSA in solution was 1 mg ml^{-1}.

does not depend on the nature of the solubilizing agent. It should be noted that the course of the curve V/V_0 versus [BF] remains unchanged by alterations in the solubilizing agent concentration. In order to estimate the inhibitory ability of BF, the inhibition constant, K_i, was calculated as follows:

$$\frac{V}{V_0} = \frac{1}{1 + [I]/K_i}$$

where [I] is the concentration of BF. The value of K_i appeared to be 0.29 + 0.02 mM. It should be noted that the sensitivity of D-amino-acid oxidase entrapped in hydrated reversed micelles to the inhibition of BF is essentially higher: K_i = 4.5 × 10–5 M at ($[H_2O]$/[aerosol-OT] = 23). This is because, in reversed micelles, localized concentration of BF in the vicinity of enzyme takes place as a result of entrapment of BF on the surface of the micelle.

Thus, the solubilization efficiency of BSA conjugates with polyalkylene oxides (PEG and pluronics) for BF was studied. Conjugates of BSA with pluronics were shown to have greater solubilizing activity for this substance than the free protein. Solubilized forms of BF are able to inhibit NADPH-dependent POL in rat liver microsomes. The inhibitory action of BF does not depend on the type of solubilizing agent. The inhibitory properties were demonstrated by a study of the influence of solubilized forms of BF on the enzyme activity of D-amino-acid oxidase from pig kidney. Thus conjugates of BSA with pluronics are a new type of carrier for the delivery of biologically active water-insoluble compounds and their screening of biological systems.

Conclusion

The design of conjugates on the basis of proteins and block co-polymers of ethylene oxide and propylene oxide allows one to mimic biological functions of membranotropic proteins and is of great importance for progress in enzymotherapy, for creation of new formulations in the pharmaceutical industry, and for production of organosoluble biocatalysts in biotechnology. Also, these compounds are promising tools in biochemical studies of water-insoluble biologically active agents and mechanisms of action of membrane-bound enzymes.

Acknowledgements

This work was supported by Grant MPE000 of the International Science Foundation. We would like to thank N.V. Khvorov for help in obtaining the DSC data and E.M. Sorokina for technical assistance.

References

ABUCHOWSKI, A., VAN ES, T., PALCZUK, N.C. AND DAVIS, F.F. (1977a). Alteration of immunological properties of bovine serum albumin by covalent attachment of polyethylene glycol. *Journal of Biological Chemistry* **252**, 3578–3581.

ABUCHOWSKI, A., MCCOY, I.R., PALCZUK, N.C. AND DAVIS, F.F. (1977b). Effect of covalent attachment of polyethylene glycol on immumogenicity and circulating life of bovine liver catalase. *Journal of Biological Chemistry* **252**, 3582–3586.

AVAKUMOV, V.M., KOVLER, M.A. AND KRUGLIKOVA-L'VOVA, R.P. (1992). Drug formulations of metabolic therapy on the basis of vitamins and coferments. *Voprosy Medizinskoi Khimii* **38**(4), 14–21.

BAILEY, F.E., JR AND KOLESKE, J.V. (1966). Configuration and hydrodynamic properties of the polyoxyethylene chain in solution. In *Nonionic Surfactants* (M.J. Schick, Ed.), pp. 794–821. Marcel Dekker, New York.

BECKMAN, J.S., MINOR, R.L., WHITE, C.W., REPINE, I.E., ROSEN, G.M. AND FREEMAN, B.A. (1988). Superoxide dismutase and catalase conjugated to polyethylene glycol increases endothelial enzyme activity and oxidant resistance. *Journal of Biological Chemistry* **263**, 6884–6992.

BIRKTOFT, J.J. AND BLOW, D.M. (1972). Structure of crystalline α-chymotrypsin. V. Atomic structure of tosyl α-chymotrypsin at 2 Å resolution. *Journal of Molecular Biology* **68**, 187–240.

BOCCU, E., LARGAJOLLI, R. AND VERONESE, F.M. (1983). Coupling of monomethoxy PEG to proteins via active esters. *Zeitschrift Naturforschung* **38**, 94–99.

EFREMOVA, N.V. AND TOPCHIEVA, N.V. (1993). Supramolecular structures based on protein conjugates with polyalkylene oxides: β-cyclodextrin complexes. *Biokhimiya* **58**(7), 753–757 (in Russian).

EFREMOVA, N.V., MOZHAEV, V.V. AND TOPCHIEVA, I.N. (1992). Thermal stability of conjugates of α-chymotrypsin with water-soluble poly(alkylene oxide)s. *Biokhimiya* **57**(3), 342–346 (in Russian).

FIELDS, R. (1971). The measurement of amino groups in proteins and peptides. *Biochemical Journal* **124**, 581–590.

GORSKAYA, I.A., BUSHUEVA, M.V., KURGANOV, B.I., TOPCHIEVA, I.N., IVANOVA, V.P., PANOVA, I.G. AND RUDAKOVA, I.P. (1991). Conjugates of bovine serum albumin with poly(alkylene oxide)s as a new type of carriers: transport of riboflavin esters to mitochondria. *Doklady Academii Nauk SSSR* **319**, 1008–1011 (in russian).

HU, I.-S., OLSON, I. AND OLSON, E.N. (1986). Protein fatty acylation: a novel mechanism for association of proteins with membrane and its membrane regulatory pathways. *Biofactors* **1**, 219–226.

KHMELNITSKY, YU. L., BELOVA, A.B., LEVASHOV, A.V. AND MOZHAEV, V.V. (1991). Relationship between surface hydrophilicity of a protein and its stability against denaturation by organic solvents. *FEBS Letters* **284**, 267–269.

KIRILLOVA, G.P., MOCHOVA, E.N., DEDUCHOVA, V.L., TARAKANOVA, A.N., IVANOVA, V.P., EFREMOVA, N.V. AND TOPCHIEVA, I.N. (1993). The influence of pluronics and their conjugates with proteins on the rate of oxygen consumption by liver mitochondria and thymus lymphocytes. *Biotechnology and Applied Biochemistry* **18**, 329–339.

KJELLANDER, R. AND FLORIN, E. (1981). Water structure and changes in thermal stability of the system poly(ethylene oxide) – water. *Journal of the Chemical Society, Faraday Transactions* **77**, 2053–2077.

KOCHETOV, G.A. (1980). Prakticheskoye rukovodstvo po enzymologii, p. 222. High School, Moscow (in Russian).

KRIGBAUM, W.R. AND GODWIN, R.W. (1968). Molecular conformation of chymotrypsinogen and chymotrypsin by low-angle X-ray diffraction. *Biochemistry* **7**, 3126–3140.

KURGANOV, B.I. AND TOPCHIEVA, I.N. (1991). Conjugates of proteins with poly (ethylene glycol) (review). *Uspekhi biologitcheskoi khimii* **32**, 63–86.

LAKIN-THOMAS, P.L. AND BRAND, M.D. (1988). Stimulation of respiration by mitogens in rat thymocytes is independent of mitochondrial calcium. *Biochemical Journal* **256**, 167–173.

LARIONOVA, N.I., GLAGYSHEVA, I.P., TOPCHIEVA, I.N. AND KAZANSKAYA, N.F. (1993). Conjugation of the classical soybean Bauman–Birk inhibitor with a copolymer of ethylene oxide and propylene oxide. *Biokhimiya* **58**, 1658–64 (in Russian).

LUNDSTED, L.G. AND SCHMOLKA, I.R. (1976). Synthesis and properties of block copolymer surfactants. In *Block and Graft Copolymerization* (R.J. Ceresa, Ed.), pp. 174–205. John Wiley, New York.

MARSHALL, J.J. AND RABINOWITZ, M.L. (1975). Modification of amino groups of α-amilaze with dextran. *Archives of Biochemistry and Biophysics* **117**(2), 777–779.

MATSUSHIMA, A., OKADA, M. AND INADA, Y. (1984). Chymotrypsin modified with polyethylene glycol catalyzes peptide synthesis reaction in benzene. *FEBS Letters* **178**(2), 275–277.

MOZHAEV, V.V. AND MARTINEK, K. AND BEREZIN, I.V. (1988). Structure–stability relationship in proteins: fundamental tasks and strategy for the development of stabilized enzyme catalysts for biotechnology. *CRC Critical Reviews in Biochemistry* **23**, 235–275.

OSIPOVA, S.V. (1990). Associative properties of block copolymers of ethylene oxide and propylene oxide in aqueous solutions. PH.D. Thesis, p. 52. MGU im. M.V. Lomonosova, Moscow.

OSIPOVA, S.V., TOPCHIEVA, I.N. AND KASAIKIN, V.A. (1990). Influence of temperature on the structure formation of block copolymers of ethylene oxide and propylene oxide in aqueous solutions. *Vysokomolekulyarnyie soedineniya* **32**(B), 240–242 (in Russian).

POLYAKOV, V.A., KOLOMNIKOVA, E.L., TOPCHIEVA, I.N. AND KABANOV, V.A. (1993). NMR Investigation of the complexation between β-cyclodextrin and poly(ethylene oxide)–poly(propylene oxide) block copolymers. *Polymer Science* **5**, 719–721.

ROTH, M. (1971). Fluorescence reaction for amino acids. *Analytical Chemistry* **43**, 880–882.

TAKAHASHI, K., KODERA, Y., YOSHIMOTO, T., AJIMA, A., MATSUSHIMA, A. AND INADA, Y. (1985). Ester-exchange catalyzed by lipase catalyzed with PEG. *Biochemical and Biophysical Research Communications* **131**, 532–536.

TAKOI, K., YAMAGATA, Y., NAKAJIMA, T. AND ICHISHIMA, E. (1989). Functional changes of polyethylene glycol-modified serine proteinase from *Aspergillus sojae*. *Agricultural and Biological Chemistry* **53**(8), 2063–2071.

TEPLOVA, M.V., KURGANOV, B.I., GORSKAYA, I.A., TOPCHIEVA, I.N., EFREMOVA, N.V. AND BUSHUEVA, T.L. (1993). Structure-functional properties of protein conjugates with polyalkylene oxides: a fluorescent study. *Biokhimiya* **58**(12), 1929–1935.

TISCHENKO, V.M., TIKTOPULO, E.L. AND PRIVALOV, P.L. (1974). Calorimetrical study of conformational transcriptions in α-chymotrypsin. *Molekularnaya Biophyzika* **19**(3), 400–404.

TOPCHIEVA, I.N., OSIPOVA, S.V., BANATSKAYA, M.I. AND VALKOVA, I.A. (1989). Membranotropic properties of ethylene and propylene oxide. *Doklady Academii Nauk SSSR* **308**, 910–913 (in russian).

TOPCHIEVA, I.N., MOMOT, I.G., KURGANOV, B.I. AND CHEBOTARYOVA, N.A. (1991). Solubilizing efficiency of protein conjugates with water soluble polyethers. *Doklady Academii Nauk SSSR* **316**, 242–246 (in Russian).

TOPCHIEVA, I.N., KOLOMNIKOVA, E.L., BANATZKAYA, M.I. AND KABANOV, V.A. (1993a). Complexation between β-cyclodextrins and poly(ethylene oxide)–poly(propylene oxide) block copolymers. *Polymer Science* **4**, 464–466.

TOPCHIEVA, I.N., KURGANOV, B.I., TEPLOVA, M.V., GORSKAYA, I.N. AND RUDAKOVA, I.P. (1993b). Use of conjugates of bovine serum albumin with poly(alkylene oxide)s for solubilization of riboflavin ester. *Biotechnology and Applied Biochemistry* **17**, 337–348.

TOPCHIEVA, I.N., MOKHOVA, E.N., KIRILLOVA, G.P. AND EFREMOVA, N.V. (1994a). Membranotropic properties of protein conjugates with block copolymers of ethylene oxide and propylene oxide. *Biochemistry (Moscow)* **59**, 11–15.

TOPCHIEVA, I.N., BLUMENFELD, A.L., KLYAMKIN, A.A., POLYACOV, V.A. AND KABANOV, V.A. (1994b). Supramolecular structures based on poly(ethylene oxide)–poly(propylene oxide) block copolymers and cyclodextrins. *Polymer Science* **36**(2), 221–227.

VERONESE, F.M., CALICETI, P., PASTORINO, A., SCHIAVON, O. AND SARTORE, L. (1989). Preparation, physico-chemical and pharmacokinetic characterization of monomethoxypoly(ethylene glycol)-derivatized superoxide dismutase. *Journal of Controlled Release* **10**, 145–154.

VON SPECHT, B.U. AND BRENDEL, W. (1977). Preparation and properties of trypsin and chymotrypsin coupled covalently to poly(*N*-vinylpyrrolidone). *Biochimica et Biophysica Acta* **484**(1), 109–114.

YOSHIMOTO, T., TAKAHASHI, K., NISHIMURA, H., AJIMA, A., TAMAURA, Y. AND INADA, Y. (1984). Modified lipase having high stability and various enzymatic activities in benzene, and its re-use by recovering from benzene solution. *Biotechnology Letters* **6**(6), 337–340.

11
Liposomes in Drug Targeting

MUKUL KUMAR BASU

Biomembrane Division, Indian Institute of Chemical Biology, Calcutta, India

Introduction

Liposomes are artificial vesicles composed of one or more concentric bi-layers which enclose an aqueous space. Water-soluble drugs or biomolecules are known to be accommodated in the inner aqueous space of liposomes, whereas lipophilic substances remain associated in the hydrophilic bilayer. Initially, the use of liposomes was mainly confined to model membrane systems, but more recently their use as vehicles for the transfer of genetic and other materials into cells in culture or for drug delivery to desired cellular and subcellular sites in a living system has been the subject of much discussion (Berg and Boman, 1973; Stahl *et al.*, 1978; Poste and Papahadjopoulos, 1978; Gregoriadis, 1980a). It appears that different applications require different types of liposomes. Although the preparation of various kinds of liposomes has recently been reviewed (Szoka and Papahadjopoulos, 1981), it seems that two main types of liposomes can be distinguished, i.e. large uni-or multilamellar liposomes, which are cleared rapidly from the circulation, and small unilamellar liposomes, which remain in circulation for longer periods of time. Obviously, large multilamellar vesicles (Bangham *et al.*, 1965), which are produced by the deposition of a thin lipid film from an organic solvent on the walls of the container followed by the agitation with an aqueous solution of the material to be encapsulated, were found to be the best for drug delivery purposes. The advantage of liposomes over other delivery systems is their lack of accumulation in the body. Moreover, liposomes are poorly immunogenic and the lipids used are biodegradable. In addition, the liposome-encapsulated drugs are protected from the bioenvironment and are reported to have reduced toxicity. Because of improved rehydration and dehydration techniques (Kirby and Gregoriadis, 1984), modern drug-encapsulated liposomes can be stored as freeze-dried powder and constituted just before use.

Abbreviations: AIDS, acquired immunodeficiency syndrome; AM, alveolar macrophages; Chol, cholesterol; DCP, dicetyl phosphate; ddITP, dideoxy-inosine triphosphate; DPPE, dipalmitoyl phosphatidyl ethanolamine; HIV, human immunodeficiency virus; LDL, low-density lipoprotein; MDP, muramyl dipeptide; MLV, multilamellar vesicles; MTP-PE, muramyl tripeptide phosphatidyl ethanolamine; PC, phosphatidyl choline; PEG, polyethylene glycol; PMN, polymorphonuclear neutrophils; SM, sphingomyelin; SPC, soya phosphatidyl choline; SUV, small unilamellar vesicles.

Biotechnology and Genetic Engineering Reviews — Vol. 12, December 1994
0264–8725/94/12/383–408 $20.00 + $0.00 © Intercept Ltd, P.O. Box 716, Andover, Hampshire SP10 1YG, UK

However, the common problem for liposome therapy is that liposomes are taken up readily by the cells of the reticuloendothelial system. Even though reversible endothelial blockade can be imposed with relative ease, it is not accompanied by any significant increase in liposome uptake by tissues outside of the mononuclear phagocyte system. The retention and uptake of circulating liposomes by reticulo-endothelial cells and blood monocytes is a major obstacle to experimental efforts to 'target' liposomes to other cell types. SUV (small unilamellar vesicle) liposomes are cleared from the circulation more slowly than larger MLV (multilamellar vesicle) liposomes. However, there is no evidence that SUV liposomes or liposomes with a prolonged half-life in the circulation accumulate to any significant degree in sites other than the liver, spleen and bone marrow. Moreover, reported intravenous injection of liposomes at regular intervals is accompanied by long-term paralysis of reticulo-endothelial function. Since many of the proposed clinical applications involve multiple-dose treatment protocols, this problem represents a potentially serious drawback. Besides, there are no appreciable methods either for targeting liposomes to specific sites *in vivo* or for ensuring the delivery of liposomal contents to target cells. Although considerable amount of work has been reported on passive targeting to the reticulo-endothelial system, not much is known about active targeting to specific subsets of circulating blood cells or to vascular endothelium. Thus, efficient delivery systems suitable for site-specific targeting are being sought.

In site-specific targeting, a recognition marker (or ligand) is usually attached to the surface of liposomes. An ideal target for such liposomes would be a cell or tissue that uniquely expresses a receptor specific to the attached ligand. Surolia, Ahmed and Bachhawat (1975) reported the potential use of glycolipids as ligands in a search for receptor sites in animal tissues. Liposomes with attached sugar residues are reported to be more effective in transportation of their contents to specific organs due to the selective binding of the sugars with their specific receptors (Ghosh and Bachhawat, 1980). Recently (Das *et al*., 1987; Medda *et al*., 1990) plant glycosides of varying terminal sugar residues have been incorporated on to the bilayer membrane of liposomes, which have been tested for their tissue specificity or for their increased uptake to specific cellular or subcellular sites. As knowledge of the receptor status of target cells undergoing infection is a very important prerequisite for designing targeted delivery vehicles, different sugar-coated liposomes have been prepared to test the feasibility of their application not only as delivery vehicles but also as efficient ligands for the estimation of specific sugar receptor status of macrophages in normal as well as in infected states (Dutta, Bandyopadhyay and Basu, 1994).

Moreover, tetrapeptide (tuftsin)-bearing liposomes (Tuft-lip) have been shown to be quite effective in activating host mcrophages (Singh, Chabbra and Srivastava, 1992), raising the possibility of their application as targeted delivery vehicles against macrophage-associated disorders. Similarly, the increased uptake of antibody-coated liposomes by erythrocytes is suggestive of their possible use as targeted drug carriers against erythrocytes undergoing infection (Singhal and Gupta, 1986). The successful targeting of amylopectin-incorporated liposomes to lung (Takada *et al*., 1984) also raised the possibility of their use as targeted drug carriers against lung associated disorders, although not much success was achieved in reducing the uptake of these liposomes by the cells of the reticulo-endothelial system.

Designing liposomal vehicles for site-specific delivery

The major emphasis during the past decade has been on designing and targeting liposomes to specific sites, in an attempt to show increased drug efficacy and reduced drug toxicity. Efforts were made to produce stable liposomes with larger circulation time, glycolipid/glycoside-bearing liposomes or sugar-coated liposomes, and peptide-or antibody-coated liposomes.

STABLE LIPOSOMES WITH LONGER CIRCULATION TIME

It was found that vesicle size and surface charge override the state of membrane permeability in determining rates of vesicle clearance (Senior, Crawley and Gregoriadis, 1985). Small unilamellar vesicles were cleared less rapidly than were large unilamellar ones. Moreover, the neutral and positively charged unilamellar liposomes were cleared less rapidly than were the unilamellar negatively charged ones (Juliano and Stamp, 1975).

The effect of cholesterol content on the stability of small unilamellar liposomes was examined *in vivo* and *in vitro* by Kirby, Clarke and Gregoriadis (1980). Cholesterol-rich liposomes, regardless of surface charge, remained stable in the blood of intravenously injected animals for up to at least 400 min. In addition, stability of cholesterol-rich liposomes was largely maintained *in vitro* in the presence of whole blood, plasma or serum for at least 90 min. But the stability of liposomes was found to be influenced not only by the biological environment with which they come into contact but also by their structural characteristics and those of the associated drugs. For instance, liposomes containing charged lipids retain certain drugs through electrostatic bonding (Gregoriadis, Neerunjun and Hunt, 1977), and lipophilic drugs can be anchored into the liposomal lipid phase (Gregoriadis, 1973). Liposome formulations with long circulation half-lives were reported by Allen, Hansen and Rutledge (1989). Liposomes with the longest circulation half-lives had compositions that mimicked the outer leaflet of red blood cell membranes (egg phosphatidyl choline/spingomyelin/cholesterol/ganglioside GM_1 in the molar ratio 1:1:1:0.14). Several other gangliosides and glycolipids were examined, but none could substitute for GM_1 in their ability to prolong circulation half-lives. In a recent report (Allen *et al.*, 1991), a novel synthetic lipid derivative of polyethylene glycol (PEG-PE), was found to decrease greatly liposome uptake by bone marrow macrophages in a concentration-dependent manner. Lipids which are known to increase the microviscosity of biological membraes, e.g. sphingomyelin (SM) and cholesterol (Chol), were found to decrease both *in vivo* and *in vitro* uptake of liposomes by bone-marrow macrophages. Moreover, the stability of liposomes in circulation was found to be markedly enhanced by structural modification of their phospholipid component (Bali, Dhawan and Gupta, 1983) or by appropriate tailoring of the phospholipid structure (Agarwal, Bali and Gupta, 1986b). Apart from the increased survival, liposome uptake by the liver was found to decrease by replacing natural phosphatidyl choline (PC) with carbamyl PC in liposomes (Gupta, 1983).

GLYCOLIPID/GLYCOSIDE-BEARING LIPOSOMES OR SUGAR-COATED LIPOSOMES

The finding that different cells can recognize different sugar residues (Krantz *et al.*,

1976) led to the idea that by modification of liposomes, by incorporating various glycolipids or glycosides having different terminal sugars, it may be possible to direct the liposomes to various tissues. Specific galactose-binding receptors were reported to be present exclusively on the parenchymal cells of liver (Gregoriades *et al.*, 1970; Surolia and Bachhawat, 1977; Hoekstra, Tomasini and Scherphof, 1980). In addition to natural glycolipids, small molecular weight synthetic glycolipids were found to be equally effective for the introduction of saccharide determinants on to the surface of liposomes. The liposomes prepared from lactosyl-phosphatidyl ethanol amine (Ghosh, Bachhawat and Surolia, 1981) or from β-D-galactosyl-phosphatidyl ethanolamine (Ghosh and Bachhawat, 1980) were found to be taken up readily by hepatocytes. Specific receptors were demonstrated on the non-parenchymal or Kupffer cells, and were found to recognize the terminal mannose residue of glycoside-bearing liposomes (Ghosh and Bachhawat, 1980) or of biopolymers (Stahl *et al.*, 1978). Ganglioside-and other glycolipid-containing liposomes were found to bind specifically to lectins (Surolia, Bachhawat and Poddar, 1975; Juliano and Stamp, 1976; Redwood and Polefka, 1976; Boldt *et al.*, 1977), and lectin-mediated agglutination of liposomes with variable hydrophilic spacers was noted by Slama and Rando (1980).

A series of investigations on the tissue distribution of liposomes bearing different glycosides isolated from indigenous plants was carried out in our laboratory (Mahato, Sahu and Pal, 1978; Mahato, Sahu and Ganguly, 1980). The plant glycosides selected for the purpose were of variable sugar chains having different sugars at the terminal end. *Figure 1* shows two of the plant glycosides isolated and used for incorporation into liposomes. Because of their interesting structure – a hydrophilic moiety was attached to a hydrophobic aglycone – the plant glycosides were very suitable for incorporation into the liposomes. A diagram of plant-glycoside-bearing liposomes and a schematic representation of synthetic glycoside-grafted liposomes are shown in *Figure 2*. Studies with two plant glycosides (gracillin with glucose as end-sugar, and fluoribundasaponin with rhamnose as end-sugar) incorporated into liposomes encapsulated with [^{125}I]IgG demonstrated the presence of anomeric-blind glucose receptors on the non-parenchymal fraction of liver cells (Das *et al.*, 1987). The rates of uptake of [^{125}I]IgG, encapsulated in different types of liposomes, by the liver (% injected dose) are presented in *Figure 3*. The uptake of gracillin liposomes by the liver was found to be 72% of the injected dose in the case of negatively charged liposomes as against 65% for the neutral ones. The uptake of liposomes in the presence of either glycogen (α-glucose) or cellobiose (β-glucose) was 48% for both, and was almost the same as that found for the control liposomes (Das *et al.*, 1987). In another report, the hepatic uptake of liposomes made from egg lecithin, cholesterol and dicetyl phosphate and either of two plant glycosides – corchorusin D with glucose as end-sugar (Mahato and Pal, 1987), and asiaticoside (Mahato *et al.*, 1987) with rhamnose as end-sugar – was compared. The hepatic uptake of asiaticoside-bearing liposomes was reduced, whereas that of corchorusin-D-bearing liposomes was enhanced and was found to be specific for glucose (Medda *et al.*, 1993). Liver perfusion followed by cell separation showed that the uptake was mostly by the non-parenchymal fraction of liver cells (*Table 1*). When the subcellular distribution pattern was examined, the corchorusin-D-bearing liposomes were found most frequently in the lysosomal fraction of the non-parenchymal cells of liver (*Table 2*). Moreover, based on this observation, an attempt was made to deliver an enzyme of plant origin, glucose

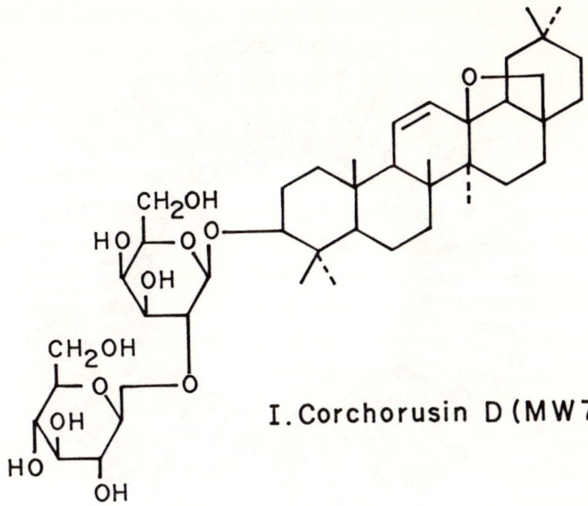

I. Corchorusin D (MW780)

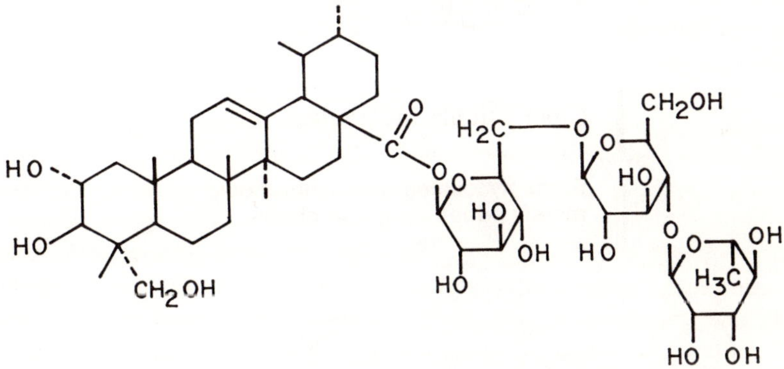

II. Asiaticoside (MW972)

Figure 1. Structures of glycosides isolated from indigenous plants. A hydrophilic sugar moiety is seen to be attached to a hydrophobic aglycone in both of the structures.

oxidase, to lysosomes using corchorusin-D-bearing liposomes as the delivery system (Medda *et al.*, 1993). As the drug action, if any of these glycosides, when incorporated in liposomes, is expected to be effective in small doses because of enhanced cell–liposome interactions, this delivery system was designed to test the selectivity and enhancement of efficacy of some indigenous medicines containing plant glycosides. Besides the plant glycosides, the synthetic glycosides with mannose as end-sugar were also incorporated into the liposomes and were successfully used to target blood–brain barrier cells or the glial cells of the brain (Umezawa and Eto, 1988).

PEPTIDE- OR ANTIBODY-COATED LIPOSOMES

Antibody-mediated targeting was done with actively endocytotic cells. Gregoriadis and Neerunjun (1975) formed large multilamellar liposomes in the presence of IgG specific for surface antigens on the target cells. Antibodies were assumed to associate

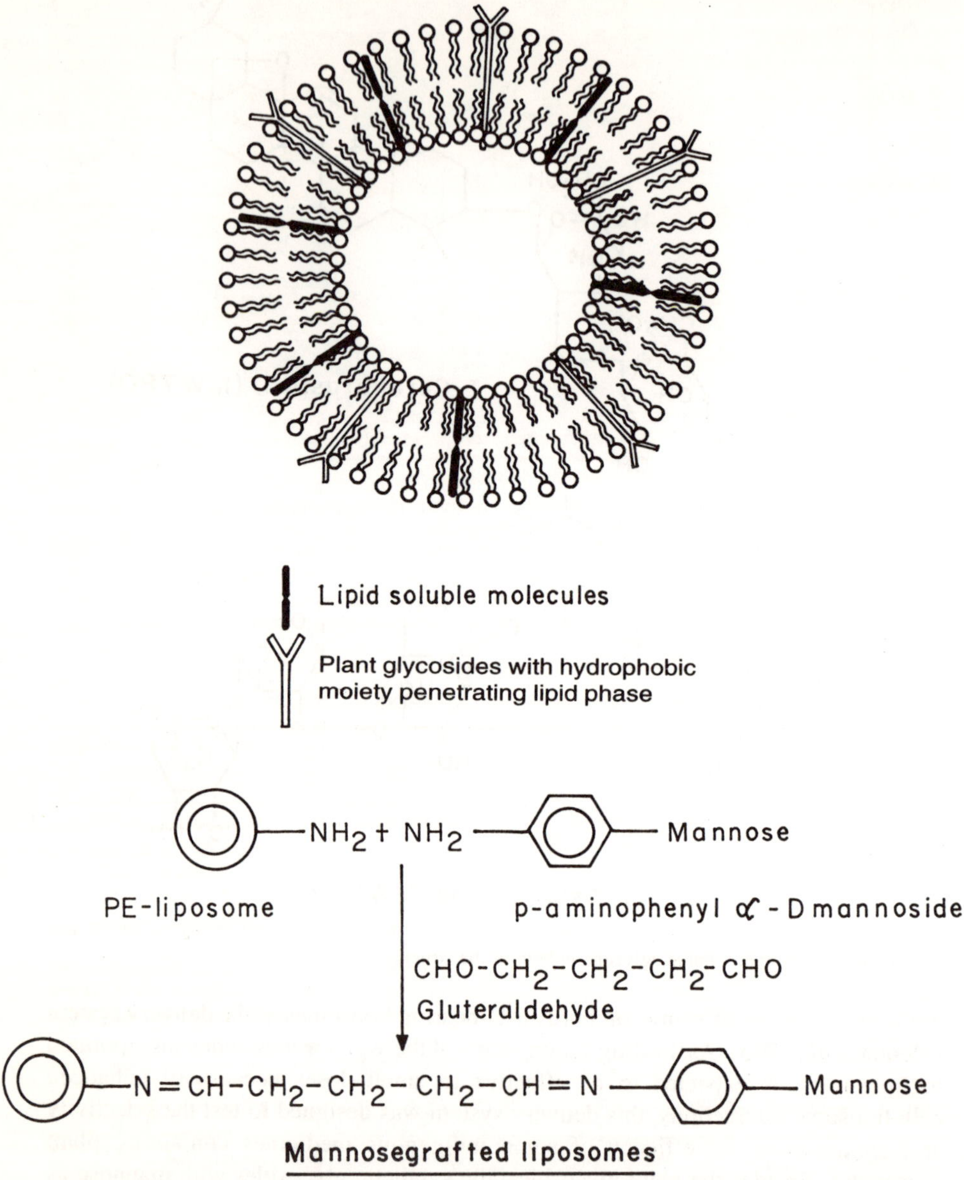

Figure 2. Diagrammatic representation of a liposome incorporating plant glycosides (top), and a schematic representation of the formation of synthetic glycoside-grafted liposomes (bottom).

with liposomes by their F_c portions, and the resulting liposomes showed preferential binding to the corresponding cells. When the cells were fractionated, most of the liposome-entrapped drug was recovered from the lysosomal fraction, consistent with the predominance of endocytosis. Weissman *et al.* (1975) formed multilamellar liposomes in the presence of heat-aggregated immunoglobulin and stimulated phagocytosis via the F_c receptor. Again, uptake was largely into the lysosomal fraction

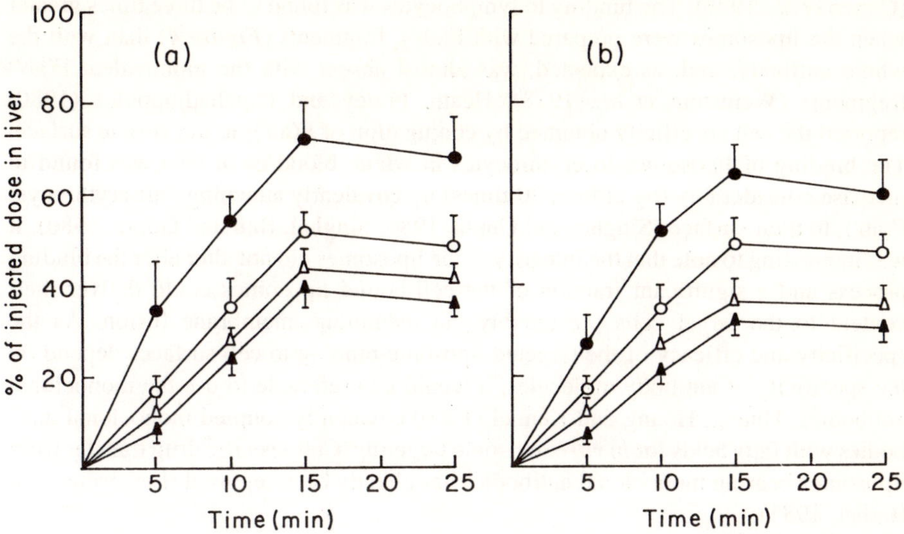

Figure 3. Rate of uptake by liver of [^{125}I]human IgG entrapped in (a) negatively charged and (b) neutral liposomes. The different liposomes used were: ●, gracillin; O, fluoribundasaponin; Δ, control with aglycone; and ▲, control without aglycone. The differences between the uptake of the gracillin liposomes, both negatively charged and neutral, and their corresponding controls are significant at $p <$ 0.001.

Table 1. Differential uptake of [^{125}I]human-IgG-entrapped liposomes by liver cell types

Liposomes	Parenchymal cells (% uptake)	Non-parenchymal cells (% uptake)
Egg PC + Chol + DCP + corchorusin D (A)	20.0 ± 4.7 (3)	69.3 ± 2.3 (3)
(A) + Glycogen	41.8 ± 0.3 (2)	42.1 ± 2.2 (2)
(A) + Mannan	19.7 ± 0.3 (2)	62.9 ± 2.5 (2)
(A) + Asialofetuin	14.5 ± 4.5 (2)	76.0 ± 3.1 (2)

The livers of experimental rats were perfused by collagenase treatment. The parenchymal and non-parenchymal cells were separated by differential centrifugation. The uptake of liposomes in both these fractions presented as a percentage of the total radioactivity in the liver cell suspension before the separation of the cells. Results are expressed as mean ± SD. For each experiment duplicate analyses were made and the number of independent experiments conducted is indicated in parentheses.

Table 2. Subcellular distribution of [^{125}I]human-IgG-entrapped liposomes by liver cell types

Liposomes	Parenchymal cells[a]			Non-parenchymal cells[a]		
	Nuclear fraction (P_1)	Mitochondrial/ lysosomal fraction (P_2)	Plasma and nuclear membrane fraction (P_3)	Nuclear fraction (P_1)	Mitochondrial/ lysomal fraction (P_2)	Plasma and nuclear membrane fraction (P_3)
Egg PC + Chol + DCP+Corchorusin	18.0 ± 2.0	32.0 ± 2.0	28.0 ± 3.0	12.0 ± 5.0	50.2 ± 4.0	18.1 ± 2.1

[a] Percentage of the total count in isolated cell fractions.
Results are shown as mean ± SD ($N = 3$).
P_2 fractions of both the parenchymal and non-parenchymal cells were analysed for the marker enzyme aryl sulphatase. The activity of the enzyme (*p*-nitrocatechol liberated/45 min/mg protein) was found to be around 0.156 ± 0.006 μmoles.

(Cohen *et al.*, 1976). The binding to lymphocytes was found to be three times greater when the liposomes were prepared with F(ab')$_2$ fragments (*Figure 4*) than with the whole antibody, and, as expected, was almost absent with the monovalent F(ab') fragments (Weinstein *et al.*, 1978). Heath, Fraley and Papahadjopoulos (1980) reported the cell specificity obtained by conjugation of F(ab')$_2$ to the vesicle surface. The binding of liposomes to erythrocytes in whole blood or *in vivo* was found to increase considerably (by at least 20 times) by covalently attaching anti-erythrocyte F(ab')$_2$ to their surface (Singhal and Gupta, 1986; Singhal, Bali and Gupta, 1986). It was interesting to note that the integrity of the liposomes did not alter after the binding process and a significant fraction of the cell-bound liposomes could deliver their content to the target cells presumably via membrane–membrane fusion. As the specificity and efficacy of the targeted liposome binding to cell surfaces depend on the specificity of antibody molecules, it would be preferable to use the monoclonal antibodies. Huang, Huang and Kennel (1980) covalently coupled monoclonal antibodies with fatty acids for *in vitro* liposome targeting. Cell-specific drug transfer from liposomes bearing monoclonal antibodies has already been reported (Lesserman and Barbet, 1981).

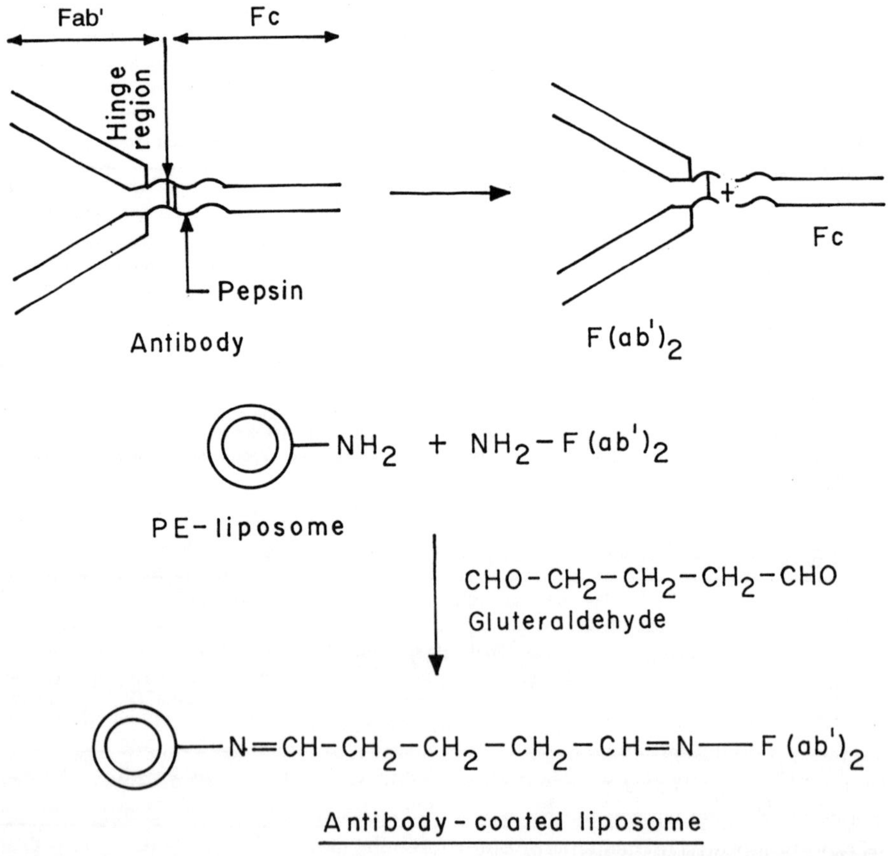

Figure 4. Schematic representation of the formation of antibody-coated, or more specifically F(ab')$_2$-fragment-coated, liposomes.

Synthetic peptides (e.g. *N*-fMet–Leu–Phe, *N*-fMet–Leu–Phe–Phe, *N*-fMet–Nle–Leu–Phe) that are reported to be chemotactic to macrophages and monocytes (Schiffmann, Corcoran and Wahl, 1975; Niedel, Wilkinson and Cuatrecasas, 1979; Snyderman and Fudman, 1980; Snyderman and Goetzel, 1981) are being utilized in our laboratory for coupling to the liposomes. The peptide-coated liposomes have already been tested against both normal and *Leishmania donovani* infected macrophages *in vitro* (Banerjee and Basu, unpublished observation). The possibility of using peptide-coated liposomes *in vivo* encapsulated with antileishmanial drug is now being explored. Tuftsin, a tetrapeptide (Thr–Lys–Pro–Arg) known to bid specifically to macrophages, monocytes and polymorphonuclear neutrophils (PMN) leucocytes and also to potentiate the natural killer activity of these cells, has successfully been used for grafting on the liposome surface (Singhal *et al.*, 1984). Subsequent studies with tuftsin-bearing liposomes revealed that the pretreatment of animals could render them resistant to malaria and leishmania infections (Gupta *et al.*, 1986; Guru *et al.*, 1989). Tuftsin-bearing liposome-intercalated Amp B has been reported to be used against experimental aspergillosis (Owais *et al.*, 1993). Liposomes bearing surface-attached antibody (L-Ab) were prepared to deliver dideoxy-inosine triphosphate (ddITP) to human monocyte/macrophages in an attempt to suppress the replication of human immunodeficiency virus (HIV), which causes the acquired immunodeficiency syndrome (AIDS), by inhibiting the reverse transcriptase of viral RNA (Betageri, Jenkins and Ravis, 1993).

Application of liposomal delivery systems in chemotherapy

AS A CARRIER OF ENZYME AND PROTEIN

The use of liposomes as enzyme carriers was reported for the first time in the treatment of lysosomal storage diseases (Gregoriades, Leathwood and Ryman, 1971; Gregoriadis and Ryman, 1972a). After administration of liposome-entrapped β-fructofuranosidases the enzyme activity was found to concentrate mainly in the lysosome-enriched fractions of liver and spleen, indicating the uptake by endocytosis (Gregoriadis and Ryman, 1972b). Gregoriadis and Neerunjun (1974) incorporated gangliosides into liposomes and showed the hepatic uptake and clearance from the circulation by measuring radioactivity of the entrapped [^{125}I]albumin or invertase. The uptake by hepatic cells of monosialo-ganglioside-bearing liposomes encapsulated with invertase was studied by Surolia and Bachhawat (1977). It was observed that model hepatitis, induced by D-galactosamine, could be prevented by administration of asialo-GM_1 liposomes encapsulated with uridine, prior to the administration of galactosamine (Mathias, Dorai and Bachhawat, 1977).

The use of liposomes for enzyme replacement therapy in Gaucher's disease has already been discussed, the missing enzyme being glucocerebrosidase. Mannose-grafted liposome-entrapped [^{125}I]γ-globulin was reported to be found in Kupffer and endothelial cells, and a potential decrease in the uptake of those liposomes was noticed when mannan was used as inhibitor (Ghosh and Bacchawat, 1980). A number of reports appeared in the literature confirming the fact that β-galactosidase liposomes are specifically taken up by the liver (Spanger and Scherphof, 1983; Szoka and Mayhew, 1983). Incorporation of D-glucose oxidase through the blood–brain barrier into the brain by means of liposomes was achieved by Naoi and Yagi (1980). Using liposomes bearing plant glycosides, our laboratory succeeded in delivering [^{125}I]glucose oxidase to the lysosomal fraction of non-parenchymal cells (Medda *et al.*, 1990).

AS A CARRIER OF CYTOTOXIC AND ANTICANCER DRUGS

Cytotoxic drugs are known for their indiscriminate action on both diseased and normal cells. Attempts have been made to direct the cytotoxicity specially to the diseased area of the body. The mechanism by which liposomal cytotoxic agents act *in vivo* was studied in partially hepatechtomized rats injected intravenously with liposomes containing actinomycin D in their lipid phase. Upon injection of actinomycin-D-or 5-fluorouracil-entrapped liposomes, it was found that cancerous tissues preferentially take up high amounts of these drugs, and there was little uptake of the drugs by bone marrow, which showed no side-effects (Gregoriadis, 1976). Several anticancer agents were incorporated in liposomes and their therapeutic efficacy was evaluated (Rahman *et al.*, 1974a; Mayhew *et al.*, 1976). The liposome-encapsulated anticancer drugs (e.g. methotrexate or 1-β-D-arabinofuranosyl-cytosine) were found to be more potent than the free drug in mice (Kosloski *et al.*, 1978; Rustum *et al.*, 1979; Kimbelberg, 1980). Many studies on the use of liposomes containing antitumour drugs for probing experimental tumours in animal models have already been reported (Kaye, 1981). There have been reports about the increased tumoricidal property of muramyl dipeptide (MDP) or its lipophilic derivatives after their incorporation into liposomes (Sone and Fidler, 1980; Kleinerman *et al.*, 1983). It was thought possible that MDP or its derivatives are protected from extracellular degradation before reaching the target cells, resulting in enhanced tumoricidal activity. Enhancement of the activity of bleomycin on Ehrlich ascites carcinoma cells by liposomal encapsulation has been reported (Sur, Hazra and Roy, 1984). Moreover induction or suppression of the immune response was found to be effected by incorporating drugs into liposomes with surface-coupled antigen (Shek, Lopez and Heath, 1986). However, no definite case was shown for which liposomes could be looked upon as one of the best drug delivery systems in cancer treatment. In solid tumours the malignant cells are known to be situated in the extravascular compartment. The inability of liposomes to extravasate from the microcirculation is obviously relevant to this problem. This must be considered as a possible factor limiting the effectiveness of liposomes in treating metastatic disease. A second major factor limiting the effectiveness of liposome-encapsulated antitumour drugs in cancer treatment originates from the problem of phenotypic heterogeneity among tumour cells (Poste, Krish and Koestler, 1984). It has been pointed out that malignant tumours are not uniform entities populated by cells with identical phenotypes. They would contain multiple subpopulations of tumour cells which could possibly exhibit notable phenotypic diversity for a wide range of cellular properties (Poste and Fidler, 1980; Poste, 1982). It was argued that the heterogeneous response of different tumour cell subpopulations to cytostatic and cytotoxic drugs seems responsible for the majority of failures in cancer chemotherapy (Poste, Krish and Koestler, 1984). Very recently, liposomal monencin has been tried in an attempt to eliminate selectively the cancer cells, and was found to potentiate the cytotoxicity of immunotoxins or hormonotoxins (Madan, Vasandani and Ghosh, 1993; Vasandani, Madan and Ghosh, 1993). However, considering all the limitations of liposomes to treat certain disorders, they have been exploited successfully in the treatment of several diseases. Also, 5-fluorouracil in combination with liposome-encapsulated MDP has been used for the detection of chemo-immunotherapy of murine liver metastases (Daemen *et al.*, 1993). Muramyl tripeptide phosphatidyl ethanolamine (MTP-PE) liposomes have been reported to act as biological response modifiers in the treatment of cancer current status (Frost, 1992). An antitumour effect of liposomes containing taxol has been reported in a taxol-resistant murine tumour model system (Sharma, Mayhew and Straubinger, 1993). Based on recent reports (Kubo *et al.*, 1993), indium III-labelled liposomes have been used successfully for dosimetry and tumour detection in patients with cancer. On the other hand, lung cancer has been shown to be eradicated by antibody-targeted delivery of doxorubicin entrapped in sterically stabilized liposomes (Ahmad *et al.*, 1993). Recent human trials in the USA have reached their second phase for the treatment of infections in cancer patients, using liposomes containing antibiotics.

AS A CARRIER OF ANTILEISHMANIAL DRUGS

The therapy of leishmaniasis poses some problems. The drugs that are often used are known to be very toxic. In experimental infections in hamsters, the toxic dose of the standard pentavalent antimonial drug was found to be equivalent to the therapeutically effective dose (Alving *et al.*, 1978). However, the liposome-encapsulated drugs were found to be much safer and more efficacious in the treatment of experimental leishmaniasis (Alving and Steck, 1979). The therapeutic dose was found to be 200–800 times higher for the unencapsulated compared to the encapsulated drug (Black, Watson and Ward, 1977; Alving *et al.*, 1978; New *et al.*, 1978). But, the requirement of dose, even for liposome-encapsulated drug, could be higher, depending on the degree of infection or the stage of the disease (Alving, 1982). The therapeutic effects of antimonial drugs, e.g. sodium stibogluconate and meglumine antimoniate, were compared with those of lepidines, e.g. 8-aminoquinolines (Kinnamon *et al.*, 1978) and tetracycline (Alving *et al.*, 1980). The efficacies of a series of liposome-entrapped drugs in the treatment of both experimental cutaneous and experimental visceral leishmaniasis were critically analysed and reported by Chance and New (1980). In addition to the visceral leishmaniasis, New, Chance and Heath (1981a) extended their observations to show that liposomes also enhance the activity of sodium stibogluconate against experimental cutaneous leishmaiasis, where the parasite is located in macrophages in peripheral tissues, rather than in the liver. The scope of their investigation was further widened to cover other compounds, e.g. griseofulvin, 5-fluorocytosine and amphotericin B, all of which were known to exhibit antifungal activity (New, Chance and Heath, 1981b). A new approach to the problem of localizing intracellular liposomes and their entrapped agents utilizing the heavy metal properties of the entrapped antileishmanial compound, sodium antimony gluconate (Pentostam), and the non-therapeutically active intracellular marker, chromium, were reported by Heath, Chance and New (1984). Besides the critical evaluation of various antileishmanial drugs, the critical evaluation of liposomes and other drug delivery systems in the treatment of leishmaniasis was also published (Croft, Neal and Rao, 1989).

Antileishmanial activity of liposome-encapsulated amphotericin B in hamsters and

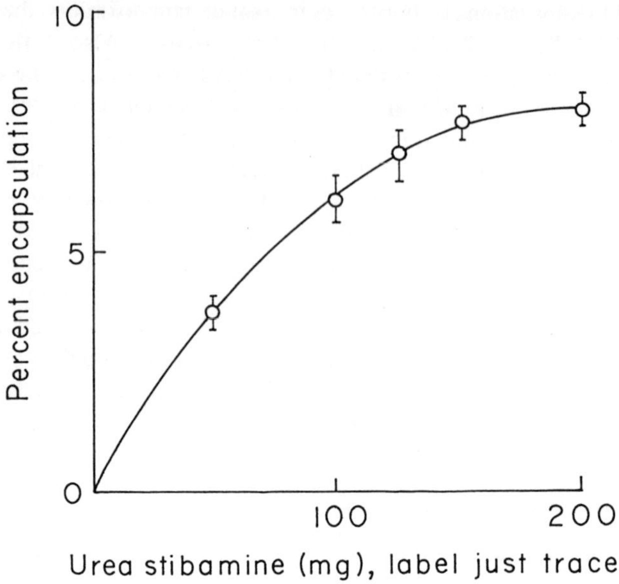

Figure 5. The relationship between percentage encapsulation and varying amounts of urea stibamine, with a trace of label used for entrapment within the liposomes. The composition and lipid concentration of the liposomes were constant throughout. The results are plotted as means ± SD (N = 3).

monkeys was reported by Berman *et al.* (1986). Successful use of liposomal amphotericin B in the treatment of drug-related visceral leishmaniasis has also been reported (Ahmad *et al.*, 1991; Davidson *et al.*, 1991). Croft, Davidson and Thornton (1991) used liposomal amphotericin B (AmBisome) in the treatment of patients suffering from multidrug-resistant visceral leishmaniasis. As part of a non-empirical approach to antileishmanial drug design and delivery, the lipoprotein-mediated antileishmanial chemotherapy was reported by Hart (1987). The idea for utilization of a specific receptor for low-density lipoprotein (LDL) and acetylated LDL on leishmania-infected macrophages was exploited *in vitro* to deliver the antileishmanial drug, Adriamycin, selectively (Hart, 1987). Tuftsin-coated Pentostam-encapsulated liposomes were reported to be used for antileishmanial activity (Guru *et al.*, 1989).

Our laboratory has used the other classical antimonial drug urea stibamine to probe experimental leishmaniasis in the hamster model (Das *et al.*, 1990). Taking advantage of the presence of mannosyl fucosyl receptors on the macrophage surface, mannose-bearing liposomal delivery was designed and planned to be used with entrapped urea stibamine for site-specific delivery (Das *et al.*, 1990). *Figure 5* shows the percentage encapsulation of urea stibamine as a function of drug concentration. The efficacies of urea stibamine – free, liposome-encapsulated and mannose-bearing liposome-encapsulated – have been compared using a 30-day infected hamster model (*Table 3*). The results demonstrate that liposome encapsulation of urea stibamine enhanced its effectiveness, an effect that was greater when mannose-bearing liposomes were used.

Very recently, different sugar-coated liposomal delivery systems were designed, either by incorporating plant glycosides or by grafting synthetic glycosides, and their specificity towards macrophages was tested *in vitro* using appropriate inhibitors (Medda *et al.*, 1993). The neoglycosylated liposomes so formed were also used as

Table 3. Effect of urea stibamine on a 30-day infected hamster model of *Leishmania donovani*

Treatment	Liver weight (g) (mean ± SD)	Spleen weight (mg) (mean ± SD)	Parasite load in the spleen (mean ± SD × 10^7)	% Suppression of parasite load in the spleen
No drug (control)	4.02 ± 0.32	267 ± 14	3.41 ± 0.34	–
Free drug	3.62 ± 0.28	238 ± 10	2.53 ± 0.56	26.0
Empty liposomes	3.24 ± 0.20	219 ± 16	2.89 ± 0.15	15.0
Empty mannose-bearing liposomes	3.22 ± 0.22	215 ± 19	2.67 ± 0.15	22.0
Liposome-encapsulated drug	2.46 ± 0.24	150 ± 12	1.54 ± 0.18[a]	55.0
Mannose-bearing liposome-encapsulated drug	2.09 ± 0.26	108 ± 12	0.97 ± 0.11[a]	72.0

Results are shown as mean ± SD ($N = 4$).
[a] $p < 0.001$ compared to no drug (control); the drug used was urea stibamine.

efficient ligands for the estimation of specific sugar receptor status of macrophages in health and in experimental leishmaniasis (Dutta, Bandyopadhyay and Basu, 1994). The efficacy and the toxicity of the drug, urea stabamine, were critically compared in a 60-day infected hamster model, using both glucose-bearing and mannose-bearing delivery systems (Medda *et al.*, 1993). The drug encapsulated in mannose-bearing delivery systems was found to be more efficient and less toxic than that encapsulated in the glucose-bearing systems. The non-antimonial classical drugs, e.g. pentamidine isothionate, and their analogues were also examined *in vitro* for antileishmanial activities (Mahato *et al.*, 1993; Nandi *et al.*, 1993) and their therapeutic efficacy and the resultant toxicity were critically analysed when encapsulated in mannose-bearing liposomes (Banerjee *et al.*, submitted). Moreover, hamycin, a polyene antibiotic, was examined in our laboratory for antileishmanial activity. When tested *in vivo* against artificial leishmaniasis in a hamster model, the liposomal hamycin and mannose-bearing liposomal hamycin were found to be more potent than the regular hamycin; the mannose-bearing liposomal hamycin being the most effective in reducing the spleen parasite load. Toxicity of the drug was reduced in the liposomal forms as judged from haemoglobin level and the level of specific enzymes related to normal liver function (*Table 4*) but was reduced further when sterol-rich liposomes were used as a delivery system (Banerjee, Bhaduri and Basu, 1994).

Table 4. Specific enzyme levels in sera of hamsters undergoing experimental leishmaniasis

Groups	Alkaline phosphatase[a]	Serum glutamate pyruvate transaminase (SGPT)[b]
Normal	8.2 ± 2.5	42.0 ± 6.0
Infected	10.2 ± 1.4	64.5 ± 18.9
Infected + free drug treated	17.6 ± 3.4	116.7 ± 21.8
Infected + liposome-intercalated drug treated	15.1 ± 3.3	103.5 ± 15.0
Infected + mannose-bearing liposome-intercalated drug treated	6.7 ± 3.0	72.5 ± 15.0

[a] μmol of *p*-nitrophenol released min^{-1} dl^{-1} serum.
[b] μmol of sodium pyruvate released min^{-1} l^{-1} serum.
Results are shown as mean ± SD ($N = 6$). Three animals per group were examined in two separate sets of experiments. The drug used was hamycin.

Besides liposomes, use of the non-ionic surfactant vesicles, 'niosomes', was proposed (Baillie *et al.*, 1986) and the therapeutic efficacy of the antimonial drug, sodium stibogluconate, encapsulated either in 'niosomes' or in liposomes, was compared *in vivo* against artificial leishmaniasis in hamsters and found to be equiactive (Hunter *et al.*, 1988). In our laboratory, 'niosomes' were prepared using Tween-20 and the therapeutic efficacy as well as the toxicity of the other antimonial drug, urea stibamine, were compared *in vivo* against artificial leishmanial infection (Medda and Basu, 1994). Although the drug toxicity at equivalent drug concentration was comparable in both the vesicular systems, the therapeutic efficacy of hamycin in the 'niosomal' form was found to be better than that in the liposomal form (*Table 5*). The non-ionic surfactants being cheaper, less toxic, biodegradable and non-immunogenic, the 'niosomes' might replace liposomes as drug carriers in the future.

Table 5. Effect of hamycin on a 60-day infected hamster model of *Leishmania donovani*

Treatment	10^{-7} x Parasite load in the spleen	Suppression of parasite load in the spleen (%)
No drug (control)	7.2 ± 0.8	–
Free drug	4.9 ± 0.4	32
Empty liposomes	6.5 ± 0.8	10
Liposomal drug	3.3 ± 0.3	54
Empty niosomes	6.2 ± 0.8	14
Niosomal drug	2.8 ± 0.2	65

Results are means ± SD ($N = 3$).
The same equivalent drug concentration was used for free, liposomal and niosomal treatments.

AS A CARRIER OF ANTIFUNGAL DRUGS

The fungal infections, e.g. aspergillosis and candidiasis, are very common among immunosuppressed patients. Such infections can be severe, sometimes causing death. Some polyene antibiotics, e.g. amphotericin B, hamycin and nystatin, are very toxic but are the only available drugs for combating the diseases related to fungal infections. Because of their highly toxic nature, the use of such drugs in high dosage, necessary for therapeutic efficacy, is restricted. Attempts to reduce drug toxicity but simultaneously to preserve therapeutic efficacy were made without success until liposomes were used as drug carriers for fungal disease treatment. Liposomes were reported to give a reduction of the toxicity of amphotericin B with complete retention of therapeutic efficacy, in a case of candidiasis (Lopez-Berestein *et al.*, 1983). Although some reports (Szoka, Milholland and Barza, 1987) on the toxicity and *in vitro* fungicidal activity of liposome-intercalated amphotericin B were published, until 1988 no systematic study on the toxicity and therapeutic efficacy of liposome-intercalated amphotericin B in an animal model was published. In 1989, while comparing the toxicity of free amphotericin B and liposomal amphotericin B in experimental aspergillosis in BALB/c mice, Ahmad, Sarkar and Bachhawat (1989) found a several-fold increase in LD_{50} of amphotericin B when associated in liposomes. Although inclusion of cholesterol in neutral liposomes drastically reduced the drug toxicity, no effect on the toxicity was shown when both positively and negatively charged liposomes were used as delivery systems (Ahmad, Sarkar and Bachhawat,

1990). Tuftsin-bearing liposomes with cholesterol were reported as a vehicle for amphotericin B in the treatment of experimental aspergillosis in mice. The drug efficacy was significantly increased by pretreating the animals with drug-free tuftsin-bearing liposomes, demonstrating that macrophage activation (possibly by tuftsin) can considerably enhance the therapeutic efficacy of antifungal drugs such as amphotericin B (Owais *et al.*, 1993). The relative toxicity and the therapeutic efficacy of the other polyene antibiotic, hamycin, intercalated in liposomes, were also tested against experimental aspergillosis in mice, although the toxicity study was concentrated only on the determination of LD_{50} of the drug intercalated in various liposomal formulations (Moonis, Ahmad and Bachhawat, 1992). The toxicity was found to be reduced when negatively charged liposomes, with increasing cholesterol composition, were used as the delivery systems (Moonis, Ahmad and Bachhawat, 1993b). Moreover, when hamycin was intercalated in mannose-bearing liposomes, the toxicity of the drug was further reduced, showing enhanced antifungal activity. The LD_{50} increased by almost twice with the inclusion of sterol, even in the neutral liposomal composition, from 1.4 (mg kg^{-1}) for SPC/DPPE-Man (molar ratio 9:1) liposomes to 2.8 (mg kg^{-1}) for SPC/Chol/DPPE-Man (molar ratio 4:5:1) liposomes (Moonis, Ahmad and Bachhawat, 1993a). Liposomal amphotericin B (AmBisome) was found to have a role in the treatment of patients with systemic fungal infections (Lopez-Berestein and Rosenblum, 1992) and in successful prophylactic use against fungal infections in bone marrow transplant recipients (Tollemar *et al.*, 1993).

AS A CARRIER OF DRUGS FOR VARIOUS OTHER DISORDERS

Alving *et al.* (1979) reported that liposomes containing neutral glycolipids with a terminal glucose or galactose, when injected intravenously, prevented the appearance of erythrocytic forms of malaria (*Plasmodium berghei*) in mice, even in the absence of antimalarial drugs. Very recently, chloroquine delivery to erythrocytes in *Plasmodium berghei* infected mice, using antibody-bearing liposomes as drug vehicles (Chandra, Agarwal and Gupta, 1991), has been reported.

In a number of storage diseases there is an inherited enzyme deficiency that leads to the accumulation of the relevant substrates. Thus treatment of such diseases with liposomes containing the appropriate enzymes seems feasible. Macrophages or fibroblasts, previously loaded with sucrose in their lysosomes, were exposed to invertase encapsulated in liposomes. Hydrolysis of sucrose to glucose and fructose occurred, and the sugars diffused out through the lysosomal membranes (Gregoriadis and Buckland, 1977). When dextran-loaded rats were treated with liposome-entrapped dextranase, a considerable loss of dextran was noted from the liver.

Attempts have been made to deliver uridine to the liver, using liposomes as carriers in model hepatitis (Mathias, Dorai and Bachhawat, 1977). The effect of modification of liposomal lipid composition on the regeneration of hepatitic liver revealed that the entrapped uridine was effective in the following order: asialoganglioside liposome > phosphatidic acid liposome > dicetylphosphate liposome.

Liposomes raise new possibilities for the oral therapeutic use of proteins, and probably other substances, which are otherwise not suitable for oral administration. Attempts to introduce liposome-encapsulated insulin into the blood circulation by the oral route have met with considerable success (Patel and Ryman, 1976). Insulin entrapped in egg phosphatidyl choline/cholesterol liposomes was less effective than in liposomes containing dipalmitoyl lecithin/cholesterol.

Application of liposomes has been extended to the treatment of rheumatoid arthritis, and it was found that cortisol palmitate-containing liposomes are stable in rheumatoid synovial fluid at 37°C. The level of the liposomal steroid in the tissue was inversely related to the chronicity of inflammation; the degree and the duration of its anti-inflammatory activity was greatest in the initial acute phase of the inflammation. These promising results have led to a preliminary clinical trial in patients with rheumatoid arthritis (de Silva *et al.*, 1979); intra-articular treatment with liposomal cortisol palmitate resulted in an improvement in both subjective and objective indices of inflammation in the affected joint. Very recently, liposome-entrapped methotrexate, injected intra-articularly, has been reported to be used in the treatment of antigen-induced arthritis in rabbits (Foong and Green, 1993). For treatment of various brain diseases, attempts have been made to target liposomes to rat brain via the internal carotid artery (Sakamoto and Ido, 1993).

Liposomes bearing asiaticoside, a plant glycoside, have been reported as used successfully *in vitro* against *Mycobacterium leprae* and *M. tuberculosis* (Medda *et al.*, submitted). Liposomes bearing tuftsin as rifampicin vehicles have been used *in vivo* in the treatment of tuberculosis in mice (Agarwal *et al.*, 1994). There have been reports on the modification of reticuloendothelial functions by muramyl dipeptide-encapsulated liposomes in jaundiced rats treated with biliary decomposition (Ding *et al.*, 1993). On the other hand, liposome-encapsulated muramyl tripeptide phosphatidyl ethanolamide was used successfully in the treatment of experimental *Klebsiella* pneumonial infection (Melissen *et al.*, 1992). Liposomes seem particularly appropriate for delivery of drugs to the lungs as they can be prepared from materials endogenous to the lung as components of lung surfactant (Schreier, Gonzalez-Rothi and Stecenko, 1993). In fact, liposomes were administered to the lung both as a means of delivering phospholipids to the alveolar surface (Gonzalez-Rothi *et al.*, 1991) and to modulate the duration of active or pulmonary absorption of pharmacologically active agents (Taylor and Newton, 1992; Taylor and Farr, 1993). In the earliest studies of pulmonary administration of liposomes for therapeutic purposes, liposomes were given as an exogeneous lung surfactant in the treatment of the neonatal respiratory distress syndrome (Ivey, Roth and Kattwinkel, 1977). Subsequently a range of liposome-associated drugs were administered to the lungs of humans and animals. These include cytotoxic drugs (McCullough and Juliano, 1979), antiasthma drugs (McCalden, Abra and Mihalko, 1989; Taylor *et al.*, 1989), antimicrobial compounds (Debs *et al.*, 1987; Gilbert *et al.*, 1988; Mihalko, Schrier and Abra, 1988) and drugs delivered for systemic action (Mihalko, Schrier and Abra, 1988; Shek *et al.*, 1988; Meisner, Pringle and Mezei, 1989).

Toxicity of liposomes

The idea of using liposomes as a convenient drug carrier might be jeopardized by their toxic effects, if any. Toxic effects could originate from liposomes themselves, from the active agents they contain or by synergistic action of the carrier and its contents. Because of the versatile nature of the use of liposomes in a drug delivery mode, acceptance of a liposomal preparation for clinical use will eventually depend on whether the overall effect is favourable to the patient. Adverse reaction may occur as a result of certain structural features; the lipid component of liposomes (Gregoriadis, 1980b); altered liposomal drug distribution; and the patient's reaction to the liposomal carrier. There have been reports that liposomes above a certain size may block lung capillaries (Rahman *et al*., 1974) and that a positively charged surface may bind circulating α_2-macroglobulin (Black and Gregoriadis, 1976) upsetting coagulation processes. But, a patient when injected intravenously with about 100 mg lipid in the form of negatively charged liposomes showed no decrease in α_2-macroglobulin levels for up to 3 days after treatment. Besides several other minor reports, another potential liposomal toxicity was observed by Chen and Keenan (1979) who noticed that egg phosphatidyl choline liposomes inhibit lymphocyte activation by attracting, and thus reducing, cellular cholesterol.

Among the reports on toxicity as a result of lipid components, a direct *in vivo* toxicity in the form of epileptic seizures or cerebral tissue necrosis has been noticed after the intracerebral injection of large amounts of liposomes containing either dicetyl phosphate or stearylamine. But, at comparable doses, the liposomes containing stearylamine were found to be non-toxic by other groups (Mayhew *et al*., 1976; Kimbelberg and Mayhew, 1978) for long-term intraperitoneal treatment. Systematic studies on liposomal toxicity were carried out in rats and beagle dogs who received intravenously, on a daily basis, sonicated soya phosphatidyl choline for several weeks. The results in behavioural changes, urine biochemistry and histology were negative (Gregoriadis, 1978). Repeatedly, clinical experience with a small number of cancer patients injected with liposomes, composed of egg lecithin, cholesterol and phosphatidic acid (molar ratio 7:2:1), has been uneventful (Gregoriadis *et al*., 1974). Thus, the toxicity of liposomes should not be a great obstacle in opening new avenues for their chemotherapeutic use.

The chronic inhalation of soya phosphatidyl choline (SPC) liposome in an attempt to target drugs to the lung, specifically to pulmonary alveolar macrophages (AM), showed no histological changes in the lung or untoward effects on the general health or survival of animals. The properties of alveolar macrophages such as phagocytic function, intracellular killing and fatty acid composition were not affected. Transmission electron microscopy and morphometry (computerized image analysis) of alveolar macrophages showed no alterations as a result of the treatment. This finding validates the concept of alveolar macrophage-directed delivery of liposome-encapsulated agents to the lung via inhalation (Myers *et al*., 1993). These data confirm and extend prior findings regarding the functional and morphological interactions of liposomes with alveolar macrophages *in vitro* (Gonzalez-Rothi *et al*., 1991).

Summary and perspectives

The popularity of liposome research is based on two attractive features of the system, namely its versatility and its similarity to natural membranes. Liposomes can vary widely in size, composition and surface characteristics, and can be made to accommodate a remarkable array of pharmacologically active substances, including antitumour and antimicrobial drugs, enzymes, hormones and vaccines. For site-specific delivery, specialized liposomes are prepared nowadays, and this has led to considerable interest in the possibility of therapeutic use of liposomes. But for liposomes, crossing barriers to the extravascular space still remains a problem. The liposomes can cross the membrane, but because of their limitations in minimal size, it is obvious that not all membranes will be crossed by even the smallest liposomes. Until drug specificity is achieved, liposomes, in spite of their limitations, should play an important role in optimizing drug action (Gregoriadis, 1980a). However, reported studies of liposome-mediated therapy are as yet too fragmentary to determine whether current investigations will lead to useful methods of treatment. Probably our incubation period for taking a positive approach towards their application in chemotherapy is still not over.

However, at present, with a much clearer understanding of liposome technology's limitations (as well as a better technical understanding of design and manufacture of these phospholipid-based microspheres) companies are putting an impressive number of liposome-encapsulated drugs into human trials, initiating a technology move from the laboratory into the clinic. In fact, more than 15 liposome products were scheduled to be in US clinics during 1988 (*Table 6*). The companies (e.g. The Liposome Company, Princeton, NJ and Vestar, Pasadana, CA) believe that there is potential in the use of liposomes for critical-care patients with cancer and infectious diseases. Vester has been able to increase the shelf-life of its liposomes from weeks to years and can now manufacture 'targeted' liposomes that congregate at tumours without having to rely on the body's macrophages to escort them. Vester's targeted daunorubicin formulation is already scheduled to go into clinical trials for the treatment of cancer, fungal infections and post-surgical healing. Technology Unlimited, Wooster, OH, had two topical liposome-based preparations in the clinic already in 1988. They have been able to coat the exterior of liposomes with sialic acid residues, thereby allowing the microspheres to avoid the reticuloendothelial system, thus raising the possibility of using these liposomes, perhaps with the help of covalently attached monoclonal antibodies, to target areas outside this system (Klausner, 1988).

Table 6. Liposome products likely to be used in human chemotherapy (after Klausner, 1988)

Liposome products	Producers	Applications
Doxorubicin	The Liposome Co.	Cancer
Daunorubicin	Vester	Cancer
Amphotericin B	Squibb/The Liposome Co.	Fungal infections
Amphotericin B	Lypho Med/Vester	Fungal infections
Insulin	Technology Unlimited	Diabetes
Miconazole	Ortho/The Liposome Co.	Vaginal infections
Gentamicin	The Liposome Co.	Gram-negative infections
Anti-inflammatory drug (non-steroidal)	Ethicon/Vester	Post-surgical healing

Acknowledgements

Financial support from the United Nations Development Programme, the Department of Science and Technology, Government of India and the Indian Council of Medical Research is gratefully acknowledged.

References

AGARWAL, K., BALI, A. AND GUPTA, C.M. (1986a). Influence of phospholipid structure on the stability of liposomes in serum. *Biochimica et Biophysica Acta* **856**, 36–40.

AGARWAL, K., BALI, A. AND GUPTA, C.M. (1986b). Effect of phospholipid structure on stability and survival times of liposomes in circulation. *Biochimica et Biophysica Acta* **883**, 468–475.

AGARWAL, A., KANDPAL, H., GUPTA, H.P., SINGH, N.B. AND GUPTA, C.M. (1994). Tuftsin bearing liposomes as rifampicin vehicles in the treatment of tuberculosis in mice. *Antimicrobial Agents and Chemotherapy* **38**, 000–000.

AHMAD, I., SARKAR, A.K. AND BACHHAWAT, B.K. (1989). Liposomal Amphotericin B in the control of experimental aspergillosis in mice: Part I – relative therapeutic efficacy of free and liposomal Amphotericin B. *Indian Journal of Biochemistry and Biophysics* **26**, 351–356.

AHMAD, I., SARKAR, A.K. AND BACHHAWAT, B.K. (1990). Effect of cholesterol in various liposomal compositions on the *in vivo* toxicity, therapeutic efficacy and tissue distribution of Amphotericin B. *Biotechnology and Applied Biochemistry* **12**, 550–556.

AHMAD, I., AGARWAL, A., PAL, A., GURU, P.Y., BACHHAWAT, B.K. AND GUPTA, C.M. (1991). Tissue distribution and antileishmanial activity of liposomised amphotericin B in Balb/c mice. *Journal of Biosciences* **16**, 217–221.

AHMAD, I., LONGEREEKER, M., SAMUEL, J. AND ALLEN, T.M. (1993). Antibody targeted delivery of dexorubicin entrapped in sterically stabilized liposomes can eradicate lung cancer in mice. *Cancer Research* **53**, 1484–1490.

ALLEN, T.M., HANSEN, C. AND RUTLEDGE, J. (1989). Liposomes with prolonged circulation times: factors affecting uptake by reticuloendothelial and other tissues. *Biochemica et Biophysica Acta* **981**, 27–35.

ALLEN, T.M., AUSTIN, G.A., CHONN, A., LIN, L. AND LEE, K.C. (1991). Uptake of liposomes by cultured mouse bone marrow macrophages: influence of liposome composition and size. *Biochimica et Biophysica Acta* **1061**, 56–64.

ALVING, C.R (1982). Therapeutic potential of liposomes as carriers in leishmaniasis, malaria and vaccines. In *Targeting of Drugs* (G. Gregoriadis, J. Senior and A. Trouet, Eds), pp. 37–347. Plenum Press, New York.

ALVING, C.R. AND STECK, E.A. (1979). The use of liposome-encapsulated drugs in leishmaniasis. *Trends in Biochemical Sciences* **4**(8), N175.

ALVING, C.R., STECK, E.A., CHAPMAN, W.L., JR, WAITS, V.B., HENDRICKS, L.D., SWARTZ, G.M., JR AND HANSON, W.L. (1978). Therapy of leishmaniasis: superior efficacies of liposome-encapsulated drugs. *Proceedings of the National Academy of Sciences, USA* **75**, 2959–2963.

ALVING, C.R., SCHNLEIDER, I., SWART, G.M., JR AND STECK, E.A. (1979). Sporozoite-induced malaria: therapeutic effects of glycolipids in liposomes. *Science* **205**, 1142–1144.

ALVING, C.R., STECK, E.A., CHAPMAN, W.L., JR, WAITS, V.B., HENDRICKS, L.D., SWARTZ, G.M., JR AND HANSON, W.L. (1980). Liposomes in leishmaniasis: therapeutic effects of antimonial drugs, 8-aminoquinolines and tetracycline. *Life Science* **26**, 2231–2236.

BAILLIE, A.J., COOMBS, G.H., DOLAN, T.F. AND LAURIE, J. (1986). Non-ionic surfactant vesicles niosomes, as a delivery system for the antileishmanial drug, sodium stibogluconate. *Journal of Pharmacy and Pharmacology* **38**, 502–505.

BALI, A., DHAWAN, S. AND GUPTA, C.M. (1983). Stability of liposomes in circulation is markedly enhanced by the structural modifications of their phospholipid component. *FEBS Letters* **154**, 373–377.

BANERJEE, G., BHADURI, A.N. AND BASU, M.K. (1994). Mannose-coated liposomal hamycin in the treatment of experimental leishmaniasis in hamsters. *Biochemical Medicine and Metabolic Biology* **53**, 1–7.

BANERJEE, G., NANDI, G., MAHATO, S.B., PAKRASHI, A. AND BASU, M.K. Drug delivery systems: targeting of pentamidines to specific sites in model leishmaniasis using sugar coated liposomes. *Antimicrobial Agents and Chemotherapy* (submitted).

BANGHAM, A.D., STANDISH, M.M. AND WATKINS, J.C. (1965). Diffusion of univalent ions across the lamellae of swollen phospholipids. *Journal of Molecular Biology* **13**, 238–252.

BERG, T. AND BOMAN, D. (1973). Distribution of lysosomal enzymes between parenchymal and Kupffer cells of rat liver. *Biochimica et Biophysica Acta* **321**, 585–596.

BERMAN, J.D., HANSON, W.L., CHAPMAN, W.L., ALVING, C.R. AND LOPEZ-BERESTEIN, G. (1986). Antileismanial activity of liposome-encapsulated amphotericin B in hamsters and monkeys. *Antimicrobial Agents and Chemotherapy* **30**, 847–851.

BETAGERI, G.V., JENKINS, S.A. AND RAVIS, W.R. (1993). Drug delivery using antibody–liposome conjugates. *Drug Development and Industrial Pharmacy* **19**, 2109–2116.

BLACK, C.D.V. AND GREGORIADIS, G. (1976). Interaction of liposomes with blood plasma proteins. *Biochemical Society Transactions* **4**, 253–256.

BLACK, C.D.V., WATSON, G.J. AND WARD, R.J. (1977). The use of pentostam liposomes in the chemotherapy of experimental leishmaniasis. *Transactions of the Royal Society for Tropical Medicine and Hygiene* **71**, 550–556.

BOLDT, D.H., SPECKART, S.F., RICHARDS, R.L. AND ALVING, C.R. (1977). Interactions of plant lectins with glycolipids in liposomes. *Biochemical and Biophysical Research Communications* **74**, 208–214.

CHANCE, M.L. AND NEW, R.R.C. (1980). The use of liposome entrapped drugs in the treatment of experimental and cutaneous leishmaniasis. In *The Host Invader Interplay* (H. van den Bossche, Ed.), pp. 659–662. Elsevier/North-Holland, Biomedical Press, Amsterdam.

CHANDRA, S., AGARWAL, A.K. AND GUPTA, C.M. (1991). Chloroquine delivery to erythrocytes in *Plasmodium berghei* infected mice using antibody bearing liposomes as drug vehicles. *Journal of Biosciences* **16**, 137–144.

CHEN, S.S.-H. AND KEENAN, R.M. (1979). Effect of phosphatidyl-choline liposomes on the mitogen-stimulated lymphocyte activation. *Biochemical and Biophysical Research Communications* **79**, 852–858.

COHEN, C.M., WEISSMAN, G., HOFFSTEIN, S., AWASTHI, Y.C. AND SRIVASTAVA, S.K. (1976). Introduction of purified hexosaminidase-A into Tay-Sachs leukocytes by means of immunoglobulin-coated liposomes. *Biochemistry* **15**, 452–460.

CROFT, S.L., DAVIDSON, R.N. AND THORNTON, E.A. (1991). Liposomal amphotericin in the treatment of visceral leismaniasis. *Antimicrobial Chemotherapy* **28**, 111–118.

CROFT, S.L., NEAL, R.A. AND RAO, L.S. (1989). Liposomes and other drug delivery systems in the treatment of leishmaniasis. In *Leishmaniasis: the Current Status and New Strategies for Control* (D.T. Hart, Ed.), pp. 783–792. Plenum Press, New York.

DAEMAN, T., DONTJE, B.H.J., REGTS, J. AND SCHERPHOF, G.L. (1993). Chemoimmunotherapy of murine liver metastases with 5-fluorouracil in combination with liposome-encapsulated muramyl dipeptide. *Journal of Immunotherapy* **13**, 31–38.

DAS, N., BACHHAWAT, B.K., MAHATO, S.B. AND BASU, M.K. (1987). Plant glycosides in a liposomal drug delivery system. *Biochemical Journal* **247**, 359–361.

DAS, N., MAHATO, S.B., NASKAR, K., GHOSH, D.K. AND BASU, M.K. (1990). Targeting of urea stibamine encapsulated in liposomes to reticuloendothelial system for the treatment of experimental leishmaniasis. *Biochemical Medicine and Metabolic Biology* **43**, 133–139.

DAVIDSON, R.N., CROFT, S.L., SCOTT, A., MAINI, M., MOODY, A.H. AND BRYCESON, A.D.M. (1991). Successful use of liposomal amphotericine B in the treatment of drug-related visceral leishmaniasis. *Lancet* **337**, 1061–1062.

DEBS, R.L., STRAUBINGER, R.M., BRUNETTE, E.N., LIN, J.M., LIN, E.J. AND MONTGOMERY, A.B. (1987). Selective enhancement of pentamidine uptake in the lung by aerosolization and delivery in liposomes. *American Review of Respiratory Diseases* **135**, 731–737.

DE SILVA, M., HAZLEMAN, B.L., PAGE-THOMAS, D.P. AND WRAIGHT, P. (1979). Liposomes in arthritis: a new approach. *Lancet* **1**, 1320–1322.

DING, J.W., ANDERSSON, R., HULTBERT, B., SOLTESZ, V. AND BENGMARK, S. (1993). Modification of reticuloendothelial function by muramyl dipeptide-encapsulated liposomes in jaundiced rats treated with biliary decompression. *Scandinavian Journal of Gastroenterology* **28**, 53–62.

DUTTA, M., BANDYOPADHYAY, R. AND BASU, M.K. (1994). Neoglycosylated liposomes as efficient ligands for the estimation of specific sugar receptor status of macrophages in health and in experimental leishmaniasis. *Parasitology* **109**, 139–147.

FOONG, W.C. AND GREEN, K.L. (1993). Treatment of antigen induced arthritis in rabbits with liposome-entrapped methotrexate injected intracardially. *Journal of Pharmacy and Pharmacology* **45**, 204–209.

FROST, H. (1992). MTP-PE in liposomes as a biological response modifier in the treatment of cancer current status. *Biotherapy* **4**, 199–303.

GHOSH, P. AND BACHHAWAT, B.K. (1980). Grafting of different glycosides on the surface of liposomes and its effect on the tissue distribution of ^{125}I-labeled r-globulin encapsulated in liposomes. *Biochimica et Biophysica Acta* **632**, 562–572.

GHOSH, P.K., BACHHAWAT, B.K. AND SUROLIA, A. (1981). Synthetic glycolipids: interaction with galactose-binding lectin and hepatic cells. *Archives of Biochemistry and Biophysics* **206**, 454–457.

GILBERT, B.E., SIX, H.R., WILSON, S.Z., WYDE, P.R. AND KNIGHT, V. (1988). Small particle aerosols of enviroxime-containing liposomes. *Antivirus Research* **9**, 355–365.

GONZALEZ-ROTHI, R.J., STRAUB, L., CARCASE, J.L. AND SCHREIER, H. (1991). Liposomes and pulmonary alveolar macrophages: functional and morphologic interactions. *Experimental Lung Research* **17**, 687–705.

GREGORIADIS, G. (1973). Drug entrapment in liposomes. *FEBS Letters* **36**, 292–296.

GREGORIADIS, G. (1976). The carrier potential of liposomes in biology and medicine. *New England Journal of Medicine* **295**, 704–710.

GREGORIADIS, G. (1978). Liposomes in therapeutic and preventive medicine: the development of the drug-carrier concept. *Annals of the New York Academy of Sciences* **308**, 343–370.

GREGORIADIS, G. (1980a). The liposome drug-carrier concept: its development and future. In *Liposomes in Biological Systems* (G. Gregoriadis and A.C. Allison, Eds), pp. 25–86. John Wiley & Sons, New York.

GREGORIADIS, G. (1980b). Tailoring liposome structure. *Nature (London)* **283**, 814–815.

GREGORIADIS, G. AND BUCKLAND, R.A. (1977). Enzyme-containing liposomes alleviate a model for storage diseases. *Nature* **244**, 170–172.

GREGORIADIS, G. AND NEERUNJUN, D.E. (1974). Control of the rate of hepatic uptake and catabolism of liposome-entrapped proteins injected into rats. Possible therapeutic applications. *European Journal of Biochemistry* **47**, 179–185.

GREGORIADIS, G. AND NEERUNJUN, D.E. (1975). Homing of liposomes to target cells. *Biochemical and Biophysical Research Communications* **65**, 537–544.

GREGORIADIS, G. AND RYMAN, B.E. (1972a). Fate of protein containing liposomes injected into rats. *European Journal of Biochemistry* **24**, 485–491.

GREGORIADIS, G. AND RYMAN, B.E. (1972b). Lysosomal localization of β-fructofuranosidase-containing liposomes injected into rats. *Biochemical Journal* **129**, 123–133.

GREGORIADIS, G., LEATHWOOD, P.D. AND RYMAN, B.E. (1971). Enzyme entrapment in liposomes. *FEBS Letters* **14**, 95–99.

GREGORIADIS, G., NEERUNJUN, E.D. AND HUNT, R. (1977). Fate of liposome-associated agent injected into normal and tumor-bearing rodents. Attempts to improve localization in tumor tissues. *Life Science* **21**, 357–369.

GREGORIADIS, G., MORELL, A.G., STERNLIEB, I. AND SCHEINBERG, I.H. (1970). Catabolism of desialylated ceruloplasmin in the liver. *Journal of Biological Chemistry* **245**, 5833–5837.

GREGORIADIS, G., SWAIN, C.P., WILLIS, E.J. AND TAVILL, A.S. (1974). Drug-carrier potential of liposomes in cancer chemotherapy. *Lancet* **2**, 1313–1316.

GUPTA, C.M. (1983). Modified liposomes: their stability as drug carriers. *Annals of the National Academy of Medical Sciences (India)* **19**, 22–30.

GUPTA, C.M., PURI, A., JAIN, R.K., BALI, A. AND ANAND, N. (1986). Protection of mice against *Plasmodium berghei* infection by tuftsin derivative. *FEBS Letters* **205**, 351–354.

GURU, P.Y., AGARWAL, A.K., SINGHA, U.K., SINGHAL, A. AND GUPTA, C.M. (1989). Drug targeting in *Leishmania donovani* infections using tuftsin bearing liposomes as drug vehicles. *FEBS Letters* **245**, 204–208.

HEATH, S., CHANCE, M.L. AND NEW, R.R.C. (1984). Quantitative and ultrastructural studies on the uptake of drug-loaded liposomes by mononuclear phagocytes infected with *Leishmania donovani*. *Molecular Biochemistry and Parasitology* **12**, 49–60.

HEATH, T.D., FRALEY, R.T. AND PAPAHADJOPOULOS, D. (1980). Antibody targeting of liposomes: cell-specificity obtained by conjugation of $F(ab')_2$ to vesicle surface. *Science* **210**, 539–543.

HOEKSTRA, D., TOMASINI, R. AND SCHERPHOF, G. (1980). Interaction of phospholipid vesicles with rat hepatocytes *in vitro*: influence of vesicle-incorporated glycolipids. *Biochimica et Biophysica Acta* **603**, 336–346.

HUANG, A., HUANG, L. AND KENNEL, S.J. (1980). Monoclonal antibody covalently coupled with fatty acid: a reagent for *in vitro* liposome targeting. *Journal of Biological Chemistry* **255**, 8015–8022.

HART, D.T. (1987). Leishmania host–parasite interactions: the development of chemotherapeutic targets and specific drug delivery systems. I. Lipoprotein mediated antileishmanial activity. In *Host–parasite Cellular and Molecular Interactions in Protozoal Infections* (K.-P. Chang and D. Suary, Eds), pp. 203–211. Springer-Verlag, Berlin.

HUNTER, C.A., DOLAN, T.F., COOMBS, G.H. AND BAILLIE, A.J. (1988). Vesicular systems (niosomes and liposomes) for delivery of sodium stibogluconate in experimental visceral leishmaniasis. *Journal of Pharmacy and Pharmacology* **40**, 161–165.

IVEY, H., ROTH, S. AND KATTWINKEL, J. (1977). Nebulization of sonicated phospholipid for treatment of respiratory distress syndrome (RDS) of infancy. *Pediatric Research* **11**, 573.

JULIANO, R.L. AND STAMP, D. (1975). The effect of particle size and charge on the clearance rates of liposomes and liposome encapsulated drugs. *Biochemical and Biophysical Research Communications* **63**, 651–658.

JULIANO, R.L. AND STAMP, D. (1976). Lectin mediated attachment of glycoprotein-bearing liposomes to cells. *Nature (London)* **261**, 235–238.

KAYE, S.B. 91981). Liposomes – problems and promise as selective drug carriers. *Cancer Treatment Review* **8**, 27–37.

KIMBELBERG, H.K. (1980). Liposomes as carriers for methotrexate. In *Liposomes in Biological Systems* (G. Gregoriadis and A.C. Allison, Eds), pp. 219–248. John Wiley & Sons, New York.

KIMBELBERG, H.K. AND MAYHEW, E. (1978). In *Critical Reviews in Toxicology* (L. Golberg, Ed.), pp. 25–79. CRC-Press, Boca Raton, Florida.

KINNAMON, K.E., STECK, E.A., LOIZEANX, P.S., HANSON, W.L., CHAPMAN, W.L., JR AND WAITS, V.B. (1978). The antileishmanial activity of lepidines. *American Journal of Tropical Medicine and Hygiene* **27**, 751–757.

KIRBY, C. AND GREGORIADIS, G. (1984). Dehydration rehydration vesicles (DRV): a new method for high yield drug entrapment in liposomes. *Biotechnology* **2**, 979–984.

KIRBY, C., CLARKE, J. AND GREGORIADIS, G. (1980). Effect of the cholesterol content of small unilamellar liposomes on their stability *in vivo* and *in vitro*. *Biochemical Journal* **186**, 591–598.

KLAUSNER, A. (1988). Will 1988 be the year of the liposome? *Biotechnology* **6**, 20.

KLEINERMAN, E.S., ERICKSON, K.L., SCHROIT, A.J., FODGLER, W.E. AND FIDLER, I.J. (1983). Activation of tumoricidal properties in human blood monocytes by liposomes containing lipophilic muramyl tripeptide. *Cancer Research* **43**, 2010–2014.

KOSLOSKI, M.J., ROSEN, F., MILHOLLAND, R.J. AND PAPAHADJOPOULOS, D. (1978). Effect of lipid vesicles (liposome) encapsulation of methotrexate on chemotherapeutic efficacy in solid rodent tumor. *Cancer Research* **38**, 2848–2857.

KRANTZ, M.L., HOLTZMAN, N.A., STOWELL, C.P. AND LEE, Y.C. (1976). Attachment of thioglycosides to proteins: enhancement of liver membrane binding. *Biochemistry* **15**, 3962–3968.

KUBO, A., NAKAMURA, K., SAMMIYA, T., KATAYAMA, N., HAHIMOTO, T., HASINOTO, S., KOBAYASHI, H. AND TERAMOTO, T. (1993). Indium-III labelled liposomes – dosimetry and tumor detection in patients with cancer. *European Journal of Nuclear Medicine* **20**, 107–112.

LESSERMAN, M.P. AND BARBET, J. (1981). Cell-specific drug transfer from liposomes bearing monoclonal antibodies. *Nature (London)* **293**, 226–228.

LOPEZ-BERESTEIN, G. AND ROSENBLUM, M.G. (1992). Treatment and pharmacokinetics of liposomal amphotericin B patients with systemic fungal infections. In *Combination Therapies: Biological Response Modifiers in the Treatment of Cancer and Infectious*

Diseases (A.L. Goldstein and E. Garaci, Eds), pp. 105–112. Plenum, New York.

LOPEZ-BERESTEIN, G., MEHTA, R., HOPFER, R.L., MILLS, K.L., MEHTA, K., FAINSTEIN, M.L., HERSH, E.M. AND JULIANO, R. (1983). Treatment and prophylaxis of disseminated infection due to *Candida albicans* in mice with liposome encapsulated amphotericin B. *Journal of Infectious Diseases* **147**, 939–945.

MCCALDEN, T.A., ABRA, R.M. AND MIHALKO, P.J. (1989). Bronchodilator efficacy of liposome formulations of metaproteranol sulfate in the anesthetized guinea pig. *Journal of Liposome Research* **1**, 211–222.

MCCULLOUGH, H.N. AND JULIANO, R.L. (1979). Organ-selective action of an antitumour drug: pharmacologic studies of liposome encapsulated cytosine arabinoside administered via the respiratory system of the rat. *Journal of the National Cancer Institute* **63**, 727–731.

MADAN, S., VASANDANI, V.M. AND GHOSH, P.C. (1993). Potentiation of ricin cytotoxicity by liposomal monensin under *in vitro* and *in vivo* conditions. *Indian Journal of Biochemistry and Biophysics* **30**, 405–410.

MAHATO, S.B. AND PAL, B.C. (1987). Triterpenoid glycosides of *Corchorus acutangulns* Lam. *Journal of the Chemical Society, Perkin Transactions* **1**, 629–634.

MAHATO, S.B., SAHU, N.P. AND GANGULY, A.N. (1980). Carbon-13 NMR spectra of dioscin and gracillin isolated from *Costus speciosus. Indian Journal of Chemistry* **19B**, 817–818.

MAHATO, S.B., SAHU, N.P. AND PAL, B.C. (1978). New steroidal saponins from *Dioscorea floribunda*: structure of floribundasaponins, C, D, E and F. *Indian Journal of Chemistry* **16B**, 350–354.

MAHATO, S.B., SAHU, N.P., LUGER, P. AND MULLER, E. (1987). Stereochemistry of a triterpenoid trisaccharide from *Centella asiatica.* X-ray determination of the strucutre of asiaticoside. *Journal of Chemical Society, Perkin Transactions* **2**, 1509–1515.

MAHATO, S.B., NANDI, G., MUKHERJEE, S. AND BASU, M.K. (1993). Chemotherapy of leishmaniasis: urea stibamine and some pentamidine analogues. In *Current Trends in Leishmania Research* (A.N. Bhaduri, M.K. Basu, A.K. Sen and S. Kumar, Eds), pp. 246–253. Publication and Information Directorate, Council of Scientific and Industrial Research, New Delhi.

MATHIAS, C., DURAI, D.T. AND BACHHAWAT, B.K. (1977) Role of uridine entrapped in liposomes in galactosamine-induced hepatitis. *Indian Journal of Biochemistry and Biophysics* **14**, 142–146.

MAYHEW, E., PAPAHADJOPOULUS, D., RUSTUM, Y.M. AND DAVE, C. (1976). Inhibition of tumor cell growth *in vitro* and *in vivo* by 1-β-D-arabino furanosyl cytosine entrapped within phospholipid vesicles. *Cancer Research* **36**, 4406–4411.

MEDDA, S. AND BASU, M.K. Vesicular systems (niosomes and liposomes): critical evaluation of hamycin in two different carriers against experimental leishmaniasis. *Journal of Pharmacy and Pharmacology* (submitted).

MEDDA, S., DAS, N., BACHHAWAT, B.K., MAHATO, S.B. AND BASU, M.K. (1990). Targeting of plant glycoside-bearing liposomes to specific cellular and subcellular sites. *Biotechnology and Applied Biochemistry* **12**, 537–543.

MEDDA, S., MUKHERJEE, S., DAS, N., NASKAR, K., MAHATO, S.B. AND BASU, M.K. (1993). Sugar-coated liposomes: a novel delivery system for increased drug efficacy and reduced drug toxicity. *Biotechnology and Applied Biochemistry* **17**, 37–47.

MEDDA, S., MAHATO, S.B., MAHADEVAN, P.R. AND BASU, M.K. Glycoside bearing liposomal delivery systems against macrophage associated disorders involving *M. leprae* and *M. tuberculosis. Indian Journal of Biochemistry and Biophysics* (submitted).

MEISNER, D., PRINGLE, J. AND MEZEI, M. (1989). Liposomal pulmonary delivery I. *In vivo* disposition of atropine base in solution and liposomal form following endotracheal instillation to the rabbit lung. *Journal of Microencapsulation* **6**, 379–387.

MELISSEN, P.M.B., VANVIANEN, W., RIJSBERGEN, Y. AND BAKKER-WONDENBERG, I.A.J.M. (1992). Free versus liposome-encapsulated muramyl tripeptide phosphatidyl ethanolamide in treatment of experimental *Klebsiella pneumoniae* infection. *Infection and Immunity* **60**, 95–100.

MIHALKO, P.J., SCHRIER, H. AND ABRA, R.M. (1988). Liposomes: a pulmonary perspective. In *Liposomes as Drug Carriers* (G. Gregoriadis, Ed.), pp. 679–694. Wiley, New York.

MOONIS, M., AHMAD, I. AND BACHHAWAT, B.K. (1992). LIposomal hamycin in the control of experimental aspergillosis in mice. Relative toxicity, therapeutic efficacy and tissue distribution of free and liposomal hamycin. *Indian Journal of Biochemistry and Biophysics* **29**, 339–345.

MOONIS, M., AHMAD, I. AND BACHHAWAT, B.K. (1993a). Mannosylated liposomes as carriers for hamycin in the treatment of experimental aspergillosis in BALB/c mice. *Journal of Drug Targeting* **1**, 147–155.

MOONIS, M., AHMAD, I. AND BACHHAWAT, B.K. (1993b). Liposomal hamycin in the control of experimental aspergillosis in mice: effect of phosphatidic acid with and without cholesterol. *Journal of Antimicrobial Chemotherapy* (in press).

MYERS, M.A., THOMAS, D.A., STRAUB, L., SONCY, D.W., NEVIN, R.W., KALTENBACH, M., HOOD, C.I., SCHREIER, H. AND GONZALEZ-ROTHI, R.J. (1993). Pulmonary effects of chronic exposure to liposome aerosols in mice. *Experimental Lung Research* **19**, 1–19.

NANDI, G., MUKHERJEE, S., BASU, M.K. AND MAHATO, S.B. (1993). Synthesis, spectroscopic properties and antileishmanial activity of pentamidine analogues. *Journal of the Indian Chemical Society* **70**, 527–531.

NAOI, M. AND YAGI, K. (1980). Incorporation of enzyme through blood–brain barrier into the brain by means of liposomes. *Biochemistry International* **1**, 591–596.

NEW, R.R.C., CHANCE, M.L. AND HEATH, S. (1981a). The treatment of experimental cutaneous leishmaniasis with liposome-entrapped pentostam. *Parasitology* **83**, 519–527.

NEW, R.R.C., CHANCE, M.L. AND HEATH, S. (1981b). Antileishmanial activity of amphotericin and other antifungal agents entrapped in liposomes. *Journal of Antimicrobial Chemotherapy* **8**, 371–381.

NEW, R.R.C., CHANCE, M.L., THOMAS, S.C. AND PETERS, W. (1978). Antileishmanial activity of antimonials entrapped in liposomes. *Nature (London)* **272**, 55–56.

NIEDEL, J., WILKINSON, S. AND CUATRECASAS, P. (1979). Receptor mediated uptake and degradation of ^{125}I-chemotactic peptide by human neutrophils. *Journal of Biological Chemistry* **254**, 10700–10706.

OWAIS, M., AHMAD, I., KRISHNAKUMAR, B., JAIN, R.K., BACHHAWAT, B.K. AND GUPTA, C.M. (1993). Tuftsin-bearing liposomes as drug vehicles in the treatment of experimental aspergillosis. *Federation of the European Biochemical Society* **326**, 56–58.

PATEL, H.M. AND RYMAN, B.E. (1976). Oral administration of insulin by encapsulation within the liposomes. *FEBS Letters* **62**, 60–63.

POSTE, G. (1982). Experimental system for analysis of the malignant phenotype. *Cancer Metabolic Review* **1**, 141–145.

POSTE, G. AND FIDLER, I.J. (1980). The pathogenesis of cancer metastasis. *Nature (London)* **283**, 139–146.

POSTE, G. AND PAPAHADJOPOULOS, D. (1978). The influence of vesicle membrane properties on the interaction of lipid vesicles with cultured cells. *Annals of the New York Academy of Sciences* **308**, 164–184.

POSTE, G., KRISH, R. AND KOESTLER, T. (1984). The challenge of liposome targeting *in vivo*. In *Liposome Technology* (G. Gregoriadis, Ed.), Vol. III, pp. 2–28. CRC Press, Florida.

RAHMAN, Y.E., CERNEY, E.A., TOLLAKSON, S.L., WRIGHT, B.J., NANCE, S.L. AND THOMSON, S.F. (1974a). Liposome-encapsulated actinomycin D: potential in cancer chemotherapy. *Proceedings of the Society for Experimental Biology and Medicine* **146**, 1173–1176.

RAHMAN, Y.E., ROSENTHAL, M.W., CERNY, E.A. AND MORELLI, E.S. (1974b). Preparation and prolonged tissue retnetion of liposome-encapsulated chelating agents. *Journal of Laboratory and Chemical Medicine* **83**, 640–647.

REDWOOD, W.R. AND POLEFKA, T.G. (1976). Interaction of wheatgerm agglutinin with phosphatidyl choline liposomes containing incorporated monosialoganglioside. *Biochimica et Biophysica Acta* **455**, 631–643.

RUSTUM, Y.M., DAVE, C., MAYHEW, E. AND PAPAHADJOPOULOS, D. (1979). Role of liposome type and route of administration in the antitumor activity of liposome entrapped 1-β-D-arabinofuranosyl-cytosine against mouse L 1210 leukemia. *Cancer Research* **39**, 1390–1395.

SAKAMOTO, A. AND IDO, T. (1993). Liposome targeting to rat brain: effect of osmotic opening of the blood brain barrier. *Brain Research* **629**, 171–175.

SCHIFFMANN, E., CORCORAN, B.A. AND WAHL, S.M. (1975). *N*-formyl-methionyl peptides as chemoattractants for leukocytes. *Proceedings of the National Academy of Sciences, USA* **72**, 1059–1062.

SCHREIER, H., GONZALEZ-ROTHI, R.J. AND STECENKO, C.A. (1993). Pulmonary delivery of liposomes. *Journal of Controlled Release* **24**, 209–223.

SENIOR, J., CRAWLEY, J.C.W. AND GREGORIADIS, G. (1985). Tissue distribution of liposomes exhibiting long half-lives in the circulation after intravenous injection. *Biochimica et Biophysica Acta* **839**, 1–8.

SHARMA, A., MAYHEW, E. AND STRAUBINGER, R.M. (1993). Antitumor effect of taxol containSing liposomes in a taxol-resistant murine tumor liver model. *Cancer Research* **53**, 5877–5882.

SHEK, P.N., LOPEZ, N.G. AND HEATH, T.D. (1986). Immune response mediated by liposome-associated protein antigens. IV. Modulation of antibody formation by vesicle encapsulated methotrexate. *Immunology* **57**, 153–157.

SHEK, P.N., JURIMA-ROMET, M., BARBER, R.F. AND DEMEESTER, J. (1988). Liposomes: potential for inhalation prophylaxis and therapy. *Journal of Aerosol Medicine* **1**, 257–258.

SINGH, S.P., CHABBRA, R. AND SRIVASTAVA, V.M.L. (1992). Respiratory burst in peritoneal exudate cells in response to a modified tuftsin. *Experientia* **48**, 994–996.

SINGHAL, A. AND GUPTA, C.M. (1986). Antibody mediated targeting of red cells *in vivo*. *FEBS Letters* **201**, 321–326.

SINGHAL, A., BALI, A. AND GUPTA, C.M. (1986). Antibody mediated targeting of liposomes to erythrocytes in whole blood. *Biochimica et Biophysica Acta* **880**, 72–77.

SINGHAL, A., BALI, A., JAIN, R.K. AND GUPTA, C.M. (1984). Specific interactions of liposomes with PMN leukocytes upon incorporating tuftsin in their bilayers. *FEBS Letters* **178**, 109–113.

SLAMA, J.S. AND RANDO, R.R. (1980). Lectin mediated aggregation of liposomes containing glycolipids with variable hydrophilic spacer arms. *Biochemistry* **19**, 4595–4600.

SNYDERMAN, R. AND FUDMAN, E.J. (1980). Demonstration of a chemotactic factor receptor on macrophages. *Journal of Immunology* **124**, 791–796.

SNYDERMAN, R. AND GOETZEL, E.J. (1981). Molecular and cellular mechanism of leukocyte chemotaxis. *Science* **213**, 830–837.

SONE, S. AND FIDLER, I.J. (1980). Synergestic activation by lymphokines and muramyl dipeptide of tumoricidal properties in rat alveolar macrophages. *Journal of Immunology* **125**, 2454–2460.

SPANGER, H.H. AND SCHERPHOF, G.L. (1983). Targeting of lactosyl-ceramide-containing liposomes to hepatocytes *in vivo*. *Biochimica et Biophysica Acta* **737**, 40–47.

STAHL, P.D., RODMAN, J.S., MILLER, M.J. AND SCHLERINGER, P.H. (1978). Evidence for receptor-mediated binding of glycoproteins, glycoconjugates, and lysosomal glycosidases by alveolar macrophages. *Proceedings of the National Academy of Sciences, USA* **75**, 1399–1403.

SUR, P., HAZRA, B. AND ROY, D.K. (1984). Enhancement of the activity of bleomycin on Ehrlich ascites carcinoma cells by liposome encapsulation. *Indian Journal of Experimental Biology* **22**, 115–119.

SUROLIA, A. AND BACHHAWAT, B.K. (1977). Monosialoganglioside liposome-entrapped enzyme uptake by hepatic cells. *Biochimica et Biophysica Acta* **497**, 760–765.

SUROLIA, A., AHMED, A. AND BACHHAWAT, B.K. (1975). Affinity chromatography of galactose containing biopolymers using covalently coupled *Ricinus communis* lectin to sepharose 4B. *Biochimica et Biophysica Acta* **404**, 83–92.

SUROLIA, A., BACHHAWAT, B.K. AND PODDAR, S.K. (1975). Interaction between lectin from *Ricinus cummunis* and liposome containing glycoside. *Nature (London)* **257**, 802–804.

SZOKA, F.C. AND MAYHEW, E. (1983). Alteration of liposome deposition *in vivo* by bilayer situated carbohydrates. *Biochemical and Biophysical Research Communications* **110**, 140–146.

SZOKA, F. AND PAPAHADJOPOULOS, D. (1981). Liposomes: preparation and characterization. In *Liposomes: from physical structure to therapeutic applications* (C.G. Knight, Ed.), Chapter 3. Elsevier/North-Holland, Amsterdam.

SZOKA, F.C., JR, MILHOLLAND, P. AND BARZA, M. (1987). Effect of lipid composition and liposome size on toxicity and *in vitro* fungicidal activity of liposome-intercalated amphotericin B. *Antimicrobial Agents and Chemotherapy* **34**, 421–429.

TAKADA, M., YUZURIHA, T., KATAYAMAA, K., IWAMOTO, K. AND SUNAMOTO, J. (1984). Increased lung uptake of liposomes coated with polysaccharides. *Biochemica et Biophysica Acta* **802**, 237–244.

TAYLOR, K.M.G. AND FARR, S.J. (1993). Liposomes for drug delivery to the respiratory tract. *Drug Development and Industrial Pharmacy* **19** (1 & 2), 123–142.

TAYLOR, K.M.G. AND NEWTON, J.M. (1992). Liposomes for controlled delivery of drugs to the lung. *Thorax* **47**, 257–259.

TAYLOR, K.M.G., TAYLOR, G., KELLAWAY, I.W. AND STEVENS, J. (1989). The influence of liposomal encapsulation on sodium cromoglycate pharmacokinetics in man. *Pharmaceutical Research* **6**, 633–636.

TOLLEMAR, J., RINGDEN, O., ANDERSSON, S., SUNDBERG, B., LJUNGMAN, P., SPARRELID, B. AND TYDEN, G. (1993). Prophylactic use of liposomal amphotericin B (ambisome) against fungal infections – a randomized trial in bone marrow transplant recipients. *Transplant Proceedings* **25**, 1495–1500.

UMEZAWA, F. AND ETO, Y. (1988). Liposomes targeting to mouse brain: mannose as a recognition marker. *Biochemical and Biophysical Research Communications* **153**, 1038–1044.

VASANDANI, Y.M., MADAN, S. AND GHOSH, P.C. (1992). *In vivo* potentiation of ricin toxicity by monensin delivered through liposomes. *Biochimica et Biophysica Acta* **1116**, 315–323.

WEINSTEIN, J.N., BLUMENTHAL, R., SHARROW, S.O. AND HENKART, P.A. (1978). Antibody mediated targeting of liposomes: binding to lymphocytes does not ensure incorporation of vesicle contents into the cells. *Biochimica et Biophysica Acta* **509**, 272–288.

WEISSMAN, G., BLOOMGARDEN, D., KAPLAN, R., COHEN, C., HOFFSTEIN, S., COLLINS, T., GOTLIEB, A. AND NAGLE, D. (1975). A general method for the introduction of enzymes by means of immunoglobulin-coated liposomes, into lysosomes of deficient cells. *Proceedings of the National Academy of Sciences, USA* **72**, 88–92.

12
Immobilization of Polysaccharide-degrading Enzymes

KENNETH B. STOREY[1] AND DORIS Y. SCHAFHAUSER-SMITH[2]

[1] *Department of Biology, Carleton University, Ottawa, Ontario, Canada and* [2] *Department of Environmental Science, Virginia Institute of Marine Science, The College of William and Mary, Gloucester Point, Virginia, USA*

Introduction

The development of biomass conversion processes for the commercial use of complex plant polysaccharides is a major focus of biotechnology research. Lignocellulosics, which comprise the cell wall materials of all woody plants, grasses and cereals, constitute the largest single biomass resource on Earth (Tan, Wong and Saddler, 1985; Tan *et al.*, 1985). Similarly, starch, the principle food reserve polysaccharide in the plant kingdom, is the next most abundant renewable biopolymer, and forms the primary source of carbohydrates in the human diet (Galliard, 1985). Polysaccharides derived from these sources continuously accumulate in large quantities as by-products of the forestry, pulp and paper, agricultural, food processing, sweetener, brewing and textile industries (Biely, 1985; Wood, 1985). Technologies that could make use of these biopolymers have several advantages including a reduction of waste biomass, an enhancement of the economic competitiveness of bioconversion technologies through the use of abundant and inexpensive starting materials, and the potential development of new products from renewable resources (Saddler and Brownell, 1982). The application of enzymatic degradations is a highly efficient means of conducting biomass conversion processes to create specific, high-quality and valuable products from waste biomass.

At present, the most commercially important enzymes are those that hydrolyse starch (amyloglucosidase, α-amylase, β-amylase, pullulanase), whereas those that degrade plant fibres, including cellulose (cellulases), xylan (xylanases) and pectin

---

Abbreviations: BSTR, batch stirred tank reactor; CMC, carboxymethylcellulose; CNBr, cyanogen bromide; ConA, concanavalin A; CSTR, continuous stirred tank reactor; DEAA, diethylaminoacetyl; DEAHP, *N, N*-diethylamino-2-hydroxypropyl cellulose; HFCS, high-fructose corn syrup; HFS, high-fructose syrup; HPLC, high-pressure liquid chromatography; PBR, packed bed reactor; PE, pectin esterase; PEG, polyethylene glycol; PEI, polyethyleneimine; PG, polygalacturonase; PMGL, endopolymethylgalacturonate lyase; PNPG, *p*-nitrophenyl glucosides; PVA, polyvinylalcohol; STR, stirred tank reactor.

Biotechnology and Genetic Engineering Reviews — Vol. 12, December 1994
0264–8725/94/12/409–465 $20.00 + $0.00 © Intercept Ltd, P.O. Box 716, Andover, Hampshire SP10 1YG, UK

(pectinases), or other plant polysaccharides, such as dextran (dextranase) and inulin (inulinase), continue to gain in industrial importance for the production of specific products. Glucose is the most commercially successful product of the enzymatic hydrolysis of both cellulose and starch and has a vast number of uses, including as a feedstock additive for single-cell protein processes, as a fermentable compound in the brewing industry, and in the production of chemical reagents and fuels, such as ethanol (Woodward and Zachry, 1982; Cheetham, 1985; Woodward, 1989). Furthermore, when cellulose or starch degradations are combined with the use of glucose isomerase the ultimate product is fructose syrup, which has wide uses in the baking, soft drink and confectionery industries (Fogarty, 1983). Glucose/fructose syrups can also be produced in other ways. For example, the hydrolysis of inulin-rich plants, such as the Jerusalem artichoke, by inulinases offers a direct production of fructose syrups, whereas the hydrolysis of sucrose or raffinose by invertase is another commercial method for the production of glucose/fructose syrups (Goodwin and Mercer, 1983).

Pectin degradation by pectic enzymes constitutes a common industrial practice for improving the quality and yield of fruit and vegetable juices during processing (Rexová-Benková and Mracková-Dobrotová, 1981; Borrego *et al.*, 1989). Xylanase enzymes have been applied to the pulp and paper industry (Dekker, 1983) and to fruit and vegetable juice production (Biely, 1985), and the hydrolysed products of xylan can be used for the production of fermented chemical reagents (Roy, Roy and Duke, 1984; Yu and Saddler, 1985). Dextranases have direct application in the production of molasses and sucrose syrups from sugar cane and sugar beet (Margaritis and Pace, 1985).

The enzymatic degradation of biopolymers offers several advantages to a commercial process:

1. enzymatic methods are highly efficient and environmentally safe procedures for biomass conversion;
2. enzymes exhibit excellent specificity with respect to substrates used and products created; and
3. enzymatic conversions can be carried out under conditions of moderate pH and temperature (Chakrabarti and Storey, 1988).

By contrast, acid-catalysed breakdown methods frequently result in the formation of undesirable by-products (Coker and Venkatasubramanian, 1985).

However, enzymatic degradation still has some disadvantages, chiefly problems of cost and reusability of catalysts. The production of purified enzymes for bioconversion processes is a major expenditure. Furthermore, reuse of enzyme catalysts is impractical in many systems because they are difficult and uneconomical to recover from a reaction mixture (Hartmeier, 1985). The economics of commercial biomass conversion processes demand that optimum conditions be determined for enzyme production and use (Schafhauser and Storey, 1992a). Two lines of research are aimed at lowering the costs of enzymatic procedures through:

1. the development of new strains of microbial mutants with higher specific activities of enzymes than those currently in use; and
2. the development of immobilization methods for the attachment of enzymes to solid supports to allow for their repeated use (Takeuchi and Makino, 1987; Woodward, 1987).

This latter topic is the subject of the present chapter.

IMMOBILIZED ENZYMES

The application of immobilized enzyme technology has great potential for improving the industrial uses of enzymes. The overall cost of an enzyme-catalysed industrial process can be greatly reduced if the enzymes can be easily recovered and reused, usually by simple filtration or centrifugation (Ohba *et al.*, 1978; Cheetham, 1985). Immobilization can also reduce the conversion time because very high enzyme concentrations can be used and recovered (Pitcher, 1980). Kinetic properties and pH or temperature optima of immobilized enzymes can also frequently be manipulated to favour specific applications, and enzyme stability is often enhanced upon immobilization (Klibanov, 1979). Enzymes within the same family, as well as those than can operate under similar conditions of pH and temperature, can also be co-immobilized to create multienzyme systems or pathways. The advantages of such systems are that faster reaction rates and more efficient substrate hydrolysis can often be achieved by a co-immobilized enzyme system compared with the same enzymes free in solution (Cheetham, 1985).

An estimation of the commerical potential of a particular immobilized enzyme system requires a thorough knowledge of certain parameters affecting its behaviour under operational conditions (Ohba *et al.*, 1978). The activities and properties of an enzyme can change quite dramatically upon immobilization, and can be attributed to conformational effects, microenvironmental effects and diffusional or mass-transfer effects (Kobayashi and Laidler, 1972). During immobilization enzymes generally experience some conformational changes which can result in changes in substrate affinity or allosteric properties, subunit dissociation or even enzyme inactivation (Klibanov, 1979; Cheetham, 1985). However, the immobilization step, which frequently involves multipoint binding of the catalyst molecule to a support, typically results in a more stable enzyme. Thus, immobilized enzymes are generally more resistant to unfolding and deactivation in response to shifts in pH or temperature, or exposure to chemicals and impurities in the biopolymer substrate mixture, and are less accessible to microbial attack relative to the free enzymes (Klibanov, 1979).

Once immobilized, enzymes come under the influence of the microenvironment of the support which may be drastically different from that which the enzyme experiences free in solution (Klibanov, 1979). Substrates and supports of opposite charge can result in attractive forces which may enhance mass transfer and increase the local concentrations of substrates (Kobayashi and Laidler, 1972). Electrostatic effects and shifts in pH optima also occur when enzymes are immobilized on to charged surfaces (Klibanov, 1979). A negatively charged matrix usually results in an alkaline shift in the pH optimum, due to the creation of a microenvironment of H^+ ions surrounding the matrix, which lowers the pH of the microenvironment surrounding the enzyme compared to that of the bulk solution (where the pH is measured). The opposite effect is created by the use of a positively charged matrix (Zaborsky, 1973). The same is true for non-polyelectrolytic supports if the partition coefficient for H^+ ions in the microenvironment and the bulk solution is not unity (Klibanov, 1979). The microenvironment of the immobilized enzyme can also possess buffering properties that can stabilize the enzyme against pH changes (Klibanov, 1979).

Since enzymes are frequently immobilized within porous bodies or retained by semipermeable membranes, mass-transfer limitation problems, particularly with substrates of large molecular weight, usually occur. Mass-transfer effects can limit the efficiency of the catalytic action of the enzyme as well as change its apparent kinetic behaviour (D.D. Lee *et al.*, 1976). Strongly diffusion-limited systems can also give the impression of stability (Caldwell *et al.*, 1976a,b). Diffusion limitation results in the creation of an unstirred layer of solvent called the Nernst layer, which surrounds water-insoluble particles, creating a concentration gradient (Zaborsky, 1973). As a result, the immobilized enzymes will usually become saturated with substrates at a higher concentration than that normally required for saturation of the free enzyme, giving rise to higher K_m values (Zaborsky, 1973). Diffusion limitation affects initial rate assays and, thus, distorts the standard plots used to estimate kinetic constants (Pitcher, 1980). Diffusion limitation effects are more apparent at high enzyme activities, and their presence in a system can be most easily detected with certain types of assays, such as via analysis of activity versus assay temperature (Arrhenius plots) (Pitcher, 1980).

External mass-transfer resistance depends primarily on particle size and fluid velocity around the particle, which is influenced by mixing conditions in stirred tanks and fluid velocity in mixed beds. Increases in stirrer speed or flow rate have been shown to minimize the problem (Bucholz and Klein, 1987). Internal mass transfer (or pore diffusion effects) is another phenomenon affecting the efficiency of an enzyme when immobilized on to porous supports. However, the apparent activity of such immobilized enzymes can be increased by either using smaller support particles, thus decreasing the diffusion path length, or by reducing the activity of the enzyme through lower enzyme loadings or lower reaction temperatures (Pitcher, 1980).

ENZYME IMMOBILIZATION METHODS AND BIOREACTOR TYPES

The economics of immobilized enzyme technology are dependent on several factors that should be considered when choosing an immobilization technique, including the cost of the support and any necessary chemicals, and whether the method is simple, reproducible, mild, cheap and safe. The resulting immobilized enzyme is worthwhile provided that the preparation remains firmly bound and stable over a wide range of conditions and with repeated uses (Kennedy, 1985; Sórensen and Emborg, 1989). Immobilization represents an added expense and operation for industries. Therefore, it must be shown that the extra costs involved can be justified by the potential savings of such a system after it is successfully implemented.

Several methods of enzyme immobilization have been considered:

1. adsorption on to solid supports;
2. covalent attachment to water-soluble or water-insoluble supports;
3. covalent cross-linking to bifunctional reagents;
4. entrapment in polymeric gels; and
5. encapsulation in microcapsules or hollow fibres (Klibanov, 1979).

Each enzyme immobilization method has its own advantages and disadvantages (*Table 1*).

Table 1. Summary of immobilization methods for polysaccharide-degrading enzymes

Method	Advantages	Disadvantages	Types of supports
Adsorption	Fast, easy, inexpensive No chemical reactants Easily regenerated and reusable Supports are often natural products	Enzymes easily desorbed due to changes in pH, ionic strength and substrate concentration	Collagen, chitin, bone, cotton, cellulose, dextran, ion-exchange resins, hydrophilic membranes, polyethylene glycol
Covalent attachment	Firm and permanent binding	Frequently causes enzyme inactivation and/or alters catalytic activity Chemical reagents often unsuitable for food industry	Nylon, silica, alumina, celite, bentonite, clay, titania, magnetite, hydrophilic membranes, polyvinyl alcohols
Covalent cross-linking	Firm and permanent binding Enhanced binding and stability compared with above methods	Immobilization procedures are long Added expense of chemical reagents Can alter catalytic activity	Glutaraldehyde, concanavalin A, polyethyleneimine, formaldehyde, aminopropyltriethoxysilane, epichlorohydrin
Entrapment	Gentle, simple No chemical modification of enzyme	Harsh polymerizing agents Unsuitable for enzymes that hydrolyse high molecular weight polymers	Gelatin, agarose, polystyrene, polyurethane, acrylamide, calcium alginate
Encapsulation in microcapsules, hollow fibres	No chemical modification of enzyme Easily scaled-up	Viscous solutions can clog reactors Substrate concentration gradients at support surface	Gelatin, nylon, polysulphone, calcium alginate, methyacrylate, polyvinyl alcohol

Immobilization by physical adsorption and ionic binding are the easiest and fastest methods (Pitcher, 1980). Since no reactive species are involved there is little or no conformational change to the enzyme following immobilization (Kennedy, 1985). However, because the binding forces involved are not permanent, changes in ionic strength, pH or substrate concentration can cause enzyme desorption (Solomon, 1978; Kennedy, 1985). In spite of this, such supports are popular, for they are usually easily regenerated and reused (Stanley *et al.*, 1976; Antrim and Auterinen, 1985).

Covalently bound enzymes give enhanced stability due to the strength and permanence of the bonds created (Solomon, 1978; Storey, Duncan and Chakrabarti, 1990). However, the conditions for achieving immobilization are more complicated and harsh, making enzymes more prone to inactivation (Kennedy, 1985). Furthermore, covalent binding methods are not used extensively in some fields, such as food processing, because the chemical reagents used for binding are not allowed in the final product (Pitcher, 1980).

Entrapment methods are usually gentle and simple, resulting in high concentrations of contained enzyme without chemical modifications or bindings that may lead to activity loss (Mazid, 1993). However, due to diffusion limitations, such methods are often unsuitable for the immobilization of enzymes that hydrolyse large molecular weight substrates (Solomon, 1978; Storey, Duncan and Chakrabarti, 1990).

A variety of bioreactor types are available for use in immobilized enzyme processes, depending on the requirements of a particular system. These include the stirred tank reactor (STR), continuous stirred tank reactor (CSTR), batch stirred tank reactor (BSTR), packed bed reactor (PBR), and membrane reactor (Gemeiner, Stefuca and Báles, 1993). The high recirculation rates of the STR minimize both the destabilization of the enzyme caused by any localized changes in reaction parameters (e.g. pH) and film diffusion resistance (Roy *et al.*, 1989). BSTR operations are broadly applicable, flexible and relatively inexpensive, whereas CSTR processes, by comparison, demand greater initial capital output for reactor design and control automation, but are more catalytically efficient processes, producing a more reliable product in a shorter reaction time (Celebi, Tsai and Tsao, 1991; Mazid, 1993). The effective use of a PBR depends on the fluidity of the reactor liquid. High molecular weight, viscous solutions can clog the column, resulting in a pressure drop across the bed (Ohba and Ueda, 1980; Pieter, Bardeletti and Coulet, 1992; Gemiener, Stefuca and Báles, 1993). Alternatively, substrate concentration gradients can develop due to internal and external diffusion limitations, which can in turn influence the catalytic efficiency (Chen and Wu, 1987). Similarly, enzymes immobilized on to small particles can also lead to pressure build-up when used in a PBR (Y.Y. Lee *et al.*, 1976).

Membrane reactors, such as the hollow fibre bioreactor and the membrane recycle bioreactor, exploit the differences in size between the enzyme and substrate trapped on one side of the membrane and the smaller molecular weight products which are freely diffusible and can be removed continuously from the reaction mixture (Darnoko, Cheryan and Artz, 1988). Although membrane bioreactors generally show higher productivities than batch processes, the catalytic efficiency can be limited due to the formation of a 'gel layer' on a membrane surface and a resultant decrease in overall permeation (Darnoko, Cheryan and Artz, 1988).

TYPES OF IMMOBILIZATION SUPPORTS

In order to be considered for use as enzyme immobilization supports, compounds must possess an adequately sized porous surface for binding, be strong, inert, and chemically and physically stable (Caldwell *et al.*, 1976a,b). A variety of diffusion supports are currently known (*Table 1*), and new ones are continually being developed and applied.

Organic compounds, including natural polymers (e.g. collagen, chitin, cellulose, bone and gelatin) and synthetic polymers (e.g. nylon, ion-exchange resins and polyurethanes) have been used in immobilized enzyme reactors. Granular chicken bone (Biobone™) and chitin are ideal supports, easily obtained from meat and seafood processing wastes, and are abundant, inexpensive, non-toxic and food grade (Findlay, Parkin and Yada, 1986; Emneus, 1993). Cellulosic materials are also inexpensive and available in many different physical forms (e.g. fibres, microcrystals, beads and membranes) (Gemeiner, Stefuca and Báles, 1993), but exhibit high hydrodynamic resistance due to compaction and clogging when used in a column (Sharma and Yamazaki, 1984). Cotton flannel exhibits high protein adsorption and shows good flow-through properties (Yamazaki, Cheok and Fraser, 1984). Gelatin is a good support matrix for entrapping enzymes, having a marked swelling power and low resistance to mass transfer (De Alteriis *et al.*, 1985).

Polyurethane foams offer a stable and homogeneous distribution of enzymes throughout the matrix and are flexible and non-reactive once polymerized (Chakrabarti and Storey, 1988). Nylon is an inexpensive support that is also available in different physical (e.g. films, membranes and powder) and chemically modified forms (Iborra *et al.*, 1992). Hydrophilic membranes, traditionally used for nucleic acid blotting and hybridization techniques, have a high protein binding capacity and good flow-through properties and represent another potential enzyme support matrix (D.Y. Schafhauser-Smith and K.B. Storey, unpublished results).

Enzymes covalently bound to polymeric carriers (e.g. styrene, acrylamide) are the most commonly used supports in continuous column reactions (Woodward, 1987). However, such polymers are relatively soft and thus prone to column plugging when operating at high flow rates and high substrate concentrations (Martensson, 1974a; Nithiandam *et al.*, 1981).

Inorganic supports (e.g. silica, alumina, celite, bentonite) have been used as carriers for the immobilization of numerous enzymes and are often preferred over organic supports due to their availability in a wide range of controlled particle sizes and porosities, their reduced susceptibility to microbial attack, and their overall mechanical and chemical stability (Kobarzewski and Paszczynski, 1983; Bajpai and Margaritis, 1987; Shimizu and Ishihara, 1987). Inorganic magnetic supports have additional advantages in that they allow for easy separation of immobilized enzymes from STR hydrolyslates through the application of a magnetic field (Dekker, 1990).

Derivatization, synthetic and chemical transformation methods are often used in an effort to make carrier surfaces more suitable for enzyme binding. Such modifications typically result in the incorporation of reactive amino and carbonyl groups on to the support surface hydroxyl groups (Weetall, 1993). Such measures may also serve to coat and thus strengthen the walls of the support, resulting in greater physical, chemical and hydrodynamic properties (Lopez-Santin *et al.*, 1983). Derivatization can also introduce spacer groups, which increase the local surface area of the support and consequently reduce the steric hindrance in the immediate vicinity of the enzyme (Abdel-Naby, 1993). However, multiple fixation of an enzyme to the more reactive matrix surface can cause a decrease in catalytic activity owing to a decrease in the flexibility of the enzyme.

One of the most common means of introducing free amino groups on to a support is by a silanization reaction involving aminopropyltriethoxsilane (Wojick *et al.*, 1987; Weetall, 1993). Cross-linking with glutaraldehyde will generate reactive surface aldehyde groups (Wojcik *et al.*, 1987). Polyethyleneimine (PEI), a polybasic aliphatic amine, is routinely used either as a cross-linking agent or as a carrier in a number of immobilized biosystems (Bahulekar, Ayyangar and Ponrathnam, 1991). Concanavalin A (ConA) is a useful ligand for the immobilization of such industrially important glycoproteins as glucose oxidase, invertase, cellulase and amyloglucosidase (Saleemuddin and Husain, 1991). ConA-linked matrices usually exhibit high activity retention and enhanced stability against various types of inactivation (Saleemuddin and Husain, 1991).

The parameters used for the coupling of an enzyme to a particular matrix are usually optimized to allow for the best binding conditions for each particular enzyme (Martensson, 1974a). However, in order to be cost-effective, the ideal immobilization procedure should be a fast and simple one. Enzyme adsorption onto granular chicken bone is a firm and immediate reaction, with activity being retained after only a 20 min binding period (Schafhauser and Storey, 1992a,b). Enzyme immobilization onto hydrophilic polyurethane foam is also fast, requires no special catalysts, and results in irreversible binding (Chakrabarti and Storey, 1988, 1989; Storey, Duncan and Chakrabarti, 1990). However, enzyme immobilization onto alkylamine derivatives of metal-activated glass beads crosslinked with glutaraldehyde is a very long, multistep process (Cabral, Norais and Cardoso, 1981). The preparation of PEI cloth, together with the enzyme immobilization step, is also a long process (approximately 19 h) (Yamazaki, Cheok and Fraser, 1984). The use of methacrylate co-polymers as supports for enzyme immobilization required intensive carrier preparation (Nithiandam *et al.*, 1981; Svec and Kalal, 1978) as well as very long immobilization times (Svec and Kalal, 1978).

IMMOBILIZED BIOCATALYSTS

A number of processes utilizing immobilized biocatalysts are now well established on a commercial level. Immobilized enzymes have been used in affinity chromatography and synthetic chemistry as well as for the production of enzyme electrodes, thermistors and biosensors (Cheetham, 1985). Biosensors, which use O_2 or H_2O_2 electrodes in conjunction with an immobilized oxidase have found use in the food and drink industry. The 'Model 2700 Industrial Analyzer', marketed by Yellow Springs Instruments is one such analyser, which provides a method for determining levels of glucose and sucrose in the sugar, molasses and confectionery industries (Luong, Groom and Male, 1991). Units are also available for the detection of total starch, fructose and dextrose (Mazid, 1993). However, other than glucose isomerase, few of the enzymes with applications for plant biomass conversion processes are presently used in immobilized forms on a commercial scale (Jensen and Rugh, 1987).

Although immobilized enzymes, in competing with free enzymes, may have the advantage of increased stability and catalyst reuse, there are several reasons why immobilized enzymes are still not used in a number of traditional enzyme applications. Where the free enzyme is cheap and the process is already developed, a change to an immobilized system may not be financially beneficial (Hartmeier, 1985). Furthermore, immobilized enzyme preparations have been plagued by high capital costs arising from the use of expensive support materials and from enzyme activity losses caused by harsh immobilization conditions, enzyme instability or reduction in catalytic efficiency once bound (Klibanov, 1979; Kennedy, 1985).

Presented in this review is a compilation of the status of immobilized enzyme technology involving many of the industrially important polysaccharide-degrading enzymes. The review highlights the different immobilization methods that have been used for each enzyme and evaluates the usefulness of immobilized preparations compared with the corresponding free enzyme, and with respect to such production characteristics as pH and temperature optima, operational and storage stability, catalytic function, and the effect of diffusion limitation on productivity. Discussion will centre on practical ideas, with emphasis on the newest literature, and general considerations behind choosing a matrix and coupling a ligand, along with some of the more common problems often encountered with their implementation. It is beyond the scope of this chapter to review extensively all the published work on this subject. Several recent and excellent reviews are available in the areas of immobilized enzyme theory (Zaborsky, 1973; Cheetham, 1985; Kennedy, 1985; Buchholz and Klein, 1987), immobilized enzyme supports and coupling methods (Hartmeier, 1985; Bahulekar, Ayyangar and Ponrathnam, 1991; Saleemuddin and Husain, 1991; Weetall, 1993), and immobilized enzyme uses (Coker and Venkatasubramanian, 1985; Heath and Belfor, 1990; Chang and Furusaki, 1991; Luong, Groom and Male, 1991; Gemeiner, Stefuca and Báles, 1993; Mazid, 1993). Several excellent reviews also focus on specific polysaccharide-degrading enzymes: cellulases (Kyosov and Consultant, 1986; Knowles, Lehtovaara and Teeri, 1987; Woodward, 1987, 1989; Klyosov, 1990; Pitson, Seviour and McDougall, 1993), hemicellulases (Biely, 1985), amylases (Fogarty, 1983; Ward, 1985; Wiseman, 1985), glucose isomerases (Chen, 1980; Verhoff *et al.*, 1985; Jensen and Rugh, 1987), or their substrates: starch (Solomon, 1978; Banks and Muir, 1980; Kandler and Hopf, 1980; Galliard, 1985; Finn, 1987; Kennedy *et al.*, 1987), and lignocelluloses (Wood, 1985; Lewis and Paice, 1989; Coughlan and Hazlewood, 1993a).

The immobilization of starch-degrading enzymes

The main commercial sources of starch are corn in the US, wheat in Canada and Australia, and grain (wheat and barley) or root (potato) crops in Europe (Kennedy *et al.*, 1987). Cassava is one of the most efficient producers of carbohydrates, and is the fourth most important source of food energy in tropical countries after rice, cane sugar and corn (Pitcher, 1980). Sweet potatoes, taro and tannia are also good sources of starch (Klyosov and Consultant, 1986).

Starch is the same regardless of source, but starch granules are of a size and shape that is characteristic of individual species. The site of the main storage area of starch varies from plant to plant, for example, being the seed in corn, the tubers in potato and the pith in sago. Starch granules are typically 25% amylose and 75% amylopectin (Banks and Muir, 1980). The characteristic features of amylose and amylopectin are their linear chains of (1-4)-linked α-*D*-glucopyranosyl residues; in addition, amylopectin has 1,4,6-tri-O-substituted residues which act as branch points (Kennedy *et al.*, 1987). Significant portions of *D*-glucose-6-phosphate and lipid material have been found in starches. Up to 1 residue in 6 in amylopectin is phosphorylated whereas the lipid material is believed to be complexed with the amylose structure (Kennedy *et al.*, 1987). Amylose is a linear molecule in the molecular weight range of 1.6×10^5 to 7.1×10^5 and varies with the source and maturity of the starch (Kennedy *et al.*, 1987). In neutral solutions amylose normally exists as a random coil but in the presence of complexing agents (e.g. iodine, fats, polar organic solvents) it forms a regular helical structure fixed via hydrogen bonding. Amylopectin has a highly branched structure with a molecular weight of about 10^8. Branches are believed to be arranged in either a comb-like or a tree-like fashion (Kennedy *et al.*, 1987). Natural starches suffer from poor solubility, inconsistent viscosity and susceptibility to acid degradation. For industrial use this can cause problems; for example, the properties of a food product containing natural starch can change on storage. For this reason, most industries make extensive use of modified starches. Most processes rely on heat and/or acid treatment which results in starches with lower viscosity and an increased number of branched chains (Finn, 1987). As a result, modified starches are more soluble and have a lower tendency towards 'set back' or gelling on standing.

AMYLOGLUCOSIDASE

Amyloglucosidase (or glucoamylase, EC 3.2.1.3.) is an exo-splitting enzyme that catalyses, by a multichain mechanism, the stepwise hydrolysis of α-1,4 linkages and, to a lesser extent, α-1,6 linkages from the non-reducing ends of starch (and glycogen) (Fogarty, 1983). Preparations may be obtained easily from culture filtrates of several fungal species of *Aspergillus* and *Rhizopus*, and from certain yeasts and bacteria (Fogarty, 1983). Amyloglucosidase is one of the most industrially important enzymes, and is used in the large-scale production of high-glucose syrups (96–98% *D*-glucose) from starch (Fogarty, 1983). High-glucose syrups are used either for the production of crystalline *D*-glucose or as a starting material for the production of high-fructose syrups which are extensively used in the baking, soft drink and confectionery industries (Fogarty, 1983).

Compared with the large-scale degradation of starch to glucose by the soluble enzyme (typical conditions are pH 4.5, 55–65°C for 48–72 h), immobilization of amyloglucosidase reduces the conversion time because very high enzyme concentrations can be used and recovered (Pitcher, 1980). The immobilization of amyloglucosidase on to different inert supports by entrapment, adsorption or covalent binding systems has been widely studied in the hopes of attaining a more controlled reusable system that can be successfully used commercially. However, immobilized amyloglucosidase has not yet found commercial success because of the difficulty in obtaining similar or higher glucose yields than with the free enzyme.

To be useful, an immobilized enzyme must remain stable and active. Schafhauser and Storey (1992b) demonstrated that 14% of added amyloglucosidase remained stable and firmly bound to granular chicken bone following repeated washes, maintained maximum activity following 5 cycles of operation both at 23°C and 55°C, and showed a storage half-life of 144 days at 4°C. Unlike the free enzyme, immobilized amyloglucosidase was protected from the inhibitory effects of Triton X-100, ethanol, KCl and pH extremes (Schafhauser and Storey, 1992a). The high hydrophilicity of the collagen present in bone helps to prevent enzyme denaturation (Coulet, 1974). Storey, Duncan and Chakrabarti (1990) cross-linked 25% of added amyloglucosidase on to the polyurethane prepolymer Hypol^{R}FHP2002 and showed improved enzyme stability under several denaturing conditions (NaCl, urea, ethanol, 95°C), and an improved storage stability of the bound (70%) over the free (50%) enzyme after 2 months at 4°C. Amyloglucosidase immobilized on to acid-hydrolysed starch-g-polyacrylonitrile resulted in a stable and active enzyme, with glucose productivities reaching 2.7 g g^{-1} carrier h^{-1} and half-lives up to 1800 h in a continuous stirred tank reactor maintained at 50°C, pH 4, and using maltose as the substrate (Slinginger, Fanta and Abbott, 1988). Amyloglucosidase bound to chitin particles packed into a 180 ml two-phase bench scale reactor (liquid expanded bed) fed with hydrolysed manoic starch (15% w/v), operated continuously for 20 days, obtaining 96% conversions of starch to glucose with a hydraulic residence time of 4 h (Freire and Sánt'Anna, 1990). In another study, 19% of added amyloglucosidase bound ionically on to co-polymers of methacrylate and maintained good storage stability with no loss of activity during 47 days of storage (Nithiandam *et al.*, 1981). Glutaraldehyde-cross-linked silanized glass has been shown repeatedly to be a good enzyme support matrix. Cabral, Novais and Cardoso (1981) bound 12% of added activity on to glutaraldehyde-cross-linked metal-activated glass beads and demonstrated a long reaction half-life (1500 h) in a draw-and-fill basket reactor using 1% w/v soluble starch with no leakage. Other authors obtained a glucose productivity of 2.6 g g^{-1} carrier h^{-1} and a half-life of 2550 h (at 50°C and pH 4.5) with amyloglucosidase bound to glutaraldehyde-cross-linked glass beads (D.D. Lee *et al.*, 1976). Amyloglucosidase immobilized on to magnetic microparticles made of PEI-coated magnetite cross-linked with glutaraldehyde continuously hydrolysed a 30% w/v maltodextrin solution at 50°C and pH 4.5 in a fluidized bed reactor; continuous two week hydrolysis resulted in 70% conversions with only 5% of initial immobilized amyloglucosidase activity lost (Pieter, Bardeletti and Coulet, 1992). This system had the added advantage that the particles were easily settled because of their high density (5 g ml^{-1}) and the loss of ultrafines due to washout was prevented by surrounding the top of the bed with an electromagnet (Peiter, Bardeletti and Coulet, 1992). Amyloglucosidase immobilized on to silanized alkylamine glass beads in a packed bed reactor maintained 80% of maximal operational activity after 30 days of continuous use with 1% soluble starch, while showing only a 20% drop in storage stability (Kennedy, 1985). Amyloglucosidase coupled to anion exchange resin resulted in a system that showed no activity loss when stored at 4°C for 6–8 months, was reusable for at least 10 cycles, showed an increased stability towards urea and metal ions, and could convert high concentrations of starch hydrolysate (32% w/w) into 96–98% w/w glucose in 24 h at 50°C and pH 4.5 (Nehete *et al.*, 1987). Amyloglucosidase immobilized in an ultrafiltration reactor (YM-5 flat sheet membrane) for the continuous hydrolysis of cassava starch, showed conversions to glucose of 64% at 22°C and 97% at 55°C in 2 h (Darnoko, Cheryan and Artz, 1988).

Many researchers, in an effort to enhance and retain enzyme binding to their chosen support material, have taken the extra step of using chemical cross-linking reagents such as carbodiimide and glutaraldehyde in an effort to enhance enzyme binding (Caldwell *et al.*, 1976b; D.D. Lee *et al.*, 1976; Svec and Kalal, 1978; Kobarzewski and Paszczynski, 1983; Nehete *et al.*, 1987; Freire and Sánt'Anna, 1990; Pieter, Bardeletti and Coulet, 1992; Emneus, 1993). Amyloglucosidase immobilized on to granular chicken bone cross-linked with silanized glutaraldehyde, silanized carbodiimide, and DEAE showed a 10–40% increase in the amount of activity bound over non-derivatized bone (Schafhauser and Storey, 1992a). Enhanced binding of amyloglucosidase to chitin was also obtained following pretreatments of the support with 25% glutaraldehyde (D.Y. Schafhauser-Smith and K.B. Storey, unpublished results).

Some enzyme systems are very unstable once immobilized or are easily inactivated or desorbed from the support, due to unfavourable conformational changes of the bound enzyme or due to the formation of weak bonds that are easily disrupted by environmental changes (pH, ionic strength of buffer) (Kennedy, 1985). Svec and Kalal (1978), using amyloglucosidase immobilized on to poly[(glycidyl methacrylate) co(ethylene dimethacrylate)] carriers, found a drop in temperature stability and a 70% loss of bound activity after 20 h, partly due to inactivation, partly to further coupling reactions of the enzyme. Amyloglucosidase bound to DEAE-cellulose-activated cyanuric chloride retained 62% of added activity and showed starch conversions of 56%; however, there was constant enzyme leakage with successive buffer washings at pH 3–6.6, a drop in activity in successive batch type processes, and a loss of temperature stability (Tomar and Prabhu, 1985). Amyloglucosidase immobilization on to the gel-forming polymer, Hypol^{R}8190H, resulted in only 1–4% of active bound activity, and a high loss (>68%) of total enzyme activity added (Storey, Duncan and Chakrabarti, 1990). The immobilization of amyloglucosidase on to hydrophilic membranes retained 46 to 51% of added activity, but resulted in higher rates of temperature deactivation, as well as a 30–40% loss of bound activity after nine cycles of use (D.Y. Schafhauser-Smith and K.B. Storey, unpublished results).

A suitable immobilized enzyme system should show functionality within the optimal industrial working range for amyloglucosidase (pH 4.5, 60°C), or show an enhanced range of productivities. The immobilization of amyloglucosidase usually results in pH stability equal to that of the free enzyme (pH 4.5) (Stanley *et al.*, 1978; Slinginger, Fanta and Abbott, 1988; Storey, Duncan and Chakrabarti, 1990; Schafhauser and Storey, 1992a), and maintains an optimal temperature of 50–60°C or better (Svec and Kalal, 1978; Nithiandam *et al.*, 1981; Storey, Duncan and Chakrabarti, 1990; Schafhauser and Storey, 1992a). However, Slinginger, Fanta and Abbott (1988) found catalyst decay at 60°C with enzyme immobilized on to starch graft co-polymers but showed that by operating at 40°C instead of 60°C the half-life of the enzyme could be increased sixteen-fold while decreasing the productivity by a factor of only 4. Thus, the major considerations that influence the choice of the most profitable operating temperature include the required production rate based on demand, capital availability for investment in reactor volume, the length of downtime and the expense of recharging the reactor (Slinginger, Fanta and Abbott, 1988).

The net charge on a carrier can have a profound influence on the pH activity behaviour of an immobilized enzyme due to the influence of the microenvironment surrounding it (Nithiandam *et al.*, 1981; Slinginger, Fanta and Abbott, 1988). An acidic shift in pH optimum from 5.5 to 4.6 occurred upon amyloglucosidase immobilization on to 2-dimethylaminoethyl methacrylate (Nithiandam *et al.*, 1981). As this carrier has a net positive charge, the shift to the lower pH optimum upon immobilization can be explained by a depletion of H^+ ions in the vicinity of the enzyme compared with H^+ concentration in the bulk solution (Nithiandam *et al.*, 1981). Amyloglucosidase immobilized to DEAE-cellulose showed a narrowing of the pH optimum from 3.8–5.2 to 3.6 (Tomar and Prabhu, 1985). The optimum pH range also became narrower with immobilization on to poly[(glycidyl methacrylate) co(ethylene dimethacrylate)] carriers (Svec and Kalal, 1978). However, Nehete *et al.* (1987) found that immobilization on to polystyrene anion-exchange resin did not alter the pH optimum of the enzyme, indicating an absence of partitioning effects of the substrate in the microenvironment of the immobilized enzyme.

Researchers frequently report changes in enzyme kinetic parameters upon immobilization, usually as the result of harsh immobilization procedures, retention by semipermeable membranes, or immobilization within porous bodies. Active site modification due to multiple point attachment of enzymes on to matrix surfaces may cause a decrease in substrate affinity (Zaborsky, 1973). Such problems are usually compounded by the use of high molecular weight substrates which cause mass transfer limitation problems. Starch, being a macromolecule, frequently shows diffusion limitation not only with respect to enzyme bound within the porous interior of a carrier, but also with surface-bound enzyme. As a result, the catalytic action and apparent kinetic behaviour of enzymes can be compromised (D.D. Lee *et al.*, 1976).

Some immobilized enzyme systems suffer from a high diffusion resistance as well as an inefficient coupling method (Nithiandam *et al.*, 1981). Darnoko, Cheryan and Artz (1988) determined that the viscosity of even a 1% w/w cassava slurry was too high, causing stirring and fouling of the membrane and excessive pressure build-up. Amyloglucosidase immobilized on to co-polymers of methacrylate showed a ten-fold increase in K_m, as well as a two-fold increase in specific activity (Nithiandam *et al.*, 1981). Amyloglucosidase immobilized on to poly[(glycidyl methacrylate) co(ethylene dimethacrylate)] carriers showed 1.8- to 2.3-fold increases in K_m values, using 1% starch (Svec and Kalal, 1978). The immobilized amyloglucosidase preparation of Tomar and Prabhu (1985) showed a 1.3-fold increase in K_m, and Lineweaver–Burk plots which indicated substrate inhibition. Amyloglucosidase immobilized on to polystyrene anion-exchange resin showed a 1.7-fold increase in K_m using a 32% w/w starch solution (Nehete *et al.*, 1987). Amyloglucosidase bound to hexyl-Sepharose, using soluble starch substrate, also showed a marginal increase in the apparent K_m (Caldwell *et al.*, 1976b).

Amyloglucosidase bound to granular chicken bone showed no significant change in its K_m values, but a 1.7-fold decrease in specific activity occurred with either Lintner or potato starches as substrates (Schafhauser and Storey, 1992a). Other kinetic data from Lineweaver–Burk plots and Arrhenius plots indicated that there were no diffusional limitations imposed by the binding to bone. However, tests using solutions of whole starch indicated that the free enzyme was 23% more effective at hydrolysing the insoluble portions of starch. Vigorous mixing helped to alleviate the problems of external mass transfer resistance, since a 3.3-fold increase in starch hydrolysis was apparent in mixed, compared with unmixed, preparations (Schafhauser and Storey, 1992a).

D.D. Lee *et al.* (1976) immobilized amyloglucosidase on to porous silica (400 Å pore diameter) packed into a recirculating batch reactor operated at high flow rates to eliminate diffusion resistance. Nehete *et al.* (1987) also found that increasing the stirring speed and decreasing the matrix particle size did not change the activity of the immobilized enzyme, thus indicating no diffusion limitation problems. The size of the glass beads (125 μm) used by Toldra, Jansen and Tsao (1992) for the immobilization of amyloglucosidase was small enough to avoid diffusion resistance.

Amyloglucosidase immobilized on Celite R649 (mean pore size 0.25 μm) was as efficient as the soluble enzyme in catalysing the hydrolysis of maltose, but less efficient in hydrolysing 30% w/v maltodextrin. The immobilization of amyloglucosidase on porous glass fibres showed 98.8% conversion of a 28% maltose solution and 93.5% conversion of a 39% dextrin solution (Toldra, Jansen and Tsao, 1992). Not surprisingly, Slinginger, Fanta and Abbott (1988) found K_m values for maltose largely unchanged when amyloglucosidase was immobilized on to starch graft polyacrylonitrile. Thus, lower product yields can occur when high molecular weight substrates are used, for they are either excluded from small carrier pores or diffuse into these pores at a much slower rate than lower molecular weight substrates. Also, glucose can form reversion products if it is slow in diffusing out of the carrier pores (Slinginger, Fanta and Abbott, 1988; Celebi, Tsai and Tsao, 1991; Emneus, 1993). This is not the case with free amyloglucosidase, where no decrease in yield occurred when high molecular weight substrates were used (D.D. Lee *et al.*, 1976; Storey, Duncan and Chakrabarti, 1990; Schafhauser and Storey, 1992a; Toldra, Jansen and Tsao, 1992). Solutions to the problem of maintaining high yields with real feeds can include increasing enzyme reactor residence times or pre-hydrolysing feeds with α-amylase to hasten the process.

Diffusion limitation can also affect the binding of the enzyme itself on to a support. The yield and activity of bound amyloglucosidase was higher when the enzyme was immobilized on to supports with larger pores (Wojcik *et al.*, 1987; Emneus, 1993). Gels with pore diameters of 50, 100, 130 Å showed 3.2, 12.4 and 16.9 units of bound amyloglucosidase per gram of gel, respectively.

Reduced catalyst activity of immobilized enzymes is not always the result of diffusion limitation. Inefficient coupling methods can also have the same result. For example, when amyloglucosidase was immobilized on to silanized alkylamine glass beads linked by carbodiimide coupling, by a modified enzyme carbohydrate, or by glutaraldehyde cross-linking, the percentage of added protein that remained bound was 29, 56.8 and 64.8%, respectively, whereas activity retention was only 11.5, 10.4 and 11.5% (Kobarzewski and Paszczynski, 1983). Amyloglucosidase immobilized on to the DEAE-, carbodiimide-or glutaraldehyde-derivatized granular chicken bone showed two-to sixfold increases in K_m compared with enzyme bound to underivatized bone (D.Y. Schafhauser-Smith and K.B. Storey, unpublished results). Immobilization of amyloglucosidase on to polyurethane foam resulted in a small, 1.1-fold, increase in K_m (Storey, Duncan and Chakrabarti, 1990). However, since high molecular weight compounds such as blue dextran could readily penetrate the foam, the large substrates (>200 kDa) were hydrolysed as effectively as smaller ones, it appeared that the K_m change was not caused by diffusion resistance in this case.

Those immobilization supports that can load high activities of enzyme per gram of support are ideal for scale-up purposes (Svec and Kalal, 1978; Kobarzewski and

Paszczynski, 1983; Storey, Duncan and Chakrabarti, 1990; Pieter, Bardeletti and Coulet, 1992; Schafhauser and Storey, 1992a). The shorter residence times that would result would decrease the number of side reactions and thus lower refining costs (Fogarty, 1983). However, Kobarzewski and Paszczynski (1983) found that the immobilization of amyloglucosidase on to glass beads in amounts greater than 10 mg protein g^{-1} beads showed only a slight increase in substrate hydrolysis, consequently decreasing the practical and economical value of the product. Enzyme loading must be optimized by considering the tradeoff between less efficient enzyme use at high loadings, and the need for greater carrier and reactor size and consequently greater immobilization cost at low enzyme loadings (Pitcher, 1980). Inexpensive supports can help to offset such extra costs (D.D. Lee *et al.*, 1976; Stanley *et al.*, 1978; Cabral, Novais and Cardoso, 1981; Freire and Sánt'Anna, 1990; Storey, Duncan and Chakrabarti, 1990; Schafhauser and Storey, 1992a; Toldra, Jansen and Tsao, 1992).

α-AMYLASE

α-Amylase ([1-4]-α-*D*-glucan amyloglucosidase, EC 3.2.1.1) is a calcium metalloprotease that is widely distributed in nature and found in animal, plant and microbial species (Solomon, 1978; Kennedy *et al.*, 1987). α-Amylase hydrolyses the α-*D*-(1,4) links between the α-*D*-glucopyranosyl residues of starch in a random endo-fashion, but is unable to hydrolyse the α-*D*-(1,6) bonds of amylopectin (Kennedy *et al.*, 1987). The end products are glucose, maltose and α-limit dextrins containing α-(1,6)-glycosidic bonds (Solomon, 1978). α-Amylases have application in the food processing industry for the production of dextrins and glucose syrups for the brewing and baking industries (Solomon, 1978; Kennedy *et al.*, 1987). The microbial α-amylases, in particular, are widely used commercially in the production of specific oligosaccharides. The enzyme from each species produces primarily a single specific oligosaccharide; for example, the major hydrolysis products of the enzymes from *Bacillus lichenformis, B. amyloliquefaciens* and *B. subtilis* are maltopentaose, maltoheptaose and maltohexaose, respectively (Kennedy *et al.*, 1987). The pH and temperature optima for maximum enzyme activities are also species-dependent, but generally occur within a pH range of 4.8–6.5 and a temperature range of 70–80°C (Kennedy *et al.*, 1987).

Sadhukhan *et al.* (1993) immobilized α-amylase from *Myceliophthora thermophila* by covalent binding to cyanogen bromide (CNBr)-activated Sepharose and by entrapment within polyacrylamide gels or calcium alginate beads in order to determine the best carrier for use in the saccharification of starch. Immobilization to calcium alginate beads gave the best improvement in α-amylase stability; compared with the free enzyme, immobilized α-amylase showed a substantial increase in thermal stability (optimal temperature rose from 60 to 65°C), a broadening of the optimal pH range (pH 4–8 versus pH 5.6) and an increase in storage stability (a half-life of 110 days versus 5 days at 28°C). However, 60% of activity was lost after six cycles of reuse (with 1% starch substrate, pH 5.5, 60°C); this was due to enzyme leakage as the result of changes in the gel structure caused by the continuous agitation.

De Cordt *et al.* (1992) analysed the reaction kinetics of the thermostable α-amylase from *B. lichenformis* in free form and when covalently bound to glutaraldehyde-activated glass beads, for application as a possible time–temperature integrator for evaluation of heat processes. The rate of thermal inactivation (D-value) was shown to be faster for the free enzyme (6–157 min) than the immobilized α-amylase (17–620 min) at 95°C, with the range of values reflecting the variety of different environmental reaction conditions tested (pH, ionic strength, Ca^{2+} concentration). Further studies analysed the performances of three different linking agents (glutaric dialdehyde, benzoquinone, s-trichlorotriazine) for α-amylase immobilization on to porous glass beads (de Cordt, 1994); all three linkers gave a similar improvement in immobilization yield and improved enzyme resistance to heat inactivation to a similar extent.

α-Amylase immobilized by attachment to soluble amino-S-triazinyl derivatives of dextran 2000, DEAE-dextran 2000 and CM-cellulose retained up to 65%, 90% and 19% of added activity, but specific activities decreased by 33%, 67% and 40%, respectively, compared to the free enzyme (Wykes, Dunnill and Lilly, 1971). All bound enzymes were stabilized in the presence of Ca^{2+} and starch, with a marked thermal stability of CM-cellulose-bound α-amylase. Indeed, the CM-cellulose-bound enzyme had a 2.6-fold greater operational half-life than the free enzyme in continuous hydrolysis reactions of soluble corn starch in an ultrafiltration reactor at 70°C (Wykes, Dunnill and Lilly, 1971).

β-AMYLASE

β-Amylase (α-1,4-glucan maltohyrolase, EC 3.2.1.2) hydrolyses α-1,4-glycosidic linkages from the non-reducing ends of starch in an exo-fashion, releasing β-maltose units (Fogarty, 1983). The enzyme occurs widely in higher plants and micro-organisms (Kennedy *et al.*, 1987). Maltose and high-maltose syrups have a variety of commercial uses in the confectionery, baking and brewing industries, due to their low viscosity in solution, low hygroscopicity, resistance to crystallization, low sweetness, heat stability and lack of colour formation (Fogarty, 1983); the syrups are also used in the manufacture of vaccines and antibiotics (Ohba, Shibata and Ueda, 1979).

The high production costs associated with the continuous use of fresh enzyme in industrial saccharification processes limits the commercial potential of the process and such has been the case to date with β-amylase. Furthermore, the β-amylases derived from plants are expensive and tend to be unstable, whereas β-amylases from microbial sources are even less active and thermostable (Saha and Shen, 1987). An immobilized β-amylase preparation that could retain high activity and functional stability would, therefore, have considerable industrial potential.

Many attempts have been made at producing an active and stable immobilized β-amylase, and a number of immobilization methods have resulted in improved functional characteristics over that of the native enzyme. The covalent coupling of β-amylase to soluble conjugated polysaccharides (dextran, amylose) made the bound enzyme as accessible as the native enzyme to macromolecular starch (Lenders and Crichton, 1984). The immobilized β-amylase also showed greater temperature stability, and could hydrolyse starch at a temperature 20°C higher than the native enzyme. CNBr-activated epichlorohydrin-cross-linked Sepharose covalently bound 35% of added barley β-amylase activity, and resulted in an enhanced temperature stability as well as increased storage and operational stabilities compared with the native enzyme (Vretblad and Axen, 1973).

β-Amylase immobilized on to polystyrene cation-exchange resin (IR-120 A13+)

would not dissociate following treatments with high concentrations of buffers, salts or EDTA solutions and retained up to 60% of activity after 3 months; furthermore, the bound enzyme showed a broader temperature optimum compared with the free enzyme (Roy and Hegde, 1987). Maeda, Tsao and Chen (1978) immobilized soybean β-amylase on to porous cellulose beads, retaining 59–69% of added activity; the immobilized enzyme showed characteristics similar to that of the native enzyme with respect to starch conversion rates, and temperature and pH optima and stabilities. Operational and storage half-lives were 24 days at 50°C and 40 days at 40°C, respectively, in packed bed reactors (Maeda, Tsao and Chen, 1978). Barley β-amylase bound firmly to granular chicken bone (21% of added activity) and, compared with the free enzyme, was less affected by treatments with detergents, ethanol, high salt and pH extremes (Schafhauser and Storey, 1994). Although pH stability did not improve, bone-immobilized β-amylase showed an increase in high-temperature stability. Sweet potato β-amylase remained bound to *p*-aminobenzylcellulose in the presence of high salt concentrations, showed improved temperature stability, and maintained full activity when stored in a dry state for 6 months (Ohba, Shibata and Ueda, 1979). Ramesh and Singh (1981) immobilized 36% of added sweet potato β-amylase activity on to alkylamine glass and showed a 1.4-fold drop in activation energy and an increased thermal stability over that of the native enzyme. β-Amylase immobilized on to arylamine glass was less stable, showing enhanced pH and thermal stability with 48% of added activity remaining bound, but a 1.5-fold drop in activation energy (Ramesh and Singh, 1981).

Although many immobilized enzyme systems have been tested with β-amylase, the enzyme has, in most cases, proven to be too unstable and immobilization has not been enough of a stabilizing factor to maintain acceptable enzyme activities over an extended period of time. The low binding ability of some supports (3–7% of added β-amylase) could be as much the result of the unstable enzyme as of other factors, such as the blocking of active sites upon immobilization or a destabilizing effect of the support on the enzyme (Roy and Hegde, 1987; Germain and Crichton, 1988). Long-term stability of β-amylase bound to Biobone was poor, with 50% of activity lost after overnight storage at 23°C (Schafhauser and Storey, 1994). However, this instability was not due to enzyme leaching from bone itself, for assays conducted in the presence versus absence of bone washings showed similar rates of substrate hydrolysis. Caldwell *et al.* (1976a) showed that the continuous use of sweet potato β-amylase immobilized on to epichlorohydrin-cross-linked hexyl-Sepharose resulted in activity leakage at a rate of 0.5% h^{-1}. The enzyme also showed poor storage and thermal stability once bound, losing 50% of bound activity after 9 days' storage at 25°C, with even greater losses at storage temperatures of 35°C and 45°C. Caldwell *et al.* (1976a) concluded that β-amylase, once immobilized, can undergo unfavourable conformational changes that can ultimately lead to activity loss, a situation that is enhanced with increasing thermal motion at higher temperatures.

The parameters affecting the coupling of enzyme to matrix can be varied to give the best possible conditions for binding and operational activities. Takeda and Hizukuri (1975) showed a 160% increase in β-amylase activity in presence of Triton X-100. Martensson (1974a), who immobilized barley β-amylase on to a carbodiimide-activated co-polymer of acrylamide–acrylic acid, demonstrated the protective effects of reduced glutathione and serum albumin for stabilizing β-amylase activity and thereby enhancing conversion rates. As β-amylase is very sensitive to the oxidation of some thiol groups necessary for enzymatic activity, reduced glutathione would impart stability through an antioxidant effect (Martensson, 1974a). the presence of serum albumin could offer stability through increased protein–protein interactions due to higher protein concentrations (Martensson, 1974a). This study also demonstrated increased operational (5.5 and 4 days versus 2 days) and storage (110 and 90 versus 18 hours) half-lives for glutathione- and albumin-protected immobilized β-amylase, compared with the unprotected bound enzyme, respectively. By contrast, the stability of β-amylase immobilized on to granular chicken bone was not improved when both immobilization and storage were conducted in the presence of bovine serum albumin, Triton X-100, dithiothreitol or mercaptoethanol (Schafhauser and Storey, 1994). Overnight stability was achieved only when the immobilized enzyme was stored in a substrate solution or when the enzyme was immobilized to derivatized bone in the presence of 1 M sorbitol or sucrose. Sugars and polyols are known protein stabilizers, which function by preferentially hydrating protein surfaces and, thereby, enhancing protein associations (Arakawa and Timasheff, 1982). Under these conditions, a storage half-life of 65 days was achieved, but only 30% of activity was maintained after five cycles of use at 45°C (Schafhauser and Storey, 1994).

Reticulating agents which serve to rigidify protein structure are valuable for increasing enzyme stability, and Germain and Crichton (1988) increased the stability of sweet potato β-amylase immobilized to porous silica by prior treatments involving enzyme attachment to soluble dextran or cross-linking with glutaraldehyde. Porous silica retained 45% of added dextran- and glutaraldehyde-modified β-amylase, but only 4% of added native enzyme. The modified-immobilized enzyme also showed greater operational and thermal stabilities compared with those of the native immobilized enzyme, for the temperatures at which 50% of activity remained bound after 10 minutes were 65°C for native β-amylase, 66°C for glutaraldehyde-cross-linked β-amylase, and 74°C for dextran-modified β-amylase (Germain and Crichton, 1988).

The immobilization of β-amylase also leads frequently to reduced catalytic activity. This may be caused by conformational changes that alter enzyme substrate affinity, or by slow intraparticle substrate diffusion or film diffusion, brought on by the large molecular weight and highly viscous starch substrates that are not conducive to rapid transport (Caldwell *et al.*, 1976a; Hon and Reilly, 1979). β-Amylase showed a six-fold increase in K_m upon immobilization on to cation-exchange resin (Roy and Hegde, 1987) and an eight-fold increase following its immobilization on to hexyl-Sepharose (Caldwell *et al.*, 1976a). The enzyme showed 1.7-and two-fold increases in K_m as well as 1.5- and two-fold decreases in specific activity upon immobilization on to alkylamine and arylamine glass (Ramesh and Singh, 1981). A 67-fold increase in K_m following the immobilization of β-amylase on to porous cellulose beads was substantially reduced after the immobilized preparation was ground into powder form to greatly increase the surface area (Maeda, Tsao and Chen, 1978). Film diffusion at flow rates below a linear velocity of 3 cm min^{-1} was also apparent (Maeda, Tsao and Chen, 1978).

The K_m values for native, glutaraldehyde-cross-linked and dextran-conjugated β-amylase, using soluble starch as the substrate, increased 33-, 3.5- and 1.4-fold, respectively, following their immobilization on to porous silica (Germain and Crichton, 1988). While the modified enzymes showed significant improvements to enzyme catalytic activity following immobilization, the systems still suffered from steric hindrance and/or diffusion limitation that could not be resolved by an increased stirring rate of the mixture. The authors also suggested that the modified β-amylase may be too big to penetrate into the porous support. By contrast, β-amylase bound to granular chicken bone showed a 1.2-fold decrease in the K_m for soluble starch, and both Lineweaver–Burk and Arrhenius plots were similar for the bound and free enzymes, all these results indicating that immobilization did not hinder the approach of high molecular weight substrates to this support (Schafhauser and Storey, 1994). However, the increase in surface area achieved by powdering the bone resulted in a 1.5-fold increase in enzyme activity.

PULLULANASE

Pullulanase (pullulan 6-glucanohydrolase, EC 3.2.1.41), also known as the 'debranching enzyme', catalyses the hydrolysis of α-*D*(1,6) and to a lesser extent the α-*D*(1,3) and α-*D*(1,4) glycosidic bonds of pullulan, amylopectin and glycogen in a random endo-fashion (Solomon, 1978; Ram and Venkatasubramanian, 1982; Yang and Coleman, 1987). The end products are maltotriose and maltoriosyl oligosaccharide side-chains of varying length (Fogarty, 1983). Useful preparations of pullulanases are obtained from a variety of micro-organisms, including *Aerobacter aerogenes, Escherichia intermedia* and *Streptomyces flavochromogenes* (Fogarty, 1983). Pullulan, a linear polymer of maltotriose linked by α-(1,6) glucosidic bonds, and waxy maize, a starch containing 80% α-(1,6) glycosidic links, are the common substrates used for pullulanase analysis (Solomon, 1978; Ram and Venkatasubramanian, 1982).

Pullulanase is widely used as an analytical tool for elucidating the structures of polysaccharides and in the determination of chain length of glycogen and amylopectin, which is important in the diagnosis of two types of glycogen storage disease (Martensson and Mosbach, 1972). For industrial use in bioconversion, the addition of pullulanase in combination with α-amylase, β-amylase or amyloglucosidase can greatly increase the purity of high-maltose and high-glucose syrups produced from starch (Yang and Coleman, 1987). Because the enzyme is of limited use on its own, only a few studies have analysed pullulanase immobilization alone, but other studies, considered later in this review, have assessed the improvements in product output brought about by the co-immobilization of pullulanase along with starch- or cellulose-degrading enzymes.

A key effort of investigators developing immobilized pullulanase systems has been the improvement of functional and thermal stability. Pullulanase from *Klebsiella aerogenes* covalently bound to azide-linked bovine-hide collagen membranes, showed improved thermal and operational stability compared with that of the native enzyme (from 2 to 7 days at 50°C) (Ram and Venkatasubramanian, 1982). *Streptomyces flavochromogenes* pullulanase pretreated with tannic acid to facilitate binding on to TEAE-cellulose, showed an operational half-life of 15 days at 50°C, and remained stable following 6 months of storage (Ohba *et al.*, 1978). The bound enzyme was also greatly affected by the presence of calcium salts, showing a substantial increase in temperature stability at low levels of Ca^{2+} (2–5 mM), whereas amounts greater than 10 mM resulted in enzyme desorption.

Lenders and Crichton (1984), who conjugated pullulanase to soluble polysaccharides (dextrans, amylose), showed a direct correlation between enzyme coupling to polymers of increasing molecular weight and greater thermal stability, resistance to urea denaturation and lower K_m values; such enhanced stabilities were attributed to the formation of multiple covalent linkages between the enzyme and the polysaccharides. Pullulanase from *Aerobacter aerogenes*, cross-linked to a co-polymer of acrylamide–acrylic acid (Biogel CM-100), retained 43% of added activity, but showed no improvement in stability over that of the free enzyme, with a half-life less than 30 min at 55°C (Martensson and Mosbach, 1972). However, coupling in the presence of substrates resulted in a fivefold increase in activity. Aware of the enzyme's sensitivity to heat denaturation, Martensson (1974b) studied the operational stability of pullulanase immobilized to a co-polymer of acrylamide–acrylic acid, at three different substrate concentrations and at three different temperatures. A stabilizing effect of high substrate concentrations was apparent, but half-lives were still low, ranging from 3 to 9 days at 40°C and 2 days at 50°C.

The immobilization of fibre-degrading enzymes

CELLULASE

Cellulases are complex multicomponent enzyme systems produced by micro-organisms, plants and perhaps some animals, that can function synergistically to hydrolyse cellulosic materials into glucose (Klyosov and Consultant, 1986; Woodward, 1989). Wood-decaying fungi far outnumber bacteria with respect to the number of cellulolytic organisms identified so far, primarily because they produce and secrete more cellulases (Saddler, 1986). Fungi such as *Trichoderma reesei* and *Aspergillus niger* are the most common commercial sources of extracellular cellulases, having specific activities greater than cellulases from other sources (Mandels *et al.*, 1981). The cellulase system can be divided into three classes, including exoglucanases such as cellobiohydrolase (1,4-β-*D*-glucan cellobiohydrolase, EC 3.2.1.91), endoglucanases (1,4-β-*D*-glucan glucanhydrolase, EC 3.2.1.4) and β-glucosidase (β-*D*-glucoside glucohydrolase, EC 3.2.1.21) (Cheetham, 1985; Woodward, 1989; Klyosov, 1990). The exo- and endoglucanases adsorb on to and act synergistically on crystalline cellulose to produce cellobiose and *D*-glucose, whereas β-glucosidase catalyses the hydrolysis of cellobiose to *D*-glucose (Woodward and Zachry, 1982). The ability of cellobiohydrolases and endoglucanases to degrade cellulose is directly correlated with the degree to which they are adsorbed to its surface, with a greater degree of adsorption resulting in a higher reaction rate and greater glucose yield (Klyosov and Consultant, 1986). Enzymes isolated from different sources can occur as isozymes and can differ in their molecular characteristics, mode of action, absorptive ability, catalytic activity and substrate specificity (Klyosov, 1990). As only one of each of the constituent enzymes is required for effective hydrolysis, the variety and/or effectiveness of cellulase complexes from different cellulase sources is quite vast (Woodward, 1989; Klyosov, 1990).

Cellulose is a major component of plant biomass and is a renewable resource, abundantly available in the form of hardwood and softwood trees, agricultural and citrus crop residues, and municipal wastes (Jain and Wilkins, 1987; Woodward, 1987; Grohmann and Baldwin, 1992). Natural cellulose is a large and complex glucose polymer. It can consist of up to 10 000 glucose units and can contain both highly ordered (crystalline) regions and more randomized (amorphous) regions, as well as forming associations with lignin and hemicellulose (Knowles, Lehtovaara and Teeri, 1987). In deference to these inconsistencies, a wide variety of cellulose-related substrates have been used for cellulase characterization studies, including soluble substrates, such as carboxymethylcellulose (CMC), cellobiose and *p*-nitrophenyl glucosides (PNPG), and insoluble substrates, such as cotton, filter paper, solka floc, amorphous cellulose and microcrystalline cellulose (Avicel) (Knowles, Lehtovaara and Teeri, 1987).

Potentially, the glucose derived from the breakdown of cellulose can be used for a vast number of processes, including as a feedstock additive for single-cell protein processes, a fermentable sugar for the production of chemical precursors and fuel, and a precursor for fructose syrups for the sweetener industry (Woodward and Zachry, 1982; Cheetham, 1985). However, the enzymatic production of glucose from cellulose has not attained commercial success. Cellulase enzymes have low specific activities for insoluble cellulosics and, thus, a high ratio of enzyme to substrate is necessary to achieve productive hydrolysis in a reasonable period of time (Takeuchi and Makino, 1987). Optimizing the conditions to maximize the activities of a multienzyme catalyst, which can involve potentially inhibitory substrate and product levels, as well as a variety of soluble and insoluble substrates and products, is a complex process (Saddler, 1986). Enzyme costs are also an important feature in the economy of enzymatic saccharification of cellulosic materials, comprising, for example, nearly 50% of the outlay required to produce sugar from corn stover (Woodward, 1987). Thus, the use of soluble cellulases for the production of inexpensive products is an expensive undertaking.

Efforts to retrieve and recycle free cellulases from fermentor reactors have not been successful (Mes-Hartree *et al.*, 1987). However, free cellulase activity was successfully recovered and reused from a spent reactor hydrosylate of steam-exploded aspen wood by a few minutes' contact with fresh substrate (Lamed *et al.*, 1991). The development of reusable cellulase systems has been an area of active study as it is seen as a practical solution for overcoming waste and high costs.

The catalytic ability of an immobilized cellulase (particularly one immobilized by entrapment or on to an insoluble support) hydrolysing an insoluble substrate should be greatly impaired, for the hydrolysis of insoluble cellulose requires both cellobiohydrolases and endoglucanases to adsorb to the surface of the substrate to initiate depolymerization (Woodward, 1987, 1989). For example, it has been shown that up to 50% of initial soluble cellulase activity can remain bound to lignocellulosic substrates following a hydrolysis reaction (Mes-Hartree *et al.*, 1987). In spite of this, immobilized cellulases can successfully hydrolyse cellulose substrates, albeit not to the same degree as the native enzyme with respect to insoluble substrates. However, as a compensation, many immobilized cellulase preparations have resulted in firmly bound, reusable enzymes, showing stability properties better than the free enzyme systems. *Trichoderma reesei* cellulase adsorbed to concanavalin A cyanogen bromide (ConA-CNBr)-activated Sepharose showed no enzyme leakage during normal operations or following treatment with urea (Woodward and Zachry, 1982). *Aspergillus niger* cellulase bound to poly-*L*-glutamic acid showed improved temperature and pH stability compared with the native enzyme, and could maintain 90% of its activity after 20 days storage (Takeuchi and Makino, 1987). Cellulases covalently bound to polyurethane particles maintained 100% of activity following more than 15 repeated uses (Kumakura, 1985). CMC cellulases immobilized on to polyvinylalcohol (PVA) and polyethylene glycol (PEG) showed no activity loss or leakage after 3 and 12 cycles of use, respectively (Mishra, Deshpande and Rao, 1983; Kumakura and Kaetsu, 1985). Cellulase from *T. reesei* covalently bound to polyurethane foam retained 33% of added activity, and showed a half-life greater than 6 weeks (assessed with 1% CMC at 23°C, pH 4.5) as well as improved pH and thermal stabilities (Chakrabarti and Storey, 1988). Cellulases immobilized on to a collagen fibril matrix saccharified a 0.33% solution of Avicel to better than 80% completion in a fluidized bed reactor at 30°C (Karube *et al.*, 1977). Cellulase enzymes from *Macrophomina phasedina* entrapped in 5% acrylamide could hydrolyse filter paper and cotton to 41% and 6.75%, respectively, over 120 h, and remained stable after 25 repeated uses (Roy, Roy and Duke, 1984).

In an effort to improve the limited degree of hydrolysis seen with immobilized enzymes against insoluble celluloses, some investigators have introduced spacer molecules between the immobilized enzyme and the support matrix, ideally allowing for increased mobility of the enzyme and, thus, a more effective interaction between the insoluble cellulose and insoluble enzyme. Cellulase immobilized on to CNBr-Sepharose and CH-Sepharose gels containing hydrocarbon spacers showed higher activity towards cellobiose, acid-swollen cellulose and cellulose powder, compared with cellulase immobilized directly on to gels (Chim-anage *et al.*, 1986). Similarly, cellulases immobilized on to CNBr-activated Sepharose, ConA-Sepharose and CNBr-activated glass beads hydrolysed alfalfa, wheat straw and pine needles to a greater extent than did the free enzyme (Fadda *et al.*, 1984). In particular, ConA-Sepharose showed 1.5- to 1.7-fold greater activity than the native enzyme against those substrates (Fadda *et al.*, 1984). The cellulase complex from *T. reesei* immobilized on to CNBr-activated ConA-Sepharose retained full activity against CMC, but only 9% and 29% activities against Avicel and cellobiose, respectively (Woodward and Zachry, 1982).

However, cellulase enzymes that are required to adsorb to the cellulose surface for proper function may not be bound strongly enough to their supports to resist desorption tendencies. At the very least, enzymes immobilized in such a way as to restrict the movement required for adsorption and desorption on to the substrate could suffer activity loss and loss of catalytic properties.

Trichoderma reesei cellulase bound to glutaraldehyde-activated nylon retained 60% of initial activity and became more thermally stable (Jain and Wilkins, 1987). However, bound activity was lost after five cycles of use, and the bound enzyme hydrolysed only 33% as much Avicel as the native enzyme over 24 h at 50°C (Jain and Wilkins, 1987). ConA-Sepharose cellulase lost between 30 and 50% of activity after only five repeated uses (Fadda *et al.*, 1984) and *T. reesei* cellulase entrapped in acrylate polymers suffered from leakage over long periods of time (Higa, Del Mastro and Castagnet, 1986). This desorption, however, could be prevented by the addition of silica gel or PEG 600 to the preparation.

Cellulases have also been immobilized on to water-soluble supports with the assumption that, as long as recovery was possible by filtration or centrifugation, such immobilized enzymes would offer greater interaction between the constituent enzyme and insoluble cellulosic substrates (Woodward, 1989), and would be more efficient catalysts compared with the free enzyme. However, no significant improvement of cellulase function was found for immobilization on to soluble versus insoluble supports. *Aspergillus niger* cellulase immobilized on to poly-*L*-glutamic acid could be recovered easily from the product solution and reused, being soluble in neutral and alkaline conditions, and insoluble at acidic pHs (Takeuchi and Makino, 1987). The immobilization of the cellulase complex to polyethylene glycols (PEG) successfully hydrolysed cellulose powder at 40°C (Kumakura and Kaetsu, 1985). Enzyme activity was also shown to be affected by an increase in the molecular weight of PEG (and the corresponding viscosity increase of the carrier); thus, cellulase bound to PEG 200 retained 60% of its original activity but only 40% of activity was retained when PEG 600 was the carrier. Cellulase from *Penicillium funiculosum* bound to carbodiimide-linked PVA showed enhanced stability, converting two-fold more alkali-treated bagasse than the native enzyme after 48 h (Mishra, Deshpande and Rao, 1983).

An understanding of the complex process of cellulase hydrolysis of cellulosics is further complicated when one considers the number of catalytic changes that could occur to the component enzymes upon immobilization. Furthermore, one would expect that any steric hindrances between the immobilized enzymes and the substrate, soluble or not, would prevent the usual binding of cellulase to substrate and thus affect enzyme catalytic properties. This subject has been addressed by surprisingly few investigators. The cellulase complex entrapped within 5% acrylamide showed K_m values only slightly higher than the native enzyme with respect to the hydrolysis of CMC and cellobiose, with enzyme activities increasing with decreasing particle size (Roy, Roy and Duke, 1984). Cellulase immobilized on to polyurethane particles also experienced an increase in activity as a function of decreasing particle diameter (Kumakura, 1985).

Not only do some immobilized cellulases utilize some insoluble forms of cellulose, but some make more efficient use of them when compared to free cellulases, thus resulting in more efficient catalysis (Fadda *et al.*, 1984; Chakrabarti and Storey, 1988). Polyurethane foam apparently presented little or no diffusional barrier to penetration by high molecular weight cellulose substrates, for compared with their respective activities against soluble CMC, the relative activities against microcrystalline cellulose were 20% for the immobilized enzyme and only 16% for the native enzyme (Chakrabarti and Storey, 1988). In addition, a substantial 7.5-fold decrease in K_m CMC occurred following cellulase immobilization on to foam. Furthermore, ConA-Sepharose-, CNBr-Sepharose- and CNBr-glass bead-immobilized cellulase showed only slight increases in K_m values (1.1-, 1.3- and 1.3-fold, respectively) compared with the free enzyme. The immobilized enzymes were also more resistant to inhibition by glucose and cellobiose, suggesting higher β-glucosidase and cellobiase activities (Fadda *et al.*, 1984). Cellulase immobilized to ConA-Sepharose could carry out long-term high rates of hydrolysis of microcrystalline cellulose to a greater extent than equal amounts of free enzyme (17 and 4 μmol ml^{-1} *D*-glucose, respectively) (Woodward and Zachry, 1982). Immobilization may have altered the conformation of the active site or regulatory and effector sites of one or more of the enzymes in the multienzyme complex, resulting in one or more effects such as increased affinity of endoglucanase for cellulose, improved enzyme efficiency, reduced feedback inhibition of cellobiohydrolase by cellobiose, or improved function of β-glucosidase that reduces cellobiose accumulation (Klyosov and Consultant, 1986; Chakrabarti and Storey, 1988).

β-GLUCOSIDASE

β-Glucosidase, which hydrolyses cellobiose to produce glucose, is essential to the complete enzymatic conversion of cellulose to glucose and is, thus, a subject of research interest in its own right. β-Glucosidase is subject to thermal inactivation and can be limited by glucose inhibition; it is considered to be the rate-limiting factor in the multienzyme process that converts cellulosics into glucose (Sundstrom *et al.*, 1981; Matteau and Saddler, 1982; Roy *et al.*, 1989). Furthermore, cellulase complexes often contain insufficient concentrations of β-glucosidase to prevent the accumulation of cellobiose, and high cellobiose levels inhibit the endoglucanase components of most cellulase systems (Sundstrom *et al.*, 1981; Woodward, 1989). Thus, supplementation of the cellulase mixture with additional β-glucosidase can lower or prevent the accumulation of inhibitory cellobiose, and enhance the rate and extent of saccharification (Sundstrom *et al.*, 1981; Roy *et al.*, 1989). β-Glucosidases are also useful in the preparation of aromatic oils as additives in the food industry (Iborra *et al.*, 1992).

The addition of purified β-glucosidase to a soluble cellulase system (Cellulosome) can result in the hydrolysis of accumulated cellobiose with a resulting ten-fold increase in cellulose degradation (Lamed *et al.*, 1991). The use of an immobilized β-glucosidase would be a further improvement, imparting all of the associated benefits of an immobilized enzyme system, including enhanced stability, reusability and reduced enzyme cost. Thus, a commercial process could become economically viable if an immobilized β-glucosidase was used. Numerous methods, of varying utility, have been developed for immobilizing β-glucosidase.

Mycelial-associated β-glucosidase encapsulated in calcium alginate beads hydrolysed cellobiose and salicin (a cellobiose analogue) to 47% and 30% conversions, respectively, in a packed bed reactor at 50°C (Matteau and Saddler, 1982). Although rapid thermal inactivation occurred at temperatures higher than 60°C, the reactor showed half-lives greater than 1000 h when operating continuously at 50°C (Matteau and Saddler, 1982). Nylon powder retained 67% of added *Penicillium funiculosum* β-glucosidase (Aguado, Romero and Rodriguez, 1993) whereas approximately 50% of added β-glucosidase remained active when immobilized to a series of (hydrophobic) alkyl and aryl derivatives of cotton cloth (Sharma and Yamazaki, 1984). Sweet-almond β-glucosidase maintained stability for 1 h at 80°C immobilized to an ultrafiltration membrane, unlike the free enzyme, which showed a rapid loss of activity under similar conditions (Sarkar and Burns, 1983). β-Glucosidase incorporated into thin films by gelation with α-casein, retained all bound activity following 10 cycles of hydrolysis of PNPG, and could be stored at 5°C for at least 1 year without activity loss or leakage (Motoki *et al.*, 1987). Woodward and Capps (1992) treated *A. niger* β-glucosidase with glutaraldehyde prior to entrapment within calcium alginate gel spheres to prevent a previously encountered enzyme leakage problem. This treatment resulted in bound enzyme activities similar to the native enzyme, as well as enhanced thermal stability (Woodward and Capps, 1992). The enzyme from *Alcaligenes faecalis* retained 25% of added activity on PNPG-CNBr-activated cellulose and showed a slight increase in temperature stability (Srinivasan and Bumm, 1974). β-Glucosidase linked to controlled pore alumina retained 50–60% of added activity, losing only 10% of activity after 500 h continuous use at 50°C against cellobiose (Sundstrom *et al.*, 1981). Polyurethane foam retained 73% of added β-glucosidase activity, and maintained essentially 100% activity after 100 h of continuous use (Chakrabarti and Storey, 1988).

Barley-meal β-glucosidase immobilized on to Sepharose, glutaraldehyde-cross-linked glass beads or phenolic resin (Duolite) retained 35%, 21% and 42% of added activity, respectively (Simos and Georgatsos, 1990). However, only Sepharose-linked β-glucosidase showed a significant increase in thermostability, whereas the cross-linked supports showed poor stability in general (Srinivasan and Bumm, 1974). β-Glucosidase bound to CNBr-activated Sepharose and entrapped within polyacrylamide gels retained 33% and 51% of added soluble enzyme activity, and showed improved temperature and storage stabilities (half-lives of 70 and 90 days, respectively) compared to the free enzyme (half-life of 4 days) but both lost approximately 30% of initial activity after four cycles of use (Roy *et al.*, 1989). *Aspergillus niger* β-glucosidase bound to a polyethyleneimine glutaraldehyde-activated magnetic support retained only 15–27% of added activity, with the continual loss of bound activity attributed to physical loss during both washing and recovery steps (Dekker, 1990). However, the immobilized enzyme could be recovered easily from lignocellulosic solids suspension in a stirred batch reactor by applying a magnetic field (Dekker, 1990). Cassava-leaf β-glucosidase bound to photo-cross-linkable prepolymer resins could be reused repeatedly and stored at 4°C without any significant loss of activity for up to 2 years (Yeoh, 1992). β-Glucosidase immobilized on to hydrazine-derivatized nylon required a long and arduous support preparation and immobilization procedure (over 16 h) (Iborra *et al.*, 1992). Although retaining 30% of added activity, the bound enzyme showed storage and operational half-lives of only 30.7 and 20.1 days, and no improvement in pH or thermostability. However, the bound enzyme showed improved catalysis over the native enzyme, with a 20% decrease in K_m for picrocrocin (the glycoside precursor of safranal oil) and an 80% decrease in K_m for glucose.

Although immobilization usually imposes steric and diffusional limitations on enzyme activity, an immobilized β-glucosidase, since it is hydrolysing a soluble substrate, should experience relatively few alterations in catalytic properties. However, β-glucosidase immobilized on to controlled pore alumina was apparently subject to diffusion limitation, showing increasing activity (from 50% to 90% of initial bound activity) with increasing liquid velocity through the packed bed (Sundstrom *et al.*, 1981). A 3.6-fold increase in K_m following immobilization was also observed, but activities of both native and bound enzyme followed a linear Arrhenius relationship with respect to temperature (Sundstrom *et al.*, 1981). Beta-glucosidase immobilized on to Sepharose, duolite or glass also showed slight increases in K_m values with respect to the free enzyme (by 10%, 40% and 80%, respectively) (Simos and Georgatsos, 1990). The 10% and 20% increases in K_m values seen following immobilization of β-glucosidase on to polyacrylamide gel and CNBr-activated Sepharose were also attributed to mass transfer limitation (Roy *et al.*, 1989). In addition, cellulose bound β-glucosidase showed an Arrhenius plot unlike that of the native enzyme, with a higher activation energy at temperatures greater than 45°C. A fivefold increase in K_m also occurred, but this was attributed to enzyme hydrolysis occurring on both the added substrate and the β-1,4 glucosidic linkages of the cellulose matrix (Srinivasan and Bumm, 1974).

XYLANASES

Lignocellulosics, which comprise the cell wall materials of all woody plants, grasses and cereals, constitute the largest single biomass resource on Earth (Tan, Wong and Saddler, 1985). They are a complex matrix principally composed of cellulose (40–60%), hemicellulose (20–30%) and lignin (15–30%) that is highly resistant to breakdown and hydrolysis (Dekker, 1983; Kluepfel *et al.*, 1990). Lignin, an aromatic polymer composed of phenylpropanoid subunits (Woodward, 1984), is not saccharide-based and, therefore, will not be discussed further in this review. Cellulose, essentially composed of non-branching polymers of *D*-glucose, retains the same structure in all plants (Biely, 1985). However, xylans, the major component of hemicelluloses, have, by comparison, much more complex structures that can vary from one plant species to another (Biely, 1985). Xylans are composed of a backbone of β-1,4-linked xylopyranose residues with side-chains of acetate, *L*-arabinofuranose, *D*-glucuronic and 4-*O*-methyl-*D*-glucuronic acid at key branch points (Biely, 1985; Wong, Tan and Saddler, 1986; Christov and Prior, 1993). Hetero-1,4-β-*D*-xylans constitute the major hemicellulosic component of the Gramineae (grasses and cereals) and angiosperms (hardwoods), whereas the hetero-1,4-β-*D*-mannans are more abundant in gymnosperms (softwoods) (Dekker, 1983).

Celluloses and xylans, the two most abundant renewable polysaccharides in nature, continuously accumulate in large quantities as waste products from the forestry, pulp and paper, agricultural, and food processing industries (Biely, 1985; Wood, 1985). Most of the biotechnological research concerned with the utilization of these waste materials has dealt with the cellulose portion for the production of liquid fuel and other cehmical products (Saddler and Brownell, 1982; Biely, 1985). The utilization of the xylan content of lignocellulosics has equal biological potential, and the development of new technologies that utilize xylans are important in several ways, including: the reduction of waste biomass, the enhancement of economic competitiveness of bioconversion technologies through the use of these abundant, inexpensive and presently under-utilized starting materials, and the development of new products from these renewable resources (Saddler and Brownell, 1982; Dekker, 1983; Yu *et al.*, 1987).

In the pulp and paper industry, xylanases can be used for the removal of contaminating hemicellulose components from high-grade cellulose pulps, for the manufacturing of dissolving pulp (Dekker, 1983; Senior *et al.*, 1988), for the processing of plant fibre sources such as flax and hemp (Biely, 1985), for the production of construction materials and as potential bleaching agents (Kluepfel *et al.*, 1990). Xylanases can also be used in the manufacture of liquid coffee, for the adjustment of wine characteristics, for the clarification of juices and for liquifying fruits and vegetables (Biely, 1985; Lewis and Paice, 1989). The hydrolysed products of xylan, namely xylose and xylo-oligosaccharides, can be used for the production of such fermentation products as ethanol, butanol and acetone (Roy, Roy and Duke, 1984; Yu and Saddler, 1985), as a supplement for microbial feedstocks or for the production of xylitol sweeteners (Abdel-Naby, 1993).

The structural rigidity of lignocelluloses necessitates that they be pretreated before being used in specific processes. Pretreatment methods include steam explosion, ball milling, swelling (with concentrated acidic or basic solutions) or enzymatic treatment (using brown- or white-rot fungi) to enhance their susceptibility to hydrolysis (Saddler and Brownell, 1982). Such preprocessing helps to increase the accessibility of cellulose and xylan by loosening these components from the binding constraints of lignin, disrupting crystalline and cell wall structures, and enhancing substrate solubility (Saddler and Brownell, 1982).

Xylans can be converted to monosaccharides by acid or enzymatic hydrolysis. However, like most acid hydrolysis reactions, the formation of toxic compounds is problematic and can hinder further microbial fermentation processes (Biely, 1985). Xylan hydrolysis by enzymatic means would allow for more efficient and controlled substraate use and product formation. Due to the level of complexity and variability that exists among xylans from different sources, their complete hydrolysis requires the synergistic action of a multitude of enzymes (Asther and Meunier, 1993; Coughlan and Hazlewood, 1993b). The enzymes involved in the hydrolysis of xylan are classified as either esterases, which rupture the ester linkages found throughout the xylan polymer, or hydrolases, which hydrolyse the glycosidic bonds of xylan (Lewis and Paice, 1989; Coughlan and Hazlewood, 1993a). The hydrolase class of enzymes includes: endoxylanases (EC 3.2.1.8), which randomly dismember the xylan backbone into shorter xylo-oligosaccharides; β-xylosidases (EC 3.2.1.37), which cleave the xylo-oligosaccharides in an exo-manner, producing xylose; and α-arabinofuranosidases (EC 3.2.1.55) and α-glucuronidases (EC 3.2.1.1), which remove the arabinose and 4-O-methylglucuronic acid substituents, respectively, from the xylan backbone (Biely, 1985; Dekker, 1983). As the result of the multifarious nature of xylans and the enzymes involved in its degradation, the analysis and characterization of xylan-degrading enzymes has proceeded slowly, compared with the cellulases, both with respect to obtaining suitable substrates for the analysis of enzyme activity, and with respect to the purification and characterization of the individual enzyme components. However, an excellent review of the subject is newly available (Coughlan and Hazlewood, 1993a).

Most biotechnological approaches to xylanolytic enzyme use have focused on the xylanases and β-xylosidases, the primary enzymes involved in xylan hydrolysis, with relatively little information available on the xylan-debranching enzymes (Christov and Prior, 1993). The role of β-xylosidase is analogous to that of β-glucosidase in the breakdown of cellulose by cellulases (Dekker, 1983). Both xylanase and β-xylosidase production are inducible when micro-organisms are grown on substrates such as xylans, or methyl β-*D*-xylopyranosides, respectively (Dekker, 1983). While xylan-degrading enzymes are produced in many organisms, those of technological interest arise from fungal sources, such as *Aspergillus* (Shimizu and Ishihara, 1987; Stålbrand *et al.*, 1992; Abdel-Naby, 1993), *Trichoderma* (Saddler and Brownell, 1982; Shimizu and Ishihara, 1987; Royer and Nakas, 1989), *Zymononas* (Saddler and Brownell, 1982), *Streptomyces* (Kleupfel *et al.*, 1990), *Thermoascus* (Senior, Mayers and Saddler, 1989) and *Aureobasidium* (Leathers, 1989) or bacterial sources such as *Cryptococcus* (Stålbrand *et al.*, 1992), *Streptomyces* (Kleupfel *et al.*, 1990) and *Bacillus* (Senior, Mayers and Saddler, 1989). Their high yields and relative ease of production make these systems ideal for commercial exploitation.

In spite of this, the costs associated with enzyme production for bioconversion processes are a major expenditure. Economics demand that optimum conditions be determined for the production and use of enzymes (Senior, Mayers and Saddler, 1989). The use of immobilized xylanolytic enzymes could improve enzyme productivity, due to the ease of recovery and reuse. To date, however, there have been relatively few attempts to immobilize xylanolytic enzymes and, of these, most have been limited to xylanases and β-xylosidases.

Shimizu and Ishihara (1987) coupled the hemicellulases of *T. reesei* and *A. niger* to a variety of supports, including porous silica, alumina and titania, with $TiCl_4$ (Titanium tetrachloride) and glutaraldehyde as the binding agents, showing enzyme loadings in the range 10–50 mg enzyme g^{-1} carrier. Using the hemicellulose-derived substrates xylan, aryl-B-xyloside, 4-O-methylglucuronoxylan and *p*-nitrophenyl-β-*D*-xylopyranoside, the immobilized activities were found to be diffusion limited, showing only 3–53% of the activity of the free native enzymes against the same substrates (Shimizu and Ishihara, 1987). However, the enzyme preparations from both sources remained stable upon immobilization, maintaining their temperature optima and showing stable rates of xylan conversions of 36–64% over 48 h.

Roy, Roy and Duke (1984) successfully immobilized both the xylanase and β-xylosidase enzymes from *Macrophomina phaseolina* by entrapment within acrylamide polymers, maintaining full activity against xylan and *o*-nitrophenol-β-*D*-xylopyranoside, respectively, after 25–27 reuses at 38°C. Xylanase showed greater temperature stability following immobilization, maintaining activity at rates as high as that of the free enzyme after 25 days at both 38° and 50°C, whereas the gel-bound β-xylosidase showed substantial reductions in activity after 4 days at 50°C. K_m values increased following immobilization by 1.8- and 1.6-fold for β-xylosidase and xylanase activities, respectively; these changes were apparently influenced by diffusion limitations, for when gel-bound β-xylosidase was ground into smaller segments using a mortar and pestle, the K_m increase was only 1.3-fold (Roy, Roy and Duke, 1984).

Oguntimein and Reilly (1980) immobilized *A. niger* β-xylosidase on to 10 different support carriers in the hope of obtaining a stable enzyme preparation; $TiCl_4$-linked alumina proved to be the best support, binding 70% of added enzyme activity. Abdel-Naby (1993) conducted a thorough study involving the immobilization of *A. niger* xylanase and β-xylosidase on to an assortment of carriers, including tannin-chitosan, Dowex-50W, glutaraldehyde-cross-linked chitosan and polyacrylamide gel beads. Both enzymes maintained the greatest stability and activity when immobilized by cross-linking or entrapment. The absorbed and ionically bound enzymes showed a decrease in thermal stability, whereas the opposite was true for the covalently bound and entrapped enzymes. Polyacrylamide retained the highest amount of added xylanase (15–25%) and β-xylosidase (16–33%) activities, and maintained continuous (albeit low) rates of larchwood xylan and *p*-nitrophenyl-β-*D*-xylopyranoside hydrolysis for over 40 days in packed bed reactors. Similarly, xylanase covalently bound to chitosan lost only 19% of its initial rate after continuous operation for 40 days, whereas the enzymes immobilized by adsorption and ionic forces lost 82% and 67%, respectively. In general, when immobilized by these methods, both xylanase and β-xylosidase showed decreases in specific activities to only 15–24% and 13–32% of the values for the corresponding soluble enzymes, respectively, as well as corresponding 23–42% and 7.5–36% increases in K_m values for substrates, respectively.

Such deviations from the free enzyme values were attributed to structural changes resulting from immobilization, as well as problems with diffusion resistance and steric hindrance (Abdel-Naby, 1993).

Stålbrand *et al.* (1992) exploited the occurrence of spontaneous mycelial pellet formation in three strains of *Aspergillus* as a possible source of immobilized β-xylosidase. Such self-immobilized enzyme systems can occur in shake flask cultures after inoculation with conidia. In general, the mycelial pellets of *A. phoenicis* had the highest pellet-bound β-xylosidase activities (280 U g^{-1} dry weight of mycelia) and maintained stability after storage at 50°C for 333 h. However, the mycelial pellets of *A. phoenicis* QM 329 and ATCC 13157 suffered from enzyme leakage, and the mycelial pellets of *A. terreus* QM 1991 were limited by both leakage and instability.

Clearly, much more research still needs to be done in the area of immobilized xylanolytic enzymes in order to improve the competitive edge of potential commercial bioconversion processes. Several areas need to be explored. For example, the co-immobilization of xylanase and cellulase enzymes could allow the efficient use of both the xylan and cellulose components of substrates (Yu and Saddler, 1985). The co-immobilization of several xylanase enzymes would also allow for a more complete degradation of the xylan. Wong, Tan and Saddler (1986) found that by using a mixture of the three soluble xylanases from *T. harzianum*, the overall hydrolysis of aspenwood xylan was increased by 30%. Furthermore, the co-immobilization of enzymes which liberate the side-chain sugars from the xylan backbone, such as α-*L*-arabino-furanoside and α-glucuronidase, or the enzymes involved in the degradation of acetyl xylans, would create new substrate sites for further xylanase activity (Biely, 1985). The results of such work would allow for the direct conversion of xylan to useful products by simple and inexpensive methods. Such steps are necessary, for a critical limitation in the commercialization of xylanolytic systems is the incomplete utilization of xylans (Biely, 1985).

PECTINASES

Pectins are a class of complex acidic polysaccharides of high molecular weight (25–360 kDa) that are present in varying amounts in higher plants. Their presence affects the consistency, turbidity and appearance of fruit juices (Fogarty and Kelly, 1983; Ward, 1985; Borrego *et al.*, 1989). Pectins consist of α-1,4-*D*-galacturonic acid polymers of varying methyl ester content, which are often complexed with *L*-araban and β-1,4-*D*-galactan polymers (Dorfman, 1955; Ward, 1985). The methoxy content of pectin varies among plant sources and changes during the development and ripening stages of fruits and vegetables. The degree of esterification of the pectins of fruits ranges from 90–95% for apples to 60–70% for oranges and only 30–35% for grapes (Ward, 1985). The two major groups of pectolytic enzymes include the pectin esterases (PE) (pectinhydrolase, EC 3.1.1.11), which hydrolyse the methyl groups that are esterified to galacturonic acid residues, and the polygalacturonases (PG) (poly[1,4-α-*D*-galacuronide]glycanohydrolase, EC 3.2.1.15), which are depolymerizing enzymes capable of hydrolysing α-1,4-galacturonosyl linkages (Dorfman, 1955; Ward, 1985; Romero *et al.*, 1989). Although PE is not a polysaccharide-degrading enzyme, PG function is dependent on the action of PE, since PG can only hydrolyse pectic substrates with a 60% or lower degree of esterification (Borrego *et al.*, 1989). Another important group is the pectic lyases (EC 4.2.2.2) such as endopolymethylgalacturonate lyase (PMGL), which cleaves the α-1,4-galacturonosyl linkages by elimination of (Δ)-4,5-*D*-galacturonate residues at the non-reducing side of the bond being split (Ward, 1985). Enzymes of this class are not hydrolytic, but act in concert with the other pectinases for the complete breakdown of pectins. Pectolytic enzymes originate from a range of plant and microbial species. Pectin esterases are commonly found in tomato (Borrego *et al.*, 1989; Romero *et al.*, 1989), *Fusarium oxysporum* and *Clostridium multifermentans*, whereas the polygalacturonases have been isolated from a variety of fungi (*Aspergillus, Fusarium, Rhizopus*) (Romero *et al.*, 1989) and bacteria (*Erwinia aroideae, Bacillus*) (Ward, 1985). Pectic lyases also originate from fungal (*Fusarium, Aspergillus, Penicillium*) and bacterial (*Bacillus, Erwinia, Clostridium, Streptomyces*) sources (Ward, 1985).

Pectin, which can be present in either soluble or insoluble (protopectin) forms, causes the cloudy and highly viscous properties of fruit and vegetable extracts, and these undesirable properties can be reduced or dissipated by treatments with pectic enzymes (Ward, 1985). Indeed, pectin degradation by pectic enzymes is a common industrial practice for improving the quality and yield of fruit and vegetable juices during processing (Ward, 1985; Borrego *et al.*, 1989). Pectinases have similar uses in the treatment of coarse fruit pulp and peel in the preparation of natural cloudifiers for use in the beverage industry, as well as in the preparation of citrus oils, the production of baby foods from macerated fruits and vegetables, and in wine production for increased yields and juice clarification (Ward, 1985).

The use of immobilized pectinolytic enzymes for continuous use in such applications would be beneficial were it not for the highly viscous substrates, which could effectively reduce the catalytic ability of an immobilized enzyme (Romero *et al.*, 1989). Thus, a successful and productive immobilized pectic enzyme reactor must be able to combat the hydrodynamic limitations of the substrate and remain stably bound to allow for high rates of conversion over long periods of operation (Lozano *et al.*, 1990).

Lozano *et al.* (1990) immobilized a commercial pectolytic enzyme preparation (Pectinol D) on to nylon membrane for use in a cross-flow packed bed reactor for the hydrolysis of 0.5% citrus pectin at pH 4 and 40°C. The cross-flow reactor design was superior to a packed bed reactor model, showing higher rates of pectin hydrolysis and higher recycling flow rates. This, in turn, effectively reduced the external film diffusion limitations on the system due to the viscous substrate. However, the system was compromised because the high flow rates resulted in enzyme deactivation due to shearing stress.

Tomato pectin esterase (PE) covalently bound to porous glass, retained only 0.32% of added activity and suffered from diffusion limitations when using citrus pectin as the substrate (Weibel *et al.*, 1975). Severe internal mass-transfer restrictions were highlighted by the twelve-fold increase in PE activity that occurred after the glass-bound enzyme was ground into smaller fragments. Citrus pectin was separated into low and high molecular weight fractions and used for kinetic studies; the kinetic behaviour of soluble PE was unaffected by the molecular weight of the substrate but the immobilized enzyme showed a five-fold increase in K_m for the high molecular weight fraction, as compared with the low molecular weight fraction (Weibel *et al.*, 1975). Immobilized PE also showed a reduction in thermal and storage stability, compared with the native enzyme, with a half-life of only 2 weeks at 25°C (Weibel *et al.*, 1975).

Borrego *et al.* (1989) immobilized tomato pectin esterase on to glyceryl-coated controlled-pore glass by four different coupling methods, in order to determine the most efficient enzyme system for the hydrolysis of citrus pectin. Although pectin esterase coupled to the thiol-derivatized glass showed the highest activity yield of 3.2%, this low value, together with the very low specific activity obtained, suggested that interactions between the enzyme and the high molecular weight substrates were severely limited by steric restrictions. Pectinesterase bound to 20 nm pore glass showed the highest rate of demethylating activity, whereas the use of 46 nm pore sized glass resulted in a decrease in activity. This was attributed to internal diffusion limitations of the pectin (MW 290 kDa) to pectin esterase immobilized within the 46 nm glass pores, whereas the 20 nm pored sized glass was inaccessible to both enzyme and substrate. A slightly higher catalytic efficiency, together with a 1.9-fold drop in K_m for citrus pectin following enzyme immobilization on to the 20 nm pore sized glass, implied that a positive conformational change had occurred. Indeed, fluorescence spectra revealed that the immobilized enzyme had undergone a partial unfolding of its tertiary structure. The immobilized enzyme remained fully stable in a continuous stirred tank reactor (0.5% pectin, pH 7, 50 nM NaCl) for 10 h, after which a steady deterioration of the support and enzyme activity occurred as the result of continuous stirring (Borrego *et al.*, 1989).

In an extension of these studies, Romero *et al.* (1989) immobilized endo-*D*-polygalacturonase from *Rhizopus* spp. on to the same porous glass support modified by the four different coupling methods. Enzyme activity, which was measured as a decrease in the viscosity of a commercial methyoxylated pectin solution at 40°C, pH 5 and 50 mM NaCl, was highest when immobilization was on to aminoaryl-derivatized glass. That greater activity was achieved against low MW pectin (MW 34 kDa) over high MW citrus pectin (MW 280 kDa) also implied limitation by steric hindrances. As seen with immobilized pectin esterase, the highest specific activity was achieved at a 20 nm pore size, and the decrease in activity seen at 46 nm pore size was the result of a higher degree of diffusion limitation (Romero *et al.*, 1989).

Van Houdenhoven, de Wit and Visser (1974) demonstrated that the continuous dydrolysis of 1% sodium pectate by a Sepharose 4B column containing immobilized *Saccharomyces fragilis* polygalacturonase was suitable for the preparation of pectic oligomers. The production of a series of such oligomers, which could be useful for the analysis of the kinetic properties of pectin-degrading enzymes, was found to vary with reaction pH. In addition, the degree of substrate degradation, as determined by thin layer chromatography, was dependent on the methyl ester and foreign sugar content and the largest number of oligomers was produced at the slowest elution flow rate (9.5 ml h^{-1}).

Aspergillus niger endo-*D*-galacturonase, covalently bound to continuous flow columns of methacrylate gel by a series of different spacer groups, showed a decrease in enzyme activity against sodium pectate in all instances, compared with the free enzyme (Rexová-Benková and Mracková-Dobrotová, 1981). Although the K_m values for sodium pectate were not affected by either immobilization or by the length of the spacer groups, the relative enzyme activity increased with increasing spacer group length, due to a progressively decreased effect of steric hindrance and diffusion limitation. Analysis of the products of immobilized and soluble enzyme hydrolysis by gel permeation chromatography showed that the randomness of the degradation by the bound enzyme was reduced because of the restriction of enzyme action to the peripheral areas of the substrate. Furthermore, analysis of immobilized enzyme activity against oligo-*D*-galactosiduronic acids of various sizes indicated that, at a common spacer length, activity was inversely proportional to substrate size, due to the limited access of bulkier molecules to the immobilized enzyme.

Endopolymethylgalacturonate lyase (PMGL), immobilized on to titanium-linked DEAE-cellulose and titanium-linked porous glass, retained only 0.46% and 9.2% of added activity, respectively (Hanisch, Rickad and Nyo, 1978). Most notable was the change in pH optimum from 5.3 for free PMGL to pH 3.7–4.7 when bound to DEAE-cellulose, and to pH 6.2–6.9 when bound to porous glass; these are typical effects for binding to polycationic versus polyanionic supports, respectively. The charged supports also influenced the catalytic activity of the bound enzyme. The attractive forces between pectin, which is negatively charged, and the polycationic DEAE-cellulose resulted in a 43% decrease in K_m whereas the repulsive forces between the pectin and the polyanionic glass support were manifest in a doubling of K_m.

The immobilization of other polysaccharide-degrading enzymes

DEXTRANASE

Dextran is a class of α-*D*-glucose branched polymers, in which the majority (95%) of the residues are linked by α-1,6-glucosidic bonds, but also some (5%) α-1,3-glucosidic linkages (Ward, 1985). Produced in abundant quantities by infection with *Leuconostoc mesenteroides* or related species in mechanically or weather-damaged sugar cane and sugar beet, dextrans can increase the viscosity of juices and thus cause problems during the clarification and evaporation of juices and in the concentration of molasses (Margaritis and Pace, 1985; Ward, 1985). However, these problems can be overcome by treating contaminated juices with dextranases for 15 min at 50–55°C, between the extraction and clarification stages of sugar production (Ward, 1985). Apart from their use in the sugar industry, other successful applications of dextranases include the production of controlled pore molecular sieves composed of dextran, and use as a toothpaste additive to help remove dental plaque (Margaritis and Pace, 1985; Wiseman, 1985).

Dextranses (1,6-α-glucan-6-glucanohydrolyases, EC 3.2.1.11) are most active in the pH range 5–7 and at temperatures up to 60°C (Ward, 1985; Wiseman, 1985). The major end products of dextran hydrolysis by dextranases are isomaltose and isomaltotriose (Wiseman, 1985). Dextranases that function in an endo-fashion are produced from strains of *Penicillium* and *Flavobacterium*, as well as *Klebsiella aerogenes, Lipomyces starkeyi* and *Fusarium monoliformi* (Smiley, Boundy and Hensley, 1982; Prabhu and Prabhu, 1985; Ward, 1985; Koenig and Day, 1989), whereas exo-dextranases are produced by *Bacillus coagulans* and *Arthrobacter globiformis* (Wiseman, 1985). Immobilization methods for dextranases have not been well documented but a few recent papers on the subject exist.

Prabhu and Prabhu (1985) covalently bound 50% of added dextranase activity from *P. aculeatum* on to glutaraldehyde-treated bentonite for the production of a stable clarifying agent for sugar cane juices. When immobilized, the enzyme optimum pH range for activity narrowed, the K_m for dextran increased by 1.3-fold, and heat stability was reduced. However, in the presence of 20% sucrose, immobilized dextranase showed enhanced thermal stability, maintained high operational activity following six cycles of repeated use, and showed a 60% hydrolysis of dextran at 50°C in 1 h (Prabhu and Prabhu, 1985).

Smiley, Boundy and Hensley (1982) showed that the action pattersn of endo-dextranases from *P. funiculosam* and *P. lilacinum* are altered when covalently bound to silanized and glutaraldehyde-treated silica beads. The active immobilized enzyme produced a substantial amount of reducing sugars but its action had little effect on the viscosity of a high molecular weight dextran (2000 kDa) at pH 5.1 and 40°C. By contrast, soluble dextranase only slightly increased the level of reducing sugars but produced a rapid decrease in substrate viscosity. However, the hydrolysis of low molecular weight dextrans (150 kDa) resulted in similar hydrolysis products for bound and free enzyme. The authors suggested that immobilized dextranase can hydrolyse relatively few internal bonds of the high molecular weight dextrans, and, as a result, the bound dextranase converts from an endo- to an exo-type of action and produces high levels of reducing sugar (Smiley, Boundy and Hensley, 1982).

INULINASE

The use of fructose as an alternative sweetener to sucrose for the food and beverage industry has gained in popularity in recent years due to its more beneficial medicinal and nutritive properties, as well as its high sweetness level (Gupta *et al.*, 1990). Apart from sucrose and starch, there are several other potential commercial sources of fructose, namely, the β-*D*-fructofuranose polymers. These fructose polymers occur in two forms, both containing a terminal sucrose residue (Goodwin and Mercer, 1983). Fructose units linked by β-β-(2,1) glycosidic bonds belong to the inulin class, whereas those linked via β-(2,6) bonds are classified as phlean (Goodwin and Mercer, 1983). Inulin is the most common fructose storage polysaccharide in nature. It is a sizeable compound consisting of 30–35 straight-chain fructose units with a molecular weight of approximately 5000 (Goodwin and Mercer, 1983; Bajpai and Margaritis, 1985). Inulins are found primarily in the Compositae family (e.g. dahlia, Jerusalem artichoke, chickory) (Goodwin and Mercer, 1983; Uhm *et al.*, 1987) and are inexpensive, high-yielding sources of fructose that are generating considerable commercial interest as a carbohydrate source for the production of foodstuffs for humans and livestock, as well as for many other industrial processes (Kosaric *et al.*, 1989). In particular, the Jerusalem artichoke tuber is approximately 80% inulin by weight (Kim and Rhee, 1989) and this inulin source has stimulated considerable experimental interest.

Inulinases (2,1-β-*D*-fructan fructanohydrolase, EC 3.2.1.7) occur in a variety of plants, particularly species with inulin-containing roots and tubers, as well as in various yeasts (*Kluyveromyces, Debaryomyces, Candida*) (Bagsvaerd, 1981; Guiraud, Demeulle and Galzy, 1981; Kim, Byun and Uhm, 1982; Bajpai and Margaritis, 1985, 1987) and fungi (*Aspergillus, Penicillium, Fusarium*) (Bagsvaerd, 1981; Kim and Rhee, 1989; Gupta *et al.*, 1990). Three functional types are known: exo-inulinases, the most common, hydrolyse the terminal β-2,1- and β-2,6-fructofuranoside linkages of inulin; endo-fructan hydrolases hydrolyse inulin to fructo-oligomers, and fructotransferases split difructose anydride off the inulin chain to produce a difructose dianhydride (Beck and Praznik, 1986).

Conventional methods of industrial high-fructose syrup production involve a multistep process employing several amylolytic enzymes as well as glucose isomerase, for the net conversion of starch to syrups typically containing 42–55% fructose (Kosaric *et al.*, 1989). By contrast, the direct production of fructose from inulin by inulinase is a single-step reaction that can result in yields of up to 95% fructose (Kim and Rhee, 1989). The potential commercial benefit of an inulin–inulinase system for fructose production is obvious. The inulin-based substrates, simple straight-chain polymers of fructose, are ideally suited for hydrolysis. To date, however, inulinases have been considered too thermally unstable for industrial use, where high operational temperatures are required to prevent microbial contamination (Bagsvaerd, 1981). Inulinases typically have temperature optima of 45–55°C and they show a rapid loss of activity at 60°C. In an effort to overcome this problem, Bagsvaerd (1981) discovered strains of *Aspergillus* that produce more thermally stable inulinases; these maintained 95% of activity after 20 min reactions at 60–65°C, and showed activity optima at 60°C (Kosaric *et al.*, 1989). Another alternative for overcoming the problem of low thermostability is enzyme immobilization, since immobilization is cost-effective and typically results in improved operational parameters. However, despite the potential advantages of immobilized inulinases, there are only a few in-depth studies on this subject.

DEAE-cellulose retained as much as 75% of added *Debaryomyces phaffii* inulinase activity, and the immobilized enzyme showed an increase in temperature optimum and thermal stability, and an operational half-life that was five times longer than that of the free enzyme at 50°C (Guiraud, Demeulle and Galzy, 1981). However, the immobilized enzyme could not be used to its fullest potential, for the optimum pH for the immobilization step (pH 5) and the optimum operational pH (pH 2.5) did not coincide; this resulted in either low activity yields or enzyme desorption. Kim and Rhee (1989) immobilized 23% of added *A. ficuum* inulinase on to glutaraldehyde-cross-linked chitin, and produced fructose from Jerusalem artichoke tubers by continuous and batch reactor processes. The immobilized enzyme demonstrated high operational stability and a storage half-life of 560 h in 10% inulin, as well as greater thermostability than the free enzyme. Kim, Byun and Uhm (1982) found that 53% of added *K. fragilis* inulinase activity was retained on aminoethylcellulose-linked glutaraldehyde, and showed that inulin had protective effects in enhancing thermal and operational stabilities. The immobilized enzyme activity was maintained for 4 h at 55°C in the presence of inulin, whereas a rapid loss of activity occurred at temperatures greater than 50°C in the absence of substrate. Also, the bound enzyme showed operational half-lives of 17 and 14 days in the presence of a Jerusalem artichoke tuber extract or a 7% (w/v) inulin solution, respectively. Using metal-link chelation, Gupta (1994) immobilized 40% of purified *K. fragilis* inulinase on cellulose; the half-life of the immobilized enzyme was 5 days at 25°C and the optimal temperature for enzyme function (55°C) was not changed by immobilization.

Fusarium oxysporum inulinase immobilized by entrapment in polyacrylamide gel retained 45% of added activity, showed an enhanced temperature optimum (45°C) compared to the free enzyme (37°C), and a storage half life of approximately 5 days at 25°C (Gupta *et al.*, 1990). Ninety per cent of immobilized enzyme was maintained after storage for 24 h at 45°C in the presence of 1% chickory root inulin, but without substrate all activity was lost. The immobilization of *Kluyveromyces marxianus* yeast cells by entrapment in gelatin also gave a very stable inulinase preparation which showed only a 2% loss of activity after continuous operation for 240 h and maintained full activity after 10 repeated uses in 7 h batch cycles at 50°C (Bajpai and Margaritis, 1985). Bajpai and Margaritis (1987) also immobilized 72% of added inulinase activity on to titanium chloride activated molecular sieves, resulting in an enzyme preparation that could function optimally at a higher temperature, was more thermostable than the free enzyme, and showed storage and operational half-lives of 83 and 30 days, respectively.

Inulinase immobilized on to glutaraldehyde-cross-linked chitin showed high conversion yields of 90 and 55% after 10 h in batch reactors and continuous packed bed column reactors, respectively, with productivities as high as 61 g l^{-1} h^{-1} (Kim and Rhee, 1989). The continuous reactor operation of gelatin-entrapped yeast cells demonstrated maximum productivities of 100 g l^{-1} h^{-1} and inulin conversions of 80% (Bajpai and Margaritis, 1985). High-fructose syrups were also produced in batch reactors with the gelatin-entrapped cells, resulting in 93% and 90% hydrolysis of Jerusalem artichoke tuber extracts or commercial inulin substrates, respectively (Bajpai and Margaritis, 1985). Inulinase adsorbed on to DEAE-cellulose maintained a constant K_m value, and the amount of inulin hydrolysis could be increased by decreasing substrate concentration and flow rate (Guiraud, Demeulle and Galzy, 1981). Total hydrolysis of the substrate was obtained using an inulin concentration of 5 g l^{-1} and a flow rate of 48 ml h^{-1}.

Immobilization of *K. marxianus* inulinase on to molecular sieves via titanium chloride activation, resulted in a two-fold increase in the K_m value for chickory root inulin (Bajpai and Margaritis, 1987). Aminoethylcellulose-immobilized inulinase showed both batch and column reactor conversions of 90% for Jerusalem artichoke extract, and high productivities of 102 mmol l^{-1} h^{-1} at 40°C and pH 5 (Kim, Byun and Uhm, 1982). An 8% increase in K_m for inulin upon immobilization was attributed to modest diffusion limitation due to an external film layer, which could be removed by increasing the reactor flow rate.

The immobilization of mono- and disaccharide-utilizing enzymes

GLUCOSE ISOMERASE

Glucose isomerase or xylose isomerase (EC 5.3.1.5) catalyses the reversible isomerization of xylose to xylulose and, of greater commercial importance, the isomerization of glucose to fructose for the production of high-fructose syrups (HFS or HFCS when corn is the substrate) (Bucke, 1981; Antrim and Auterinen, 1985). Glucose isomerase is an intracellular enzyme produced by microbial organisms from a range of genera, principally *Streptomyces, Bacillus, Pseudomonas, Lactobacillus* and *Anthrobacter* (Bucke, 1981; Al-Tai, Ali and Abdul-Razzak, 1987).

Before 1970, the principle source of food sweeteners was sugar cane and sugar beets (Verhoff *et al.*, 1985). However, with a 1.7-fold greater sweetening ability, HFCS was found to be a good alternative to sucrose for use as a food sweetener in the soft drink and candy industries (Chen, 1980; Bucke, 1981). Industrial processes treat corn starch first with amylases to produce glucose syrups and then use glucose isomerase to produce syrup with a significant fructose content (42% at equilibrium) (Verhoff *et al*., 1985). Soluble glucose isomerases have been used by corn-syrup producers for decades (Jensen and Rugh, 1987). Unfortunately, the cost of the isomerization step was huge due to the expense of using fresh glucose isomerase each time (Verhoff *et al*., 1985). Alkaline isomerization is an alternative, but produces excessive colour and undesirable side-products (Cheetham, 1985). Thus, to make this process economically feasible, glucose isomerase is now used in an immobilized form (Jensen and Rugh, 1987).

The commercial processes for glucose isomerase immobilization are of two types, cell-free enzyme immobilization and whole-cell immobilization (Jensen and Rugh, 1987). In 1969, Takasaki patented the first whole-cell immobilized glucose isomerase. This batch isomerization process, which used heat-fixed *Streptomyces wedmorensis* cells containing glucose isomerase, produced a 48% fructose product (70°C, pH 7.5, 10 mM Mg^{2+}) in 24 h with an operational half-life of approxiamtely 170 h (Jensen and Rugh, 1987). Other methods of micro-organism immobilization include flocculation, entrapment, adsorption and cross-linking (Chen, 1980; Jensen and Rugh, 1987). Although immobilized cell systems do not require enzyme isolation or purification steps, they usually produce less stable glucose isomerase preparations and can suffer from product contamination by cell debris, as compared with methods that immobilize pure enzyme (Chen, 1980; Jensen and Rugh, 1987).

Critical to the expansion of the sweetener industry has been the considerable effort devoted to the development of reusable, pure enzyme-immobilized glucose isomerase systems. Indeed, the enzymatic conversion of glucose to fructose by immobilized glucose isomerase on an industrial scale was the first successful application of immobilized enzyme biotechnology (Chen, 1980). The first reusable, pure enzyme-immobilized glucose isomerase, patented in 1970 by the Clinton Corn Company, used adsorption on to DEAE-cellulose and gave 50% isomerization (pH 6.5, 60°C) and an activity half-life of approximately 200 h. However, these early systems were very susceptible to changes in process conditions (Jensen and Rugh, 1987). The recent literature on the immobilization of glucose isomerase is vast and cannot be mentioned fully here. However, good comprehensive reviews of the subject are provided by Jensen and Rugh (1987) and Chen (1980).

Immobilized glucose isomerase is available commercially. Glucose isomerase immobilized on to SiO_2 beads can be purchased under the trade name Optisweet 22 from Kali-Chemie and the enzyme bound to derivatized alumina particles can be obtained from UOP Inc. Gist Brocades and Miles Laboratories Inc. immobilize whole cells by entrapment in gelatin treated with glutaraldehyde (Maxazyme) or by flocculation of whole cells treated with glutaraldehyde (Taka sweet) (Gaikwad and Deshpande, 1992). The following discussion highlights only a fraction of the papers published on the subject of glucose isomerase immobilization. We have chosen to focus on some of the different methods of glucose isomerase immobilization as well as the various factors involved in the successful function of an immobilized glucose isomerase bioreactor, including the maintenance of high enzyme activity, stability and reusability.

Glucose isomerase has a characteristically high reaction temperature and thermal stability (Bucke, 1981) and immobilized glucose isomerase usually maintains the pH (operational range pH 7–9) and temperature stabilities (70–90°C) similar to the free enzyme (Y.Y. Lee *et al*., 1976; Huitron and Limon-Lason, 1978; Gaikwad and Deshpande, 1992; Schafhauser and Storey, 1992b; Deshmukh, Choudhury and Shankar, 1993). Antrim and Auterinen (1985) demonstrated the successful immobilization of glucose isomerase on to DEAE-cellulose, with industrial-sized columns showing a high immobilized activity of 1500 U g^{-1}, a productivity greater than 9 metric tons of 42% fructose syrup solids per kilogram of immobilized enzyme, and a half-life of 70 days. Glutaraldehyde-activated Indion 48-R retained 30–40% of added glucose isomerase activity and could convert 36% glucose syrups into 18.5% fructose in 5 h (Deshmukh, Choudhury and Shankar, 1993). Glutaraldehyde-cross-linked porous glass particles bound 56% of added glucose isomerase activity, with an enzyme loading of 48 mg g^{-1} glass and a half-life reaching 240 days at 50°C (Y.Y. Lee *et al*., 1976). A silanized and glutaraldehyde-treated silicon oxide based carrier bound as much as 30 units g^{-1} of support material without a decrease in activity (Sórensen and Emborg, 1989). Granular chicken bone bound up to 32% of added enzyme and binding was stable even after exposure to a wide variety of chemical and physical conditions, including treatments with detergents, organic solvents and high salt solutions (Schafhauser and Storey, 1992b). Enzyme activity was also stable after five cycles of operation but the operational half-life was only 168 h at 55–60°C. Glucose isomerase immobilized on to polyurethane foam retained 55% of added activity but the storage half-life was only 160 h at 55°C (Storey and Chakrabarti, 1990).

Increases in K_m values for glucose of 1.6- to 3-fold following immobilization are not unusual for glucose isomerase (Huitron and Limon-Lason, 1978; Volkin and Klibanov, 1989; Storey and Chakrabarti, 1990; Schafhauser and Storey, 1992b). Such changes could be due to partial inactivation due to immobilization in an unfavourable configuration, or to steric hindrance from overcrowding of the immobilized enzyme (Kennedy, 1985). However, effects of immobilization on glucose isomerase kinetic properties are fairly inconsequential compared with the effects of immobilization on many other enzyme systems.

Immobilized glucose isomerase is rarely affected by pore diffusion limitation or external mass-transfer resistance problems, as both the substrate and product molecules are small (Bucke, 1981). Experiments that typically identify diffusion-limitation problems, such as comparisons of the Arrhenius and Lineweaver–Burk plots for free versus immobilized enzymes, or measurement of reaction rates for the enzyme bound to supports of varying pore or particle sizes, have concluded that no appreciable diffusion limitation exists (Sórensen and Emborg, 1989; Storey and Chakrabarti, 1990; Schafhauser and Storey, 1992b). Similarly, experiments conducted to assess the presence of external mass-transfer limitations, such as increasing the agitation speed of the bioreactor, had minimal effect on the apparent kinetic properties of the immobilized glucose isomerase and, therefore, showed no restrictions on the catalytic efficiency of the immobilized enzyme (Huitron and Limon-Lason, 1978; Schafhauser and Storey, 1992b). However, Y.Y. Lee *et al*. (1987), in comparing the reaction rates of immobilized glucose isomerase at changing flow rates in a recirculated packed bed reactor, observed significant resistance to substrate diffusion through the external liquid film when the flow rate was below a certain velocity. Upon comparing the reaction rates of two support sizes (310 v. 20 μm diameter), the authors showed that the reaction rate differences above 60°C were the result of a suppressed reaction rate caused by diffusion resistance in the micropores of the enzyme support.

Enzyme stability is one of the principal factors affecting reactor productivity for free or immobilized glucose isomerase (Chen and Wu, 1987). The primary causes for activity loss are poor binding stability, thermal inactivation, impurities, and the generation of chemical inactivators at high temperature and at pH extremes (Y.Y. Lee *et al.*, 1976; Huitron and Limon-Lason, 1978; Klibanov, 1979; Chen, 1980; Bucke, 1981; Antrim and Auterinen, 1985; Chen and Wu, 1987; Jensen and Rugh, 1987; Volkin and Klibanov, 1989; Schafhauser and Storey, 1992b). Thus, few immobilized enzyme systems have the properties necessary for economically feasible, commercial production of HFCS (Verhoff *et al.*, 1985; Gaikwad and Deshpande, 1992). But due to its commercial importance, extensive research has focused on new methods of immobilization that could result in a more stable glucose isomerase, for even a modest increase in the half-life of the enzyme would substantially reduce the production cost of HFCS (Klibanov, 1979; Volkin and Klibanov, 1989).

Poor binding stability is a common problem in immobilized enzyme technology. Immobilization by absorption tends to be weaker, although this is not always the case. Glucose isomerase immobilization on to an anion-exchange resin showed a 35% loss of enzyme activity after seven cycles of reuse (Gaikwad and Deshpande, 1992) whereas glucose isomerase bound to glutaraldehyde-activated Indion 48-R lost approximately 40% of initial bound activity after five cycles of use (Deshmukh, Choudhury and Shankar, 1993). Immobilization by entrapment in polyacrylamide (Stranberg and Smiley, 1971) also resulted in very unstable preparations. Chitin cross-linked with glutaraldehyde retained 40% of added glucose isomerase activity, but showed a decreasing half-life with increasing flow rates at 60°C (7 days at 7.8 ml h^{-1} and 16.5 days at 5 ml h^{-1}) (Stanley *et al.*, 1976). These supports, however, can be regenerated and reused (Antrim and Auterinen, 1985; Sórensen and Emborg, 1989).

Glucose isomerase is a metalloenzyme, containing distinct metal-binding sites which require a combination of cobalt, magnesium and manganese for maximum activity (Verhoff *et al.*, 1985; Marg and Clark, 1990). Glucose isomerase immobilized on to porous glass particles showed 39% and 40% increases in activity in the presence of 10 mM Mg^{2+} and 0.1 mM Co^{2+}, respectively (Y.Y. Lee *et al.*, 1976). Co^{2+} also stabilized glucose isomerase against heat denaturation and increased the enzyme half-life ten-fold. However, Co^{2+} is very toxic and must be removed when the reaction is complete in order to meet government standards for food products (Verhoff *et al.*, 1985; Jensen and Rugh, 1987). This problem can be alleviated by adding 1.5–50 mM Mg^{2+}, which allows full expression of free and immobilized glucose isomerase activity in the absence of cobalt (Huitron and Limon-Lason, 1978; Antrim and Auterinen, 1985; Chen and Wu, 1987; Sórensen and Emborg, 1989; Volkin and Klibanov, 1989; Storey and Chakrabarti, 1990; Schafhauser and Storey, 1992b). The addition of bisulphate in the range of 0.1–2 mM also helps to maintain enzyme stability and prevent oxidation (Antrim and Auterinen, 1985; Sórensen and Emborg, 1989; Storey and Chakrabarti, 1990; Schafhauser and Storey, 1992b).

The degree of stability given to glucose isomerase by the presence of ligands (glucose, fructose, HFCS) has also been extensively studied. Substrate protection of an enzyme may be due to the accompanying stability of the tertiary protein structure when the enzyme combines with the substrate to yield a complex form (Chen and Wu, 1987). A knowledge of enzyme protection by a substrate may be used in selecting the operating conditions of enzyme reactors, optimizing reactor performance and enzyme replacement policy (Y.Y. Lee *et al.*, 1976; Chen and Wu, 1987). Indeed Y.Y. Lee *et al.* (1976) observed an eight-fold increase in the half-life for glucose isomerase immobilized on to porous glass beads in the presence of 1.5 M fructose. Substrates or products can also stabilize enzymes against heat-induced unfolding by forming intermediate complexes which have a slower thermal decay rate compared with that of the uncomplexed enzyme (Klibanov, 1979; Ahern and Koibanov, 1987; Chen and Wu, 1987; Volkin and Klibanov, 1989). Glucose isomerase immobilized on to glutaraldehyde-cross-linked glass beads showed a half-life of 37 min at 90°C in the presence of HFCS, but less than 1 min its absence (Volkin and Klibanov, 1989). Immobilized glucose isomerase also showed an increasing half-life from 120 to 190 hours with increasing glucose concentrations of up to 3 M at 70°C (Chen and Wu, 1987).

Although most glucose isomerases exhibit high temperature optima, industrial bioreactors carry out the reaction at 55–57°C, as the enzyme is subject to thermal inactivation with long-term use at high temperatures. For example, glucose isomerase covalently bound to glutaraldehyde-cross-linked porous glass beads showed a functional half-life of only 120 min at 80°C (Volkin and Klibanov, 1989). At any given temperature, irreversible thermoinactivation of an enzyme is caused by a combination of covalent and conformational processes. Volkin and Klibanov (1989) provided a mechanistic description of the irreversible thermoinactivation of glucose isomerase under normal operating conditions. Thermoinactivation of free or immobilized glucose isomerase at 90°C in the presence of HFCS was shown to be the result of deamidation of the enzyme's asparagine/glutamine residues. The authors also showed that the time-dependent decay of enzyme activity in industrial bioreactors is brought about by oxidation of the enzyme's cysteine residues (a significant problem since total exclusion of O_2 in a bioreactor is not feasible) and/or heat-induced deleterious reactions with HFCS or its impurities.

The purity of the substrates and products are also critical factors for glucose isomerase activity and stability. Saccharified corn syrups usually contain particulate matter and soluble impurities such as amino acids, peptides, lipids and metals which can cause enzyme inactivation or inhibition if not removed; decreases in productivity of as much as 50% have been recorded (Jensen and Rugh, 1987).

Jensen and Rugh (1987) described various types of bioreactors used for HFCS production and their optimization. Batch processes typically function in the presence of Co^{2+} and at low pH values, in order to stabilize the enzyme and to limit by-product formation during the long residence times of the reaction. Plug flow (fixed bed) processes are more advantageous, as much higher rates of productivity and shorter residence times can be achieved; these both eliminate the need for Co^{2+} and considerably reduce by-product formation. Production capacity can also be adapted according to demand by adjusting isomerization temperature and age of the enzyme. When a high syrup production capacity is needed, the isomerization temperature is increased and old enzyme is continuously replaced by new, high-activity enzyme. The lowest cost of conversion is obtained by running at a low temperature (e.g. 56–57°C) and by utilizing the enzyme until only 10% residual activity remains. Syrups containing 55, 60 and 90% fructose can now be manufactured from corn syrup, but at much lower quantities than the 42% fructose syrups (MacAllister, 1980).

INVERTASE

Invertase (β-*D*-fructofuranosidase, EC 3.2.1.26) is an enzyme, typically found in baker's yeast (*Saccharomyces cerevisiae*), that can hydrolyse disaccharide compounds with unsubstituted β-*D*-fructofuranosyl residues, such as sucrose and raffinose, into their constituent monosaccharides (Goldstein and Lampen, 1975; Lopez-Santin *et al.*, 1983). The invert sugar syrups that result find use in the confectionery, beverage and bakery industries (Mansfeld, Schellenberger and Römbach, 1992). Invertase also exhibits transferase activity at high substrate concentrations (e.g. 1 M sucrose), and is capable of transferring the β-*D*-fructofuranosyl residue to primary alcohols (Goldstein and Lampen, 1975). Notably, the very first study of an immobilized enzyme occurred in 1916 and involved the stable immobilization of invertase on to activated charcoal (Nelson and Griffin, 1916). Immobilized invertase preparations are now commercially available from several companies, including Gist-Brocades Inc. (Maxinvert), Rohm GmbH (Plexazym IN) and Tate and Lyle Inc. (charcoal-bound invertase). However, although yeast invertase is used commercially for the production of glucose/fructose syrups, immobilized invertase has not found widespread use (Yamazaki, Cheok and Fraser, 1984), possibly because most invertase preparations suffer from substrate (sucrose) inhibition (Dickensheets, Chen and Tsao, 1977).

Yeast invertase is a glycoprotein; 50% of its molecular weight is carbohydrate, mainly glucan and mannan (Lampen, 1971). The immobilization of glycoprotein enzymes via their carbohydrate moieties is advantageous, since the bonds always occur away from the enzyme active site, thus allowing the enzyme to maintain all of its original activity (Marek, Valentová and Kás, 1984). The covalent coupling of yeast invertase on to *Ocimum basilcum* seeds through its carbohydrate moieties resulted in an alkaline shift in pH optimum (from pH 3.8 to 4.4) and an increase in thermostability with no appreciable change in temperature optimum (Melo and D'Souza, 1992). The immobilized enzyme could be reused more than 10 times in batch process without appreciable loss of activity.

Marek, Valentová and Kás (1984) compared the pH optima and kinetic constants of invertase immobilized on to various solid supports, in an effort to determine the ideal conditions for attraction and binding of invertase via its carbohydrate moiety. All of the supports tested contained groups that reacted with the invertase aldehyde groups activated by periodate oxidation. Enzyme conjugates prepared by direct reaction to glycidyl methacrylate, styrene-divinyl benzene or bead cellulose had higher pH optima and higher K_m values for sucrose (2.8-, 3.8- and 2.0-fold higher, respectively), compared with the native enzyme. By comparison, invertase immobilized on to the same supports by the modified Ugi method (addition of 20 ml cyclohexyl isocyanide and 40 ml 0.5 M acetic acid to the enzyme/support mixture followed by incubation with shaking at 4°C for 6 h), showed overall higher enzyme activities, higher pH optima and lower K_m values than the corresponding immobilized enzymes prepared by direct action.

Yeast invertase immobilized on to ConA-Sepharose through its carbohydrate moiety also retained high enzyme activity (70%) and showed enhanced stability at 60°C and in the presence of 3 M urea, compared to the native enzyme (Iqbal and Saleemuddin, 1983, 1985). Stability was further improved when the immobilized enzyme was cross-linked with glutaraldehyde. The continuous column hydrolysis of 1 M sucrose by ConA-Sepharose-bound invertase resulted in over 80% sucrose conversions at 30°C and at a flow rate of 100 ml h^{-1}, with the column retaining over 80% of bound activity after 60 days (Iqbal and Saleemuddin, 1983, 1985).

Baker's yeast invertase immobilized on to microcrystalline cellulose, DEAE-or CM-Sephadex, and insolubilized ConA retained 46, 41, 70 and 73% of added activity, respectively (Woodward and Wiseman, 1978). ConA- and microcrystalline-cellulose-bound invertase were more thermally stable, particularly in the presence of sucrose, and showed broader pH optima (pH 2.5–6) and unaltered K_m values, compared with the native enzyme. By contrast, DEAE- and CM-Sephadex bound invertase both showed a 3.6-fold increase in the K_m value for sucrose and had poor thermal stability, losing all activity after 1 min of heating at 65°C. However, stability was restored in the presence of 20 mM succinic acid and 20 mM ethylenediamine, respectively, suggesting that new stabilizing salt-linkages had formed between the enzymes and these compounds to replace those that were broken in the enzyme during the immobilization step. These results, together with additional chemical modification experiments, led the authors to conclude that the alteration of either amino or carboxyl groups would result in the loss of conformational stability, unless these stabilizing salt-linkages were preserved.

Invertase immobilized on to *N,N*-diethylamino-2-hydroxypropyl (DEAHP)-cellulose beads remained firmly bound, with a similar temperature optimum, a slightly higher thermal stability and a more acidic pH optimum, compared with the native enzyme (Hradil and Svec, 1981). Furthermore, the enzyme retained the same activation energy but showed a slight increase (20%) in the K_m value upon immobilization. The amount of enzyme adsorbed on to DEAHP-cellulose was proportional to the porosity and the exchange capacity of the beads, with activities in a stirred tank reactor exceeding 8000 mmol min^{-1} g^{-1} in 10% sucrose at 30°C and pH 5.

Candida utilis invertase immobilized on to porous cellulose beads by ionic-guanidino attachment showed a broader pH optimum (from pH 4 to 5.4) compared with the native enzyme (pH 4.1), a 1.8-fold lower activation energy and, unlike native invertase, remained active at temperatures above 70°C (Dickensheets, Chen and Tsao, 1977). However, the 2.8-fold higher K_m value, together with a 1.7-fold higher K_i value for sucrose, suggested that immobilized invertase activity was limited by diffusion effects. This was verified by the fact that a minimum column flow rate of 18 ml min^{-1} was required to eliminate the effect of external film diffusion on the reaction rate of bound invertase. Furthermore, a comparison of the reaction rates of invertase bound to cellulose beads of three different mesh sizes revealed pore diffusion effects with the smallest mesh bead size (Dickensheets, Chen and Tsao, 1977). Diethylaminoacetyl (DEAA)-cellulose retained 30–45% of added *C. utilis* invertase activity by ionic interactions, and the immobilized enzyme showed similar K_m values, storage stability, heat stability and temperature optimum to native invertase (Maeda and Suzuki, 1973). An immobilized column preparation catalysed continuous inversion of 50% sucrose for 9 days at 30°C.

Yeast invertase covalently bound to glutaraldehyde-cross-linked polystyrene anion-exchange resin showed continuous sucrose hydrolysis in both 3 ml packed bed reactors and 50 litre pilot-scale reactors, without loss of efficiency or operational stability (Mansfeld, Schellenberger and Römbach, 1992). Studies of the dependence of productivity on substrate concentration, temperature and catalyst stability also found that the loss of reactor productivity due to product inhibition at sucrose concentrations higher than 2.1 M could be overcome by increasing the reactor temperature from 40 to 50°C (Mansfeld, Schellenberger and Römbach, 1992). The 60% decrease in half-life of the catalyst that resulted from the higher temperature could be overcome by using a higher sucrose concentration (2.4 M). The half-life of the catalyst increased from 15 to 131 days at higher substrate concentrations, due to substrate protection. As a result, the productivity of the 50 litre immobilized invertase reactor was 0.76 kg sucrose hydrolysed h^{-1} l^{-1} at 40°C and a substrate concentration of 2.4–2.6 M. Sucrose conversions of 92% could be maintained for 1 year under these conditions. This demonstrates the value of searching for the optimal compromise between different operational parameters in order to achieve optimal results with respect to economy and catalyst exploitation (Mansfeld, Schellenberger and Römbach, 1992).

Stefuca, Gemeiner and Báles (1988) compared the reaction kinetics and stability of chlorotriazinyl cellulose-bound invertase in a stirred tank reactor (STR) and a packed bed reactor (PBR). Invertase in the STR, when maintained at an agitation speed of 300 rev min^{-1}, showed no change in K_m values and, thus, was not limited by external diffusion. In the PBR, the reaction rate and K_m value for invertase was affected by external mass transfer but only when the flow rate dropped from 4 to 0.5 ml min^{-1}. However, the beads in both reactors were deformed or partially destroyed under the reactor conditions. Under optimal storage conditions (in the presence of formaldehyde) the immobilized invertase maintained 100% activity after 40 days at 4°C. Monsan and Combes (1992) compared the activities of invertase immobilized on to corn grits and operating in either a continuous stirred tank reactor (CSTR) or a packed bed reactor (PBR) of 100 ml volumes. The productivity of the PBR was 30% and 40% higher than that of the CSTR when using either low (0.4 M) or high (2.44 M) sucrose concentrations, respectively. The PBR was successfully scaled up to 17.6 litres, achieving productivities of 0.45 kg sucrose hydrolysed h^{-1} l^{-1} or 93% sucrose conversions at 2.5 M substrate and 50°C.

Invertase entrapped in gelatin discs hardened with formaldehyde was limited by sucrose resistance to diffusion into the insoluble matrix (de Alteriis *et al.*, 1985). The 39% enzyme activity yield that was obtained under conditions of pH 5 and 30°C was nearly doubled after immobilized invertase was crushed into smaller segments (de Alteriis *et al.*, 1985). Invertase covalently bound to glutaraldehyde-cross-linked polyethyleneimine-derivatized cotton cloth showed in a decreased optimal pH range with significant activity loss above pH 5 (Yamazaki, Cheok and Fraser, 1984). However, the enzyme preparation maintained full activity after 3 months' storage at 30°C in 1 M sucrose. López-Santin *et al.* (1983) immobilized 20–30% of added invertase on to cyanuric chloride-activated clay; the immobilized enzyme showed a broader pH optimum than its free counterpart but kinetic parameters were similar in assays at 25°C and 0.3 M sucrose. However, immobilized enzyme batch reactors were still limited by extensive substrate and production inhibition as seen with the free enzyme at 5% sucrose, and also showed significant levels of thermal deactivation above 30°C.

Co-immobilization studies

A new generation of enzyme technology has explored the concept of co-immobilizing enzymes on to a single support, for the creation of a conversion system that can degrade complex substrates into monomer sugars in a single process. Reactions catalysed by a co-immobilized enzyme system could offer a variety of advantages over a comparable series of individual soluble enzyme reactions, including:

1. reduction of the diffusion distance between enzymes for substrates and products;
2. prevention of the accumulation of inhibitory concentrations of intermediates and products in reactors;
3. a potential reduction in the overall amount of enzyme required;
4. an improved sugar yield and/or shorter residence times;
5. a potential increase in the half-lives of constituent enzymes due to the stability imparted by the close proximity of multiple enzymes on a single matrix; and
6. a potentially greater cost-effectiveness with respect to energy and equipment use (Mårtensson, 1974a,b; Hägerdal, 1980; Chakrabarti and Storey, 1990; Xiao *et al*., 1990).

Such advantages are in addition to those basic properties already associated with the use of immobilized enzymes.

The greatest potential for success occurs when the enzymes involved have similar characteristics (optimum pH and temperature, heat stabilities) and when they are co-immobilized in a ratio that has been optimized to give the highest enzyme conversion rates possible without overcrowding the support (Chakrabarti and Storey, 1989; Schafhauser and Storey, 1993). Although relatively few immobilized multienzyme systems have been developed, such processes show great potential as a more advanced and efficient method of polymer degradation (especially starch and cellulose) for the production of high-sugar syrups.

CO-IMMOBILIZED ENZYMES FOR STARCH DEGRADATION

β-Amylase can hydrolyse the α-1,4-glycosidic links, but is unable to bypass the α-1,6-glycosidic linkages in the amylopectin portions of starch and does not attack maltotriose; hence, the use of β-amylase alone results in incomplete degradation of branched polymers. The end products typically resulting from β-amylase hydrolysis of liquified starch are 55% maltose, 15% maltotriose and 30% β-limit dextrins (Finnsugar Co.Ltd. production information). Realizing the limitations imposed by the specificity of β-amylase, researchers have focused on developing multienzyme systems that combine pullulanase and β-amylase to produce a high-maltose product from starch.

Ohba and Ueda (1980) combined a pullulanase complex immobilized to TEAE-cellulose with sweet potato β-amylase bound to *p*-aminobenzyl cellulose in the same reactor in order to produce a functioning multienzyme complex. Unlike the immobilized β-amylase alone, the reactor containing the mixture of both enzymes could completely convert starch to maltose at 50°C and showed a long operational half-life (20 days) (Ohba and Ueda, 1980). Måartensson (1974b) co-immobilized β-amylase and pullulanase on to acrylamide–acrylic acid co-polymers but the system showed intraparticle/film diffusion resistance and steric hindrance despite vigorous shaking of the mixtures. Although activation energies and Arrhenius plots for the free and co-immobilized enzyme systems were similar, product yields were strongly influenced by flow rate, and the K_m value for starch increased by 5.8-fold with immobilization. Furthermore, the stability of β-amylase and pullulanase, even when co-immobilized, was still dependent on substrate concentration and remained sensitive to temperature. Operational stability improved with increasing substrate concentration for the co-immobilized enzymes, but half-lives decreased from 20 days at 40°C to 1–5.6 days at 50°C (Mårtensson, 1974b).

Amyloglucosidase can attack both α-(1,4) and α-(1,6) bonds in starch to produce glucose. However, the efficiency of conversion is increased if the debranching enzyme, pullulanase, is added (Galliard, 1985). The enzymes function interdependently. Pullulanase, by debranching starch, makes it more accessible to amyloglucosidase which has higher reaction rates on linear glucose polymers (Ram and Venkatasubramanian, 1982). The amyloglucosidase–pullulanase complex covalently coupled to polyurethane foam retained 32% of combined added activity (compared with a soluble mixture of the two enzymes) and showed no activity loss after 4 months storage at 4°C (Chakrabarti and Storey, 1990). The similar pH and temperature optima of the individual enzymes allowed for enhanced activities when co-immobilized and doubled the glucose yield over immobilized amyloglucosidase alone (Chakrabarti and Storey, 1990).

The co-immobilization of amyloglucosidase and pullulanase on to azide-linked bovine hide collagen membranes resulted in enhanced levels of substrate hydrolysis (4–5% dextrose equivalent) over that of immobilized amyloglucosidase alone; pullulanase was especially effective in enhancing activity in the final stages of long-term hydrolysis (e.g. 160 h) (Ram and Venkatasubramanian, 1982). Co-immobilization also greatly improvd the stability of pullulanase, which remained active for at least 6 days compared with only 1 day for the free enzyme. At 50°C, pH 5.0 was considered to be the optimal operational pH. Although immobilized pullulanase would show greater activity and stability at pH 6 or 7, these higher pHs had an adverse effect on the activity of amyloglucosidase (Ram and Venkatasubramanian, 1982).

The addition of pullulanase to amyloglucosidase immobilized on to granular chicken bone imparted a certain degree of instability not seen when amyloglucosidase was immobilized alone; storage half-life decreased (70 days versus 144 days, respectively), activity was lost with continuous reuse (24% lost versus 100% remaining after five cycles) and a significant activity loss occurred in the presence of 0.2 and 0.5 M KCl (Schafhauser and Storey, 1993). However, the co-immobilization of pullulanase with amyloglucosidase maintained high temperature optima and resulted in higher product yields than those produced by immobilized amyloglucosidase alone, particularly in the later stages of a hydrolysis reaction (4–24 h). Furthermore, the co-immobilized enzymes maintained a broader pH range of activity (from pH 3 to 6.5) which was particularly different from the free-enzyme mixture on the alkaline side of the range, the free-enzyme mixture retaining only about 60% of maximal activity at pH 6.5.

With near-optimal activities of the bone-immobilized amyloglucosidase/pullulanase still retained at pH 6.5, the possibility of adding glucose isomerase into the immobilized enzyme system became feasible. Such an enzyme complex would allow the complete conversion of starch to fructose. A one-step conversion process would be ideal for, at present, the processes of starch hydrolysis and glucose isomerization using soluble enzymes must, of necessity, be conducted in two separate stages using very different reaction conditions (particularly of pH) (Verhoff *et al.*, 1985). Furthermore, the enriched glucose product afforded by co-immobilized amyloglucosidase and pullulanase activity would lead to a higher conversion efficiency and reduced reaction time for the isomerization step, for high-fructose syrups are more easily produced by starting with as high a glucose concentration as economy allows (Antrim and Auterinen, 1985).

The co-immobilization of amyloglucosidase, pullulanase and glucose isomerase on to granular chicken bone resulted in a sufficient broadening of the operational parameters that the three enzymes could successfully catalyse a one-step conversion process that converted glycogen to fructose (Schafhauser and Storey, 1992c). Since the immobilized complex could not hydrolyse insoluble starch in any appreciable amount, glycogen was used as the substrate for experimental optimization of the system. The reaction pH used with the three-enzyme co-immobilized system strongly influenced the yield and ratio of the hexose sugars produced. Lowering the reaction pH increased the glucose yield by increasing the activity of amyloglucosidase and pullulanase, but fructose production dropped due to the inhibition of glucose isomerase. This is a positive result, however, for it demonstrates that the product ratios can be altered readily by a simple manipulation of pH in order to suit different goals of HFCS producers. The greatest fructose yield achieved was 20% at the optimal pH of 6.5–7 of the co-immobilized system, versus 30% at pH 7.5–8 for a mixture of the free enzymes. Reactor pH could also be favourably manipulated during the hydrolysis reaction to increase fructose output; a neutral pH in the early stages of the reaction favoured glucose production, whereas a more alkaline pH in the later stages favoured fructose production, resulting in a net 25% fructose product solution.

Xiao *et al.* (1990) separately immobilized amyloglucosidase on to anilinosulphonic-linked polystyrene beads and glucose isomerase on to triethanolamine-linked polystyrene beads and then combined the two in the same reactor. The mixture carried out simultaneous saccharification and isomerization reactions successfully at pH 6.6 and 60°C. The charged supports altered the pH profiles of the two enzymes such that immobilized amyloglucosidase and glucose isomerase retained high activities (90% and 55%, respectively) at neutral pH. By contrast, soluble amyloglucosidase and glucose isomerase showed much lower activities (40–50%) due to their very different pH optima (pH 4.5 and 9.5, respectively).

Wheat β-amylase and *A. niger* amyloglucosidase were successfully co-immobilized on to glutaraldehyde-cross-linked silica beads due to the similarities in their pH optima (Bohnenkamp and Reilly, 1980). By co-immobilizing different ratios of β-amylase and amyloglucosidase in batch reactors with increasing process times, syrup products of a variety of compositions were produced (Bohnenkamp and Reilly, 1980). Hägerdal (1980) successfully co-immobilized sweet almond β-amylase and *Saccharomyces cerevisiae* cells for the production of ethanol from cellobiose. At 22°C the complex, which was entrapped into calcium alginate gel, showed approximately 50% conversion of cellobiose to glucose, with 40% of the resulting glucose fermented to ethanol.

CO-IMMOBILIZED ENZYMES FOR CELLULOSE DEGRADATION

Natural cellulases from most fungal sources are limited in their overall activity by low amounts of β-glucosidase (Klyosov and Consultant, 1986) and by cellobiose inhibition of cellobiohydrolase (Sundstrom *et al.*, 1981). Although advantageous to the growth of the fungus, this is a limiting factor with respect to their industrial application for cellulose processing. Improved rates of cellulose hydrolysis and improved glucose yield can be achieved under lab or industrial conditions by combining additional β-glucosidase with cellulases (Sundstrom *et al.*, 1981).

The co-immobilization of *Trichoderma reesei* cellulase and *A. niger* β-glucosidase on to polyurethane foam resulted in a 2.5- and 4-fold greater rate of glucose production for soluble cellulose (CMC) and insoluble cellulose (microcrystalline), respectively, compared with immobilized cellulase alone (Chakrabarti and Storey, 1989). A ratio of 30:2 units of immobilized cellulase:β-glucosidase was optimal for maximum cellulose hydrolysis, which represents a five-fold increase in β-glucosidase content compared with the natural content of the fungal cellulase complex (Chakrabarti and Storey, 1989). Furthermore, the long operational half-life (approximately 500 h) and the capacity for long-term linear rates of hydrolysis of microcrystalline cellulose, implied that the co-immobilization of β-glucosidase with cellulase resulted in a stable complex which eliminates the problem of cellobiose feedback inhibition of the cellulase complex.

Co-immobilization of cellulase and β-glucosidase on to polyurethane foam also resulted in an alkaline shift in the pH optimum from 5 to 6 with a significant amount of activity still maintained at pH 7 and 8 (Chakrabarti and Storey, 1989). This, plus the fact that foam-immobilized glucose isomerase also showed significant activity at a pH as low as 6.8, justified experiments on the co-immobilization of all three enzymes in an attempt to create a single-process conversion of cellulose to fructose. Such a one-step conversion process is not possible with the soluble enzymes under similar reaction conditions. The formation of an operational immobilized multienzyme complex would allow for greater rates of cellulose conversion, for the ultimate product fructose, unlike glucose, is a poor feedback inhibitor of β-glucosidase (Woodward and Arnold, 1981).

Like the multienzyme system designed for the one-step conversion of starch to fructose (Schafhauser and Storey, 1992c), the specific reaction pH used with the co-immobilized complex of cellulase, β-glucosidase and glucose isomerase on polyurethane foam strongly influcenced the yield and ratio of the hexose sugars produced (Storey and Chakrabarti, 1990). Lowering the pH increased the total yield of hexose sugars by increasing the activity of cellulase and β-glucosidase, but reduced the fructose yield by inhibiting glucose isomerase activity. Increasing the pH had the opposite effect. However, unlike the complex of amyloglucosidase, pullulanase and glucose isomerase, the immobilized complex of cellulase, β-glucosidase and glucose isomerase became a better catalyst than a mixture of the free enzymes. The co-immobilized system showed a fructose yield as high as 30% from insoluble as well as soluble cellulose substrates at its pH optimum of 7 (Storey and Chakrabarti, 1990).

CO-IMMOBILIZED ENZYMES FOR PECTIN DEGRADATION

Compared with the individually immobilized enzymes, a co-immobilized system including both pectin esterase (PE) and polygalacturonase (PG) should represent a logical and beneficial advance for industrial processes such as decreasing the viscosity of pectin-containing solutions and enhancing the rate of juice clarification. The depolymerizing activity of PG is dependent on the demethoxylating activity of PE, such that PG cannot hydrolyse the glycosidic linkages of *D*-galacturonic acid until they have been sufficiently demethoxylated (i.e. by about 45%).

Romero *et al.* (1989) co-immobilized tomato PE with a *Rhizopus* spp. PG on to an aminoaryl-derivatized porous glass support, and showed that the co-immobilized enzymes were much more effective than the individual immobilized enzymes in reducing the viscosity of both high molecular weight citrus pectin and low molecular weight methoxylated pectin. The pH optima for stability and activity of the two enzymes overlapped sufficiently (for PE pH 5–8 and for PG pH 3.5–6.5) to result in significant activities of both enzymes when the co-immobilized system was run at pH 5.9. Co-immobilized PE and PG showed better activity when immobilized on to 46 nm pore size glass as compared with 20 nm pore size glass. The 46 nm pore size glass had a larger and more accessible surface area (both internal and external) for enzyme binding. The pores of 20 nm pore sized glass, however, were too small to allow for enzyme entrance and binding and, as a result, the enzymes were limited to the external surface of the support for binding.

Manjón *et al.* (1992) also co-immobilized the two enzymes on to porous glass and showed that although PG acts as an activity enhancer for PE activity, both soluble and immobilized PE was competitively inhibited by polygalacturonic acid, the substrate of PG. Whereas PE showed a decrease in temperature optimum with co-immobilization (from 60 to 45°C), the temperature optimum (rising from 40 to 60°C) and the temperature stability of PG increased substantially in the co-immobilized system. However, the co-immobilized enzymes suffered from diffusional limitations, with K_m values increasing by 1.3-fold for PE and by 1.7-fold for PG. Both enzymes showed greater operational stability after co-immobilization, with an increase in half-life during continuous operation of 1.2-fold for PE and 1.6-fold for PG.

INVERTASE CO-IMMOBILIZATION WITH OTHER ENZYMES

Various studies have also attempted to immobilize invertase along with other enzymes to achieve the production of specific sugar products. Inman and Hornby (1974) co-immobilized yeast invertase and glucose oxidase on to the inner surface of nylon tubes in an attempt to create a one-step sucrose analyser; however, the system showed no functional improvements as compared with the combined activities of the individually immobilized enzymes. It was suggested that α-*D*-glucose, the hydrolysed product of sucrose, must first undergo mutarotation to β-*D*-glucose before it can be oxidized by glucose oxidase. Thus, the co-immobilization of the two enzymes offered no improvements in the system, as mutarotation was the rate-limiting step for the reaction.

Yeast invertase and *Penicillium chrysogenum* glucose oxidase were co-immobilized on to urea- and formaldehyde-activated cheesecloth for the detection of saccharose in solution (Krysteva and Yotova, 1992). The immobilized enzyme preparation was used in conjunction with a 'liquid' dextran catalase, and saccharose concentration was measured indirectly by means of an amperometric oxygen electrode. The co-immobilized invertase and glucose oxidase maintained constant pH optima (pH 5.0 and 5.8, respectively) and showed slight increases in their temperature optima (60° and 35°C, respectively). The immobilized preparation remained highly stable and active, maintaining reproducible readings with short response times (2–3 min) over a 30 day period, and showing no detectable change in enzyme activity when stored at 4°C for 1 year. Kiba (1993) immobilized invertase and pyranose oxidase on to porous glass beads for use as column reactors in a HPLC system for the detection of sucrose and glucose; the hydrogen peroxide produced was monitored chemiluminometrically. The system was highly sensitive, with detection limits of 2–3 $\times 10^{-7}$ M and a linear response up to 5×10^{-4} M.

Ro and Kim (1991) co-immobilized invertase bound to glutaraldehyde-cross-linked chitin with *Zymomonas mobilis* cells entrapped in calcium alginate beads for the continuous production of gluconic acid and sorbitol from sucrose. The optimum pH for the oxidoreductase enzyme of immobilized *Z. mobilis* (pH 6.2) and for immobilized invertase (pH 4.5) were close enough that the co-immobilized system could function at a compromise pH of 6.0. Immobilized invertase was less sensitive to substrate inhibition than the free enzyme (i.e. inhibition was first seen at 20% sucrose v. 5% sucrose, respectively), possibly due to the formation of a concentration gradient between the bulk solution and the pores of the alginate beads. The co-immobilized system operating in a continuous stirred tank reactor showed maximum productivities of 4.6 g l^{-1} h^{-1} with 20% sucrose as substrate, but the operational half-life was only 170 h due to abrasion of the beads. A packed bed reactor, which was subsequently designed to overcome this problem, showed maximum productivities of 5.2 g l^{-1} h^{-1} for 250 h with 93% substrate conversion after 22 h (Ro and Kim, 1991).

References

ABDEL-NABY, M.A. (1993). Immobilization of *Aspergillus niger* NRC 107 xylanase and β-xylosidase, and properties of the immobilized enzymes. *Applied Biochemistry and Biotechnology* **38**, 69–81.

AGUADO, J., ROMERO, M.D. AND RODRIGUEZ, L. (1993). Immobilization of β-glucosidase from *Penicillium funiculosum* on nylon powder. *Biotechnology and Applied Biochemistry* **17**, 49–55.

AHERN, T.J. AND KLIBANOV, A.M. (1987). Thermal inactivation of proteins. *Methods of Biochemical Analysis* **33**, 91–127.

AL-TAI, A.M., ALI, Y. AND ABDUL-RAZZAK, S.H. (1987). Isomerization of glucose to fructose: production and some properties of glucose isomerase from *Streptomyces* sp. *Enzyme and Microbial Technology* **9**, 632–634.

ANTRIM, R.L. AND AUTERINEN, A.-L. (1985). A new regenerable, immobilized glucose isomerase. Oral paper at the 36th Detmond Starch Convention (Finnsugar Biochemics Inc.).

ARAKAWA, T. AND TIMASHEFF, S.N. (1982). Preferential interactions of proteins with salts in concentrated solutions. *Biochemistry* **21**, 6545–6552.

ASTHER, M. AND MEUNIER, J.-C. (1993). Immobilization as a tool for the stabilization of lignin peroxidase produced by *Phanerochaete chrysosporium* INA-12. *Applied Biochemistry and Biotechnology* **38**, 57–67.

BAGSVAERD, L.Z. (1981). Enzymatic hydrolysis of inulin – an alternative way to fructose production. *Starch* **11**, 373–377.

BAHULEKAR, R., AYYANGAR, N.R. AND PONRATHNAM, S. (1991). Polyethyleneimine in immobilization of biocatalysts. *Enzyme and Microbial Technology* **13**, 858–868.

BAJPAI, P. AND MARGARITIS, A. (1985). Immobilization of *Kluyveromyces marxianus* cells containing inulinase activity in open pore gelatin matrix: 2. application for high fructose syrup production. *Enzyme and Microbial Technology* **7**, 459–461.

BAJPAI, P. AND MARGARITIS, A. (1987). Characterization of molecular-sieve-bound inulinase. *Journal of Fermentation Technology* **65**, 239–242.

BANKS, W. AND MUIR, D.D. (1980). Structure and chemistry of the starch granule. In *Carbohydrates: Structure and Function*, volume 3 *The Biochemistry of Plants* (J. Preiss, Ed.), pp. 321–369. Academic Press, New York.

BECK, R.H.F. AND PRAZNIK, W. (1986). Determination of inulase isoenzymes on polyacrylamide gels by activity staining. *Journal of Chromatography* **369**, 240–243.

BIELY, P. (1985). Microbial xylanolytic systems. *Trends in Biotechnology* **3**, 286–290.

BOHNENKAMP, C.G. AND REILLY, P.J. (1980). Use of immobilized glucoamylase–β-amylase and glucoamylase–fungal amylase mixtures to produce high-maltose syrups. *Biotechnology and Bioengineering* **22**, 1753–1758.

BORREGO, F., TARI, M., MANJON, A. AND IBORRA, J.L. (1989). Properties of pectinesterase immobilized on glycophase-coated controlled-pore glass. *Applied Biochemistry and Biotechnology* **22**, 129–140.

BUCHOLZ, K. AND KLEIN, J. (1987). Characterization of immobilized biocatalysts. *Methods in Enzymology* **135**, 3–29.

BUCKE, C. (1981). Glucose transforming enzymes. In *Enzymes and Food Processing* (G.G. Birch, N. Blakebrough and K.J. Parker, Eds), pp. 51–72. Applied Science Publishers, London.

CABRAL, M.S., NOVAIS, J.M. AND CARDOSO, J.P. (1981). Immobilization of amyloglucosidase on alkylamine derivatives of metal-link-activated inorganic supports. *Biotechnology and Bioengineering* **23**, 2083–2092.

CALDWELL, K.D., AXON, R., BERGWALL, M. AND PORATH, J. (1976a). Immobilization of enzymes based on hydrophobic interaction. I. Preparation and properties of a β-amylase adsorbate. *Biotechnology and Bioengineering* **18**, 1573–1588.

CALDWELL, K.D., AXEN, R., BERGWALL, M. AND PORATH, J. (1976b). Immobilization of enzymes based on hydrophobic interaction. II. Preparation and properties of an amyloglucosidase adsorbate. *Biotechnology and Bioengineering* **18**, 1589–1604.

CELEBI, S.S., TSAI, G.-J. AND TSAO, G.T. (1991). Characterization of glucoamylase immobilized on celite. *Applied Biochemistry and Biotechnology* **27**, 163–171.

CHAKRABARTI, A.C. AND STOREY, K.B. (1988). Immobilization of cellulase using polyurethane foam. *Applied Biochemistry and Biotechnology* **19**, 189–207.

CHAKRABARTI, A.C. AND STOREY, K.B. (1989). Enhanced glucose production from cellulose using coimmobilized cellulase and β-glucosidase. *Applied Biochemistry and Biotechnology* **22**, 263–278.

CHAKRABARTI, A.C. AND STOREY, K.B. (1990). Co-immobilization of amyloglucosidase and pullulanase for enhanced starch hydrolysis. *Applied Biochemistry and Biotechnology* **33**, 48–50.

CHANG, H.N. AND FURUSAKI, S. (1991). Membrane bioreactors: present and prospects. *Advances in Biochemical Engineering/Biotechnology* **44**, 27–64.

CHEETHAM, P.S.J. (1985). Principles of industrial enzymology: the basis of utilization of soluble and immobilized enzymes in industrial processes. In *Handbook of Enzyme Biotechnology* (A. Wiseman, Ed.), pp. 54–146. Ellis Horwood, Chichester.

CHEN, K.-C. AND WU, J.-Y. (1987). Substrate protection of immobilized glucose isomerase. *Biotechnology and Bioengineering* **30**, 817–824.

CHEN, W.-P. (1980). Glucose isomerase (a review). *Process Biochemistry* June/July, pp. 30–35.

CHIM-ANAGE, P., KASHIWAGI, Y., MAGAE, Y., OHTA, T. AND SASAKI, T. (1986). Properties of cellulase immobilized on agarose gel with spacer. *Biotechnology and Bioengineering* **28**, 1876–1878.

CHRISTOV, L.P. AND PRIOR, B.A. (1993). Esterases of xylan-degrading microorganisms: production, properties, and significance. *Enzyme and Microbial Technology* **15**, 460–465.

COKER, L.E. AND VENKATASUBRAMANIAN, K. (1985). Starch conversion processes. In *Comprehensive Biotechnology* (M. Moo-Young, Ed.), volume 3, pp. 777–787. Pergamon Press, Oxford.

COUGHLAN, M.P. AND HAZLEWOOD, G.P. (Eds) (1993a). *Hemicellulose and Hemicellulases.* Portland Press, London.

COUGHLAN, M.P. AND HAZLEWOOD, G.P. (1993b). β-D-Xylan-degrading enzyme systems: biochemistry, molecular biology and applications. *Biotechnology and Applied Biochemistry* **17**, 259–289.

COULET, P.R. (1974). A mild method of general use for covalent coupling of enzymes to chemically activated collagen films. *Biotechnology and Bioengineering* **16**, 1055–1068.

DARNOKO, D., CHERYAN, M. AND ARTZ, W.E. (1988). Saccharification of cassava starch in an ultrafiltration reactor. *Enzyme and Microbial Technology* **11**, 154–159.

DE ALTERIIS, H., PARASCANDOLA, P., SALVADORE, S. AND SCARDI, V. (1985). Enzyme immobilization within insolubilized gelatin. *Journal of Chemical Technology and Biotechnology* **B35**, 60–64.

DE CORDT, S.F. (1994). Convenience of immobilized *Bacillus lichenformis* α-amylase as time–temperature integrator. *Journal of Chemical Technology and Biotechnology* **59**, 193–199.

DE CORDT, S.F., VANHOOF, K., HU, J., MAESMANS, G., HENDRICKS, M. AND TOBBACK, P. (1992). Thermostability of soluble and immobilized α-amylase from *Bacillus lichenformis. Biotechnology and Bioengineering* **40**, 396–402.

DEKKER, R.F.H. (1983). Bioconversion of hemicellulose: aspects of hemicellulase production by *Trichoderma reesei* QM 9414 and enzymatic saccharification of hemicellulose. *Biotechnology and Bioengineering* **25**, 1127–1146.

DEKKER, R.F.H. 1990). Application of magnetic immobilized β-glucosidase in the enzymatic saccharification of steam-exploded lignocellulosic residues. *Applied Biochemistry and Biotechnology* **23**, 25–39.

DESHMUKH, S.S., CHOUDHURY, M.D. AND SHANKAR, V. (1993). Preparation and properties of glucose isomerase immobilized on Indion 48-R. *Applied Biochemistry and Biotechnology* **42**, 95–104.

DICKENSHEETS, P.A., CHEN, L.F. AND TSAO, G.T. (1977). Characteristics of yeast invertase immobilized on porous cellulose beads. *Biotechnology and Bioengineering* **19**, 365–375.

DORFMAN, A. (1955). Enzymes of carbohydrate metabolism: pectic enzymes. *Methods in Enzymology* **1**, 158.

EMNEUS, J. (1993). Comparison between different inorganic supports for the immobilization of amyloglucosidase and alpha-amylase to be used in enzyme reactors in flow-injection systems. Hydrolysis of maltose, maltooligosaccharides and soluble starch. *Analytica Chimica Acta* **276**, 303–318.

FADDA, M.B., DESSI, M.R., MAURICI, R., RINALDI, A. AND SATTA, G. (1984). Highly efficient solubilization of natural lignocellulosic materials by a commercial cellulase immobilized on various solid supports. *Applied Microbiology and Biotechnology* **19**, 306–311.

FINDLAY, C.J., PARKIN, K.L. AND YADA, R.Y. (1986). Bone as a solid support for the immobilization of enzymes. *Biotechnology Letters* **8**, 649–652.

FINN, R.K. (1987). Conversion of starch to liquid sugar and ethanol. *Biotechnology* **1**, 101–107.

FOGARTY, W.M. (1983). *Microbial Enzymes and Biotechnology.* Applied Science Co., New York.

FOGARTY, W.M. AND KELLY, C.T. (1983). Pectic enzymes. In *Microbial Enzymes and Biotechnology* (W.M. Fogarty, Ed.), pp. 131–182. Applied Science Publishers, London.

FREIRE, D.M.G. AND SANT'ANNA, G.L., JR (1990). Hydrolysis of starch with immobilized glucoamylase: a comparison between two types of expanded bed reactors. *Applied Biochemistry and Biotechnology* **26**, 23–34.

GAIKWAD, S.M. AND DESHPANDE, V.V. (1992). Immobilization of glucose isomerase on Indion 48-R. *Enzyme and Microbial Technology* **14**, 855–858.

GALLIARD, T. (1985). Bulk chemicals from plants: starch and starch derived products. In *Plant Products and the New Technology* (K.W. Fuller and J.R. Gallon, Eds). *Annual Proceedings of the Phytochemical Society of Europe*, volume 26, pp. 103–115. Clarendon Press, Oxford.

GEMEINER, P., STEFUCA, V. AND BALES, V. (1993). Biochemical engineering of biocatalysts immobilized on cellulosic materials. *Enzyme and Microbial Technology* **15**, 551–566.

GERMAIN, P. AND CRICHTON, R.R. (1988). Characterization of a chemically modified β-amylase immobilized on porous silica. *Journal of Chemical Biotechnology* **38**, 297–315.

GOLDSTEIN, A. AND LAMPEN, J.O. (1975). β-D-Fructofuranoside fructohydrolase from yeast. *Methods in Enzymology* **42**, 504–511.

GOODWIN, T.W. AND MERCER, E.I. (1983). *Introduction to Plant Biochemistry*. Pergamon Press, Oxford.

GROHMANN, K. AND BALDWIN, E.A. (1992). Hydrolysis of orange peel with pectinase and cellulase enzymes. *Biotechnology Letters* **14**, 1169–1174.

GUIRAUD, J.P., DEMEULLE, S. AND GALZY, P. (1981). Inulin hydrolysis by the *Debaryomyces phaffii* inulinase immobilized on DEAE cellulose. *Biotechnology Letters* **3**, 683–688.

GUPTA, A.K. (1994). Production, purification and immobilization of inulinase from *Kluyveromyces fragilis*. *Journal of Chemical Technology and Biotechnology* **59**, 377–385.

GUPTA, A.K., RATHORE, P., KAUR, N. AND SINGH, R. (1990). Production, thermal stability and immobilization of inulinase from *Fusarium oxysporum*. *Journal of Chemical Technology and Biotechnology* **47**, 245–257.

HÄGERDAL, B. (1980). Enzymes co-immobilized with microorganisms for the microbial conversion of non-metabolizable substrates. *Acta Chemica Scandinavica* **B34**, 611–613.

HANISCH, W.H., RICKARD, P.A.D. AND NYO, S. (1978). Poly(methoxygalaturonide) lyase immobilized via titanium onto solid supports. *Biotechnology and Bioengineering* **20**, 95–106.

HARTMEIER, W. (1985). Immobilized biocatalysts – from simple to complex systems. *Trends in Biotechnology* **3**, 149–153.

HEATH, C. AND BELFOR, G. (1990). Membranes and bioreactors. *International Journal of Biochemistry* **22**, 823–835.

HIGA, O.Z., DEL MASTRO, N.L. AND CASTAGNET, A.C. (1986). Immobilization of cellulase and cellobiase by radiation-induced polymerization. *Radiation Physical Chemistry* **27**, 311–316.

HON, C.C. AND REILLY, P.J. (1979). Properties of β-amylase immobilized to alkylamine porous silica. *Biotechnology and Bioengineering* **21**, 505–512.

HRADIL, J. AND SVEC, F. (1981). Inversion of sucrose with β-D-fructofuranosidase (invertase) immobilized on bead DEAHP-cellulose: batch process. *Enzyme and Microbial Technology* **3**, 331–335.

HUITRON, C. AND LIMON-LASON, J. (1978). Immobilization of glucose isomerase to ion-exchange materials. *Biotechnology and Bioengineering* **20**, 1377–1391.

IBORRA, J.L., CASTELLAR, M.R., CANOVAS, M. AND MANJON, A. (1992). Picrocrocin hydrolysis by immobilized β-glucosidase. *Biotechnology Letters* **14**, 475–480.

INMAN, D.J. AND HORNBY, W.E. (1974). Preparation of some immobilized linked enzyme systems and their use in the automated determination of disaccharides. *Biochemical Journal* **137**, 25–32.

IQBAL, J. AND SALEEMUDDIN, M. (1983). Activity and stability of glucose oxidase and invertase immobilized on Concanavalin A Sepharose: influence of lectin concentrations. *Biotechnology and Bioengineering* **25**, 3191–3195.

IQBAL, J. AND SALEEMUDDIN, M. (1985). Sucrose hydrolysis using invertase immobilized on concanavalin A-Sepharose. *Enzyme and Microbial Technology* **7**, 175–178.

JAIN, P. AND WILKINS, E.S. (1987). Cellulase immobilized on modified nylon for saccharification of cellulose. *Biotechnology and Bioengineering* **30**, 1057–1062.

JENSEN, V.J. AND RUGH, S. (1987). Industrial-scale production and application of immobilized glucose isomerase. *Methods in Enzmology* **136**, 356–370.

KANDLER, O. AND HOPF, H. (1980). Occurrence, metabolism and function of oligosaccharides. In *Carbohydrates: Structure and Function*, volume 3 *The Biochemistry of Plants* (J. Preiss, Ed.), pp. 221–279. Academic Press, New York.

KARUBE, I., TANAKA, S., SHIRAI, T. AND SUZUKI, S. (1977). Hydrolysis of cellulose in a cellulase-bead fluidized bed reactor. *Biotechnology and Bioengineering* **19**, 1183–1191.

KENNEDY, J.F. (1985). Principles of immobilization of enzymes. In *Handbook of Enzyme Biotechnology* (A. Wiseman, Ed.), pp. 147–207. Ellis Horwood, Chichester.

KENNEDY, J.F., CABRAL, J.M.S., SA-CORREIA, I. AND WHITE, C.A. (1987). Starch biomass: a chemical feedstock for enzyme and fermentation processes. In *Starch: Properties and Potential* (T. Galliard, Ed.), volume 13, pp. 115–130. John Wiley and Sons, New York.

KIBA, N. (1993). Determination of sucrose and glucose by HPLC using immobilized enzyme reactors and a chemiluminescence detector. *Bunseki Kagaku* **42**, 649–652.

KIM, C.H. AND RHEE, S.K. (1989). Fructose production from Jerusalem artichoke by inulinase immobilized on chitin. *Biotechnology Letters* **11**, 201–206.

KIM, W.Y., BYUN, S.M. AND UHM, T.B. (1982). Hydrolysis of inulin from Jerusalem artichoke by inulinase immobilized on aminoethylcellulose. *Enzyme and Microbial Technology* **4**, 239–244.

KLIBANOV, A.M. (1979). Enzyme stabilization by immobilization: a review. *Analytical Biochemistry* **93**, 1–25.

KLUEPFEL, D., VATS-MEHTA, S., AUMONT, F., SARECK, F. AND MOROSOLI, R. (1990). Purification and characterization of a new xylanase (xylanase B) produced by *Streptomyces lividans* 66. *Biochemical Journal* **267**, 45–50.

KLYOSOV, A.A. (1990). Trends in the biochemistry and enzymology of cellulose degradation. *Biochemistry* **29**, 10577–10585.

KLYOSOV, A.A. AND CONSULTANT, U. (1986). Enzymatic conversion of cellulosic materials to sugars and alcohol. The technology and its implications. *Applied Biochemistry and Biotechnology* **12**, 249–300.

KNOWLES, J., LEHTOVAARA, P. AND TEERI, T. (1987). Cellulase families and their genes. *TIBTECH* **5**, 255–261.

KOBARZEWSKI, J. AND PASZCZYNSKI, A. (1983). Catalytic properties of immobilized crude and pure glucoamylase from *Aspergillus niger* C. *Biotechnology and Bioengineering* **25**, 3207–3212.

KOBAYASHI, T. AND LAIDLER, K.J. (1972). Kinetic analysis for solid supported enzymes. *Biochimica et Biophysica Acta* **302**, 1–12.

KOENIG, D. AND DAY, D. (1989). The purification and characterization of a dextranase from *Lipomyces starkeyi*. *European Journal of Biochemistry* **183**, 161–167.

KOSARIC, N., WIECZOREK, A., COSENTINO, G.P. AND DUVNJAK, Z. (1989). Industrial processing and products from the Jerusalem artichoke. In *Advances in Biochemical Engineering/ Biotechnology* (A. Fiechter, Ed.), volume 32, pp. 15–21. Springer-Verlag, New York.

KRYSTEVA, M.A. AND YOTOVA, L.K. (1992). Multienzyme membranes for biosensors. *Journal of Chemical and Technical Biotechnology* **54**, 13–18.

KUMAKURA, M. (1985). Preparation and use of polyurethane enzyme particles. *Separation and Purification Methods* **14**, 109–125.

KUMAKURA, M. AND KAETSU, I. (1985). Fluid immobilized cellulase. *Biotechnology Letters* **7**, 773–778.

LAMED, R., KENIG, R., MORAG, E., CALZADA, J.F., DE MICHEO, F. AND BAYER, E.A. (1991). Efficient cellulose solubilization by a combined cellulosome–β-glucosidase system. *Applied Biochemistry and Biotechnology* **27**, 173–183.

LAMPEN, J.O. (1971). Yeast and *Neurospora* invertase. In *The Enzymes* (P.D. Boyer, Ed.), volume 5, pp. 291–306. Academic Press, New York.

LEATHERS, T.D. (1989). Purification and properties of xylanase from *Aureobasidium*. *Journal of Industrial Microbiology* **4**, 341–348.

LEE, D.D., LEE, Y.Y., REILLY, P.J., COLLINS, E.V. AND TSAO, G.T. (1976). Pilot plant production with glucoamylase immobilized to porous silica. *Biotechnology and Bioengineering* **28**, 253–267.

LEE, Y.Y., FRATZKE, A.R., WUN, K. AND TSAO, G.T. (1976). Glucose isomerase immobilized on porous glass. *Biotechnology and Bioengineering* **18**, 389–413.

LENDERS, J.-P. AND CRICHTON, R.R. (1984). Thermal stabilization of amylolytic enzymes by covalent coupling to soluble polysaccharides. *Biotechnology and Bioengineering* **26**, 1343–1351.

LEWIS, N.G. AND PAICE, M.G. (Eds) (1989). *Plant Cell Wall Polymers: Biogenesis and Biodegradation*. ACS Symposium Series No. 399, American Chemical Society, Washington.

LOPEZ-SANTIN, J., SOLA, C., PARIS, J.M. AND CAMINAL, G. (1983). Invertase immobilized on activated clay: properties and kinetic study. *Journal of Chemical and Technical Biotechnology B* **33**, 101–106.

LOZANO, P., MANJON, A., IBORRA, J.L., CANOVAS, M. AND ROMOJARO, F. (1990). Kinetic and operational study of a cross-flow reactor with immobilized pectolytic enzymes. *Enzyme and Microbial Technology* **12**, 499–505.

LUONG, J.H.T., GROOM, C.A. AND MALE, K.B. (1991). The potential role of biosensors in the food and drink industries. *Biosensors and Bioelectronics* **6**, 547–554.

MACALLISTER, R.V. (1980). Manufacture of high fructose corn syrup using immobilized glucose isomerase. In *Immobilized Enzymes for Food Processing* (W.H. Pitcher, Jr, Ed.), pp. 81–111. CRC Press, Cleveland, OH.

MAEDA, H. AND SUZUKI, H. (1973). Preparation of immobilized invertase. *Biotechnology and Bioengineering* **15**, 403–412.

MAEDA, H., TSAO, G.T. AND CHEN, L.F. (1978). Preparation of immobilized soybean β-amylase on porous cellulose beads and continuous maltose production. *Biotechnology and Bioengineering* **20**, 383–402.

MANDELS, M., MEDEIROS, J.E., ANDREOTTI, R.E. AND BISSETT, F.H. (1981). Enzymatic hydrolysis of cellulose: evaluation of cellulase culture filtrates under use conditions. *Biotechnology and Bioengineering* **23**, 2009–2026.

MANJON, A., IBORRA, J.L., ROMERO, C. AND CANOVAS, M. (1992). Properties of pectinesterase and endo-D-polygalacturonase coimmobilized in a porous glass support. *Applied Biochemistry and Biotechnology* **37**, 19–30.

MANSFELD, J., SCHELLENBERGER, A. AND RÖMBACH, J. (1992). Application of polystyrene-bound invertase to continuous sucrose hydrolysis on pilot scale. *Biotechnology and Bioengineering* **40**, 997–1003.

MAREK, M., VALENTOVA, O. AND KAS, J. (1984). Invertase immobilization via its carbohydrate moiety. *Biotechnology and Bioengineering* **26**, 1223–1226.

MARG, G.A. AND CLARK, D.S. (1990). Activation of glucose isomerase by divalent cations: evidence for two distinct metal-binding sites. *Enzyme and Microbial Technology* **12**, 367–373.

MARGARITIS, A. AND PACE, G.W. (1985). Microbial polysaccharides. In *Comprehensive Biotechnology* (M. Moo-Young, Ed.), volume 3, pp. 1005–1044. Pergamon Press, Oxford.

MÅRTENSSON, K. (1974a). Preparation of an immobilized two-enzyme system, β-amylase–pullulanase, to an acrylic copolymer for the conversion of starch to maltose. I. Preparation and stability of immobilized β-amylase. *Biotechnology and Bioengineering* **16**, 567–577.

MÅRTENSSON, K. (1974b). Preparation of an immobilized two-enzyme system, β-amylase–pullulanase, to an acrylic copolymer for the conversion of starch to maltose. III. Process kinetic studies on continuous reactors. *Biotechnology and Bioengineering* **16**, 1567–1587.

MÅRTENSSON, K. AND MOSBACH, K. (1972). Covalent coupling of pullulanase to an acrylic copolymer using a water soluble carbodi-imide. *Biotechnology and Bioengineering* **14**, 715–724.

MATTEAU, P.P. AND SADDLER, J.N. (1982). Continuous glucose production using encapsulated mycelial-associated β-glucosidase from *Trichoderma* E58. *Biotechnology Letters* **4**, 715–721.

MAZID, M.A. (1993). Biocatalysis and immobilized enzyme/cell bioreactors. *Bio/techology* **11**, 691–694.

MELO, J.S. AND D'SOUZA, S.F. (1992). Immobilization of invertase through its carbohydrate moiety on *Ocimum basilicum* seed. *Applied Biochemistry and Biotechnology* **32**, 159–170.

MES-HARTREE, M., HOGAN, C., HAYES, R.D. AND SADDLER, J.N. (1987). Recycle of enzymes and substrate following enzymatic hydrolysis of steam pre-treated aspenwood. *Biotechnology and Bioengineering* **30**, 558–564.

MISHRA, C., DESHPANDE, V. AND RAO, M. (1983). Immobilization of *Penicillium funiculosum* cellulase on a soluble polymer. *Enzyme and Microbial Technology* **5**, 342–344.

MONSAN, P. AND COMBES, D. (1992). Application of immobilized invertase to continuous hydrolysis of concentrated sucrose solutions. *Biotechnology and Bioengineering* **26**, 347–351.

MONTENECOURT, B.S., CARROLL, J.O. AND LANZILOTTA, R.P. (1985). Assay of industrial microbial enzymes. In *Comprehensive Biotechnology* (M. Moo-Young, Ed.), volume 4, pp. 329–361. Pergamon Press, Oxford.

MOTOKI, M., ASO, H., SEGURO, K. AND NIO, N. (1987). Immobilization of enzymes in protein films prepared using transglutaminase. *Agricultural Biological Chemistry* **51**, 997–1002.

NEHETE, P.N., HEDGE, M.V., REDDY, G. AND SHANKAR, V. (1987). Immobilization of amyloglucosidase on polystyrene anion exchange resin II. Kinetics and stabilities. *Biotechnology Letters* **9**, 651–654.

NELSON, J.M. AND GRIFFIN, E.G. (1916). Adsorption of invertase. *Journal of the American Chemical Society* **38**, 1109.

NITHIANDAM, V.S., SRINIVASAN, K.S.V., JOSEPH, K.T. AND SANTAPPA, M. (1981). Preparation and properties of immobilized amyloglucosidase. *Biotechnology and Bioengineering* **23**, 2273–2282.

OGUNTIMEIN, G.B. AND REILLY, P.J. (1980). Purification and immobilization of *Aspergillus niger* β-xylosidase. *Biotechnology and Bioengineering* **22**, 1127–1142.

OHBA, R. AND UEDA, S. (1980). Production of maltose and maltotriose from starch and pullulan by an immobilized multienzyme of pullulanase and β-amylase. *Biotechnology and Bioengineering* **22**, 2137–2154.

OHBA, R., SHIBATA, T. AND UEDA, S. (1979). Preparation and properties of covalently immobilized β-amylase on *p*-aminobenzyl-cellulose. *Journal of Fermentation Technology* **57**, 146–150.

OHBA, R., CHAEN, H., HAYASHI, S. AND UEDA, S. (1978). Immobilization of *Streptomyces flavochromogenes* pullulanase on tannic acid and TEAE-cellulose. *Biotechnology and Bioengineering* **20**, 665–676.

PIETER, B.R., BARDELETTI, G. AND COULET, P.R. (1992). Glucoamylase immobilization on a magnetic microparticle for the continuous hydrolysis of maltodextrin in a fluidized bed reactor. *Applied Biochemistry and Biotechnology* **32**, 37–53.

PITCHER, W.H. (1980). *Immobilized Enzymes for Food Processing*. CRC Press, Boca Raton.

PITSON, S.M., SEVIOUR, R.J. AND MCDOUGALL, B.M. (1993). Noncellulolytic fungal β-glucanases: their physiology and regulation. *Enzyme and Microbial Technology* **15**, 178–192.

PRABHU, M. AND PRABHU, K.A. (1985). Immobilization of dextranase on bentonite. *Enzyme and Microbial Technology* **7**, 279–282.

RAM, K.A. AND VENKATASUBRAMANIAN, K. (1982). Enhancement of starch conversion efficiency with free and immobilized pullulanase and α-1,4-glucosidase. *Biotechnology and Bioengineering* **24**, 355–369.

RAMESH, V. AND SINGH, C. (1981). Studies on the kinetics and activation of soluble and immobilized sweet potato β-amylase. *Journal of Molecular Catalysis* **10**, 341–355.

REXOVÁ-BENKOVÁ, L. AND MRACKOVÁ-DOBROTOVÁ, M. (1981). Effect of immobilization of *Aspergillus niger* extracellular endo-D-galacturonanase on kinetics and action pattern. *Carbohydrate Research* **98**, 115–122.

RO, H.-S. AND KIM., H.-S. (1991). Continuous production of gluconic acid and sorbitol from sucrose using invertase and an oxidoreductase of *Zymomonas mobilis*. *Enzyme and Microbial Technology* **13**, 920–924.

ROMERO, C., SANCHEZ, S., MANJON, A. AND IBORRA, J.L. (1989). Optimization of the pectinesterase/endo-D-polygalacturonase coimmobilization process. *Enzyme and Microbial Technology* **11**, 837–843.

ROSEVEAR, A. (1984). Immobilized biocatalysts – a critical review. *Journal of Chemical Biotechnology B* **34**, 127–150.

ROY, F. AND HEGDE, M.V. (1987). Immobilization of β-amylase on polystyrene/cation exchange resin equilibrated with Al^{3+} ions (IR-120 Al^{3+}). *Enzyme and Microbial Technology* **9**, 550–552.

ROY, S.K., ROY, U. AND DUKE, D.K. (1984). Immobilized cellulolytic and hemicellulolytic enzymes from *Macrophomina phaseolina*. *Journal of Chemical Technology and Biotechnology B* **34**, 165–170.

ROY, S.K., RAHA, S.K., DEY, S.K. AND CHAKRABARTY, S.L. (1989). Immobilization of β-glucosidase from *Myceliophthora thermophila. Enzyme and Microbial Technology* **11**, 431–435.

ROYER, J.C. AND NAKAS, J.P. (1989). Xylanase production by *Trichoderma longibrachiatum. Enzyme and Microbial Technology* **11**, 405–410.

SADDLER, J.N. (1986). Factors limiting the efficiency of cellulase enzymes. *Microbiological Sciences* **3**, 84–87.

SADDLER, J.N. AND BROWNELL, H.H. (1982). Pretreatment of wood cellulosics to enhance enzymatic hydrolysis to glucose. Presentation at the International Symposium on Ethanol from Biomass sponsored by the Royal Society of Canada 13–15 October, Winnipeg, Canada.

SADHUKHAN, R., ROY, S.K. AND CHAKRABARTY, S.L. (1993). Immobilization of α-amylase from *Myceliophthora thermophila* D-14 (ATCC 48104). *Enzyme and Microbial Technology* **15**, 801–804.

SAHA, B.C. AND SHEN, G.-J. (1987). Behaviour of a novel thermostable β-amylase on raw starch. *Enzyme and Microbial Technology* **9**, 598–601.

SALEEMUDDIN, M. AND HUSAIN, Q. (1991). Concanavalin A: a useful ligand for glycoenzyme immobilization – a review. *Enzyme and Microbial Technology* **13**, 290–295.

SARKAR, J.M. AND BURNS, R.G. (1983). Immobilization of β-glucosidase and β-D-glucosidase-polyphenolic complexes. *Biotechnology Letters* **5**, 619–624.

SCHAFHAUSER, D.Y. AND STOREY, K.B. (1992a). Immobilization of amyloglucosidase onto granular chicken bone. *Applied Biochemistry and Biotechnology* **32**, 89–109.

SCHAFHAUSER, D.Y. AND STOREY, K.B. (1992b). Immobilization of glucose isomerase onto granular chicken bone. *Applied Biochemistry and Biotechnology* **32**, 79–87.

SCHAFHAUSER, D.Y. AND STOREY, K.B. (1992c). Fructose production: coimmobilized amyloglucosidase, pullulanase and glucose isomerase on Biobone™. *Biotechonology and Applied Biochemistry and Biotechnology* **36**, 63–74.

SCHAFHAUSER, D.Y. AND STOREY, K.B. (1993). Co-immobilization of amyloglucosidase and pullulanase onto granular chicken bone for enhanced starch degradation. *Applied Biochemistry and Biotechnology* **17**, 103–113.

SCHAFHAUSER, D.Y. AND STOREY, K.B. Immobilization of β-amylase onto granular chicken bone. *Applied Biochemistry and Biotechnology* (Submitted).

SENIOR, D.J., MAYERS, P.R. AND SADDLER, J.N. (1989). Production and purification of xylanases. In *Plant Cell Wall Polymers: Biogenesis and Biodegradation* (N.G. Lewis and M.G. Paice, Eds) American Chemical Society Series No. 399, pp. 641–654, American Chemical Society, Washington.

SENIOR, D.J., MAYERS, P.R., MILLER, D., SUTCLIFFE, R., TAN, L. AND SADDLER, J.N. (1988). Selective solubilization of xylan in pulp using a purified xylanase from *Trichoderma harzianum. Biotechnology Letters* **10**, 907–912.

SHARMA, S. AND YAMAZAKI, H. (1984). Preparation of hydrophobic cotton cloth. *Biotechnology Letters* **6**, 301–306.

SHIMIZU, K. AND ISHIHARA, M. (1987). Immobilization of cellulolytic and hemicellulolytic enzymes on inorganic supports. *Biotechnology and Bioengineering* **29**, 236–241.

SIMOS, G. AND GEORGATSOS, J.G. (1990). Immobilization of barley β-glucosidase on solid supports – yields and properties. *Applied Microbiology and Biotechnology* **33**, 51–53.

SLINGINGER, P.J., FANTA, G.F. AND ABBOTT, T.T. (1988). Glucoamylase immobilized on acid-hydrolysed starch graft polyacrylonitrile. *Biotechnology and Bioengineering* **31**, 759–769.

SMILEY, K.L., BOUNDY, J.A. AND HENSLEY, D.E. (1982). Action patterns of immobilized dextranase. *Carbohydrate Research* **104**, 319–324.

SOLOMON, B. (1978). Starch hydrolysis by immobilized enzymes: industrial applications. In *Advances in Biochemical Engineering*, volume 10 *Immobilized Enzymes* (T.K. Ghose, A. Fiechter and N. Blakebrough, Eds), pp. 132–153. Springer-Verlag, New York.

SORENSEN, J.E. AND EMBORG, C. (1989). Influence of immobilization technique on the quality of immobilized enzyme products based on a porous SiO_2 carrier. *Enzyme and Microbial Technology* **11**, 26–28.

SRINIVASAN, V.R. AND BUMM, M.W. (1974). Isolation and immobilization of β-D-glucosidase from *Alcaligenes faecalis. Biotechnology and Bioengineering* **26**, 1413–1418.

STÅLBRAND, H., HAHN-HÄGERDAL, B., RECZEY, K. AND TJERNELD, F. (1992). Mycelia-associated β-xylosidase in pellets of *Aspergillus* ss. *Applied Biochemistry and Biotechnology* **34/35**, 261–272.

STANLEY, W.L., WATTERS, G.G., KELLY, S.H., CHAN, B.G., GARIBALDI, J.A. AND SCHADE, J.E. (1976). Immobilization of glucose isomerase on chitin with glutaraldehyde and by simple adsorption. *Biotechnology and Bioengineering* **18**, 439–443.

STANLEY, W.L., WATTERS, G.G., KELLY, S.H. AND OLSON, A.C. (1978). Glucoamylase immobilized on chitin with glutaraldehyde. *Biotechnology and Bioengineering* **20**, 135–140.

STEFUCA, V., GEMEINER, P. AND BALES, V. (1988). Study of porous cellulose beads as an enzyme carrier via simple mathematical models for the hydrolysis of saccharose using immobilized invertase reactors. *Enzyme and Microbial Technology* **10**, 306–311.

STOREY, K.B. AND CHAKRABARTI, A.C. (1990). One-step conversion of cellulose to fructose using coimmobilized cellulase, β-glucosidase and glucose isomerase. *Applied Biochemistry and Biotechnology* **23**, 139–154.

STOREY, K.B., DUNCAN, J.A. AND CHAKRABARTI, A.C. (1990). Immobilization of amyloglucosidase using two forms of polyurethane polymer. *Applied Biochemistry and Biotechnology* **23**, 221–236.

STRANBERG, G.W. AND SMILEY, K.L. (1971). Free and immobilized glucose isomerase from *Streptomyces phaeochromogenes. Applied Microbiology* **21**, 588–593.

SUNDSTROM, D.W., KLEI, H.E., COUGHLIN, R.W., BIEDERMAN, G.J. AND BROUWER, C.A. (1981). Enzymatic hydrolysis of cellulose to glucose using immobilized β-glucosidase. *Biotechnology and Bioengineering* **23**, 473–485.

SVEC, F. AND KALAL, J. (1978). Immobilization of amyloglucosidase on poly[(glycidyl methacrylate) co(ethylene dimethacrylate)] carrier and its derivatives. *Biotechnology and Bioengineering* **20**, 1319–1328.

TAKEDA, Y. AND HIZUKURI, S. (1975). Effect of Triton X-100 on sweet potato β-amylase. *Biochimica et Biophysica Acta* **268**, 175–183.

TAKEUCHI, T. AND MAKINO, K. (1987). Cellulase immobilized on poly-L-glutamic acid. *Biotechnology and Bioengineering* **29**, 160–164.

TAN, L.U.L., WONG, K.K.Y. AND SADDLER, J.N. (1985). Functional characteristics of two D-xylanases purified from *Trichoderma harzianum. Enzyme and Microbial Technology* **7**, 431–436.

TAN, L.U.L., WONG, K.K.Y., YU, E.K.C. AND SADDLER, J.N. (1985). Purification and characterization of two D-xylanases from *Trichoderma harzianum. Enzyme and Microbial Technology* **7**, 425–430.

TOLDRA, F., JANSEN, N.B. AND TSAO, G.T. (1992). Hydrolysis of maltose and cornstarch by glucoamylase immobilized in porous glass fibres and beads. *Process Biochemistry* **27**, 177–181.

TOMAR, M. AND PRABHU, K.A. (1985). Immobilization of glucoamylase on DEAE-cellulose activated with chloride compounds. *Enzyme and Microbial Technology* **7**, 557–560.

UHM, T.-B., JEON, D.-Y., BYUN, S.M., HONG, J.-S. AND GROOTWASSINK, J.W.D. (1987). Purification and properties of B-fructofuranosidase from *Aspergillus niger. Biochimica et Biophysica Acta* **926**, 119–126.

VAN HOUDENHOVEN, F.E.A., DE WIT, P.J.G.M. AND VISSER, J. (1974). Large-scale preparation of galacturonic acid oligomers by matrix-bound polygalacturonase. *Carbohydrate Research* **34**, 233–239.

VERHOFF, F.H., BOGUSLAWSKI, G., LANTERO, O.J., SCHLAGER, S.T. AND JAO, Y.C. (1985). Glucose isomerase. In *Comprehensive Biotechnology* (M. Moo-Young, Ed.), volume 3, pp. 837–859. Pergamon Press, Oxford.

VOLKIN, D.B. AND KLIBANOV, A.M. (1989). Mechanism of thermoinactivation of immobilized glucose isomerase. *Biotechnology and Bioengineering* **33**, 1104–1111.

VRETBLAD, P. AND AXEN, R. (1973). Preparation and properties of an immobilized barley β-amylase. *Biotechnology and Bioengineering* **15**, 783–794.

WARD, O.P. (1985). Hydrolytic enzymes. In *Comprehensive Biotechnology* (M. Moo-Young, Ed.), volume 3, pp. 820–835. Pergamon Press, Oxford.

WEETALL, H.H. (1993). Preparation of immobilized proteins covalently coupled through silane coupling agents to inorganic supports. *Applied Biochemistry and Biotechnology* **41**, 157–188.

WEETALL, H.H. AND PITCHER, W.H. (1986). Scaling up an immobilized enzyme system. *Science* **232**, 1396–1403.

WEIBEL, M.K., BARRIOS, R., DELOTTO, R. AND HUMPHREY, A.E. (1975). Immobilized enzymes: pectin esterase covalently coupled to porous glass particles. *Biotechnology and Bioengineering* **17**, 85–98.

WISEMAN, A. (Ed.) (1985). *Handbook of Enzyme Biotechnology*. Ellis Harwood, UK.

WOJCIK, A., LOBARZEWSKI, J., BLASZCZYNSKA, T. AND FIEDUREK, J. (1987). Silica gels activated by BC13 and aliphatic diamines as supports for glucoamylase immobilization. *Biotechnology and Bioengineering* **30**, 983–986.

WONG, K.K.Y., TAN, L.U.L. AND SADDLER, J.N. (1986). Functional interactions among three xylanases from *Trichoderma harzianum. Enzyme and Microbial Technology* **8**, 617–622.

WOOD, D.A. (1985). Useful biodegradation of lignocellulose. In *Plant Products and the New Technology* (K.W. Fuller and J.R. Gallon, Eds). *Annual Proceedings of the Phytochemical Society of Europe*, volume 26, pp. 295–309. Clarendon Press, Oxford.

WOODWARD, J. (1984). Xylanases: functions, properties and applications. *Tropics in Enzyme and Fermentation Biotechnology* **8**, 9–30.

WOODWARD, J. (1987). Utilization of cellulose as a fermentation substrate: problems and potential. In *Carbon Substrates in Biotechnology*, Special Publication of the Society for General Microbiology, volume 21, pp. 45–65. IRL Press, Oxford.

WOODWARD, J. (1989). Immobilized cellulases for cellulose utilization. *Journal of Biotechnology* **11**, 299–312.

WOODWARD, J. AND ARNOLD, S.L. (1981). The inhibition of β-glucosidase activity in *Trichoderma reesei* C30 cellulase by derivatives and isomers of glucose. *Biotechnology and Bioengineering* **23**, 1553–1562.

WOODWARD, J. AND CAPPS, K.M. (1992). Cellobiose hydrolysis by glutaraldehyde-treated β-glucosidase entrapped in propylene glycol alginate/bone gelatin spheres. *Applied Biochemistry and Biotechnology* **34–35**, 341–347.

WOODWARD, J. AND WISEMAN, A. (1978). The involvement of salt links in the stabilization of Baker's yeast invertase. *Biochimica et Biophysica Acta* **527**, 8–16.

WOODWARD, J. AND ZACHRY, G.S. (1982). Immobilization of cellulase through its carbohydrate side chains. A rationale for its recovery and reuse. *Enzyme and Microbial Technology* **4**, 245–248.

WYKES, J.R., DUNNILL, P. AND LILLY, M.D. (1971). Immobilization of α-amylase by attachment to soluble support materials. *Biochimica et Biophysica Acta* **25**, 522–529.

XIAO, C., HUI, Z., WEI, L. AND JIACONG, S. (1990). Simultaneous saccharification and isomerization by immobilized glucoamylase and glucose isomerase. *Journal of Chemical Technology and Biotechnology* **47**, 161–169.

YAMAZAKI, H., CHEOK, R.K.H. AND FRASER, D.E. (1984). Immobilization of invertase on polyethylenimine-coated cotton cloth. *Biotechnology Letters* **6**, 165–170.

YANG, S.-S. AND COLEMAN, R.D. (1987). Detection of pullulanase in polyacrylamide gels using pullulan-reactive red agar plates. *Analytical Biochemistry* **160**, 480–482.

YEOH, H.-H. (1992). Cassava β-glucosidase electrode prepared with photocrosslinkable prepolymer resins. *Biotechnology and Applied Biochemistry* **15**, 221–225.

YU, E.K.C. AND SADDLER, J.N. (1985). Biomass conversion to butanediol by simultaneous saccharification and fermentation. *Trends in Biotechnology* **3**, 100–104.

YU, E.K.C., TAN, L.U.L., CHAN, M.K.-H., DESCHATELETS, L. AND SADDLER, J.N. (1987). Production of thermostable xylanase by a thermophilic fungus, *Thermoascus aurantiacus. Enzyme and Microbial Technology* **9**, 16–24.

ZABORSKY, O. (1973). Properties of covalently bonded water-insoluble enzyme-polymer conjugates. In *Immobilized Enzymes* (R.C. Weast, Ed.), pp. 49–60. CRC Press, Cleveland.

13
Methylglyoxal and Lipid Hydroxperoxide as Endogenous Cytotoxic Molecular Species: Detoxification and Regulation of Gene Expression in Yeasts

YOSHIHARU INOUE

Research Institute for Food Science, Kyoto University, Uji, Kyoto, Japan

Introduction

In 1978, Hinnen, Hicks and Fink first reported the transformation of protoplast cells of *Saccharomyces cerevisiae* using polyethylene glycol and $CaCl_2$. Harashima, Takagi and Oshima (1984) indicated that transformants obtained by the protoplast method were sometimes diploids or polyploids, because cell fusion occurred during the transformation. On the other hand, Ito *et al.* (1983) reported that intact yeast cells of *S. cerevisiae* could be transformed by treating the cells with alkaline cations. The transformants obtained showed the same ploidy as that of the recipient (Harashima, Takagi and Oshima, 1984). Through several modifications, the alkaline cation method (lithium acetate method) was established as a convenient method for the transformation of yeasts, not only *S. cerevisiae* but also other yeast strains such as *Hansenula polymorpha* (Roggenkamp *et al.*, 1986) and *Candida biodinni* (Sakai, Kazarimoto and Tani, 1991).

As well as the method for transformation, studies of the construction of vectors have also been developed. In the transformation of yeast, the genes complementing the auxotrophic markers, such as *HIS3, LEU2* and *URA3* are used as marker genes. Several genes for resistance to drugs such as G418 (Webster and Dickson, 1983) and tunicamycin (Rine *et al.*, 1983), are also used as dominant selectable marker genes

---

Abbreviations: BHP, butyl hydroperoxide; BSO, buthionine sulphoximine; CDNB, 1-chloro-2,4-dinitrobenzene; CHAPS, 3-cholamidopropyldimethylammonio-1-propane sulphonate; EDTA, ethylenediaminetetraacetic acid; GPx, glutathione peroxidase; GSH-I, γ-glutamylcysteine synthetase; GSH-II, glutathione synthetase; GST, glutathione *S*-transferase; HPLC, high-pressure liquid chromatography; HSE, heat-shock element; HSF, heat-shock factor; HSP, heat-shock protein; IPTG, isopropyl-β-*D*-thiogalactopyranoside; MCO, metal-catalysed oxidation; mRNA, messenger RNA; PHGPx, phospholipid hydroperoxide glutathione peroxidase; Pi, inorganic phosphate; SDS-PAGE, sodium dodecyl sulphate polyacrylamide gel electrophoresis; SOD, superoxide dismutase; UAS, upstream activating sequence.

Biotechnology and Genetic Engineering Reviews — Vol. 12, December 1994
0264–8725/94/12/467–508 $20.00 + $0.00 © Intercept Ltd, P.O. Box 716, Andover, Hampshire SP10 1YG, UK

and can be used to select transformants of prototropic and industrial strains of yeast. The author has also investigated the possibility of using some stress-resistance genes as selectable markers for the transformation of yeasts.

Several environmental stresses are known to trigger intracellular alterations in micro-organisms, e.g. cellular responses against extracellular signals or the induction of some genes. On the other hand, intracellular stresses, such as waste materials in the cells or some metabolites possessing cytotoxicity, would also affect the growth or metabolism of micro-organisms. Against those extracellular and/or intracellular stresses, micro-organisms have some mechanisms for adaptation or resistance. In this chapter, I have chosen two kinds of stresses, i.e. methylglyoxal and lipid hydroperoxide as the endogenous cytotoxic molecular species.

Methylglyoxal is a typical cellular 2-oxoaldehyde. Formation of this compound is performed both enzymatically and non-enzymatically, and the major precursors of methylglyoxal are dihydroxyacetone phosphate or glyceraldehyde 3-phosphate, which are derived from glycolysis, an ubiquitous energy-generating system in organisms. Though methylglyoxal is a normal metabolite, it arrests the growth of various organisms at millimolar concentrations. Because the compound is toxic, its synthesis and degradative pathway must be strictly controlled in cells to avoid its overaccumulation.

On the other hand, lipid hydroperoxide is also one of the endogenous cytotoxic molecular species, and it affects the growth or proliferation of cells. Lipid hydroperoxide is synthesized both enzymatically and non-enzymatically from unsaturated fatty acid or lipid containing unsaturated fatty acid moieties. Phospholipids included in the cytoplasmic or mitochondrial membrane are also peroxidized. Lipid hydroperoxide is degraded in the presence of transition metal ions to yield radicals such as the alcoxy radical and the peroxy radical. These radicals are reactive oxygen species and they cause radical chain reactions together with other lipids, lipid hydroperoxide and oxygen. Thus, lipid hydroperoxide would be one of the oxidative stresses in living cells.

In this chapter, the biochemical aspects of metabolism of methylglyoxal and lipid hydroperoxide in yeast cells are described, as are the genes corresponding to the metabolism or detoxification of methylglyoxal and lipid hydroperoxide in yeast cells, and the possibility of the use of such genes as selectable marker genes for the transformation of yeast is discussed.

Glyoxalase systems in yeasts

METABOLISM OF METHYLGLYOXAL IN MICRO-ORGANISMS

Formation of methylglyoxal

The major route for formation of methylglyoxal in microbial cells is catalysed by methylglyoxal synthase (EC 4.2.99.11) (*Figure 1*). The enzyme catalyses the conversion of dihydroxyacetone phosphate to methylglyoxal without formation of Schiff base (Hopper and Cooper, 1972). Methylglyoxal synthases were purified from several micro-organisms, such as *Escherichia coli* (Hopper and Cooper, 1972), *Pseudomonas saccharophilia* (Cooper, 1974), *Proteus vulgaris* (Tsai and Gracy,

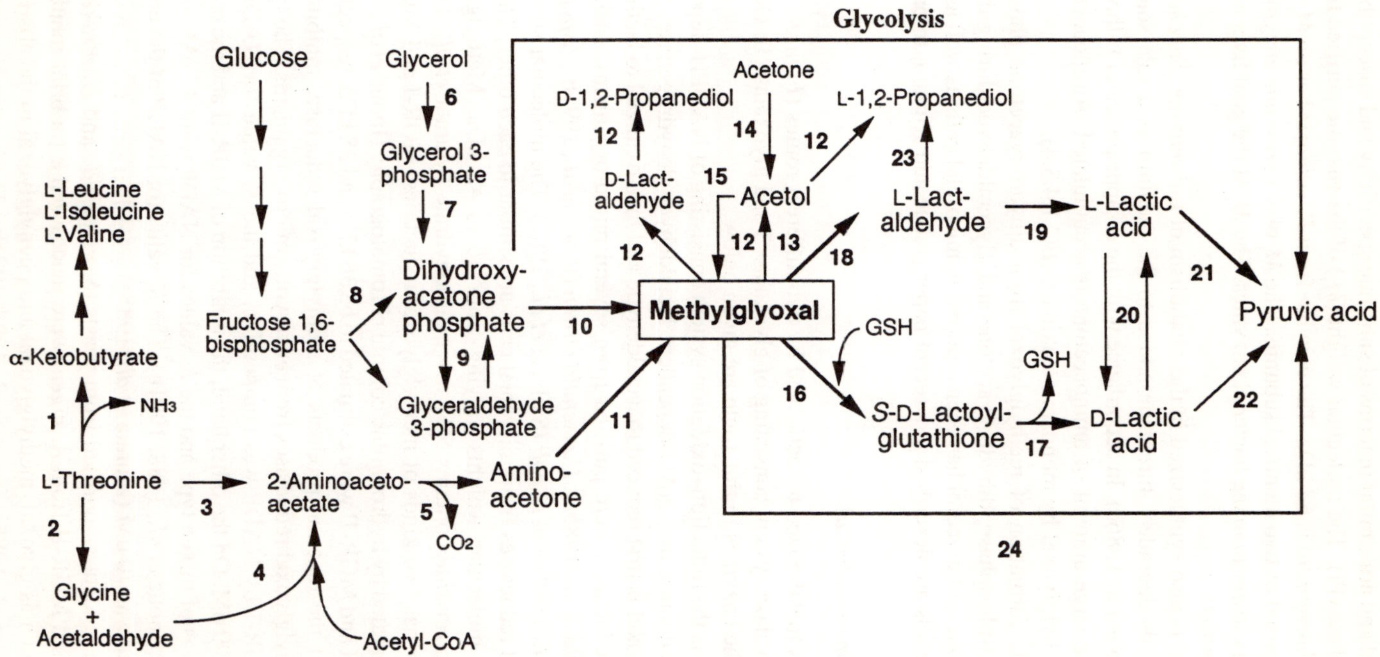

Figure 1. Metabolism of methylglyoxal. The enzymes indicated by the numbers are: 1, threonine deaminase; 2, threonine aldolase; 3, threonine dehydrogenase; 4, aminoacetone synthase; 5, spontaneous decarboxylation; 6, glycerol kinase; 7, glycerol 3-phosphate dehydrogenase; 8, fructose 1,6-bisphosphate aldolase; 9, triosephosphate isomerase; 10, methylglyoxal synthase; 11, monoamine oxidase; 12, aldolase reductase; 13, aldehyde reductase; 14, acetone mono-oxygenase; 15, acetol dehydrogenase; 16, glyoxalase I; 17, glyoxalase II; 18, methylglyoxal reductase; 19, lactaldehyde dehydrogenase; 20, lactate racemase; 21, L-lactate dehydrogenase; 22, D-lactate dehydrogenase; 23, propanediol oxidoreductase; 24, 2-oxoaldehyde dehydrogenase.

1976) and *S. cerevisiae* (Murata *et al.*, 1985b). The same enzyme activity was also reported in goat liver (Ray and Ray, 1981). The properties of these enzymes have been well documented and are similar in terms of substrate specificity and susceptibility to inorganic phosphate (Pi). The molecular weight (M_r) of the enzyme purified from *E. coli* and *P. saccharophilia* is 67 kDa. The enzyme from *P. vulgaris* has an M_r of 135 kDa and is composed of two identical subunits. The M_r of *S. cerevisiae* enzyme is 26 kDa, significantly lower than the bacterial enzymes. The M_r of the goat liver enzyme has not been reported.

Methylglyoxal is also synthesized in the catabolism of L-threonine. *Saccharomyces cerevisiae* cells degrade L-threonine added to the medium as a sole source of nitrogen (Murata *et al.*, 1986c). In accordance with the disappearance of L-threonine in the medium, a large amount of aminoacetone is accumulated. Aminoacetone is converted to methylglyoxal by monoamine oxidase (EC 1.4.3.4).

Non-enzymatic formation of methylglyoxal from dihydroxyacetone phosphate, glyceraldehyde 3-phosphase, dihydroxyacetone and glyceraldehyde during the metabolism of acetone and glycerol has been reported in microbial cells as well as those of mammals. Details are described in a recent paper (Inoue and Kimura, in press).

Degradation of methylglyoxal

Methylglyoxal is metabolized to lactic acid by two different routes (*Figure 1*). One route is the glyoxalase system consisting of glyoxalase I and glyoxalase II. Detailed descriptions of the enzymes in this route are given below.

Another route is the reduction–oxidation system consisting of NADPH-dependent methylglyoxal reductase and NAD-dependent lactaldehyde dehydrogenase. In this route, methylglyoxal is first reduced to lactaldehyde and then oxidized to lactic acid. Methylglyoxal reductases were purified from several micro-organisms, such as *S. cerevisiae* (Murata *et al.*, 1985a), *Hansenula mrakii* (Inoue *et al.*, 1991b), *Aspergillus niger* (Inoue *et al.*, 1988) and *E. coli* (Saikusa *et al.*, 1987). The molecular structures of methylglyoxal reductases from microbial cells are similar to each other. They are monomers with molecular weights varying from 36 to 43 kDa. Methylglyoxal reductase of *S. cerevisiae* contains 5% (w/w) carbohydrate (Murata *et al.*, 1985a). *Aspergillus niger* has two kinds of methylglyoxal reductases, MGR-I and MGR-II, which can be separated by hydrophobic column chromatography (Inoue *et al.*, 1988). The M_r of MGR-I and MGR-II were estimated to be 36 kDa and 38 kDa, respectively.

NADP, one of the reaction products of methylglyoxal reductase, inhibited the activities of methylglyoxal reductases from eukaryotic micro-organisms. The type of inhibition of MGR-I by NADP was competitive, and the K_i value for NADP was estimated to be 490 μM. On the other hand, the inhibition of MGR-II and the enzyme from *H. mrakii* was of mixed type and the K_i values for NADP were 45 μM and 250 μM, respectively (Inoue *et al.*, 1988, 1991b). The K_i value for NADP of the enzyme from *S. cerevisiae* was 70 μM (Murata *et al.*, 1985a).

Methylglyoxal reductases purified from yeasts (*S. cerevisiae* and *H. mrakii*) and mould (MGR-I and MGR-II from *A. niger*) were inactivated by a brief incubation with substrates such as glyoxal, methylglyoxal and phenylglyoxal in the absence of NADPH (Murata *et al.*, 1985a; Inoue *et al.*, 1988, 1991b). The inactivation was not observed when the enzyme was preincubated with NADPH prior to exposure to 2-

oxoaldehydes. Since methylglyoxal and phenylglyoxal are known to modify the arginine residue in proteins (Takahashi, 1968), methylglyoxal reductases from these sources may contain the arginine residue in the NADPH-binding site.

Lactaldehyde thus formed is further metabolized to lactic acid by lactaldehyde dehydrogenase (*Figure 1*). Purification and characterization of lactaldehyde dehydrogenase from microbial cell has been reported for *E. coli* (Sridhara and Wu, 1969) and *S. cerevisiae* (Inoue *et al.*, 1985). The M_r of the *E. coli* enzyme was estimated to be 100 kDa (Sridhara and Wu, 1969), whereas Baldoma and Aguilar (1987) reported that the M_r of the enzyme, measured by SDS-PAGE, was 55 kDa. On the other hand, the enzyme from *S. cerevisiae* was a monomer with an M_r of 40 kDa. The *E. coli* enzyme catalyses the oxidation of L-lactaldehyde to L-lactic acid in the presence of NAD, and is almost specific for L-lactaldehyde; but other aldehydes, such as D-lactaldehyde (relative activity, 5%), pyruvaldehyde (relative activity, 3%), D,L-glyceroaldehyde (relative activity, 3%) and propionaldehyde (relative activity, 12%), are also slightly oxidized by the enzyme. On the other hand, lactaldehyde dehydrogenase from *S. cerevisiae* was almost completely specific to L-lactaldehyde, i.e. utilization of D-lactaldehyde by the enzyme was only 0.2% compared with that of L-lactaldehyde. The enzyme was specific to NAD, and NADP could not substitute for NAD. Recently, Inoue and Kimura (1994) found that NADH formed by the reaction of lactaldehyde dehydrogenase is reoxidized by NAD(P)H dehydrogenase, which is closely linked with lactaldehyde dehydrogenase.

Other routes

Methylglyoxal is reduced to acetol in the presence of NADH by aldehyde reductase. The enzyme was partially purified from *H. mrakii* (Inoue *et al.*, 1992a). This enzyme activity has not been reported in *S. cerevisiae* (Murata *et al.*, 1986a). Some bacteria, such as *Bacillus subtilis* and *Pseudomonas putida* could directly oxidize methylglyoxal to pyruvic acid by the action of 2-oxoaldehyde dehydrogenase (Willetts and Turner, 1970; Rhee *et al.*, 1987). Details are described by Inoue and Kimura (in press).

GLYOXALASE SYSTEM

Glyoxalase I

Conversion of methylglyoxal to D-lactic acid *via S*-D-lactoylglutathione is catalysed by the glyoxalase system, consisting of glyoxalase I and glyoxalase II in the presence of glutathione. Methylglyoxal is non-enzymatically condensed with glutathione to give an adduct, hemi-mercapal (hemi-thioacetal), which is the intrinsic substrate for glyoxalase I (*Figure 1*). The dissociation constant (K_d) for hemi-mercaptal is 3×10^{-3} M (Vander Jagt, 1993).

Since methylglyoxal is cytotoxic, bacterial cells such as *E. coli* or *Pseudomonas putida* cannot grow in media containing 1.0–1.2 mM methylglyoxal (Egyud and Szent-Gyorgyi, 1966; Rhee, Murata and Kimura, 1987). The author and his co-workers found that the yeast *Hansenula mrakii* IFO 0895 was highly resistant to methylglyoxal and was able to grow in a medium containing up to 25 mM methylglyoxal (Inoue *et al.*, 1991a). The specific activity of glyoxalase I in cell extracts of

H. mrakii (1.48–2.11 units mg^{-1} protein) (Inoue *et al.*, 1991a) was relatively higher than those of *S. cerevisiae* (0.238 units mg^{-1} protein) (Kosugi *et al.*, 1988), *A. niger* (0.026 units mg^{-1} protein) (Inoue *et al.*, 1987), *P. putida* (0.145 units mg^{-1} protein) (Rhee, Murata and Kimura, 1986) and *E. coli* (0.012 units mg^{-1} protein) (Rhee, Murata and Kimura, 1987).

Glyoxalase I is widely distributed in organisms and has been purified from various sources, from mammals to micro-organisms. Mammalian glyoxalase Is are dimers, with molecular weights around 43–48 kDa. On the other hand, microbial glyoxalase Is are monomers and molecular weights vary from 19 kDa to 38 kDa. Douglas, Seddon and Nakagawa (1986) suggested that yeast glyoxalase I contained approximately 0.75% (w/v) carbohydrate.

Each subunit of mammalian glyoxalase Is contains one atom of zinc (Aronsson, Marmstal and Mannervik, 1978; Marmstal, Aronsson and Mannervik, 1979). The zinc ion is located in the catalytic site of the enzyme and can be replaced by other bivalent metal ions, such as Mg^{2+}, Mn^{2+}, Co^{2+} and Ni^{2+} (Uotila and Koivusalo, 1975; Han *et al.*, 1977; Aronsson *et al.*, 1981; Sellin *et al.*, 1983; Sellin and Mannervik, 1984). The metal ion in the active site is suggested to be co-ordinated to two nitrogen and to four oxygen atoms (Sellin *et al.*, 1987), although the amino acid residues involved in the co-ordination were not identified (Mannervik and Ridderstorm, 1993).

The activity of glyoxalase I purified from *S. cerevisiae* was inhibited by ethylenediaminetetraacetic acid (EDTA), and the activity of EDTA-inhibited enzyme was restored partially by Mg^{2+} or Ca^{2+}, and slightly by Mn^{2+} (Murata *et al.*, 1986d). Fe^{2+}, Ni^{2+} and Co^{2+} were all without any effect on reactivation. Glyoxalase I purified from *P. putida* (Rhee, Murata and Kimura, 1986) was not inhibited by EDTA. The enzyme from a yeast, *H. mrakii*, was also insensitive to bivalent metal ion chelators such as EDTA, 1,2-cyclohexanediaminetetraacetic acid and 8-hydroxyquinoline (Inoue *et al.*, 1991a). In analogy with other microbial glyoxalase Is, the activity of the enzyme from *A. niger* was also inhibited by Zn^{2+} (Inoue *et al.*, 1987). The enzyme was protected from inhibition by Zn^{2+} by the addition of an equimolar amount of EDTA. The activity of glyoxalase I from the mould was inhibited by EDTA, and the inhibition was found to be a competitive type against hemi-mercaptal (K_i=1.3 mM). This inhibitory effect of EDTA on the reaction might be due to the structural similarity between hemi-mercaptal and EDTA. Douglas and Shinkai (1985) suggested the model for the reaction of glyoxalase I. They indicated that two oxygen atoms of hemi-mercaptal at the position of C-1 and C-2 could easily chelate Zn^{2+} in the enzyme with the subsequent isomerization of hemi-mercaptal. On the other hand, EDTA also has a similar structure of the oxygen atoms, being able to chelate bivalent metal ions. Therefore, EDTA may compete with hemi-mercaptal at the active site of the mould glyoxalase I. Mannervik, Lindsstorm and Bartfai (1972) suggested that glyoxalase I from mammals contained Zn^{2+} and that the inhibitory effect of EDTA was due to a removal of Zn^{2+}, the EDTA-inhibited enzyme being reactivated by Zn^{2+} and/or other bivalent metal ions. The reaction using the mould glyoxalase I was inhibited by EDTA; however, the mould enzyme was not found to be inactivated by either preincubation with EDTA or dialysis against the buffer containing EDTA. The mould enzyme, as well as the bacterial (*P. putida*) and the yeast (*H. mrakii*) enzymes, may not contain Zn^{2+} in its active site, or Zn^{2+} might be tightly associated with the enzyme.

Glyoxalase I activity in the haploid cells of *S. cerevisiae* changes when the cells are exposed to the opposite-type mating factor. It seems to be regulated through the phosphorylation/dephosphorylation state of the enzyme (Inoue *et al.*, 1989, 1990a; Inoue and Kimura, 1991). The enzyme activity in *S. cerevisiae* also increases when the cells are exposed to a high-cell concentration culture. In this case, cell–cell interaction affects the amount of glyoxalase I mRNA (Inoue, Yano and Kimura, 1993d). From these observations, glyoxalase I is not only a detoxification enzyme of methylglyoxal in the cells but it may also have some significant functions relating to cell growth, proliferation and differentiation.

Glyoxalase II

Glyoxalase II is an alternative enzyme in the glyoxalase system. It catalyses the hydrolysis of *S*-D-lactoylglutathione, formed by the reaction of glyoxalase I, to glutathione and D-lactic acid (*Figure 1*). Glyoxalase IIs have been purified from various sources. Mammalian enzymes are distributed in various tissues, such as brain, liver and erythrocytes. The enzymes are found in mitochondria as well as in the cytosol, although other organelles are unlikely to contain the enzyme (Jerzykowski *et al.*, 1978; Talesa *et al.*, 1988; Uotila, 1989). The molecular weighs of glyoxalase IIs vary within the range 18–30 kDa. Specific activity of the enzyme is approximately 600–900 μmol min^{-1} mg^{-1} protein, except for yeast enzymes (*S. cerevisiae*, 1.34 μmol min^{-1} mg^{-1} protein; *H. mrakii*, 31.9 μmol min^{-1} mg^{-1} protein) (Murata *et al.*, 1986b; Inoue and Kimura, 1992b). The K_m value for *S*-D-lactoylglutathione is in the range 86–450 μM, except for an extraordinarily low K_m value of the enzyme from *S. cerevisiae* (Murata *et al.*, 1986b). The values of k_{cat} of calf brain and rat erythrocyte were estimated to be 16×10^3 min^{-1} and 17×10^3 min^{-1}, respectively, with k_{cat}/K_m values of about 10^8 M^{-1} min^{-1}; suggesting that the hydrolysis of *S*-D-lactoylglutathione by glyoxalase II is diffusion limited (Guha, Vander Jagt and Creighton, 1988).

Glyoxalase II can hydrolyse a wide variety of *S*-hydroxyacylglutathione derivatives, such as *S*-acetyl, *S*-lactoyl, *S*-glycoyl, *S*-mandelyl, *S*-glyceroyl, *S*-acetoacetylglutathiones. Glyoxalase II from *S. cerevisiae* was, however, specific to *S*-D-lactoylglutathione. *Saccharomyces cerevisiae* has other enzymes that are active to *S*-D-lactoylglutathione as well as *S*-acetylglutathione, and one of the enzymes was purified (Murata *et al.*, 1987). The enzyme (P3) showed group-transfer activity and catalysed the reaction to form acetyl-coenzyme A from *S*-acetylglutathione and coenzyme A. *S*-D-Lactoylglutathione-hydrolysing activity was not divided into more than one peak during purification from the yeast, *H. mrakii* (Inoue and Kimura, 1992b; Inoue *et al.*, 1994a).

Glyoxalase II is inhibited by many derivatives of glutathione. *S*-(Substituted-carbobenzoxy)glutathione derivatives are the strongest competitive inhibitors; suggesting the occurrence of a hydrophobic substrate-binding site in glyoxalase II (Al-Timari and Douglas, 1986; Principato *et al.*, 1989). Hemi-mercaptal, a non-enzymatic condensation product between methylglyoxal and glutathione, inhibited glyoxalase II (Uotila, 1973; Oray and Norton, 1980; Murata *et al.*, 1986b; Rae *et al.*, 1990; Rae, Board and Kuchel, 1991; Inoue and Kimura, 1992b). Yeast glyoxalase IIs were also inhibited by hemi-mercaptal, although the activity of the P3-enzyme, which hydrolysed *S*-D-lactoylglutathione, was not affected by hemi-mercaptal (Murata *et al.*, 1987).

Glyoxalase I-related genes in *Saccharomyces cerevisiae*

GLYOXALASE I-DEFICIENT MUTANTS: A COMPLEMENTARY GENE

Although methylglyoxal is a normal metabolite in living cells, it arrests the growth of several organisms at millimolar concentrations. Detoxification of methylglyoxal is mainly performed by glyoxalase I. To clone the gene for glyoxalase I from *S. cerevisiae*, the mutants highly sensitive to methylglyoxal were isolated. Several kinds of mutants which showed slow growth in a medium containing methylglyoxal were obtained, some of which showed a temperature-sensitive phenotype (M26, M35, M39, M50 and M52). When such mutants were cultured at 23°C, the cells showed almost the same growth rate compared with that of wild-type cells in the presence of 0.5 mM methylglyoxal (*Figures 2a,b*). On the other hand, when the mutants were cultured at 35°C in the presence of 0.5 mM methylglyoxal, the growth of M35 and M39 was slightly inhibited (*Figure 2e*). In the presence of 1.0 mM methylglyoxal, notable growth inhibition was observed in all mutants (*Figure 2f*). Glyoxalase I activity in the mutants was assayed and was found to be lower than that of the wild-type strain in mutants M35 and M39, whereas no difference in enzyme activity was observed in other mutants. Therefore, the decrease in glyoxalase I activity was thought to be a major factor for the growth inhibition of M35 and M39 in the presence of methylglyoxal at 35°C. Penninckx, Jespers and Legrain (1983) also reported that the growth rate of a glyoxalase I-deficient yeast mutant was slightly lower than that of the parental cells.

To clone the gene corresponding to glyoxalase I, a genomic DNA library of *S. cerevisiae* was screened for complementation to the mutation of M39. Several candidates were obtained, and the mutation was phenotypically complemented in all transformants. A recombinant plasmid was recovered from one clone and was designated p39-b. It contained 3.8 kb of *S. cerevisiae* chromosomal DNA. The nucleotide sequence of the 3.8 kb fragment was determined. It contained one open reading frame of 510 bp and the molecular weight (M_r) of the polypeptide deduced from the nucleotide sequence was calculated to be 18 875 Da (170 amino acids) (*Figure 3*). The gene has four putative TATA-like sequences at 1002 bp (TATATA), 989 bp (TATA), 434 bp (TATAA) and 133 bp (TATA) upstream, respectively, from the translational initiation codon (ATG). The polyadenylation signal (AATAAA) was located 131 bp downstream of translational termination codon (TGA).

Glyoxalase I purified from *S. cerevisiae* was digested with trypsin and the N-terminal amino acid sequences of two tryptic peptides obtained by reverse-phase high-pressure liquid chromatography (HPLC) (peaks 5 and 13) were determined. The sequence was as follows: peak 5, Phe–Tyr–Thr–Glu–His–Phe–Gly; peak 13, Asp–Pro–Asp–Gly–Tyr–Ser–Ile–Glu–Val–Val–Pro–Xaa–Gly. These sequences were not found in the amino acid sequence deduced from the nucleotide sequence of the gene, herewith sequenced. Therefore, the gene cloned and sequenced was thought not to correspond to the structural gene of yeast glyoxalase I.

GLYOXALASE I GENE AND REGULATION OF ITS EXPRESSION

Glyoxalase I activation conferring (GAC) gene from S. cerevisiae

Growth of the yeast *S. cerevisiae* was inhibited in the presence of 2 mM methylglyoxal.

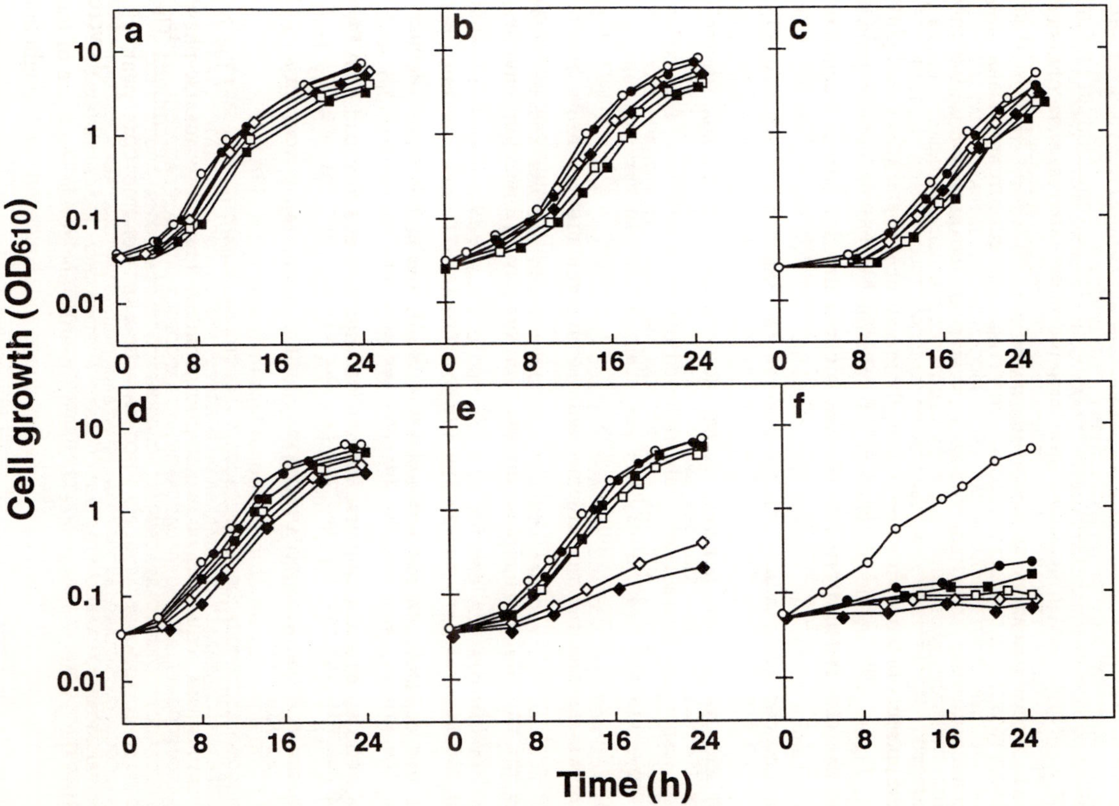

Figure 2. Growth of methylglyoxal-sensitive mutants. Cells were cultured in the synthetic dextrose (SD) minimal medium containing 0 mM (a and d), 0.5 mM (b and e), and 1.0 mM (c and f) methylglyoxal, with reciprocal shaking at 23°C (a, b and c) or 35°C (d, e and f), respectively. Cell growth was monitored by measuring the optical density of the culture at 610nm (OD_{610}). Symbols: O, wild type; ●, M26; ◊, M35; ♦, M39; □, M50; ■, M52.

```
                                                GTACTCTTCGCTGAACCGGGTTTTTT -1088
    TCTTTGCAATTTTTTTTTCGTTCTCCTAAAGCATACACAAATAAATCCTTTTTTTTATTTTCTATTT -1021
    ATTTTGTTATTTATCATCTATATAGCAATAATATACTTTGTTTTTATTCGTATTTCACACTTTTGCT -954
    TTTTCCTTATGCAGGCAGTGTAATTCATTGGGGAGGTGATTTTCATGTGCGCATATCCTTGCGTGCA -887
    AGCAGCCGGTCGGTGGCAAATCCGGCGCTTCCCCCTCAAAAAAAAAAAAAAAAAAAAAAAAAGGGAA -820
    CTCAGAACGGGGGAGGTTGAAGAGCAGGCCAAGGGAAATATTAGTTTTGACCTATGTGGGAAACAGA -753
    ATTTTCAATGAGTTATGGCAACTTGGCCGAGTGGTTAAGGCGAAAGATTAGAAATCTTTTTGGGCTT -686
    TGCCCGCGCAGGTTCGAGTCCTGCAGTTGTCGTTATTTTTCTCTTTTTTTTCAACTTTCCCTTGTTC -619
    ATCAGATCGAGGCGGTAGAAGAAACAATTACTTTTCCTAAATGGGTAAAAACTCGTGTTTTAGGAAA -552
    AAAAAGAAAAATTTGGTCAAAACTCGAAAGATAGGTTCTTAATCTTCTTTCAAGTTGAAAAGGCCTA -485
    CGCTCTTTTCCTTGAAGCATTTTCATCCTACTGCTCGTATTGAACTCCACTATAAGCGCACCAAAAA -418
    GATACAAACGTCAATTATGGAAGCTCAACGGGATTTGATGGAGACGCTACTACTTTTTTCGCTCCAG -351
    ACGCTGTGTTTGGTGACAGAGTGCGCAGATTTCAAGAGTTTTTAGATACTTTCACCTCATACAGAGA -284
    CTCTGTAAGGTCCATACAAGTTTACAACAGCAATAACGCGGCCAACTACAACGATGATCAAGATGAC -217
    GCAGACGAACGAGATTTGCTAGGTGATGACGACGGTGATGATCTTGAAAAGGAAAAGAAAGCAGCAT -150
    CGTCCACCTCATTGAATATACTCCCTCACAGGATTATCATCTCGCTTGATGACTTGAGAGAATTCGA -83
    CAGGTCGTCTGGTCGGCATTTAGTCGAACCAGCATACTTCATCCCGCCTGCCGAAAAGGCGCTTACT -16
                    +1
    GACCTAGCAGATTCC ATG GAC GAT GTT CCA CAT CCC AAT GCC TCT GCA GTA TCG +39
                    Met Asp Asp Val Pro His Pro Asn Ala Ser Ala Val Ser

    TCT CGC CAT CCT TGG AAG CTT TCG TTC AAA GGC TCA TTT GGT GCA CAC GCA +90
 14 Ser Arg His Pro Trp Lys Leu Ser Phe Lys Gly Ser Phe Gly Ala His Ala

    TTG TCT CCT CGT ACT CTA ACG GCA CAA CAT TTA AAC AAA CTG GTC TCT GTT +141
 31 Leu Ser Pro Arg Thr Leu Thr Ala Gln His Leu Asn Lys Leu Val Ser Val

    GAG GGT ATC GTA ACT AAG ACT TCG TTG GTC AGG CCA AAG CTT ATC AGA TCT +192
 48 Glu Gly Ile Val Thr Lys Thr Ser Leu Val Arg Pro Lys Leu Ile Arg Ser

    GTC CAC TAC GCG GCA AAG ACT GGT AGA TTC CAT TAC AGA GAT TAT ACA GAT +243
 65 Val His Tyr Ala Ala Lys Thr Gly Arg Phe His Tyr Arg Asp Tyr Thr Asp

    GCT ACT ACA ACT CTC ACC ACC CGC ATC CCA ACG CCT ACC ATC TAT CCA ACG +294
 82 Ala Thr Thr Thr Leu Thr Thr Arg Ile Pro Thr Pro Thr Ile Try Pro Thr

    GAG GAC ACT GAA GGT AAC AAA CTA ACC ACC GAA TAT GGG TAT AGT ACG TTC +345
 99 Glu Asp Thr Glu Gly Asn Lys Leu Thr Thr Glu Tyr Gly Tyr Ser Thr Phe

    ATA GAC CAT CAC GTA TCA CTG TGC AAG AAA TGC CCG AAA TGG CCC CCG CTG +396
116 Ile Asp His His Val Ser Leu Cys Lys Lys Cys Pro Lys Trp Pro Pro Leu

    GCC AAC TTC CCA GGT CCA TTG ACG TCA TTC TCG ATG ACG ACC TTG TGG ACA +447
133 Ala Asn Phe Pro Gly Pro Leu Thr Ser Phe Ser Met Thr The Leu Trp Thr

    AGA CCA AGC CAG GTG ACA GAG TTA ACG TTG TCG GGG TAT TCA AGT CGC TTG +498
150 Arg Pro Ser Gln Val Thr Glu Leu Thr Leu Ser Gly Tyr Ser Ser Arg Leu

    GTG CTG GTG GCA TGA ACCAGTCCAACTCTAATACATTGATCGGGTTCAAAACTCTGATCCTA +560
167 Val Leu Val Ala ***

    GGTAATACGGTGTATCCTTCCACGCCAGATCCACGGGTGTCGCTGCGAGACAAATGTTGACAGATTTC +627
    GATATAAGAAATATCAATAAACTATTCCAAAAAAAGGACATTTTCGATATCTTGTCTCAATCTTTAG +694
    CGCCTTCTATTTATGGACATGATCATATAAAGAAGCATTTATTGATCGTCATGGAGTGTGAGAAAAAT +761
    TTAGAAAATGCTCGCATTTAAGAAGTGACATCAATATCCTAATGGTGGTGATCCATCCACTGCCAAGT +828
    CCCAATTGCTAAGGTTTGTGTTGAATACAGCATCACAGGCAATTGCTACTACTGGTAGAGGTTCTTCC +895
    GGTGTCGGTTTGACCGCAGCGGTCACTACTGATAGGGAAACAGGTGAAAGAAGACTGGTGCCATGGTT +962
    CTTGCTGACCGCGGGGTTGTATGTATTGATGAATTTGATAAGATGACAGATGTGGATAGAGTCGCCAT +1029
    TCATGAAGTAATGGAACAACAAACGGTGACGATTGCCAAAGCAGGTATTCACACAAACATTAAATGCT +1096
    CGTTGTAGTGTTATTGCTGCCGAAATCCCGTTTTTGGGCAGTACGATGTCAATAGAGATC
```

Figure 3. Nucleotide sequence of a DNA fragment in p39-b. The recombinant plasmid (p39-b) contained a 3.8 kb inserted fragment and the whole sequence of the fragment was determined. The longest open reading frame and its 5'- and 3'-flanking regions are shown. Putative TATA-like sequences and the polyadenylation signal are indicated by underlining. Numbers on the right-hand side of the figure show the nucleotide position, which started from the translational initiation codon (ATG). Numbers on the left-hand side of the figure show the amino acid position, beginning from the N-terminal methionine (Met).

The addition of an equimolar amount of glutathione or L-cysteine to the culture containing methylglyoxal eliminated the toxic effect of methylglyoxal, indicating that this effect is chemically neutralized by sulphydryl compounds. However, in the presence of a large amount of methylglyoxal (20 mM) and glutathione (20 mM), growth of yeast cells was inhibited. Since a non-enzymatic adduct (hemi-mercaptal) of methylglyoxal and glutathione is an intrinsic substrate for glyoxalase I, the yeast genomic DNA library was screened for the glyoxalase I gene by selecting transformants showing rapid growth on minimum agar plates containing 20 mM methylglyoxal and 20 mM glutathione. Three candidates were obtained and the glyoxalase I activity in these clones was appreciably higher (three- to fivefold) than that of the control strain (Murata *et al.*, 1988).

Since the phenotypic characters of these clones were identical, the nucleotide sequence of the inserted DNA in a hybrid plasmid from one of the clones was determined, and an open reading frame of 318 bp was identified. The M_r of the gene product deduced from the DNA sequence was calculated to be 14 700 Da (106 amino acids). However, yeast glyoxalase I is a monomer with M_r of 32 kDa. Thus, the open reading frame found was not thought to correspond to yeast glyoxalase I. Thus, we named the gene *GAC* (glyoxalase I activation conferring) (Inoue *et al.*, 1990c). The *GAC* gene gave two transcripts with different molecular sizes in *S. cerevisiae* cells. Cells dosed with the *GAC* gene by multicopy plasmids could grow in a medium containing methylglyoxal at concentration giving arrested growth of the control strain, suggesting that the *GAC* gene product was in fact functioning in the detoxification of methylglyoxal.

To confirm that the *GAC* gene was not the structural gene for yeast glyoxalase I, an expression system of the *GAC* gene in *E. coli* was constructed. The coding region of the *GAC* gene was ligated downstream of a *trc* promoter in an expression vector (pKK233-2) of *E. coli* and the resultant plasmid was introduced into *E. coli*. Expression of the *GAC* gene was analysed by Maxi-cell method and a protein with an M_r of approximately 15 kDa was found to be synthesized. To monitor the expression of the *GAC* gene in *E. coli* easily, the *GAC* gene was fused with a *lacZ* gene encoding the β-galactosidase of *E. coli*. The fusion gene was ligated downstream of a *tac* promoter of the expression vector pKK223–3 and the replication origin (*ori*) of the plasmid DNA was replaced with that of pUC19. Although the pKK223–3 is a multicopy vector, the *ori* of the plasmid was from pBR322. The copy number of pUC19 in *E. coli* cells is much higher than that of pBR322 because of the mutation near the *ori* region of pUC19. The resultant plasmid (pUG-Lac) was introduced into *E. coli* JM109 and expression was induced by IPTG (isopropyl-β-D-thiogalactopyranoside). The GAC–LacZ fusion protein was expressed in *E. coli*, and β-galactosidase activity could be detected. The GAC–LacZ fusion protein was purified and the N-terminal amino acid sequence was determined. The sequence coincided with that deduced from the nucleotide sequence of the *GAC* gene. A full-length gene of *GAC* was then ligated downstream of the *tac* promoter using the same system of *GAC–lacZ* fusion, and the resultant plasmid (pUGAC) was introduced into *E. coli* JM109. GAC protein produced in the *E. coli* cell reached approximately 7% of the total cellular proteins, although glyoxalase I activity did not increase (Inoue and Kimura, in press). Therefore, the *GAC* gene was confirmed as not encoding the structural gene for the yeast glyoxalase I.

Molecular cloning of the structural gene for glyoxalase I from S. cerevisiae

A structural gene for yeast glyoxalase I (*GLO1*) was cloned using anti-glyoxalase I IgG as a probe from the genomic DNA library constructed with λgt11. Approximately 2 × 10^5 plaques were screened, and a positive clone was obtained. The recombinant phage (λGI-10) was purified and re-transfected to *E. coli* Y1090, and the expression of the fusion protein with β-galactosidase and yeast glyoxalase I (GloI) was examined by immunoblotting analysis. A lysate of *E. coli* Y1090 transfected with λGI-10 was found to contain a fusion protein of LacZ–GloI. The λGI-10 contained approximately 3.2 kb of insert DNA in the *Eco*RI site of the vector. The nucleotide sequence of the insert DNA downstream of *Eco*RI was determined, and the amino acid sequence deduced from the nucleotide sequence coincided with that of a peak in the peptide fractionation. From the results of immunoblotting and nucleotide sequence analysis, we concluded that a part of the *GLO1* gene was cloned on to λGI-10 (Inoue *et al.*, 1991c).

Using the *Eco*RI-*Eco*RI small fragment of λGI-10 as a probe (approximately 500 bp), the *S. cerevisiae* genomic DNA library constructed using YEp13 as a vector was screened. A recombinant plasmid (p89-1) was isolated from one of the positive clones and the insert DNA was sequenced. The amino acid sequence deduced from the nucleotide sequence had the same sequence as those of peaks 5 and 13 in peptide fractionation. Recently a cDNA of human glyoxalase I was cloned and sequenced by independent groups (Kim *et al.*, 1993; Mannervik and Riddestrom, 1993; Ranganathan *et al.*, 1993). Sequence of our peaks 5 and 13 showed homology with an amino acid sequence deduced from the cDNA sequence of human glyoxalase I.

A structural gene for glyoxalase I of *Pseudomonas putida* was also cloned and its nucleotide sequence was determined (Rhee, Murata and Kimura, 1987, 1988). Human glyoxalase I showed 51% homology at the nucleotide level and 42% homology at amino acid level with *P. putida* glyoxalase I. Mammalian glyoxalase I is a dimer (M_r=43–48 kDa) composed of identical subunits, and each subunit contains one Zn^{2+} at its catalytic site. The binding frequency of zinc atoms to the proteins was thought to be as follows: His>>Glu>Asp=Cys, and these amino acids have appropriate spacing (Valee and Auld, 1990). Ranganathan *et al.* (1993) predicted that the amino acids concerned with Zn^{2+} binding are Glu100 and His103, with one of Asp121, His127, Cys139 or Glu143. They also suggested that the Zn^{2+}-binding motif of bacterial glyoxalase I was: Glu91 and His94 with one of Asp112, His118, Cys130 or Glu132. However, Rhee, Murata and Kimura (1986) reported that *P. putida* glyoxalase I was not inhibited by an excess amount (10 mM) of EDTA and suggested no contribution of metal ions in the catalytic function of the bacterial enzyme.

Effect of the GAC gene product on the expression of the GLO1 gene in S. cerevisiae

Although the expression of the *GAC* gene in *E. coli* did not increase the glyoxalase I activity, a yeast dosed with the gene showed increased activity of glyoxalase I and then showed resistance to methylglyoxal. Therefore, the biological function of the *GAC* gene product in *S. cerevisiae* cells should be a subject of considerable interest. The mRNA level of the *GLO1* gene in *S. cerevisiae* dosed with the *GAC* gene increased in yeast cells compared with that of control cells (*Figure 4*) (Inoue *et al.*, 1991c). The result suggested that the *GAC* gene product affected the activity of yeast glyoxalase I at the mRNA level of the *GLO1* gene; i.e. transcriptional activation or enhancing the stability of *GLO1* mRNA.

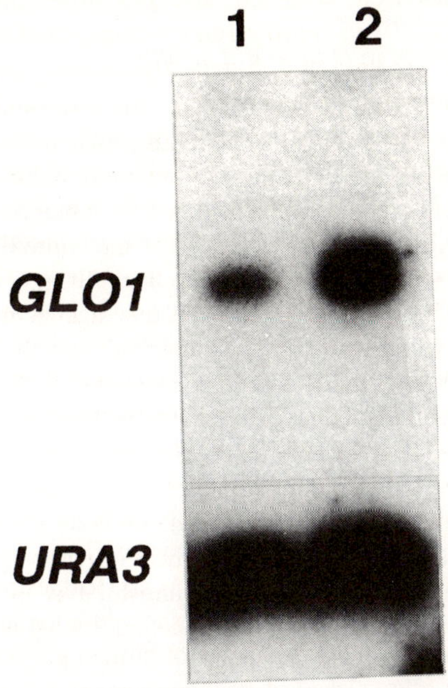

Figure 4. Effect of the *GAC* gene on the expression of the *GLO1* gene in *S. cerevisiae*. Total RNA prepared from the *S. cerevisiae* cell transformed with YEp13 (lane 1, vector alone as control) and YEp13 + *GAC* (lane 2) was subjected to electrophoresis in an agarose gel containing formaldehyde and transferred on to a nylon membrane. The level of *URA3* mRNA was used as an internal control by re-probing the membrane as indicated at the bottom.

It is known that the 5'-flanking regions of several yeast genes contain a specific DNA sequence for the binding of the *trans*-activator for transcription, designated UAS (upstream activating sequence) (Guarente, 1984, 1988; Struhl, 1987). Several proteins that bind to the UAS and activate the transcription of yeast genes have been identified (Verdier, 1990). DNA binding proteins often contain the Zn^{2+} ion in their molecule through tetrahedral co-ordination of cysteine residues, the protein being termed a 'zinc finger' (Johnston, 1987a). In this model, the DNA binding protein binds to the specific site of the 5'-flanking region of the gene and accelerates the transcription of DNA through interaction with RNA polymerase (Allison and Ingles, 1989). The Zn^{2+} co-ordinated by cysteine residues is essential for the protein to bind to DNA (Johnston, 1987b). According to amino acid sequences deduced from the DNA sequence of *GAC* gene, the gene product is expected to contain five cysteine residues near the carboxyl terminus. However, the spacing of the cysteine residues is not similar to that of the C_6 zinc fingers of GAL4 and other yeast activator proteins (Inoue *et al.*, 1990c). It bears a similarity to the C_5 zinc finger, which is characteristic of the hormone receptor superfamily.

Recent findings suggested that the GAC protein produced in *E. coli* had the capability for DNA binding; the results of a gel retardation assay indicated that GAC protein produced in *E. coli* could bind to the chromosomal DNA of *S. cerevisiae* (H. Ginya, 1991). This suggested that the *GAC* gene product may function in the yeast cell to activate the transcription of the *GLO1* gene through interaction with its 5'-flanking region.

Oxidative stress response in *Hansenula mrakii*

All aerobic organisms use molecular oxygen for respiration or the oxidation of nutrients to acquire the energy to live. The molecular oxygen is reduced to water through the acceptance of four electrons. During the reduction of molecular oxygen, several reactive oxygen species are formed; i.e. acceptance of one, two and three electrons leads to the formation respectively, of the suproxide anion radical (O_2^-), hydrogen peroxide (H_2O_2) and the hydroxyl radical (HO•) (*Figure 5*). Such reactive oxygen species have been reported as causative agents in several degenerative diseases (Ames, 1983; Cerutti, 1985). They attack almost all cell components, DNA, protein and lipid membranes, and sometimes cause lethal damage to the cells. Both prokaryotic and eukaryotic cells have defensive methanisms against such oxidative damage. *Escherichia coli* and *Salmonella typhimurium* cells have the *oxyR*-controlled regulon of hydrogen peroxide-inducible genes (Christman, Storz and Ames, 1989; Storz, Tartaglia and Ames, 1990). *Escherichia coli* cells also have a *soxRS* regulon, which is induced by the superoxide anion radical (Greenberg *et al*., 1990; Tsaneva and Weiss, 1990; Wu and Weiss, 1991; Nunoshiba *et al*., 1992). The yeast *S. cerevisiae* has cytochrome *c* peroxidase (Yonetani, 1970) as well as a superoxide dismutase-catalase system.

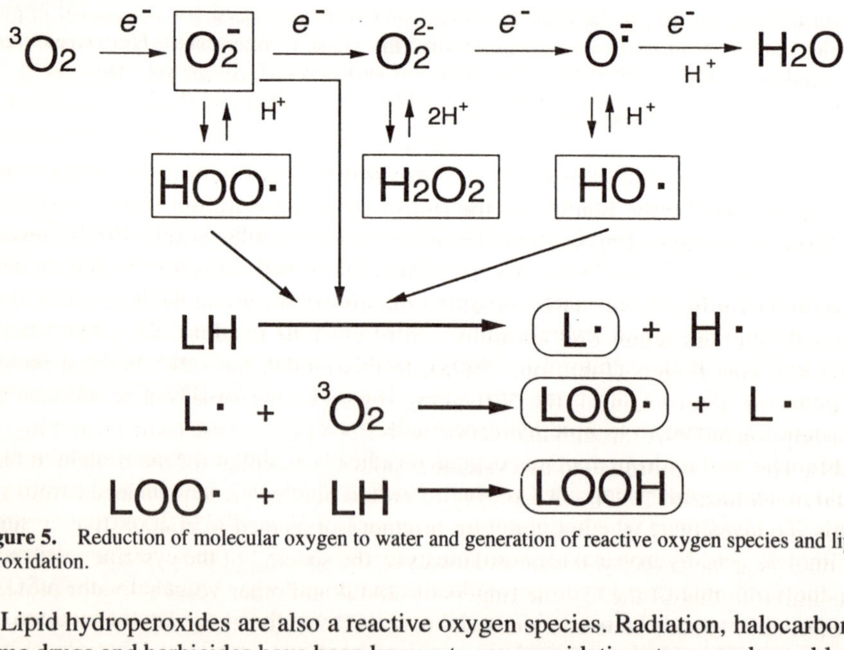

Figure 5. Reduction of molecular oxygen to water and generation of reactive oxygen species and lipid peroxidation.

Lipid hydroperoxides are also a reactive oxygen species. Radiation, halocarbons, some drugs and herbicides have been known to cause oxidative stress, and are able to peroxidize the biological membrane *in vivo*. Among the reactive oxygens, HO•, as well as the perhydroxyl radical (HOO•) can extract *bis*-allylic hydrogen atoms of unsaturated fatty acids (LH) to form lipid alkyl radicals (L•). The L• is oxidized by molecular oxygen to generate a lipid peroxy radical (LOO•), and the LOO• thus formed reacts with LH to give lipid hydroperoxide (LOOH) and L•. The radical chain reaction is then propagated (*Figure 5*). Occurrence of the lipid hydroperoxides in the biological membrane may be one of the major reasons for oxidative damage of the cells.

Contrasting with the voluminous studies of the defensive mechanism against reactive oxygen such as O_2^- and H_2O_2 in bacterial cells (Demple and Amabile Cuevas, 1991; Farr and Kogoma, 1991; Storz and Tartaglia, 1992), studies of adaptation or resistance to oxidative stress in yeast have been fewer (Collinson and Dawes, 1992; Jamieson, 1992). Furthermore, the defensive mechanism for oxidative stress caused by lipid hydroperoxide has not been studied in detail in micro-organisms (Inoue *et al.*, 1990b, 1992b). As the first step in this study, the author and his co-workers screened several yeast strains for resistance against lipid hydroperoxide, and found that *Hansenula mrakii* IFO 0895 could grow in a medium containing 4 mM linoleic acid hydroperoxide in which all other yeast strains tested could not grow (Inoue *et al.*, 1990b). The resistance was proved to be due to a membrane-bound glutathione peroxidase which was induced when the yeast was exposed to oxidative environments.

INDUCTION OF GLUTATHIONE PEROXIDASE BY LIPID HYDROPEROXIDE IN *H. MRAKII*

Screening of yeast resistant to linoleic acid hydroperoxide

Several yeast strains in laboratory collections were screened for resistance against 1 mM linoleic acid hydroperoxide. Among the yeast strains tested, *Hansenula* yeast (*H. anomala* IFO 0149, *H. californica* IFO 0800, *H. canadensis* IFO 0973, *H. canadensis* IFO 0976, *H. jadinii* IFO 0987, *H. saturnus* IFO 0117 and *H. mrakii* IFO 0895) could grow in the linoleic acid hydroperoxide-containing medium. *Hansenula mrakii* IFO 0895 could grow in the minimal medium containing 4 mM linoleic acid hydroperoxide (*Figure 6a*). Other strains, except for *Rhodotorula minuta* IFO 0387, *Zygo-saccharomyces rouxii* IFO 0487, *Metschnikowia zobellii* IFO 1680, *Candida maltosa* IFO 1975 and *Pichia kluveri* IFO 1165, could not grow in the medium containing 1 mM linoleic acid hydroperoxide. *Metschnikowia zobellii* belongs to the Spermophthoraceae and several strains in this family are plant parasites. Synthesis of lipid hydroperoxide catalysed by lipoxygenase is known as one of the defensive responses of plants against microparasites. Thus, *M. zobellii* IFO 1680 may have a resistance mechanism to lipid hydroperoxide.

Linoleic acid hydroperoxide is degraded radically in the presence of metal ions to yield many secondary degraded products, such as aldehydes, ketones and carboxylic acids. To investigate whether or not the tolerance of *H. mrakii* to the oxidative stress of linoleic acid hydroperoxide resulted from resistance to the secondary degraded products of linoleic acid hydroperoxide, the compound was incubated with $CuSO_4$ to promote the radical chain reaction, and the mixture was then added to the medium. As shown in *Figure 6b*, the cells of *H. mrakii* could grow in a medium containing 1 mM linoleic acid hydroperoxide or 5 μM $CuSO_4$ alone, but cell growth was completely inhibited in a medium containing 1 mM linoleic acid hydroperoxide together with

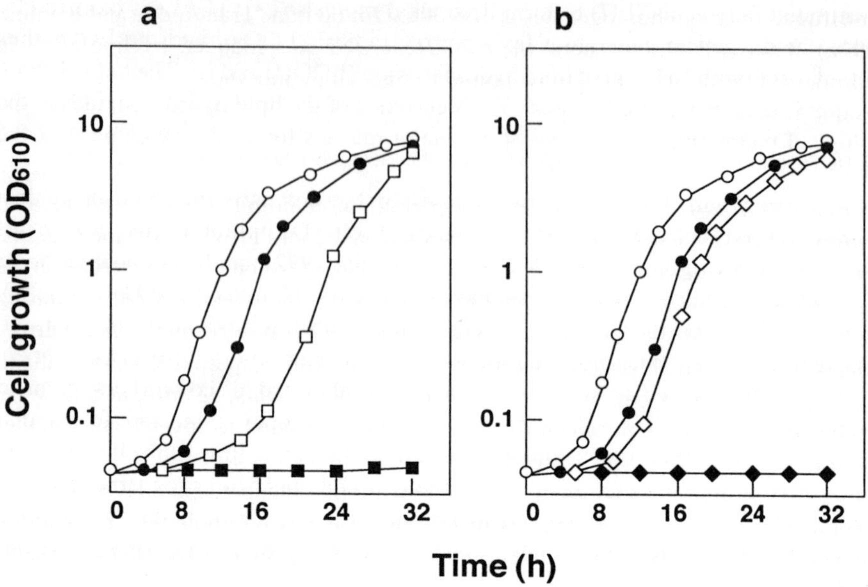

Figure 6. Effect of lipid hydroperoxide on the growth of *H. mrakii*. (a) Cells of *H. mrakii* IFO 0895 were cultured in the SD minimal medium containing O mM (O), 1 mM (●), 4 mM (□) and 10 mM (■ linoleic acid hydroperoxide at 28°C, respectively. Cell growth was monitored by measuring the OD_{610}. (b) Cells of *H. mrakii* were cultured in the absence of chemicals (O) or the presence of 1 mM linoleic acid hydroperoxide (●), 5 μM $CuSO_4$ (◊) or a mixture of 1 mM linoleic acid hydroperoxide and 5 μM $CuSO_4$ (♦). In the latter case, the mixture was preincubated at 28°C for 12 h before it was added to the medium.

5μM $CuSO_4$. This suggested that *H. mrakii* was resistant to linoleic acid hydroperoxide itself but not to the secondary degraded products of lipid hydroperoxide.

Induction of glutathione peroxidase by linoleic acid hydroperoxide

Residual linoleic acid hydroperoxide in the culture of *H. mrakii* was analysed by thin layer chromatography. The hydroperoxide moiety of the compound was found to be reduced to an alcohol moiety. Thus, the tolerance of *H. mrakii* to linoleic acid hydroperoxide may result from the reduction of the hydroperoxide moiety of the linoleic acid hydroperoxide to an alcohol moiety (Inoue *et al.*, 1990b). Reduction of the hydroperoxide moiety to alcohol is expected to be catalysed by several peroxidases. The activity of potential candidates, such as ascorbate peroxidase (Boveris *et al.*, 1980), chloride peroxidase (Hager, 1970), cytochrome *c* peroxidase (Yonetani, 1970), NAD-peroxidase (Dolin, 1957), peroxidase (Kenten and Mann, 1954) and glutathione peroxidase (Awasthi, Beutler and Srivastava, 1975) was assayed in the soluble and insoluble fractions of cell homogenates of *H. mrakii*, which was grown in the medium with or without 1 mM linoleic acid hydroperoxide. Among the enzymes assayed, only glutathione peroxidase was found to be dramatically induced when the cells of *H. mrakii* were cultued with 1 mM linoleic acid hydroperoxide (Inoue *et al.*, 1990b). Almost all of the glutathione peroxidase was recovered from the insoluble fractions using several detergents such as Triton X-100, lubrol PX (polyethylene-glycol (9) lauryl ether) and CHAPS (3-cholamidopropyldimethylammonio-1-propane sulphonate). No activity was detected in the soluble fractions obtained by ultracentrifugation of cell homogenates at 105 000 *g* for 60 min (Tran, Inoue and Kimura, 1993a). Thus, glutathione peroxidase of *H. mrakii* was induced by linoleic acid hydroperoxide and it seemed to be bound to the cell membrane.

INDUCTION OF GLUTATHIONE PEROXIDASE BY REACTIVE OXYGEN IN *H. MRAKII*

Hansenula mrakii could grow in the medium containing 4 mM linoleic acid hydroperoxide, in which no other yeast strain tested could grow. Resistance against lipid hydroperoxide was due to a glutathione peroxidase which was induced when the yeast was exposed to exogenous lipid hydroperoxide (Inoue *et al.*, 1990b). Among the rective oxygen species, the hydroxy radical (HO•) as well as the perhydroxy radical (HOO•), which is generated by protonation of the superoxide anion radical, can initiate the peroxidation of unsaturated fatty acid. Thus, *H. mrakii* cells were incubated in a mixture in which O_2^- or HO• was generated to investigate whether glutathione peroxidase was induced.

Intact cells of yeast (possessing cell walls) were first incubated in a HO•- and O_2^--generating system; however, glutathione peroxidase was not induced. The reaction rate constant (*k*) of HO• for abstraction of the *bis*-allylic hydrogen atom of unsaturated fatty acid is estimated to be $10^8\ M^{-1}\ s^{-1}$ (Barber and Thomas, 1978), although the radical is generally thought not to be involved in the peroxidation process of membrane lipid *in vivo*, because the HO• in the cytosol would react non-specifically with other compounds surrounding the radical before it reached the membrane (Pryor, 1986; Aikens and Dix, 1991). Therefore, when intact cells are used, the cell wall might prevent the radical from reaching or interacting with the cell membrane. Thus, protoplasts of *H. mrakii* were exposed to HO• and O_2^-, respectively.

When the protoplasts of *H. mrakii* were incubated with H_2O_2 alone, or Fe^{2+} + ADP, glutathione peroxidase activity was not detected in any of the cases (*Figures 7a,b,c*). However, when the protoplasts were incubated in the HO•-generating mixture ($H_2O_2 + Fe^{2+} \rightarrow HO\bullet + OH^- + Fe^{3+}$; Fenton reaction) for 1 h, the cells regained growth after a 20 h lag time, and at this time glutathione peroxidase activity was detected (*Figure 7d*).

To examine whether or not other reactive oxygen radicals, such as O_2^- can induce the synthesis of glutathione peroxidase in *H. mrakii*, the protoplasts were treated in an O_2^--generating system. When protoplasts of *H. mrakii* cells were pre-incubated with xanthine or xanthine oxidase alone (*Figures 7e,f*), glutathione peroxidase activity was not detected. However, when the protoplasts were pretreated for 1 h in a mixture generating O_2^- (xanthine + xanthine oxidase), glutathione peroxidase was induced after a 10 h time lag (*Figure 7g*). The superoxide anion radical itself is thought not to be able to withdraw the *bis*-allylic hydrogen atom of unsaturated fatty acid (Gebicki and Bielski, 1981; Halliwell and Gutteridge, 1990), although it serves as proton acceptor to yield the perhydroxyl radical ($O_2^- + H^+ \rightarrow HOO\bullet$) (Bielski, Arudi and Sutherland, 1983; Sawyer, McDowell and Yamaguchi, 1988). It has been reported that HOO• could initiate lipid peroxidation (Gebicki and Bielski, 1981; Bielski, Arudi and Sutherland, 1983; Aikens and Dix, 1991). At physiological pH (i.e. pH 6.8), only 1% of O_2^- exists as HOO• (pKa=4.88), though HOO• may be formed more abundantly near the surface of the membrane where the pH might be lower owing to the negative surface charge of the membrane (Bielski, 1978; Barber, 1980). Therefore,

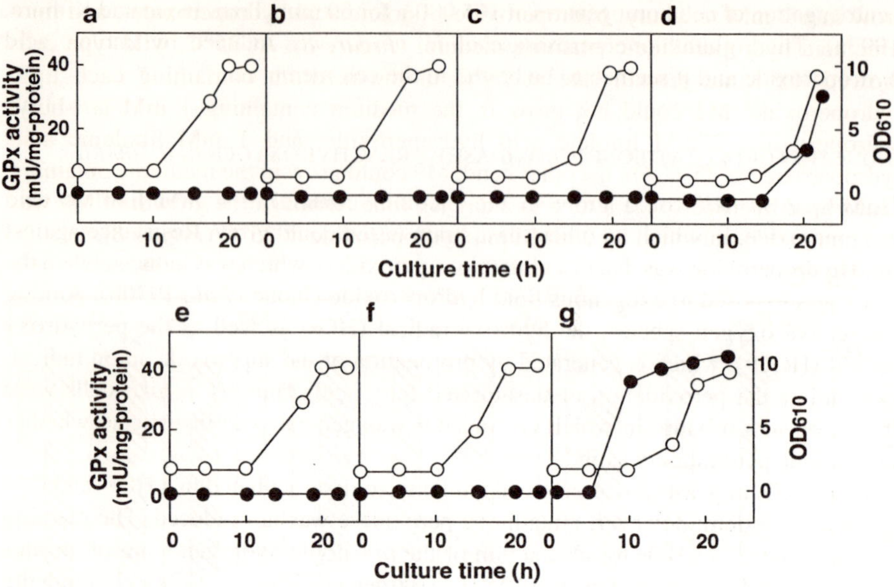

Figure 7. Induction of glutathione peroxidase by reactive oxygen. Upper panel: induction of glutathione peroxidase by HO•; protoplasts of *H. mrakii* were pretreated with (a) no chemicals, (b) 30 mM H_2O_2, (c) 0.1 mM $FeSO_4$ + 0.25 mM ADP, and (d) 0.1 mM $FeSO_4$ + 0.25 mM ADP + 30 mM H_2O_2. Lower panel: induction of glutathione peroxidase by O_2^-; protoplasts of *H. mrakii* were pretreated with (e) 0.5 mM xanthine, (f) 2 mU ml^{-1} xanthine oxidase, and (g) 0.5 mM xanthine + 2 mU ml^{-1} xanthine oxidase. Symbols: O, OD_{610}; ●, glutathione peroxidase (GPx) activity.

induction of glutathione peroxidase by the xanthine + xanthine oxidase system may be due to HOO• rather than O_2^-.

The following putative model for adaptation to oxidative stress in *H. mrakii* is suggested. Reactive oxygen species such as HO• and HOO• attack the membrane of protoplasts, and initiate the peroxidation of membrane lipid. Peroxidation of the membrane lipid constitutes damage to the cell, and it may serve as a signal for the yeast to trigger the synthesis of some apparatus for adaptation to oxidative stress.

ROLE OF GLUTATHIONE PEROXIDASE AGAINST OXIDATIVE STRESS IN *H. MRAKII*

Wild-type cells of *H. mrakii* were highly resistant to the oxidative stres caused by lipid hydroperoxide. The resistance was due to a glutathione peroxidase which was induced when the yeast was exposed to the oxidative environments (Inoue *et al.*, 1990b; Tran, Inoue and Kimura, 1993b). To investigate the role of glutathione peroxidase, the mutants sensitive to lipid hydroperoxide were isolated.

Two mutants (M1 and M9) showing small colonies on the minimal agar plate containing 2 mM *tert*-butyl hydroperoxide at 30°C were obtained. Wild-type cells of *H. mrakii* could make large colonies under the same conditions. To confirm the phenotypes of M1 and M9, each cell was cultured in a liquid medium containing *tert*-butyl hydroperoxide, linoleic acid hydroperoxide and linolenic acid hydroperoxide, respectively, with various concentrations, at 28°C and 35°C. Wild-type cells as well as M1 and M9 cells showed almost the same growth rates at 28°C, and they could grow in a 2 mM *tert*-butyl hydroperoxide-, linoleic acid hydroperoxide-, and linolenic acid hydroperoxide-containing medium (*Figure 8*). At 35°C, wild-type cells showed the same growth rate at 28°C in the medium containing each lipid hydroperoxide. M1 could not grow in the medium containing 1 mM *tert*-butyl hydroperoxide, 2 mM linoleic acid hydroperoxide, and 1 mM linolenic acid hydroperoxide at 35°C. On the other hand, M9 could grow in the medium containing 1 mM lipid hydroperoxide after a 1–2 day lag time, although the growth of M9 was also completely inhibited by 2 mM lipid hydroperoxide at 35°C (*Figure 8*).

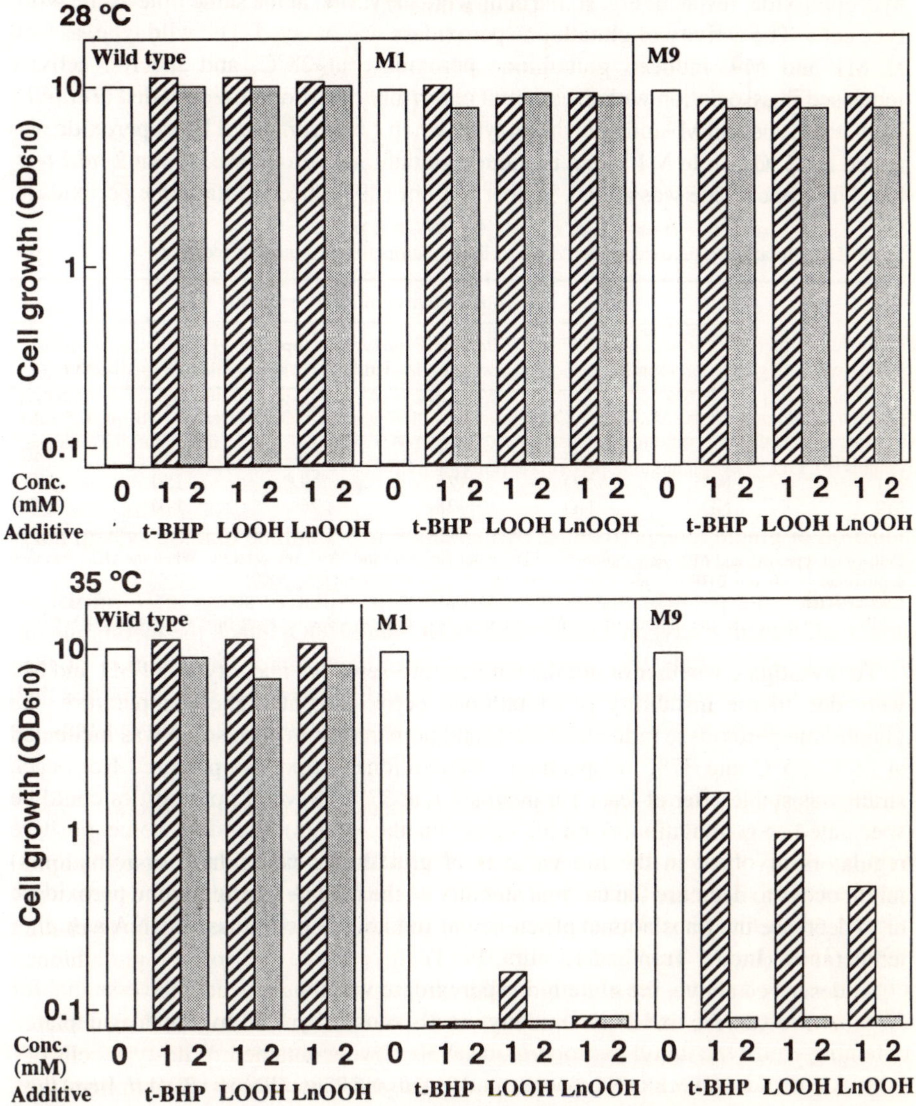

Figure 8. Effect of temperature on growth of lipid hydroperoxide-sensitive mutants. Cells (wild type, M1 and M9) were cultured in the SD medium containing 0 mM (▭), 1 mM (▬) and 2 mM (▬) lipid hydroperoxides, at 28°C (upper panel) and 35°C (lower panel), respectively. Cell growth was monitored by measuring the OD_{610}. Each bar indicates the OD_{610} of the 3 day culture. t-BHP, *tert*-butyl hydroperoxide; LOOH, linoleic acid hydroperoxide; LnOOH, linolenic acid hydroperoxide.

The resistance against lipid hydroperoxide of the wild-type cells of *H. mrakii* was due to glutathione peroxidase, which was induced when the cells were cultured in the presence of lipid hydroperoxide (Inoue *et al.*, 1990b). Hence, the activities of glutathione peroxidase in M1 and M9 were examined. Cells of the wild type, M1 and M9 were cultured in the minimal medium at 28°C and 35°C, respectively. When the OD_{610} of the medium reached approximately 1.0, *tert*-butyl hydroperoxide was added to each medium, and the cultivation was continued until the OD_{610} reached 8.0. At 35°C, M1 and M9 could not grow after the additions of 1 mM and 2 mM *tert*-butyl hydroperoxide, respectively, so the cells were harvested at the same time as the wild-type cells. The activity of glutathione peroxidase was assayed. The wild type, as well as M1 and M9, induced glutathione peroxidase at 28°C, and specific activity increased in association with the amount of *tert*-butyl hydroperoxide added (*Table 1*). Both wild type and M9 induced the enzyme when 1 mM *tert*-butyl hydroperoxide was added at 35°C, while M1 failed to induce glutathione peroxidase. When 2 mM *tert*-butyl hydroperoxide was added, neither M1 nor M9 induced glutathione peroxidase.

Table 1. Glutathione peroxidase activities in lipid hydroperoxide-sensitive mutants

Strain	GPx activity (mU mg^{-1} protein)					
	tert-BHP added into medium					
	0 mM		1 mM		2 mM	
	28°C	35°C	28°C	35°C	28°C	35°C
Wild type	ND[a]	ND	31	26	180	140
M1	ND	ND	25	ND[b]	130	ND[b]
M9	ND	ND	28	26	110	ND[b]

Cells (wild type, M1 and M9) were cultured in SD medium at 28°C and 35°C, respectively. When the OD_{610} reached approximately 1.0, *tert*-BHP was added and cultivation was carried out until OD_{610}=8.0.
[a] Not detected.
[b] Cells did not grow after the addition of *tert*-BHP, so the cells were collected at the same time as the wild-type cells.

To investigate whether or not the temperature-sensitive phenotypes of M1 and M9 were due to the instability of glutathione peroxidase at higher temperature, the glutathione peroxidase induced at 28°C and prepared from each strain was incubated at 28°C, 35°C and 37°C, respectively. Glutathione peroxidase prepared from each strain was stable after at least 1 h incubation at 37°C. Several possibilities could be speculated, e.g. mutation(s) might occur on the gene(s) encoding some positive regulators involved in the biosynthesis of glutathione peroxidase, or mutation(s) might occur to decrease the thermal stability of the mRNA of glutathione peroxidase or to decrease the translational efficiency of flutathione peroxidase mRNA at higher temperature (Inoue, Tran and Kimura, 1993).

As described above, the glutathione peroxidase was thus proved to be essential for *H. mrakii* to survive under the oxidative stress caused by lipid hydroperoxide.

PURIFICATION OF MEMBRANE-BOUND GLUTATHIONE PEROXIDASE FROM *H. MRAKII*

Purification of glutathione peroxidase from H. mrakii

As described above, the glutathione peroxidase was recovered from the insoluble fractions of the cell homogenates of *H. mrakii* (Inoue *et al.*, 1990b). Therefore, the glutathione peroxidase of *H. mrakii* appeared to occur in a membrane-bound form. Since CHAPS has a high critical micellar concentration value and small micellar size, preferable properties for the purification of an enzyme, this detergent was chosen to solubilize the glutathione peroxidase from total membrane fractions of *H. mrakii*. As shown in *Table 2*, almost all glutathione peroxidase activity remained in the precipitates after KCl-treatment; suggesting that the glutathione peroxidase was bound strongly to the biological membrane. After treatment of the precipitates with 0.5% CHAPS, the enzyme was completely solubilized. Recovery of glutathione peroxidase in the CHAPS extracts was approximately 82%. Therefore, almost all the glutathione peroxidase was believed to be bound to the biological membrane of *H. mrakii* cells.

Table 2. Recovery of glutathione peroxidase activity in membrane fractions

Step	Total activity (unit)	Recovery (%)
Homogenates	3.52	100[a]
200 000 × *g*		
Supernatant	0.22	6.30
Pellet	3.30	93.7
2 M KCl, 200 000 × *g*		
Supernatant	0.40	11.4
Pellet	2.90	82.4
0.5% CHAPS, 200 000 × *g*		
Supernatant	2.90	82.4
Pellet	ND[b]	–

[a] Total activity of homogenates was taken as 100%.
[b] Not detected.

Glutathione peroxidase was purified to the homogenous state as judged by SDS-PAGE from the total membrane fractions of *H. mrakii* (Tran, Inoue and Kimura, 1993a), and the molecular weight of the purified enzyme was estimated to be 28 kDa.

Glutathione peroxidases purified and characterized so far can be divided into two groups: a selenium-dependent enzyme (glutathione peroxidase) and a selenium-independent enzyme (glutathione *S*-transferase). The glutathione peroxidase purified from *H. mrakii* (*Hansenula*-GPx) was different from the glutathione peroxidases so far purified in substrate specificity. As shown in *Table 3*, various lipid hydroperoxides and their methyl esters were well used by *Hansenula*-GPx. Some properties of the enzyme are summarized in *Table 4* in comparison with glutathione peroxidases from other sources. Specific activity of purified *Hansenula*-GPx for *tert*-butyl hydroperoxide (625 μmol min^{-1} mg^{-1} protein) was higher than those of other glutathione peroxidases as listed in *Table 4* (human erythrocyte GPx-I, 103.3 μmol min^{-1} mg^{-1} protein (Awasthi, Beutler and Srivastava, 1975); pig heart PHGPx, 15.5 μmol min^{-1} mg^{-1} protein (Ursini, Maiorino and Gregolin, 1985); rat liver GPx-II, 3.2 μmol min^{-1} mg^{-1} protein (Prohaska and Ganther, 1977); rat liver microsome GST, 1.4 μmol min^{-1} mg^{-1} protein for cumene hydroperoxide (Morgensten and DePrierre, 1983); *Mucor hiemalis* GPx, 187.2 μmol min^{-1} mg^{-1} protein (Aisaka, Uwajima and Terada, 1983); *Euglena gracilis* GPx, 42 μmol min^{-1} mg^{-1} protein (Overbaugh and Fall, 1985)). Hydroperoxide of phosphatidylcholine, a representative of membrane phospholipid, was preferably reduced by the *Hansenula*-GPx. Usually glutathione peroxidase and glutathione *S*-transferase were not active on hydroperoxides of phosphatidylcholine and cholesterol, except for the phospholipid hydroperoxide glutathione peroxidase (PHGPx) reported by Ursini, Maiorino and Gregolin (1985) (*Table 4*). Cholesterol-5α-hydroperoxide was also reduced by the *Hansenula*-GPx.

Table 3. Substrate specificity of glutathione peroxidase from *H. mrakii*

Substrate	Relative activity (%)
Hydrogen peroxide	0
tert-Butyl hydroperoxide	100
Cumene hydroperoxide	144
Methyl ethyl ketone hydroperoxide	101
Benzoyl peroxide	59
Di-*tert* butyl peroxide	0
Dicumene peroxide	0
cis-Vaccenic acid hydroperoxide	162
Linoleic acid hydroperoxide	142
Linolenic acid hydroperoxide	109
Oleic acid hydroperoxide	98
Methyl linolate hydroperoxide	150
Methyl linolenate hydroperoxide	124
Methyl oleate hydroperoxide	123
Cholesterol 5α-hydroperoxide	120
Phosphatidylcholine hydroperoxide	120
Substrate for glutathione *S*-transferase:	
o-Dinitrobenzene	0
1-Chloro-2,4-dinitrobenzene	0
1,2-Dichloro-4-nitrobenzene	0
p-Nitrobenzyl chloride	0
p-Nitrophenethyl bromide	0
4-Nitropyridine-*N*-oxide	0
1,2-Epoxy-3-(*p*-nitrophenoxy) propane	0

Activity on *tert*-butyl hydroperoxide was taken as 100%.

The *Hansenula*-GPx could not reduce hydrogen peroxide, while the other glutathione peroxidases, except for glutathione peroxidase from *Mucor hiemalis* (Aisaka, Uwajima and Terada, 1983), can use it as a substrate. Since the glutathione *S*-transferase could not reduce hydrogen peroxide, *Hansenula*-GPx was thought to be a kind of glutathione *S*-transferase. However, as shown in *Table 3*, the *Hansenula*-GPx did not catalyse the conjugation of glutathione with electrophilic compounds such as 1-chloro-2,4-dinitrobenzene and *o*-dinitrobenzene, which are used as substrates of glutathione *S*-transferase in yeast (Habig, Pabst and Jakoby, 1974; Kumagai *et al*., 1988).

One of the characteristics of the *Hansenula*-GPx is that the enzyme is strongly bound to the biological membrane. Morgenstern and co-workers (Morgenstern *et al*., 1980; Morgenstern, Guthenberg and DePrierre, 1982; Morgenstern and DePrierre, 1983) purified the glutathione *S*-transferase from the membrane fractions of rat liver microsome, although the enzyme was not active on phospholipid hydroperoxide. PHGPx with activity on phosphatidylcholine hydroperoxide was purified from cytosolic fractions of pig heart (Ursini, Maiorino and Gregolin, 1985). The *Hansenula*-GPx has not been subjected to a direct analysis of selenium as yet because of insufficient availability of the purified enzyme. However, whether the *Hansenula*-GPx contains selenium or not, the enzyme previously purified can be distinguished

Table 4. Properties of glutathione peroxidases from different sources

Enzyme source	Subcellular localization	Molecular weight	No. of subunits	Substrate			
				H_2O_2	*tert*-BHP	PLHP	CDNB
Human erythrocyte (GPx-I)[a]	Cytosol	95 000	4	Yes	Yes	No	No
Pig heart (PHGPx)[b]	Cytosol	20 000	1	Yes	Yes	Yes	No
Rat liver (GPx-II, GST)[c]	Cytosol	47 000	2	No	Yes	No	Yes
Rat liver microsome (GST)[d]	Membrane	14 000	1	No	?[g]	No	Yes
Mucor hiemalis[e]	Cytosol	45 000	2	No	Yes	No	No
Euglena gracilis[f]	Cytosol	130 000	4	Yes	Yes	No	No
Hansenula mrakii	Membrane	28 000	?	No	Yes	Yes	No

[a] Data shown for the enzyme from human erythrocyte (Awasthi, Beutler and Srivastava, 1975).
[b] Data shown for pig heart (Ursini, Maiorino and Gregolin, 1985).
[c] Data shown for rat liver (Prohaska and Ganther, 1977).
[d] Data shown for rat liver microsome (Morgenstern, Guthenberg and DePrierre, 1982; Morgenstern and DePrierre, 1983).
[e] Data shown for *M. hiemalis* (Aisaka, Uwajima and Terada, 1983).
[f] Data shown for *E. gracilis* (Overbaugh and Fall, 1985).
[g] Cumene hydroperoxide was used instead of *tert*-butyl hydroperoxide.
Abbreviations: *tert*-BHP, *tert*-butyl hydroperoxide; PLHP, phospholipid hydroperoxide; CDNB, 1-chloro-2,4-dinitrobenzene; GPx, glutathione peroxidase; PHGPx, phospholipd hydroperoxide glutathione peroxidase; GST, glutathione *S*-transferase.

from classical types of glutathione peroxidase and glutathione *S*-transferase so far purified because of its substrate specificity and intracellular localization.

Glutathione peroxidase is thought to play an important role in the protection of cells against oxidative stress, such as peroxidation of membrane lipids. Indeed, the *Hansenula*-GPx was shown to be essential for *H. mrakii* to protect the cells against oxidative stress caused by lipid hydroperoxide by using mutants (Inoue, Tran and Kimura, 1993). Since classical glutathione peroxidases have been purified from cytosolic fractions of cell homogenates, the membrane-protective function of the enzyme was supposed to be associated with a preceding action of phospholipase (Grosmann and Wendel, 1983). Glutathione *S*-transferases have been purified from the cytoplasm as well as from membrane fractions; however, the direct reduction of peroxidized membrane lipids by glutathione *S*-transferase has been disputed (Gibson, Hornbrook and McCay, 1980; Tan *et al.*, 1984). On the other hand, PHGPx was shown to be able to reduce peroxidized membrane lipids, thus suggesting an alternative mechanism for protection of the membrane from peroxidation through direct interaction of the enzyme with peroxidized membrane lipids. Although PHGPx activity was also reported to be detected from the cell membranes, the PHGPx purified so far has been from the cytosolic fractions (Ursini, Maiorino and Gregolin, 1985; Roveri *et al.*, 1992). From this point of view, the *Hansenula*-GPx turns out to be of physiological interest. Because the enzyme was purified from the membrane fractions and was active toward hydroperoxide of phosphatidylcholine, a representative of membrane phospholipid, it can serve as evidence for the physiological function of glutathione peroxidase in the direct reduction of peroxidized membrane lipids (Inoue and Kimura, 1992c) (*Figure 9*).

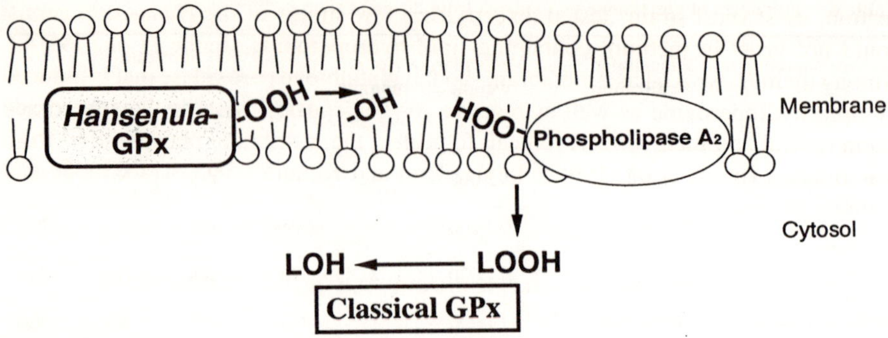

Figure 9. Possible model for protection of membrane lipid by glutathione peroxidase of *H. mrakii*.

Oxidative stress response in *Saccharomyces cerevisiae*

Several environmental stresses are known to trigger intracellular alterations in organisms. Organisms of all types show the synthesis of stress-inducible proteins, and the most advanced undertanding of the stress-inducible proteins has been obtained from the study of heat-shock protein (HSP). A sudden increase in temperature of the environment in which cells are growing induces increased synthesis of a set of heat-shock mRNAs and proteins in the cells. When *E. coli* cells are shifted from 30°C to 42°C, the intracellular concentration of σ^{32} increases 15- to 20-fold. The σ factor is one of the components of RNA polymerase of *E. coli* and substitution of a vegetative σ factor (σ^{70}) to σ^{32} changes the RNA polymerase so as specifically to recognize the HSP-encoding genes. When the yeast cells are exposed to the heat-shock stress, the heat-shock factor (HSF), which constitutively binds to the specific sequence (HSE, heat-shock element) of the genes encoding HSP, is trimerized and phosphorylated by heat shock, and the phosphorylated HSF then activates the transcription of *HSP* genes.

Intracellular stresses would also affect the expression of *hsp* genes. For example, reactive oxygen species, which are generated during respiration in aerobic organisms, are known to induce the synthesis of some stress shock proteins. *Escherichia coli* and *S. typhimurium* have *oxyR*-controlled hydrogen peroxide-inducible genes. When the bacterial cells are treated with low doses of hydrogen peroxide, synthesis of at least 30 proteins is induced. The synthesis of nine of these is positively controlled by OxyR protein. OxyR binds to the promoter regions of several genes such as *katG* (catalase), *ahpC* and *ahpF* (alkylhydroperoxide reductase) and *gorA* (glutathione reductase), whose products are concerned with the protection of the cells against oxidative stress, to activate their expression. OxyR belongs to the LysR family and it binds to its own promoter region of the *oxyR* gene to control its transcription negatively (Storz *et al.*, 1993). Several oxidative stress proteins are known to be induced by multiple types of stress such as heat shock and carbon starvation; however, the molecular mechanisms of the overlapping regulation have not been well elucidated.

Recently the author and his co-workers have started to study the mechanism for adaptation and resistance to oxidative stress caused by lipid hydroperoxide in yeasts (Inoue *et al.*, 1990b, 1992a, 1993; Inoue, Kobayashi and Kimura, 1993; Inoue, Tran and Kimura, 1993; Tran, Inoue and Kimura, 1993a,b). As described in the previous section, most yeast strains tested, except for a few strains of the *Hansenula* genus, could not grow in medium containing lipid hydroperoxide (Inoue *et al.*, 1990b). Among them *H. mrakii* has a membrane-bound glutathione peroxidase that is induced by lipid hydroperoxide as well as reactive oxygen species, such as the superoxide anion radical and the hydroxyl radical, to protect the membrane phospholipid from peroxidation (Inoue *et al.*, 1990b; Inoue, Tran and Kimura, 1993; Tran, Inoue and Kimura, 1993a,b).

MOLECULAR CLONING AND PHENOTYPIC CHARACTER OF OXIDATIVE STRESS RESISTANT GENE FROM *S. CEREVISIAE*

Cloning

A genomic DNA library of *S. cerevisiae* was screened for the gene(s) that confer resistance against lipid hydroperoxide in yeast. Several candidates were obtained by screening the transformants showing rapid growth on lipid hydroperoxide-containing medium. Finally two independent clones (designated No. 2 and No. 10) were obtained (*Figure 10a*). Growth of the control strain carrying YEp13 in the medium containing 0.75 mM *tert*-butyl hydroperoxide and 0.5 mM linoleic acid hydroperoxide was inhibited for approximately 2 days, and then the cells started to grow (*Figure 10b*). On the other hand, the transformants No. 2 and No. 10 could grow at almost same growth rate compared with growth without chemicals. By the addition of 2.0 mM *tert*-butyl hydroperoxide and 1.0 mM linoleic acid hydroperoxide, the growth of the control strain was completely inhibited. Whereas both transformants (No. 2 and No. 10) could grow in the medium containing *tert*-butyl hydroperoxide and linoleic acid hydroperoxide up to 2.0 mM and 1.0 mM, respectively (*Figure 10b*).

Plasmids isolated from strains No. 2 and No. 10 were named pYHP2 and pYHP10, respectively. The size of the inserted fragment in pYHP2 and pYHP10 was 5.9 kb and 4.0 kb, respectively. The restriction map of the inserted fragment was different in the two plasmids (Inoue *et al.*, 1993). Several kinds of genes corresponding to the resistance to oxidative stress caused by lipid hydroperoxide may exist on the genome of *S. cerevisiae*. Collinson and Dawes (1992) reported that *S. cerevisiae* cells have at least four polypeptides involved in the peroxide-stress response. Jamieson (1992) and Flattery-O'Brien, Collinson and Dawes (1993) also reported that *S. cerevisiae* has two distinct adaptive responses to hydrogen peroxide and a superoxide-generating reagent (menadione). Therefore, amplification of such a gene would allow the transformant cell to resist oxidative stress.

The transformant carrying pYHP10 could grow in the medium containing *tert*-butyl hydroperoxide, linoleic acid hydroperoxide and hydrogen peroxide up to 2.0 mM, 1.0 mM and 5.0 mM, respectively, in which the control strain could not grow. The gene was designated as *OSR*, oxidative stress resistant gene (Inoue, Kobayashi and Kimura, 1993).

Glutathione-dependent phenotype

Glutathione is synthesized by two sequential reactions catalysed by γ-glutamylcysteine synthetase (GSH-I, EC 6.3.2.2) and glutathione synthetase (GSH-II, EC 6.3.2.3) in

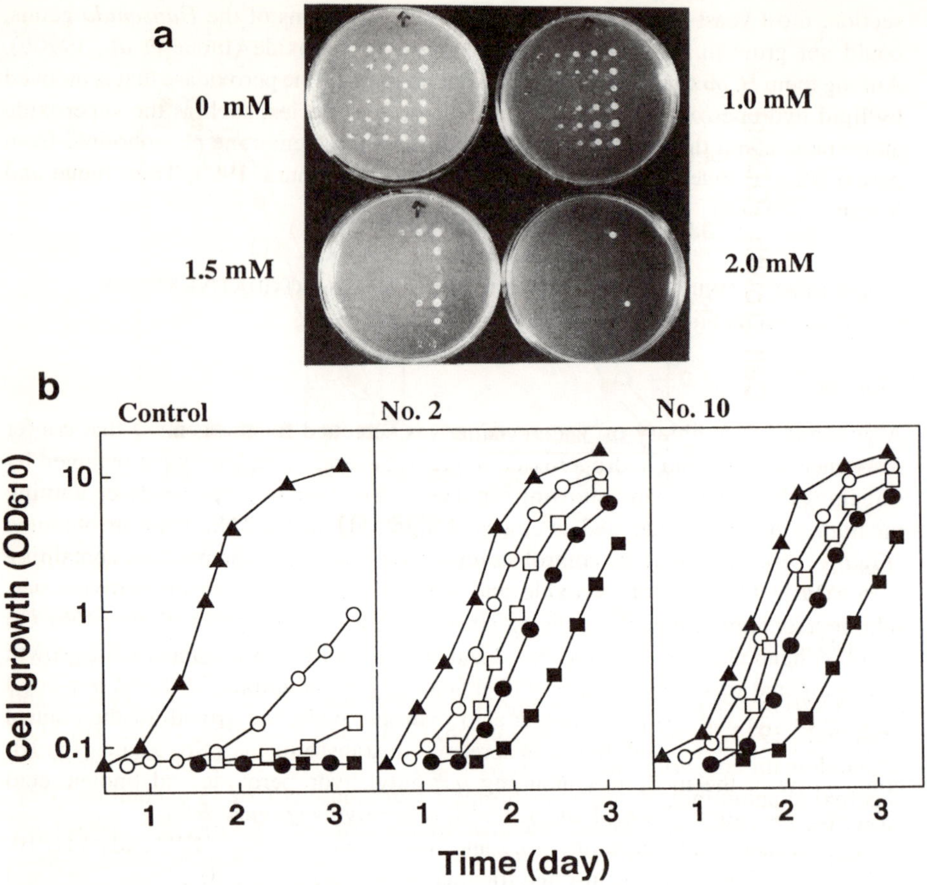

Figure 10. Growth of *S. cerevisiae* cells dosed with the *OSR* gene. (a) Screening of lipid hydroperoxide resistant clones from a genomic DNA library. Several candidates obtained by primary screening were replica plated on the SD minimal agar plate containing *tert*-butyl hydroperoxide with various concentrations as indicated. (b) Growth of yeast transformants in lipid hydroperoxide-containing medium. Symbols: ▲, without chemicals; O, 0.75 mM *tert*-butyl hydroperoxide; ●, 2.0 mM *tert*-butyl hydroperoxide; □, 0.5 mM linoleic acid hydroperoxide; ■, 1.0 mM linoleic acid hydroperoxide.

the presence of ATP. Glutathione is distributed in almost all aerobic organisms and the compound has many biological functions in the cells, e.g. detoxification of various cytotoxic compounds, maintenance of redox potential in the cells, cofactor for various enzymes, protection of SH group of various proteins, transportation of amino acids (constituting γ-amino cycle), and so on. Both prokaryotic and eukaryotic cells have a glutathione-dependent detoxification system for xenobiotics (Inoue and Kimura, in press). In order to investigate whether or not resistance of the yeast transformant carrying pYHP10 was glutathione-dependent, buthionine sulphoximine (BSO), a potent inhibitor for GSH-I, was added to the medium containing *tert*-butyl hydroperoxide. BSO alone did not affect the growth of the transformants carrying pYHP10 and YEp13, whereas when BSO and *tert*-butyl hydroperoxide were added simultaneously, the growth of both transformant was completely arrested (*Figure 11*).

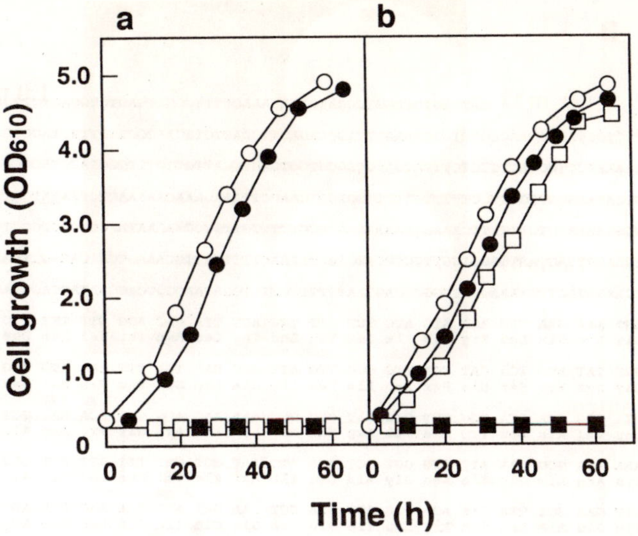

Figure 11. Effect of BSO on growth of yeast transformants. *Saccharomyces cerevisiae* carrying YEp13 (a) or pYHP10 (b) was cultured in the SD medium containing BSO and/or *tert*-butyl hydroperoxide. Symbols: O, without chemicals; ●, 0.1 mM BSO; □, 1.5 mM *tert*-butyl hydroperoxide; ■, 0.1 mM BSO + 1.5 mM *tert*-butyl hydroperoxide.

Saccharomyces cerevisiae has superoxide dismutases (Cu,Zn-SOD (Goscin and Fridovich, 1972; Bermingham-McDonogh, Gralla and Valentine, 1988), Mn-SOD (Ravindranath and Fridovich, 1975; Marres *et al.*, 1985)) for disproportion of the superoxide anion radical, and cytochrome *c* peroxidase (*CCP1* (Yonetani and Ray, 1965; Kaput, Golz and Blobel, 1982)) as well as catalases (*CTT1* (Seah, Bhatti and Kaplan, 1973; Hartig and Ruis, 1986) and *CTA1* (Seah and Kaplan, 1973; Cohen, Rapat and Ruis, 1988)) for the reduction or breakdown of hydrogen peroxide; however, glutathione-dependent enzymes for scavenging of reactive oxygen species in *S. cerevisiae* have not been identified at the genetic level. The author and his co-workers proved that a yeast, *H. mrakii*, has a membrane-bound glutathione peroxidase to protect the cell against peroxidation of membrane lipid (Inoue *et al.*, 1990b; Inoue, Tran and Kimura, 1993; Tran, Inoue and Kimura, 1993a,b). In mammalian cells, glutathione peroxidase is one of the major enzymes that scavenge hydrogen peroxide and lipid hydroperoxide. Jacobson *et al.* (1989) have proposed that alkyl hydroperoxide reductase serves as a prokaryotic equivalent to the glutathione reductase/glutathione peroxidase system in mammalian cells. The intracellular glutathione concentration of *S. cerevisiae* is comparable with that of the mammalian cell, and yeast also has glutathione reductase (Racker, 1955); thus the occurrence of some glutathione-dependent enzymatic systems for scavenging the reactive oxygen species in *S. cerevisiae* would be expected.

NUCLEOTIDE SEQUENCE AND FUNCTION OF THE *OSR* GENE IN *S. CEREVISIAE*

Nucleotide sequence of the OSR gene

The nucleotide sequence of the *OSR* gene was determined. *Figure 12* shows the 2297

```
                                        GATCGGTGCAGGTCCCATGAGATAAACTTTTGCGTTGAGTCTGGACTCTTGATGTATCTCGGA -547
CTAAAGCATTCATTCTGCTAAGGGAAGCGTCTGGCTAAGTTCAGCCAGACTTCCATTCTACAACCCTTTTTACTAGCTGAGCCTTCTTCCC -456
AGGAAGCAAAAAAAAAATCTTGCTACTTTCTCTTCATGACGCAGCGTGGTGAGTGCATTGCTCTTTCCCTACCTACGCCGAACATATATGT -365
CATATAACGGAAGTCATAAACGGTGGCTCCTTTTTGTTTTTGCTGGTCAACCTTAAAGAAACAAAAACTCTAATATGATTTCTTGGCATAG -274
ACCCACACACCCTCGGAAAACCTCATTTTCAAAAGAAAAAAAATGCGTTTTTGTTTGCGAAAAATAACGTCCCTCCCTCTGTTTTCCTTC -183
TTCAGATGGCCTTGAAGGTGAGGAATGAGGTGGTTGCTGCATCAGCTAAAACTTTTCATTGCAAAGTCATCAGAGCTCAAAGTTTCATAAA -92
AGAATATAACATTCAAGATACTTTAAAAGGCTTGGCGCAGTAATTTTTCCCTATCGATAACCGTCAGGAATACCAGGAAACTAATAGAGTC -1

    ATG ATC ACC GGT AAA GAA TTG AGA ATC ATC TCT CTT TTG ACC TTA GAC ACG GTT TTT TTC CTA TTG GAA +69
  1 Met Ile Thr Gly Lys Glu Leu Arg Ile Ile Ser Leu Leu Thr Leu Asp Thr Val Phe Phe Leu Leu Glu

    ATT ACC ATA GGT TAT ATG TCA CAT TCA TTG GCC TTG ATT GCC GAT TCA TTT CAC ATG TTG AAT GAT ATC +138
 24 Ile Thr Ile Gly Tyr Met Ser His Ser Leu Ala Leu Ile Ala Asp Ser Phe His Met Leu Asn Asp Ile

    ATC TCT CTT TTA GTG GCA CTA TGG GCT GTG GAT GTG GCC AAA AAC AGG GGT CCA GAC GCT AAA TAC ACT +207
 47 Ile Ser Leu Leu Val Ala Leu Trp Ala Val Asp Val Ala Lys Asn Arg Gly Pro Asp Ala Lys Tyr Thr

    TAT GGA TGG AAA AGA GCG GAA ATT TTG GGT GCT TTA ATC AAT GCT GTT TTT CTT ATT GCC CTG TGT TTC +276
 70 Tyr Gly Trp Lys Arg Ala Glu Ile Leu Gly Ala Leu Ile Asn Ala Val Phe Leu Ile Ala Leu Cys Phe

    TCT ATT ATG ATT GAA GCT TTA CAA AGA TTG ATT GAA CCT CAA GAA ATT CAA AAC CCA AGG TTG GTT TTA +345
 93 Ser Ile Met Ile Glu Ala Leu Gln Arg Leu Ile Glu Pro Gln Glu Ile Gln Asn Pro Arg Leu Val Leu

    TAC GTT GGT GTA GCA GGG TTA ATT TCT AAT GTC GTA GGT TTA TTT TTG TTC CAC GAT CAT GGC AGC GAT +414
116 Tyr Val Gly Val Ala Gly Leu Ile Ser Asn Val Gal Gly Leu Phe Leu Phe His Asp His Gly Ser Asp

    AGT CTG CAC TCA CAC TCT CAT GGC TCT GTG GAA AGC GGG AAT AAC GAT TTG GAC ATA GAA TCT AAT GCG +483
139 Ser Leu His Ser His Ser His Gly Ser Val Glu Ser Gly Asn Asn Asp Leu Asp Ile Glu Ser Asn Ala

    ACT CAT TCC CAC TCT CAT GCA TCT CTT CCA AAC GAT AAT TTG GCC ATC GAT GAA GAT GCT ATT TCG AGT +552
162 Thr His Ser His Ser His Ala Ser Leu Pro Asn Asp Asn Leu Ala Ile Asp Glu Asp Ala Ile Ser Ser

    CCT GGG CCC TCA GGG CAA ATT GGT GAA GTG TTG CCA CAA TCA GTA GTA AAC AGA TTA TCA AAC GAA AGC +621
185 Pro Gly Pro Ser Gly Gln Ile Gly Glu Val Leu Pro Gln Ser Val Val Asn Arg Leu Ser Asn Glu Ser

    CAA CCC TTA TTG AAC CAC GAT GAT CAT GAC CAC AGC CAT GAA TCA AAG AAA CCA GGT CAT CGT TCT TTG +690
208 Gln Pro Leu Leu Asn His Asp Asp His Asp His Ser His Glu Ser Lys Lys Pro Gly His Arg Ser Leu

    AAT ATG CAT GGT GTC TTC TTA CAT GTA CTA GGT GAT GCT CTG GGT AAT ATT GGT GTT ATT GCA GCT GCT +759
123 Asn Met His Gly Val Phe Leu His Val Leu Gly Asp Ala Leu Gly Asn Ile Gly Val Ile Ala Ala Ala

    TTG TTT ATT TGG AAA ACT GAA TAT TCT TGG AGA TAT TAC TCG GAT CCA ATC GTT TCT TTA ATC ATC ACC +828
254 Leu Phe Ile Trp Lys Thr Glu Tyr Ser Trp Arg Tyr Tyr Ser Asp Pro Ile Val Ser Leu Ile Ile Thr

    ATT ATT ATT TTC TCT TCC GCT CTG CCC TTA TCA CGT AGA GCT TCA AGA ATT TTA CTA CAG GCT ACT CCT +897
277 Ile Ile Ile Phe Ser Ser Ala Leu Pro Leu Ser Arg Arg Ala Ser Arg Ile Leu Leu Gln Ala Thr Pro

    TCT ACA ATT TCT GCT GAT CAG ATT CAA AGA GAG ATT TTG GCA GTA CCT GGC GTG ATA GCG GTC CAT GAC +966
300 Ser Thr Ile Ser Ala Asp Gln Ile Gln Arg Glu Ile Leu Ala Val Pro Gly Val Ile Ala Val His Asp

    TTC CAC GTC TGG AAC TTA ACT GAA TCA ATA TAT ATT GCA TCT ATC CAC GTT CAA ATA GAC TGT GCA CCT +1035
323 Phe His Val Trp Asn Leu Thr Glu Ser Ile Tyr Ile Ala Ser Ile His Val Gln Ile Asp Cys Ala Pro

    GAT AAA TTC ATG AGC TCC GCC AAG CTG ATA AGA AAA ATA TTC CAT CAA CAC GGT ATT CAT TCT GCA ACT +1104
346 Asp Lys Phe Met Ser Ser Ala Lys Leu Ile Arg Lys Ile Phe His Gln His Gly Ile His Ser Ala Thr

    GTT CAA CCA GAA TTT GTC TCT GGA GAT GTT AAT GAG GAT ATT CGC AGA AGA TTT TCT ATC ATA GCA GGT +1173
369 Val Gln Pro Glu Phe Val Ser Gly Asp Val Asn Glu Asp Ile Arg Arg Arg Phe Ser Ile Ile Ala Gly

    GGT TCA CCA TCT TCG TCT CAA GAA GCC TTT GAC AGC CAT GGA AAC ACT GAG CAT GGT AGA AAA AAG AGT +1242
392 Gly Ser Pro Ser Ser Ser Gln Glu Ala Phe Asp Ser His Gly Asn Thr Glu His Gly Arg Lys Lys Ser

    TCA CCT ATT GCC TAT GGT GCT ACT ACA CAT CAT CTA ATT GTA TTG TAGATGACGCTGTAAACTGCAATACTTCCAA +1318
415 Ser Pro Ile Ala Tyr Gly Ala Thr Thr His His Leu Ile Val Leu ***

    TTGCCTGTAATAAAGATATAAAGGGCTGAAGCGTCGGTGAATTTATCCAAGGTTCTACAGAGAACATAAACTATATACTATTTACATAATT +1409
    AATTCTTAGACGTATAGGAAGAAGCAGTTCTTCCTAACTCAAGAACTGCGTTATTCAATGTATGAATACGCTATCTTTTAGCTATAAATCT +1500
    CCATTAAAAATTTGTATGCAATATGTGCCACAGTCCTTCCTCATCTTTTTTTTGAGACTTTTAACCCAGACATAAAATACCCCCATACATC +1591
    CAGAATAGCGTTTTTATCTCTCAAATAGTTAACTTTTTTCAGCGCGCGCAGAGCAGGAAACAAATTAAGCTCACTTCGGTGTGGTAC
```

Figure 12. Nucleotide sequence of the *OSR* gene. Putative TATA-like sequences and the polyadenylation signal are indicated by underlining. Numbers on the right-hand side of the figure show the nucleotide position, which started from the translational initiation codon (ATG). Numbers on the left-hand side of the figure show the amino acid position beginning from the N-terminal methionine (Met).

bp nucleotide sequence and deduced amino acid sequence derived from the DNA sequence. It contained an open reading frame with 1287 bp (encoding 429 amino acids), and the molecular weight of the peptide was calculated to be 47 075 Da. The 5'-non-coding region of the *OSR* gene contained four putative TATA-like sequences at the position of –371 (TATAT), –362 (TATAA), –291 (TATGA) and –87 (TATAA).

A homology search revealed that the *OSR* gene was identical with *ZRC1*, the zinc-resistant conferring gene of *S. cerevisiae* (Kamizono *et al.*, 1989). The amino acid sequence near the C-terminus of OSR was different from that of ZRC1. The OSR protein consisted of 429 amino acids, whereas the ZRC1 protein contained 442 amino acids. The amino acid sequence near the C-terminus of OSR was ^{422}TTHHLIVL, whereas that of ZRC1 was ^{422}TTASSNCIVDDAVNCNTSNCL. This was due to a one-base deletion of the *OSR* gene. If we add 'G' 1270 bp downstream of ATG, the reading frame is shifted and the amino acid sequence thereafter completely coincides with that of ZRC1. However, we could not read 'G' at this position.

It was interesting that the *OSR* gene was identical with the *ZRC1* gene. The *OSR* gene was cloned independently using a different probe (lipid hydroperoxide resistance). Reactive oxygen species in the cells are formed by various routes. Besides the respiration of molecular oxygen, reactive oxygen species are formed by several biochemical reactions, such as xanthine+xanthine oxidase reaction to form the superoxide anion radical. On the other hand, transition metal ions also produce reactive oxygen species by metal-catalysed oxidation (MCO) systems. One possible explanation why the same gene was cloned by different methods is that both probes (zinc ion and oxidative stress) are closely related to the reactive oxygen species. Furthermore, the *ZRC1* gene was recently cloned in a study of ageing of *S. cerevisiae* by Guarente and his co-workers (M. Nishizawa, personal communication). Oxidative stress is known to be one of the factors that determine ageing (Ames and Shigenaga, 1992). Therefore, zinc ions, which may trigger the generation of reactive oxygen species, and oxidative stress, which is caused by reactive oxygen species, may cause similar effects in the cell, and may also be concerned with ageing.

Gene disruption

When the *OSR* gene was introduced into *S. cerevisiae* with the multicopy plasmid, the yeast showed higher resistance against oxidative stress caused by several hydroperoxides, such as *tert*-butyl hydroperoxide and linoleic acid hydroperoxide. To investigate the function of the *OSR* gene product in *S. cerevisiae* cells, the corresponding gene on the chromosomal DNA was disrupted by replacing with the *URA3* gene.

Growth of a knockout mutant in the presence of *tert*-butyl hydroperoxide was compared with that of wild-type cell. First of all, since we could obtain the disruptant using a haploid cell, *OSR* gene seemed not to be necessary for viability of the yeast. As shown in *Figure 13*, the knockout mutant could grow in the *tert*-butyl hydroperoxide-containing medium up to 0.5mM, although the growth rate was slightly reduced. In the presence of 0.7 mM *tert*-butyl hydroperoxide the mutant could not grow at all, whereas the wild-type cell could grow. These results suggested that the *OSR* gene product is necessary for scavenging the excess amount of oxidants produced during the normal metabolism of the cell.

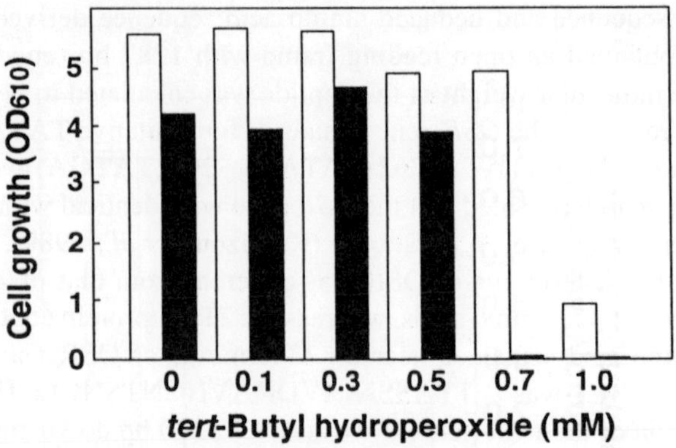

Figure 13. Growth of an *OSR* gene disruptant. *Saccharomyces cerevisiae* whose *OSR* gene was disrupted (*osr::URA3*) was cultured in the SD minimal medium containing *tert*-butyl hydroperoxide. White bars indicate the growth of wild type, and black bars indicate the growth of disruptant. Each bar indicates the OD_{610} of 3 days' culture.

Glutathione content

Previously the author proposed that the function of the *OSR* gene product seemed to be expressed in a glutathione-dependent manner (Inoue, Kobayashi and Kimura, 1993). Intracellular glutathione content was then measured in wild type, knockout mutant and transformant cells carrying YEp24 + *OSR* and YCp50 + *OSR*. The glutathione content in knockout mutants was lowered approximately 44% compared with that in wild-type cells, whereas the cells dosed with the *OSR* gene had approximately a three-fold higher glutathione content (*Figure 14a*).

In the biosynthesis of glutathione, γ-glutamylcysteine synthetase (GSH-I) is a key enzyme. The effect of the *OSR* gene on the expression of the *GSH-I* gene in *S. cerevisiae* was then investigated by quantitative slot blot hybridization (*Figure 14b*). The amount of mRNA of the *GSH-I* gene in the knockout mutant cell was lower compared with that of the wild-type cell. On the other hand, the mRNA level of the *GSH-I* gene was slightly increased in the cells harbouring YEp24 + *OSR* and YCp50 + *OSR*. Biosynthesis of glutathione is believed to be controlled at the substrate level, i.e. GSH-I is a rate-limiting enzyme and the GSH-I activity is negatively controlled by glutathione (feedback inhibition). It implies that the increase in the mRNA level of the *GSH-I* gene may not directly reflect the amount of intracellular glutathione. The results also suggested a possibility that the *OSR* gene product may stabilize the *GSH-I* mRNA, or inhibit the degradation of mRNA of the *GSH-I* gene.

Stress response of OSR gene expression

The promoter activity of the 5'-flanking region (612 bp) of the *OSR* gene was examined by constructing an *OSR*'–'*lacZ* fusion. The *Sau*3AI–*Eco*RV fragment (*Figure 15a*; 744 bp) was cloned into the *Sma*I site of pMC1871 (Shapira *et al.*, 1983) which contained the *lacZ* gene of *E. coli* without its original promoter region and N-terminal eight amino acids. The resultant fusion gene contained the 5'-non-coding region of the *OSR* gene, 45 amino acids beginning from N-terminal methionine and *lacZ* structural gene (*Figure 15a*). The fusion gene was cloned into YCp50 and

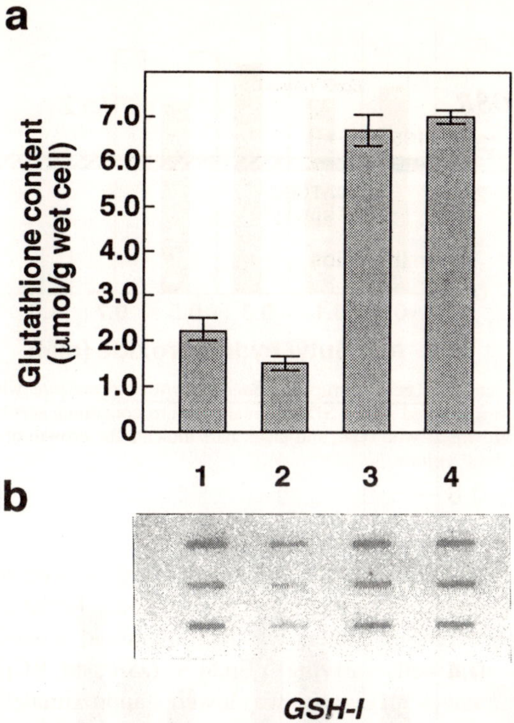

Figure 14. Effect of the *OSR* gene on the biosynthesis of glutathione. (a) Intracellular glutathione content. Lane 1, wild type; lane 2, knockout mutant; lane 3, wild type carrying YEp24 + *OSR*; lane 4, wild type carrying YCp50 + *OSR*. (b) Effect of the *OSR* gene on the mRNA level of the *GSH-I* gene in *S. cerevisiae*. Each slot corresponds to the lanes in (a). Total RNA was prepared from each cell and the amount of *GSH-I* mRNA was monitored by slot hybridization. Results of three independent experiments are shown.

YEp24, respectively, and each recombinant plasmid was transformed into *S. cerevisiae*. β-Galactosidase activity could be detected from both transformants carrying YCp50 + *OSR*'–'*lacZ* and YEp24 + *OSR'–'lacZ*; suggesting that the 5'-non-coding region of *OSR* gene has the promoter activity in *S. cerevisiae*. These observations show that the *OSR* gene is constitutively expressed in *S. cerevisiae* cells, which was also confirmed by Northern blotting analysis (Inoue, Kobayashi and Kimura, 1993). Expression of the *OSR* gene under oxidative conditions was then investigated using the *OSR*'–'*lacZ* fusion gene. The activity increased in accordance with the increased concentrations of *tert*-butyl hydroperoxide and cumene hydroperoxide added to the medium (*Figure 15b*).

The expression of many yeast genes is known to be regulated at the transcriptional level by DNA-binding proteins. To investigate whether some proteins bind to the promoter regions of the *OSR* gene, a gel retardation assay was conducted (*Figure 16a*). Band shift was observed; however, unexpectedly, the shift was also observed even when the cell extracts prepared from those cells not exposed to oxidative stress were used. Band shift was suppressed by the addition of an excess of competitor, thus some proteins specifically recognizing the 5'-flanking region of *OSR* gene might constitutively exist and bind to this region.

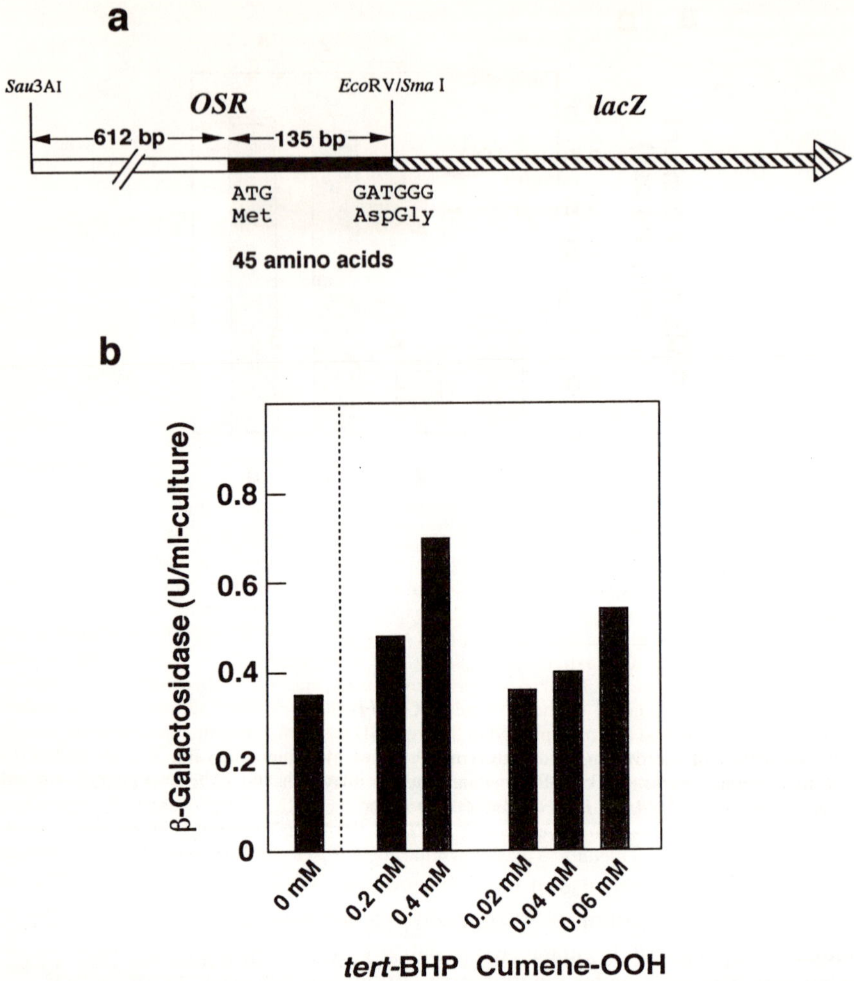

Figure 15. Effect of oxidative stress on translational efficiency. (a) Structure of *OSR*′–′*lacZ* fusion. ▭, 5′-flanking region of *OSR* gene; ▬, a part of the open rearing frame of *OSR* gene (45 amino acids); ▭, *lacZ* gene. (b) *S. cerevisiae* carrying the *OSR*′–′*lacZ* fusion gene with single copy vector (YCp50) was cultured in the SD minimal medium containing low concentrations of lipid hydroperoxide as indicated, and β-galactosidase activity was assayed. *tert*-BHP, *tert*-butyl hydroperoxide; Cumene-OOH, cumene hydroperoxide.

The amount of mRNA of the *OSR* gene after the treatment of cells with sublethal concentrations of *tert*-butyl hydroperoxide and cumene hydroperoxide was examined (*Figure 16b*). No significant changes were observed before and after the treatment. Therefore, the transcription of the *OSR* gene was not affected by the oxidative stress. On the other hand, the translational level seemed to be positively regulated under the oxidative conditions (*Figure 15b*); the region involved might be within the first 45 amino acids (within 135 bases from the initiation codon, AUG) (*Figure 15a*). The secondary structure of mRNA of this region is shown in *Figure 17*. Total secondary structure energy was calculated to be –41.2 kcal mol^{-1}. This structure is, therefore, expected to be possible. Translational efficiency of this mRNA may be

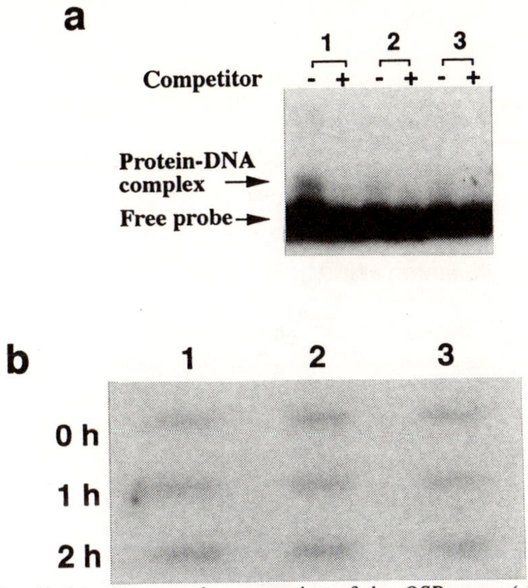

Figure 16. Effect of oxidative stress on the expression of the *OSR* gene. (a) Gel retardation assay. *Saccharomyces cerevisiae* was cultured in SD minimal medium without chemicals, 0.3 mM *tert*-butyl hydroperoxide and 0.1 mM cumene hydroperoxide, respectively, and cell extracts were prepared from each cell. The 5'-flanking region of the *OSR* gene (612 bp as indicated in *Figure 15a*) was labelled with ^{32}P and was used as a probe. Cell extracts used in each lane were prepared from the cells cultured with: lane 1, without chemicals; lane 2, 0.3 mM *tert*-butyl hydroperoxide; land 3, 0.1 mM cumene hydroperoxide, respectively. (b) Slot hybridation of the *OSR* gene. Cells of *S. cerevisiae* cultured for 16 h without oxidative stress were washed and resuspended in the fresh SD medium containing no chemicals (slot 1), 0.3 mM *tert*-butyl hydroperoxide (slot 2) or 0.1 mM cumene hydroperoxide (slot 3), and incubated at 30°C with reciprocal shaking. Total RNA was prepared after 1 h and 2 h, respectively, and slot hybridization was carried out using the *OSR* gene as a probe.

lower in the yeast cell because of its high secondary structure energy. The effect of heat shock on the translational efficiency was examined using the *OSR*'–'*lacZ* fusion, although no change of β-galactosidase activity was observed. Alteration of the translational efficiency under the oxidative conditions appears to be of interest.

UTILIZATION OF THE *OSR* GENE AS A SELECTABLE MARKER GENE FOR YEAST TRANSFORMATION

In the transformation of *S. cerevisiae*, complemental genes for the auxotrophic markers of host strains are commonly used for the selection of transformant. Several drug-resistant genes are also used if the host strains do not have appropriate auxotrophic markers. Since the amplification of the *OSR* gene has proved to make the yeast transformant resistant against oxidative stress caused by lipid hydroperoxide, the plasmid carrying the *OSR* gene (pYHP10) was introduced into *S. cerevisiae* S288C (*MAT*α *SUC2 mal mel gal2 CUP1*), LB1-10B (*MAT*α *mnn1 SUC2 mal gal2 CUP1*) and LB1-16A (*MAT*α *mnn2 SUC2 mal gal2 CUP1*), which were commonly used as laboratory strains. These yeast strains have no appropriate markers such as *leu2* and *ura3* for the detection of transformants, although transformants could be screend for resistance against 2.0 mM *tert*-butyl hydroperoxide on the minimal agar plate. By

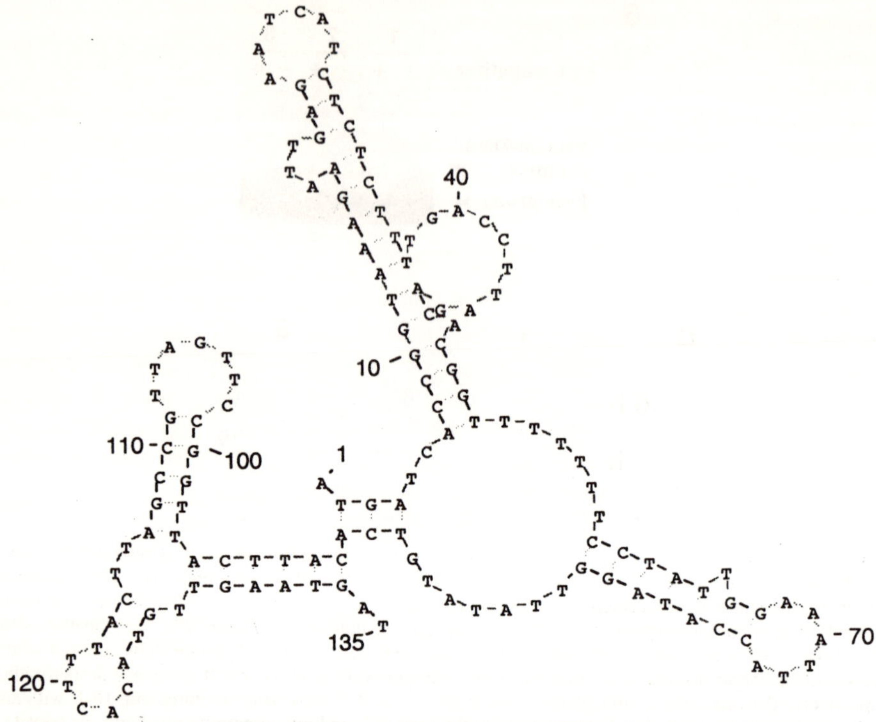

Figure 17. Secondary structure of mRNA of the *OSR* gene; the first 45 amino acid region.

Southern blotting analysis using a part of the *bla* gene in the vector (YEp13), pYHP10 was maintained in the *tert*-butyl hydroperoxide-resistant transformants. Therefore, the *OSR* gene was proved to be useful as the selection marker gene in the transformation of *S. cerevisiae* having no appropriate markers (Inoue *et al.*, 1993).

Most industrial users of yeast probably will not want to grow the transformants continuously in the presence of lipid hydroperoxide, because the chemical would cause the oxidation of the product. Therefore, the previously cloned *OSR* gene must be modified for use in conjunction with integration of transforming DNA into the genome of the host strain, with multicopy, and must confer on the transformants stable resistance against lipid hydroperoxide.

The gene product of *OSR* has not been identified yet; however, the gene could be used as a selection marker for the transformation of *S. cerevisiae* strains that do not have an appropriate auxotrophic marker. Promoter replacement study has also been started, to make it more suitable as the selectable marker gene.

Conclusions

Methylglyoxal and lipid hydroperoxide are endogenous cytotoxic molecular species. Micro-organisms as well as mammals have several defensive mechanisms against these stresses. By amplification of the genes corresponding to resistance against these stresses, breeding of useful organisms would be expected. Indeed, the author and his co-workers succeeded in producing *S*-D-lactoylglutathione, which is a reaction product of glyoxalase I and has several physiological activities, by using genetically engineered *E. coli* cells (Inoue and Kimura, 1992a). The *OSR* gene was found to be useful as a selectable marker gene for the transformation of *S. cerevisiae* strains that do not have appropriate auxotrophic markers.

It has been believed that micro-organisms do not have peroxidases that use glutathione as the electron donor. Micro-organisms are believed to use cytochrome *c* as an electron donor for the peroxidase reaction (cytochrome *c* peroxidase). In plants, ascorbate is an electron donor for ascorbate peroxidase. In mammalian systems, glutathione is used as an electron donor. However, the author and his co-workers have discovered a peroxidase that uses glutathione as an electron donor in the yeast, *H. mrakii*. Furthermore, the glutathione peroxidase is tightly bound to the biological membrane. Very recently, we found that the enzyme is bound to the cell membrane as well as both the inner and outer membranes of mitochondria, an organelle in which large amounts of reactive oxygen species are formed by respiration. Molecular characterization of glutathione peroxidase in *H. mrakii* would give us a hint not only for the elucidation of the mechanisms to protect membrane phospholipid but also for the evolution of mechanisms against oxidative stress in organisms.

References

AIKENS, J. AND DIX, T.A. (1991). Perhydroxyl radical (HOO•) initiated lipid peroxidation. The role of fatty acid hydroperoxides. *Journal of Biological Chemistry* **266**, 15091–15098.

AISAKA, K., UWAJIMA, T. AND TERADA, O. (1983). Purification and properties of glutathione peroxidase from *Mucor hiemalis*. *Agricultural and Biological Chemistry* **47**, 1107–1113.

ALLISON, L.A. AND INGLES, C.J. (1989). Mutations in RNA polymerase II enhance or suppress mutation in GAL4. *Proceedings of the National Academy of Sciences, USA* **86**, 2794–2798.

AL-TIMARI, A. AND DOUGLAS, K.T. (1986). Inhibition of mammalian glyoxalase I (lactoylglutathione lyase) by *N*-acylated *S*-blocked glutathione derivatives as a probe for the role of the *N*-site of glutathione in glyoxalase I mechanism. *Biochimica et Biophysica Acta* **870**, 160–168.

AMES, B.N. (1983). Dietary carcinogens and anticarcinogens. Oxygen radicals and degenerative diseases. *Science* **221**, 1256–1264.

AMES, B.N. AND SHIGENAGA, M.K. (1992). DNA damage by endogenous oxidants and mitogenesis as causes of aging and cancer. In *Molecular Biology of Free Radical Scavenging Systems* (J.G. Scandalios, Ed.), pp. 1–22. Cold Spring Harbor Laboratory Press, Cold Spring Harbor, New York.

ARONSSON, A.-C., MARMSTAL, E. AND MANNERVIK, B. (1978). Glyoxalase I, a zinc metalloenzyme of mammals and yeast. *Biochemical and Biophysical Research Communications* **81**, 1235–1240.

ARONSSON, A.-C., SELLIN, S., TIBBELIN, G. AND MANNERVIK, B. (1981). Probing the active site glyoxalase I from human erythrocytes by use of the strong reversible inhibitor *S-p*-bromobenzylglutathione and metal substitution. *Biochemical Journal* **197**, 67–75.

AWASTHI, Y.C., BEUTLER, E. AND SRIVASTAVA, S.K. (1975). Purification and properties of human erythrocyte glutathione peroxidase. *Journal of Biological Chemistry* **250**, 5144–5149.

BALDOMA, L. AND AGUILAR, J. (1987). Involvement of lactaldehyde dehydrogenase in several metabolic pathways of *Escherichia coli* K-12. *Journal of Biological Chemistry* **262**, 13991–13996.

BARBER, D.J.W. AND THOMAS, J.K. (1978). Reaction of radicals with lecithin bilayers. *Radiation Research* **74**, 51–65.

BARBER, J. (1980). Membrane surface charges and potentials in relation to photosynthesis. *Biochimica et Biophysica Acta* **694**, 253–308.

BERMINGHAM-MCDONOGH, O., GRALLA, E.B. AND VALENTINE, J.S. (1988). The copper, zinc-superoxide dismutase gene of *Saccharomyces cerevisiae*: cloning, sequencing and biological activity. *Proceedings of the National Academy of Sciences, USA* **85**, 4789–4793.

BIELSKI, B.H.J. (1978). Reevaluation of the spectral and kinetic properties of HO_2 and O_2^- free radicals *Photochemistry and Photobiology* **28**, 645–649.

BIELSKI, B.H.J., ARUDI, R.L. AND SUTHERLAND, M.W. (1983). A study of the reactivity of HO_2/O_2^- with unsaturated fatty acids. *Journal of Biological Chemistry* **258**, 4759–4761.

BOVERIS, A., SIES, H., MALTINO, E.E., DOCAMPO, R., TURRENS, J.F. AND SOPPANI, O.M. (1980). Deficient metabolic utilization of hydrogen peroxide in *Tryphanosoma cruzi*. *Biochemical Journal* **188**, 643–648.

CERUTTI, P.A. (1985). Peroxidant states and tumor promotion. *Journal of Biological Chemistry* **251**, 7816–7820.

CHRISTMAN, C.F., STORZ, G. AND AMES, B.N. (1989). *OxyR*, a possible regulator of hydrogen peroxide-inducible genes in *Escherichia coli* and *Salmonella typhimurium*, is homologus to a family of bacterial regulatory proteins. *Proceedings of the National Academy of Sciences, USA* **86**, 3484–3488.

COHEN, G., RAPATZ, W. AND RUIS, H. (1988). Sequence of the *Saccharomyces cerevisiae CTA1* gene and amino acid sequence of catalase A derived from it. *European Journal of Biochemistry* **176**, 159–163.

COLLINSON, L.P. AND DAWES, I. (1992). Inducibility of the response of yeast cells to peroxide stress. *Journal of General Microbiology* **138**, 329–335.

COOPER, R.A. (1974). Methylglyoxal formation during glucose catabolism by *Pseudomonas saccharophilia*. Identification of methylglyoxal synthase. *European Journal of Biochemistry* **44**, 81–86.

DEMPLE, B. AND AMABILE, CUEVAS, C.F. (1991). Redox redux: the control of oxidative stress responses. *Cell* **67**, 837–839.

DOLIN, M.I. (1957). The *Streptococcus faecalis* oxidase for reduced diphosphopyridine nucleotide. III. Isolation and properties of a flavin peroxidase for reduced diphosphopyridine nucleotide. *Journal of Biological Chemistry* **225**, 557–573.

DOUGLAS, K.T. AND SHINKAI, T. (1985). Chemische Grundlagen des Wirkungsmechanisums von Glyoxalase I, einem Zielenzym für Cancerostica. *Angewandte Chemie* **97**, 32–45.

DOUGLAS, K.T., SEDDON, A.P. AND NAKAGAWA, Y. (1986). Yeast glyoxalase I. Circular dichronic spectra and pH effects. *International Journal of Biochemistry* **18**, 549–555.

EGYUD, L.G. AND SZENT-GYORGYI, A. (1966). Cell division, SH, ketoaldehydes and cancer. *Proceedings of the National Academy of Sciences, USA* **55**, 388–393.

FARR, S.B. AND KOGOMA, T. (1991). Oxidative stress responses in *Escherichia coli* and *Salmonella typhimurium*. *Microbiological Reviews* **55**, 561–585.

FLATTERY-O'BRIEN, J., COLLISON, L.P. AND DAWES, I.W. (1993). *Saccharomyces cerevisiae* has an inducible response to menadione which differs from that to hydrogen peroxide. *Journal of General Microbiology* **139**, 501–507.

GEBICKI, J.M. AND BIELSKI, B.H.J. (1981). Comparison of the capacities of the perhydroxyl and the superoxide radicals to initiate chain oxidation of linoleic acid. *Journal of American Chemical Society* **103**, 7020–7022.

GIBSON, D.D., HORNBROOK, K.R. AND MCCAY, P.B. (1980). Glutathione-dependent inhibition of lipid peroxidation by a soluble, heat-labile factor in animal tissue. *Biochimica et Biophysica Acta* **620**, 572–582.

GINYA, H. (1991). MSc Thesis, Kyoto University.

GOSCIN, S.A. AND FRIDOVICH, I. (1972). The purification and properties of superoxide dismutase from *Saccharomyces cerevisiae*. *Biochimica et Biophysica Acta* **289**, 276–283.

GREENBERG, J.T., MONACH, P., CHOU, J.H., JOSEPHY, D. AND DEMPLE, B. (1990). Positive control of a global antioxidant defense regulon activated by superoxide-generating agents in *Escherichia coli*. *Proceedings of the National Academy of Sciences, USA* **87**, 6181–6185.

GROSMANN, A. AND WENDEL, A. (1983). Non-reactivity of selenoenzyme glutathione peroxidase with enzymatically hydroperoxized phospholipid. *European Journal of Biochemistry* **135**, 549–552.

GUARENTE, L. (1984). Yeast promoters: positive and negative elements. *Cell* **36**, 799–800.

GUARENTE, L. (1988). UASs and enhancer: common mechanism of transcriptional activation in yeast and mammals. *Cell* **52**, 303–305.

GUHA, M.K., VANDER JAGT, D.L. AND CREIGHTON, D.J. (1988). Diffusion-dependent rates for the hydrolysis reaction catalyzed by glyoxalase II from rat erythrocytes. *Biochemistry* **27**, 8818–8822.

HABIG, W.H., PABST, M.J. AND JAKOBY, W.B. (1974). Glutathione *S*-transferase: the first enzymatic step in mercaptric acid formation. *Journal of Biological Chemistry* **249**, 7130–7139.

HAGER, L.P. (1970). Iodination of tyrosine by chloroperoxidase; preparation of chloroperoxidase (*Caldariomyces fumago*). *Methods in Enzymology* **17**, 648–652.

HALLIWELL, B. AND GUTTERIDGE, J.M.C. (1990). Role of free radicals and catalytic metal ions in human disease: an overview. *Methods in Enzymology* **186**, 1–85.

HAN, L.-P.B., SCHIMANDLE, C.M., DAVISON, L.M. AND VANDER JAGT, D.L. (1977). Comparative kinetics of Mg^{2+}-, Mn^{2+}-, Co^{2+}- and Ni^{2+}-activated glyoxalase I. Evaluation of the role of the metal ion. *Biochemistry* **16**, 5478–5484.

HARASHIMA, S., TAKAGI, A. AND OSHIMA, Y. (1984). Transformation of protoplasted yeast cells is directly associated with cell fusion. *Molecular and Cellular Biology* **4**, 771–778.

HARTIG, A. AND RUIS, H. (1986). Nucleotide sequence of the *Saccharomyces cerevisiae CTT1* gene and deduced amino-acid sequence of yeast catalase T. *European Journal of Biochemistry* **160**, 487–490.

HINNEN, A., HICKS, J.B. AND FINK, G.R. (1978). Transformation of yeast. *Proceedings of the National Academy of Sciences, USA* **75**, 1929–1933.

HOPPER, D.J. AND COOPER, R.A. (1972). The purification and properties of *Escherichia coli* methylglyoxal synthase. *Biochemical Journal* **128**, 321–329.

INOUE, Y. AND KIMURA, A. (1991). Expression of glyoxalase I gene and signal transduction in *Saccharomyces cerevisiae. Bulletin of the Research Institute for Food Science, Kyoto University* **54**, 31–32.

INOUE, Y. AND KIMURA, A. (1992a). Production of physiological active materials by microorganisms. *Bulletin of the Research Institute for Food Science, Kyoto University* **55**, 33–35.

INOUE, Y. AND KIMURA, A. (1992b). Purification and characterization of glyoxalase II from *Hansenula mrakii. Journal of Fermentation and Bioengineering* **73**, 271–276.

INOUE, Y. AND KIMURA, A. (1992c). Glyoxalase system and new type of glutathione peroxidase in yeast. *Nippon Nogeikagaku Kaishi* **66**, 1505–1508.

INOUE, Y. AND KIMURA, A. Methylglyoxal and regulation of its metabolism in microorganisms. In *Advances in Microbial Physiology* (A.H. Rose and D.W. Tempest, Eds). Academic Press, London, in press.

INOUE, Y. AND KIMURA, A. (1994). Purification and some properties of NAD(P)H dehydrogenase from *Saccharomyces cerevisiae*: contribution to glycolytic methylglyoxal pathway. *Journal of Fermentation and Bioengineering* **77**, 557–561.

INOUE, Y., KOBAYASHI, S. AND KIMURA, A. (1993). Cloning and phenotypic characterization of a gene enhancing the resistance against oxidative stress in *Saccharomyces cerevisiae. Journal of Fermentation and Bioengineering* **75**, 327–331.

INOUE, Y., TRAN, L.-T. AND KIMURA, A. (1993). Role of glutathione peroxidase against oxidative stress in yeast: phenotypic character of lipid hydroperoxide-sensitive mutants derived from *Hansenula mrakii. Journal of Fermentation and Bioengineering* **75**, 229–231.

INOUE, Y., YANO, H. AND KIMURA, A. (1993). Effect of cell–cell interaction on the mRNA level of glyoxalase I gene in *Saccharomyces cerevisiae. Journal of Fermentation and Bioengineering* **76**, 388–390.

INOUE, Y., WATANABE, K., SHIMOSAKA, M., SAIKUSA, T., FUKUDA, Y., MURATA, K. AND KIMURA, A. (1985). Metabolism of 2-oxoaldehydes in yeasts. Purification and characterization of lactaldehyde dehydrogenase from *Saccharomyces cerevisiae. European Journal of Biochemistry* **153**, 243–247.

INOUE, Y., RHEE, H.-I., WATANABE, K., MURATA, K. AND KIMURA, A. (1987). Metabolism of 2-ketoaldehydes in mold: purification and characterization of glyoxalase I from *Aspergillus niger. Journal of Biochemistry* **102**, 583–589.

INOUE, Y., RHEE, H.-I., WATANABE, K., MURATA, K. AND KIMURA, A. (1988). Metabolism of

2-oxoaldehydes in mold. Purification and characterization of two methylglyoxal reductase from *Aspergillus niger. European Journal of Biochemistry* **171**, 213–218.

INOUE, Y., CHOI, B.-Y., MURATA, K. AND KIMURA, A. (1989). Sexual response in *Saccharomyces cerevisiae*: alteration of enzyme activity in the glyoxalase system by mating factor. *Biochemical and Biophysical Research Communications* **165**, 1091–1095.

INOUE, Y., CHOI, B.-Y., MURATA, K. AND KIMURA, A. (1990a). Sexual response of *Saccharomyces cerevisiae*: phosphorylation of yeast glyoxalase I by a cell extract of mating factor-treated cells. *Journal of Biochemistry* **108**, 4–6.

INOUE, Y., ICHIRYU, T., YOSHIKAWA, K., TRAN, L.-T., MURATA, K. AND KIMURA, A. (1990b). Induction of glutathione peroxidase by linoleic acid hydroperoxide in *Hansenula mrakii. Agricultural and Biological Chemistry* **54**, 3289–3293.

INOUE, Y., LING, F., CHOI, B.-Y., MURATA, K. AND KIMURA, A. (1990c). Nucleotide sequence of a gene which enhances the activity of glyoxalase I in *Saccharomyces cerevisiae. Biotechnology and Applied Biochemistry* **12**, 341–345.

INOUE, Y., TRAN, L.-T., YOSHIKAWA, K., MURATA, K. AND KIMURA, A. (1991a). Purification and characterization of glyoxalase I from *Hansenula mrakii. Journal of Fermentation and Bioengineering* **71**, 131–133.

INOUE, Y., TRAN, L.-T., YOSHIKAWA, K., MURATA, K. AND KIMURA, A. (1991b). Purification and characterization of methylglyoxal reductase from *Hansenula mrakii. Journal of Fermentation and Bioengineering* **71**, 134–136.

INOUE, Y., YANO, H., GINYA, H., TSUCHIYAMA, H., MURATA, K. AND KIMURA, A. (1991c). Positive effect of *GAC* gene product on the mRNA level of glyoxalase I gene in *Saccharomyces cerevisiae. Biotechnology and Applied Biochemistry* **14**, 391–394.

INOUE, Y., IKEMOTO, S., KITAMURA, K. AND KIMURA, A. (1992a). Occurrence of a NADH-dependent methylglyoxal reducing system: conversion of methylglyoxal to acetol by aldehyde reductase from *Hansenula mrakii. Journal of Fermentation and Bioengineering* **74**, 46–48.

INOUE, Y., TRAN, L.-T., YOSHIKAWA, K. AND KIMURA, A. (1992b). Phenotypic character of the lipid hydroperoxide-resistant mutant: positive relationship between flocculation and resistance against lipid hydroperoxide in *Saccharomyces cerevisiae. World Journal of Microbiology and Biotechnology* **8**, 296–300.

INOUE, Y., KOBAYASHI, S., YOSHIKAWA, K., TRAN, L.-T. AND KIMURA, A. (1993). Lipid hydroperoxide resistant gene in *Saccharomyces cerevisiae*: utilization as a selectable marker gene for yeast transformation. *Biotechnology and Applied Biochemistry* **17**, 305–310.

INOUE, Y., FUKUDA, K., WAKAI, Y., SUDSAI, T. AND KIMURA, A. (1994). Ester formation by a yeast *Hansenula mrakii* IFO 0895: contribution of esterase for *iso*-amyl acetate production in sake brewing. *Lebensmittel-Wissenschaft und Technologie* **27**, 189–193.

ITO, H., FUKUDA, Y., MURATA, K. AND KIMURA, A. (1983). Transformation of intact yeast cells treated with alkali cations. *Journal of Bacteriology* **153**, 163–168.

JACOBSON, F.S., MORGAN, R.W., CHRISTMAN, M.F. AND AMES, B.N. (1989). An alkyl hydroperoxide reductase from *Salmonella typhimurium* involved in the defense of DNA against oxidative damage. *Journal of Biological Chemistry* **264**, 1488–1496.

JAMIESON, D.J. (1992). *Saccharomyces cerevisiae* has distinct adaptive responses to both hydrogen peroxide and menadione. *Journal of Bacteriology* **174**, 6678–6681.

JERZYKOWSKI, T., WINTER, R., MATSUZEWSKI, W. AND PISKORSKA, D. (1978). A re-evaluation on the distribution of glyoxalases in animal and tumor tissues. *International Journal of Biochemistry* **9**, 853–860.

JOHNSTON, M. (1987a). A model fungal gene regulatory mechanism: the *GAL* genes of *Saccharomyces cerevisiae. Microbiological Reviews* **51**, 458–476.

JOHNSTON, M. (1987b). Genetic evidence that zinc is an essential co-factor in the DNA-binding domain of GAL4 protein. *Nature* **328**, 353–355.

KAMIZONO, A., NISHIZAWA, M., TERANISHI, Y., MURATA, K. AND KIMURA, A. (1989). Identification of a gene conferring resistance to zinc and cadmium ions in the yeast *Saccharomyces cerevisiae. Molecular and General Genetics* **219**, 161–167.

KAPUT, J., GOLZ, S. AND BLOBEL, G. (1982). Nucleotide sequence of the yeast nuclear gene for cytochrome *c* peroxidase precursor. Functional implications of the pre sequence for protein transport into mitochondria. *Journal of Biological Chemistry* **257**, 15054–15058.

KENTEN, R.H. AND MANN, P.J.C. (1954). A simple method for the preparation of horseradish peroxidase. *Biochemical Journal* **57**, 347–348.

KIM, N.-S., UMEZAWA, Y., OHMURA, S. AND KATO, S. (1993). Human glyoxalase I. cDNA cloning, expression and sequence similarity to glyoxalase I from *Pseudomonas putida. Journal of Biological Chemistry* **268**, 11217–11221.

KOSUGI, N., INOUE, Y., RHEE, H.-I., MURATA, K. AND KIMURA, A. (1988). Production of *S*-lactoylglutathione by glycerol-adapted *Saccharomyces cerevisiae* and genetically engineered *Escherichia coli* cells. *Applied Microbiology and Biotechnology* **28**, 263–267.

KUMAGAI, H., TAMAKI, H., KOSHINO, Y., SUZUKI, H. AND TOCHIKURA, T. (1988). Distribution, formation and stabilization of yeast glutathione *S*-transferase. *Agricultural and Biological Chemistry* **52**, 1377–1382.

MANNERVIK, B. AND RIDDERSTROM, M. (1993). Catalytic and molecular properties of glyoxalase I. *Biochemical Society Transactions* **21**, 515–517.

MANNERVIK, B., LINDSSTORM, L. AND BARTFAI, T. (1972). Partial purification and characterization of glyoxalase I from porcine erythrocytes. *European Journal of Biochemistry* **29**, 276–281.

MARMSTAL, E., ARONSSON, A.-C. AND MANNERVIK, B. (1979). Comparison of glyoxalase I purified from yeast (*Saccharomyces cerevisiae*) with the enzyme from mammalian sources. *Biochemical Journal* **183**, 23–30.

MARRES, C.A.M., VAN LOON, A.P.G.M., OUDSHOORN, P., VAN STEEG, H., GRIVELL, L.A. AND SLATER, E.C. (1985). Nucleotide sequence analysis of the nuclear gene coding for manganase superoxide dismutase of yeast mitochondria, a gene previously assumed to code for the Rieske iron–sulphur protein. *European Journal of Biochemistry* **147**, 153–161.

MORGENSTERN, R. AND DEPRIERRE, J.W. (1983). Microsomal glutathione transferase. Purification in unactivated form and further characterization of the activation process, substrate specificity and amino acid composition. *European Journal of Biochemistry* **134**, 591–597.

MORGENSTERN, R., GUTHENBERG, C. AND DEPRIERRE, J.W. (1982). Microsomal glutathione *S*-transferase. Purification, initial characterization and demonstration that it is identical to the cytosolic glutathione *S*-transferase A, B and C. *European Journal of Biochemistry* **128**, 243–248.

MORGENSTERN, R., MEIJER, J., DEPRIERRE, J.W. AND ERNSTER, L. (1980). Characterization of rat-liver microsomal glutathione *S*-transferase activity. *European Journal of Biochemistry* **104**, 167–174.

MURATA, K., FUKUDA, Y., SHIMOSAKA, M., WATANABE, K., SAIKUSA, I. AND KIMURA, A. (1985a). Metabolism of 2-oxoaldehydes in yeasts. Purification and characterization of NADPH-dependent methylglyoxal reducing enzyme from *Saccharomyces cerevisiae. European Journal of Biochemistry* **151**, 631–636.

MURATA, K., FUKUDA, Y., WATANABE, K., SAIKUSA, T., SHIMOSAKA, M. AND KIMURA, A. (1985b). Characterization of methylglyoxal synthase in *Saccharomyces cerevisiae. Biochemical and Biophysical Research Communications* **131**, 190–198.

MURATA, K., INOUE, Y., SAIKUSA, T., WATANABE, K., FUKUDA, Y., SHIMOSAKA, M. AND KIMURA, A. (1986a). Metabolism of α-ketoaldehydes in yeasts: inducible formation of methylglyoxal reductase and its relation to growth arrest of *Saccharomyces cerevisiae. Journal of Fermentation and Technology* **64**, 1–4.

MURATA, K., INOUE, Y., WATANABE, K., FUKUDA, Y., SAIKUSA, T. AND KIMURA, A. (1986b). Metabolism of α-ketoaldehydes in yeasts: purification and characterization of glyoxalase II from *Saccharomyces cerevisiae. Agricultural and Biological Chemistry* **50**, 135–142.

MURATA, K., SAIKUSA, T., FUKUDA, Y., WATANABE, K., INOUE, Y., SHIMOSAKA, M. AND KIMURA, A. (1986c). Metabolism of 2-oxoaldehydes in yeasts. Possible role of glycolytic bypath as a detoxification system in L-threonine catabolism in *Saccharomyces cerevisiae. European Journal of Biochemistry* **157**, 297.

MURATA, K., SAIKUSA, T., WATANABE, K., INOUE, Y., FUKUDA, Y., SHIMOSAKA, M. AND KIMURA, A. (1986d). Complete purification of yeast glyoxalase I and its activation by ferrous ions and polyamines. *Agricultural and Biological Chemistry* **50**, 2381–2383.

MURATA, K., SATO, N., RHEE, H.-I., WATANABE, K. AND KIMURA, K. (1987). Purification and characterization of glutathione thiolesterase from *Saccharomyces cerevisiae*. *Agricultural and Biological Chemistry* **51**, 1901–1907.

MURATA, K., SAIKUSA, T., INOUE, Y. AND KIMURA, A. (1988). Molecular cloning of a gene enhancing yeast glyoxalase I activity and its effect on yeast cell size. *Journal of Fermentation and Technology* **66**, 495–500.

NUNOSHIBA, T., HIDALGO, E., AMABILE CUEVAS, C.F. AND DEMPLE, B. (1992). Two-stage control of an oxidative stress regulon: the *Escherichia coli* SoxR protein triggers redox-inducible expression of the *soxS* regulatory gene. *Journal of Bacteriology* **174**, 6054–6060.

ORAY, B. AND NORTON, S.J. (1980). Purification and characterization of mouse liver glyoxalase II. *Biochimica et Biophysica Acta* **611**, 168–173.

OVERBAUGH, J.M. AND FALL, R. (1985). Characterization of a selenium-independent glutathione peroxidase from *Euglena gracilis*. *Plant Physiology* **77**, 437–442.

PENNINCKX, M.J., JESPERS, C.J. AND LEGRAIN, M.J. (1983). The glutathione-dependent glyoxalase pathway in yeast *Saccharomyces cerevisiae*. A vital defense line against methylglyoxal production during glycerol catabolism. *Journal of Biological Chemistry* **258**, 6030–6036.

PRINCIPATO, G.B., TALESA, V., NORTON, S.J. AND GIOVANNINI, E. (1989). Effects of some S-blocked gutathione derivatives on prevalent glyoxalase II (a form) of rat liver. *Enzyme* **41**, 175–180.

PROHASKA, J.R. AND GANTHER, H.E. (1977). Glutathione peroxidase activity of glutathione *S*-transferases purified from rat liver. *Biochemical and Biophysical Research Communications* **76**, 437–445.

PRYOR, W.A. (1986). Oxy-radicals and related species: their formation, lifetimes, and reactions. In *Annual Review of Physiology* (R.M. Berne and J.F. Hoffman, Eds), volume 48, pp. 657–667. Annual Reviews Inc., Palo Alto, California.

RACKER, E. (1955). Glutathione reductase from baker's yeast and beef liver. *Journal of Biological Chemistry* **217**, 855–865.

RAE, C., BOARD, P.G. AND KUCHEL, P.W. (1991). Glyoxalase 2 deficiency in the erythrocyte of a horse: ^{1}H NMR studies of enzyme kinetics and transport of *S*-lactoylglutathione. *Archives of Biochemistry and Biophysics* **291**, 291–299.

RAE, C., BERNERS-PRICE, S.J., BULLIMAN, B.T. AND KUCHEL, P.W. (1990). Kinetic analysis of the human erythrocyte glyoxalase system using ^{1}NMR and computer model. *European Journal of Biochemistry* **193**, 83–90.

RANGANATHAN, S., WALSH, E.S., GODWIN, A.K. AND TEW, K.D. (1993). Cloning and characterization of human colon glyoxalase-I. *Journal of Biological Chemistry* **268**, 5661–5667.

RAVINDRANATH, S.D. AND FRIDOVICH, I. (1975). Isolation and characterization of a manganase-containing superoxide dismutase from yeast. *Journal of Biological Chemistry* **250**, 6107–6112.

RAY, S. AND RAY, M. (1981). Isolation of methylglyoxal synthase from goat liver. *Journal of Biological Chemistry* **256**, 6230–6234.

RHEE, H.-I., MURATA, K. AND KIMURA, A. (1986). Purification and characterization of glyoxalase I from *Pseudomonas putida*. *Biochemical and Biophysical Research Communications* **141**, 993–999.

RHEE, H.-I., MURATA, K. AND KIMURA, A. (1987). Molecular cloning of the *Pseudomonas putida* glyoxalase I gene in *Escherichia coli*. *Biochemical and Biophysical Research Communications* **147**, 831–838.

RHEE, H.-I., MURATA, K. AND KIMURA, A. (1988). Nucleotide sequence of the glyoxalase I gene of *Pseudomonas putida*. *Agricultural and Biological Chemistry* **52**, 2243–2246.

RHEE, H.-I., WATANABE, K., MURATA, K. AND KIMURA, A. (1987). Metabolism of 2-ketoaldehydes in bacteria: oxidative conversion of methylglyoxal to pyruvate by an enzyme from *Pseudomonas putida*. *Agricultural and Biological Chemistry* **51**, 1059–1066.

RINE, J., HANSEN, W., HARDEMAN, E. AND DAVIS, R.W. (1983). Targeted selection of recombinant clones through gene dosage effects. *Proceedings of the National Academy of Sciences, USA* **80**, 6750–6754.

ROGGENKAMP, R., HANSEN, H., ECKART, M., JANOWICZ, Z. AND HOLLENGERG, C.P. (1986). Transformation of the methylotrophic yeast *Hansenula polymorpha* by autonomous replication and integration vectors. *Molecular and General Genetics* **202**, 302–308.

ROVERI, A., CASASCO, A., MAIORINO, M., DALAN, P., CALLIGARO, F. AND URSINI, F. (1992). Phospholipid hydroperoxide glutathione peroxidase of rat testis. *Journal of Biological Chemistry* **267**, 6142–6146.

SAIKUSA, T., RHEE, H.-I., WATANABE, K., MURATA, K. AND KIMURA, A. (1987). Metabolism of 2-oxoaldehydes in bacteria: purification and characterization of methylglyoxal reductase from *Escherichia coli. Agricultural and Biological Chemistry* **51**, 1893–1899.

SAKAI, Y., KAZARIMOTO, T. AND TANI, Y. (1991). Transformation system for an asporogenous methylotrophic yeast, *Candida boidinii*: cloning of the orotidine-5'-phosphate decarboxylase gene (*URA3*), isolation of uracil auxotrophic mutants, and use of the mutants for integrative transformation. *Journal of Bacteriology* **173**, 7458–7463.

SAWYER, D.T., MCDOWELL, M.S. AND YAMAGUCHI, K.S. (1988). Reactivity of perhydroxyl (HOO•) with 1,4-cyclohexadiene (model for allylic groups in biomembranes). *Chemical Research of Toxicology* **1**, 97–100.

SEAH, T.C.M. AND KAPLAN, J.G. (1973). Purification and properties of the catalase of bakers' yeast. *Journal of Biological Chemistry* **248**, 2889–2893.

SEAH, T.C.M., BHATTI, A.R. AND KAPLAN, J.G. (1973). Novel catalytic proteins of bakers' yeast. I. An atypical catalase. *Canadian Journal of Biochemistry* **51**, 1551–1555.

SELLIN, S. AND MANNERVIK, B. (1984). Metal dissociation constants for glyoxalase I reconstituted with Zn^{2+}, Co^{2+}, Mn^{2+} and Mg^{2+}. *Journal of Biological Chemistry* **259**, 11426–11429.

SELLIN, S., ERIKSSON, L.E.G. AND MANNERVIK, B. (1987). Electron paramagnetic resonance study of the active site of copper-substituted human glyoxalase I. *Biochemistry* **26**, 6779–6784.

SELLIN, S., ARONSSON, A.-C., ERIKSSON, L.E.G., LARSEN, K., TIBBELIN, G. AND MANNERVIK, B. (1983). Properties and catalytic function of glyoxalase I. In *Function of Glutathione: Biochemical, Physiological, Toxicological, and Clinical Aspects* (A. Larsson, S. Orrenius, A. Holmgren and B. Mannervik, Eds), pp. 187–197. Raven Press, New York.

SHAPIRA, S.K., CHOU, J., RICHAUD, F.V. AND CASADABAN, M.J. (1983). New versatile plasmid vectors for expression of hybrid proteins coded by a cloned gene fused to *lacZ* gene sequences encoding an enzymatically active carboxy-terminal protein of β-galactosidase. *Gene* **25**, 71–82.

SRIDHARA, S. AND WU, T.T. (1969). Purification and properties of lactaldehyde dehydrogenase from *Escherichia coli. Journal of Biological Chemistry* **244**, 5233–5238.

STORZ, G. AND TARTAGLIA, L.A. (1992). *OxyR*: a regulator of antioxidant genes. *Journal of Nutrition* **122**, 627–630.

STORZ, G., TARTAGLIA, L.A. AND AMES, B.N. (1990). Transcriptional regulator of oxidative stress-inducible genes: direct activation by oxidation. *Science* **248**, 189–194.

STORZ, G., BAIRD, P.T., TOLEDANO, M.B., ALTUVIA, S. AND KULLIK, I. (1993). Regulation of oxidative stress-inducible genes in bacteria. *Proceedings of the XIIth International Symposium on the Stress Proteins*, Fukuoka, Japan, pp. 206–210.

STRUHL, K. (1987). Promoters, activator proteins and the mechanism of transcriptional initiator in yeast. *Cell* **49**, 295–297.

TAKAHASHI, K. (1968). The reaction of phenylglyoxal with arginine residues in proteins. *Journal of Biological Chemistry* **243**, 6171–6179.

TALESA, V., UOTILA, L., KOIVUSALO, M., PRINCIPATO, G.B., GIOVANNINI, E. AND ROSI, G. (1988). Demonstration of glyoxalase II in rat liver mitochondria. Partial purification and occurrence in multiple forms. *Biochimica et Biophysica Acta* **955**, 103–110.

TAN, K.H., MEYER, D.J., BELIN, J. AND KETTERER, B. (1984). Inhibition of microsomal lipid peroxidation by glutathione and glutathione transferases B and AA. *Biochemical Journal* **220**, 243–252.

TRAN, L.-T., INOUE, Y. AND KIMURA, A. (1993a). Oxidative stress response in yeast: purification and some properties of a membrane-bound glutathione peroxidase from *Hansenula mrakii. Biochimica et Biophysica Acta* **1164**, 166–172.

TRAN, L.-T., INOUE, Y. AND KIMURA, A. (1993b). Induction of glutathione peroxidase by reactive oxygen in the yeast *Hansenula mrakii*. *Journal of Fermentation and Bioengineering* **76**, 64–66.

TSAI, P.-K. AND GRACY, R.W. (1976). Isolation and characterization of crystalline methylglyoxal synthase from *Proteus vulgaris*. *Journal of Biological Chemistry* **251**, 364–367.

TSANEVA, I.R. AND WEISS, B. (1990). *SoxR*, a locus governing a superoxide response regulon in *Escherichia coli* K-12. *Journal of Bacteriology* **172**, 4197–4205.

UOTILA, L. (1973). Purification and characterization of *S*-2-hydroxyacylglutathione hydrolase glyoxalase II from human liver. *Biochemistry* **12**, 3944–3951.

UOTILA, L. (1989). Glutathione thiol esterases. In *Glutathione: Chemical, Biochemical and Medical Aspects, Part A* (D. Dolphin, R. Poulson and O. Avramovic, Eds), pp. 767–804. Wiley-Interscience, New York.

UOTILA, L. AND KOIVUSALO, M. (1975). Purification and properties of glyoxalase I from sheep liver. *European Journal of Biochemistry* **52**, 493–503.

URSINI, F., MAIORINO, M. AND GREGOLIN, C. (1985). The selenoenzyme phospholipid hydroperoxide glutathione peroxidase. *Biochimica et Biophysica Acta* **839**, 62–70.

VALLEE, B.L. AND AULD, D.S. (1990). Active-site zinc ligands and activated H_2O of zinc enzymes. *Proceedings of the National Academy of Sciences, USA* **87**, 220–224.

VANDER JAGT, D.L. (1993). Glyoxalase II: molecular characteristics, kinetics and mechanism. *Biochemical Society Transactions* **21**, 522–527.

VERDIER, J.-M. (1990). Regulatory DNA-binding proteins in yeast: an overview. *Yeast* **6**, 271–297.

WEBSTER, T.D. AND DICKSON, R.C. (1983). Direct selection of *Saccharomyces cerevisiae* resistant to the antibiotic G418 following transformation with a DNA vector carrying the kanamycin-resistance gene of *Tn*903. *Gene* **26**, 243–252.

WILLETTS, A.J. AND TURNER, J.M. (1970). Threonine metabolism in a strain of *Bacillus subtilis*: enzymes acting on methylglyoxal. *Biochimica et Biophysica Acta* **222**, 668–670.

WU, J. AND WEISS, B. (1991). Two divergently transcribed genes, *soxR* and *soxS*, control a superoxide response regulon of *Escherichia coli*. *Journal of Bacteriology* **173**, 2864–2871.

YONETANI, T. (1970). Cytochrome *c* peroxidase. In *Advances in Enzymology and Related Areas of Molecular Biology* (F.F. Nord, Ed.), volume 33, pp. 309–335. Interscience, New York.

YONETANI, T. AND RAY, G.S. (1965). Studies on cytochrome *c* peroxidase. I. Purification and some properties. *Journal of Biological Chemistry* **240**, 4503–4508.

14
Monoclonal Antibody Production using Free-suspended and Entrapped Hybridoma Cells

GYUN MIN LEE[1] AND BERNHARD O. PALSSON[2]

[1] Department of Biotechnology, Korea Advanced Institute of Science and Technology, Taejon, Korea and [2] Department of Chemical Engineering, University of Michigan, Ann Arbor, USA

Introduction

Various cell culture methods have been used for the large-scale production of monoclonal antibodies (MAbs). Broadly one can distinguish between the systems in which the cells are entrapped (or immobilized) and those in which the cells are grown in free suspension. Many researchers have recognized the potential for MAb production using entrapped cells. An active discussion has appeared in the literature concerning the relative benefits of entrapped cell culture for large-scale MAb production (Adamson and Schmidli, 1986; Nilsson, 1987; Tyo and Spier, 1987; Heath and Belfort, 1990; Looby and Griffiths, 1990). In fact, MAb production has been carried out successfully using entrapped hybridoma cells (Nilsson *et al*., 1983, 1986; Shirai *et al*., 1987; Bugarski *et al*., 1989; Sinacore, Creswick and Buehler, 1989). The potential advantages that the entrapped cell culture offers over conventional free-suspended cell culture are as follows (Nilsson, 1987; Tyo and Spier, 1987).

First, entrapped cells can be protected from shear stress induced by mechanical agitation and/or air sparging. Hybridoma cells in stationary and decline phases of growth were more sensitive to shear force caused by mechanical agitation than the cells in the growth phase (Lee *et al*., 1988; Petersen, McIntire and Papoutsakis, 1988). Since a significant portion of MAb is produced during the stationary and decline phases of growth (Boraston *et al*., 1984; Velez *et al*., 1986; Brennan, Shevitz and Macmillan, 1987; Hu *et al*., 1987; Lee *et al*., 1988; Miller, Blanch and Wilke, 1988; Renard *et al*., 1988; Dalili and Ollis, 1989; Savinell, Lee and Palsson, 1989), it is important to maintain cell viability during the stationary and decline phases by

Abbreviations: BSA, bovine serum albumin; EDTA, ethylenediaminetetraacetic acid; FBS, fetal bovine serum; IL-6, interleukin-6; IMDM, Iscove's modified Dulbecco's medium; MAb, monoclonal antibodies; MWCO, molecular weight cut-off; NP, non-producing hybridoma population; P, producing hybridoma population; PEC, peritoneal exudate cells; SFLP, serum-free, low-protein; TGFβ_1, transforming growth factor β_1.

Biotechnology and Genetic Engineering Reviews — Vol. 12, December 1994
0264–8725/94/12/509–533 \$20.00 + \$0.00 © Intercept Ltd, P.O. Box 716, Andover, Hampshire SP10 1YG, UK

protecting the cells from shear stress. Furthermore, cells placed in low-serum and serum-free media were more shear-sensitive than cells placed in high-serum media (Murakami, Masui and Salo, 1982; Kunas and Papoutsakis, 1989, 1990; Lee, Savinell and Palsson, 1989; McQueen and Bailey, 1989; Shacter, 1989; van der Pol *et al.*, 1990). More complete reviews on the shear sensitivity of animal cells have been published (Papoutsakis, 1991a,b). The problem of free-suspended hybridoma cells regarding shear sensitivity can be overcome by entrapping cells into hydrogel beads or capsules.

Secondly, entrapped cell particles are about 50–500 times larger than free-suspended cells and are easier to use in perfusion systems because they do not clog conventional filter systems and can be easily isolated during medium changes.

Thirdly, entrapped cells may create a microenvironment that is more favourable for MAb production than the environment found in the surrounding medium. Finally, higher local and overall cell concentrations (of the order of 1×10^7 to 1×10^8 cells ml^{-1}) can be achieved, resulting in higher MAb titre, volumetric MAb productivity and reduced capital cost for a MAb production facility.

Although a few studies on entrapped hybridoma cell culture are available, most of them give very limited information on specific MAb productivity and metabolism (Nilsson *et al.*, 1983, 1986; Shirai *et al.*, 1987; Bugarski *et al.*, 1989; Sinacore, Creswick and Buehler, 1989). Furthermore, since different hybridoma cell lines were employed in different entrapped cell culture systems, it is difficult to determine the advantages of using the entrapped hybridoma cells for MAb production over free-suspended cells.

We have carried out a series of experiments to compare entrapped hybridoma cells with free-suspended hybridoma cells for MAb production. The model cell line used was the murine hybridoma cell line, S3H5/γ2bA2. In this chapter, the advantages of using entrapped cells over free-suspended cells for MAb production are discussed with respect to efficiency of serum utilization for MAb production, specific MAb productivity and stability of MAb productivity. Cells were entrapped into calcium alginate beads (Lee and Palsson, 1990) because calcium alginate can be easily dissolved by treatment with calcium chelating against such as EDTA and citrate buffer (Lim, 1980, 1988).

Efficiency of serum utilization for antibody production

One of the main economic drawbacks in the large-scale cultivation of hybridoma cells is the requirement for expensive serum that is used in growth media (Griffiths, 1986, 1987). Furthermore, animal-derived serum contains high levels of contaminants and unwanted proteins that must be separated from the MAbs during downstream processing, thus increasing production costs (Ratafia, 1989). The serum requirement for growth of a murine hybridoma cell line in batch cultivation has been reduced at high initial cell concentrations (Lee, Huard and Palsson, 1980). Therefore, one would expect high-concentration hybridoma cell cultures to be economically advantageous (Glacken, Fleischaker and Sinskey, 1983). High cell concentrations can be obtained by immobilizing the cells in gel beads. In addition, it has been reported that immobilization enhances the specific productivity of secreted products, such as ethanol by yeast (Okita, Bonham and Gainer, 1985; Doran and Bailey, 1986) and αamylse by recombinant yeast (Walls and Gainer, 1989). Immobilization may similarly enhance the specific MAb productivity (q_{MAb}) of hybridoma cells, improving the efficiency of serum utilization for MAb production.

To compare an entrapped cell culture with a free-suspended cell culture with respect to the efficiency of serum utilization for MAb production, free-suspended and calcium alginate-entrapped cell cultures were carried out in a repeated fed-batch mode, stepping down the serum concentration from 10% to 0% v/v (Lee, Varma and Palsson, 1991a). We observed two interesting phenomena in the entrapped cell culture.

First, the q_{MAb}, unlike the specific glucose uptake and lactate production rates (q_{glu} and q_{lac}), was enhanced significantly in the entrapped cell culture, and this enhancement was not a transient phenomenon (*Figure 1*). Intracellular MAb content of the entrapped cells as measured by flow cytometry was also higher than that of free-suspended cells. Secondly, in the entrapped cell culture, the reduction of serum concentration from 10% to 1% did not decrease the volumetric MAb productivity (r_{MAb}). In fact, because the q_{MAb} remained constant, only viable cell concentration directly affected the r_{MAb}; the serum concentration affected r_{MAb} only indirectly. In the free-suspended cell culture, with each decrease in serum concentration, the viable cell concentration fell, thus reducing r_{MAb}. In the entrapped cell culture, with each decrease in the serum concentration from 10% to 1%, the viable cell concentration remained constant, and thereby the r_{MAb} also remained constant. When the serum concentration was reduced further from 1% to 0%, the viable cell concentration fell along with the r_{MAb}. So the limitation to r_{MAb} is probably a direct result of some limit to viable cell concentration. In the entrapped cell culture with the viable cell concentration over 10^7 cells ml^{-1} of gel, there might be some limiting components required for cell growth other than serum. Thereby, the effect of serum concentrations on the viable cell concentrations might be masked by other limitations.

As discussed above, the entrapped cell culture system has a higher r_{MAb} than the free-suspended cell culture system because of both the enhanced q_{MAb} and the maintenance of a high viable cell concentration. In addition, because the reduction of serum concentration from 10% to 1% did not influence the viable cell concentration in the entrapped cell culture, serum concentration in medium could be decreased to 1% without any loss in MAb productivity. Therefore, the entrapped cell culture offers significant advantages in efficiency of serum utilization for MAb production when compared to the free-suspended cell culture. When 1% serum medium was used, the efficiency of serum utilization for MAb production by the entrapped cell culture was about three times higher than that by free-suspended cell culture. By using the entrapped cells with 1% serum medium, MAb yield was approximately 2.2 mg ml^{-1} of serum.

Specific MAb productivity

A high volumetric MAb productivity (r_{MAb}) which can be achieved by increasing the cell concentration and/or increasing the specific MAb productivity (q_{MAb}), is desired for economical production of MAbs. Accordingly, the enhanced q_{MAb} of calcium alginate-entrapped S3H5/γ2bA2 hybridoma cells could improve the economics of *in vitro* production of MAbs (Lee, Varma and Palsson, 1991a).

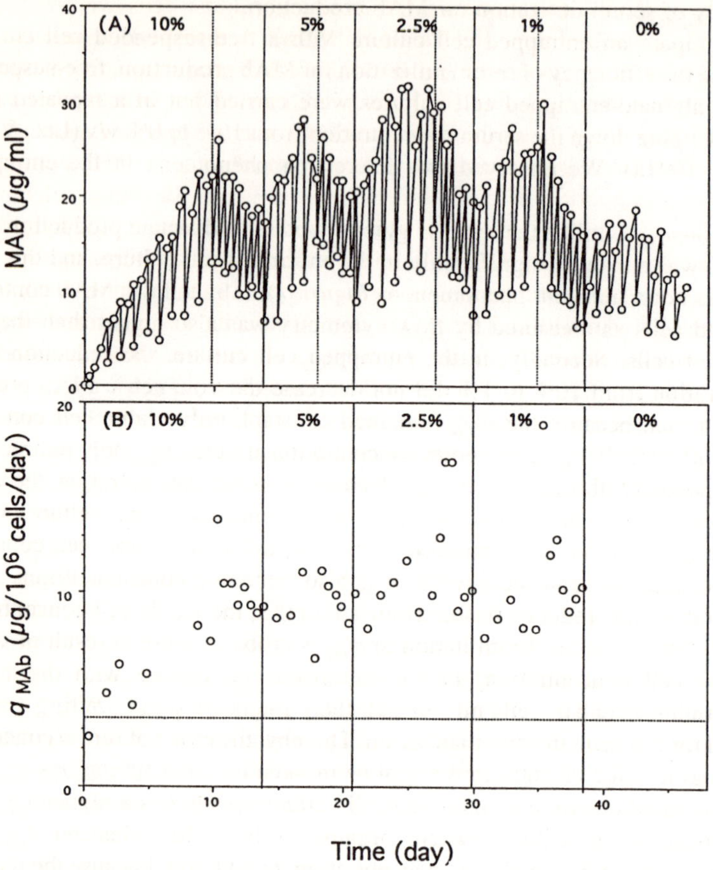

Figure 1. MAb concentration and specific MAb productivity (q_{MAb}) in the immobilized cell culture. Viable cell concentration at the beginning of this experiment was 1.8×10^6 cells ml^{-1}, and decreased to 1.0×10^6 cells ml^{-1} after 2 days of cultivation in 10% serum medium. This decrease in viable cell concentration may have occurred because of cell damage during the immobilization process. Viable cell concentration increased to 1.8×10^6 cells ml^{-1} after 8 days of cultivation in 10% serum medium. Viable cell concentration in 1% serum medium on cultivation day 38 was 1.7×10^6 cells ml^{-1} with a cell viability of 50%. The percentages in the figure represent the serum concentrations used in the media. (A) MAb concentration. Dotted and solid lines indicate times when the feeding schedule was changed. After the dotted line, 50 ml of spent medium was replaced by fresh medium twice a day. After the solid line, 60–65 ml of spent medium was replaced once a day. (B) q_{MAb} of immobilized cells. Dotted line represents an average q_{MAb} of free-suspended cells in a control experiment.

Contrary to our observation of the enhanced q_{MAb} of entrapped hybridoma cells, other researchers have reported that the q_{MAb} of entrapped hybridoma cells was comparable with that of hybridoma cells grown in suspension (Shirai *et al.*, 1987; Sinacore, Creswick and Buehler, 1989; Wohlpart, Gainer and Kirwan, 1991). These results were obtained using several hybridoma cell lines immobilized in calcium alginate beads. The differences between the results we reported and those found by others (Shirai *et al.*, 1987; Sinacore, Creswick and Buehler, 1989; Wohlpart, Gainer and Kirwan, 1991) are possibly due to the different culture conditions under which the alginate beads were carried, or due to differences in the cell lines used.

In our work, the entrapped S3H5/γ2bA2 hybridoma cells were cultivated in a spinner flask (Lee, Varma and Palsson, 1991a). A threefold increase in q_{MAb} was obtained only after viable cell concentrations in 1 mm diameter calcium alginate beads reached the maximum. Culture conditions under which other researchers repported that the q_{MAb} of the entrapped cells was comparable with that of free-suspended cells are summarized as follows. Shirai *et al.* (1987) cultivated 4H11 human–mouse hybridoma cells immobilized in 2 mm diameter calcium alginate beads in a six-well plate. Wohlpart and co-workers cultivated 63D hybridoma cells immobilized in 1 mm and 3.9 mm diameter calcium alginate beads in a spinner flask, but they did not measure the q_{MAb} of entrapped cells in the spinner flask (Wohlpart, Gainer and Kirwan, 1991). The q_{MAb} of entrapped cells was measured by incubating washed beads (removed from the spinner flask) in fresh medium in a sterile culture tube for 4 hours. Sinacore and co-workers cultivated KTI16A hybridoma cells in 1–1.5 mm diameter calcium alginate beads in a spinner vessel (Sinacore, Creswick and Buehler, 1989). The q_{MAb} was measured during the growth phase of entrapped cells in a batch culture.

Considering the importance of q_{MAb}, it is necessary to determine whether the enhanced q_{MAb} of entrapped hybridoma cells is, in fact, cell-line specific or whether the differences in the results reported are due to cell culture conditions. Thus, we cultivated S3H5/γ2bA2 hybridoma cells entrapped into calcium alginate beads with different bead sizes in both T- and spinner flasks in order to determine whether cultivation methods and bead sizes, as employed by other researchers (Shirai *et al.*, 1987; Wohlpart, Gainer and Kirwan, 1991), influence the q_{MAb}. We also carried out free-suspended cell cultures with the cells recovered from calcium alginate beads to determine whether changes in the q_{MAb} are reversible.

The cultivation method was found to influence significantly the q_{MAb} of the entrapped cells (*Table 1*). When the entrapped cells in 3 mm diameter alginate beads were cultivated in T-flasks, q_{MAb} was not enhanced by 200% as previously observed in an entrapped cell culture using 1 mm diameter alginate beads in spinner flasks. The q_{MAb} of the entrapped cells was approximately 72% higher than that of the free-suspended cells in a control experiment. Therefore, the conclusion that the enhanced q_{MAb} of entrapped cells is cell-line specific cannot be drawn. Because the degree of q_{MAb} enhancement varies with cultivation method, the q_{MAb} of entrapped 4H11 hybridoma cells may be improved by cultivating them in spinner flasks. Unlike the cultivation methods, the size of alginate beads in the range 1–3 mm did not influence significantly the q_{MAb} of the entrapped cells in T-flasks. These results are consistent with the results of Wohlpart, Gainer and Kirwan (1991), that the q_{MAb} of entrapped cells was little affected by bead size in the range of 1–3.9 mm.

When free-suspended cell cultures with the cells recovered from calcium alginate beads were carried out, the enhanced q_{MAb} as well as q_{glu} and q_{lac} decreased to the value of the free-suspended cells in a control free-suspended cell culture (*Figure 2*). Thus, the changes in the entrapped cells were reversible. In addition, the changes in the entrapped cells more rapidly reverted in spinner flasks than in T-flasks. In spinner flasks, the q_{MAb} of free-suspended cells recovered from the beads dropped significantly at the first passage (*Figure 2*) and the enhanced intracellular MAb contents

Table 1. The q_{MAb}, q_{glu}, q_{lac}, and $Y_{lac/glu}$ of S3H5/γ2bA2 hybridoma cells under various culture conditions

Cultivation method	Bead size (mm)	q_{MAb} (μg 10^6 $cell^{-1}$ day^{-1})	q_{glu} (mmol 10^9 $cells^{-1}$ day^{-1})	q_{lac} (mmol 10^9 $cells^{-1}$ day^{-1})	$Y_{lac/glu}$ (mmol $mmol^{-1}$)
Free cell culture in T-flasks[a]		3.88±0.53 (3.88±0.98)	3.32±0.42 (3.26±0.43)	5.57±0.82 (5.37±0.70)	1.69±0.18 (1.65±0.14)
Free cell culture in spinner flasks		3.70±0.25	3.83±0.46	6.85±1.06	1.78±0.07
Entrapped cell culture in T-flasks	1.1	6.13±1.27	6.90±1.07	12.49±1.64	1.83±0.26
	1.6	6.16±1.03	7.00±0.79	13.57±1.45	1.95±0.19
	3	6.66±1.78	7.14±1.36	13.76±1.38	1.98±0.42
Entrapped cell culture in spinner flasks	1	10.93±1.57	5.24±1.24	9.97±1.57	1.95±0.30
	3	15.91±1.39	8.18±1.34	14.84±1.50	1.83±0.16
Recovered free cell culture in T-flasks[b]	1.1	3.51±0.45	3.92±0.29	6.97±0.18	1.79±0.10
	1.6	2.76±0.16	4.08±0.44	6.76±0.16	1.66±0.03
	3	3.32±0.80	3.21±0.04	5.44±0.41	1.70±0.15
Recovered free cell culture in spinner flasks[b]	3	5.89±1.26	4.46±0.56	7.16±0.06	1.62±0.19

[a] For free-suspended cell culture in T-flasks, duplicate cultures (flasks A and B) were carried out in a repeated-fed batch mode.
[b] The values are average data from the last three batches.

decreased to the value of the free-suspended cells used for entrapment at the second passage. Hybridoma cells are often described as a heterogeneous population with regard to distribution of intracellular MAb content (Heath, Dilwith and Belford, 1990), surface MAb content (McKinney, Dilwith and Belfort, 1991), and chromosome number (Goding, 1980). The microenvironment in alginate beads may favour the population of a high MAb producer. We could hypothesize that the population of a high MAb producer became dominant during the entrapped cel culture, resulting in the enhanced q_{MAb} of entrapped cells. However, the results obtained here do not support this hypothesis, since the reversion of changes in the entrapped cells occurs quickly, within two passages. Thus, the enhanced q_{MAb} of entrapped cells is more likely to be due to changes in cellular behaviour caused by entrapment.

As discussed above, the cultivation method (T- or spinner flasks) was found to influence significantly the q_{MAb} of the calcium alginate-entrapped S3H5/γ2bA2 hybridoma cells, indicating that the enhanced q_{MAb} may not be cell-line specific. Thus, the remaining question to be addressed is to determine whether the enhanced q_{MAb} of entrapped hybridomas is cell-line specific. We cultivated calcium alginate-entrapped

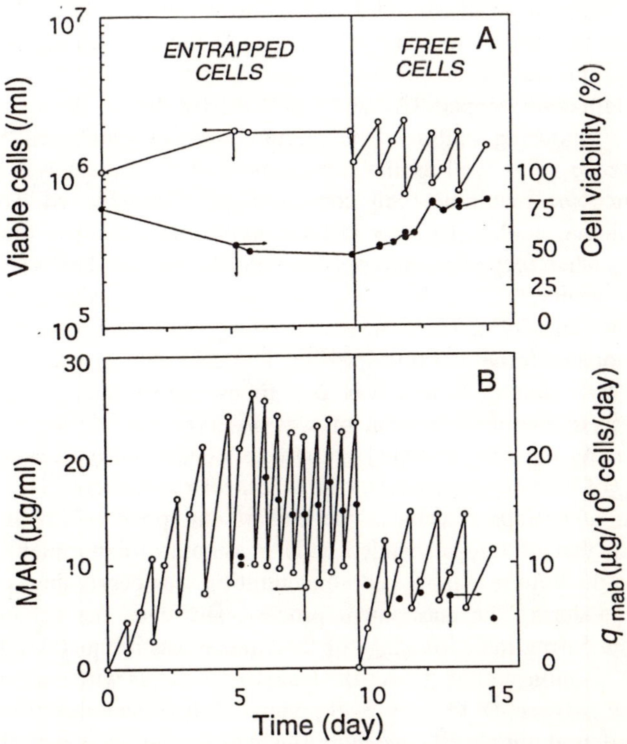

Figure 2. Alginate-entrapped (3 mm diameter) S3H5/γ2bA2 hybridoma cell culture followed by free-suspended cell culture in spinner flasks: (A) viable cell concentration and cell viability. (B) MAb concentration and q_{MAb}. The entrapped cells were cultivated in a repeated fed-batch mode. After 10 days, the alginate beads containing the cells were recovered from the spinner flasks and were dissolved in an isotonic citrate solution for 10 minutes. The cells were then centrifuged at 1500 r.p.m. for 10 minutes and then washed with a fresh medium. The recovered, free suspended cells were re-inoculated into a spinner flask. The cells were diluted 1:2 with fresh medium every day.

murine hybridomas under the conditions where we observed the significantly enhanced q_{MAb} of calcium alginate-entrapped S3H5/γ2bA2 hybridoma (Lee, Varma and Palsson, 1991) and, thereby, determined whether the enhanced q_{MAb} of entrapped hybridomas is indeed cell line-specific.

The two murine hybridomas used were 4A2 and DB9G8 hybridomas. Both hybridomas have been extensively studied in the literature (Heath, 1988; Mancuso *et al*., 1990; Hiller *et al*., 1991; McKinney, Dilwith and Belfort, 1991; Hiller, Clark and Blanch, 1993; Lee and Palsson, 1993). Since the physiological properties of hybridomas can change during passage, there might be some difficulty in comparing the results obtained here directly with those in the literature. Some of the features of hybridomas used here are as follows. The q_{MAb} of a slow-growing, high-concentration culture of 4A2 hybridoma in a hollow fibre reactor with high porosity and large pore size was almost identical to that of free-suspended cells in continuous suspension culture at low dilution rate (Mancuso *et al*., 1990). Accordingly, we may be able to compare the q_{MAb} of calcium alginate-entrapped cells indirectly with that of the cells in the hollow fibre reactor. It has been suggested that the DB9G8 hybridoma has an upper limit for MAb concentration in the culture medium as the q_{MAb} dropped to zero at around 27–36 μg ml^{-1} antibody (McKinney, Dilwith and Belfort, 1991). It may thus be informative to monitor the changes in q_{MAb} of the calcium alginate-entrapped DB9G8 hybridoma during the culture.

Calcium alginate-entrapped 4A2 and DB9G8 hybridomas showed very similar changes in q_{MAb} during culture. For calcium alginate-entrapped S3H5/γ2bA2 hybridoma, the q_{MAb} as well as the intracellular MAb content was significantly enhanced since the maximum cell concentration was achieved. Unlike S3H5/γ2bA2 hybridoma, neither 4A2 nor DB9G8 hybridomas showed the persistently enhanced q_{MAb} when they were entrapped in calcium alginate beads (*Figures 3* and *4*). A similar observation has been reported for the 63D3 hybridoma (Wohlpart, Gainer and Kirwan, 1991). Thus, the enhanced q_{MAb} of calcium alginate-entrapped hybridomas appears to be cell-line specific. The enhanced q_{MAb} of entrapped 4A2 and DB9G8 hybridomas, which was 2–3 times higher than the q_{MAb} of free-suspended cells in a control experiment, was observed only during the early stage of the culture. As the cells resumed growth, this enhanced q_{MAb} decreased to the level of q_{MAb} of free-suspended cells in a control experiment. The intracellular MAb content of entrapped cells at the end of entrapped cell culture was also comparable to that of free-suspended cells in a control experiment. During the early stage of the culture, viable cell concentration was decreasing, probably due to cell damage during the entrapment process. However, the enhanced q_{MAb} of entrapped cells during the early stage of the culture was not just due to a low cell concentration, as indicated by the results from experiments with varying initial cell concentrations (*Figure 4*). On day 5, the viable cell concentration in the experiments with different initial cell concentrations was similar. However, the entrapped cells with an initial cell concentration of 1.0×10^6 cells ml^{-1} were actively growing while the entrapped cells with an initial cell concentration of 2.0×10^6 cells ml^{-1} were still in the lag phase. The q_{MAb} of entrapped cells with an initial cell concentration of 1.0×10^6 cells ml^{-1} was 28.29 μg 10^6 cells^{-1} day^{-1} while the q_{MAb} of entrapped cells with an initial cell concentration of 2.0×10^6cells ml^{-1} was 64.24 μg 10^6 cells^{-1} day^{-1}. Thus, the enhanced q_{MAb} of entrapped cells at the beginning of

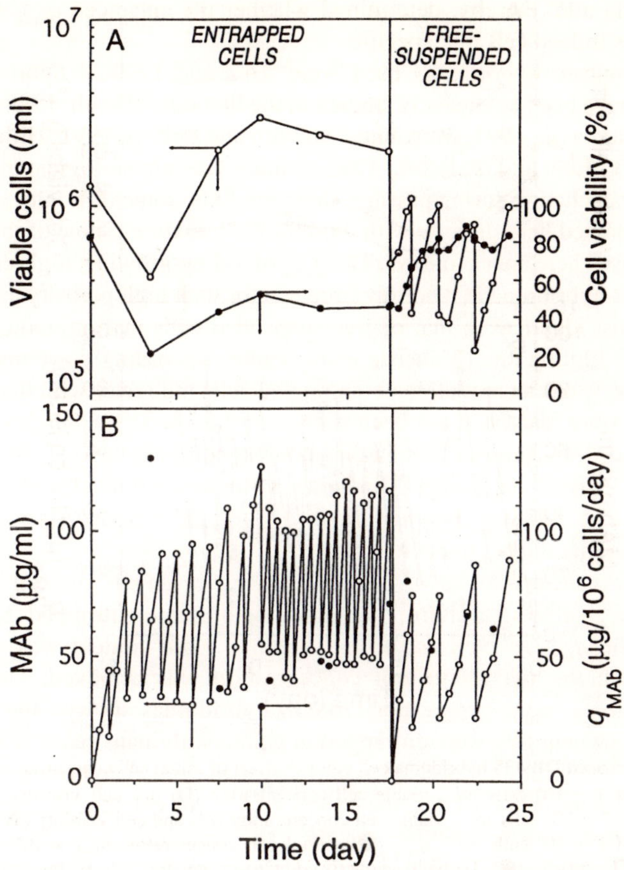

Figure 3. Entrapped 4A2 hybridoma cell culture followed by free-suspended cell culture in spinner flasks: (A) viable cell concentration and cell viability. (B) MAb concentration and q_{MAb}.

the culture was not just due to the low cell concentration. Furthermore, released antibody from dead cells did not account for the enhanced q_{MAb} because the antibody contained in viable and dead cells together accounted for less than 10% of the total extracellular antibody (Meilhoc, Wittrup and Bailey, 1989; Passini and Goochee, 1989; Reddy, Bauer and Miller, 1992). Accordingly, the enhanced q_{MAb} of entrapped cells was likely to be due to the stressful culture conditions induced by entrapment. After the cells adapted to a new microenvironment in the beads, the entrapment might not provide the stressful culture condition to the cells any longer. Thus, when the cells resumed growth, the enhanced q_{MAb} of entrapped cells might disappear.

The enhancement of q_{MAb} in immobilized cell culture could be attributed to a number of causes: slowed growth rate, stressful culture conditions and presence of autocrine growth factors. The specific growth rate (μ) of immobilized cells was almost zero after the cell concentration reached saturation. Many reports show that

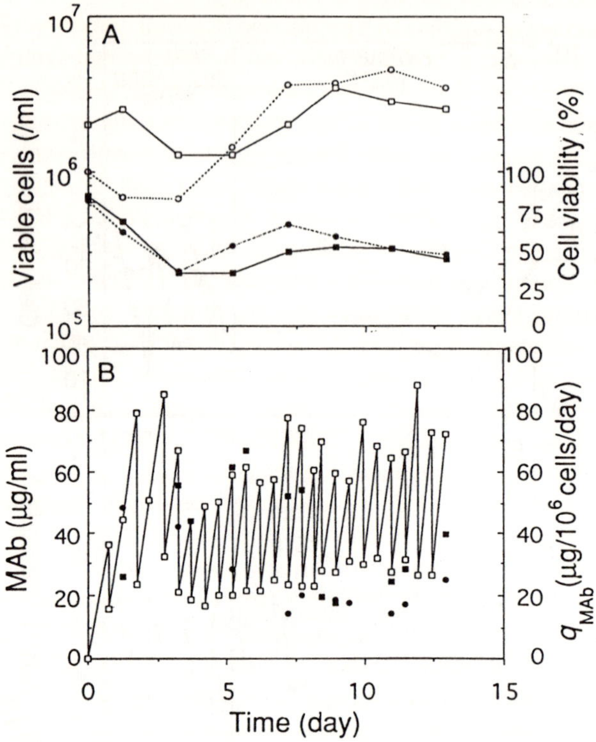

Figure 4. Entrapped DB9G8 hybridoma cell culture: effect of initial cell concentration. (A) Initial cell concentration of 1×10^6 cells ml^{-1}, viable cell concentration (O) and cell viability (●); initial cell concentration of 2×10^6 cells ml^{-1}, viable cell concentration (□) and cell viability (■). (B) Initial cell concentration of 1×10^6 cells ml^{-1}, q_{MAb} (●); initial cell concentration of 2×10^6 cells ml^{-1}, MAb concentration (□) and q_{MAb} (■). To avoid an overly complex figure, only q_{MAb} for the entrapped cells with an initial cell concentration of 1.0×10^6 cells ml^{-1} has been plotted.

q_{MAb} is inversely related to the µ of hybridoma cells (Reuveny *et al.*, 1986; Dean *et al.*, 1987; Low, Harbour and Barford, 1987; Heath, 1988; Miller, Blanch and Wilke, 1988; Hagedorn and Kargi, 1990; van der Pol *et al.*, 1990), although there are some exceptions (Shirai *et al.*, 1987; Glacken, Adema and Sinskey, 1988; Dalili and Ollis, 1989). One structured model, proposed by Suzuki and Ollis (1990), shows that the q_{MAb} can be increased by 3.5 times by reducing the µ to zero. Stressful culture conditions would also result in slowed cell growth rate. The q_{MAb} of the S3H5/γ2bA2 hybridoma was elevated at slowed growth rates in free-suspended cell cultures (Lee, Huard and Palsson, 1989). However, when the entrapped S3H5/γ2bA2 hybridoma showed slowed growth rates and was under stressful conditions, the q_{MAb} was not always enhanced (Lee, Chuck and Palsson, 1993). Cultivation methods (T-flask and spinner flask) determined the extent of enhancement of q_{MAb} of entrapped S3H5/γ2bA2 hybridoma. Thus, slowed growth rates and stressful culture conditions of the entrapped S3H5/γ2bA2 hybridoma do not directly account for its enhanced q_{MAb}.

The difference in the q_{MAb} of entrapped S3H5/γ2bA2 hybridoma might be explained by different mass transfer characteristics of autocrine growth factors. Increasing evidence exists to show that hybridoma cells produce autocrine growth factors which either stimulate or inhibit MAb production. Kidwell and co-workers observed that NOT human–human hybridoma cells and human normal lymphocytes grew better in hollow-fibre cartridges with low molecular weight cut-offs (3000 MWCO) (Kidwell, Knazek and Wu, 1990). Direct evidence for the presence of growth-stimulating factor accumulation in the hollow-fibre reactor with 3000 MWCO fibres was obtained by recovering growth-stimulating activity from the extracapillary space of the bioreactor cartridge (Kidwell, Knazeck and Wu, 1990). Interleukin-6 (IL-6, molecular weight approximately 26 kDa) is an autocrine growth factor known to stimulate immonoglobulin synthesis in lymphoid cells (Hirano *et al.*, 1986; Kawano *et al.*, 1988; Aarden, 1989). Makishima and co-workers found high IL-6 levels in active culture supernatants of peritoneal exudate cells (PEC), which may be responsible for the enhanced q_{MAb} *in vivo* (Makishima, Mikami and Terada, 1990). They also found that recombinant human IL-6 significantly enhanced the q_{MAb} of 2E3 hybridoma cells *in vitro* and it did not augment proliferation of the cells (Makishima, Mikami and Terada, 1990; Terada, Makishima and Takamatsu, 1992).

On the other hand, END and Col12 hybridomas are found to produce inhibitory factors, one of which has been identified as transforming growth factor β_1 (TGFβ_1). TGFβ_1 inhibits MAb production as well as cell growth (Kidwell, 1989). Its specific productivity was also found to be cell-line dependent (Kidwell, 1989). Thus, when compared with ultrafiltration membrane-based hollow-fibre cartridges with low molecular cut-offs (4000 MWCO), yield of hybridoma cells and MAb production were significantly higher in bioreactors based on high MWCO hollow fibres (0.5 μm pore) that allow exchange of very large macromolecules across their walls (Kidwell, 1989).

Hagedorn and Kargi (1990) made a similar observation in flat-sheet membrane bioreactors with several different MWCO membranes. When HDP1 hybridoma cells were cultivated in a flat sheet membrane bioreactor, the q_{MAb} was significantly enhanced to different degrees depending on the MWCO of the membrane used. The q_{MAb} in a 50 000 MWCO membrane system was 3.5 times greater than the q_{MAb} in a batch reactor without a membrane, whereas the q_{MAb} in a 100 000 MWCO membrane system was five times greater than the q_{MAb} in a batch reactor without a membrane. Accordingly, inhibitors such as TGFβ_1 may not be removed rapidly enough or may be unable to diffuse out of the 50 000 MWCO membrane system, resulting in the lower q_{MAb}. TGFβ_1 has a molecular weight of approximately 25 000, but is often recovered from cell-conditioned media as a high molecular weight complex (Kidwell, Mohanam and Salomon, 1987).

The results from the experiment employing varying bead size in both T- and spinner flasks indicate that, in the bead size in the range of 1–3 mm, external mass-transfer resistance is more influential on the q_{MAb} of S3H5/γ2bA2 hybridoma than internal mass-transfer resistance. Thus, the accumulation of inhibitors around the beads in T-flasks may hamper the inhibitors diffusing out of the beads and, thereby, reduce the q_{MAb}.

The enhanced q_{MAb} of the entrapped cells was not unique to calcium alginate-entrapped S3H5/γ2bA2 hybridoma cells in spinner flasks. A significant enhancement of q_{MAb} has been observed for SPO1 hybridoma cells entrapped in alginate/agarose composite gel in a packed bed perfusion reactor (Shen, Reid and Greenfield, 1992). When the medium was exchanged three to four times a day, the q_{MAb} of the entrapped cells was three to eight times greater than that of free-suspended cells in a control suspension culture. The q_{MAb} of hybridoma cells entrapped in collagen microspheres in a fluidized bed perfusion bioreactor was higher than that of free-suspended cells in a continuous stirred tank bioreactor (Ray *et al.*, 1990). Because the medium was recirculated very rapidly in both packed bed perfusion and fluidized bed perfusion bioreactors (Ray *et al.*, 1990; Shen, Reid and Greenfield, 1992), the external mass-transfer resistance around the beads may have been insignificant.

In contrast, the enhancement in q_{MAb} of free-suspended 9.2.27 hybridoma cells was not observed in the total cell recycling suspension reactor where a high cell concentration with a very slow hybridoma growth was achieved (Flickinger, Goebel and Bohn, 1990). In addition, the q_{MAb} of a slow-growing, high-concentration culture of X-D hybridoma cells in a hollow fibre with a high porosity and large pore size, was almost identical to that of free-suspended cells in continuous suspension culture at a low dilution rate (Mancuso *et al.*, 1990). When S3H5/γ2bA2 hybridoma cells used in this study were cultivated in a hollow-fibre bioreactor, the intracellular MAb content of the cells, unlike the calcium alginate-entrapped cells, did not increase (Halberstadt, 1991). Accordingly, mass transfer of possible autocrine growth factors, as well as high cell concentration with slowed growth rate, may be an important factor to enhance q_{MAb}.

As discussed earlier, the transiently enhanced q_{MAb} of 4A2 and DB9G8 hybridomas is likely due to stressful conditions. Thus, even when the q_{MAb} is enhanced by entrapment, the reason for the enhanced q_{MAb} of entrapped cells appears to be different among cell lines. Different responses of the cells to entrapment are probably because of the nature of hybridoma cells. Since hybridomas are the randomly fused cells between immortalized myeloma and antibody-producing lymphocytes, their properties are expected to be very different. For example, 4A2 and DB9G8 hybridomas showed similar responses to entrapment with regard to their MAb production. However, they showed different responses to stress induced by freezing and thawing concerning their MAb production (Lee, Kim and Palsson, in press). Reddy and Miller (1994) showed that the effect of hyperosmotic stress on q_{MAb} is cell-line dependent. Furthermore, the production of autocrine growth factors and inhibitory factors was also found to be cell-line dependent (Kidwell, 1989). Accordingly, as shown in this study, it is unlikely for all the hybridomas to show a similar response to cell entrapment with respect to their MAb production.

The results reported here may be compared with those in the literature (Mancuso *et al.*, 1990; McKinney, Dilwith and Belfort, 1991). For 4A2 hybridoma cells, like the q_{MAb} of a slow-growing, high-concentration culture of 4A2 hybridoma cells in a hollow fibre reactor with high porosity and large pore size, the q_{MAb} of calcium alginate-entrapped cells was similar to that of free-suspended cells in a control experiment. Thus, calcium alginate beads do not provide 4A2 hybridomas with a unique environment to improve the q_{MAb}. For DB9G8 hybridoma cells, we did not find any MAb feedback inhibition. However, as we mentioned earlier, direct comparison may be difficult because culture history was different. In addition, a different medium was used here.

In conclusion, unlike the S3H5/γ2bA2 hybridoma, neither 4A2 nor DB9G8 hybridomas showed the persistently enhanced q_{MAb} when they were entrapped in calcium alginate beads. Thus, the enhanced q_{MAb} of calcium alginate-entrapped hybridomas appears to be cell-line specific. Although calcium alginate entrapment did not enhance the q_{MAb} of 4A2 and DB9G8 hybridomas, it did enhance the r_{MAb} because of high cell concentration. For both cell lines, the r_{MAb} in entrapped cell culture was approximately 3–4 times higher than that in free-suspended cell culture. As long as entrapment does not deteriorate the q_{MAb}, MAb production using entrapped cells may be feasible.

Stability of MAb production

Loss of MAb productivity of hybridoma cells in long-term cultivation has become a major concern to those interested in large-scale production of MAbs. This loss of MAb productivity is often attributed to the appearance of a non-producing population of hybridomas (NP) in the culture which has a growth advantage over a producing population of hybridomas (P) (Frame and Hu, 1990, 1991a,b; Lee and Palsson, 1990; Ozturk and Palsson, 1990; Chuck and Palsson, 1992; Merritt and Palsson, 1993). When an NP appears in the culture, it does not always take over the whole culture (Westwoudt, Naipal and Harrisson, 1984; Gardner *et al.*, 1985; Chuck and Palsson, 1992). Populations of P and NP may be balanced under certain culture conditions, depending on the relative growth rate of producing and non-producing cells and the rate of loss of MAb productivity (Lee, Varma and Palsson, 1991b). The appearance of an NP is thought to be due to mutations or loss of genes associated with antibody regulation as well as antibody synthesis ability because the losses in MAb productivity are usually reversible (Galfré *et al.*, 1980; Wilde and Milstein, 1980; Gardner *et al.*, 1985). Several hybridoma cell lines have been reported to lose their MAb productivity at a rate of 10^{-2}–10^{-5} cells/generation (Cotton, Secher and Milstein, 1973; Galfré *et al.*, 1980; Clark *et al.*, 1983; Gardner *et al.*, 1985). Although the tendency to lose MAb productivity is particular to each cell line, cells in low-serum or serum-free medium are often found to be more prone to lose MAb productivity than cells in high-serum medium (Lee and Palsson, 1990; Ozturk and Palsson, 1990).

Immobilization has been suggested to improve plasmid stability in genetically modified bacteria in bioreactors (Kumar *et al.*, 1991; Flickinger and Rouse, 1993). The observation that plasmid copy number can be stabilized in cells immobilized in gel beads and biofilms has been made and can be predicted mathematically (de Taxis du Poët *et al.*, 1986, 1987; Nasri *et al.*, 1987; Sayadi *et al.*, 1988, 1989; Barbotin, Nava Saucedo and Thomasset, 1990; Kumar and Schügerl, 1990), although the mechanism for this is unknown (Huang, Peretti and Bryers, 1993). Immobilization may similarly improve the stability of the hybridoma cells' MAb productivity.

We examined the hypothesis that the instability of S3H5/γ2bA2 hybridoma cells' MAb productivity in low-serum or serum-free medium may be overcome by entrapping the cells in gel beads (Lee and Palsson, 1990). First, free-suspended cell cultures with low-serum and serum-free media were carried out for several months to assess cell growth and loss of MAb productivity. Cell culture was performed using spinner flasks in a repeated fed-batch mode. Cells were cultivated in 1% serum media and then passed to serum-free media followed by serum-free, low-protein (SFLP) media. The specific growth rates of the cells in 1% serum media were in the range 0.6–1.2

day^{-1}. Cells in serum-free and SFLP media grew as fast as in 1% serum media for 2 months. MAb productivity was maintained in 1% serum media. However, cells in serum-free media showed a decrease in MAb productivity, and it completely disappeared in IMDM-based SFLP media (*Figure 5*). Flow cytometric study showed that the loss of MAb productivity was due to the occurrence of an NP which lost its MAb productivity. A single peak of P was found in 1% serum medium. However, in serum-free media, a second peak of NP appeared. As the passage number increased, the NP increased, and after 15–20 passages, the NP took over the whole culture, resulting in a complete loss of MAb productivity. Next, cells immobilized in alginate beads were also cultivated in serum-free media to test whether immobilization can improve the stability of the cell's MAb productivity. As shown in *Figure 6*, calcium alginate-entrapped hybridoma cells in SFLP medium were producing MAb over a month, indicating that immobilization can improve the stability of the cell's MAb-producing ability. The intracellular MAb content of free-suspended cells is compared with immobilized cells in *Figure 7*. The fluorescent intensity of immobilized cells in SFLP, which indicates the amount of intracellular MAb, was approximately three times higher than that of free-suspended cells in 10% serum medium. The q_{MAb} of immobilized cells in SFLP medium was determined to be approximately 9.6 μg 10^6 cells^{-1} day^{-1}, which was also approximately three times higher than that of free-suspended cells in 10% serum medium. Consequently, we found that immobilization not only improved the stability of S3H5/γ2bA2 hybridoma cells, but also enhanced their q_{MAb}.

Several lines of reasoning argue for this improved stability of MAb productivity of calcium alginate-entraped S3H5/γ2bA2 hybridoma cells. First, a microenvironment, created by a high local cell concentration in the calcium alginate beads, may keep the cells from a genetic drift or instability. Secondly, entrapped cells may have limited

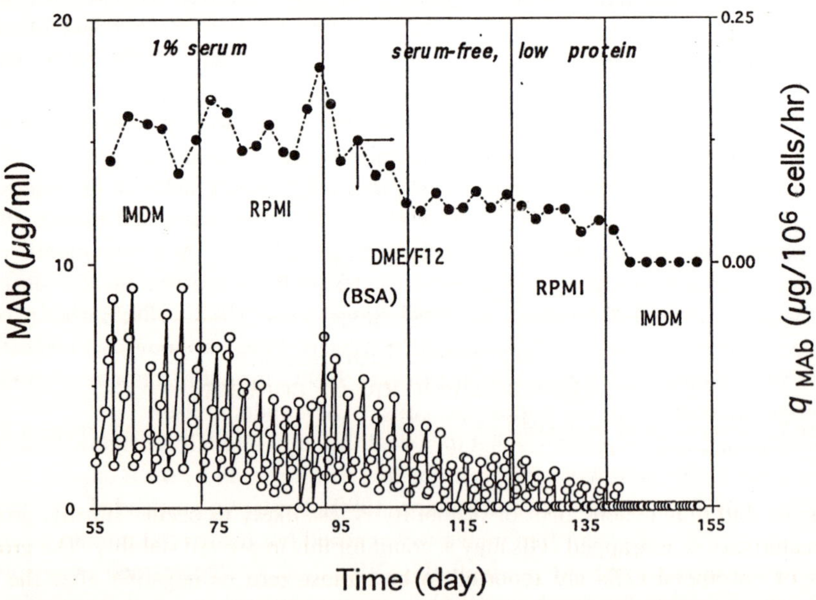

Figure 5. MAb production of free-suspended S3H5/γ2bA2 hybridoma cells in low-serum and serum-free media. The figure indicates how the media composition was changed with time. BSA, bovine serum albumin; DME/F12, IMDM and RPMI are the basal media used.

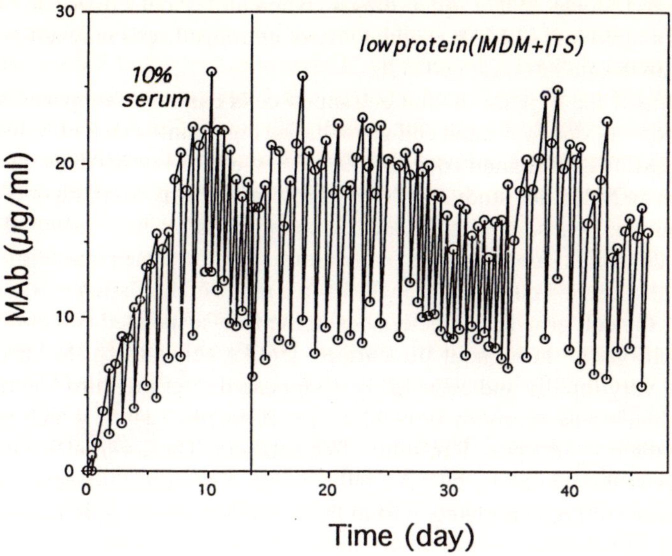

Figure 6. MAb concentrations in entrapped S3H5/γ2bA2 hybridoma cell culture. The feeding schedule was as follows: days 0–10, 60 ml/24 h; days 10–14, 50 ml/12 h; days 14–26, 60 ml/24 h; days 26–34, 50 ml/12 h; day 34 onwards, 60 ml/24 h.

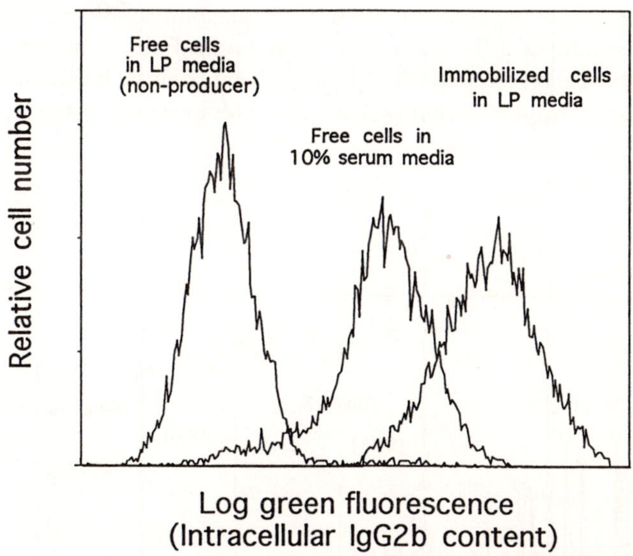

Figure 7. Comparison of intracellular MAb contents between free-suspended and entrapped S3H5/γ2bA2 hybridoma cells.

division, and thus genetic drift or instability is less likely to occur. Thirdly, growth characteristics of entrapped cells may account for this improved stability. The growth rates of entrapped cells are reported to be almost zero or negative after the cell concentration reaches its maximum (Nilsson *et al.*, 1986; Shirai *et al.*, 1987; Lee and Palsson, 1990; Lee, Varma and Palsson, 1991; Overgaard *et al.*, 1991; Wohlpart, Gainer and Kirwan, 1991; de la Broise *et al.*, 1992; Ogata *et al.*, 1992; Shirai,

Hashimoto and Steele, 1992). Accordingly, when an NP outgrowing a P appears in the culture, a rapid loss of MAb productivity of entrapped cells may not occur due to the slow growth rate of entrapped cells.

We examined the hypothesis that entrapped cells can have improved stability of MAb productivity due to limited cell growth. During maintenance of hybridoma cell line 4A2 in IMDM supplemented with 10% FBS, the cells lost MAb productivity and were used as an NP in this study. An NP outgrowing a P was co-entrapped with a P in calcium alginate beads, then cultivated in spinner flasks in a repeated fed-batch mode. As a control, a free-suspended cell culture with the same population of a P and an NP was also carried out in spinner flasks in a repeated fed-batch mode. As shown in *Figure 8*, the cell growth characteristics in the free-suspended cell culture did not change significantly throughout the culture. Unlike cell growth, MAb production deteriorated very rapidly and completely disappeared after the third batch. The q_{MAb} in the first batch was approximately 11.23 μg 106 cells^{-1} day^{-1}, which is less than one-fifth of the q_{MAb} of a pure P in serum-free medium. Thus, a significant portion of an inoculum is thought to be NP. An NP present in an inoculum appeared to take over the whole culture completely within three batches, resulting in a rapid decrease in the q_{MAb}.

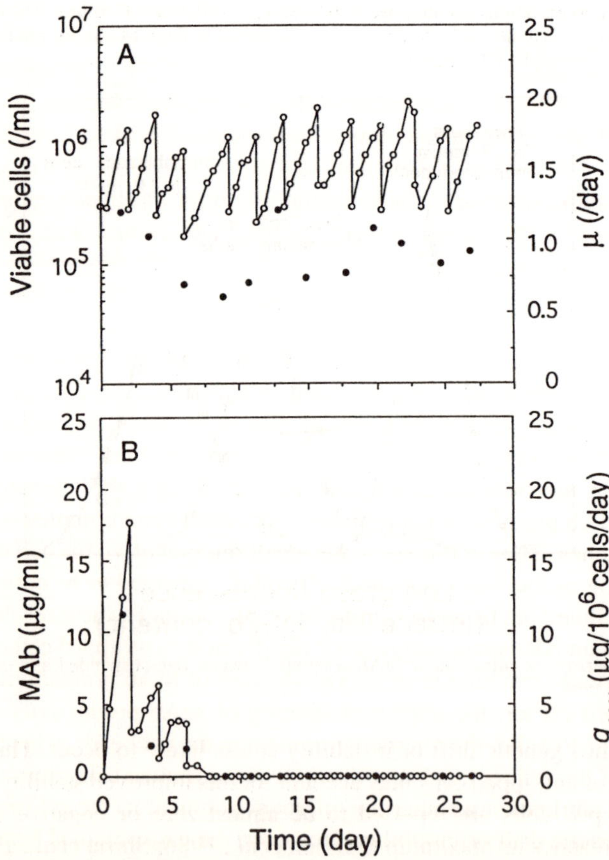

Figure 8. Free-suspended 4A2 hybridoma cell culture: (A) viable cell concentration (O) and μ (●). (B) MAb concentration (O) and q_{MAb} (●).

To examine whether entrapped cells can have improved stability of MAb productivity due to limited cell growth, two sets of entrapped cell cultures were carried out in a repeated fed-batch mode. Since the same inoculum used for free-suspended cell culture was also used for cell entrapment, a significant portion of the calcium alginate-entrapped cells at the beginning of culture belonged to an NP. Unlike the free-suspended cell culture, MAb production in the entrapped cell culture was maintained until the end of the culture, though a little deteriorated (*Figure 9*). The MAb concentration dropped significantly on day 8 because medium was exchanged twice a day after 8 days of cultivation. Accordingly, the r_{MAb} did not drop significantly on day 8. The r_{MAb} initially increased and gradually decreased in accordance with the viable cell concentration. However, the changes in q_{MAb} differed considerably from those in r_{MAb}. The q_{MAb} of entrapped cells at the beginning of the culture was approximately 11.58 μg 106 cells^{-1} day^{-1}, which is close to that of the first batch of free-suspended cell culture (*Figure 8*). The q_{MAb} dropped significantly during the growth phase of the entrapped cells and thereafter remained almost constant at 2.76±0.45 μg 10^6 cells^{-1} day^{-1}. Thus, like free-suspended cell culture, an NP appeared to outgrow a P in entrapped cell culture during the growth phase. Because the estimation of q_{MAb} is based on total viable cell concentration including an NP, the rapid outgrowth of an NP in entrapped cell culture results in a significant decrease in q_{MAb}. However, when the cell growth is limited, an NP cannot outgrow a P in a short period of time, as indicated by the stable value of q_{MAb} after 8 days of cultivation. Assuming that the q_{MAb} of the entrapped cells was equal to that of the free-suspended cells in serum-free medium, less than 5% of total viable cell concentration in entrapped cells at the end of the culture belonged to a P.

Next, to examine whether maintenance of a P in the entrapped cells is due to the limited cell growth, the second set of entrapped cell culture was carried out in a repeated fed-batch mode. After 19 days of cultivation, the alginate bead-entrapped cells were recovered aseptically from the spinner flask and were dissolved in a sterile isotonic citrate solution. The recovered, free-suspended cells were cultivated in a repeated fed-batch mode so that the NP could freely compete against the P. As shown in *Figure 10*, the cell growth, viability and MAb production characteristics of the entrapped cells were similar to those in *Figure 9*. Thus, we assumed that the conditions of the entrapped cells in two sets of entrapped cell cultures were almost identical. When the free-suspended cells recovered from the alginate beads were subcultured, MAb production deteriorated very rapidly and completely disappeared within two batches. Thus, a P present at a small fraction of viable cell concentration in the beginning of the free-suspended cell culture appeared to be outgrown rapidly by an NP and diluted out from the spinner flask. Accordingly, these results support the hypothesis that an NP cannot outgrow a P rapidly in the entrapped cell culture due to limited cell growth, resulting in improved stability of MAb productivity.

As mentioned earlier, the improved stability of MAb productivity of entrapped cells may be due to either the microenvironment in the gel beads, where genetic drift or instability of the cells is less likely to occur, or the limited cell growth of entrapped cells. It is still not known whether the microenvironment in the gel beads keeps the cells from a genetic drift or instability. However, the data obtained here suggest that even if an NP occurs in the long-term culture of entrapped cells, it cannot take over the whole culture in a short period of time due to the limited growth of these cells.

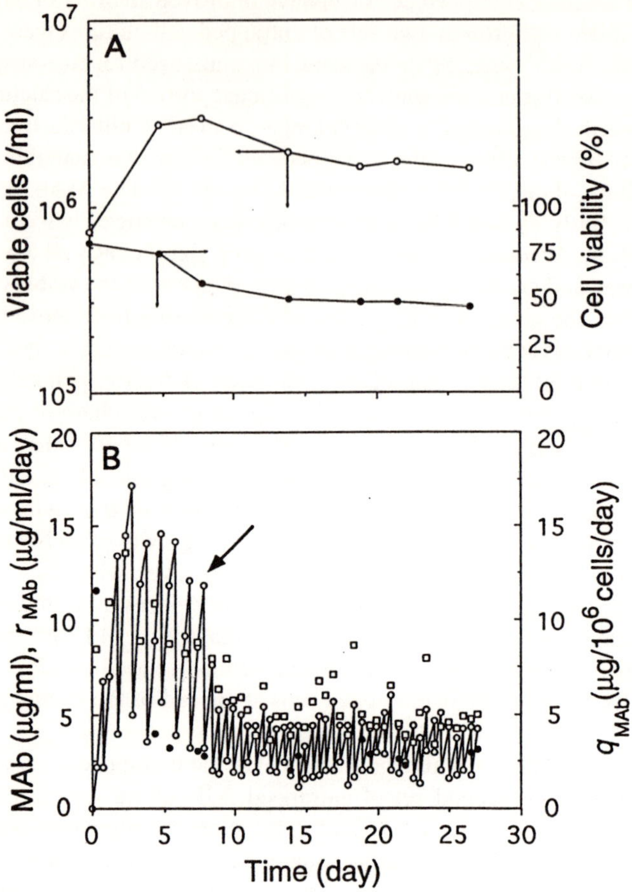

Figure 9. Entrapped 4A2 hybridoma cell culture: (A) viable cell concentration (O) and cell viability (●); (B) MAb concentration (O), r_{MAb} (□), and q_{MAb} (●). The arrow indicates changes in the feeding schedule.

Furthermore, even if an NP occurs in the exponential phase of growth in entrapped cell culture, the q_{MAb} can be decreased, but not significantly, due to the relatively short duration of the exponential growth phase. When an initial viable cell concentration in entrapped cell culture is approximately 1–2 × 10^6 cells ml^{-1} of medium, a maximum cell concentration is often achieved in less than 10 days (Lee and Palsson, 1990; Lee, Varma and Palsson, 1991a). In conclusion, due to the characteristics of growth of entrapped cells, the stability of MAb productivity of the cells can be improved using cell entrapment in gel beads.

Conclusion

Taken together, the results from questions addressed in this chapter show that immobilized cell culture is more efficient for MAb production than free-suspended cell culture with respect to the efficiency of serum utilization for MAb production, volumetric MAb productivity and stability of MAb production.

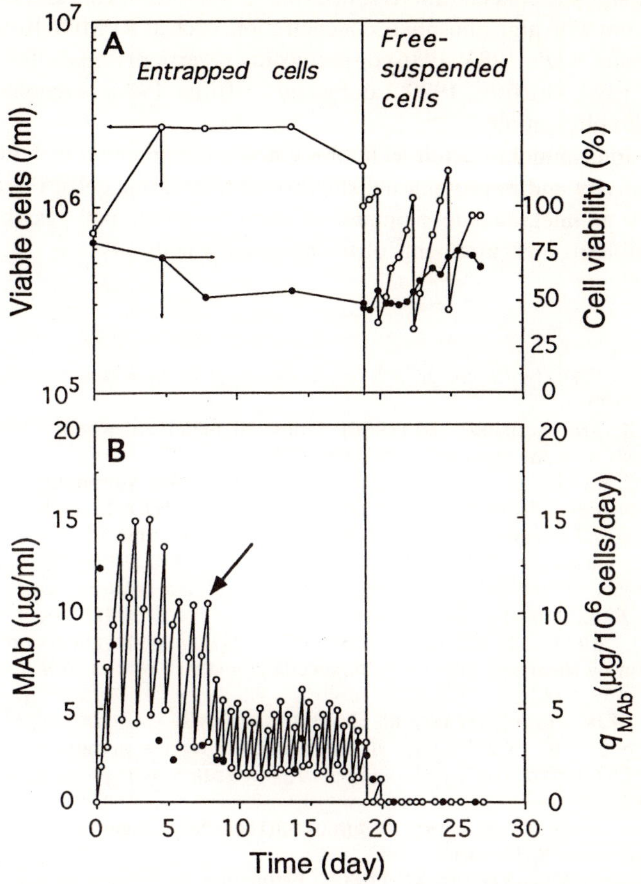

Figure 10. Entrapped 4A2 hybridoma cell culture followed by free-suspended cell culture: (A) viable cell concentration (O) and cell viability (●); (B) MAb concentration (O) and q_{MAb} (●). The arrow indicates changes in the feeding schedule.

The possible disadvantages of an entrapped cell culture system are the material cost, the potential for contamination during the immobilization process, the lack of compatibility of some cell types with the matrix components, and the potential for oxygen limitation within beads or microcapsules. However, some of these potential disadvantages can be overcome by choosing the proper immobilization method. Sodium alginate, which was used here for a gel material, is inexpensive and biocompatible with most hybridoma cell lines. The risk of contamination during immobilization is not significant because the process of making calcium alginate beads is a one-step process. The typical radius of alginate beads is in the range 400–500 μm, so that oxygen transport may not be limited. Typical penetration depth of oxygen for immobilized animal cells is approximately 500–1000 μm (Chang and Moo-Young, 1988). The main drawback of alginate beads is the possible diffusional limitation with respect to a large protein molecule like IgM and the difficulty of using media containing high phosphate concentrations. However, it was observed that IgM produced by the entrapped cells could diffuse out of the calcium alginate beads regardless of alginate concentrations tested (0.8–2.5%) (Kim, Han and Lee, in press). When a medium with high phosphate concentration, such as RPMI1640 must be used, agarose (Nilsson *et al.*, 1983; 1986) or porous microcarrier (Runstadler and Cernek, 1988; Cahn, 1990; Griffiths, 1990; Looby and Griffiths, 1990) is recommended as a matrix for immobilization.

In conclusion, immobilization technology holds the potential of both increasing MAb productivity and improving the stability of hybridoma cells. Future developments will see an integrated development of immobilization methods, the bioreactor and the physiological requirements of the hybridoma cells.

References

AARDEN, L.A. (1989). Hybridoma growth factor. *Annals of the New York Academy of Sciences* **557**, 192–199.

ADAMSON, S.R. AND SCHMIDLI, B. (1986). Industrial mammalian cell culture. *Canadian Journal of Chemical Engineering* **64**, 531–539.

BARBOTIN, J.N., NAVA SAUCEDO, J.E. AND THOMASSET, B. (1990). Morphological observations on immobilized cells. In *Physiology of Immobilized Cells* (J.A.M. de Bont, J. Visser, B. Mattiasson and J. Tramper, Eds), pp. 487–498. Elsevier, Amsterdam.

BORASTON, R., THOMPSON, P.W., GARLAND, S. AND BIRCH, J.R. (1984). Growth and oxygen requirements of antibody producing mouse hybridoma cells in suspension culture. *Develop. Biol. Standard.* **55**, 103–111.

BRENNAN, A.J., SHEVITZ, J. AND MACMILLAN, J.D. (1987). A perfusion system for antibody production by shear-sensitive hybridoma cells in a stirred reactor. *Biotechnology Techniques* **1**, 169–174.

BUGARSKI, B., KING, G.A., JOVANOVIC, G., DAUGULIS, A.J. AND GOOSEN, M.F.A. (1989). Performance of an external loop air-lift bioreactor for the production of monoclonal antibodies by immobilized hybridoma cells. *Applied Microbiology and Biotechnology* **30**, 264–269.

CAHN, F. (1990). Biomaterials aspects of porous microcarriers for animal cell culture. *Trends in Biotechnology* **8**, 131–136.

CHANG, H.N. AND MOO-YOUNG, M. (1988). Estimation of oxygen penetration depth in immobilized cells. *Applied Microbiology and Biotechnology* **29**, 107–112.

CHUCK, A.S. AND PALSSON, B.O. (1992). Population balance between producing and nonproducing hybridoma clones is very sensitive to serum level, state of inoculum, and medium composition. *Biotechnology and Bioengineering* **39**, 354–360.

CLARK, M., COBBOLD, S., HALE, G. AND WALDMANN, H. (1983). Advantages of rat monoclonal antibodies. *Immunology Today* **4**, 100–101.

COCO-MARTIN, J.M., OBERINK, J.W., BRUNINK, F., VAN DER VELDEN-DE GROOT, T.A.M. AND BEUVERY, E.C. (1992). Instability of a hybridoma cell line in a homogeneous continuous perfusion culture system. *Hybridoma* **11**, 653–665.

COTTON, R.G.H., SECHER, D.S. AND MILSTEIN, C. (1973). Somatic mutation and the origin of antibody diversity. Clonal variability of the immunoglobulin produced by MOPC 21 cells in culture. *European Journal of Immunology* **3**, 135–140.

DALILI, M. AND OLLIS, D.F. (1989). Transient kinetics of hybridoma growth and monoclonal antibody production in serum-limited cultures. *Biotechnology and Bioengineering* **33**, 984–990.

DEAN, R.C., JR, KARKARE, S.B., RAY, N.G., RUNSTADLER, P.W., JR, AND VENKATASUBRAMANIAN, K. (1987). Large-scale culture of hybridoma and mammalian cells in fluidized bed bioreactors. *Annals of the New York Academy of Sciences* **506**, 129–146.

DE LA BROISE, D., NOISEUX, M., MASSIE, B. AND LEMIEUX, R. (1992). Hybridoma perfusion systems: a comparison study. *Biotechnology and Bioengineering* **40**, 25–32.

DE TAXIS DU POËT, P., DHULSTER, P., BARBOTIN, J.-N. AND THOMAS, D. (1986). Plasmid inheritability and biomass production: comparison between free and immobilized cell cultures of *Escherichia coli* BZ18(pTG201) without selection pressure. *Journal of Bacteriology* **165**, 871–877.

DE TAXIS DU POËT, P., ARCAND, Y., BERNIER, R., JR, BARBOTIN, J.-N. AND THOMAS, D. (1987). Plasmid stability in immobilized and free recombinant *Escherichia coli* JM105 (pKK223-200): importance of oxygen diffusion, growth rate, and plasmid copy number. *Applied and Environmental Microbiology* **53**, 1548–1555.

DORAN, P.M. AND BAILEY, J.E. (1986). Effects of immobilization on growth, fermentation properties, and macromolecular composition of *Saccharomyces cerevisiae* attached to gelatin. *Biotechnology and Bioengineering* **28**, 73–87.

FLICKINGER, M.C. AND ROUSE, M.P. (1993). Sustaining protein synthesis in the absence of a rapid cell division: an investigation of plasmid-encoded protein expression in *Escherichia coli* during very slow growth. *Biotechnology Progress* **9**, 555–572.

FLICKINGER, M.C., GOEBEL, N.K. AND BOHN, M.A. (1990). Determination of specific monoclonal antibody secretion rate during very slow hybridoma growth. *Bioprocess Engineering* **5**, 155–164.

FRAME, K.K. AND HU, W.S. (1990). The loss of antibody productivity in continuous culture of hybridoma cells. *Biotechnology and Bioengineering* **35**, 469–476.

FRAME, K.K. AND HU, W.S. (1991a). Kinetic study of hybridoma cell growth in continuous culture: I. a model for non-producing cells. *Biotechnology and Bioengineering* **37**, 55–64.

FRAME, K.K. AND HU, W.S. (1991b). Kinetic study of hybridoma cell growth in continuous culture: II. behavior of producers and comparison to nonproducers. *Biotechnology and Bioengineering* **38**, 1020–1028.

GALFRÉ, G., BUTCHER, G.W., HOWARD, J.C., WILDE, C.D. AND MILSTEIN, C. (1980). Clonal competition and stability of hybrid myelomas of mouse and rat origin. *Transplantation Proceedings* **12**, 371–375.

GARDNER, J.S., CHIU, A.L.H., MAKI, N.E. AND HARRIS, J.F. (1985). Quantitative stability analysis of human monoclonal antibody production by heteromyeloma hybridomas, using an immunofluorescent technique. *Journal of Immunological Methods* **85**, 335–346.

GLACKEN, M.W., ADEMA, E. AND SINSKEY, A.J. (1988). Mathematical description of hybridoma culture kinetics: I. initial metabolic rates. *Biotechnology and Bioengineering* **32**, 491–506.

GLACKEN, M.W., FLEISCHAKER, R.J. AND SINSKEY, A.J. (1983). Mammalian cell culture: engineering principles and scale-up. *Trends in Biotechnology* **1**, 102–108.

GODING, J.W. (1980). Antibody production by hybridomas. *Journal of Immunological Methods* **39**, 285–308.

GRIFFITHS, J.B. (1986). Can cell culture medium costs be reduced? Strategies and possibilities. *Trends in Biotechnology* **4**, 268–272.

GRIFFITHS, J.B. (1987). Serum and growth factors in cell culture media – an introductory review. *Develop. Biol. Standard.* **66**, 155–160.

GRIFFITHS, J.B. (1990). Advances in animal cell immobilization technology. In *Animal Cell Biotechnology* (R.E. Spier and J.B. Griffiths, Eds), volume 4, pp. 149–166. Academic Press, London.

HAGEDORN, J. AND KARGI, F. (1990). Production of monoclonal antibodies by hybridoma cells in a flat sheet membrane bioreactor. *Biotechnology Progress* **6**, 220–224.

HALBERSTADT, C.R. (1991). Design, implementation and modeling of a transtubular bioreactor for the growth of mammalian cells. Ph.D. thesis, University of Michigan, Ann Arbor, MI, USA.

HEATH, C. (1988). Engineering aspects of improved antibody production by hybridomas. Ph.D. thesis, Rensselaer Polytechnic Institute, Troy, NY, USA.

HEATH, C. AND BELFORT, G. (1990). Membranes and bioreactors. *International Journal of Biochemistry* **22**, 823–825.

HEATH, C., DILWITH, R. AND BELFORT, G. (1990). Methods for increasing monoclonal antibody production in suspension and entrapped cell cultures: biochemical and flow cytometric analysis as a function of medium serum content. *Journal of Biotechnology* **15**, 71–90.

HENGARTNER, H., MEO, T. AND MÜLLER, E. (1978). Assignment of genes for immunoglobulin κ and heavy chains to chromosomes 6 and 12 in mouse. *Proceedings of the National Academy of Sciences, USA* **75**, 4494–4498.

HILLER, G.W., CLARK, D.S. AND BLANCH, H.W. (1993). Cell retention-chemostat studies of hybridoma cells – analysis of hybridoma growth and metabolism in continuous suspension culture on serum-free medium. *Biotechnology and Bioengineering* **42**, 185–195.

HILLER, G.W., AESCHLIMANN, A.D., CLARK, D.S. AND BLANCH, H.W. (1991). A kinetic analysis of hybridoma growth and metabolism in continuous suspension culture on serum-free medium. *Biotechnology and Bioengineering* **38**, 733–741.

HIRANO, T., YASUKAWA, K., HARADA, H., TAGA, T., WATANABE, Y., MATSUDA, T., KASHIWAMURA, S., NAKAJIMA, K., KOYAMA, K., IWAMATSU, A., TSUNASAWA, S., SAKIYAMA, F., MATSUI, H., TAKAHARA, Y., TANIGUCHI, T. AND KISHIMOTO, T. (1986). Complementary DNA for a novel human interleukin (BSF-2) that induces B lymphocytes to produce immunoglobulin. *Nature* **324**, 73–76.

HU, W.S., DODGE, T.C., FRAME, K.K. AND HIMES, V.B. (1987). Effect of glucose on the cultivation of mammalian cells. *Develop. Biol. Standard.* **66**, 279–290.

HUANG, C.-T., PERETTI, S.W. AND BRYERS, J.D. (1993). Plasmid retention and gene expression in suspended and biofilm cultures of recombinant *Escherichia coli* DH5α(pMJR1750). *Biotechnology and Bioengineering* **41**, 211–220.

KAWANO, M., HIRANO, T., MATSUDA, T., TAGA, T., HORII, Y., IWATO, K., ASAOKU, H., TANG, B., TANABE, O., TANAKA, H., KURAMOTO, A., KISHIMOTO, T. (1988). Autocrine generation and requirement of BSF-2/IL-6 for human multiple myelomas. *Nature* **332**, 83–85.

KIDWELL, W.R. (1989). Filtering out inhibition. *Bio/Technology* **7**, 462–463.

KIDWELL, W.R., KNAZEK, R. AND WU, Y. (1990). Effect of fiber pore size on performance of cells in hollow fiber bioreactors. In *Trends in Animal Cell Culture Technology* (H. Murakami, Ed.), pp. 29–33. Kodansha, Tokyo.

KIDWELL, W.R., MOHANAM, S. AND SALOMON, D.S. (1987). Growth factor production by mammary tumor cells. In *Cellular and Molecular Biology of Breast Cancer* (D. Medina, Ed.), pp. 239–252. Plenum Press, New York.

KIM, S.J., HAN, B.K. AND LEE, G.M. Feasibility study on the use of calcium alginate-entrapped hybridoma cells for IgM production. *Bioprocess Engineering*, in press.

KUMAR, P.K.R. AND SCHÜGERL, K. (1990). Immobilization of genetically engineered cells: a new strategy for higher stability. *Journal of Biotechnology* **14**, 255–272.

KUMAR, P.K.R., MASCHKE, H.-E., FRIEHS, K. AND SCHÜGERL, K. (1991). Strategies for improving plasmid stability in genetically modified bacteria in bioreactors. *Trends in Biotechnology* **9**, 279–284.

KUNAS, K.T. AND PAPOUTSAKIS, E.T. (1989). Increasing serum concentrations decrease cell death and allow growth of hybridoma cells at higher agitation rates. *Biotechnology Letters* **11**, 525–530.

KUNAS, K.T. AND PAPOUTSAKIS, E.T. (1990). The protective effect serum against hydrodynamic damage of hybridoma cells in agitated and surface aerated bioreactors. *Journal of Biotechnology* **15**, 57–70.

LEE, G.M. AND PALSSON, B.O. (1990). Immobilization can improve the stability of hybridoma antibody productivity in serum-free media. *Biotechnology and Bioengineering* **36**, 1049–1055.

LEE, G.M. AND PALSSON, B.O. (1993). Stability of antibody productivity is improved when hybridoma cells are entrapped in calcium alginate beads. *Biotechnology and Bioengineering* **42**, 1131–1135.

LEE, G.M., CHUCK, A.S. AND PALSSON, B.O. (1993). Cell culture conditions determine the enhancement of specific monoclonal antibody productivity of calcium alginate-entrapped S3H5/γ2bA2 hybridoma cells. *Biotechnology and Bioengineering* **41**, 330–340.

LEE, G.M., HUARD, T.K. AND PALSSON, B.O. (1989). Effect of serum concentration on hybridoma cell growth and monoclonal antibody production at various initial cell densities. *Hybridoma* **8**, 369–375.

LEE, G.M., KIM, S.J. AND PALSSON, B.O. Enhanced specific antibody productivity of calcium alginate-entrapped hybridoma is cell line-specific. *Cytotechnology*, in press.

LEE, G.M., SAVINELL, J.M. AND PALSSON, B.O. (1989). Serum can act as a shear protecting agent in agitated hybridoma cell cultures. *Hybridoma* **8**, 639–645.

LEE, G.M., VARMA, A. AND PALSSON, B.O. (1991a). Production of monoclonal antibody using free-suspended and immobilized hybridoma cells: effect of serum. *Biotechnology and Bioengineering* **38**, 821–830.

LEE, G.M., VARMA, A. AND PALSSON, B.O. (1991b). Application of population balance model to the loss of hybridoma antibody productivity. *Biotechnology Progress* **7**, 72–75.

LEE, G.M., HUARD. T.K., KAMINSKI, M.S. AND PALSSON, B.O. (1988). Effect of mechanical agitation on hybridoma cell growth. *Biotechnology Letters* **10**, 625–628.

LIM, F. (1980). Microencapsulation of living cells and tissues-theory and practice. In *Biomedical Application of Microencapsulation* (F. Lim, Ed.), pp. 137–154. CRC Press, Boca Raton.

LIM, F. (1988). Microencapsulation of living mammalian cells. In *Upstream Processes: Equipment and Techniques* (A. Mizrahi, Ed.), pp. 185–197. Alan R. Liss, New York.

LOOBY, D. AND GRIFFITHS, G. (1990). Immobilization of animal cells in porous carrier culture. *Trends in Biotechnology* **8**, 204–209.

LOW, K.S., HARBOUR, C. AND BARFORD, J.P. (1987). A study of hybridoma cell growth and antibody production kinetics in continuous culture. *Biotechnology Techniques* **1**, 239–244.

MCKINNEY, K.L., DILWITH, R. AND BELFORT, G. (1991). Manipulation of heterogeneous hybridoma cultures for overproduction of monoclonal antibodies. *Biotechnology Progress* **7**, 445–454.

MCQUEEN, A. AND BAILEY, J.E. (1989). Influence of serum level, cell line, flow type and viscosity on flow-induced lysis of suspended mammalian cells. *Biotechnology Letters* **11**, 531–536.

MAKISHIMA, F., MIKAMI, T. AND TERADA, S. (1990). Effect of some cytokines on antibody productivity of a murine B-cell hybridoma. In *Trends in Animal Cell Culture Technology* (H. Murakami, Ed.), pp. 299–302. Kodansha, Tokyo.

MANCUSO, A., FERNANDEZ, E.J., BLANCH, H.W. AND CLARK, D.S. (1990). A nuclear magnetic resonance technique for determining hybridoma cell concentration in hollow fiber bioreactors. *Bio/Technology* **8**, 1282–1285.

MANO, T., MITSUDA, S., KUMAZAWA, E. AND TAKESHITA, Y. (1992). New immobilization methods of mammalian cells using alginate and polyacrylate. *Journal of Fermentation Bioengineering* **73**, 486–489.

MEILHOC, E., WITTRUP, K.D. AND BAILEY, J.E. (1989). Application of flow cytometric measurement of surface IgG in kinetic analysis of monoclonal antibody synthesis and secretion by murine hybridoma cells. *Journal of Immunological Methods* **121**, 167–174.

MERRITT, S.E. AND PALSSON, B.O. (1993). Loss of antibody productivity is highly reproducible in multiple hybridoma subclones. *Biotechnology and Bioengineering* **42**, 247–250.

MILLER, W.M., BLANCH, H.W. AND WILKE, C.R. (1988). A kinetic analysis of hybridoma growth and metabolism in batch and continuous suspension culture: effect of nutrient concentration, dilution rate, and pH. *Biotechnology and Bioengineering* **32**, 947–965.

MILLER, W.M., WILKE, C.R. AND BLANCH, H.W. (1987). Effects of dissolved oxygen concentration on hybridoma cell growth and metabolism in continuous culture. *Journal of Cellular Physiology* **132**, 524–530.

MURAKAMI, H., MASUI, H. AND SATO, G.H. (1982). Suspension culture of hybridoma cells in serum-free medium: soybean phospholipids as the essential components. In *Growth of Cells in Hormonally Defined Media* (G.H. Sato, A.B. Pardee and D.A. Sirbasku, Eds), pp. 711–715. Cold Spring Harbor Laboratories, Cold Spring Harbor, New York.

NASRI, M., SAYADI, S., BARBOTIN, J.-N., DHULSTER, P. AND THOMAS, D. (1987). Influence of immobilization on the stability of pTG201 recombinant plasmid in some strains of *Escherichia coli. Applied Environmental Microbiology* **53**, 740–744.

NILSSON, K. (1987). Methods for immobilizing animal cells. *Trends in Biotechnology* **5**, 73–78.

NILSSON, K., SCHEIRER, W., MERTEN, O.W., ÖSTBERG, L., LIEHL, E., KATINGER, H.W.D. AND MOSBACH, K. (1983). Entrapment of animal cells for production of monoclonal antibodies and other biomolecules. *Nature* **302**, 629–630.

NILSSON, K., SCHEIRER, W., KATINGER, H.W.D. AND MOSBACH, K. (1986). Production of monoclonal antibodies by agarose-entrapped hybridoma cells. *Methods in Enzymology* **121**, 352–360.

OGATA, M., ARIKAWA, K., MATSUKURA, H., YAMAZAKI, Y. AND SUZUKI, A. (1992). Large-scale mammalian cell cultures by application of macroporous cellulose carriers. In *Animal Cell Technology: Basic and Applied Aspects* (H. Murakami, S. Shirahata and H. Tachibana, Eds), pp. 201–207. Kluwer Academic Publishers, Dordrecht.

OKITA, W.B., BONHAM, D.B. AND GAINER, J.L. (1985). Covalent coupling of microorganisms to a cellulosic support. *Biotechnology and Bioengineering* **27**, 632–637.

OVERGAARD, S., SCHARER, J.M., MOO-YOUNG, M. AND BOLS, N.C. (1991). Immobilization of hybridoma cells in chitosan alginate beads. *Canadian Journal of Chemical Engineering* **69**, 439–443.

ØYAAS, K., BERG, T.M., BAKKE, O. AND LEVINE, D.W. (1989). Hybridoma growth and antibody production under conditions of hyperosomotic stress. In *Advances in Animal Cell Biology and Technology for Bioprocesses* (R.E. Spier, J.B. Griffiths, J. Stephenne and P.J. Crooy, Eds), pp. 212–218. Butterworths, London.

OZTURK, S.S. (1990). Kinetic characterization of hybridoma growth, metabolism, and monoclonal antibody production rates. Ph.D. thesis, University of Michigan, Ann Arbor, MI, USA.

OZTURK, S.S. AND PALSSON, B.O. (1990). Loss of antibody productivity during long-term cultivation of a hybridoma cell line in low serum and serum-free media. *Hybridoma* **9**, 167–175.

PAPOUTSAKIS, E.T. (1991a). Media additives for protecting freely suspended animal cells against agitation and aeration damage. *Trends in Biotechnology* **9**, 316–324.

PAPOUTSAKIS, E.T. (1991b). Fluid-mechanical damage of animal cells in bioreactors. *Trends in Biotechnology* **9**, 427–437.

PASSINI, C.A. AND GOOCHEE, C.F. (1989). Response of a mouse hybridoma cell line to heat shock, agitation, and sparging. *Biotechnology Progress* **5**, 175–188.

PETERSEN, J.F., MCINTIRE, L.V. AND PAPOUTSAKIS, E.T. (1988). Shear sensitivity of cultured hybridoma cells (CRL-8018) depends on mode of growth, culture age and metabolite concentration. *Journal of Biotechnology* **7**, 229–246.

RATAFIA, M. (1989). Serum-free media markets. *Bio/Technology* **7**, 574–575.

RAY, N.G., VOURNAKIS, J.N., RUNSTADLER, P.W., JR, TUNG, A.S. AND VENKATASUBRAMANIAN, K. (1990). Physiology of hybridoma and recombinant Chinese hamster ovary cells immobilized in collagen matrix. In *Physiology of Immobilized Cells* (J.A.M. de Bont, J. Visser, B. Mattiasson and J. Tramper, Eds), pp. 699–711. Elsevier, Amsterdam.

REDDY, S. AND MILLER, W.M. (1994). Effects of abrupt and gradual osmotic stress on antibody production and content in hybridoma cells that differ in production kinetics. *Biotechnology Progress* **10**, 165–173.

REDDY, S., BAUER, K.D. AND MILLER, W.M. (1992). Determination of antibody content in live versus dead hybridoma cells: analysis of antibody production in osmotically stressed cultures. *Biotechnology and Bioengineering* **40**, 947–964.

RENARD, J.M., SPAGNOLI, R., MAZIER, C., SALLES, M.F. AND MANDINE, E. (1988). Evidence that monoclonal antibody production kinetics is related to the integral of viable cells curve in batch systems. *Biotechnology Letters* **10**, 91–96.

REUVENY, S., VELEZ, D., MILLER, L. AND MACMILLAN, J.D. (1986). Comparison of cell propagation methods for their effect on monoclonal antibody yield in fermentors. *Journal of Immunological Methods* **86**, 61–69.

RUNSTADLER, P.W., JR, AND CERNEK, S.R. (1988). Large-scale, fluidized-bed, immobilized cultivation of animal cells at high densities. In *Animal Cell Biotechnology* (R.E. Spier and J.B. Griffiths, Eds), volume 3, pp. 305–320. Academic Press, London.

SAVINELL, J.M., LEE, G.M. AND PALSSON, B.O. (1989). On the orders of magnitude of epigenic dynamics and monoclonal antibody production. *Bioprocess Engineering* **4**, 231–234.

SAYADI, S., BERRY, F., NASRI, M., BARBOTIN, J.N. AND THOMAS, D. (1988). Increased stability of pBR322-related plasmids in *Escherichia coli* W3101 grown in carrageenan gel beads. *FEMS Microbiology Letters* **56**, 307–312.

SAYADI, S., NASRI, M., BARBOTIN, J.N. AND THOMAS, D. (1989). Effect of environmental growth conditions on plasmid stability, plasmid copy number, and catechol 2,3-dioxygenase activity in free and immobilized *Escherichia coli* cells. *Biotechnology and Bioengineering* **33**, 801–808.

SHACTER, E. (1989). Serum-free media for bulk culture of hybridoma cells and the preparation of monoclonal antibodies. *Trends in Biotechnology* **7**, 248–253.

SHEN, B.Q., REID, S. AND GREENFIELD, P.F. (1992). Continuous monoclonal antibody production by a composite gel perfusion in a protein-free medium. In *Animal Cell Technology: Basic and Applied Aspects* (H. Murakami, S. Shirahata and H. Tachibana, Eds), pp. 173–178. Kluwer Academic Publishers, Dordrecht.

SHIRAI, Y., HASHIMOTO, K. AND STEELE, L.M. (1992). Change in monoclonal antibody productivity in immobilized hybridoma cells at low dissolved oxygen concentrations. In *Animal Cell Technology: Basic and Applied Aspects* (H. Murakami, S. Shirahata and H. Tachibana, Eds), pp. 383–389. Kluwer Academic Publishers, Dordrecht.

SHIRAI, Y., HASHIMOTO, K., YAMAJI, H. AND TOKASHIKI, M. (1987). Continuous production of monoclonal antibody with immobilized hybridoma cells in an expanded bed fermentor. *Applied Microbiology and Biotechnology* **26**, 495–499.

SINACORE, M.S., CRESWICK, B.C. AND BUEHLER, R. (1989). Entrapment and growth of murine hybridoma cells in calcium alginate gel microbeads. *Bio/Technology* **7**, 1275–1279.

SUZUKI, E. AND OLLIS, D.F. (1990). Enhanced antibody production at slowed growth rates: experimental demonstration and a simple structured model. *Biotechnology Progress* **6**, 231–236.

TERADA, S., MAKISHIMA, F. AND TAKAMATSU, H. (1992). Enhancing effect of interleukin-6 on antibody productivity of a hybridoma cell line. In *Animal Cell Technology: Basic and Applied Aspects* (H. Murakami, S. Shirahata and H. Tachibana, Eds), pp. 413–418. Kluwer Academic Publishers, Dordrecht.

TYO, M.A. AND SPIER, R.E. (1987). Dense cultures of animal cells at the industrial scale. *Enzyme and Microbial Technology* **9**, 514–520.

VAN DER POL, L., ZIJLSTRA, G., THALEN, M. AND TRAMPER, J. (1990). Effect of serum concentration on production of monoclonal antibodies and on shear sensitivity of a hybridoma. *Bioprocess Engineering* **5**, 241–245.

VELEZ, D., REUVENY, S., MILLER, L. AND MACMILLAN, J.D. (1986). Kinetics of monoclonal antibody production in low serum growth medium. *Journal of Immunological Methods* **86**, 45–52.

VOURNAKIS, J.N. AND RUNSTADLER, P.W. (1989). Microenvironment: the key to improved cell culture products. *Bio/Technology* **7**, 143–145.

WALLS, E.L. AND GAINER, J.L. (1989). Retention of plasmid bearing cells by immobilization. *Biotechnology and Bioengineering* **34**, 717–724.

WESTERWOUDT, R.J., NAIPAL, A.M. AND HARRISSON, C.M.H. (1984). Improved fusion technique. II. stability and purity of hybrid clones. *Journal of Immunological Methods* **68**, 89–101.

WILDE, C.D. AND MILSTEIN, C. (1980). Analysis of immunoglobulin chain secretion using hybrid myelomas. *European Journal of Immunology* **10**, 462–467.

WOHLPART, D., GAINER, J. AND KIRWAN, D. (1991). Oxygen uptake by entrapped hybridoma cells. *Biotechnology and Bioengineering* **37**, 1050–1053.

15
High Cell Density Growth of Micro-organisms

ALAN W. BUNCH

Research School of Biosciences, Biological Laboratory, The University, Canterbury, Kent, UK

What is high cell density growth? The scope of this chapter

It is difficult to define exactly what is meant by high cell density growth because it is a relative term. Thus when working with protozoa a cell density of 2.2×10^7 and 54 g dry weight l^{-1} has been considered as growth at high cell density, whereas with bacteria and fungi this would be only a moderate growth density (Chang and Furusaki, 1991; Kiy and Tiedtke, 1992). The highest biomass levels ever recorded in a bioreactor were those described by Chung, Chang and Kim (1987). In these experiments a *Nocardia* strain was grown in a dual hollow-fibre bioreactor, reaching densities of 550 g dry weight l^{-1}. No record of the cell number was given, but 10^{12} *Escherichia coli* cells ml^{-1} of accessible reactor space has been reported, again for a hollow-fibre based system (Inloes *et al.*, 1983). The common factor in most of these reports is the comparison of a system with batch culture in a shake flask, or a stirred tank reactor. Therefore I will use this approach when discussing the alternatives available for increasing the culture cell density of micro-organisms. As far as possible, all types of micro-organisms will be covered, although inevitably the information available relates mainly to a relatively narrow range of bacteria and fungi.

It is rare that figures are given for both the dry weight and cell numbers present in a particular reactor, yet the two are not always proportional. Therefore the culture conditions are sometimes given in more detail than may seem necessary to allow readers to get as good an impression as possible about the physiological state of the microbes in the systems described. The demand for high cell density growth is mainly to improve the volumetric productivity of a process, although there is growing evidence that it can also be a strategy for stabilizing recombinant organisms, slowing down the loss of plasmid-encoded genes (Huang, Perelti and Bryers, 1993). Interest in high cell density culture is also growing because of a need to obtain increased biomass levels of microbes that are reputed to be difficult to cultivate, for biochemical

Abbreviations: ATP, adenosine triphosphate; GRAS, generally regarded as safe.

Biotechnology and Genetic Engineering Reviews — Vol. 12, December 1994
0264–8725/94/12/535–561 $20.00 + $0.00 © Intercept Ltd, P.O. Box 716, Andover, Hampshire SP10 1YG, UK

as well as commercial studies. A good example of this is the protozoan *Tetrahymena thermophila* (Kiy and Tiedtke, 1992). It is therefore likely that interest in developing such systems will continue for some time.

Factors that limit culture cell density

Microbial cells are complex entities. *Table 1* shows the elemental composition of a dried mass of *Escherichia coli*, which has to be supplied to the cell from the culture medium. This sort of balance is seen in most types of microbial cell. Neidhardt, Ingraham and Schaechter (1990) illustrate nicely the task required of a micro-organism during growth where, again using *Escherichia coli* as an example and referring only to macromolecules, over 25 million molecules representing more than 1550 different types need to be replicated in order to construct a new cell. Even under conditions allowing for slow growth there is a significant need for cell structure repair and turnover of molecules.

Table 1. The approximate elemental composition of dried *Escherichia coli* biomass

Element	Composition
Carbon	50–70%
Nitrogen	15–20%
Oxygen	10–20%
Hydrogen	6–8%
Phosphorus	2–4%
Sulphur	1–3%
Potassium	1–2%
Iron	0.1–0.3%
Calcium	0.01–0.05%
Magnesium	0.01–0.05%
Chlorine	0.02–0.03%
Trace elements	0.2–0.5%

Growth conditions, both nutritional and physical, will affect the chemical composition of the cells and the metabolic needs of the bacterium. Similar elemental compositions are found in other micro-organisms.

The driving force for molecular synthesis has to be provided from the components present in the growth media. When concentrating on adenosine triphosphate (ATP) synthesis, one of the most important molecules used by living cells in this context, it is apparent that different micro-organisms have developed alternative strategies for providing enough synthetic capacity. The most efficient way of making ATP is via oxidative phosphorylation, using oxygen as the terminal electron acceptor. If electron transport based synthesis is not possible, recourse is made to substrate-level phosphorylation. It is no accident that many biotechnological processes have developed around microbial strains whose biochemistry matches the efficient synthesis of the desired product. Thus *Clostridium* has been used for the synthesis of acetone and butanol, yeast for ethanol production and *Acetobacter* for vinegar manufacture, a strictly aerobic process. Problems usually arise when the physiological characteristics of a microbe are in conflict with the intensity of the process needed for a commercially viable operation.

One of the biggest difficulties is found with the supply of oxygen to cells. Tempest, Neijssel and Zewenboon (1983) recorded the fact that at 30°C the concentration of oxygen in water/air equilibria is only 0.23 mM. This figure quickly decreases as the temperature is increased and/or the salt concentration rises. *Alcaligenes faecalis* consumes 140–170 nmol O_2 min^{-1} (mg protein)$^{-1}$ in an aerobic denitrification process (van Niel *et al.*, 1992) and *Saccharomyces cerevisiae* during aerobic growth needs 2.5 mmol (g dry weight)$^{-1}$ h^{-1} (Verdwyn *et al.*, 1992), which are fairly typical values. It can be seen therefore that oxygen can quickly become the main growth rate or cell maintenance limiting nutrient. Transfer of oxygen into microbial growth media is a continuing problem in the development of biotechnological processes that demand a constant supply. A pragmatic solution to this problem is to select micro-organisms that can grow well under oxygen-limited conditions. *Escherichia coli*, a facultative anaerobe, has proven an excellent choice where high-density growth is required but under conditions where oxygen supply will be limited. The lactic acid bacteria, whose biochemistry allows them to grow almost equally well with or without oxygen, are another group of microbes used under these conditions (Linton *et al.*, 1987). However, in both cases significant changes can occur in the metabolic profile of these bacteria grown under oxygen sufficient or deficient conditions (Harrison and Loveliss, 1971; Tseng and Montville, 1992). Nevertheless many clever solutions have been found to supply oxygen to high cell density cultures with some success. These will be reviewed later.

The supply of other nutrients to high cell density cultures can also be a problem and requires good mixing regimes in order to avoid the development of cultures with heterogeneity of physiological states. Again, many of the systems described later have found solutions to this difficulty, but in some bioreactors it cannot be avoided. It is often difficult to ascertain from reports on high cell density systems which nutrient is limiting the growth and/or the biotransformation potential of the microbial cells. It is, however, important to know the answer to this facet of bioreactor operation because the quality of the cells performing the desired function will, in most cases, be affected radically by the growth-limiting nutrient.

Earlier studies on the effects of culture conditions on the composition and size of microbial cells showed that merely recording the biomass levels in terms of dry weight does not always give an accurate indication of the useful biomass present. Herbert (1961) showed that growth rate dramatically influenced the chemical composition of *Aerobacter aerogenes*. At a growth rate of 0.8 h^{-1} this bacterium, under the conditions used, contained nearly three times more RNA than when grown at 0.1 h^{-1}. Under nitrogen-limited conditions Herbert showed that *Escherichia coli* had nearly five times as much cellular glycogen at low growth rates, than when grown at 0.5 h^{-1}. The mean cell volume can be affected by both the growth environment and the growth rate. The review by Tempest *et al.* (1983) recorded many examples of how microbial cells perform in laboratory culture. For example, van Liere (1979) showed that the cyanobacterium *Oscillatoria agardlia* changes size from 10μm^3 at a growth rate of 0.005 h^{-1} with 5 pg dry weight/cell, to 35 μm^3 at a growth rate of 0.025 h^{-1} with 21 pg dry weight/cell. More recently, Cooper (1991) has reported that bacterial cell density, within the bounds of experimental error, stays constant under a wide range of growth conditions. Finally, the viability of cells is also influenced by the growth-limiting nutrient. Postgate and Hunter (1962) showed that the death rate of *Aerobacter aerogenes* was unaffected by the oxygen in the atmosphere over the range of 2–100%. Populations at cell densities greater than 20 g dry weight l^{-1} survived longer than lower-density cultures, and in cultures where the growth-limiting nutrient was carbon, phosphorus or sulphur the viability declined linearly, whereas with nitrogen limitation the decline was sigmoidal.

These observations show how variable a high cell density community can be in comparison to the same cells grown under less extreme conditions. It is also likely that the gross changes in cell composition indicate significant changes in the biochemistry of the cells and so reveal the complex task of understanding the rationale behind productivity changes in high- and low-density cultures of a given micro-organism. In many systems the classical techniques of microbiology employed to study changes in the biochemistry of microbial cells under different growth conditions cannot be used. This is usually because the recovery of cells from the bioreactor imposes stresses that change the cell. Physical techniques, such as nuclear magnetic resonance and microcalorimetry, have been used on a limited basis to follow microbial metabolism in such bioreactors. However, changes in the product profile in cultures of well-characterized microbes probably still gives the best indication of the metabolic status of cultures. Examples of such studies will be given in the following sections.

Bioreactors for high cell density growth

Many cultivation methods have been employed in the quest for increased cell density. They fall into two groups: those that are a modification of the stirred tank reactor and those that are developed around a very different concept. The former type of reactor is attractive to most investigators because it employs equipment that, for the main part, is already available in a standard microbial cultivation facility and therefore this type will be dealt with first.

FED BATCH BIOREACTORS

This type of bioreactor has proved to be one of the most successful for increasing the volumetric productivity of a standard batch fermentation. In the basic system the micro-organism is added to the culture medium, usually in a stirred tank reactor when volumes greater than 500 ml are involved, and allowed to grow until one or more of the original nutrients becomes growth limiting. This is an excellent way of producing growth-associated products, but does have limitations when attempting the production of products which are partially or non-growth associated. In the fed-batch process the first stage is that which would occur in a standard batch process. This part of the fed-batch procedure is used to develop the majority of the biocatalytic potential used in the second part of the process. When the culture has reached a point where nutrient limitation begins, the growth-limiting nutrient (and often other metabolites) is fed to the culture to maximize the biomass development and/or product formation (see *Figure 1*). Penicillin synthesis is a good example of one of the most thoroughly designed fed-batch systems and shows how this type of bioreactor can be used to increase catalytic density in addition to directing metabolism to product formation (Swartz, 1985). It has also found extensive use in the processes, rapidly coming on-line, that are the result of recombinant gene technology. Most of these systems are still based on plasmid-encoded genes and fed-batch technology has proved to be reasonably useful in maintaining the population in a productive state, overcoming some of the problems related to segregational and structural instability problems associated with many plasmids (Chang *et al.*, 1991). Many examples illustrate the use of this type of bioreactor for enhancing productivity, with enhanced cell density being a major factor, but the examples cited here have been selected to show the range of applications rather than to provide an exhaustive review.

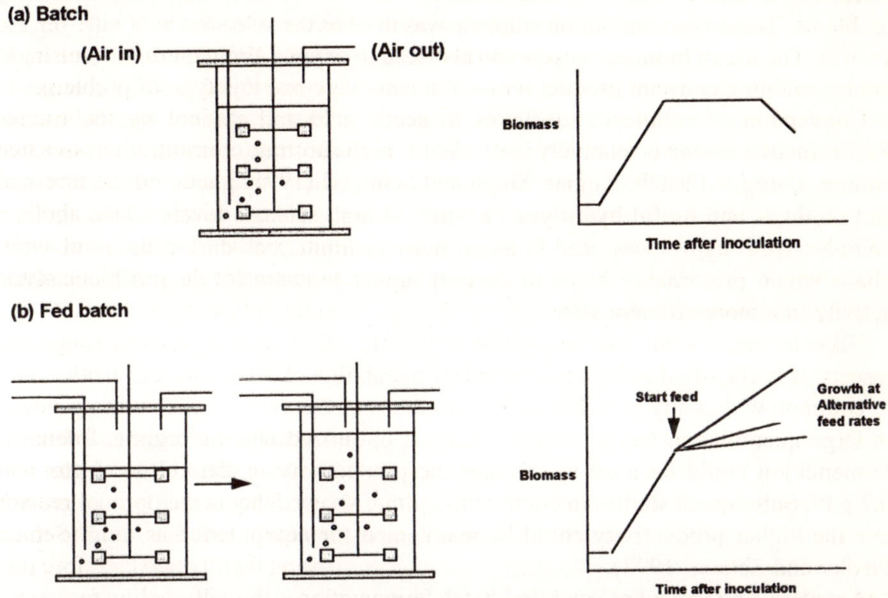

Figure 1. Reactor conformation and biomass development in batch and fed-batch culture. A stirred tank reactor is used to illustrate the stylized representation of batch and fed-batch culture. The design and biomass production profiles are usually more complex than the illustrations indicate. Air supply is an option depending on the micro-organism being grown, inert gases can sometimes be used to remove toxic gaseous end products.

The most straightforward application for fed-batch reactors is the synthesis of growth-associated natural products of microbial metabolism. Lactic acid is a major end product of the metabolism of many bacteria. As well as being produced during the manufacture of dairy products, it is an important chemical in its own right (Vickroy, 1985). Frequently the exploitation of this microbial metabolite is limited by the availability of a commercially viable manufacturing procedure. One of the biggest problems is that focused on by Ohara, Hiyama and Yoshida (1992) and involves inhibition of growth and substrate uptake by the end product, lactic acid. By using the information gained from this study the authors were able to predict suitable operating conditions for maximizing productivity. Substrate inhibition causing limited bioreactor operation is also a problem that can be partially overcome by using fed-batch systems. This can be illustrated by reference to the catabolism of the toxic molecule salicylic acid by *Pseudomonas cepacia* (Tocaj *et al.*, 1993). When grown in batch culture with salicylic acid as the sole source of carbon and energy, concentrations greater than 2 g l^{-1} are inhibitory (Kamin, White-Stevens and Presswood, 1978). At concentrations greater than 10 g l^{-1} no growth occurs. By carefully controlling the feed rate to the reactor, using a monitor that measured the salicylate concentration in the reactor, a much higher conversion rate to biomass was recorded. However, it is interesting to note that modifications had to be made to the culture medium due to the nature of fed-batch fermentation. Previous studies on the system in batch and cell recycle fermentation used potassium nitrate as the nitrogen source for growth (Kamin, White-Stevens and Presswood, 1978; Berg, Holst and Mattiasson, 1989). In the fed-batch reactor, where the elevated biomass conentrations led to oxygen limitation, the nitrate was converted to nitrite (due to its use as an electron acceptor), which caused toxicity problems. Therefore ammonium sulphate was used as the sole source of nitrogen for growth. The use of biomass recycle can also lead to oxygen-limited growth (see later) but by enabling constant product removal it can overcome this type of problem.

Conversion of cellulosic substrates to acetic acid and ethanol by the fungus *Fusarium oxysporum* is relatively inefficient at high substrate concentrations in batch culture. Using fed batch, Kumar, Singh and Schugerl (1991) increased the levels of end products and useful hydrolytic enzymes at high substate levels. Once again, a complex feed regime was used to avoid nutrient limitations during the production phase which presumably helps to support higher biomass levels and biocatalytic activity in a more efficient state.

Nikkomycin is a nucleoside-peptide antibiotic active against various fungi and insects. It is classified as being a secondary metabolite of *Stretptomyces tendae* and, in common with many useful microbial products of this type, is difficult to produce in large quantities in batch culture. Using an optimized feeding regime, fed-batch fermentation could be used to increase the productivity of the culture from 3 to 6.7 g l^{-1}. Subsequent studies in continuous culture showed that at the optimal growth rate the higher productivity could be maintained for nearly twice as long (Schüz, Fiedler and Zahner, 1993).

An interesting variant of batch/fed-batch fermentation is the self-cycling fermenter systems described by van Walsum and Cooper (1993). In this process the metabolic activity of the culture is monitored (in this case oxygen consumption) and the cycle time matched to the doubling time of the organism. *Bacillus subtilis* and *Acinetobacter calcoaceticus* were used in this study. When a cycle has reached the end point, half of the reactor volume is removed and the volume in the reactor immediately restored with fresh growth medium. Using a holding tank for the removed medium, the culture is allowed a further period of incubation before harvesting. The products investigated were surfactants, classified as being secondary products of the organisms concerned. The additional incubation time was shown to markedly improve the yield from the fermentation.

In these first examples of the use of batch culture the enhancement of cell density in the reactors is often not stated. They were selected to illustrate that although there is often some implication of increased cell densities, increased productivity is a combination of providing the right physiological environment and increasing the biomass level. The fact that these two facets of fed-batch operation are important is often understated and yet both are crucial for maximizing the yield of recombinant products as well as the types of natural product illustrated above.

Recently work relating to recombinant DNA technology was published in the *Annals of the New York Academy of Sciences* (Prokop and Bajpai, 1991). The majority of papers were concerned with the construction of new producer strains. A much smaller number of papers was concerned with the actual production of the recombinant products. Fed-batch fermentation methods were highlighted as a means of improving productivity. *Escherichia coli* was the recombinant host in two of the papers concerned. Concentrating on the high cell density that can be achieved, Knorre *et al.* (1991) stated that this has the advantages of reducing reactor volume, improving space–time yield, and media costs and simplifying downstream processing. They showed how optimizing the growth rate in the fed stage and designing the process to avoid oxygen limitation can improve the productivity of fermentations where the cell density reaches 110 g dry weight l^{-1}. Chang *et al.* (1991) alluded to the complexity of bacterial cells mentioned earlier. They showed how the nature of the selection pressure for the retention of plasmid-containing cells can dramatically influence the proportion of the population capable of synthesizing the recombinant protein, porcine somatotropin, in the production phase. Recombinant protein synthesis was induced by heat shock, the efficiency of which was affected by the cell density at the time of induction. Their studies showed how media design can reduce the catalytic development phase of the fed-batch culture. An interesting observation was why the location of the temperature-sensitive repressor on a plasmid, rather than the chromosome, determined the factors that affect expression of the recombinant protein. The growth rate at which the culture was switched to production is also important in optimizing expression. At the higher growth rates tested expression was greatest, but the final growth yield was lowest. A compromise was needed to maximize productivity. The authors showed that computer control to maintain the optimum nutrient balance in the fermentation could improve productivity further.

Yeasts have proved to be an almost equally popular host for recombinant protein synthesis as *Escherichia coli*, where their GRAS status (generally regarded as safe) is an attractive feature. Increasing numbers of species are employed in this context but reference to *Saccharomyces cerevisiae* illustrates the results obtained in fed-batch cultivation. Once again, good productivity has been obtained in cultures with cell densities in excess of 100 g dry weight l^{-1} (Alberghina *et al.*, 1991). Nurmianto, Greenfield and Lee (1994) showed that, as with the bacteria, there is often an optimum growth rate, supported by optimum nutrient feed rate, for maximal stable recombinant protein synthesis. A recurring observation that is made when reviewing the literature in this area of biotechnology, is the variety of variable factors that can determine the productivity of high cell density cultures. For example, Ibba *et al.* (1993) have shown that the culture medium can affect the plasmid content and stability in recombinant *Saccharomyces cerevisiae*, but cells with the highest copy number are not always most productive. There is clearly a need in many systems to gain a better understanding of the interaction of plasmid copy number, gene expression and protein synthesis. It was also reported that in a similar study repeated fed batch was a better way of producing recombinant protein than continuous culture (see the case of nikkomycin production above). Blondeau *et al.* (1994), from their studies on the yeast *Kluyveromyces lactis*, also remind us that it is possible to increase cell density over that seen in batch culture by using either fed batch or continuous culture, while the physiological conditions need to be optimized for optimal expression of heterologous proteins.

Fed-batch bioreactors need fairly minor modifications to conventional batch reactors. The modifications needed to allow continuous operation of such reactors are also often straightforward. For commercial production of high-value products, continuous culture has never been employed extensively. Reasons for this relate both to the working practices of industry and to the biological properties of micro-organisms, which can often show unstable productivity in such systems. Nevertheless continuous culture, in its various operation modes, remains a useful tool for studying the behaviour of cells growing at high cell densities and may become more attractive as a production method in the future.

CONTINUOUS CULTURE SYSTEMS FOR ATTAINING HIGH CELL DENSITIES

It has already been shown that continuous culture systems of the sort shown in *Figure 2* can be designed to attain moderately high cell densities. In chemostat culture, growth is usually limited by a single nutrient. Ethanol production by *Saccharomyces cerevisiae* has been compared in batch and continuous systems (Gupta and Chand, 1994). It was shown that some of the best productivities were obtained as the system was perturbed under conditions of glucose limitation. In addition, changes in cell density that resulted from 'shift-up' or 'shift-down' did not immediately affect the productivity of the system. The rates of synthesis in the continuous processes were competitive with the best batch process reported by these authors. It is therefore not surprising that optimization of continuous culture has allowed big improvements in the yields of some microbes that grow comparatively poorly in batch culture. An extreme example of this is the growth of the hyperthermophilic *Pyrococcus furiosus*. Raven *et al.* (1992) showed that growth of the archaeum was limited by the build-up of fermentation products, especially hydrogen. Choosing the best gas to remove this metabolic end product continuously, cell concentrations in the bioreactor could be increased ten-fold to 3×10^9 cells ml^{-1}. In a modified type of continuous culture, using a perfusion device, *Tetrahymena thermophila* could be grown to cell densities 30–40 times that attainable in standard batch reactors. Total cell retention was used in this system and the study revealed that culture growth was apparently limited by both the glucose and protein content of the medium, although the production of excreted protein by the cells when grown at high cell densities gave problems in determining the exact limiting factors (Kiy and Tiedtke, 1992a,b). Many further modifications can be performed on the basic chemostat reactor. One interesting variation briefly reported by Cubarsi and Villaverde (1993) uses a two-stage system for growing and then inducing cells for product formation. However, the most interesting evolutions of the basic continuous culture design are those that incorporate artificial membranes

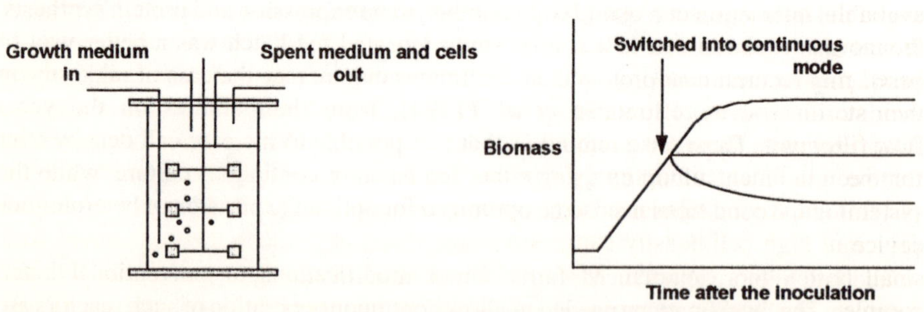

Figure 2. Continuous culture fermentation. The basic design is again illustrated by using a stirred tank reactor. In reality the steady state obtained after switching the reactor into continuous operation can often last for a very short time before mutants appear that affect both the biomass levels and/or product synthesis.

for the continuous separation of products from the microbial catalyst. At this point it is worth reviewing briefly the technology that is available for such separations.

Many techniques can be used to separate cells from culture media. Immobilization of micro-organisms has been used widely where cell retention or product removal on a continuous basis is required. There is, however, often a problem in using such a technique. Few immobilization methods can maintain long-term retention of microbial cells. The systems frequently underutilize the potential biocatalyst present (e.g. due to mass-transfer problems) and it is often difficult to monitor the physiological state of the immobilized cells. In the latter case it causes difficulties in rational process development. Flocculation of microbial cells has been used successfully, but is restricted to microbial species which either readily form natural workable aggregates, or aggregate with the limited range of chemical flocculants available. Alternative physical methods, such as centrifugation, can be too expensive for the application envisaged. From an operational point of view artificial membranes have offered great potential. The difficulty usually arises when long-term continuous operation at high cell densities is envisaged. Frequently the materials available are not robust enough, too expensive, or in a device that quickly fouls during operation. Chang and Furusaki (1991) have comprehensively reviewed the types of reactors that have been constructed incorporating artificial membranes. Heath and Belfort (1992) give details of many of the membranes available and the materials from which they can be constructed. Cross-flow filtration is the most widely applicable separation mode used in bioreactor design and I will limit this discussion to developments in this area of filtration, given their potential for improvement of high cell density growth in continuous culture.

Broadly speaking, the materials available for the construction of filtration units fall into two categories, plastic or inorganic based. The ever increasing range of plastics allows the construction of membranes in a wide variety of conformations, ranging from flat sheet structures to hollow fibres. In addition, the surface properties vary with the polymer used, and many polymer surfaces can be chemically modified to increase the range of surface properties. This facility can be extremely important in separation processes. The inorganic supports available are in some respects less varied, but are still available in a number of conformations, usually in the form of tubular structures. Ceramic membranes have been the main types used in the applications relevant to the topic of this chapter, but newer materials such as zirconium–carbon structures are also available; both types can often have their surfaces modified. The main advantage of the inorganic membranes is that they are usually much more robust than the plastic-based membranes, allowing them to be used under harsher conditions and making their sterilization more straightforward. *Figure 3* shows the basic details of a cross-flow filter unit. There are guidelines for the design of such units, which are essential for the attainment of optimal filtration fluxes. This is particularly important for systems employing tubular structures where increased flow rates through a separation device of high cell density cultures will not result in efficient media removal with small bore filters (Warren, Macdonald and Hill, 1991). It is fair to say that the complete mechanism of cross-flow filtration has not been fully worked out (Patel, Mehaia and Cheryan, 1987; Riesmier, Kroner and Kula, 1989, 1990). Tanaka *et al.* (1993) have investigated the filtration of *Saccharomyces cerevisiae* in cross-flow systems employing a flat-sheet membrane. Careful selection of flow rates at low cell

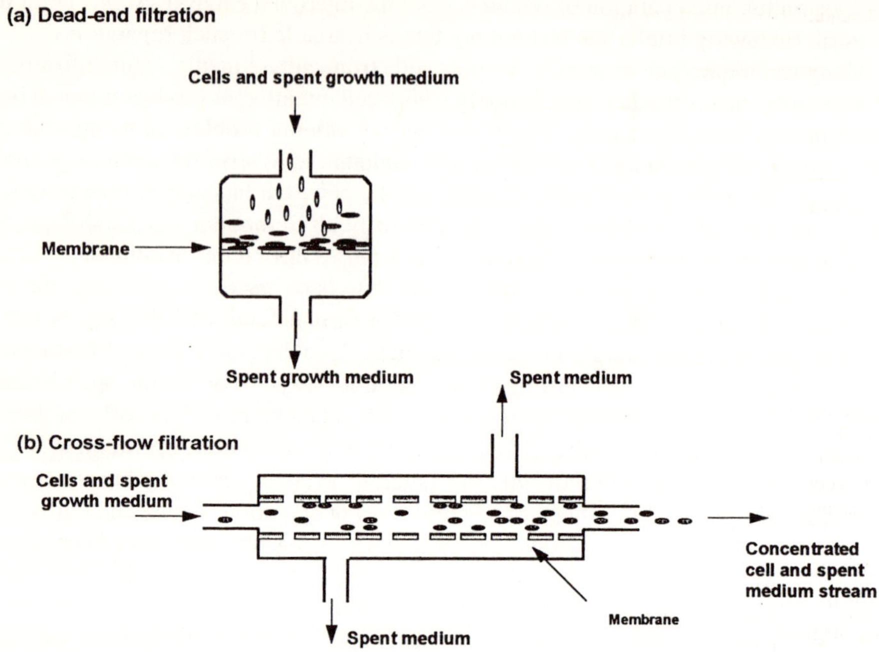

Figure 3. A comparison between dead-end and cross-flow filtration. Cross-flow filtration systems can be constructed from a wide variety of membrane types in both flat-sheet and tubular configurations.

densities allows efficient operation of the separator. As the cell density increases to that often encountered in high cell density culture systems fluxes dramatically decrease. Backflushing of such systems is possible but usually becomes unsuitable when continued long-term filtration is needed in a bioreactor system. There have been several attempts to increase the flow instability in cross-flow systems with the objective of reducing membrane fouling (Heath and Belfort, 1992). Some systems allow the use of ceramics. The rotating ceramic membrane unit described by Nakano, Matsumara and Kataoka (1993) has been shown to allow high cell densities of *Corynebacterium glutamicum* (120 g dry weight l^{-1}) and *Propionobacterium freudenreichii* (53 g dry weight l^{-1}) to be obtained in perfusion culture lasting several hundred hours. By using flat-sheet dimpled membranes and an oscillating flow, Millward *et al.* (1994) have described a cross-flow filtration device that is extremely efficient. Current studies employing this device have shown that it is capable of operation for over 100 h while operating in a cell-recycle continuous culture device with either yeast or lactic acid bacteria (Ruanglek *et al.*, unpublished observations). The membrane arrangement and surface detail are shown in *Figure 4*. Further details, including gas transfer characteristics, have also been described, illustrating the potential operational flexibility of the device (Bellhouse *et al.*, 1973; Beeton *et al.*, 1991, 1994). Thus advances in membrane technology are likely to increase the attractiveness of continuous systems using filters to achieve high cell densities. Of these systems, continuous culture with cell recycle probably offers the greatest potential when fitted with an efficient separations device.

In his classic text, Pirt (1975) outlined the mathematical properties predicted for

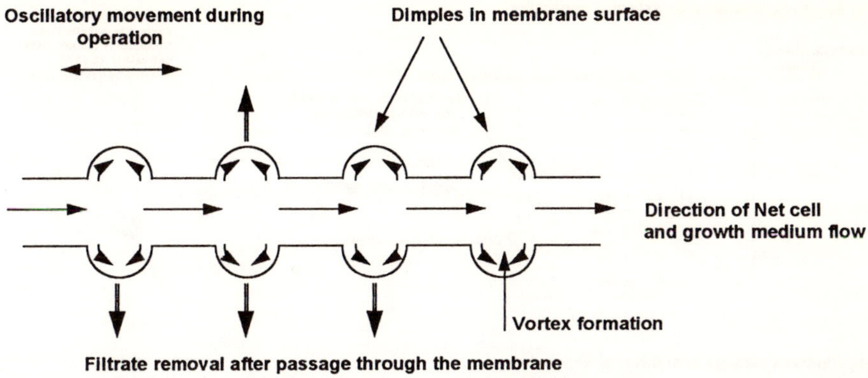

Figure 4. Detail of the twin membrane arrangement in cross-flow filtration with vortex wave formation. The polysulphone membranes are moulded to produce a regular dimpled structure over the surfaces of the two membranes used in this filtration device. During operation the bi-membrane unit oscillates in the direction shown, causing the formation of vortexes, which delay the formation of polarization layers and enhance the filtration rate. Full details can be found in Millward *et al.* (1994).

continuous culture with cell recycle and some of the advantages of using such a system. An example of the basic nature of this type of reactor is given in *Figure 5*. Although the illustration shows an external separation device, internal designs have also been used. It is not appropriate in a review of this sort to expound the potential of these systems in detail. They have great applicability to the fundamental study of microbial physiology as well as to the manufacture of microbial products. The cell-recycle ratio (R) is the proportion of biomass loss from a continuous culture that is recycled back into the main reactor. From chemostat theory it is possible to define the specific growth rate (μ) of a microbe in such a device as:

$$\mu = (1-R)D$$

where D is the dilution rate of the culture. Thus the specific growth rate can be altered by either changing the dilution rate of the system or the recycle ratio. With full recycle one has zero growth rate at steady state, a sort of perfusion culture. It was very difficult in early types of cross-flow filtration to achieve 90–100% cell recycle while attaining a steady state. Thus, as we will see, most of the recent reports in the literature refer to cultures that are not at steady state under conditions of 100% cell retention. These processes are analogous to those used for the attainment of high cell density animal cell cultures using the perfusion technique (Zhang, Handa-Corrigan and Spier, 1993). Some of the predicted properties of the system are, in any case, clearly unattainable. For example, this theory would predict that the steady-state biomass concentration at 100% cell recycle will be infinite. But other properties of the system merit continued study. These systems, by allowing high biomass densities and delaying washout at a given dilution rate, offer attractive prospects for dealing with poorly water-soluble growth substrates, toxic products and the enhancement of microbial products, negatively correlated with, or independent of, the growth rate of the microbe. A few selected examples will be used to illustrate the current status of this technology in microbial cell cultivation, as Chang and Furasaki (1991) have reviewed many earlier cell-recycle systems.

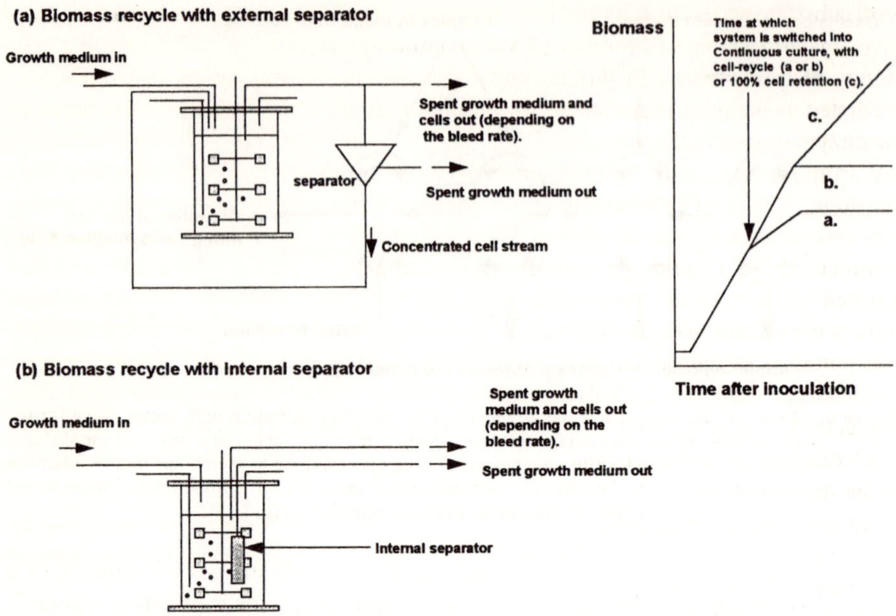

Figure 5. Continuous culture with cell recycle using either external or internal separators. A wide range of devices are available for separating cells from spent growth media. External devices can include centrifugation as well as filtration, and with both internal and external set-ups it is possible to use cell flocculation followed by sedimentation to retain or separate biomass produced during the fermentation. The biomass profiles shown indicate that where 100% cell retention is used no steady state is obtained. With partial cell-recycle the attainment of a steady state can take some time (graph a).

Tin and Mawson (1993) investigated the production of ethanol from whey by *Kluveromyces marxianus* growing in continuous culture with cell recycle, in an attempt to overcome the problems relating to the poor conversion rates observed in batch and continuous culture systems without cell recycle. From an application standpoint it was important to avoid, as far as possible, high sugar levels in the product stream, which in commercial processes would lead to problems of waste disposal. Both centrifugation and flocculation have been used in similar processes to achieve cell recycle. (Howell, 1981; Barry, 1982; Mawson, 1987). Tin and Mawson (1993) used tubular ceramic membranes in the filtration device used to separate the yeast cells from ethanol and unused substrates. Setting the recycle ratio to either 0 or 100%, the effects of increased lactose concentration and increased dilution rate on productivity were examined. The cell densities employed in this study were fairly low, explaining why filter blockage was apparently not a problem. Productivity was increased under cell-recycle conditions and predictably the residual substrate concentrations at a given dilution rate were lower. Others have shown that increasing the biomass density in recycle systems allows higher dilution rates to be used while keeping sugar levels low in the effluent from the reactor (Cheryan and Mehaia, 1983; Mehaia and Cheryan, 1990).

As mentioned previously, lactic acid is a commercially important microbial product. The conversion of starch to lactic acid in batch fermenters is often limited by the low concentration of microbes in the system. Immobilization methods have been used but have often failed to achieve the productivities expected, due to problems with substrate access to cells and loss of cell activity. Zhang and Cheryan (1994) have investigated the use of biomass recycle continuous culture for improving the efficiency of this process. In this example selection of the membrane properties was predicted as being as important as the filtration unit design. This is because many of the enzymes used in the biotransformation are extracellular and would be continually lost from the reactor if the membrane pore size is too large. Using a hollow-fibre cartridge, operating at relatively high transmembrane pressures, a workable laboratory reactor system was developed. Redesign of this system would be necessary for commercial use. Continuous culture studies identified the range of dilution rates that resulted in low residual starch levels in the outflow from the reactor. Full retention of cells in the reactor was maintained and, more surprisingly, so was the biotransformation potential of the system even when membranes with molecular weight cut-offs much larger than the hydrolytic enzymes likely to be involved were used. The productivity under optimal conditions was nearly twice that observed in comparable batch reactors. A cell density of 38.8 g dry weight l^{-1} was calculated towards the end of the fermentation.

An interesting application of cell-retention continuous culture is that described by Kang, Lee and Chang (1993). *Bacillus thuringiensis* has been used for many years as a source of spores for use as insecticides. It is a fermentation that needs optimal productivity to allow the small profit margins to be maximized. Two systems were compared. The first involved spore production in the cell-retention system operating on its own, the second coupled this reactor to a second-stage continuous culture process with no cell-recycle. Earlier experiments in glucose-limited fed-batch culture obtained a maximum of 5.5×10^9 spores ml^{-1} (Kang, Lee and Chang, 1992). This process was limited by the volume of the reactor. In the total cell retention system the spore count rose to over 1×10^9 spores ml^{-1} but with three times the number of colony-forming units present at the same time. Converting the remaining cells to spores could not be achieved by switching the reactor into batch mode at the time of maximum spore counts. High levels of glucose in the feed to cells in the total cell retention system prevented spore formation. In this case when the reactor was subsequently switched into batch mode spore formation was quickly observed and reached 7.2×10^9 spores ml^{-1}, with over 70% of cells converted to spores. To further decrease the stress on cells in the culture and improve their subsequent conversion to spores, higher levels of glucose in the total cell retention system were used. The best productivity was observed at a glucose feed concentration of 50 g l^{-1}. The reactors that used a continuous culture rather than batch culture in the second stage were an attempt to obtain continuous spore production. A bleed ratio of 0.33 (67% cell retention) from the first reactor was used to feed the continuous culture spore-production phase. Continuous production of spores could be achieved, but at lower spore concentrations in the continuous culture reactor, than were in the experiments mentioned earlier. Problems arose due to the almost inevitable generation of asporogenous mutants. Increasing the bleed ratio led to a gradual decline in productivity. At lower bleed ratios the internal ceramic filter device caused problems, presumably due to the membrane fouling. A maximum cell mass of 82.2 g l^{-1} was recorded in this device and the authors reported that the spore yields under optimum operating conditions were the highest recorded so far for this bacterium.

Using a similar reactor design, fitted with an internal stainless-steel membrane device, Chang, Lee and Kim (1993) investigated the ability of *Saccharomyces cerevisiae* to produce ethanol. The membrane pore sizes were much larger than usually used and were either 2 or 10 μm. Filter fouling could be reduced by increasing the agitation speed used to mix the components present in the reactor, although there are both operational and cost limits (for commercial production) to using this approach. Cell densities in these reactors reached a maximum of 157 g dry weight l^{-1}. For best productivities, slightly lower cell densities were preferable, which could be controlled by increasing the cell bleed rate (reducing cell retention). The use of co-culture of the yeast *Saccharomyces diastaticus* and *Pichia stipitis* in a cell retention bioreactor, using flocculation to effect cell/media separation, yielded less impressive cell densities but still an efficient process for converting glucose and xylose to ethanol (Laplace *et al.*, 1993). Uribelarrea *et al.* (1993) have investigated the carbon and energy balances in cultures of *Schizosaccharomyces pombe* growing in cell-recycle fermenters at different bleed and glucose feed rates. In previous studies where total cell retention was employed, it was shown that growth, under conditions of glucose excess, was limited by the inhibitory effects of ethanol (De Queiroz, Uribelarrea and Pareilleux, 1991). With partial cell recycle at different growth rates under glucose limitation, little ethanol was produced and the maximum cell density recorded was 32.7 g dry weight l^{-1}. Over a similar range of growth rates with glucose excess, over 40 g l^{-1} ethanol and 165 g dry weight l^{-1} biomass were recorded, the highest values at the lowest growth rates. It was shown that a variety of maintenance requirements were greatest when glucose was in excess. The use of recycle fermentation to examine effects on the specific growth rate independent of dilution rate was highlighted as being of great value in assessing the influence of environmental parameters.

To complete this section three processes have been selected to illustrate the range of variations in fermentation design that can be employed. Babu and Panda (1991) have illustrated the use of fermentation broth recycling in overcoming waste generation problems for the production of penicillin amidase by *Escherichia coli*. In this system the carbon and energy source (phenylacetic acid) has a complex influence on enzyme synthesis and cell growth rate. Centrifugation was used to separate cells from product and the medium nutrient composition analysed and restored to the starting concentrations after cells were removed. It was shown that the process water could be successfully re-used, but some of the products that accumulate influence penicillin amidase synthesis.

Itaconic acid production by *Aspergillus terreus* has been studied in many bioreactor systems. The use of cell immobilization has resulted in disappointing productivities, identified as being caused by oxygen limitation. Using an air-lift reactor in three modes of operation Park *et al.* (1994) showed that repeated batch culture, without cell recycling, gave the most reproducible activities. When cell recycling was attempted, at the end of each batch run, using either centrifugation or filtration to separate the cells from the culture medium, problems arose due to the increased mycelium concentration.

The synthesis of glutamic acid by *Brevibacterium flavum* is an example of a biotechnological process that has been optimized from a variety of approaches ranging from genetic manipulation to growth medium optimization (Beppu, 1986; Minoda, 1986). There is still room, however, for further improvements in productivity by redesign of the bioreactors employed. Glutamate causes end-product inhibition of its synthesis. The continued removal of this amino acid should therefore help improve the productivity of the fed-batch process routinely used for its manufacture. Glutamic acid can be separated from culture media by electrodialysis. This requires prior separation of cells from the media for maximal efficiency. Ishazaki, Takasaki and Furuka (1993) have designed a total cell recycle system using filtration, to both perform cell separation and maintain an adequate oxygen supply while the cells are in the recycle loop. Their results showed that a separation device incorporating vortex flow (by filtration unit rotation) maintained good productivity, even when the unit was not used to supply air to the cells in the filtration loop.

Many variations of reactor design have been developed to improve the continuous production of a wide range of microbial products. Some of the improvements have been achieved solely by increasing the biomass density in the reactors. However, in most cases it has been appreciated that this strategy is most successful when the undesirable problems associated with extreme biomass densities are avoided, thus avoiding the generation of biomass in a form that is not able to make the desired product because of its location or physiological state. The final type of reactors covered has been shown to have impressive abilities in generating extremely high cell densities, but has much less potential for modulating the accessibility of substrate to cells.

HIGH-DENSITY CULTURE IN IMMOBILIZED CELL BIOREACTORS

Some of the earliest attempts to solve the problem of toxic product removal on a continuous basis involved the use of dialysis culture (Chang and Furusaki, 1991). Technically this sort of bioreactor generates high cell densities with only a minority of the cells attached to the reactor structure, the rest remaining in suspension. However, in many immobilization systems a similar type of culture develops and therefore some reference to this type of cultivation is relevant. Because of the materials available to construct dialysis culture reactors, systems have tended to be restricted to the laboratory scale. Using a two-chamber device, with the growth chamber constructed from a cylindrical dialysis membrane, Märkl *et al.* (1993) grew *Escherichia coli* to densities in excess of 170 g dry weight l^{-1}. They calculated, making generalizations on bacterial cell composition, that the theoretical maximum cell density was between 160 and 200 g dry weight l^{-1}. Growth of this bacterium was obtained by using glycerol as the carbon and energy source. It was shown that for maximal biomass development both low and high concentrations of glycerol should be avoided. The authors attributed the capacity to feed a larger proportion of the culture evenly and without large fluctuations in the nutritional environment, plus continued product removal, to the attainment of such high cell densities when compared to techniques such as fed-batch culture. They also highlighted the fact that high substrate concentrations in the outer part of the reactor (which have to be maintained to allow adequate supply to the bacterial cells) is a problem with the system. Ogbonna and Märkl (1993) showed that adding glycerol directly to the dialysis chamber of a reactor, in which *Escherichia coli* was growing, could reduce nutrient loss while maintaining the high biomass concentrations mentioned above. Fitting an external dialysis chamber to the reactor resulted in much lower biomass densities being recorded. It was proposed that the changes in parameters, including oxygen tension, in the external dialysis loop were responsible for the lower biomass levels.

Dialysis culture of micro-organisms has many similarities to cultures of microbes growing immobilized in hollow-fibre bioreactors, the mode of cultivation credited with the record for the highest cell density ever reported. A schematic view of this type of bioreactor is shown in *Figure 6*. Past applications of this technique have been reviewed previously (Bunch, 1988; Chang and Furusaki, 1991). Although impressive biomass densities have been reported in these reactors, the individual cell productivities are usually low. This is mainly because mass-transfer limitations prevent all the cells in such reactors from participating in the process involved. It is possible to improve the supply of nutrients to the immobilized cells by changes in the way the bioreactor is constructed or operated (see below). One of the biggest difficulties with these reactors is knowing what factors are limiting productivitity. the first problem is how to determine the biomass levels during operation. This has to be performed by indirect means. Linton *et al.* (1989) used a pressure sensor to follow biomass development. It measured the back pressure as the increasing number of cells restricted the flow of media through the reactor. A calibration graph could be constructed over a fairly wide range of operating cell densities to predict the biomass levels at any stage of the incubation. The results confirmed the fact that in many of these systems the cells grew in a way similar to that observed in dialysis culture. A second problem relates to the physiological state of cells growing in hollow-fibre bioreactors at high cell densities. Linton *et al.* (1987) used a species of lactic acid bacteria, *Streptococcus faecalis*, as a biological sensor. Other bacteria in which the relationship between excreted end-product synthesis and environmental stress has been investigated could be used. Lloyd and Bunch (unpublished observations) using *Escherichia coli* in this way have shown that it is possible to optimize reactor design using this technique for the production of non-growth-associated products, including secondary metabolites and recombinant proteins. In these and other cases it has been shown that, as expected, the supply of oxygen to cells in hollow-fibre bioreactors is most often the major problem. By operating hollow-fibre bioreactors in transverse rather than direct mode (see *Figure 7*), it is possible to improve the delivery of nutrients to cells immobilized on or around the fibres. However, in many cases this can cause fibre rupture to occur (Inloes *et al.*, 1983). Roberson and Kim (1983) designed a dual hollow-fibre system which used one type of fibre for oxygen supply and carbon dioxide removal, the other acting as the immobilization support. Chang and Furusaki (1991) have shown that antibiotics and citric acid can be produced successfully in these reactors and that the productivity can be over 30 times that seen in batch reactors. Other membranes can also be used to replace hollow fibres. Higton *et al.* (1992) compared several types of membrane for the growth of *Streptococcus faecalis*, showing that ceramic membranes, but not glass or metal membranes, gave good results and had the advantage of increased robustness.

A variant of the artificial membrane immobilization method has been reported by Tiltscher and Storr (1993) for high-density growth of the slime mould *Dictyostelium discoideum*, when expressing recombinant human antithrombin III. In this study cells were loaded into inorganic porous aggregates. Up to 15 times the biomass density could be achieved using this method as compared to reactors employing suspended cells. It was shown that although cells growing in these porous structures had a longer

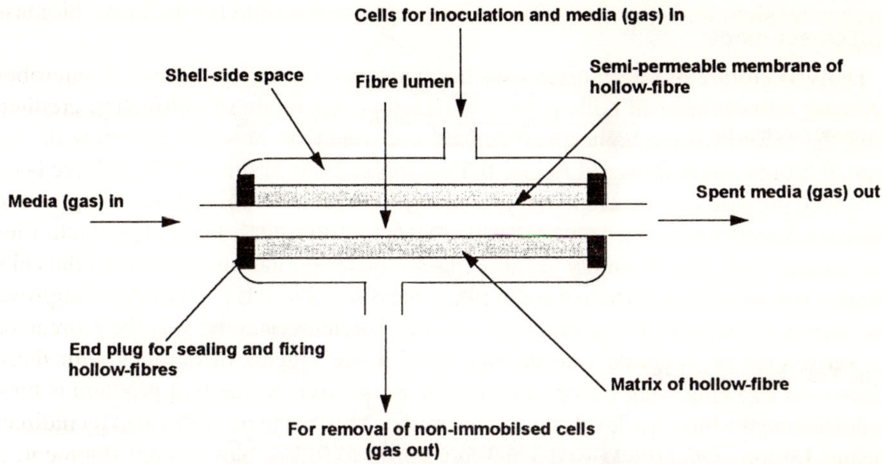

Figure 6. A schematic representation of a hollow-fibre bioreactor for the immobilization of microbial cells. For convenience a single-fibre reactor is illustrated. Although there are reports of such reactors being used for microbial and enzyme immobilization, most reactors will have between 8 and 20 fibres to provide a reasonable immobilization surface plus even media flow at laboratory scale. If anisotropic hollow fibres are used, cells may penetrate into the matrix structure. In most systems cell growth is associated with the surface of the fibre and its surroundings. The semipermeable membrane can be selected to retain all molecules greater than 1 kDa or only retain particles several microns in diameter. Examples of fibre structures and the range of membranes available can be found in Bunch (1988). In dual hollow-fibre reactors a mixture of fibres is employed to allow enhanced gas supply (Chung, Chang and Kim, 1987).

lag phase than suspension batch cultures or cultures with cells absorbed on to the porous aggregates, the cells could be regenerated after storage for use in subsequent experiments. Repeated batch production of the recombinant protein could be achieved using this method. Using a packed bed reactor, operating in an qnalogous way to 100% biomass retention systems, it was possible to obtain continuous production of antithrombin III at levels nearly 20 times higher than that obtained in suspension culture.

A large range of immobilization techniques has been used to enhance biomass densities in reactors. The use of a popular method, gel entrapment, usually leads to a restricted proportion of the biomass being active in a given process. Once again, oxygen limitation is a recurring problem. Although techniques such as co-immobilization of different cell types or the use of agents that can liberate oxygen at the necessary sites have been attempted (Omar, 1993), these systems are likely to have restricted uses in biotechnology. This is particularly true because in microbial cells making natural or recombinant products any loss of competitive advantage between producing and non-producing mutants (or strains that have lost a plasmid containing a recombinant gene) will contribute to the reduction of the cultures' production stability. Using engineered clones of *Escherichia coli*, Varma and Palsson (1994) have shown that careful maintenance of the nutritional environment, not only with respect to the oxygen supply, can stabilize productivity during screening and fermentation procedures. This paper is fascinating as it emphasizes the way that metabolism must be integrated with environment to achieve optimum results. It is an encouraging fact that more and more emphasis is being placed on gaining a better understanding of how microbial metabolism functions under different growth condi-

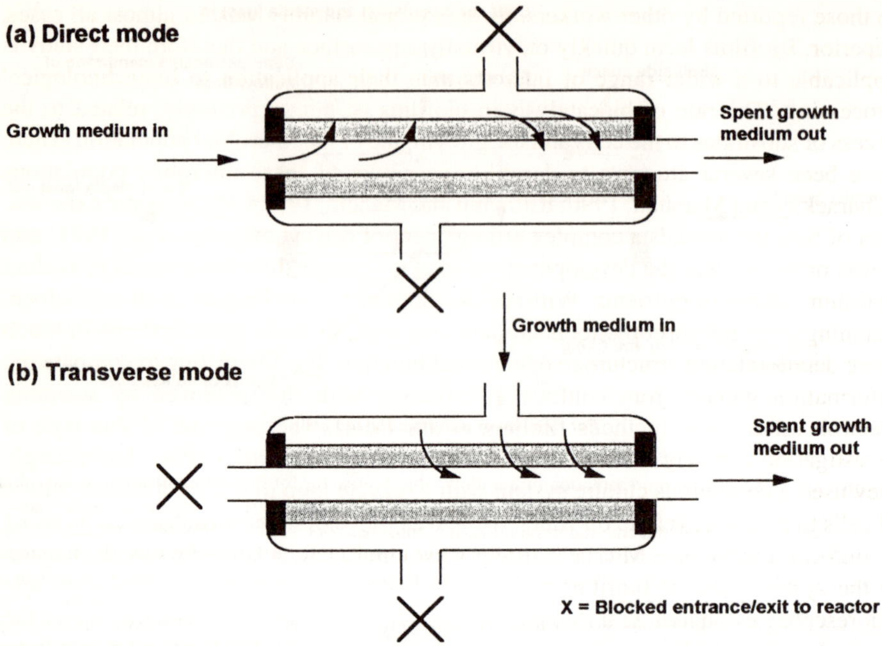

Figure 7. Operational modes for hollow-fibre bioreactor operation. In both operation modes cells are inoculated into the shell side space where they grow on or around the fibres. When using the direct mode, growth medium is supplied via the fibre lumen, where it diffuses to the cells. Products diffuse out into the fibre lumen where they are removed in the outflow from the reactor. Several reports show that this mode of operation can lead to relatively fast fibre breakdown in comparison to the transverse mode systems (Bunch, 1988). Gasses can be supplied or removed via the shell side space if required. In transverse mode the growth medium is passed directly to the cells. The spent medium and products are collected after filtration through the semipermeable part of the hollow-fibre structure. Gasses can be supplied via either the fibre lumen or with the growth medium. To avoid operational problems other solutions to gas supply have been found (see text).

tions, coupled with defining the limits within which it is possible to engineer rationally new metabolic functions (Majewski and Domach, 1990; Bailey, 1991; Stephanopoulos and Vallino, 1991; Varma and Palsson, 1994). In the meantime one of the most interesting ways of holding microbial cells in a form which reduces mass-transfer problems, biofilms, is worthy of some mention as a method of avoiding loss of biocatalytic potential in bioreactors. Tyagi, Gupta and Chand (1992) compared the ethanol productivity of cells immobilized in alginate beads and on the surface of raw and activated bagasse (the latter positively charged with an amino hexyl group in an attempt to gain better establishment of cells). Oxygen and carbon dioxide uptake rates were highest in the 'bagasse'-based reactors where high cell density biofilms were generated. More cells could be loaded on to the activated bagasse support than was the case with the raw material. This compared well with other types of support, such as ion-exchange resins and ceramics. Cell retention of 0.410 g cells/g carrier was reported for the activated bagasse. Loss of cells from the biofilms present in raw bagasse was higher than in the activated bagasse systems. In both cases when operating in a packed bed reactor the flow rates used influenced the rate of cell loss from the system. Although the ethanol productivity of these reactors was comparable to those reported by other workers, its operational stability was, in almost all cases, superior. Biofilms form quickly on virtually any surface and therefore their study is applicable to a wider range of interests than their application to biotechnological processing. The rate of biocatalysis in biofilms is, not surprisingly, related to the access of substrates to the cells and the growth rate of the microbial population. There have been several attempts to describe the nature of these microbial populations (Characklis and Marshall, 1990; Rittmann and Manem, 1992). Microscopic examination of biofilms reveals a complex arrangement of cells (Lawrence *et al.*, 1991) and it was proposed that the development of such systems is directed towards providing optimum access of nutrients. With the development of techniques such as confocal scanning laser microscopy, tools are now available to study these systems in much more detail, relating structure to operational function. It is interesting to compare the information gained from confocal microscopy with that achieved by scanning electron microscopy methods (Zellner *et al.*, 1994). The outcome of this type of investigation is nicely illustrated by the work of de Beer *et al.* (1994). Interestingly they used a continuous culture system with a recycle loop to deliver a mixed culture of cells to a glass coverslip on which the biofilm development was observed by using a confocal microscope. Microelectrodes were employed to study the oxygen tension in the system. Mature biofilms were developed over a period of several days and fluorescence exclusion with fluorescein was used to enhance the visualization of the internal structures. Penetration of the biofilm was investigated by using fluorescent microspheres of 0.87 μm diameter. Cell clusters 80 μm thick and 100–300 μm in diameter were a key feature in these biofilms, separated from each other by open spaces 50–100 μm wide. The clusters covered approximately 50% of the slide surface. Subsequent development of the biofilm increased the thickness of the cell clusters but not the percentage of the surface covered. Penetrability studies showed that the larger clusters were not closely attached to the substratum. At the point when observation using confocal microscopy becomes severely limited (600 μm), it was shown that an interconnecting channel system had developed in the biofilm. In an elegant study these authors showed that in the channels running from the surface to the substratum the oxygen concentration was always much higher than in the cell clumps and was maintained at the level seen in the supply medium until close to the coverslip surface. Thus it shows that there is potential for 'self-design' of high cell density biofilms to limit mass-transfer problems. The use of this type of microbial culture therefore has the potential to overcome many of the disadvantages of standard immobilization procedures while allowing high biomass densities to be employed.

Weuster-Botz (1993) described the application of biofilm technology to the continuous production of ethanol using *Zymomonas mobilis* growing on porous glass beads in a fluidized bed reactor. The productivity of this system is amongst the highest so far reported, and with the addition of a second reactor stage high productivity with almost total substrate conversion is possible. Demirci, Pometto and Johnson (1993a) have used a biofilm-based system for another product employed as a chemical feedstock, with the same objective of reducing production costs to allow greater industrial use; a mixed culture of *Streptomyces viridosporus* and *Lactobacillus casei* was employed.

Polypropylene chips were used as the support and were blended with a range of agricultural materials, employed as carbon and nitrogen sources on the support surfaces. *Streptomyces* was used to generate an extensive biofilm in which *Lactobacillus* was immobilized. This process gave a three- to five-fold increase in production rates when compared to suspension culture, and using a mixed culture system improved production over reactors operating with pure culture. As with all biofilm systems, some microbial cells were found in the product stream. In a subsequent paper by Demirci, Pometto and Johnson (1993b) alternative mixed cultures were constructed using a variety of biofilm generators and lactic acid producers. Once again the best production was observed with the mixed cultures when compared to pure cultures of lactic acid producers.

Product removal in biofilms is another feature of these systems that needs careful investigation. It may be envisaged that in biotechnology, in addition to the treatment of environmental wastes, removal of products that inhibit biofilm activity will be of great importance. There have been several attempts to model the consequences of toxic substrates on biofilm activity (van Ede, Bollen and Beenackers, 1993). Continued effort towards toxic product removal is critical for full exploitation of these systems. An extreme example of the sort of possible application of biofilm technology is the biotransformation of waste sulphuric acid by sulphate-reducing bacteria (Stucki, Hanselmann and Hurzeler, 1993). These authors illustrate how such a process can be integrated with physical and chemical systems for the bioremediation of wastes, a rapidly growing concept in environmental technology. The isolation of sulphate-reducing bacteria capable of growth on cheap and abundant waste products is relatively recent (Widdel and Pfennig, 1977). Processes using fixed biofilms show that the key parameter to monitor and control is the production of hydrogen sulphide, which is inhibitory to even those bacteria that synthesize it as a natural product. Using acetate as a co-substrate the process can be summarized as follows:

$$SO_4^{-2} + CH_3COO^- + 3H^+ \rightarrow H_2S + 2CO_2 + 2H_2O$$

Thus the products, as gases, should be easily removed, achieved in this case by incorporating a gas stripping column in the recycle loop. This example illustrates how a knowledge of biofilm development and operation allows clear guidance for the overal design of the bioreactor.

Some reference should be made to solid state fermentations. In some respects these can be considered as the ultimate type of biofilm reactor, employing localized high cell densities. It is a specialized, but important, area of biotechnology. A relatively narrow range of substrates can be used for this type of microbial cultivation. The substrates become heterologous during the fermentation, and excessive heat build-up can cause problems and, once again, monitoring what is happening is extremely difficult (Sargantanis *et al.*, 1993). The latter authors showed how very high biomass densities can occur in these systems, requiring careful process design to avoid premature loss of reactor operation. Scale-up strategies have been discussed by Lonsane *et al.* (1992). Using sugar-cane bagasse impregnated with growth nutrients, Soccol *et al.* (1994) have shown that this type of reactor can be used as a 'low-tech' method of producing lactic acid. Their process used the fungus *Rhizopos oryzae*. An alternative system using bacteria has been described by Xavier and Lonsane (1994), while Sarangbin, Kirimura and Usami (1993) described a process for the manufacture of citric acid using the fungus *Aspergillus niger*. Both accounts again highlight the fact that although the technology is simple, the design of an efficient process needs careful consideration.

The future of high cell density cultivation of microbial cells

In many examples, investigators have been driven by curiosity about how much biomass can be developed in a given type of reactor. There have been some surprises. Generation of over 500 g dry weight biomass l–1 in dual hollow-fibre reactors (Chung, Chang and Kim, 1987) is much higher than would be predicted using the available data on microbial cell composition. What is more surprising is that these cultures were still productive. With more information about the way microbial cultures establish in these systems, some of the productivities observed in high cell density culture become more understandable. A study of biofilms using confocal laser microscopy is a good illustration of how mass-transfer problems, for example, can be reduced by the way colonies develop.

The operation of few microbial processes has been compared in a wide range of different bioreactors. Chang and Furusaki (1991) have performed such an exercise for citric acid production by micro-organisms; Bunch (1988) for lactic acid synthesis. Using our examples, a limited comparison is possible for the production of ethanol. However, even for this small range of microbial products it is not surprising to learn that the generation of high biomass densities is not the only way of further improving a fermentation process. Some of the most efficient ways of synthesizing lactic acid use low cell density continuous culture, although the highest reactor volumetric productivity is obtained in a hollow-fibre system. Whether the benefits associated with high-density cultivation are worthwhile can only be assessed in the light of factors such as substrate costs, substrate conversion rates, fermentation run-time, reactor stability and process integration with working practices.

Given the fact that enhancing biomass density will still remain part of the approach to process intensification in the future, it is encouraging to see continued development of materials for use as reactors or as separation devices in reactors. As such systems become more robust their application to cultivation problems is likely to grow rapidly. For example, cell recycle/cell retention systems are likely to find extensive use for growing micro-organisms which generate only low biomass densities in batch and chemostat culture. Several examples have been cited which show how reactor selection can lead to useful cell yields of micro-organisms, such as protozoa and extremophiles. New techniques for improving cell separation continue to appear. An ingenious example is that used by Trampler *et al.* (1994) to increase the cell density of hybridoma cultures. Using an acoustic flow-through device they caused aggregation of cells to take place, allowing separation by sedimentation to occur. The result was a 70-fold increase in productivity with 99% cell retention. It is possible therefore to predict that the development of new types of filtration devices will enable further improvements to high cell density systems.

The diversity of reactor types shows that there is great potential for applying high cell density technology. The use of solid-state fermentations shows that relatively low-tech processes can be attractive for the generation of microbial products. Ohno, Ano and Shoda (1993) have shown that such systems can be used with aerobes as well as anaerobes or facultative anaerobes. They show that the productivity of *Bacillus subtilis* cultures making the antifungal peptide antibiotic, iturin, growing on soybean curd, can be enhanced 6–8 times over that achieved in cell suspension reactors. Nevertheless, it is inevitable that the best increases in productivity will use high cell density culture in combination with other approaches for improving process efficiency. It is not possible to review all these other methods comprehensively here. Nevertheless, two will be mentioned because they reflect the way that process optimization is being approached.

At the beginning of this chapter the complexity of microbial cells was highlighted. We are still a long way from being able to comprehend fully the intricate way that the metabolic processes of micro-organisms are integrated, especially in the vast range of physiological environments provided in bioreactors. One approach is that taken by Varma, Boesch and Palsson (1993). They remind us that only in the case of the red blood cell is it possible to characterize the dynamic capabilities of a cell. Using *Escherichia coli* they performed a flux balance analysis to calculate the production capability of this bacterium in respect of selected amino acids and related products. Their paper reveals how much further this approach has to go before it can be generally applied to microbial fermentations, but gives an insight as to how increasing information on microbial metabolism can be used to determine rationally the capabilities of a given reactor system. This will allow guidelines to be developed to evaluate the potential of metabolic engineering to overcome productivity limitations relating to a single enzyme activity (or lack of activity). It is likely in the near future that larger changes in the natural metabolic capability of a microbial cell could be made.

The use of artificial neural networks is the second approach to process design. In a lucid account of this technique Zhang *et al.* (1994) examined its application to *Bacillus thuringiensis* fermentations. These authors identified three major problems in the development of reliable automated control systems for microbial fermentations: lack of reliable mathematical models to describe cell growth and product formation, a general lack of appropriate on-line sensors and the complicated regulatory systems found within microbial cells. They stated that the use of neural network analysis is well matched to processes where there is much data but poor system knowledge. The best features of this approach are that: it allows deductions without a precise mathematical model of the process, works with incomplete information and it can handle large amounts of information quickly through parallel processing. A neural network solves problems by adapting ('training' in the terminology of this subject) to the input data; in this example, five parameters that affect culture optical density. They showed by selecting one of a variety of different networks that they could model accurately optical density changes in the culture during its development. This technique takes advantage of computer power that is now available, and has been applied to many problems.

All these developments in the drive towards more productive and efficient microbial product synthesis make it possible to predict that limitations of existing cultivation methods will be overcome, to the mutual benefit of those who are seeking to understand more about how micro-organisms function and those who desire to design a commercially viable process.

References

ALBERGHINA, L., PORRO, D., MARTEGANI, E. AND RANZI, B.M. (1991). Efficient production of recombinant DNA proteins in *Saccharomyces cerevisiae* by controlled high-cell density fermentation. *Biotechnology and Applied Biochemistry* **14**, 81–92.

BABU, P.S.R. AND PANDA, T. (1991). Effects of recycling fermentation broth for the production of penicillin amidase. *Process Biochemistry* **26**, 7–14.

BAILEY, J.E. (1991). Toward a science of metabolic engineering. *Science* **252**, 1668–1675.

BARRY, J.A. (1982). Alcohol production from cheese whey. *Dairy Industry International* **47**, 19–22.

BEETON, S., MILLWARD, H.R., BELLHOUSE, B.J., NICHOLSON, A.M., JENKINS, N.J. AND KNOWLES, C.J. (1991). Gas transfer characteristics of a novel membrane bioreactor. *Biotechnology and Bioengineering* **38**, 1233–1238.

BEETON, S., BELLHOUSE, B.J., KNOWLES, C.J., MILLWARD, H.R., NICHOLSON, A.M. AND WYATT, J.R. (1994). A novel membrane bioreactor for microbial growth. *Applied Microbiology and Biotechnology* **40**, 812–817.

BELLHOUSE, B.J., BELLHOUSE, F.H., CURL, C.M., MACMILLAN, T.I., GUNNING, A.J., SPRATT, E.H., MACMURRAY, S.B. AND NELEMS, J.M. (1973). A high efficiency membrane oxygenator and pulsatile pumping system and its application to animal trials. *Transactions of the American Society of ARtificial Internal Organs* **34**, 747–754.

BEPPU, T. (1986). Breeding of amino acid-producing strains of coryneform bacteria by recombinant DNA technology and protoplast fusion. In *Biotechnology of Amino Acid Production* (K. Aida, I. Chibata, K. Nakayama, K. Takinami and H. Yamada, Eds), pp. 29–33. Kodansha-Elsevier, Tokyo.

BERG, A.-C., HOLST, O. AND MATTIASSON, B. (1989). Continuous culture with complete cell-recycle cultivation of *Pseudomonas cepacia* ATCC 29351 on salicylate for production of salicylate hydroxylase. *Applied Microbiology and Biotechnology* **30**, 1–4.

BLONDEAU, K., BOUTUR, O., BOZE, H., JUNG, G., MOULIN, G. AND GALZY, P. (1994). Development of high-cell-density fermentation for heterologous interleukin 1β production in *Kluyveromyces lactis* controlled by the PH05 promoter. *Applied Microbiology and Biotechnology* **41**, 324–329.

BUNCH, A.W. (1988). The uses and future potential of microbial hollow-fibre bioreactors. *Journal of Microbiological Methods* **8**, 103–119.

CHANG, H.N. AND FURUSAKI, S. (1991). Membrane bioreactors: present and prospects. *Advances in Biochemical Engineering/Biotechnology* **44**, 28–64.

CHANG, H.W., LEE, W.G. AND KIM, B.S. (1993). Cell retention culture with an internal filter module. Continuous ethanol fermentation. *Biotechnology and Bioengineering* **41**, 677–681.

CHANG, L.-L., HWANG, L.Y., HWANG, C.-F. AND MOU, D.-G. (1991). Study of high density *Escherichia coli* fermentation for production of porcine somatotropin protein. *Annals of the New York Academy of Sciences* **646**, 259–272.

CHARACKLIS, W.G. AND MARSHALL, K.C. (1990). Biofilms: a basis for an interdisciplinary approach. In *Biofilm* (W.G. Characklis and K.C. Marshall, Eds), pp. 3–15. Wiley, New York.

CHERYAN, M. AND MEHAIA, M.A. (1983). A high-performance membrane bioreactor for continuous fermentation of lactose to ethanol. *Biotechnology Letters* **8**, 519–524.

CHUNG, B.H., CHANG, H.N. AND KIM, I.H. (1987). Rifamycin production by *Nocardia mediterrenei* in a dual hollow-fibre bioreactor. *Enzyme and Microbial Technology* **9**, 345–349.

COOPER, S. (1991). *Bacterial Growth and Division: Biochemistry and Regulation of Prokaryotic and Eukaryotic Division Cycles*. Academic Press, New York.

CUBARSI, R. AND VILLAVERDE, A. (1993). Conditions for a continuous production of transient microbial products in a two-stage fermentation system. *Biotechnology Letters* **15**, 827–832.

DE BEER, D., STOODLEY, P., ROE, F. AND LEWANCHOWSKI, Z. (1994). Effects of biofilm structures on oxygen distribution and mass transport. *Biotechnology and Bioengineering* **43**, 1131–1138.

DEMIRCI, A., POMETTO, A.L. AND JOHNSON, K.E. (1993a). Evaluation of biofilm reactor solid support for mixed-culture lactic acid production. *Applied Microbiology and Biotechnology* **38**, 728–733.

DEMIRCI, A., POMETTO, A.L. AND JOHNSON, K.E. (1993b). Lactic acid production in a mixed culture biofilm reactor. *Applied and Environmental Microbiology* **59**, 203–207.

DE QUEIROZ, J.H., URIBELARREA, J.L. AND PAREILLEUX, A. (1991). High cell density cultures of *Schizosaccharomyces pombe* in a cell-recycle reactor. Growth kinetics and metabolic status. *Applied Biochemistry and Biotechnology* **30**, 283–295.

GUPTA, S.K. AND CHAND, S. (1994). Bioconversion of sugars to ethanol in a chemostat employing *S. cerevisiae* – dynamic response to perturbations in process parameters. *Process Biochemistry* **29**, 343–354.

HARRISON, D.E.F. AND LOVELISS, J.E. (1971). Effect of growth conditions on respiratory activity and growth efficiency in facultative anaerobes grown in chemostat culture. *Journal of General Microbiology* **68**, 35–43.

HEATH, C.A. AND BELFORT, G. (1992). Synthetic membranes in biotechnology: realities and possibilities. *Advances in Biochemical Engineering* **47**, 46–85.

HERBERT, D. (1961). The chemical composition of microorganisms as a function of their environment. *Society of General Microbiology Symposia* **10**, 391–416.

HIGTON, G., BALL, J., BUNCH, A.W. AND KNOWLES, C.J. (1992). A method for growth and immobilisation of *Streptococcus faecalis* var. *zymogenes* in bioreactors containing hollow-fibre, ceramic, metallic or glass-fibre membranes. *Journal of Microbiological Methods* **16**, 1–11.

HOWELL, M.J. (1981). High-grade ethanol from whey. *Chemical Industry (N2)* **15**, 4–6.

HUANG, C.-T., PERELTI, S.W. AND BRYERS, J.D. (1993). Plasmid retention and gene expression in suspended and biofilm cultures of recombinant *Escherichia coli* DH5α(pMJR1750). *Biotechnology and Bioengineering* **41**, 211–220.

IBBA, M., BONARIUS, D., KUHLA, J., SMITH, A. AND KUENZI, M. (1993). Mode of cultivation is critical for the optimal expression of recombinant hirudin by *Saccharomyces cerevisiae*. *Biotechnology Letters* **15**, 667–672.

INLOES, D.S., SMITH, W.J., TAYLOR, D.P., COHEN, S.N., MICHAELS, A.S. AND ROBERTSON, C.R. (1983). Hollow-fibre membrane bioreactors using immobilised *Escherichia coli* for protein synthesis. *Biotechnology and Bioengineering* **25**, 2653–2681.

ISHIZAKI, A., TAKASAKI, S. AND FURUKA, Y. (1993). Cell-recycled fermentation of glutamate using a novel cross-flow filtration system with constant air supply. *Journal of Fermentation and Bioengineering* **76**, 316–320.

KAMIN, H., WHITE-STEVENS, R.H. AND PRESSWOOD, R.P. (1978). Salicylate hydroxylase. *Methods in Enzymology* **53**, 527–543.

KANG, B.C., LEE, S.Y. AND CHANG, H.N. (1992). Enhanced production of spores of *Bacillus thuringiensis* by fed-batch culture. *Biotechnology Letters* **14**, 721–726.

KANG, B.C., LEE, S.Y. AND CHANG, H.N. (1993). Production of *Bacillus thuringiensis* spores in total cell retention culture and two-stage continuous culture using an internal ceramic filter system. *Biotechnology and Bioengineering* **42**, 1107–1112.

KIY, T. AND TIEDTKE, A. (1992a). Continuous high cell density fermentation of the ciliated protozoon *Tetrahymena*: I. A perfused bioreactor. *Applied Microbiology and Biotechnology* **38**, 141–146.

KIY, T. AND TIEDTKE, A. (1992b). Mass cultivation of *Tetrahymena thermophila* yielding high cell densities and short generation times. *Applied Microbiology and Biotechnology* **37**, 576–579.

KNORRE, W.A., DECKWER, W.-D., KORZ, D., POHL, H.-D., RIESENBERG, D., ROSS, A., SANDERS, E. AND SCHULZ, V. (1991). High cell density fermentation of recombinant *Escherichia coli* with computer-controlled optimal growth rate. *Annals of the New York Academy of Sciences* **646**, 301–321.

KUMAR, P.K.R., SINGH, A. AND SCHUGERL, K. (1991). Fed-batch culture for the direct conversion of cellulosic substrates to acetic acid/ethanol by *Fusarium oxysporum*. *Process Biochemistry* **26**, 209–216.

LAPLACE, J.M., DELGENES, J.P., MOLELTA, R. AND NAVARRO, J.M. (1993). Ethanol production from glucose and xylose by separated and co-culture processes using high-cell density systems. *Process Biochemistry* **28**, 519–525.

LAWRENCE, J.R., KORBER, D.-R., HOYLE, B.D., COSTERTON, J.W. AND CALDWELL, D.E. (1991). Optical sectioning of microbial biofilms. *Journal of Bacteriology* **273**, 6558–6567.

LINTON, E.A., KNOWLES, C.J., BUNCH, A.W. AND HIGTON, G. (1987). The biotechnological potential of microbial hollow-fibre bioreactors. In *Bioreactors and Biotransformations* (G.W. Moody and P.B. Baker, Eds), pp. 299–309. Elsevier Applied Science, Amsterdam.

LINTON, E.A., HIGTON, G., KNOWLES, C.J. AND BUNCH, A.W. (1989). Monitoring microbial growth in a hollow-fibre reactor using an electronic pressure sensor. *Enzyme and Microbial Technology* **11**, 257–320.

LONSANE, B.K., SAUCEDO-CASTANEDA, G., RAIMBAULT, M., ROUSSOS, S., VINIEGRA-GONZALEZ, G., GHILDYAL, N.P., RAMAKRISHNA, M. AND KRISHNAIAH, M.M. (1992). Scale-up strategies for solid state fermentation systems. *Process Biochemistry* **27**, 259–273.

MAJEWSKI,. R.A. AND DOMACH, M.M. (1990). Simple constrained-optimization view of acetate overflow in *Escherichia coli. Biotechnology and Bioengineering* **35**, 732–738.

MÄRKL, H., ZENNECK, C., DUBACH, C.H. AND OGBONNA, J.C. (1993). Cultivation of *Escherichia coli* to high cell densities in a dialysis reactor. *Applied Microbiology and Biotechnology* **39**, 48–52.

MAWSON, A.J. (1987). Ethanol production from whey in New Zealand. *Australian Journal of Biotechnology* **1**, 64–66.

MEHAIA, M.A. AND CHERYAN, M. (1990). Ethanol from hydrolysed whey permeate using *Saccharomyces cerevisiae* in a membrane recycle bioreactor. *Bioprocess Engineering* **5**, 57–61.

MILLWARD, H.R., BELLHOUSE, B.J., NICHOLSON, A.M., BEETON, S., JENKINS, N. AND KNOWLES, C.J. (1994). Mammalian cell damage in a novel membrane bioreactor. *Biotechnology and Bioengineering* **43**, 899–906.

MINODA, Y. (1986). Raw materials for amino acid fermentation. In *Biotechnology of Amino Acid Production* (K. Aida, I. Chibata, K. Nakayama, K. Takinami and H. Yamada, Eds), pp. 51–66. Kodansha-Elsevier, Tokyo.

NAKANO, K., MATSUMARA, M. AND KATAOKA, H. (1993). Application of a rotating ceramic membrane to dense cell culture. *Journal of Fermentation and Bioengineering* **76**, 49–54.

NEIDHARDT, F.C., INGRAHAM, J.L. AND SCHAECHTER, M. (1990). *Physiology of the Bacterial Cell: A Molecular Approach.* Sinawer Associates, Massachusetts.

NURMIANTO, L., GREENFIELD, P.F. AND LEE, P.L. (1994). Optimal feeding policy for recombinant protein production via the yeast *Saccharomyces cerevisiae* in fed-batch culture. *Process Biochemistry* **29**, 55–68.

OGBONNA, J.C. AND MÄRKL, H. (1993). Nutrient-split feeding strategy for dialysis cultivation of *Escherichia coli. Biotechnology and Bioengineering* **41**, 1092–1100.

OHARA, H., HIYAMA, K. AND YOSHIDA, T. (1992). Kinetics of growth and lactic acid production in continuous culture and batch culture. *Applied Microbiology and Biotechnology* **37**, 544–548.

OHNO, A., ANO, T. AND SHODA, M. (1993). Production of the antifungal peptide antibiotic iturin by *Bacillus subtilis* NB22 in solid state fermentation. *Journal of Fermentation and Bioengineering* **75**, 23–27.

OMAR, S.H. (1993). Oxygen diffusion through gels employed for immobilization. *Applied Microbiology and Biotechnology* **40**, 173–181.

PARK, Y.S., ITIDA, M., OHTA, N. AND OKABE, M. (1994). Itaconic acid production using an air-lift bioreactor in repeated batch culture of *Aspergillus terreus. Journal of Fermentation and Bioengineering* **77**, 329–331.

PATEL, P.N., MEHAIA, M.A. AND CHERYAN, M. (1987). Cross-flow membrane filtration of yeast suspensions. *Journal of Biotechnology* **5**, 1–16.

PIRT, S.J. (1975). *Principles of Microbe and Cell Cultivation*, pp. 42–57. Blackwell Scientific Publications, Oxford.

POSTGATE, J.R. AND HUNTER, J.R. (1962). Survival of starved bacteria. *Journal of General Microbiology* **29**, 233–263.

PROKOP, A. AND BAJPAI, R.K. (1991). Recombinan DNA Technology I. *Annals of the New York Academy of Sciences* **646**, whole volume.

RAVEN, N., LADWA, N., COSSAR, D. AND SHARP, R. (1992). Continuous culture of the hyperthermophilic archaeum *Pyrococcus furiosus*. *Applied Microbiology and Biotechnology* **38**, 263–267.

RIESMEIR, B., KRONER, K.H. AND KULA, M.-R. (1989). Tangential filtration of microbial suspensions: filtration resistances and model development. *Journal of Biotechnology* **12**, 153–171.

RIESMEIR, B., KRONER, K.H. AND KULA, M.-R. (1990). Harvest of microbial suspensions by microfiltration. *Desatination* **77**, 219–233.

RITTMAN, B.E. AND MANEM, J.A. (1992). Development and experimental evaluation of a steady-state multispecies biofilm model. *Biotechnology and Bioengineering* **39**, 914–922.

ROBERSON, C.R. AND KIM, I.H. (1983). Dual aerobic hollow-fibre bioreactor for cultivation of *Streptomyces aureofaciens*. *Biotechnology and Bioengineering* **37**, 1012–1020.

SARANGBIN, S., KIRIMURA, K. AND USAMI, S. (1993). Citric acid production from cellobiose by 2-deoxyglucose-resistant mutant strains of *Aspergillus niger* in semi-solid culture. *Applied Microbiology and Biotechnology* **40**, 206–210.

SARGANTANIS, J., KARIM, M.N., MURPHY, V.G. AND RYOO, D. (1993). Effect of operating conditions on solid substrate fermentation. *Biotechnology and Bioengineering* **42**, 149–158.

SCHÜZ, T.C., FIEDLER, H.-P. AND ZAHNER, H. (1993). Optimised nikkomycin production by fed-batch and continuous fermentation. *Applied Microbiology and Biotechnology* **39**, 433–437.

SOCCOL, C.R., MARIN, B., RAIMBAULT, M. AND LEBEAULT, J.-M. (1994). Potential of solid state fermentation for production of L-(+)-lactic acid by *Rhizopus oryzae*. *Applied Microbiology and Biotechnology* **41**, 286–290.

STEPHANOPOULOS, G. AND VALLINO, J.J. (1991). Network rigidity and metabolic engineering in metabolite over production. *Science* **252**, 1675–1681.

STUCKI, G., HANSELMANN, K.W. AND HURZELER, R.A. (1993). Biological sulphuric acid transformation: Reactor design and process optimization. *Biotechnology and Bioengineering* **41**, 303–315.

SWARTZ, R.W. (1985). Penicillins. In *Comprehensive Biotechnology* (M. Moo-Young, Ed.), volume 3, pp. 8–43. Pergamon Press, Oxford.

TANAKA, T., KAMIMURA, R., ITOH, K. AND NAKANISHI, K. (1993). Factors affecting the performance of cross flow filtration of yeast cell suspension. *Biotechnology and Bioengineering* **41**, 617–624.

TEMPEST, D.W., NEIJSSEL, O.M. AND ZEWENBOON, W. (1983). Properties and performance of microorganisms in labaratory culture: their relevance to growth in natural ecosystems. In *Microbes in their Natural Environment* Society for General Microbiology Symposium. Cambridge University Press **34**, 119–153.

TILTSCHER, H. AND STORR, M. (1993). Immobilisation of the slime mould *Dictoyostelium discoideum* for the continuous production of recombinant human antithrombin III. *Applied Microbiology and Biotechnology* **40**, 246–250.

TIN, C.S.F. AND MAWSON, A.J. (1993). Ethanol production from whey in a membrane recycle bioreactor. *Process Biochemistry* **28**, 217–221.

TOCAIJ, A., HOF, A., HAGANDER, P. AND HOLST, O. (1993). Fed batch cultivation of *Pseudomonas cepacia* with on-line control of the toxic substrate salicylate. *Applied Microbiology and Biotechnology* **38**, 463–466.

TRAMPLER, F., SONDERHOFF, S.A., PUI, P.W.S., KILBURN, D.G. AND PINET, J.M. (1994). Acoustic cell filter for high density perfusion culture of hybridoma cells. *Biotechnology* **12**, 281–284.

TSENG, C.-P. AND MONTVILLE, T.J. (1992). Enzymatic regulation of glucose catabolism by *Lactobacillus plantarum* in an aerobic chemostat. *Biotechnological Progress* **8**, 126–131.

TYAGI, R.D., GUPTA, S.K. AND CHAND, S. (1992). Process engineering studies on continuous ethanol production by immobilised *S. cerevisiae*. *Process Biochemistry* **27**, 23–32.

URIBELARREA, J.L., DE QUEIROZ, H., GOMA, G. AND PAREILLEUX, A. (1993). Carbon and energy balances in cell-recycle cultures of *Schizosaccharomyces pombe*. *Biotechnology and Bioengineering* **42**, 729–736.

VAN EDE, C.J., BOLLEN, A.M. AND BEENACKERS, A.A.C.M. (1993). Analytical effectiveness calculations concerning the degradation of an inhibitive substrate by a steady state biofilm. *Biotechnology and Bioengineering* **42**, 267–278.

VAN LIERE, L. (1979). On *Oscillatoria agardhii* gomont; experimental ecology and physiology of a nuisance bloom-forming cyanobacterium. Ph.D. thesis, University of Amsterdam, The Netherlands.

VAN NIEL, E.W., BRABER, K.J., ROBERTSON, L.A. AND KWENEN, J.G. (1992). Heterotrophic nitrification and aerobic denitrification in *Alcaligenes faecalis* strain Two. *Antonie van Leeuwenhoek* **62**, 231–237.

VAN WALSUM, G.P. AND COOPER, P.G. (1993). Self-cycling fermentation in a stirred tank reactor. *Biotechnology and Bioengineering* **42**, 1175–1180.

VARMA, A. AND PALSSON, B.O. (1994). Predictions for oxygen supply control to enhance population stability of engineered production strains. *Biotechnology and Bioengineering* **43**, 275–285.

VARMA, A. BOESCH, B.W. AND PALSSON, B.O. (1993). Biochemical production capabilities of *Escherichia coli. Biotechnology and Bioengineering* **42**, 59–73.

VERDWYN, C., POSTMA, E., SCHEFFERS, W.A. AND VAN DIYKEN, J.R. (1992). Effect of benzoic acid on metabolic fluxes in yeasts: a continuous culture study on the regulation of respiration and alcoholic fermentation. *Yeast* **8**, 501–517.

VICKROY, T.B. (1985). Lactic acid. In *Comprehensive Biotechnology* (M. Moo-Young, Ed.), volume 3, pp. 8–43. Pergamon Press, Oxford.

WARREN, R.K., MACDONALD, D.G. AND HILL, G.A. (1991). Cross-flow microfiltration of *Saccharomyces cerevisiae. Process Biochemistry* **26**, 337–342.

WEUSTER-BOTZ, D. (1993). Continuous ethanol production by *Zymomonas mobilis* in a fluidized bed reactor. Part I. Kinetic studies of immobilisation in macroporous glass beads. *Applied Microbiology and Biotechnology* **39**, 679–684.

WIDDEL, F. AND PFENNIG, W. (1977). A new anaerobic, sporing, acetate-oxidising, sulphate-reducing bacterium, *Desulphotomaculum (emend) acetoxidans. Archives of Microbiology* **112**, 119–122.

XAVIER, S. AND LONSANE, B.K. (1994). Sugar-cane press mulch as a novel and inexpensive substrate for production of lactic acid in a solid-state fermentation system. *Applied Microbiology and Biotechnology* **41**, 291–295.

ZELLNER, G., DIEKMANN, H., AUSTERMANN-HAUN, U. AND SEYFRIED, C.F. (1994). Scanning electron microscopy of biofilm development in anaerobic fixed-bed reactors: influence of the inoculum. *Biotechnology Letters* **16**, 315–320.

ZHANG, D.X. AND CHERYAN, M. (1994). Starch lactic acid in a continuous membrane bioreactor. *Process Biochemistry* **29**, 145–150.

ZHANG, Q., REID, J.F., LITCHFIELD, J.B., REN, J. AND CHANG, S.-W. (1994). A prototype neural network supervised control system for *Bacillus thuringiensis* fermentations. *Biotechnology and Bioengineering* **43**, 483–489.

ZHANG, S., HANDA-CORRIGAN, A. AND SPIER, R.E. (1993). A comparison of oxygenation methods for high-density perfusion cultures of animal cells. *Biotechnology and Bioengineering* **41**, 685–692.

Index

Biotechnology & Genetic Engineering Reviews VOLUME 4 (September 1986); ISBN 0 946707 07 3; ISSN 0264–8725; vii + 436 pp; 234 x 156 mm; illustrated; comprehensive index; hardback.

CONTENTS

Biotechnology & Genetic Engineering Reviews VOLUME 5 (September 1987); ISBN 0 946707 10 3; ISSN 0264–8725; x + 419 pp; 234 x 156 mm; illustrated; comprehensive index; hardback. Retail Price £80.00.

CONTENTS

Biotechnology & Genetic Engineering Reviews VOLUME 6 (September 1988); ISBN 0 946707 12 X; ISSN 0264–8725; x + 457 pp; 234 x 156 mm; illustrated; comprehensive index; hardback.

CONTENTS

Biotechnology & Genetic Engineering Reviews VOLUME 7 (December 1989); ISBN 0 946707 22 7; ISSN 0264–8725; xii + 324 pp; 234 x 156 mm; illustrated; comprehensive index; hardback.

CONTENTS

Biotechnology & Genetic Engineering Reviews VOLUME 8 (November 1990); ISBN 0 946707 32 4; ISSN 0264–8725; xiii + 410 pp; 234 x 156 mm; illustrated; comprehensive index; hardback.

CONTENTS

Biotechnology & Genetic Engineering Reviews VOLUME 9 (December 1991); ISBN 0 946707 39 1; ISSN 0264–8725; x + 425 pp; 234 x 156 mm; illustrated; comprehensive index; hardback.

CONTENTS

Biotechnology & Genetic Engineering Reviews VOLUME 10 (December 1992); ISBN 0 946707 39 1; ISSN 0264–8725; xi + 510 pp; 234 x 156 mm; illustrated; comprehensive index; hardback.

CONTENTS

Biotechnology & Genetic Engineering Reviews VOLUME 11 (December 1993); ISBN 0 946707 77 4; ISSN 0264–8725; xi + 381 pp; 234 x 156 mm; illustrated; comprehensive index; hardback.

CONTENTS